ENGINEERING ELECTRODYNAMICS

Electric Machine, Transformer, and Power Equipment Design

Janusz Turowski
Marek Turowski

ENGINEERING ELECTRODYNAMICS

Electric Machine, Transformer, and Power Equipment Design

CRC Press
Taylor & Francis Group
Boca Raton London New York

CRC Press is an imprint of the
Taylor & Francis Group, an **informa** business

CRC Press
Taylor & Francis Group
6000 Broken Sound Parkway NW, Suite 300
Boca Raton, FL 33487-2742

Printed on acid-free paper
Version Date: 20130716

International Standard Book Number-13: 978-1-4665-8931-5 (Hardback)

Library of Congress Cataloging-in-Publication Data

Turowski, J.
[Elektrodynamika techniczna. English]
Engineering electrodynamics : electric machine, transformer, and power equipment design / Janusz Turowski and Marek Turowski.
pages cm
Includes bibliographical references and index.
ISBN 978-1-4665-8931-5 (alk. paper)
1. Electric machinery. 2. Electrodynamics. I. Turowski, Marek. II. Title.

TK2211.T8617 2013
621.31'042--dc23 2013026049

Visit the Taylor & Francis Web site at
http://www.taylorandfrancis.com

and the CRC Press Web site at
http://www.crcpress.com

To Our Family

Contents

Preface

Modern industrial engineering, especially its scientific management, is based on principles of mechatronics (J. Turowski [1.20]). These principles consist mainly of simple tools of rapid design and development, based on sophisticated, comprehensive fundamental researches. One such tool is the expert approach, which among others consists of a knowledge base and software. The knowledge base is created by a knowledge expert in a given field of knowledge and software is prepared by the knowledge engineer.

The more knowledge that is implemented into the knowledge base, the simpler, faster, and more economical is the resulting design tool. This book presents theoretical and practical materials for the creation and use of an effective knowledge base for design in electrical engineering.

Generally, in engineering, there are two main paths of design:

1. Development and building of machines or systems
2. Motion and dynamics of the system

For the development and building of machines and systems, the theories different physical fields and material science are basic scientific tools.

Nowadays, the design of motion and dynamics is supported by software packages, such as Saber (http://www.synopsys.com/Systems/Saber/Pages/default.aspx, J. Turowski [1.20]).

However, as A. H. Jasinski [1.4] rightly said: "A considerable interest is now ... to ensure shortening of 'time to market' for new product development ... to reduce time lag ... and minimize development costs." It is briefly expressed as "*rapid design*" (J. Turowski [1.4], p. 192 and Fig. 1.1).

There are a number of computer methods of modeling and simulation of design processes. However, the most user-friendly, fast and low-priced have proven to be two basic methods:

1. For development and building—a flow network method, FNM-3D, confirmed by the worldwide success of the authors' RNM-3D (reluctance network method, three-dimensional) software package, implemented and used broadly in over 40 transformer industry institutions all over the world.
2. For motion—the author's "Hamilton" package [1.20], based on the variational Hamilton's principle of least action, a generalized theory of electromechanical energy conversion, the Euler–Lagrange equation, and the Lagrange state function.

Both these design methods can provide a solution within seconds for each design variant, whereas other sophisticated numerical packages (e.g., finite element method, FEM-3D) require much more time and effort for almost the same solution. All this can be included in a mechatronics philosophy and technology.

To deal with the rapidly developing modern theory of electrical engineering, current engineers need to be good in mathematics as well as electrical engineering. They should understand well the physical phenomena and be able to use this knowledge in their professional work. The technical universities in many countries adapt their curricula to meet these requirements.

One such discipline, especially interesting for research, development, and design engineers, is the applied, macroscopic Maxwell's theory of electromagnetic fields. There still exists a gap between advanced electrical theories and their practical application in industry. The objective of engineering electrodynamics, which is considered a scientific discipline, is to reduce this gap.

Electrodynamics is the science describing the motion of matter, that is, fields, energy, charges, bodies, and media, under the influence of forces acting in electric and magnetic fields.

Classical electrodynamics (Jackson [2.10]) is a theory of electrodynamics. It is a part of physics, based on pure basic laws, and somewhat abstract mathematical assumptions.

Engineering electrodynamics (J. Turowski [1.15]), on the other hand, is the science of the practical industrial application of classical electrodynamics and physics. It takes into account physical properties of materials, magnetic and thermal nonlinearities, forces and motion, and so on. Semi-empirical, analytical–experimental, rationally reduced models and simplified computer programs are often used to accelerate and simplify obtaining the results and provide practical tools of rapid design.*

Frequently, very useful for rapid design are the achievements of previous generations of engineers, in the form of parametric formulas derived analytically, quite often with practical correction coefficients.

> "Mathematicians resolve what is resolvable, whereas engineer has to resolve … everything", is the author's (JT) instruction for his students, to illustrate the tasks and challenges of a busy practical engineer. Easily available computer software and computer skills of students are not enough. Experienced industrial scientists rightly confirmed (*Coulson* [7.6]) that … there is a risk that the need for engineers with good physical insight is overlooked. Physical insight is essential when the practical problem is translated into a form which can be solved mathematically … it is important to ensure that there is an adequate supply of engineers with the right outlook …

A contemporary designer who has to deal with various materials, including conductors, semiconductors, and dielectrics, of constant and variable parameters, operating within temperatures from almost absolute zero to several thousands of degrees centigrade, at weak or very strong fields, has to combine his skills in classical electrodynamics with the knowledge of materials' microstructure and their physical and processing properties. This book attempts to address the practical aspects of these issues.

* The best design programs or tools should be *fool-proof* (German: *idiotensicher*), that is, as simple as, for example, driving a modern sophisticated automobile using the same steering-wheel like it was 100 years ago. For electrodynamics, such easy steering tools are the well-known Ohm's and Kirchhoff's laws. See, for example, the authors' *RNM-3D* packages.

To a large extent, this book is based on the authors' vast industrial experience and scientific works published from 1956 to the present days in various Polish and international books, scientific journals, and industrial conferences. This book also contains the results of research works that so far have not been published, as well as references to numerous publications of other specialists. Since the book has partly the character of a handbook, a review of basic laws and formulae of electrodynamics, as well as necessary information in the form of tables, examples, and working formulae, has been introduced.

In the world of the engineering community, interest in applied electromagnetism is permanently growing. It is mainly due to the continuous and significant increase in the capacity of larger power equipment, in which the concentration of electromagnetic fields sometimes reaches rates far higher than earlier values. As a consequence, excessive power losses caused by eddy currents in constructional elements, locally exceeding permissible values, and electrodynamic forces endangering the reliability of the whole equipment as well as the entire power system connected with it may appear.

To understand these phenomena and to overcome the problems, it is necessary to continuously enhance the understanding and knowledge of electrodynamic processes, and their positive and negative effects, primarily due to the growing concern about the reliability of large and increasingly sophisticated power equipment.

Since the first Polish edition of this book (Warsaw 1968), followed by an improved Russian edition (Moscow 1974) and a third edition (Warsaw 1993), many years have passed, but the problems mentioned above are still of immense interest. Almost 150 years have passed since the 1857 publication of Urbanski's book [1.21] in Lwów (Lviv), at that time under Austrian occupation and after the Soviet invasion (1939). It was one of the first Polish and international works on electrodynamics even though it was 25 years after Faraday's (1832) discovery, 16 years before Maxwell's theory (1873), and 27 years before the famous Poynting's work (1884) [3.3]. Since that time, especially since the 1970s, electrodynamics has become one of the main subjects of Polish specialization. Engineering electrodynamics was introduced into the teaching curricula at the Technical University of Lodz (TUL) as well as many other technical universities.

In spite of the tremendous progress in computer modeling and numerical computation, particularly in the fast calculation of three-dimensional (3-D) coupled fields, the basic principles and methodologies are still the same. On the other hand, significant progress has taken place in the area of magnetic and superconducting materials.

For faster development and popularization of the methods of electromagnetic engineering in Poland, a state-sponsored project "Analysis and Synthesis of Electromagnetic Fields" (1980–1990) was undertaken under the author's direction (J. Turowski [1.17]). The results of that research were presented at subsequent national and international conferences on applied electromagnetism (Uniejów 1974, Jadwisin 1982, Rydzyna 1983). In particular, the renowned International Symposium on Electromagnetic Fields in Electrical Engineering (ISEF) was founded by the author J. Turowski and his respected colleagues, and has continued for many years (Lodz 1979, Warsaw 1985, Pavia-Italy 1987, Lodz 1989, Southampton 1991, Warsaw 1993, Thessaloniki-Greece 1995, Gdansk 1997, Pavia 1999, Cracow 2001,

Maribor-Slovenia 2003, Vigo-Spain 2005, Prague 2007, etc., to date; ISEF'2013 in Macedonia [1.53]), organized biennially by the TUL-Poland and chaired by the author for many years (then, since 2001 by Professor S. Wiak). The ISEF Symposia delivered new important materials. The technical universities of Lodz, Warsaw, Szczecin, Poznan, and the like are the strongest Polish research centers in this field. All these experiences have had some influence on this book. Worldwide, if one were to consider the leading conferences, like ISEF, Compumag, Intermag, ICEM, and BISEF-ICEF, over 100 publications on electromagnetism appear every year.

Some of the most important research subjects are related to the rapid solution of coupled, nonlinear, and 3-D fields, with the application of interactive methods of design and optimization.

Recent inventions in materials, such as ferromagnetic amorphous strips (METGLAS), high-temperature superconductors, permanent magnets with rare earths, MHD, semiconductors, mechatronics, and so on, which revolutionize electrical engineering technology, are also covered in this book.

Critical and sensible choices from many offered computational methods, such as ANM, RNM, FDM, FEM, BEM, and other simplified calculation methods are often crucial. For example, some 3-D industrial problems can be solved easier and faster with *Fourier's* series, *Ampere's and Biot–Savart* laws, or RNM3D than with other sophisticated numerical methods. The economic advantage of solution speed-up for *rapid design* could be as much as computation time below 1 s (RNM3D) for each design variant versus much longer and cumbersome model preparations and computations (FEM-3D).

As a professor at the TUL in the Department of Intelligent Information Systems at AHE-Lodz, the author (*J. Turowski*) has taught these topics for many years in his lectures on electric machines and transformers, electromechanical components of automatics and mechatronics, technological electrodynamics, and electromagnetic components of power electronics, innovation management, and new technologies. This book is, in principle, about creating physical and computer models, whereas the other book *Electromagnetic Calculation of Elements of Electric Machines and Devices* by J. Turowski [1.16] covered practical design solutions. Subsequent books by the author and coauthors, *Analysis and Synthesis of Electromagnetic Fields* (J. Turowski [1.17/1]) and its extended English translation *Computational Magnetics* [1.17/2], *Modern Electrical Drives* (Ertan,..., J. Turowski, et al. [1.3]), *Fundamentals of Mechatronics* [1.20], *Transformers. Analysis, Design, and Measurement* (Lopez Fernández, Ertan, J. Turowski [1.8]), present many other scientific and practical aspects.

Optimization of the design and characteristics of modern machines and devices is not possible without electromagnetic modeling and calculation. It was, among others, the subject of joint publications of Polish–Italian–Indian–Spanish–Portuguese–Mexican teams including K. Komęza, A. Pelikant, J. Turowski, M. Turowski, S. Wiak (Poland), M. Rizzo, A. Savini, P. Di Barba (Italy), S. V. Kulkarni (India), X. Lopez-Fernandez (Spain/Portugal), J. Gieras (USA), J. Sykulski (UK), and others.

A similar fundamental work was conducted during industrial verifications of the authors' method of equivalent reluctance network (RNM-3D), for fast 3-D modeling and calculation of stray fields and losses in large power transformers, carried out by the Polish–Indian team of M. Kopeć, D. A. Koppikar, S. V. Kulkarni, J. Turowski and M. Turowski and supported by the transformer works at ABB-ELTA in Lodz-Poland, Crompton Greaves Ltd. in Mumbai-India, EMCO Transformers in Thane-India, Chinese Transformer Works Shenyang (Mr. Xiu), and Baoding (Dr. Cheng), EFACEC-Porto, Portugal, as well as transformer works in the United States, the United Kingdom, Australia, Iran, Russia, and others. The authors also express their gratitude to their colleagues at universities in Italy—Pavia and Palermo; in Japan—Okayama (Professor T. Nakata) and Kanazawa (Professor K. Bessho); in Turkey—Ankara (Professor B. Ertan); in China—Shenyang (Professor T. Reunian), Hubei, and Huazhong (Professor Zhou Ke Ding); Poland—Electrotechnical Institute in Lodz and Warsaw; and others.

In this book, the authors used not only their own experiences, but also the results of joint publications and valuable discussions with many colleagues and specialists from Poland and abroad, PhD and diploma students, as well as many industry specialists. All of them have influenced the content of this book to a certain extent. Therefore, the authors express their sincere gratitude and acknowledgments to the publishers and authors for permission to quote the cited works.

When discussing modern applications of the industrial electrodynamics, one cannot forget the importance of the new interdisciplinary domain—“mechatronics.” It is changing the traditional concurrent engineering approach to modern design and manufacturing as a result of the recent incredible progress in software, design tools [1.20], [4.35], and advances in integrated circuits and electronic control systems.

Authors

Janusz Turowski, DSc 1963, PhD 1958, MSc, electr. eng. 1951, Honorary Doctor (*Honoris Causa*) University of Pavia (Italy) 1998, worked as a full professor during 1964–2003 (retired 2003) in electrical machines, applied electromagnetics, and mechatronics. Since 1949, he has been with the Institute of Electrical Machines and Transformers (now Mechatronics and Information Systems 2003–), and as vice dean (1964–1969) and first vice rector (1990–1996) of the Technical University of Lodz (TUL) Poland. He also was as full professor and dean (1999–2010) in the Department of Intelligent Information Systems, Academy of Humanities and Economics (AHE)—Lodz. He is a full member of the International Academy of Electrotechnical Sciences (1999–), member of the Ukrainian Academy of Science, senior member of IEEE (1988–2001), and member of CIGRE-Paris (1964–2001).

Dr. Turowski graduated from TUL with an MSc degree and worked as an electrical engineer in 1951 and later was upgraded to a full professor (1978–). During the years 1973–1992, he was the director of the Institute of Electrical Machine and Transfer.

Dr. Turowski was president (1976–1982) of the Polish Association of Theoretical and Applied Electrotechnics-PTETiS, as well as a member (1978–1909) and the vice chairman (1999–2003) of the Electrical Committee Polish Academy of Science. He was an all-country coordinator of fundamental research in analysis and synthesis of electromagnetic fields (1980–1990), and the chairman (1979–2001) and honorary chairman (2001–) of biennial international symposia on electromagnetic fields in electrical engineering—ISEF. Professor Turowski was a member of the Polish Main Council for University Education (1990–1996) and honorary president (2004–) of the Polish UNESCO-UNISPAR Working Group Society.

He was also elected a member of the Council for European Education by the Minister of National Education (1996–1999) and Science Council of Technology Agency (1996–1997) and a member of the EC/PHARE Higher Education Steering Committee (1996–1997). In addition, he was a member of SEFI (1992–1996), "Alliance Universities for Democracy"—USA (1992–1996), "International Compumag Society" (1996–2003), The New York Academy of Science (1994–1995), and UATI Administrative Board, Paris (1999–2002).

Dr. Turowski is the president of the Polish UNISPAR Working Group Society (1996–), and was coordinator of the EuroUNISPAR Network (2000–2005) and joint Baltic Sea Innovation Network BASIN (1999–2005), past editor in chief of the journal *Polish Innovation Market*, besides holding other positions.

Professor Turowski is the author or coauthor of about 310 publications, including 20 books and 35 chapters in joint books in Poland, the United Kingdom, the United States, France, Germany, Russia, China, India, Japan, NATO, and other countries; his works have been cited by other authors about a 1000 times, with more than 210 from abroad.

Turkowski is the author of *Technical Electrodynamics* (Warsaw 1968, 1993, Moscow 1974), *Electromagnetic Calculations of Elements of Electrical Machines and Devices* (Warsaw 1982, Moscow 1986), and *Fundamentals of Mechatronics* (Lodz, 2008).

He was the joint author and editor of *Analysis and Synthesis of Electromagnetic Fields* (Polish Academy of Sciences 1990), *Electromagnetic Fields in Electrical Engineering* (New York 1989, London 1990, 1992, 1994), joint author of *Computational Magnetics* (London 1994), *Modern Electric Drives* (NATO Science Series), *Electric Machines*, and others. He is also one of the main designers of big ELTA (ABB) Transformer Works.

Dr. Turowski is a consultant in transformers and electric machine works in Poland, China, India, Australia, the United States, Canada, and other countries. He has lectured in universities and companies in Italy, the United States, the United Kingdom, Japan, Russia, China, India, Spain, Portugal, Australia, Germany, Brazil, and other countries. He has supervised 18 PhD and 3 DSc candidates successfully and authored 450 referee opinions.

Turkowski was a member of approximately 30 international steering committees of conferences in China 3, Japan 2, Turkey 3, India 2, Russia 4, the United Kingdom 2, Italy 2, Romania 2, Hungary, Australia, PHARE, Greece 2, Algeria, Poland 10, and Spain ARWtr (many times).

Turkowski was the chairman and as well as a member of the editorial boards of *Rozprawy El.* (1975–1987) and a member of the *Archives of Electrical Engineering* 1988–1996 of the Polish Academy of Science; international scientific journals *Electromotion* and *Australasian Journal of Engineering Education*; and of the ARWtr4 International Technical Committee and ICEM4 editorial board.

Turkowski is the recipient of State Crosses "Polonia Restituta" (1973) and a higher rank (1991). He is also a honorary recipient of "Medaglia Teresiana" (1998) and "Volta Silver Medal" from the University of Pavia, Italy (1999). Turkowski is an honorary member of the Polish Association PTETiS (1986), honorary chairman of ISEF (2001), and he has received 20 scientific and organizational awards from the Minister of Higher Education. He received the 2009 State Cross of the Exiles of Siberia (from the former President Lech Kaczynski) Nr 9-2009-157. Professor Turowski received the 2010 medal for 30-year participation in the action in the Independent Self-Governing Trade Union "Solidarity" and fidelity to his ideals and others. He received the 1st Award of Polish Federation Technical Societies, the "Technology Master '2001" award for Program RNM-3D applied in over 44 transformer works and research centers all over the world. He received medals from Polish Electricians Association (SEP) for merits in the domain of Transformers and Power Systems. In 2004, he received the Silver Medal from the International Academy of Electrotechnical Sciences "For a prominent contribution to the development of world science, education technology and international cooperation."

Marek Turowski earned his MSc and PhD in electronics engineering from the Technical University of Lodz, Poland, in 1983 and 1992, respectively. Dr. Turowski has over 30 years of experience in modeling and computer simulation (numerical methods, algorithms, models, and simulation techniques) for multi-domain (electrical, electromagnetic, optical, thermal, and mechanical) analysis, semiconductor physics, design, and optimization of electronic devices and integrated circuits (ICs), electromagnetic fields, photonic devices, optoelectronic systems, high-frequency ICs, and micro-electro-mechanical systems (MEMS). He carried out a part of his doctoral research at LAAS-CNRS in Toulouse, France, on physics-based modeling of semiconductor devices. In 1992–1993, he was a postdoctoral research fellow at the University of Sydney, Australia, working on 3D modeling of electromagnetic fields in electric machines and transformers. He collaborated with international industry and academia on modeling of electromagnetic fields and losses in power transformers, for example, in Poland, Italy, Spain, Portugal, India, and the United States. In 1996–1998, he was with the Department of Electrical Engineering, University of Kentucky in Lexington, Kentucky, USA, developing novel models for semiconductor devices. From September 1998 to April 2013, he was a full-time employee of CFD Research Corporation (CFDRC) in Huntsville, Alabama, USA, where later he was the director of Nano Electronics & Plasma Technologies. Since May 2013 he has been with Robust Chip Inc. in Pleasanton, California, a company devoted to the reliability of electronic integrated circuits.

Dr. Turowski has been an author or coauthor of over 50 journal papers and book chapters and over 155 conference papers and presentations. He is a senior member of IEEE, the Electron Device Society, and Nuclear and Plasma Sciences Society. He has been a reviewer for *IEEE Transactions on Device and Materials Reliability, IEEE Transactions on Computer-Aided Design (CAD), on Nuclear Science,* and *Materials Science and Engineering B.* He has led numerous projects in modeling and CAD for electronic semiconductor devices, optoelectronics and photonics (solar cells, sensors, photodetectors, VCSELs), electromagnetics in RF ICs (DARPA NeoCAD), thermal issues in electronics, MEMS, novel physical design tools for 3D/stacked ICs (DARPA 3DIC Program, 2002–2005), and radiation effects in modern submicron semiconductor devices, ICs, and solar cells (AFRL 1999–2004; MDA 2003; DARPA/Mission Research 2004–2005; DTRA 2005–2011; AFRL 2005–2006; NASA projects on rad-hard space electronics & photonics (quantum dot solar cells) 2005–2009; DOE SBIR on radiation detectors, 2006–2007; DTRA project on radiation effects in high-speed compound semiconductor microelectronics (2008–2011); NASA projects on radiation hardening of space electronics for extreme environments (2007–2013); AFOSR and DTRA projects on radiation effects in solar cells and photodetectors (2008–2013) and radiation hardening of the advanced ultra-deep submicron semiconductor technologies, devices, and ICs (2012–2014).

List of Symbols

$\boldsymbol{A}$: magnetic vector potential (definition: $\boldsymbol{B} = \nabla \times \boldsymbol{A}$ or $\boldsymbol{B}$ = curl $\boldsymbol{A}$ or $\boldsymbol{B}$ = rot $\boldsymbol{A}$), instantaneous value; A: modulus (Wb/m)

A: surface area, d$\boldsymbol{A}$: vector of area element perpendicular to the surface (m^2)

A_1: specific electric loading = linear current density, instantaneous value; A_{m1}: maximum value (A/m)

a_1, a_2: width of HV and LV transformer winding, respectively (m)

$\boldsymbol{B}$: magnetic flux density vector, instantaneous value; B: modulus of vector; $\underline{B}_m = B_m\, e^{j\omega t}$, where $B_m = |B_m|e^{j\psi}$: complex value; B^*: conjugate value
B_m: maximum value in time,
$B_{avr,m}$: average value in space of maximum flux density in time;
B_t: tangential component; B_n: normal component; B_s: surface value;
B_{sat}: saturation flux density (T = Wb/m²)

b, a: linear dimensions (m)

C: capacitance of capacitor (F); constant

c: specific heat [J/(kg · °K)]

c_o: speed of light (m/s)

$\boldsymbol{D}$: vector of electric flux density, electric displacement (C/m²);
D: internal stator diameter (m)

d: thickness of sheet or screen (m)

e: instantaneous value of electromotive force EMF (V); base of a natural logarithm (e = 2.718); e = 1.60219×10^{-19} C electron charge

$\boldsymbol{E}$: vector of electric field intensity, instantaneous value; $\boldsymbol{E}_m$: maximum value (V/m); other symbols as in case of $\boldsymbol{B}$; $\boldsymbol{E}_{ext}$: external (e.g., thermoelectric) field intensity

E_u: electromotive force EMF, rms value; E_{um}: maximum value (V); $\boldsymbol{F}$: force vector (N)

F: current flow, magnetomotive force MMF, ampere-turns; F_m: maximum flow (A)

f: frequency (Hz)

$\boldsymbol{H}$: vector of magnetic field intensity, instantaneous value; $\boldsymbol{H}_m$: maximum value (A/m); $\hat{H}_m$: maximum in time and space; other symbols as in case of $\boldsymbol{B}$

h: linear dimension, height (m)

i: current, instantaneous value; I: rms value; I_m: maximum value (A); $i = \int \boldsymbol{J} \cdot \mathrm{d}\boldsymbol{s}$

i, j, k or $\mathbf{1}_x$, $\mathbf{1}_y$, $\mathbf{1}_z$: unit vectors in rectangular (*Cartesian*) coordinates

$\boldsymbol{J}$: vector of electric current density, instantaneous value; $\boldsymbol{J}_m$: maximum value (A/m²); other symbols as in case of $\boldsymbol{B}$;

$j = \sqrt{-1}$: imaginary (complex) unit, $j^2 = -1$

$k = \sqrt{\dfrac{\omega\mu\sigma}{2}}$: attenuation constant of electromagnetic wave in metal (1/m)

l: length; d$\boldsymbol{l}$: vector of length element (m)

L, M: self-inductance and mutual inductance, respectively (H)

M, m: coefficients of mirror images

n: rotational (angular) speed, $n_s = (f/2p)$(revolutions/s = 1/s); $n_s = (3000/p)$(rev/min)-synchronous speed
N: number of turns
P: active power (W); $P_s = P + jQ$: apparent power (V · A); P_1: value per unit surface (W/m^2)
p: screening factor; number of poles pairs
Q: reactive power (var); Q_1: per unit surface, length or volume
Q: electric charge (C); q: volumetric electric charge density (C/m^3); q_s: surface charge density (C/m^2)
R: electric resistance (Ω)
R_m: reluctance (magnetic resistance), $\underline{R}_m = R_m + R_{mr}$: complex reluctance (magnetic impedance) (1/H)
$\boldsymbol{r}$: radius vector, r: radius (m); $\mathbf{1_r}$: unit vector in radial direction
$\boldsymbol{S} = \boldsymbol{E} \times \boldsymbol{H}$: Poynting's vector; $\underline{S}_s = S_p + jS_q$: vector modulus of complex Poynting's vector; S_p: active component (W/m^2); S_q: reactive component (var/m^2)
$\boldsymbol{s}_i = \mathbf{i}s_{xi} + \mathbf{j}s_{yi} + \mathbf{k}s_{zi}$: vector in direction of wave propagation
T: period (s); thermodynamic temperature (K)
t: time (s); temperature (°C)
u or v: voltage, instantaneous value; U or V: rms value; V_m: maximum value (V); $V_m = \int \boldsymbol{E} \cdot d\boldsymbol{l}$
$\boldsymbol{v}$: linear velocity vector (m/s)
W: energy (J); w: energy per unit volume (J/m^3)
X: reactance; X_L: leakage reactance (Ω)
$Z = R + jX$: complex impedance; $Z^* = R - jX$; $|Z| = \sqrt{R^2+X^2}$ (Ω); Z_1, Z_2, Z_3: wave impedances of media 1, 2, and 3
($Z_{air} = 377\ \Omega \gg |Z_{Steel}| \approx 2.4 \times 10^{-4}\ \Omega \gg |Z_{Cu}| = 2.7 \times 10^{-6}\ \Omega$)
$\alpha = (1 + j)k$: propagation constant in space, for solid metal (1/m); $\alpha^* = (1 - j)k$; $|\alpha| = \sqrt{2}k = \sqrt{\omega\mu\sigma}$; α: thermal coefficient of resistivity (1/K)
α': coefficient of heat dissipation (W/m^2)
$\Gamma = \sqrt{j\omega\mu(\sigma+j\omega\varepsilon)} = \alpha'' + j\beta''$: propagation constant in space
α'': attenuation constant; β'': phase constant
$\delta = (1/k) = \sqrt{(2/\omega\mu\sigma)}$: equivalent depth of field penetration into conducting half-space (m)
δ_{air}: interwinding insulation gap in transformer; $\delta' = \delta_{air} + (a_1 + a_2/3)$: equivalent leakage gap
$\varepsilon = \varepsilon_r \varepsilon_o$: dielectric permittivity; ε_r: relative permittivity
$\varepsilon_o \cong 8.854 \times 10^{-12}$ F/m: absolute dielectric permittivity = electric constant (permittivity of free space)
$\Theta = t - t_{ambient}$: temperature rise (–°K)
Λ: permeance: magnetic conductivity (H)
λ: length of wave (m); thermal conductivity [W/(m · °K)]
$\mu = \mu_r \mu_o$: magnetic permeability; μ_r: relative permeability; $\mu_o = 4\pi \times 10^{-7}$ H/m: absolute magnetic permeability = magnetic constant (permeability of free space)
$\nu = 1/\mu$: reluctivity (m/H)
$\rho = 1/\sigma$: resistivity = specific resistance (Ω/m)

$\sigma = 1/\rho$: conductivity = specific conductance (S/m)

ρ: volume density of space charge (C/m^3)

ρ_m: mass density (kg/m^3)

$\tau = \dfrac{\pi D}{2p}$: pole pitch of electric machine (m)

$\Phi = \int \boldsymbol{B} \cdot \mathrm{d}\boldsymbol{s}$: magnetic flux, instantaneous value; Φ_m: maximum value (Wb)

Φ_L: leakage flux; φ: angle of phase shift (rad)

χ_m: magnetic susceptibility, $\mu_r + \chi_m$ (*pure number*) = a measure of magnetic polarization

$\Psi = \int \boldsymbol{D} \cdot \mathrm{d}\mathbf{s}$: electric flux (C)

$\omega = 2\pi f$: pulsation, angular frequency (rad/s)

$W = \mathbf{A} \cdot \boldsymbol{B} = AB \cos \alpha = A_x B_x + A_y B_y + A_z B_z$: scalar product; div $\boldsymbol{A}$ (or $\equiv \nabla \cdot \boldsymbol{A}$)

$\boldsymbol{F} = \boldsymbol{A} \times \boldsymbol{B} = \boldsymbol{n}\ AB\ \sin\ \alpha = \mathbf{i}(A_y B_z - A_z B_y) + \mathbf{j}(A_z B_x - A_x B_z) + \mathbf{k}(A_x B_y - A_y B_x)$: vector product; curl $\boldsymbol{A} \equiv$ rot $\boldsymbol{A} \equiv \nabla \times \boldsymbol{A}$

$\nabla^2 \boldsymbol{A} \equiv \mathrm{Lap}\, \boldsymbol{A} \equiv \Delta \boldsymbol{A}$

Abbreviations of Names of Computer Methods

ANM	analytically numerical method
BEM	boundary element method
FDM	finite difference method
FEM	finite element method
FNM	flow (resistant) network method
MIE	method of integral equations
RNM	reluctance network method

1 Methods of Investigation and Constructional Materials

1.1 METHODS OF INVESTIGATIONS

The solutions to engineering tasks by applying the methods of industrial electrodynamics can be divided into several stages (Figure 1.1a):

1. *Formulating mathematical equations* and finding a *function*, which describes the electromagnetic field and its properties in the investigated region, considering the constant or variable characteristics of media (air, copper, steel, etc.) in this region
2. *Determining the limiting conditions*, that is, *boundary conditions* and initial conditions on the surface of the investigated region, imposed by the type and configuration of sources in the investigated field (configuration of conductors, coils or magnetic cores, type of current, etc.) and the border surfaces of adjacent media
3. Selecting *constants* and *parameters* of equations in such a way that satisfies the boundary and initial conditions, that is, finding a final mathematical solution
4. *Experimental verification* of the assumptions, adequacy of a computation models, intermediate simplifications, and final results
5. *Demonstrating* the obtained results in a form of *simple* formulae, user-friendly programs, tables, and/or diagrams, facilitating optimal use of the results of the object being investigated

Formulating adequate, that is, being in agreement with reality, equations (stage 1) that correctly describe an object, its phenomena, and solution is a difficult task. Often, we have to limit the calculation to simplified mathematical models, based on one of the laws or group of laws of physics and ignore others. Examples of formulation and solutions of mathematical equations based on the fundamental equations of electrodynamics are given in Chapter 2. Stages 1 through 4 (Figure 1.1a) belong to the *analysis* of the problem, in which investigations of the physical properties of materials play an important role. This is discussed in Section 1.2.

The objective of industrial or engineering electrodynamics, after all, is the design, that is, the creation of new structures, (the *synthesis*). Therefore, the stage of *analysis* should be limited to a minimum to avoid making the design process too long and too expensive. An absolutely necessary element of a full solution is the *experimental*

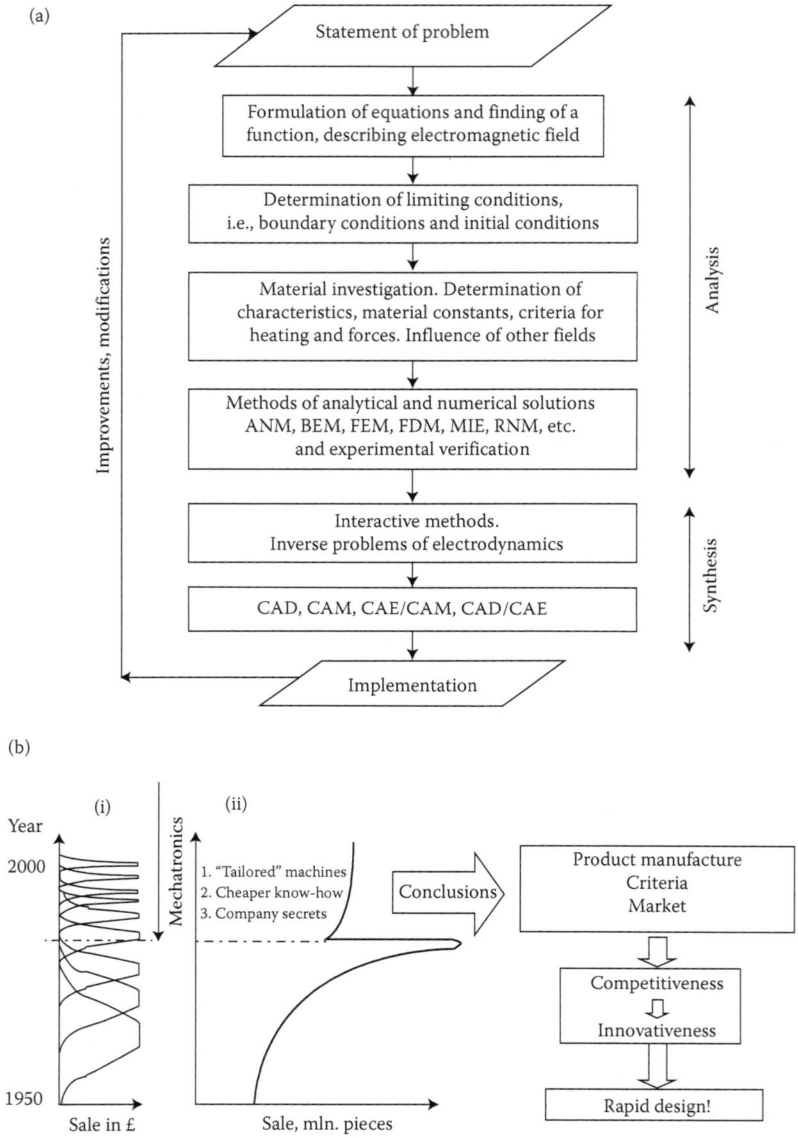

FIGURE 1.1 (a) Classification of modeling, computational, and research tasks in engineering electrodynamics and electromechanics. Process of design—see (b) through (d). (b) Impact of mechatronics upon (i) "time to market" and (ii) sale of small catalog machines in the United Kingdom (W. Wood 1990) [1.20]. (c) Block diagram of an expert system for designing machines: 1—large portion of introduced knowledge and experience = simple, inexpensive and rapid solution, for example, 1 s; 2—small portion of knowledge and experience = difficult, expensive, labor-consuming solution. (Adapted from Turowski J.: *Fundamentals of Mechatronics* (in Polish). AHE-Lodz, 2008.) (d) RNM-3D interactive design in less than 1 s design cycle for one constructional variant (Adapted from Turowski J.: *Fundamentals of Mechatronics* (in Polish). AHE-Lodz, 2008.)

(c)

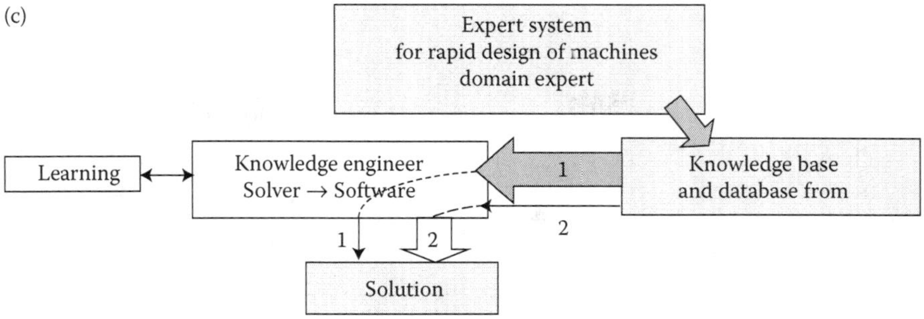

(d)

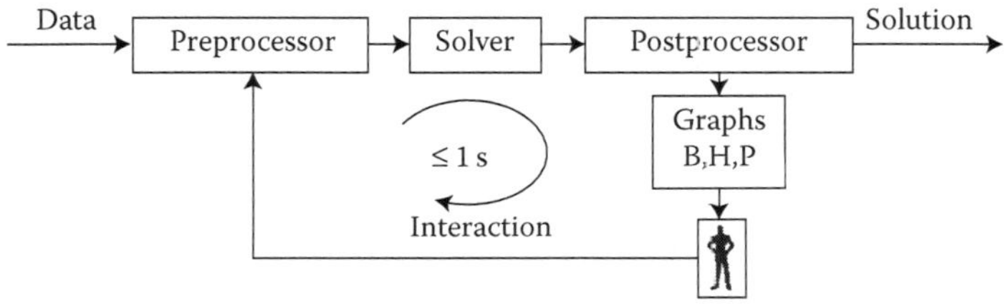

FIGURE 1.1 (continued) (a) Classification of modeling, computational, and research tasks in engineering electrodynamics and electromechanics. Process of design—see (b) through (d). (b) Impact of mechatronics upon (i) "time to market" and (ii) sale of small catalog machines in the United Kingdom (W. Wood 1990) [1.20]. (c) Block diagram of an expert system for designing machines: 1—large portion of introduced knowledge and experience = simple, inexpensive and rapid solution, for example, 1 s; 2—small portion of knowledge and experience = difficult, expensive, labor-consuming solution. (Adapted from Turowski J.: *Fundamentals of Mechatronics* (in Polish). AHE-Lodz, 2008.) (d) RNM-3D interactive design in less than 1 s design cycle for one constructional variant (Adapted from Turowski J.: *Fundamentals of Mechatronics* (in Polish). AHE-Lodz, 2008.)

verification of the results of calculations (stage 4). It is especially important today when field problems are resolved with the help of sophisticated commercial computer programs. Oftentimes, authors are the only ones who know the structure of such programs and applied assumptions.

Synthesis, that is, assembling elements of the analysis into a new product, based earlier on the trial-and-error method, has recently gained the following tools:

- Interactive methods of design, which are a higher-level and faster trial-and-error method [1.20]
- CAD and Auto-CAD (computer-aided design), mainly for design and graphics
- CAM (computer-aided manufacturing) systems, to assist the production process
- CAE (computer-aided engineering), which is a combination of the systems mentioned above, where a physical model (prototype) is substituted by a computer model and its characteristics are evaluated and improved by the computer simulation, including the manufacturing process itself

Automated CAD/CAE systems revolutionize the design and manufacturing processes of many electromagnetic devices and machines, but will never obviate the necessity of human control and physical insight into phenomena.

Therefore, it is impossible to resolve an electrodynamic problem without at least a simplified consideration of the structure and physical properties of the materials.

The new discipline of *mechatronics* (J. Turowski [1.20]), which emerged in 1970s–1980s, as the synergistic combination of the *mechanical engineering, electronic control, engineering electromagnetics,* and system thinking, exerts serious impact on the modern design of products and manufacturing processes.*

The *principles of mechatronics* can be listed as (1) system approach, (2) rapid design (Figure 1.1b), (3) employment of artificial intelligence, (4) substitution of concurrent engineering by mechatronic engineering, (5) collective work, (6) simple methods based on comprehensive fundamental research, (7) accuracy relevant to the need, (8) analytical methods wherever possible, (9) linearization of nonlinear parameters, (10) short, interactive design cycles (Figure 1.1d), (11) simple machines with sophisticated control systems, (12) employment of the ISO9000, SWOT, and outsourcing rules, and (13) employment of expert systems, which are different for (a) building (Figure 1.1b) and (b) motion.

The more the knowledge implemented into the knowledge base, the less the time to success!

One of the main objectives of this book is to help designers and researchers to employ the above-mentioned principles into the machine design and to reduce the still existing gap between the theory and industrial practice. The process of design should be as rapid as possible. Figure 1.1d shows a practical design of a hybrid and semi-intelligent software package for rapid simulation and the design of the leakage region screening of stray fields in large power transformers for reduction of additional losses in tank and windings, excessive local heating hazard, and crushing short-circuit forces in windings.

The authors of this book wish to express their gratitude to multiple transformer works on all continents, and colleagues at different universities, for their cooperation in the industrial implementation of this methodology.

1.2 CONSTRUCTIONAL MATERIALS

1.2.1 Structure and Physical Properties of Metals

Among the many conducting materials, the most important, from the design point of view, are the solid-state materials and among them—the metals. A designer is interested, first of all, in electromagnetic, thermal, and mechanical properties of metals.

The most important *electromagnetic properties* include

- Electric conductivity, σ†, and/or its inverse—resistivity, ρ

* J. Turowski, X.M. Lopez-Fernandez, A. Soto, D. Souto: "Stray Losses Control in Core- and Shell Type Transformers. Part I: Upgrading of Energy-Saving and Reliability of Large Transformers" and "Part II: Method of Three-Dimensional Reluctance Network Solution (RNM-3Dshell)." Chapters in the book *International Advanced Research Workshop on Transformers—ARWtr2007*, Baiona, Spain, 2007.

† In some older publications, it is signed by γ.

- Temperature coefficient of resistivity and thermal limit of linearity (e.g., metal melting point, superconducting transition temperature)
- Magnetic properties, such as magnetization curves, specific per-unit (p.u.) power losses, magnetizability, and limits of linearity (Curie point)
- Other specific properties, such as thermal electromotive force in joining with another metal (usually Cu), electronic work function, and so on

The most important *thermal properties* include

- Coefficient of thermal conductivity, λ
- Coefficients of thermal dissipation by convection and radiation
- Specific heat
- Thermal elongation coefficient
- Curie point
- Melting temperature

The most important *mechanical properties* include

- Tensile strength limit and liquidity limit
- Relative elongation and modulus of elasticity (Young's modulus) under tension

The biggest computational difficulties are encountered while accounting for the nonlinear (magnetic and thermal) properties of metal. In order to select the proper computational methods and to avoid these difficulties, the knowledge of the basic material structure and its properties is necessary.

1.2.1.1 Atomic Structure

Metals, like all elements, have an atomic structure. Around the positive-charged nucleus, the negative-charged electrons* circulate in different orbits. The electron mass equals 9.1095×10^{-28} g (Ashcroft [1.1]), that is, 0.000551 of the mass of the smallest of all atoms—the atom of hydrogen. In normal conditions, the number of electrons in the atom equals the number of positive-charged protons in the nucleus. Due to this, the atom as a whole is a neutral particle. According to de Broglie hypothesis (1924) about the wave properties of elementary particles, the electron has the definite wavelength λ depending on its velocity. The particle with the momentum $p = mv$ corresponds to the wavelength of $\lambda = h/p$ (where $h = 6.6262 \times 10^{-34}$ Js is the Planck constant). In the case of the simplest atom—the hydrogen atom, which has a circular electron orbit—the length $2\pi r$ of the orbit should be a multiple of the wavelength ($2\pi r = n\lambda$); otherwise, an interferential quenching of electron waves would occur (Figure 1.2c).

Hence, electrons can occupy only strictly defined orbits of discretely changing diameters. The regions between the permitted orbits are the "prohibited" zones for electrons. This phenomenon is called *quantization of orbits*, where the integer value of $n = 1, 2, 3, \ldots, \infty$ is called the *main quantum number*.

* Discovered in 1897 by J. J. Thomson ([1.1], p. 22).

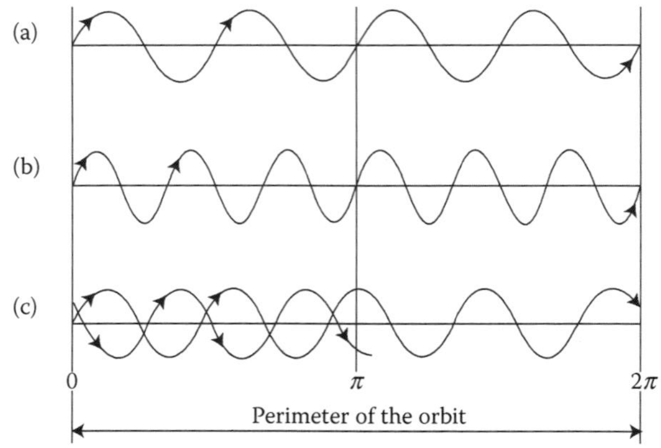

FIGURE 1.2 Scheme of the quantization of orbits: (a) four waves, (b) six waves, (c) interferential quenching of electron waves at the fragmentary (noninteger) number of waves on the orbit.

In reality, in a multielectron atom, the electrons and nucleus are subordinated to complex influences of a Coulomb and centrifugal forces as well as an external (e.g., terrestrial) magnetic field. Due to that, they move on more complex orbits, which have forms of ellipses relocating in the space. The electrons themselves, however, are in a rotating motion with an angular momentum, called *spin*.

Therefore, the full description of the electron state in the orbit requires a definition of a set of four quantum numbers:

1. The *main quantum number*, $n = 1, 2, 3, \ldots, \infty$, which defines the large axis of elliptic orbit, that is, *the main energy level* or the so-called *electronic shell* (electronic layer). It has been proven by investigations using cathode rays. These shells are typically denoted with the letters K, L, M, N, O, P, Q (Figure 1.3).
2. *Azimuthal quantum number*, also called the secondary quantum number $l = 0, 1, 2, \ldots, (n - 1)$, which separates the electrons of each shell into *n subshells*, each having slightly different energy. This number defines the small axis of the elliptic orbit. These are the *energy sublevels*, which create sets of trajectories in the frame of every main energy level. They are marked with the letters *s, p, d, f* (Figure 1.3).
3. *Magnetic quantum number*, also called the third quantum number, $m_l = 0, \pm 1, \pm 2, \ldots, \pm l$, which defines the spatial quantization of the plane elliptic orbits. It is connected with the existence of magnetic moment of an electron, which causes directional orientation of the atom in external (e.g., terrestrial) magnetic field.
4. *Spin magnetic quantum number*, $m_s = \pm 1/2$, which defines the orientation of the vector (axis) of the spin (levorotatory or dextrorotatory).

Both the magnetic quantum numbers m_l and m_s define a number of electrons in a subgroup.

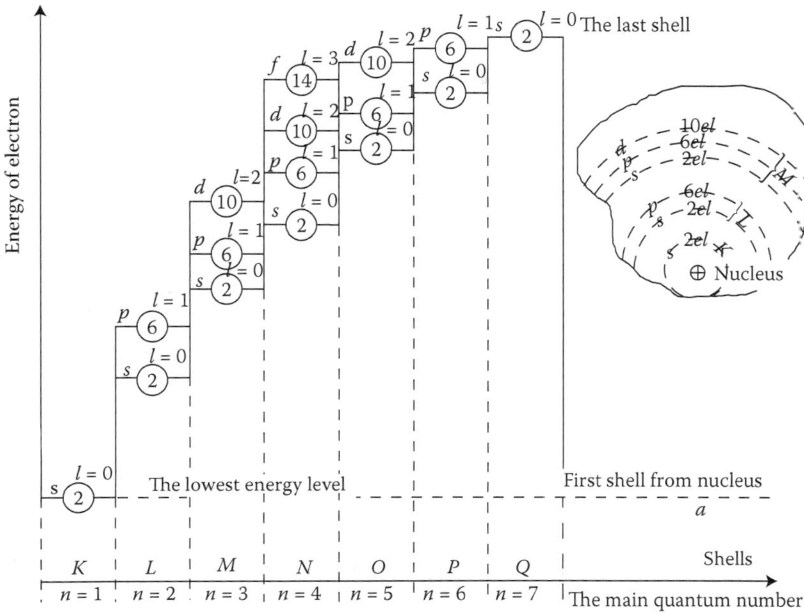

FIGURE 1.3 Diagram of permissible energy levels of electrons in a multielectron atom (with no scale regards): *l*—Azimuthal quantum number; *s, p, d, f*—subshells; ②, ⑥, and so on—maximal permitted number of electrons in a given subshell.

According to the principle called Pauli exclusion principle (1925), none of the electrons in the atom can have the same set of values of the quantum numbers mentioned above like any other electron. Since all these quantum numbers are strictly connected to each other, it means that in a given shell and subshell, there can exist only a strictly defined number of electrons "filling" the given energy level (Figure 1.3).

For instance, for $n = 1$, there can only be the following possible numbers $l = 0$, $m_l = 0$, $m_s = \pm 1/2$. It means that in the first shell, only up to two electrons can exist. If there is only one electron, it is a chemically active atom of hydrogen. If there are two electrons, it is a chemically inactive helium atom with the complete shell. At the shell $n = 2$, two subgroups $l = 0$ and $l = 1$ are possible, with the numbers $m_l = 0$, 0 and ±1 and $m_s = +1/2$ and $-1/2$; which means together 8 electrons in the shell, and so on. In this way, all sublevels *s* can hold at the most 2 electrons, subshells *p*–6 electrons, *d*–10 electrons, and *f*–14 electrons. In the normal state of an atom, electrons successively occupy the lowest (nearest to nucleus) energy levels.

After completing a given shell, it starts building a new shell. However, not all sublevels, as shown in Figure 1.3, are fully filled by electrons. Sometimes, in multielectron atoms, the repulsing forces from other electrons in the atom cause that farther subshells (Figure 1.3) are more advantageous, from the energy point of view, for a new electron, despite the fact that the subgroup nearer the nucleus is not yet completed (see groups *M* and *N* in Figures 1.3 and 1.17 [later in the chapter]). The elements with such a structured atom are called *transitory*. To this class belong all elements with ferromagnetic properties, such as Fe, Co, Ni, and others (Figure 1.4).

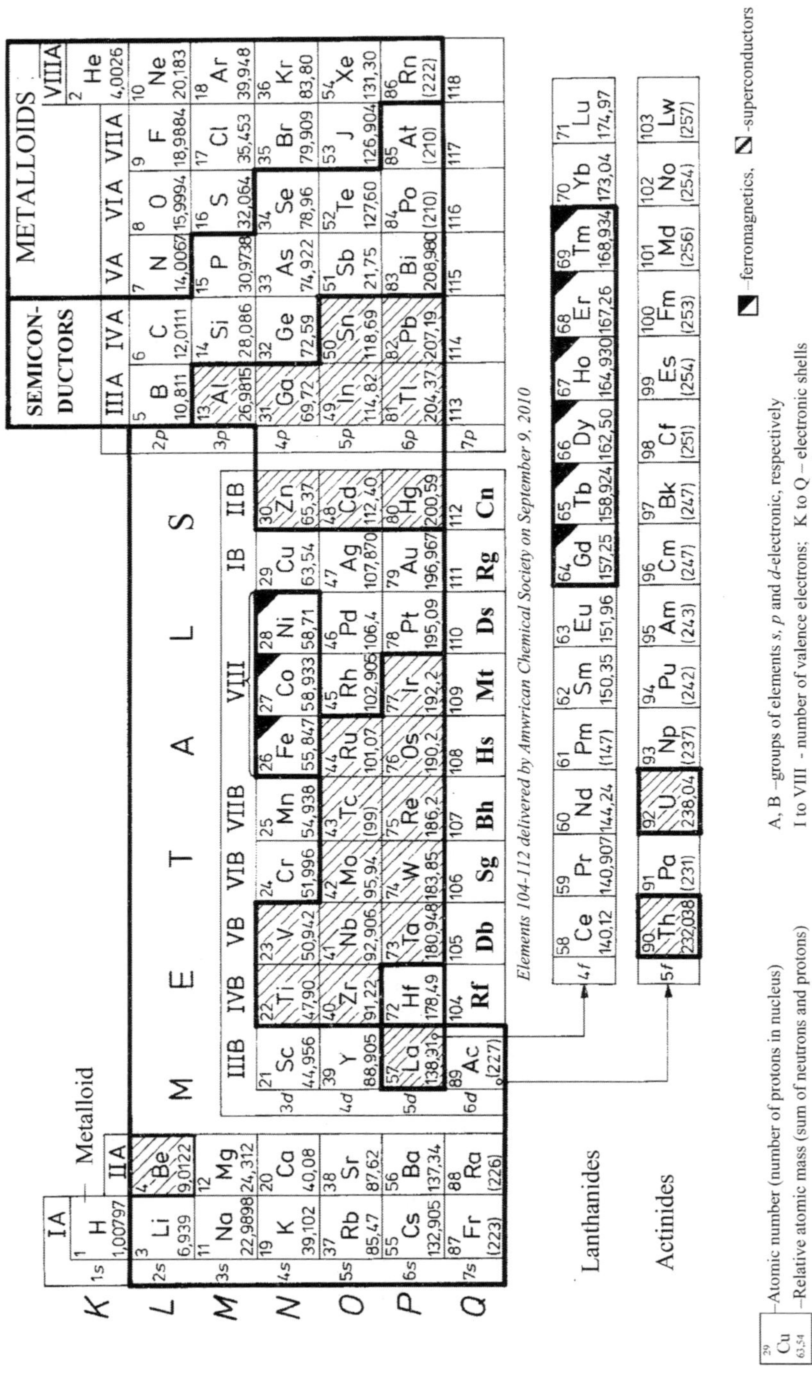

FIGURE 1.4 Periodic classification of the elements with division on electronic blocks *s*, *p*, *d*, *f* and metals, semiconductors, nonmetals.

The number of electrons in the extreme, outer shell of a given element predominantly decides the chemical properties of a metal. Therefore, these properties show periodicity moving from the atomic number $Z = 1–109$, corresponding to the number of electrons in the atom. The electrons placed at the outer main orbit (outer shell) are called *valence electrons.* In the theory of metals, these electrons are called *conduction electrons* (or *charge carriers*) and these electrons decide the *electric conductivity* of a body. According to the simplified P. Drude's model (1900) of electrical and thermal conductivity ([1.1], p. 23), "metal atoms that assemble into a solid body get rid of (lose) its valence electrons." These electrons can move freely within the metal creating the so-called *electron gas*, whereas ions of metal remain unchanged and immovable.

1.2.1.2 Ionization

In order to lift an electron to a higher energy level than what it occupies in the normal state of the atom, it is necessary to deliver to the electron an additional energy, such as a *photon*, with a certain defined portion of energy (e.g., of light), equal to 1 *energy quantum.* We then say that the atom has become *excited.* The easiest to excite are the valence electrons, because the nearest higher energy level is always open for them, and the distances between outer energy levels are smaller than between internal levels.

The excitation of an electron can lead to its full separation from the atom. It causes *ionization of the atom.* In such a case, a single-positive-charged ion is created. It is possible to create double- and triple-charged ions. In another case, when an atom "catches" an additional electron onto its orbit, a *negative ion* is created. The ionization potential and the excitation potential (*resonant potential*) are expressed in electron-volts (eV).

The energy of 1 eV *equals the energy that one electron gains while shifted between points of potential different by* 1 V. Examples of ionization energy values for various atoms are: 13.5 eV for hydrogen, 24.5 eV for helium, 13.6 eV for oxygen, 21.5 eV for neon, 10.4 eV for mercury, and 14.5 eV for nitrogen. A totally or partly ionized gas is called *plasma.* The degree of ionization of plasma changes with the change of temperature. Electric conductivity of plasma in the presence of a magnetic field is a tensor quantity (anisotropy).

1.2.1.3 Crystal Structure of Metals

Metals have crystal structure, that is, ions of metals are distributed in space in an ordered mode. Among 14 possible combinations of distribution of ions in the space, the most important in metals are three types of elementary grids presented in Figure 1.5.

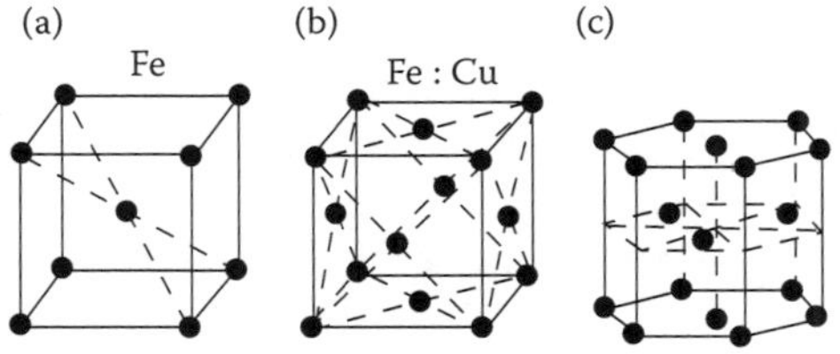

FIGURE 1.5 Types of crystallographic grids taking place most often in metals: (a) regular space-centered, (b) regular face-centered; (c) hexagonal grid with the densest packing in space.

The metals sodium, vanadium, chromium, niobium, and wolfram (tungsten) crystallize according to the *regular space-centered grid* (Figure 1.5a). The metals copper, silver, gold, nickel, and aluminum crystallize according to the *regular face-centered* crystallographic grid (Figure 1.5b).

Iron, however, can occur in two different crystallographic forms. At normal temperatures, it has the regular space-centered grid (Figure 1.5a), whereas at temperatures higher than 906°C, it adopts the regular face-centered crystallographic grid (Figure 1.5b).

About 30 elements, including α-cobalt, magnesium, neodymium, platinum, titanium, zinc, and zirconium, crystallize according to the *hexagonal grid structure* (Figure 1.5c) with the densest packing in space [1.1].

Properties of crystalline bodies depend essentially on the spacing of ions in crystal. The spacing of ions along one crystallographic axis can be different than the spacing along another axis. It results in different physical properties of the body in different directions, that is, *anisotropy of crystals.* Physical properties of metals in solid state depend mainly on their crystallographic structure. For instance, the metal mass density at any temperature can be calculated on the basis of knowing the structure and characteristics of the elementary space grid of the metal.

In reality, metals do not have an ideal crystallographic structure. Their grid is deformed by admixtures, unoccupied nodes, thermal motion of ions around the equilibrium position, and so on. As a result of these deviations, "the electric conductivity of metals is not infinite" ([1.1], p. 167).

1.2.1.4 Electrical Conductivity and Resistivity of Metals

The proximity of particular atoms in a solid body against each other, and especially in crystal, causes mutual penetration of electrons from one atom to another, creating considerable forces between interacting atoms and splitting of stable energy levels into a big number of intermediate energy levels, close to each other, which are permitted for the motion of electrons. This effect obviously appears strongest on the external surface. This mutual interaction of atoms (ions) in the case of metals is so large that for external valence electrons it creates a practically continuous zone of permitted energy levels adherent to each other, called energy *band.*

In metals, only a part of the allowed energy levels in the valence zone is occupied by electrons. The remaining higher energy levels, which also create a continuous zone (*energy band*), are free. Therefore, while it is necessary to supply defined discrete (quantum) portions of energy in single atoms for the excitation of a valence electron to a higher energy level, in metal, the continuity of permitted energy levels makes it possible to lift the valence electron to a higher free level by means of any arbitrary small amount of energy. Therefore, if in a sample of metal we produce an electric field, the interaction force of this field causes the excitation of electrons to higher free energy levels of this zone (called *conduction band*) as well as their motion toward the direction of higher potential of the imposed field. This motion of electrons is the *electric current*, and the described phenomenon is called *electron conductivity.*

Hence, any continuous zone of partially filled permitted energy levels is called *conduction band.* Electrons in fully filled energy levels (as in the *valence band*) cannot therefore in principle participate in the electron conductivity. This is due to the lack of sufficiently near, free energy level that could be occupied by electron after receiving an additional small portion of energy from the electric field.

Generally, we say that valence electrons in metal are in an "unbounded" state, creating a specific *electron gas* filling the space created by regularly disposed positive ions. This gas can move in ionic lattice under the influence of imposed electric field. Unbounded electrons, colliding with the ions of the lattice, recoil from them in an elastic way, but cannot leave the metal, because it is prevented by the difference of potentials on metal–vacuum (or metal–dielectric) boundary. This potential difference is related to performing a certain work called *electronic work function* ($L = eV$). A number of electrons, which grows with increasing temperature of metal, acquires sufficiently large energy required for an electron to leave the metal. This effect is called *electron thermoemission.*

Despite the fact that densities of electron gas are a thousand times higher than that of classical gases, in the Drude's model (1900) of electrical and thermal conductivity ([1.1], p. 23), the electron gas follows the normal kinetic theory of gases, with small modifications. According to this theory, the free electrons, moving with a constant speed under the influence of applied voltage, convey their energy by colliding with the crystal lattice and are subject to scattering. This scattering, appearing as a "*phonon*" *resistivity* (or *collision resistivity*), ρ_i(T), occurs on irregularities of lattice, caused mainly by thermal vibration of ions (Wyatt [1.22], p. 551). By analogy to the quanta of radiation field—*photons*, the quanta of field of displacement of vibrating ions are called *phonons* (Ashcroft [1.1], p. 540). Owing to the described collisions, the electron gas brings body to the status of thermal equilibrium with the ambience. It is because the hotter the body region (bigger ion vibrations) where collisions occur, the faster the electron will abandon it. This causes the strong dependence of the resistivity $\rho_i(T)$ on temperature (Figure 1.6).

In the temperature range occurring in electric machines and power equipment, exactly this type of *resistivity of electron–phonon interaction* dominates. In this range of temperatures, the resistivity is approximated in the known way, with a straight line:

$$\rho_i(T) = \rho_t = \rho_o\,(1 + \alpha t) \tag{1.1}$$

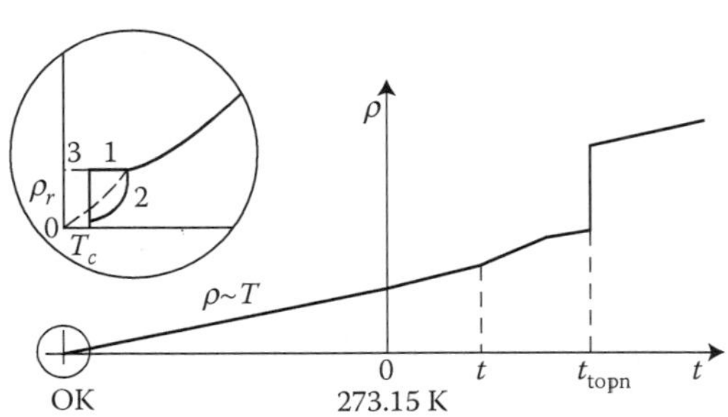

FIGURE 1.6 Typical characteristics of the metal resistivity ρ versus temperature: ρ_r—residual resistivity, T_C—absolute temperature of transition in superconductive status in K, t_{melt}—melting point; 1—superconductor, 2—pure metal, 3—normal metal. (Adapted from Smolinski S.: *Superconductivity.* (in Polish) Warsaw: WNT 1983.)

At significant reduction of the temperature T, the dispersion of phonons lessens—at the beginning linearly to about one-third of the Debye temperature (344 K for Cu), and next according to the fifth power of the temperature, up to the steady-state *residual value* ρ_r. In this way, the next components of resistance come to light, and as a result, in accordance with the Matthiessen's rule (1864), the *total resistance* ρ of metal can be expressed (Handbook [1.13], p. 26); ([1.22], p. 555]) as

$$\rho(T) = \rho_r + \rho_i(T) + \rho_H(T) + \rho_l(T) + \rho_{\text{eddy}}(T) \tag{1.2}$$

where ρ_r is the *residual resistance* in low (helium) temperature, to a large extent dependent on lattice defects and even on trace impurities whose influence at room temperature can be completely neglected.

The $\rho_i(T)$ component of the electron–phonon interaction, as per Bloch and Grüneisen (Smolinski [1.12], p. 27), is expressed as the *ideal resistivity*:

- At high temperatures

$$\rho_i = \frac{A}{MT_D}\left(\frac{T}{T_D}\right)^5 \tag{1.3}$$

- At very low temperatures

$$\rho_i = \frac{A}{MT_D}\left(\frac{T}{T_D}\right)^5 \int_0^{T/T_D} \frac{z^5}{(e^z - 1)(1 - e^z)}\, dz \approx \begin{cases} aT & (T > T_D) \\ bT^5 \left(T < \frac{T_D}{10}\right) \end{cases} \tag{1.4}$$

where T_D is the Debye temperature (426 K for Al, 1160 K for Be) [1.12], M is the atom mass of metals (26.97 for Al, 9.013 for Be), and A is the Bloch's coefficient.

Since aluminum and beryllium have large T_D and small M and A, only these metals, but not, for example, Cu (344 K; 63.54), should be applied as conductors for operation at low temperatures. Furthermore, they should be very pure metals.

$\rho_H(T)$—the so-called *magnetoresistance*, related to the Hall effect described below; it is strong in cadmium, weak in aluminum, and negligible in many others metals [1.12].

$\rho_l(T)$—resistivity related to the so-called *dimensional effect* that appears when dimensions of wire, foil, grains, and so on, are comparable with the mean free path of electrons (e.g., about 0.1 mm or more); in such a case, the resistivity increase is inversely proportional to the smallest dimension [1.12].

$\rho_{\text{eddy}}(T)$—a component resulting from the *skin effect* and eddy currents at alternating currents (ACs); in engineering, it is taken into account with the help of the coefficient k_{ad} of additional losses:

$$\rho_{\text{eddy}} = (k_{\text{ad}} - 1)\,\rho_i$$

In the design of a cryoelectric apparatus, where conductors made of high-purity metals and of thickness no more than 50 μm are applied, a compromise should be found between the "dimensional effect" and the "skin effect."

The flow of electric current of density $\boldsymbol{J}$ under the influence of an external electric field $\boldsymbol{E}$ is described by Ohm's law:

$$\boldsymbol{E} = \rho \boldsymbol{J} \quad \text{or} \quad \boldsymbol{J} = \sigma \boldsymbol{E} \tag{1.5}$$

where the coefficient of proportionality $\sigma = 1/\rho$ is called *conductivity*.*

If N is the number of free electrons of charge e, per unit volume of conductor, and v is the net velocity of electrons as an effect of operation of the electric field E and retarding collisions with atoms oscillating due to heat, the electric current flowing through the surface A will be

$$i = NevA \tag{1.6}$$

After the introduction of *mobility of electrons* $\mu_e = v/E$, we obtain from Ohm's law $\boldsymbol{E} = \boldsymbol{J}\,\rho$ the important formula for the *resistivity of metal*:

$$\rho = \frac{1}{Ne\mu_e} = \frac{m}{Ne^2\tau} = \frac{1}{\sigma} \tag{1.7}$$

where $m = 9.1095 \times 10^{-31}$ kg is the rest mass of electron; $\tau = m/(\rho N e^2)$ is the *relaxation time*, that is, the average time of free motion of electron between the succeeding collisions, or otherwise time constant, of convection of electron; and $v_u(t) = (|e|E)/m\,(\tau(1 - e^{-t/\tau}))$ (Kozlowski L. [1.7]). The τ values, in the temperature range from 273 to 373 K, vary in the limits: $\tau_{\text{Cu}} = (2.7 \ldots 1.9) \times 10^{-14}$ s and $\tau_{\text{Fe}} = (0.24 \ldots 0.14) \times 10^{-14}$ s.

It means that up to the voltage frequency of the order $f = 10^8$ MHz, the electron transit time can be neglected, as well as their velocity, which "at the strongest fields appearing in metals is on the order of 0.01 mm/s" ([1.7], p. 187).

1.2.1.5 Influence of Ingredients on Resistivity of Metals

Centers of dissipation of oscillations in the concentration $n_{\text{osc}} = rT$ and the corresponding conductivity σ_{osc}, plus defects (n_d, σ_d) and impurities (n_{imp}, σ_{imp}), reduce the mean free path of electrons, which causes an increase of metal resistivity, according to the formula ([1.7], p. 188)

$$\rho = \frac{1}{\sigma} = \frac{m^* v_F}{ne^2}(rT\sigma_{\text{osc}} + n_d\sigma_d + n_{\text{imp}}\sigma_{\text{imp}}) \tag{1.8}$$

where m^* is the effective mass of the electron, v_F is the Fermi velocity, and r is the coefficient of proportionality to absolute temperature T of lattice oscillation concentration.

* In new standards, the symbol of conductivity is σ, instead of the former γ.

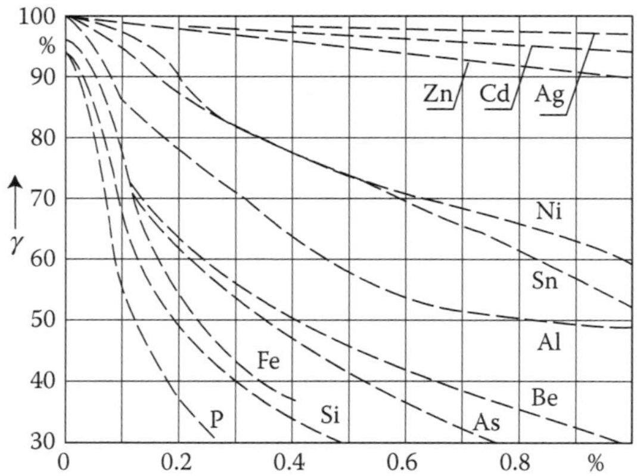

FIGURE 1.7 Dependence of copper (Cu) conductivity σ on weight contents of different admixtures in relation to the conductivity of pure (100%) copper. (Adapted from *Handbook of Electrical Materials*. (in Russian) Vol. 2, Moscow: Gosenergoizdat, 1960.)

The conductivity σ of materials used in electrical engineering is usually expressed in percents of conductivity of the *international standard* of annealed copper ([1.22], p. 562), IACS (International Annealed Copper Standard), of the value $\sigma = 58.824 \times 10^6$ S/m, at 20°C (100% IACS).* At present, there is available a copper with conductivity $\sigma = 103\%$ IACS. Pure silver has $\sigma = 106\%$ IACS and pure aluminum has $\sigma = 60\%$ IACS. The resistivity of conductors significantly depends on ingredients and impurities. For example, the resistivity of alloy Cu–Ni is $\rho \approx (1.5 + 1.35 \times \delta_{\mathrm{Ni}\%}) \times 10^{-8}\ \Omega$m at the nickel content $\delta_{\mathrm{Ni}\%}$ from 0 to 3.32% ([1.22], p. 556).

As per J. Linde (Ann. Phys. 1932), at admixture contents 1% dissolved in copper, silver, or gold, the resistivity increase is proportional to $(\Delta z)^2$, where Δz is the difference between the valence of the material dissolved and the solvent.

Pure metals show a regular crystalline structure and have a small resistivity. Plastic deformation and presence of admixtures, even in small amounts, cause the deformation of the crystal lattice and an increase in metal resistivity.

At *recrystallization* by annealing, the resistivity increased due to plastic processing can be reduced back to the initial value. In Figures 1.7 through 1.9, the influence of different admixtures on the conductivity of copper, aluminum, and iron is shown.

1.2.1.6 Resistivity at Higher Temperatures

At temperatures higher than room temperature, up to about 100°C, the resistivity of bronze, Cu, and Al maintains linear dependence (Figure 1.10), whereas the resistivity of steel increases rapidly.

At transition from the solid to the liquid state in most metals, a jump in resistivity occurs (Figure 1.6). The ratio of this increase is: for mercury, 3.2; for tin and zinc, 2.1; for copper, 2.07; for silver, 1.90; for aluminum, 1.64; and for sodium, 1.45 [1.13].

* More exactly, it is 1 m of Cu wire of 1 g mass and resistance 0.15328 Ω at 20°C.

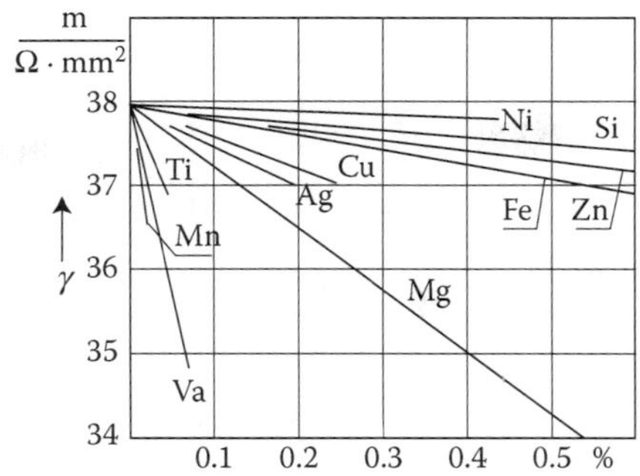

FIGURE 1.8 Dependence of the conductivity ($\gamma \equiv \sigma$) of annealed aluminum (Al) on contents of different admixtures. (Adapted from *Handbook of Electrical Materials*. (in Russian) Vol. 2, Moscow: Gosenergoizdat, 1960.)

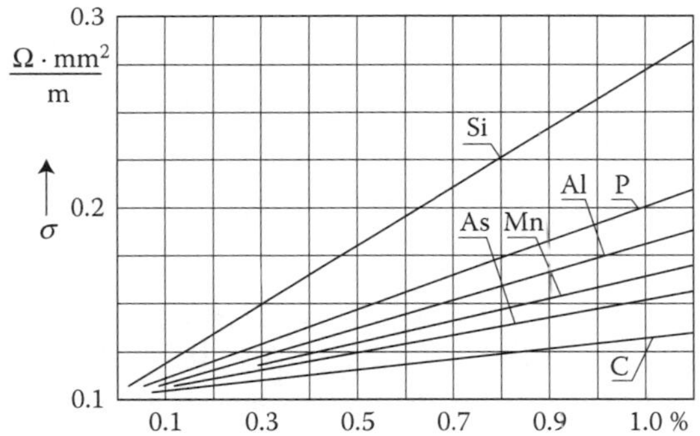

FIGURE 1.9 Dependence of resistivity of steel on contents of different admixtures. (Adapted from *Handbook of Electrical Materials*. (in Russian) Vol. 2, Moscow: Gosenergoizdat, 1960.)

1.2.1.7 Thermoelectricity

When two different metals come into contact, there appears between them a difference of potentials, which is caused by different values of *electron work functions* (of exit from metal) and also because the numbers of free electrons in different metals are not equal. Hence, the pressure of electronic "gas" in different metals may be different. This contact difference of potentials for different pairs of metals is in a range from a few tenth (fraction) of volt to several volts. If temperatures of contacts are equal, the sum of potential differences in a closed circuit, consisting of two different conductors, equals zero. If, however, one of the welds in this circuit has a higher temperature T_2 than the second one T_1, in the circuit, a resultant *thermoelectric force* is created (the Seebeck effect, 1821), equal to

$$E_u = \frac{k}{e}(T_1 - T_2)\ln\frac{n_A}{n_B} = C(T_1 - T_2) \tag{1.9}$$

where k is the Boltzmann constant, e is the charge of the electron, and n_A and n_B are the number of electrons per unit volume in metals A and B.

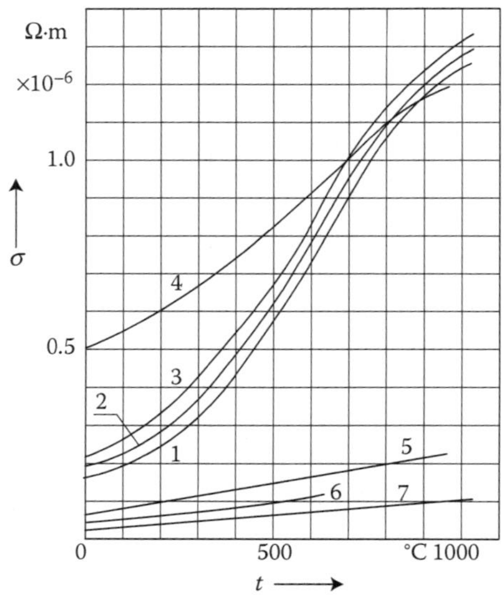

FIGURE 1.10 Resistivity ρ of metals at higher temperatures: 1—carbon steel 0.11%C; 2—carbon steel 0.5%C; 3—carbon steel 1%C; 4—stainless and acid-resistant steels; 5—brass 60%Cu; 6—aluminum; 7—copper. (From W. Liwinski, WNT 1968.)

This effect is used for the measurement of temperature with *thermoelements* (thermocouples). Keeping the temperature of one of the junctions at 0°C, we use the electromotive force E_u of the circuit to read the temperature of the second junction, after corresponding calibration.

For measurements of temperature in different ranges, one uses different pairs of metals [1.22]: from –200°C to 400°C copper–constantan (60%Cu + 40%Ni); from 0°C to 1000°C chromel (90%Ni + 10%Cr)–alumel (90%Ni + 5%Al); up to 1700°C platinium + platinium + 13%rodium; and in very low temperatures, for example, Au + 0.03%Fe (10^{-5} V/K) ([1.22], p. 560).

In measurement systems and calibration resistors, one tries to use metals with a thermoelectric force as small as possible with respect to copper, in order to prevent the introduction of an additional error. Such an alloy with very small thermoelectric force in relation to copper (about 1 μV/°C) is *manganin*, as opposed to *constantan* (4.05×10^4 μV/°C) used for thermocouples.

Thermoelectric forces E_u of different metals in relation to platinum are given in Table 1.1. In order to calculate the thermoelectric force of a circuit created by two metals, we subtract from each other two corresponding values from Table 1.1. For example, a copper–constantan thermoelement, with the weld temperature of 100°C and its ends at 0°C, produces $E_u = 0.75 - (-3.5) = 4.25$ mV, and the copper has positive polarity. Thermoelectric forces also appear in uniform conductors if a drop in temperature exists along their length. It means the gradient of the temperature is accompanied by the gradient of the electrical potential (the Thomson effect, 1854).

TABLE 1.1
Average Thermoelectric Force E_u of Metals with Respect to Platinum in the Temperature Range from 0°C to 100°C

Metal	E_u (mV/100°C)	Metal	E_u (mV/100°C)
Bismuth	−6.5	Aluminum	0.4
Constantan	−3.5	Manganin	0.6
Nickel	−1.5	Silver	0.7
Palladium	−0.5	Zinc	0.7
Mercury	0.0	Gold	0.7
Platinum	0.0	Copper	0.75
Tin	0.4	Brass	1.1
Lead	0.4	Iron	1.8

The phenomena described above are reversible, that is, an opposite to the Seebeck effect is the Peltier effect (1834): when a current is made to flow through a junction composed of materials A and B, heat is generated at one junction and absorbed at the other junction. The Peltier effect, due to its small efficiency, is used to cool only small volumes, utilizing semiconductor thermoelements, which demonstrate higher thermoelectric forces and lower thermal conductivity. The thermoelectricity of semiconductors is utilized, among others, in temperature sensors and other devices.

1.2.1.8 Thermal Properties

The high thermal conductivity of metal conductors is connected with their electrical conductivity. It is because heat transfer takes place mainly with the help of free conduction electrons, that is, the electron gas. Between Ohm's law and Fourier's equation of thermal conductivity exists a formal analogy.

According to the experimental Wiedemann-Franz law (1853), at a given temperature the thermal conductivity of metal λ is proportional to the electric conductivity σ, that is,

$$\frac{\lambda}{\sigma} = aT \tag{1.10}$$

where the constant a, called the Lorentz number, at 373 K equals $a_{Fe} = 2.88 \times 10^{-8}$ V²/K², $a_{Cu} = 2.29 \times 10^{-8}$ V²/K² [1.1] and is approximately constant for most of the metals, T is the absolute temperature (in K), λ is the thermal conductivity (in W/(K m)), and σ is the electrical conductivity (in S/m). Pure metals have higher thermal conductivity than alloys. Thermal conductivity of dielectrics is much lower because of the absence of free conductive electrons and the thermal energy in dielectrics is transferred only by elastic oscillations of ions. In metals, both types of thermal conductivity exist, but the first of them plays a decisive role.

TABLE 1.2
Important Electrical, Thermal, and Chemical Properties of Copper

Properties	Values	Remarks
Electrical conductivity of purest electrolytic copper at 20°C	59.77×10^6 S/m	See Figure 1.7 for the influence of admixtures
Electrical conductivity of standard, annealed copper at 20°C, after international standard and Polish Standard PN/E-4	58×10^6 S/m	Adopted as reference 100% for evaluation of conductivity of other metals
Temperature coefficient of resistivity at temperature 0–150°C, after international standard and Polish Standard PN/E-4	0.00393 1/K	For purest Cu, 0.0043 1/K
Electron work functions	4.07–2.61 eV	
Thermal conductivity λ at 20°C	385–394 W/(m K)	
Coefficient of thermal linear expansion at temperature 20–100°C	16.5×10^{-6} 1/K	Standard copper as per PN/E-4, 17×10^{-6} 1/K
Specific heat at temperature 20°C	385 J/(kg K)	
Melting point temperature	1083 ± 0.1°C	
Casting temperature	1150–1200°C	
Recrystallization temperature	200–300°C	
Annealing temperature	500–700°C	
In dry and humid air, in water, in hydrochloric or sulfuric acid of concentration less than 80%	Not oxidized and not active	
In dry air at temperature 100°C	Creates protective oxide layer	
In dry air at temperature ≥200°C	Oxidizes (color coating)	
In oxygen	Oxidizes at room temperature	
With sulfur at heating	Creates Cu_2S not protecting against corrosion	
With nitrogen	Creates compounds Cu_3N, CuN_3, Cu_6N_2 obtained with indirect methods	
In hydrochloric and sulfuric acid, in ammonia at heating, in nitrogen acid	Dilutes	

Source: Adapted from *Handbook of Electrical Materials*. (in Russian) Vol. 2, Moscow: Gosenergoizdat, 1960; Wesolowski K.: *Materials Science*. Vols. I, II, III. (in Polish) Warsaw: WNT, 1966.

The electrons moving under the influence of a temperature gradient generate at the ends of an open metal conductor the thermoelectromotive force (the Seebeck effect).

In constructional problems, an important role is played by the material's *coefficient of thermal expansion* (CTE), especially at cooperation of parts made of different metals, for instance, copper rods in iron slots of electric machines. In Table 1.2, some important electrical, thermal, and chemical properties of copper are collected.

1.2.1.9 Mechanical Properties

The mechanical properties of metals to a large extent depend on their crystallographic structure and temperature (Figure 1.11). Among the metals used in electrical construction, a special attention is paid to *copper*. Its mechanical properties significantly depend on thermal and plastic processing and on the contents of admixtures.

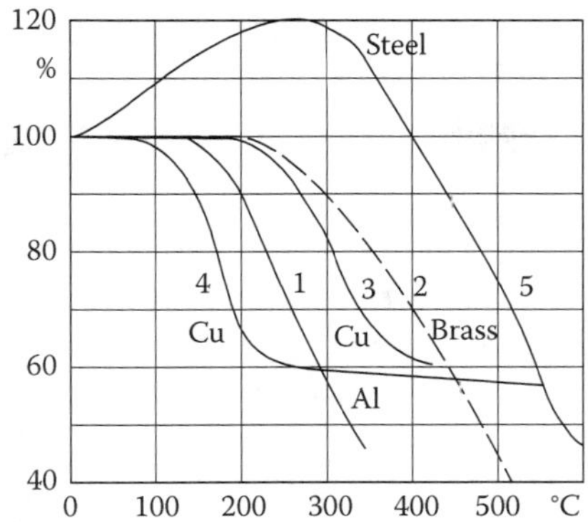

FIGURE 1.11 Dependence of mechanical strength of metals on temperature (*per Babikov*): 1—aluminum, 2—brass, 3—hard copper or at short-duration heating, 4—electrolytic copper or at long-lasting heating, 5—steel. (Adapted from Kozłowski L.: *Elements of Atomic Physics and Solid*. Cracow: AGH, 1972.)

In the case of copper, the relation of the extension force P versus elongation Δl does not show an evident border of plasticity P_{pl}, in contrast to the analogical graph for steel (Figure 1.12). Up to the limit of P_s, the elongation is exclusively elastic, and above that limit—it is elastic and plastic. The point P_H (limit of proportionality) is where the curve begins deviation from the straight line. The force P_r represents the limit of strain endurance to lengthening. *Modulus of elasticity* (Young's modulus) is the ratio of stress, below the limit of elasticity, to the relative elongation (strain), that is, the slope of the linear portion of the stress–strain curve.

Plastic processing of copper at cold temperatures (*squeeze*), appearing during draw-out, causes its hardening with a significant increase of endurance to split (up to 400–500 N/mm^2), and a small elongation at split (1–2%). Such a wire is strongly elastic at bending and is not suitable for winding.

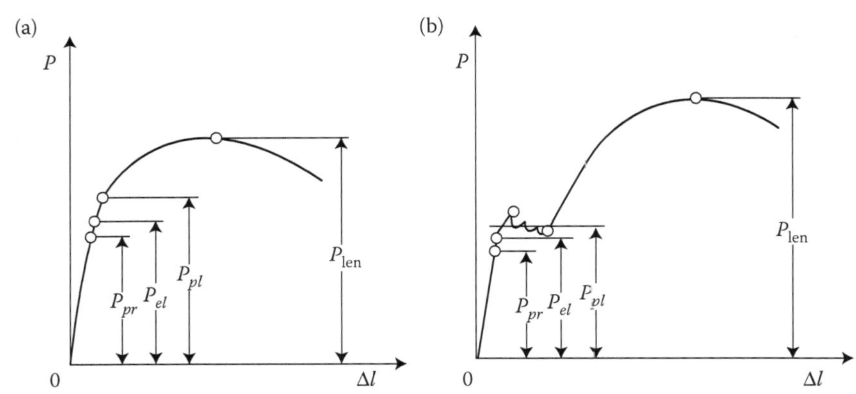

FIGURE 1.12 Tensile forces versus elongation of a sample for (a) copper and (b) steel.

Annealing, which consists of heating copper to a temperature of several hundred degrees Celsius (above the recrystallization temperature) and then rapid cooling, causes significant softening of copper. Along with the softening, a remarkable reduction of endurance to split takes place (up to about 200 N/mm^2) as well as an increase of plasticity. It also lessens the resistivity, by 2–3%.

The *recrystallization temperature* is the temperature at which the removal of internal stresses of plastic deformation of crystals appears, by increasing the size of some crystals at the cost of others. The recrystallization temperature varies in the range between 280°C and 400°C, for different sorts of copper and for different deformations, and is accompanied by a sudden fall of strain strength and hardness and at the same time an increase of elongation. In the case of pure copper, the initial temperature of recrystallization is about 180°C [1.23]. For most metals, the higher the deformation, the lower the recrystallization temperature. Any impurities and ingredients in copper increase its recrystallization temperature.

In turbine generators of older construction, rotor windings were made of soft electrolytic copper with the modulus of elasticity of 10^5 N/mm^2, limit of elasticity of 42 N/mm^2, and CTE of 17×10^{-6} K^{-1}. At such small strengths of copper, any faster starting or stopping of large turbogenerators was accompanied by a permanent deformation of conductors in the slots, caused by reciprocal interaction of thermal expansion and the friction of the conductors on slots. It resulted in damages of winding, insulation, or clampings. Application of copper along with the addition of 0.07–0.1% of silver, with cold press treatment, increased its limit of elasticity to 150 N/mm^2, which reduced the risk of such damages. Also, for windings of large transformers, sometimes a copper with silver additions is used, which, contrary to normal copper, does not lose its increased elasticity obtained by plastic treatment during exploitation. Due to similar reasons, for the rotor windings, sometimes a conducting aluminum alloy is used, for example, Cond-Al (Latek [1.35]) with a elasticity limit of 11 deca-newton (daN)/mm^2 and a thermal coefficient of expansion of 13.1×10^{-6} 1/K. In Table 1.3, selected important mechanical properties of copper are collected.

For commutator bars of electric machines, sometimes an abrasion-resistant, sufficiently hard copper is used, with a hardness of at least 75 daN/mm^2.

Because components of electric circuits and material for construction elements (clamping plates, consoles) of electric machines and transformers are under hazard from strong magnetic leakage fields, various bronzes are used occasionally. These bronzes contain, apart from copper, tin, beryllium, chromium, magnesium, zinc, cadmium, silicon, and other metals. At properly selected composition, these alloys can reach endurance to split in the order of 80–100 daN/mm^2, and even more. However, they have a higher resistivity than copper. For instance, cadmium bronzes with contents of 0.9% Cd have after broaching the conductivity of 83–90% of the conductivity of copper and tensile strength up to 73 daN/mm^2. They are used for manufacturing slipper conductors and corresponding contact elements (commutators) due to their high abrasion resistance. Beryllium bronzes (2.25% Be) reach a strength of 110 daN/mm^2 at a conductivity of 30% [1.23].

Aluminum, often used as a material replacing copper, is about 3.5 times lighter than copper, and its resistivity is 1.65 times higher than the resistivity of copper.

TABLE 1.3
Basic Mechanical Properties of Copper at Temperature 20°C

Properties	Unit	Casting	Plastic Processing Soft	Plastic Processing Hard	Dependence on Increase of Annealing Temperature
Limit of tensile strength	daN/mm²	18–22	20–28	25–50	Decreases
Limit of proportionality	daN/mm²	—	2.2–3	14–20	
Limit of plasticity at residual elongation 0.2%	daN/mm²		6–7	23–38	
Relative elongation before disruption	%	18	18–50	0.5–5	Grows
Relative narrowing	%	23	To 75	To 55	
Modulus of elasticity					
Static	daN/mm²	—	11,700	12,200–13,200	
Dynamic	daN/mm²	—	7400	11,200	
Brinell hardness	daN/mm²	40	35–37.5	65–120	Decreases
Limit of compressive strength	daN/mm²	157	—	—	
Settlement at compression	%	65			
Specific striking viscosity	daN/mm²	5.3	15.6		
Limit of shear strength	daN/mm²	—	19	43	
Limit of fatigue strength at torsion	daN/mm²	—	2.8	4.3	
Limit of fatigue strength at bending	daN/mm²	—	—	11	
Mass density at temperature 20°C	g/cm³	8–8.9	8.87–8.89	8.85	

Source: After *Handbook of Electrical Materials.* (in Russian) Vol. 2, Moscow: Gosenergoizdat, 1960.

Aluminum conductors of the same length and resistance as that of copper will be about 2 times lighter than the copper conductor. However, its diameter must be 1.28 times bigger than the diameter of the copper conductor of the same resistance. It is additionally difficult in the case of replacing copper conductors by aluminum conductors, when the space is limited.

In Table 1.4, selected important technical properties of aluminum are presented.

Pure aluminum has a remarkably (2–3 times) lower mechanical strength than copper, but its alloys with magnesium (Mg), silicon (Si), iron (Fe), and so on have much better mechanical properties. For instance, the alloy called *aldrey* (0.3–0.5% Mg; 0.4–0.7 Si, and 0.2–0.3% Fe) has the mass density and conductivity almost the same as aluminum, and the tensile strength (35 daN/mm²) close to the strength of copper. This material is used for the conductors of overhead power lines.

In the air, aluminum is always covered by a thin layer of oxide (about 0.001 mm) which protects the metal against further corrosion, but also makes it difficult to join aluminum conductors with each other. Welding aluminum elements or aluminum with copper requires a special technology. On a humid contact of Al–Cu appears an

TABLE 1.4
Selected Important Electric, Mechanical, Thermal, and Chemical Properties of Aluminum

Properties	Numerical Values	Comments
Conductivity at temperature 20°C of purest annealed aluminum (99.997% Al)	38×10^6 S/m	Influence of admixtures, Figure 1.8
Conductivity at temperature 20°C of aluminum conductors	$(33–35) \times 10^6$ S/m	Polish standards PN/E-103 and 106
Thermal coefficient of resistivity at temperature 0–150°C	0.004 1/K	
Mass density at temperature 20°C	2.70 g/cm^3	
Limit of tensile strength		
Soft Al	8–9 daN/mm^2	
Hard Al	15–17 daN/mm^2	
Limit of proportionality		
Soft Al	2.5 daN/mm^2	
Hard Al	7–8 daN/mm^2	
Yield point		
Soft Al	3.5–4.5 daN/mm^2	
Hard Al	10–13 daN/mm^2	
Coefficient of thermal conductivity at temperature 20°C	209 W/(m K)	
Temperature coefficient of linear expansion at temperature 20°C	24×10^{-6} 1/K	
Melting point	600–657°C	
Annealing temperature	350–400°C	
Water, water vapor, CO, CO_2, nitric acid	Does not affect aluminum at normal temperature	
In nitric and sulfur acids at heating	Easy dissolves	
In bases (even weak)	Dissolves	
On a humid contacts with copper	It creates cell that causes strong corrosion of aluminum	

Source: Adapted from *Handbook of Electrical Materials*. (in Russian) Vol. 2, Moscow: Gosenergoizdat, 1960.

electric cell with a current flow from aluminum to copper, which causes a considerable corrosion of the aluminum conductor.

1.2.1.10 Hall Effect and Magnetoresistivity of Metals

In order to calculate the resistivity of metal (1.7), one has to know the *density* of free electrons of valence and *mobility of electrons*. In one of the method of determination of these values, the Hall effect (1879) is used. If a current i flows in the x direction through a sample of metal placed in a uniform magnetic field of flux density B directed along the z axis (Figure 1.13), then the Lorentz force that acts on the electrons moving in the metal

$$F_m = -e\,(v \times B) \tag{1.11}$$

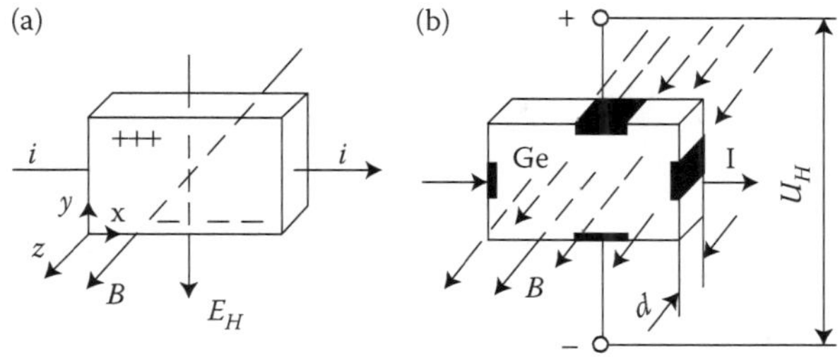

FIGURE 1.13 Illustration of the Hall effect: (a) vector relations; (b) scalar values.

directed along the y axis in the negative direction. This force presses the electrons to the lower side of the conductor (Figure 1.13a). This results in a nonuniform distribution of electric charges, which causes the appearance of a transverse electric field E_H (Hall field) directed along the negative direction of the y axis and opposing the gathering of electrons in the lower part of the conductor. This phenomenon is called the Hall effect.

The electromotive force $F = eU_H$ acting on electrons due to the Hall effect should, in equilibrium state, compensate the magnetic force (1.11), which means

$$eU_H = evB \tag{1.12}$$

Introducing, according to Equation 1.6, the current density $J = Nev$, we obtain

$$eU_H = \frac{JB}{N} \tag{1.13}$$

The value

$$R_H = -\frac{1}{Ne} = -\frac{U_H}{JB} \tag{1.14}$$

characterizing the body properties is called the Hall constant. Equation 1.14 allows the experimental determination of this constant, as well as the density of free electrons N in the conductor. Knowing the Hall constant and resistivity of the metal, it is possible then to determine, on the basis of Equation 1.7, the *electron mobility* μ_e. Considering the dimensions of the plate sample (Figure 1.13b), we can express Equation 1.14 in a more convenient form:

$$U_H = R_H \frac{IB}{d} \tag{1.15}$$

The resistivity of the metal in the direction of the current flow

$$\rho_H(H) = \frac{U_x}{J} \tag{1.16}$$

is called *magnetoresistivity*. This value, in magnetic fields of flux density B not higher than about 10 T, grows with an increase of the magnetic flux intensity H ([1.22], p. 564) according to the dependence

$$\frac{\Delta\rho}{\rho} = aH^2(a \approx 10^{-6}\mathrm{m}^2/\mathrm{A}^2) \tag{1.17}$$

and at stronger fields, this growth is directly proportional to H.

Since in most metals the density of free electrons N is on the order of 10^{29} electrons/m^3, the Hall constant R_H is not large. At room temperature, R_H equals, for example, 2.5×10^{-10} Vm^3/(A · Wb) for Na and 0.55×10^{-10} Vm^3/(A · Wb) for Cu (Wilkes [1.24], p. 272). In spite of that, the Hall effect in metals has been used for building direct current (DC) generators in the form of a copper disk rotating between magnet poles ([1.22], p. 563). Electrons in it are shifted to the edge of the rotating disk, which causes a creation of voltage between the axis and edge of the disk. This system is reversible and can also operate as a motor. The Hall effect is also utilized for building magnetohydrodynamic generators (Figure 2.6), or pumps of liquid metals, and other equipment. The Hall effect, even more distinctively than in metals, appears in semiconductors (Section 1.2.4).

1.2.2 Superconductivity

Superconductivity is referred to as a state in which a body sufficiently cooled down loses its resistivity, and an electric current—once excited—can flow in such conductors without losses, for instance, for several years. Research work on liquefaction of gases opened up the way to superconductivity. The first liquefaction of SO_2 gas (around 1780) was achieved by French scientists J. Clouet and G. Monge.*

However, the first liquefaction of permanent gases (air, oxygen, carbon monoxide, and nitrogen in static state, as well as hydrogen in hazy state) was achieved in 1883 by Poles K. Olszewski and Z. Wróblewski from Cracow Jagiellonian University. From that moment, rapid development of the physics of low temperature followed, the so-called *cryogenics*.

In 1898, Scotsman J. Dewar liquefied hydrogen again. In 1908, Dutchman H. Cammerlingh-Onnes liquefied helium, reaching temperature 4.2 K, and in 1911, he discovered the *superconductivity effect* (of mercury). It consists in it that in proximity of temperature of absolute zero (–273.16°C), the resistivity ρ of metal begins to dramatically decrease zero, proportionally to the fifth power of the absolute temperature ($\rho \sim T^{-5}$) of the body (Figure 1.6). The vanishing of resistance occurs almost suddenly in the range of temperature differences of 0.01 K. Many metals can transition into the superconductivity state at temperatures higher than absolute zero. Such metals are called *superconductors*. They include about 27 pure metals and over 2000 known compounds and alloys. They are usually inferior metallic conductors, in which the free conductance electrons are able to join into pairs (*Cooper*

* After "*General Encyclopedia*," Warsaw, 1976.

pairs—1957) as a result of electron–phonon interaction at certain temperatures (*transition temperature*). It happens in such a way that the resistivity caused by a collision of one electron with ions of crystal lattice is reduced exactly to zero due to rebound of a second counterpart electron, without loss of energy. Copper and ferromagnetic materials are not superconductors. Superconductors lose superconducting properties when the disruption of *Cooper* pairs occurs due to the surpassing of specific temperature T_c, called the *critical temperature*, or when the magnetic field intensity on the superconductor surface exceeds the *critical value* H_c (or $B_c = \mu_0 H_c$) defined for given metal and given temperature, or when the *critical current density* J_c is exceeded in a superconductor. This current density is closely connected with the critical field intensity H_c.

According to the Silsbee hypothesis from 1916 ([1.22], p. 566), for a conductor of radius r, the critical current is

$$I_c = \frac{r}{2} H_c \tag{1.18}$$

The critical field, per Ref. [1.22], is related to temperature by the dependence:

$$\frac{H}{H_r(0)} = 1 - \left(\frac{T}{T_c}\right)^2 \tag{1.19}$$

The transition temperature at $H_c = 0$ and at $J_c = 0$ is called the *critical temperature* T_c. Above the critical parameters, the electron pairs mentioned above disintegrate and the resistivity of the metal returns to its normal value (Figure 1.14).

Superconductors are divided into two main classes:

1. *Type I superconductors, ideal* or "soft" (pure metals—Figure 1.15a) with strong diamagnetic properties, caused by the fact that the superconductivity current can flow in them merely in a thin surface layer

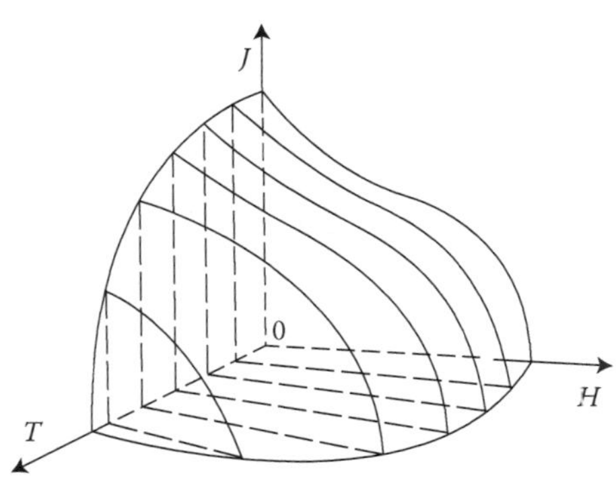

FIGURE 1.14 The critical surface *T–H–J* embracing superconductivity region. (Adapted from Smolinski S.: *Superconductivity*. (in Polish) Warsaw: WNT 1983.)

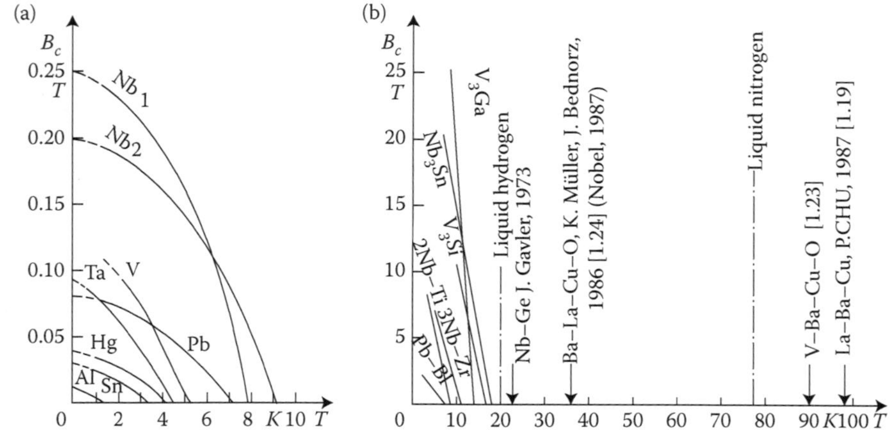

FIGURE 1.15 Dependence of critical flux density B_c on the absolute temperature T, after different sources: (a) ideal, "soft" superconductors; (b) nonideal, "hard" superconductors (Kunzler 1962) and ceramic superconductors (30–100 K).

2. *Type II superconductors, nonideal*, so-called "hard" (alloys and intermetallic compounds—Figure 1.15b) contain superconducting filaments distributed in the whole mass of metal—thanks to the permanent magnetic field and current that occur in the entire superconductor cross section. These filaments can, similarly to miniature superconducting rings, "catch" the field, causing irreversible *hysteresis* effects (Figure 1.16) in superconducting properties and two critical fields (Figure 1.16b). The hard superconductors do not have a clear transition border, like soft superconductors (curves B_C in Figure 1.15)

Type I superconductors (elements) behave in the superconducting state like ideal diamagnetics.

Type II superconductors (alloys and intermetallic compounds) at fields lower than the first *critical value* ($H < H_{c1}$) behave similarly to Type I. At $H_{c1} < H < H_{c2}$, the field successively penetrates into the superconductor and due to that, in one sample, there can be *superconducting zones* and *normal zones* with different sets of values T, H, J (mixed state—Figure 1.16b). After exceeding the *second critical value*, $H > H_{c2}$, the superconductor passes to the *normal state*. For explanation of the phenomena that occur in superconductors, the so-called *two-liquid model* of electric conductivity is used. In such a model, one liquid represents the normal electrons of conductivity, and the second liquid represents the created doublet electrons of superconductivity (*Cooper pairs*). As long as the electrons of superconductivity exist, the electric field in a metal does not exist and normal electrons do not participate in the conduction of the current.

An evidence of quantum character of superconductivity is presented by the Josephson's phenomenon (Nobel Prize, 1973) occurring on junctions SIS or SNS (S—superconductor, I—insulator, N—normal metal) in which through a thin barrier

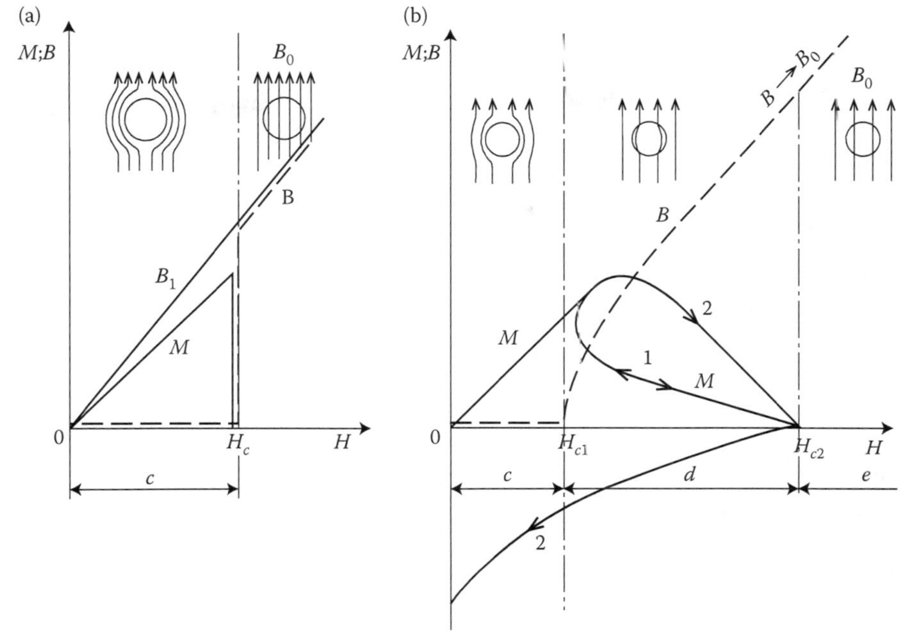

FIGURE 1.16 Magnetization M and flux density B in superconductors: (a) Type I—soft. (b) Type II—hard (Adapted from Ashcroft N.W. and Mermin N.D.: *Solid State Physics*. Brooks/Cole, 1976; Smolinski S.: *Superconductivity*. (in Polish). Warsaw: WNT, 1983; Wyatt O.H. and Dew-Hughes D.: *Metals, Ceramics & Polymers: An Introduction to the Structure and Properties of Engineering Materials*. Cambridge University Press, 1974.) c—expulsion of field, *Meissner* effect ($B = 0$); d—progressive penetration of field, mixed state; e—normal state ($H > H_{c2}$); 1—reversible superconductor; 2—nonreversible superconductor (Adapted from Wyatt O.H. and Dew-Hughes D.: *Metals, Ceramics & Polymers: An Introduction to the Structure and Properties of Engineering Materials*. Cambridge University Press, 1974.); B—magnetization characteristic of superconductor (B) and normal metal (B_1); with $\mu \approx$ const.

of thickness $d \approx 10^{-7}$ cm can penetrate both normal electrons and *Cooper* pairs, giving the superconducting state a strong nonlinearity $i = f(u)$. These junctions have found applications in metrology of low voltages (e.g., 10^{-14} V) and in microelectronics [1.12].

Although soft superconductors have been known for about 100 years, they have not found a broad application because of the low value of their critical magnetic field intensity (Figure 1.15a). Only the discovery and investigation of fibrous (hard) superconductors in 1960–1961, with the critical magnetic field intensity H_c exceeding 8 MA/m ($B_c = 10$ T) and critical current density up to $10^4 \ldots 10^5$ A/cm^2, caused more broad development of practical applications of superconductivity. It is discussed later, in Section 2.6.

Until lately, the highest observed temperatures and critical fields were reached by compounds of Nb_3Sn, V_3Ga, and NbAlGe (T_c about 20 K). They are, however, fragile and expensive. Therefore, the broadest application found so far has been the alloy of niobium with zirconium Nb–Zr25% easily undergoing plastic working.

TABLE 1.5
Features of Cryogenic Liquids

Properties	Nitrogen	Hydrogen	Helium	Neon
Chemical symbol	N_2	H_2	He	Ne
Molecular mass	28.2	2.02	4.00	20.18
Inflammability at contents in air in %	Incombustible	4–74	Incombustible	27.25
Mass of 1 L of liquid in gram	815	64.8	125	24.70
Boiling point in K at 1 atm ≈ 0.1 MPa	77.32	20.37	4.216	
Freezing point in K at 1 atm	63.15	13.96	—	
Cubical expansion at 1 atm in boiling point	697	777	682	
Latent heat of vaporization in cal/kg	47.7	107.5	5.72	

The main difficulty in the broad application of superconducting technology, apart from the low critical parameters of superconductors known until the 1960s–1980s, was the high cost of cryogenic equipment and especially liquefiers of helium, whose costs reach tens or even hundreds of thousands of US$. Moreover, a high input power required to evacuate 1 W of power from the region of demanded temperature (0.6–3 kW/W) and the high price of helium were additional hindrances.

Achieving sufficiently low temperatures in the range of liquid helium (Table 1.5) was therefore very expensive and technically cumbersome (low heat of evaporation, materials, leak of welds, gaskets, etc.) Approaching the temperatures of liquid hydrogen (Figure 1.15) in the 1970s promised significant cost reduction, but increased the risk of explosion.

Suddenly, in 1986, there appeared the long expected discovery of the so-called *high-temperature superconductors* (30–70 K), when the Polish-German scientist J. Bednorz [1.32] "advanced to the level of 30 K" on the basis of *ceramic superconductors*. In April 1987, Bednorz together with Swiss K. A. Müller observed superconducting transition in oxide compounds based on *rare earth* materials (Y, La) of the type Ba–La–Cu–O, at 35 K. It was recognized as the opening of *new era* in the field of superconductivity,* by awarding both research with the Nobel Prize in Physics for the year 1987.

Following this approach caused an avalanche development of new discoveries and popularization of the superconductivity research (also in the Technical University of Lodz—J. Turowski, J. Jackowski, and others).

In the Y–Ba–Cu–O system, superconductivity was determined at $T_c \approx 90$ K [1.32]. In February 1987, Paul Chu (Houston, USA) discovered superconductivity in the composite of La–Ba–Cu oxide at 98 K [1.26]. Research of 100 K has been conducted [1.32] and the discovery of superconductors at transition temperature 240 K and even higher have been anticipated [1.38], [1.12]. Recently, there has been more and more talk of a discovery of superconductivity at room temperature and higher

* Yet, until 1978, some authors wrote ([1.22], p. 573): "Nowadays, it seems very improbable increasing critical temperature over 25 K." It reminds the famous theoretical testimonies from the 1950s on "impossibility to break away from Earth to the outer space..."

in organic material. See http://www.physorg.com/news134828104.html.* Exceeding the temperature of liquid nitrogen (77 K) is in fact a real technical revolution, because this fluid is easily available and safe. A worse situation is with the critical current in which oxide superconductors based on yttrium Y and lanthanum La are very small in comparison with the NbTi or Nb_3Sn superconductors [1.38]. Therefore, some years will pass until new high-temperature superconductors will be widely introduced into industrial practice. But first, the 240 MVA grid autotransformer was designed (Sykulski [1.54]).

Presently, for building experimental synchronous generators, wires made of Nb–Ti are used as superconductors.

Casual or operational changes of temperature, field, or current can cause sudden, uncontrolled transitions of the superconductor into a normal state, what can create a hazard of explosion from the energy accumulated in the magnetic field and hence, the destruction of the whole device. That is why a fundamental problem of designing superconducting devices is the assurance of *stability of superconductors*. As per S. Smolinski ([1.12, p. 94]), a superconducting wire is internally stable when its thickness or diameter does not exceed

$$x \leq \sqrt{\frac{(3/\mu_0)C_{sc}\,J_c}{-(dJ_c/\mathrm{d}T)}} \tag{1.20}$$

where J_c is the critical current density (A/m²) and C_{sc} is the volume thermal capacity (J/(K m²)). For instance, in Nb–Ti, $x < 50$ μm.

Technically, the stability of conductors in superconducting windings (*cryogenic stability*) is assured by fusion in, rolling in, or splicing of the thread of superconductors into copper surrounding. This hazard is especially high at AC currents, due to the hysteresis losses in superconductor and eddy-current losses in the stabilization envelope. In a case when a step change of flux destroys superconductivity in some section, the conduction of current is taken over by a surrounding normal conductor (Cu) with a sufficiently high cross section. The Cu stabilizer is in turn seated in a material with much higher resistivity, such as Cu–Ni.

In order to limit magnetic couplings between particular fibers, the pitch of the twist of conductors must not be bigger than the permissible value ([1.12], p. 96).

In the rotor of the cryogenerator of Mitsubishi company (Ueda [1.44]), there were used, for instance, conductors in the form of a multicore cable (17 wires) with the dimensions of the conductors 2.1 × 9.3 mm, Cu/Cu–Si/Nb–Ti = 2/0.4/1, superconducting threads of diameter 14 μm with pitch of twist 15 mm and wires of diameter 1.12 mm and length of lay 70 mm. The critical current was 5000 A at the flux density 6.5 T [1.44]. See also Sykulski [1.54].

The later developed multicore conductors for AC 50/60 Hz superconducting coils (Tanaka [1.40]) have the diameter of 0.1 mm, twist pitch of 0.9 mm, threads of superconductors Nb–Ti of diameter ca. 0.5 μm, in a Cu–Ni envelope, the current 31.8 or 100 A, at the critical magnetic flux density 0.8 to 0.5 T (see Figures 1.14 and 1.15).

* Room temperature superconductivity: One step closer to the Holy Grail of physics. *Nature*, 9 July 2008.

S. Smolinski ([1.12], p. 97) divides technical superconducting conductors in two classes, from the point of view of their crystalline structure:

Class 1: Alloys of a regular space-centered structure (Figure 1.5a), like Nb–Zr and Nb–Ti, with good ductility and parameters: $T_c = 10$ K, $B_c(T = 4\text{ K}) = 10$ T; $J_c(B = 0) = (4 \text{ to } 6) \times 10^9$ A/m^2.

Class 2: Compounds of a special structure, for example, Nb_3, Sn, V, Ga, fragile, used mainly for spraying of strips made of Cu or Al, with parameters: $T_c = 18$ K, $B_c(T = 4\text{ K}) = 22$ T; $J_c(B = 0) = 5 \times 10^{10}$ A/m^2.

A broader description of the processing and technical data of superconducting conductors is given in the works of Smolinski [1.12] and Sykulski [1.54]. However, that latest discoveries may have significantly changed some of this information.

It is necessary to mention that besides superconductors there are also the so-called *cryoconductive conductors* ([1.12], p. 168), working over the transition temperature. For instance, a cable made of pure Al in temperatures of 20–30 K has a resistivity 250–100 times lower than resistivity of copper at room temperature. It can also be copper, pure beryllium in temperature of liquid nitrogen, and sodium (Na).

More recently, superconductors have found broader application in superconducting coils for the production of strong magnetic field [10.15]. However, the eddy-current concentrators of AC field (Bessho [1.29])* still compete with them to obtain 60 Hz flux densities even up to 16 T.

1.2.2.1 Superconductor Era in Electric Machine Industry

As it was mentioned, Poles K. Olszewski and Z. Wróblewski (1883), by liquefaction of permanent gases (air, oxygen, carbon monoxide, and nitrogen in static state as well as hydrogen in hazy state) initiated *cryogenics*. 25 years later, Dutchman H. Kammerlingh-Onnes liquefied helium and discovered superconductivity at a temperature of 4.2 K. It was too low and too expensive to be used in industrial technology. A new way was opened with the discovery of the so-called *high-temperature superconductors* by J. Bednorz and K. A. Müller (Nobel Prize, 1987). It was revolutionary for cryotechnology of liquid nitrogen (N_2) (77.3 K) and hydrogen (H_2). Both gases are very cheap and sufficient to start building modern superconducting machines and transformers.

For example:

According to S.P. Mecht i N. Avers from Waukesha Electric System and M.S. Walker of Intermagnetics General Corp., dimensions of an average distribution transformer of 30 MVA and 138/13.8 kV lessen at the cooling: from (1) conventional (49 ton and 22 800 L of oil), to (2) cryogenic (24 ton), to (3) liquid nitrogen in open cycle LN2 (16 ton).

"The cryo-cooled HTS transformer operates at closed cycle and all refrigeration is self-contained. The open cycle unit has an internal supply of liquid nitrogen refrigerant, which is automatically replenished, periodically, from remotely located liquefiers or storage vessels. Oil-filled transformers in urban location, which

* By the way, Professor K. Bessho (Japan) told the author (JT) during a conversation in Kanazawa in 1987 that his eddy-current concentrator was inspired by J. Turowski's logo of ISEF'87: .

superconducting transformers may replace, are often surrounded by sprinkler systems and oil containment structures" (IEEE Spectrum, July 1997, p. 43).

On the other hand, according to R.D. Blaucher from the U.S. National Renewable Energy Laboratory, power losses in two large 300 MVA generators amount to (1) over 5 MW in a conventional construction ($\eta = 98.6\%$), and (2) about 2 MW in a superconducting construction ($\eta = 99.4\%$).

"Conventional and low-temperature superconducting (LTS) generators having 100–600 MVA ratings have similar loss profiles but different efficiencies. A 300-MVA LTS generator's losses total 2 MW or so—almost the same as just the resistive losses with a conventional rotor. The total includes refrigeration power offset by reduced exciter losses. Assume that 50 W of heat is removed by a liquid helium refrigerator, requiring 50 kW for the compressor (room temperature power): refrigerator loss thus represents about 3 percent of the entire LTS machine losses and only 0.02 percent of its efficiency.

In effect, once the rotor is rid of resistive losses, as in this LTS application, the added refrigerating losses barely make a dent in total machine efficiency—at least, for generators with ratings in excess of 100 MW" (IEEE Spectrum, July 1997).

1.2.3 Magnetic Properties of Bodies (Ferromagnetism)

Advances in ferromagnetic materials technology, together with developments in the applied electromagnetic field theory and computation techniques, are the main sources of continuous progress observed in the design of electromechanical energy converters, since the mid-nineteenth century.

Ferromagnetic materials have been known from prehistoric times. As per the work of W. Gilbert (1600) on magnetism, the first theoretical models of elementary magnets, molecular current loops (orbital and spins), and magnetic dipoles are connected with the names of Coulomb (1736–1806), Kirwan (1733–1812), and Ampere (1775–1836), confirmed later by the electron theory (Thomson, 1897) and the domain theory (Weiss, 1907).

Magnetic properties of a body are related to the action of the magnetic field of intensity H on the body and to the internal features of atomic structure of the body, described by magnetization density, also called the density of magnetic moment (Jackson [2.10], p. 189) or magnetization (Ashcroft [1.1], p. 758):

$$\mathrm{M} = -\frac{1}{v}\frac{\partial F(H)}{\partial H} \tag{1.21}$$

where $F(H)$ is the free energy of a system placed in the field H.

1.2.3.1 Magnetic Polarization and Magnetization

Electrons in atoms, due to the rotation around the nucleus (orbital moment) and around its own axis (spin moment), act as currents flowing in closed circles and therefore produce a magnetic field. Such elementary circulating currents appear in all bodies. They can be substituted by equivalent dipoles with the magnetic moments

$$\boldsymbol{p}_m = \mu_0 i_0 \boldsymbol{s}_0 = m\boldsymbol{d} \tag{1.22}$$

where $\mu_0 = 4\pi \times 10^{-7}$ H/m is the *magnetic constant*, or the permeability of vacuum; i_0 is the elementary current of rotating electrons; $\boldsymbol{s}_0$ is the vector numerically equal to the surface encircled by this current, directed perpendicularly as per the rule of a right-handed screw; m is the magnetic flux going out from the pole, sometimes called fictitious magnetic mass; and $\boldsymbol{d}$ is the vector of distance between the poles, directed as $\boldsymbol{s}_0$.

Magnetic properties of bodies are shaped by the nature of these dipoles and their behavior in a magnetic field. Under the influence of a magnetic field, the dipoles existing in any material medium are more or less ordered. In this way, the body becomes magnetically polarized.

In order to describe the level of magnetization of a body, the vector notions of *magnetic polarization* $\boldsymbol{J}_i$ and *magnetization* $\boldsymbol{H}_i$ were introduced:

$$\boldsymbol{J}_i = \mu_0 \boldsymbol{H}_i \tag{1.23}$$

The magnetic polarization is the full *magnetic moment* of the volumetric unit of a body, which has N accordingly directed dipoles, that is

$$J_i = \frac{N \boldsymbol{p}_i}{V} \tag{1.24}$$

This is the value expressed in teslas, similar to the flux density $\boldsymbol{B} = \mu\boldsymbol{H}$.

Inside a magnetized body, the flux density from an external field ($\boldsymbol{B}_0 = \mu_0\boldsymbol{H}$) adds itself to the vector $\boldsymbol{J}_i$ from the elementary dipoles. Both values add to each other as vectors and yield the effective flux density

$$\boldsymbol{B} = \mu_0\boldsymbol{H} + \boldsymbol{J}_i = \mu_0(\boldsymbol{H} + \boldsymbol{H}_i) \tag{1.25}$$

1.2.3.2 Ferromagnetics, Paramagnetics, and Diamagnetics

In magnetically isotropic media, the magnetic polarization is proportional to the magnetic field intensity, that is

$$\boldsymbol{J}_i = \chi \mu_0 \boldsymbol{H} \tag{1.26}$$

The coefficient $\chi = (\partial \boldsymbol{H}_i / \partial \boldsymbol{H})$ is called *magnetic susceptibility*. It is the measure of changes in body magnetization under the influence of an external field. If we divide both sides of Equation 1.25 by $\boldsymbol{H}$, and consider the last expression, we shall obtain

$$\mu = \frac{\boldsymbol{B}}{\boldsymbol{H}} = \mu_0 + \frac{\boldsymbol{J}_i}{\boldsymbol{H}} = \mu_0 \mu_r \tag{1.27}$$

The coefficient $\mu_r = 1 + \chi$ is called *relative permeability*. Depending on this relationship, all materials can be divided into the following groups:

- Ferromagnetic ($\chi \gg 1$), like Fe, Co, Ni, Cd, and their alloys
- Paramagnetic ($0 < \chi < 1$), for example, steel over the Curie temperature, and antiferromagnetics (e.g., MnSe, MnTe)
- Diamagnetic ($\chi < 0$) and quasi-diamagnetic, for example, Cu, Al in AC field, or superconductors

From the technical point of view, ferromagnetics may be subdivided, on the basis of their hysteresis loops width (Figure 1.20b), into soft and hard magnetic materials.

Soft ferromagnetics have narrow hysteresis loops, with a small value of the coercive magnetic field intensity H_c (0.6–50 A/m) and rather high remnant magnetic flux density B_r (up to 1.7 T). Soft ferromagnetics are used for the laminated cores of electric machines and transformers with alternating flux, where small iron losses and high flux densities are desired.

Hard magnetic materials, on the contrary, have broad hysteresis loops, with H_c from about 8 to 200 kA/m and, for rare earth magnets, even more than 550 kA/m (Figure 1.22 [later in the chapter]). The latter materials are used for the production of permanent magnets of DC, synchronous, hysteresis, stepping, and other motors. Their energy per unit volume is high.

The excitation in electric machines is produced by electromagnets in large machines and by permanent magnets in small ones. This is due to the fact that the magnetic energy of permanent magnets is proportional to the cubic power (l^3) of the linear dimensions l, while the magnetic energy of electromagnets is proportional to the product of flux $\Phi = Bs$ and the magnetizing force $F_m = IN$, that is, to the fourth power of linear dimensions. At the same time, if it is considered that the volume of a permanent magnet in an electric machine is inversely proportional to its maximal energy $(BH)_{max}$, and that since the beginning of the twentieth century, the specific energy of permanent magnets has increased more than 30 times (as shown in Figure 1.22 [later in the chapter]), the impact on machine design and weight that has been achieved can be understood. In the same period, the per-unit power loss in laminated cores has been reduced almost 10 times, which again has improved the design parameters. It is expected that amorphous magnetic materials, high-temperature superconductivity, and silicon micromechanics will have further significant impact on modern motors and their performance.

The magnetic permeability μ of a body depends, therefore, on the number N of magnetic dipoles in a unit volume of the body and on the direction of the polarization vector $\boldsymbol{J}_i$ with regard to the vector $\boldsymbol{H}$ of magnetic field intensity.

In a general case of *anisotropic* medium, the permeability can be a *tensor* (Equation 2.82). Normally, however, the vectors are parallel or antiparallel to each other (i.e., in opposite directions).

Depending on the values and directions of the vectors, all bodies can be divided into the following groups:

- *Ferromagnetic* bodies, in which the second component in Equation 1.27 is positive and much bigger than μ_0. This group includes the bodies with clearly evident magnetic properties, that is, iron (Fe), nickel (Ni), cobalt (Co), gadolinium (Gd), and their alloys, with the permeability μ_r higher than 1.1.

- *Paramagnetic* bodies, in which the second component in Equation 1.27 is positive but much lower than μ_0. Most of these bodies have the relative permeability $\mu_r = 1.000–1.001$, not depending on the external field. They are, for instance, iron at a temperature higher than the Curie point, platinum family, sodium (Na), potassium (K), iron salts, oxygen (O), and others. In some materials, interatomic forces act in a way that magnetic moments of adjusted atoms are antiparallel against each other. This phenomenon is called *antiferromagnetism* and also shows hysteresis and *Curie* point. Since the permeability of *antiferromagnetics* is very small, they are considered as paramagnetic bodies (e.g., MnSe and MnTe).
- *Diamagnetic* bodies, in which the second component in Equation 1.27 is negative. Their relative permeability μ_r is therefore a little smaller than 1. Placed in a strong external field, they are repelled toward a weaker field. This is due to the fact that in the bodies the external field causes such a change in motion of electrons along orbits, which according to the *Lenz* rule will produce a field opposite to the excitation field. This effect exists also in ferromagnetics, but in them it is masked by a much stronger opposite action of magnetic moments of spins. This is why a body becomes diamagnetic– when the resulting magnetic moment of particle equals zero, which means that the outer electron states are full.

The differences in the above properties of particular bodies follow from different structures of atoms and particles, as well as from different crystallographic structure.

It should be mentioned that nonferromagnetic conductors placed in an alternating field also behave like diamagnetics and repel external field. It concerns especially the *superconducting* state, in which the body acts as an ideal diamagnetic. This effect is caused, of course, by the induced eddy currents and not due to the internal atomic structure of a body.

1.2.3.3 Atomic Structure of Ferromagnetics

Magnetization of ferromagnetics follows from the specifics of their atomic and crystallographic structure.

Let us consider an isolated atom of iron. Iron has the atomic number 26. It means that 26 electrons are present in the orbits of the iron atom. The same is the number of protons in its nucleus. Figure 1.17 shows a schematic distribution of these electrons in shells determined by the main quantum number n and in *subshells* determined by the *azimuthal quantum number l*. Magnetic properties of iron are determined in principle by the electrons in the subshells ($n = 3$, $l = 2$). This subshell has space for 10 electrons (see Figure 1.3), but in iron it contains only 6 electrons; it means that is filled in only partly. In spite of that, in the next shell, there are two electrons filling the subgroup of $n = 4$, $l = 0$ (Figure 1.3). In a solid body, these electrons are free and are located in the conductive zone. This is the required condition of participating electrons from a not-filled subgroup ($n = 3$, $l = 2$) in ferromagnetic phenomena.

Every electron has a spin moment of quantity of motion with regard to its axis. It is accompanied by the *magnetic moment of electron* (in $\text{Wb} \cdot \text{m}$), defined by the formula (Kozłowski [1.7], p. 230)

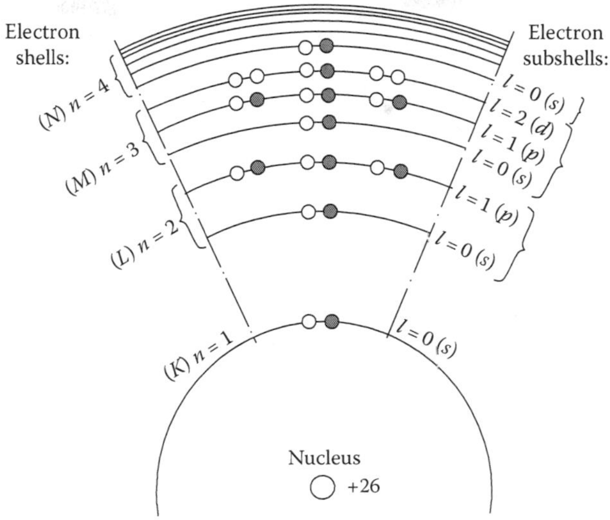

FIGURE 1.17 Distribution of 26 electrons of the atom of iron on the permitted energy levels: the filled/black circles (–) and the white circles (+) correspond to the two possible directions of the spin; *(K)* $n = 1$, $l = 0$ (s) is the lowest energy level.

$$p_s = \frac{\mu_0 eh}{4\pi m} \tag{1.28}$$

where e is the charge of the electron, μ_0 is the magnetic constant, m is the mass of the electron, and h is the Planck constant.

The value p_s/μ_0 is called Bohr magneton. The total spin magnetic moment of an atom is the sum of spin moments of particular electrons.

These moments, similar to the spins themselves, can be both parallel and antiparallel. In many bodies, the number of positive spins equals the number of negative spins. In the case of an atom of iron, the spins of particular electrons in all filled shells also compensate each other, but in the unfilled shell $n = 3$, $l = 2$, there exist four uncompensated spins, thanks to which the atom as a whole has the resultant spin magnetic moment equal to the four Bohr magnetons (Figure 1.17). Atoms can also have an uncompensated *orbital magnetic moment.* However, it is much lower than the spin moment and its participation in the total spin magnetic moment of atom in the case of solid bodies is very low.

In a solid ferromagnetic body, adjacent atoms approach each other and as result of their interaction, they change the above-described motion of electrons in particular atoms or in gases of ferromagnetic metals. At crystallization, in the resulting action of interatomic forces of exchange, of electrostatic nature, splitting and mutual overlaps of external energy levels take place. Owing to that, the external electrons are no more tied to a specific atom, but have a tendency to pass from one atom to another, as well as from the last level ($n = 4$, $l = 0$) to the one before the last ($n = 3$, $l = 2$), and vice versa. The resulting effect of these motions is a partial compensation

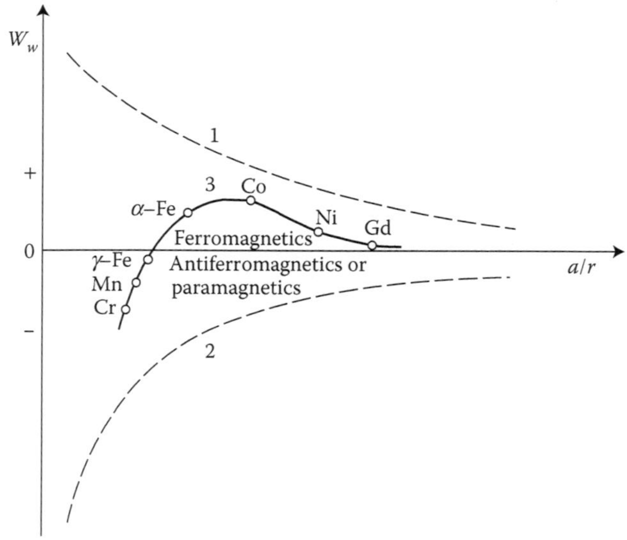

FIGURE 1.18 Dependence of the energy of exchange W_w on the relative distance between atoms. 1—The force of electrostatic attraction electron–nucleus; 2—the repulsion force as result of Pauli exclusion principle; 3—the resultant force of exchange interaction.

of nonbalanced spins, which leads to the reduction of the time-averaged magnetic moment of an atom. Consequently, the resultant spin magnetic moment is reduced at crystallization, in the case of iron from 4 to 2.22, and in the case of nickel—from 2 to 0.6 of Bohr magnetons.

According to Heisenberg quantum theory (1928), expanded by Bethe and others, the described interaction between the dipoles is an effect of the so-called *energy of exchange* W_w, which is a function of the distance (a) *between atoms* relative to the radius (r) of the unfilled shell 3d (4f for Gd); see Figure 1.18.

The *exchange forces* put dipoles in order. At big relative distances a/r, the exchange forces are weak and the material is paramagnetic. At decreasing a/r, the exchange forces are growing and cause parallel positioning of adjacent dipoles, which is a characteristic for ferromagnetics. At further reduction of a/r, the interaction becomes negative and causes the creation of antiferromagnetism or *ferrimagnetism*,* which also has antiparallel adjacent spins, but not with the same magnetic moments, which causes some unbalanced magnetization. The introduction of alloy ingredients to pure manganese (Mn), which is antiferromagnetic, increases the distance a between atoms and such alloys, for instance, MnBi or MnCuAl [1.22], can become ferromagnetic.

1.2.3.4 Zones of Spontaneous Magnetization

According to Weiss domains theory (1907), ferromagnetic materials consist of a large number of microzones of spontaneous magnetization (the so-called *domains*) that contain a significant number of atoms. Their magnetic moments (spins) are

* The name is originated from ferrites, in which this effect was first discovered.

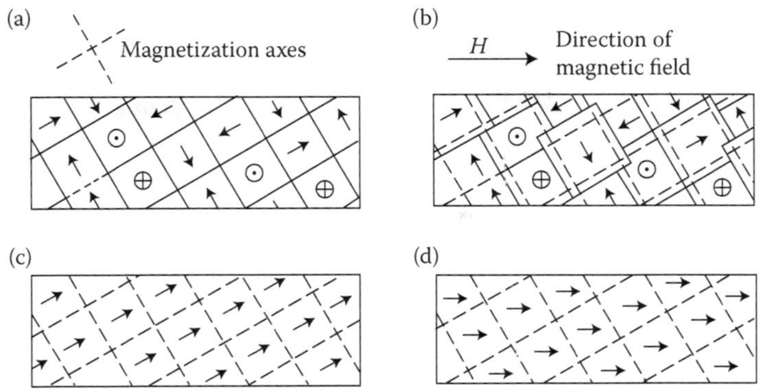

FIGURE 1.19 Scheme of change of domain structure of iron at growing magnetic field (After Bozorth R.M.: *Ferromagnetism*. New York: Van Nostrand, 1951.). (a) Demagnetized sample; (b) partial magnetization at the cost of reversible shift of borders; (c) magnetization vectors of all domains are directed identical, in result of an irreversible shifting of borders; (d) full saturation of the sample (rotation of vectors in strong fields).

oriented in parallel. In this way, these zones even in the absence of an external field are always magnetized to the saturation. Magnetic moment of domain is defined by the magnitude and direction of magnetization and the volume of the domain itself. The direction of magnetization depends on the crystallographic structure of the body. In the absence of an external field and external mechanical stresses, the vectors of magnetic polarization $\boldsymbol{J}_i$ of particular domains are directed in one of the six directions of easiest magnetizations. These directions correspond to the edges of a cube of elementary crystal lattice of iron (Figure 1.5a). In an unmagnetized sample of ferromagnetic, the vectors of magnetization of domains are directed in a chaotic, disordered way (Figure 1.19a), so that the resultant magnetic moment of the sample equals zero. Applying an external field causes, in the final phase, a rotation of the magnetic moments of every atom around their axis (Figure 1.19c and d). Inside a domain, the magnetic moments of atoms, remaining parallel to each other, are oriented in a direction closer to the direction of external field.

Dimensions of domains are big enough (from 0.1 to 0.001 mm) so that by using special methods (powder figures), one can observe them with the help of a normal optical microscope of 200× magnification. Domains can be smaller (or rarely bigger) than particular crystals within which they create something like closed magnetic circuits. Bloch (1932) proved that boundaries between particular domains are not sharp, but they occupy a zone of dimensions of the order of 100 radii of an atom, in which a successive rotation of electron spins at passage from atom to atom takes place. These zones are called *boundary layers* or Bloch walls. The magnetic moment of a domain can also change itself with no change of direction from its magnetization, but only by a shifting of the Bloch wall, causing an increase of volume of the domain (Figure 1.19b).

1.2.3.5 Form of the Magnetization Curve

The initial magnetization curve (1 in Figure 1.20a) represents the dependence of the flux density B on the magnetic field intensity H in a sample, which in the initial state

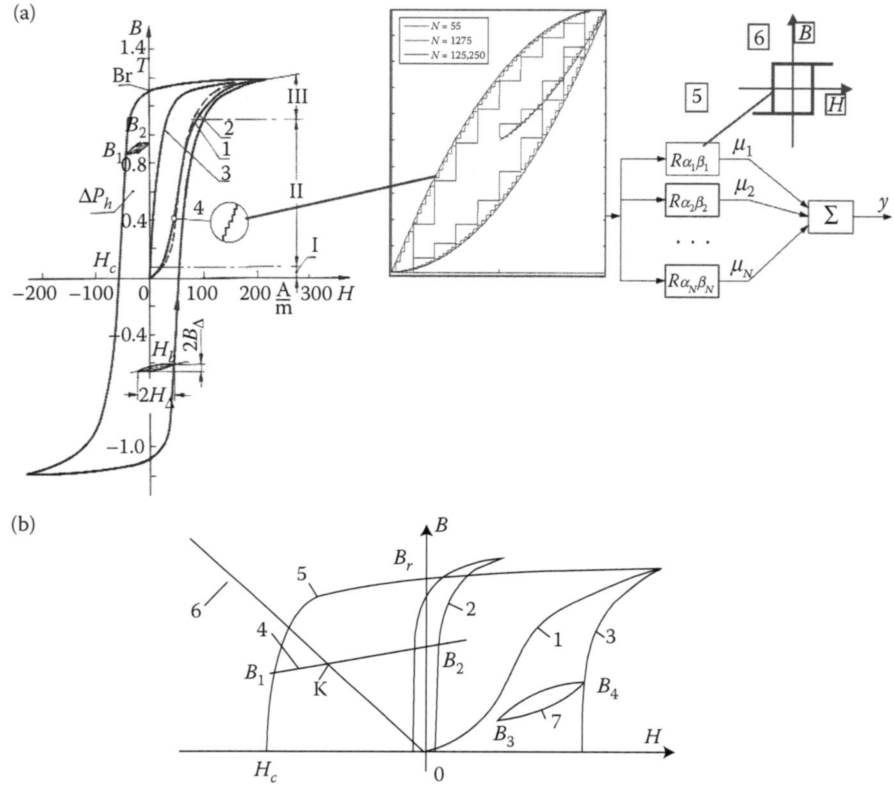

FIGURE 1.20 (a) Types of magnetization curves of annealed technical iron (After Bozorth R.M.: *Ferromagnetism*. New York: Van Nostrand, 1951.), and their characteristic points: 1—initial curve; 2—arithmetic mean of ordinates of hysteresis; 3—ideal curve; ±Br—residual flux density (remanence); 4—Barkhausen's effect (magnification of order 109 times); H_Δ—partial cycle; 5—discrete Preisach's mathematical model of elementary rectangular domains connected in parallel; 6—with accuracy depending on assumed number N of small parallel rectangular elementary domains (hysterons) (after http://en.wikipedia.org/wiki/Preisach_model_of_hysteresis). (b) Typical hysteresis loops: 1—initial line, 2—soft material, 3—hard material, 4—recoil line, 5—demagnetization line, 6—permeance line of external magnetic circuit, 7—partial cycle.

was not magnetized. The magnetization curve can be divided into three main sections (Figure 1.20a).

Section I—initial, where the curve goes out from the origin of coordinates at the angle defined by the initial permeability ($\mathrm{d}B/\mathrm{d}H = \mu_{\text{in}}$). In this section, the curve is concave toward the bottom and is subject to *Rayleigh's law* (per [1.2]):

$$B = \mu_{\text{in}} H + \upsilon H^2 \tag{1.29}$$

where $\upsilon = \mathrm{d}\mu/\mathrm{d}H$ is a constant value.

Changes of flux density in this section of the curve are, in principle, *reversible*. It means that at diminution of the magnetic field intensity, the flux density returns practically to its previous value. In section I, a reversible shifting of borders takes place, causing an increase of these domains whose direction of magnetization is close to the direction of the external field (Figure 1.19b).

Section II of the magnetization curve (Figure 1.20a) proceeds as the steepest. In the scope of this section, irreversible changes of flux density take place, which are accompanied with the irreversible shifting of borders of the spontaneous magnetization (Figure 1.19c) and, next, the irreversible jumping rotation of magnetization vectors toward the direction of easy magnetization, closer to the direction of the applied field. This process of magnetization, progressing in a jump-like manner (4 in Figure 1.20a), is called the Barkhausen effect. The permeability dB/dH in this section is the highest.

Section III, corresponding to saturation, has the smallest inclination and the permeability dB/dH of the sample. At infinite growth of the magnetic field H, the curve tends to the permeability μ_0. In a significant part of this section, the changes of flux density are reversible. It corresponds to reversible rotation of magnetization vector of the sample from the state shown in Figure 1.19c to the state of full saturation (Figure 1.19d). The polarization of the sample, $J_i = B - \mu_0 H$, at growth of the external field H is approaching to a certain constant value J_{is} (saturation). The flux density B instead, according to formula (1.25), increases further to infinity together with an increase of the field intensity H, according to the linear dependence ($B = J_{is} + \mu_0 H$). The slope of this straight line is small after all, in comparison to the course of magnetization curve of iron. Therefore, we can say here also about a constant *flux density of saturation* $B_s \approx J_{is}$. It equals about 2.16 T for pure and slightly siliconized iron, ca. 1.98 T for hot-rolled and 2.02 T for cold-rolled transformer sheets, 2 T for cast steel, and 1.5 T for cast iron (Tables 1.6 and 1.7).

By diminishing the external field, the vectors of magnetic polarization return to the nearest direction of easy magnetization, which causes the residual flux density (remanence) B_r.

The rotation of a magnetic polarization vector in phase III requires overcoming the *energy of magnetic anisotropy* in order to rotate magnetic moments from the direction of easiest magnetization toward the direction of more difficult magnetization. This energy is much higher than the energy necessary for irreversible shifting of Bloch walls. In *soft magnetic materials* (narrow hysteresis loop), the motion of Bloch walls in phase II goes without greater disturbances. Thanks to that, the rotations of magnetic polarization vectors in phase III are so small that they do not pass the "difficult" direction and due to that they are reversible.

In *hard magnetic materials*, the motions of Bloch walls are more difficult and the magnetic polarization vectors in the course of doing significant rotations are passing the direction of difficult magnetization, and for their return to the previous state an additional energy is needed, which increases the hysteresis loop.

1.2.3.6 Hysteresis

The irreversible processes occurring in section II of the magnetization curve are the cause of the *hysteresis loop*.

TABLE 1.6

Conductivity σ, Per-Unit Hysteresis Losses Δp_{h1} at 50 Hz and at the Flux Density B_m, Coefficient η in *Steinmetz* Formula (1.39), Metal Mass Density ρ_m, and Saturation Flux Density $B_{sat.}$ of Different Types of Steel

Material	Percentage Chemical Constitution in % and Working	$\sigma_{20°C}$ (10^6 S/m)	B_m (T)	Δp_{h1} (W/kg)	η ($\times 10^{-3}$)	ρ_m (10^3 kg/m^3)	$B_{sat.}$ (*T*)
Amorphous strips	METGLASS alloy	0.769	1.4	0.18		7.18	1.56
Nonmagnetic cast iron	“Metallic glass” 30–50 μm	0.98					
$\mu_{r\,max}$ =	3C; 2.8 Si;						
	For yoke beams (H. Kerr, 1964)						
Silicon iron	4 Si, hot-rolled transformer sheets	1.6–1.7	1.0	1.06		7.60	1.97
Silicon iron	4 Si, 96 Fe, annealed 800°C; after hot-rolling	1.67	1.0	0.86	17.2	7.60	1.97
Cast iron	3.5 C;						
$\mu_{r\,max} = 1400$	For yoke beams (H. Kerr, 1964)	1.82	—	—	—	—	—
Molybdenum permalloy	4 Mo, 79 Ni, 17 Fe annealed at temperature						
	1350°C in hydrogen atmosphere	1.82	0.5	0.005	—	8.58	0.87
Silicon iron [1.30]	Anisotropic transformer sheets (*Bochnia*)						
Silicon iron [1.30]	0.28 mm ET 3	2.096	1.5	0.84	—	7.65	—
Silicon iron	0.30 mm ET 4	2.096	1.5	0.89	—	7.65	—
Silicon iron	0.30 mm ET 5	2.096	1.5	0.93	—	7.65	—
	0.30 mm ET 6	2.096	1.5	1.0	—	7.65	—
	Isotropic generator sheets	—	1.5	4.26	—	—	—
	mixed samples EP 20	—	1.5	4.52	—	—	—
	EP 23	—	1.5	5.40	—	—	—
	0.5 mm (*Bochnia*) EP 26	2.2–2.9	1.0	1.68	—	7.65	—
	2.5 Si; hot-rolled transformer sheets	2.5	1.0	0.2	3.9	7.67	2.00
	3 Si, 97 Fe, annealed 1200°C, after cold-rolling						

Gray cast iron	0.5 C; (1.25–3.8) Si	3.3	—	—	—	7.2	—
Silicon cast iron	1 Si, hot-rolled transformer steel	3.6–4.2	1.0	1.9	–	7.80	—
Constructional steel	0.27 C; 0.60 Mn; 0.25 Si; 0.044 P,						
	0.024 S; tubes 107/114 mm; $\mu_{r\,max} = 570$	4.0	1.0	9.97	—	7.80	—
Constructional steel	0.35 C; 0.53 Mn; 0.19 Si; 0.018 P;	4.6	1.0	10.25	—	7.76	—
	0.025 S; tubes 95/99 mm; $\mu_{r\,max} = 660$	5.6	1.0	4.46	—	7.75	—
Constructional steel	0.08 C; 0.46 Mn; traces of Si; 0.02 P;	6.7	1.0	3.31	—	7.81	—
	0.021 S; tubes 110/111 mm; $\mu_{r\,max} = 2420$						
Constructional steel	0.06 C; 0.36 Mn; traces of Si; 0.033 P;						
	0.050 S; tubes 76.5/82 mm; $\mu_{r\,max} = 2000$						
Cast steel	0.3 C; 0.3 Si; 0.5 Mn;	6.7	—	—	—	7.8	—
Constructional steel	0.06 C; 0.40 Mn; traces of Si; 0.008 P;						
	0.025 S; tubes 73/75 mm; $\mu_{r\,max} = 1330$	7.3	1.0	4.62	—	7.50	—
Constructional steel	0.05 C; 0.40 Mn; traces of Si; 0.079 P;	7.6	1.0	4.48	—	7.74	—
	0.036 S; tubes 44/48 mm; $\mu_{r\,max} = 1480$						
Soft cast iron	0.2 C; annealed at 950°C	10	–	–	–	7.8	—
Iron	99.9 Fe; cold-rolled; cold work 50%	10	1.0	6.35	127	7.88	—
	99.9 Fe; annealed at 900°C;	10	1.0	1.9	38	7.88	2.25
	99.9 Fe; annealed at 1200°C in H_2	10	1.4	0.12	1.4	7.88	2.15

Source: Adapted from Bozorth R.M.: *Ferromagnetism*. New York: Van Nostrand, 1951; Jezierski E.: *Transformers. Theory*. WNT Warsaw, 1975 (see also [5.2]); Jezierski E. et al., *Transformers. Construction and Design*. (in Polish) Warsaw: WNT 1963; Neiman L. R. and Zaicev I. A.: Experimental verification of the surface effect in tubular steel bars. (in Russian) *Elektrichestvo*, 2, 1950, 3–8, and others.

TABLE 1.7
Magnetic Materials Used in Electronics and Power Electronics

Name	Composition (Rest Fe) %	Mass Density (mg/m³)	H_C (A/m)	B_r (T)	$B_{sat.}$ (T)	μ_{init} (×1000)	μ_m (×1000)	Application
"Armco" iron	0.025 C	7.88	70	1.3	2.17	0.25	4	Magnetic amplifiers 10^{-2}–10^6 W
Silicon iron	0.035 O_2	7.65	10–45	1.4	2.0	0.2–1	3–8	EFA 0.8–6.15 kW
	2–5 Si							
Permalloys with rectangular	50 Ni		24	1.4	1.5	—	—	Magnetic amplifiers 10^{-3}–10^3 W
hysteresis loop			24	1.4	1.5	—	—	
Daltamax			24	1.45	1.5	—	—	
Orthonol								
Hypernik								
Permalloys with narrow								
hysteresis loop								
Molybdenum permalloy	4 Mo; 79 Ni	8.58	2.4	0.52	0.68	—	—	Magnetic amplifiers 10^{-13}–10 W
Supermalloy	5 Mo; 79 Ni	—	2.4	0.52	0.68	—	—	Magnetic amplifiers 10^{-13}–10 W EFA 10–300 W
Supermalloy	5 Mo; 79 Ni	8.76	0.6	0.42	0.8	80	900	—
Permiwar 45	25 Co, 45 Ni	8.6	96	—	1.55	0.4	2	—
Izoperm 50	50 Ni	8.25	480	—	1.60	011	0.12	—

Ferrite with	—	—	8–20	—	0.2–0.43	—	—	Magnetic amplifiers
Narrow hysteresis loops	—	—	32–100	0.22–0.26	0.24–0.28	—	—	Magnetic amplifiers
Rectangular hysteresis loops								
Ferrite Ni–Zn	—	4	2–6	—	0.25	0.01–2	0.03–8	—
Ferrite Mn–Zn	—	5	4–30	—	0.4	0.7–6	2–12	—
Amorphous strips METGLASS 2605 S-2	$Fe_{78}B_{13}Si_9$ annealed	7.18	2.4	1.30	1.56	—	—	Transformers and reactors of low frequency (50–60 Hz)
Amorphous strips METGLASS 2605 SC with rectangular hysteresis loops	$Fe_{81,5}B_{13,5}Si_{3,5}C_2$ annealed	7.32	3.2	1.42	1.61	—	—	Magnetic amplify, power and impulse transformers, current converters from low frequency to very high frequency

Source: After *Handbook of Automation Engineer* (in Polish), WNT 1973, and others.

One of the best known methods of a mathematic description of the complicated magnetization processes is the *Preisach model of hysteresis* from 1935 (published in *Zeitschrift für Physik*, 1938) in which a ferromagnetic material is represented as a collection of small elementary domains (*hysterons*) with parallel-connected rectangular hysteresis loops (Figure 1.20a, right). Each *hysterons* is magnetized to a value of either h or $-h$. They interact with each other and create a stepped graph with accuracy, depending on the number of elementary loops, N. The Preisach model has been followed and improved by other researchers too (e.g., Atherton [1.28]).

At small values of H_m, the symmetric branches of the hysteresis loops are parabolic. When H_m increases, the loops become longer, reaching the shape of the letter S and their dimensions approach a certain boundary shape called *boundary hysteresis loop* (Figure 1.21). At yet bigger values of H_m, only the "moustaches" of hysteresis loop lengthen. They run along the normal magnetization curve and tend to the saturation flux density $B_{sat} = J_{is} + \mu H \approx J_{is} = \text{const}$ (Table 1.6).

Vertexes of hysteresis loop create the so-called *commutation (vertex) magnetization curve*, usually identified with the *initial magnetization curve.*

The *boundary hysteresis loop* in the points of crossing of coordinate axes defines two characteristic parameters: *residual (remanence) flux density*, $\pm B_r$, and *coercive magnetic field intensity*, $\pm H_c$ (Figure 1.21). The H_c determines the external field necessary for the full demagnetization of a sample. The hysteresis loop shape is the basis for the division of magnetic materials into *soft* with a narrow hysteresis loop (cores of transformers and electric machines—Figures 1.21 through 1.24) and *hard* with a broad hysteresis loop (permanent magnets—Figure 1.22). The surface area inside the hysteresis loop, expressing the energy necessary for the remagnetization of the sample, in J/m³

$$W_s = \oint H\,\mathrm{d}B \tag{1.30}$$

is at the same time equal to the power loss for the hysteresis Δp_{h1} during one (1) period. A hysteresis loop measured oscillographically at an alternating current AC (called *dynamic*) is broader (Figure 1.21) than the loop measured at direct current DC. This is caused by additional power losses from eddy currents induced by changes of magnetization. In soft magnetic materials, one pursues to reach the hysteresis loop as narrow as possible. Such are the so-called *silicon steels*. Addition of silicon, without changing any other conditions, causes a significant reduction of power losses caused by hysteresis loop phenomena. In addition, the hysteresis losses depend on many other factors, of which some more important ones are ingredients, manufacturing and plastic working, mechanical stresses, and heat treatment (annealing).

There were great improvements to the working process of iron. For instance, transition from what was widely used until the beginning of the 1960s, *hot-rolled* (4% Si) transformer sheets to *anisotropic, cold-rolled* (2–3% Si) transformer sheets, caused 4 times reduction of hysteresis losses (at $B_m = 1$ T) (Table 1.6; Figure 1.23). Next, the discovery and application (around 1980) of the rapidly cooled *amorphous strips* (Figure 1.24) caused another 4 times reduction of the iron loss in comparison to cold-rolled sheets.

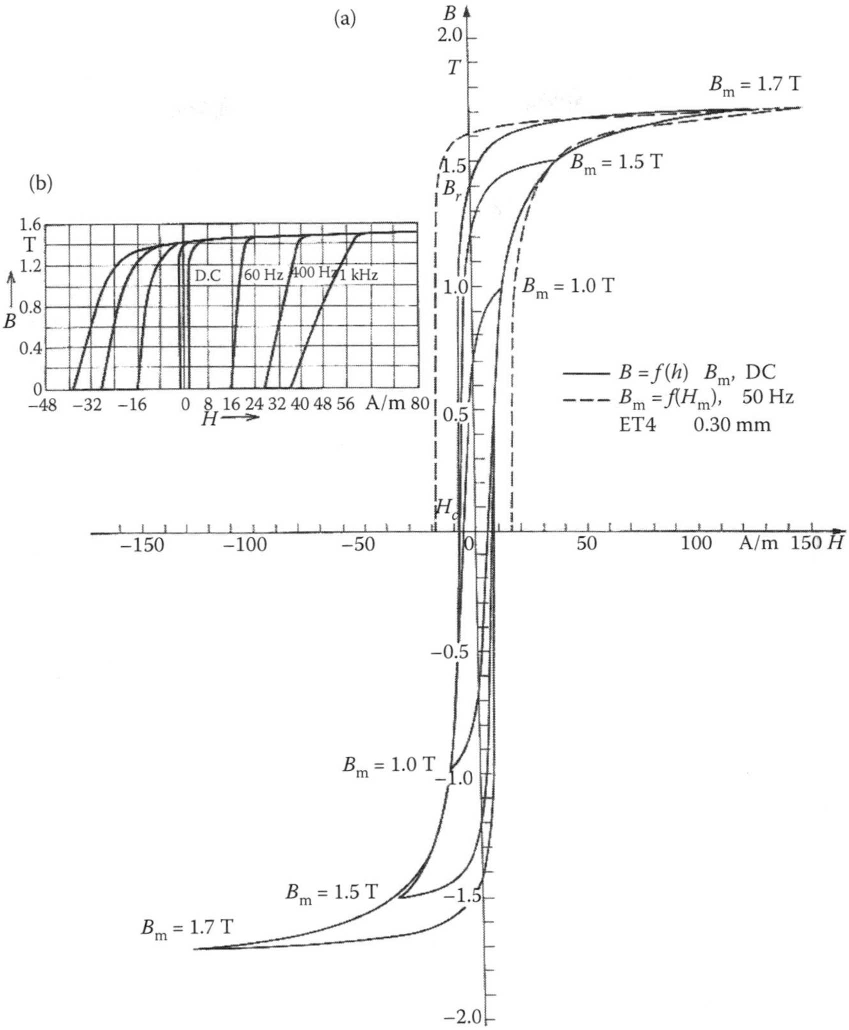

FIGURE 1.21 Family of symmetric hysteresis loops: (a) anisotropic transformer sheets (ET4, 0.3 mm, *Bochnia Works* [1.30]). (b) Amorphic sheets METGLASS Alloy 2605 S-2 with "dynamic" hysteresis loops. (After METGLAS Electromagnetic Alloys. Allied Corporation, USA 1981.)

A *soft* magnetic material of small hysteresis losses should be very pure, with uniform structure, free of internal deformations, and with small anisotropy, so that motions of Bloch walls and the rotation of magnetization vectors can be executed with as little obstacles as possible.

On the other hand, a *hard* magnetic material (with a high coercion) should have maximally hindered motions of Bloch walls and the rotation of magnetic polarization vectors. This is supported by strong stresses inside crystal lattices, big anisotropy, presence of other phases, and grainy or powdery material structure. Fine crystals of

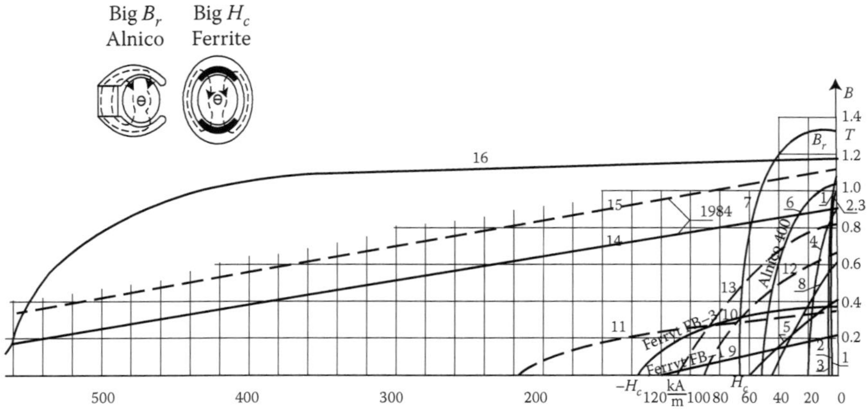

FIGURE 1.22 Demagnetization characteristics of permanent magnets. 1—chromium steel 3.5% Cr; 2—wolfram steel 6%W; 3—cobalt steel 2% Co; 4—cobalt steel 36% Co; 5—cast alnisi; 6—cast alnico 400; 7—anisotropic alloy type magnico cast with column structure "Columex" (UK 1956); 8—isotropic alnico 160 sintering; 9—barium ferrite isotropic FB-1; 10—barium ferrite anisotropic FB-3; 11—ferrite anisotropic sintered*; 12—alnico isotropic sintered/casted*; 13—alnico anisotropic sintered/casted*; 14—SmCo anisotropic sintered*; 15—NdFe2 anisotropic sintered*; 16—Nd–Fe–B, after IIM Warsaw University of Technology 1987 (*H.P. Kreuth. Bull. SEV 1984); 14 to 16–rare earth *permanent magnets*.

powder may not have Bloch walls at all and therefore magnetization of most of them can be carried out solely by nonreversible rotation of a polarization vector, which requires very strong fields. That is the reason for the large coercive intensity of powdered magnetic materials.

In electronics and power electronics engineering (energoelectronics), the alloys most often used are that of ferromagnetic metals, Fe, Ni, and Co, as well as ferrites.

Ferrites are semiconductors (σ = from 1 to 10^{-9} S/m) created from complex compounds of the ferric oxide (Fe_2O_3) with oxides of other bivalent metals with the general formula $MeO \cdot Fe_2O_3$, where symbol Me means Ni, Mn, Fe, Co, Li, Mg, Zn, or Cu. They are manufactured by a onefold or twofold burning of mixture of these components in temperatures from 900°C to 1400°C. There are distinguishable *soft magnetic ferrites* (cores of induction coils and transformers of high frequency) and *hard magnetic ferrites* (permanent magnets). Ferrites with rectangular hysteresis loops have been used for manufacturing memory elements for computers, magnetic amplifiers, and so on. In magnetic amplifiers (Table 1.7), small power transformers, instrument transformers, and HF coils, apart from silicon sheets, amorphous strips, and ferrites, also so-called *permalloys*, are used. The permalloys are nickel–iron alloys, with a high relative magnetic permeability and contents of 35–85% of Ni, and the rest is Fe, Mo (molybdenum permalloy), Mn, Cr, or Cu. The great success of the 1980s was the previously mentioned *amorphous strips* of type METGLASS, so-called "metallic glass" (Table 1.7; Figure 1.24) used for magnetic cores, from low (50 Hz) to high frequency,

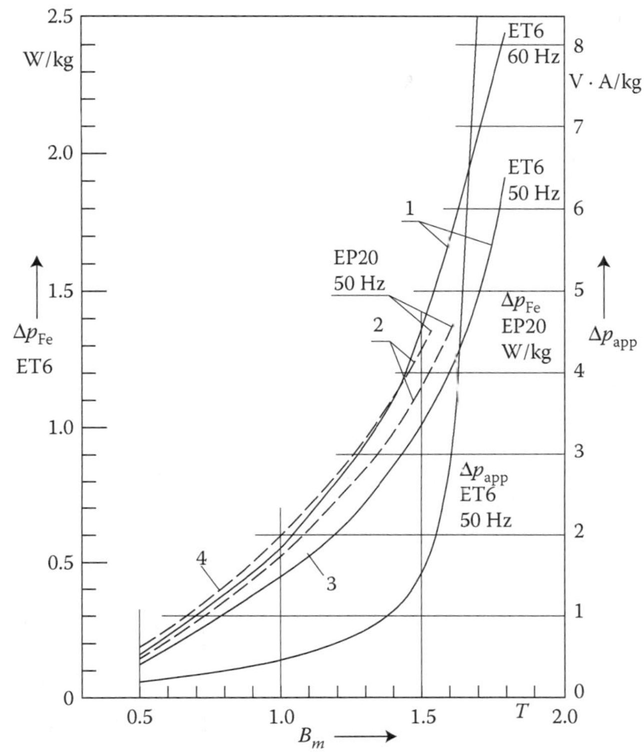

FIGURE 1.23 Plots of the per-unit iron losses ΔpFe (in W/kg), at 50 Hz: 1—anisotropic transformer sheets ET6, 0.30 mm; 2—isotropic generator sheets EP20, 0.50 mm, 50 Hz; 3—longwise; 43—longwise; mixed; Δpapp—curve of apparent per-unit losses in VA/kg consumed for core excitation. (Adapted from *Silicon Electrical Sheets*. Bochnia: Catalogue of Metallurgical Processing Plant, 1982.)

for example, 100 kHz, and for pulse and signal transformers (alloys 2605 SC, 2605 S-3) for space vehicles of 400 Hz (2605 CO), and others [1.36].

Hard magnetic materials for permanent magnets are characterized by the part of the hysteresis loop situated in the second quadrant, between B_r and H_C (Figures 1.21 and 1.22), called *demagnetization curves*. The smaller the volume and mass of the permanent magnet, V_m, with other conditions being the same, the bigger is the magnet energy ($BH/2$) (in J/m^3) and is expressed by the dependence (Turowski [1.18]):

$$V_m = \frac{\Phi \Sigma U_m}{BH} \tag{1.31}$$

where Φ is the magnetic flux of the magnet and ΣU_m is the sum of magnetic voltages in the magnetic circuit. Therefore, dimensions of the circuit are selected so that the operation of the system be executed near the maximum energy $(BH)_{max}$, called *energy factor* of magnets, or its *specific energy*.

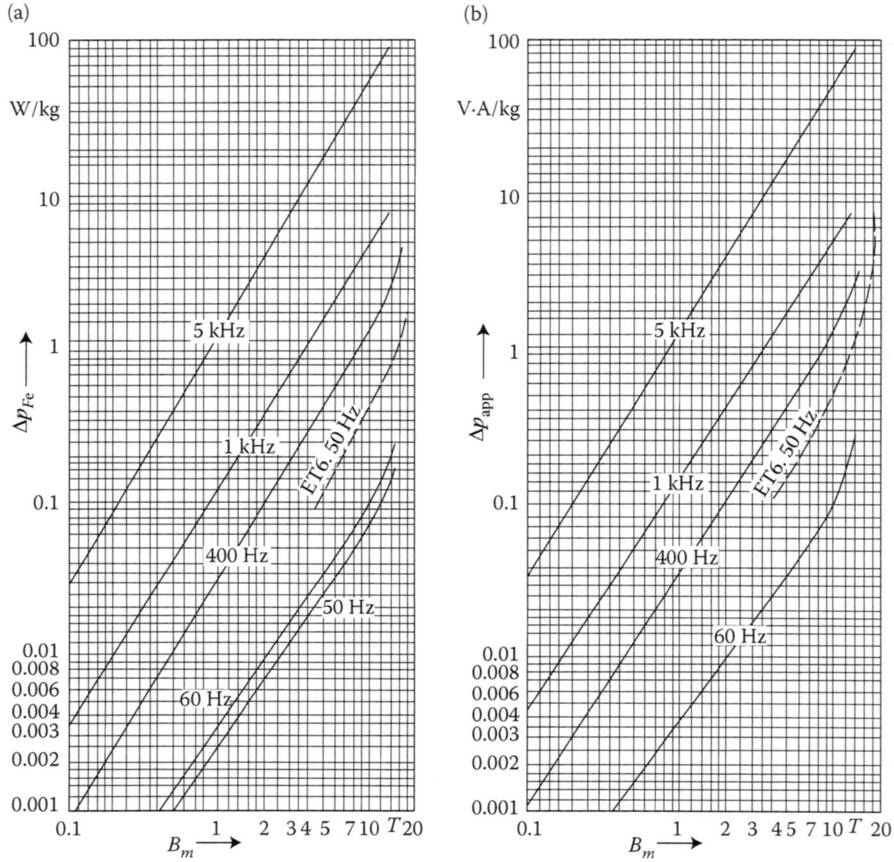

FIGURE 1.24 Characteristics of the "metallic glass" amorphous strips for transformer cores—METGLASS Alloy 2605 (Adapted from METGLAS Elactromagnetic Alloys. What if? Allied Corporation, USA 1981; Partyga S. and Turowski J.: Current problems of exploitation and construction of transformer. (in Polish). *Przegląd Elektrotechniczny*, No. 8–9, 1982, 234–236.): (a) per-unit loss Δp_{Fe} (W/kg); (b) apparent loss per-unit Δp_{app} (VA/kg) (courtesy of Allied Corporation): – – anisotropic transformer sheets ET6, as in Figure 1.23.

A typical shape of a demagnetization curve is a sloping line from the point $(B_r, 0)$ to $(0, H_C)$* (e.g., ALNICO alloys—Figure 1.22) or to a straight line (e.g., ferrites). In the former case, Bozorth ([1.2], p. 277) recommends an analytical approximation of the magnetization curve, after the so-called Frölich–Kenelly law from 1881:

$$\frac{H}{B} = \frac{1}{\mu} = \mathrm{a+b}H \tag{1.32}$$

* In the rare earth permanent magnets with high H_C (500...1000 kA/m, and more), two notions are distinguished: traditional B^{Hc} and J^{Bc}, determined by the vector of polarization $J_i = B - \mu_0 H$ (1.25), below which the magnet loses magnetization completely.

which leads to the formula

$$\frac{B}{B_r} = \frac{1 + H/H_C}{1 + (H/H_C)(B_r/B_{\text{sat.}})} \tag{1.32a}$$

Other approximation formulae of magnetization curves and their evaluations are given in Section 7.1.

The shape of a magnetization curve is characterized with the help of the so-called *shape factor* or *convexity of the curve*

$$\gamma = \frac{(BH)_{\max}}{B_r H_C} \tag{1.33}$$

For the curve approximated with hyperbola (1.32), as per Bozorth [1.2]

$$\gamma = \left(\frac{1 - \sqrt{1 - H_C/H_{\text{sat.}}}}{H_C/H_{\text{sat.}}}\right)^2 = \left(\frac{1 - \sqrt{1 - B_r/B_{\text{sat.}}}}{B_r/B_{\text{sat.}}}\right)^2 \tag{1.34}$$

Theoretically, the γ range is between 0.25 (rectilinear magnetization curve) and 1.0 (rectangular hysteresis loop). Practically, the γ value lies between 0.25 (barium ferrite FB1) and 0.65 (ALNICO) [1.2].

After many years of exploiting permanent magnets made of alloy or ferrites, in the 1970s–1980s, there appeared revolutionary discoveries and then quick implementations of new magnetic materials with contents of rare earths *neodymium* (Nd) and *samarium* (Sm) (Figure 1.22). In 1986, in Poland, the *rare earth permanent magnets* (REPM) of type Nd–Fe–B were manufactured and put into practice by the Institute of Material Technology (IMT) of the Warsaw University of Technology.

A relatively low cost of manufacturing was achieved, which promised a broad implementation into industrial practice.

The *energy factor* $(BH)_{\max}$ increased immediately from 5–18 kJ/m^3 to 200–225 kJ/m^3 (for the Nd–Fe–B magnets from IMT, Warsaw University of Technology), which changed the whole concept of design, especially of small electric machines with permanent magnets.

The largest resources of rare earths (5 times more than the rest of the world) are in China (Zhang [1.47]). Thanks to the large production of neodymium oxide (200–300 tons a year), the price of neodymium fell in 1986 and 1987 by about 20–30% a year. Magnets of $(BH)_{\max}$ up to 45 MGOe* = 358.2 kJ/m^3 have already been reached. Magnets of Nd–Fe–B have been used in China for manufacturing of thin-dimensional loudspeakers of aerophones, for magnetic faces, magneto-mechanical devices, motors, generators, motors for window wipers, wind generators, DC tachometric generators, servomotors, and stepping and synchronous motors. In 1987, Zhang Xi [1.47] stated that the production of neodymium was higher than the demand for it,

* The units still used by some physicists, albeit they do not belong to the legal units (M.P. No 4. 1978—in Polish).

and there was a need to accelerate the exploitation of the Nd–Fe–B magnets. Some important drawbacks of REPMs are their low Curie points and strong sensitivity to changes of temperature and corrosion. The industrial application of REPM is discussed in the book chapter (J. Turowski, S. Wiak et al. in [1.3], pp. 19–49).

Power losses in core iron, P_{Fe}, are one of the fundamental characteristics of magnetic materials for laminated cores of AC electromagnetic equipment. Traditionally, we divide them into those *from eddy currents* (*eddy loss*) P_{eddy} and those *from hysteresis loops* (*hysteresis loss*) P_{h}.

$$P_{\mathrm{Fe}} = P_{\mathrm{eddy}} + P_h \tag{1.35}$$

The eddy-current losses are further subdivided into so-called *classic* $P_{\mathrm{ed.cl}}$—connected with the thickness and resistivity of iron sheets—which are calculated with the methods of electrodynamics (Section 6.2), and the so-called losses from *eddy-current anomalies* (Turowski [2.34] and Hammond [2.8], p. 224) $P_{\mathrm{ed.an}}$ depending on the crystallographic structure of the sheet, which means

$$P_{\mathrm{eddy}} = P_{\mathrm{ed.cl}} + P_{\mathrm{ed.an}} \tag{1.36}$$

where as per Equation 6.16, the "classic" *eddy-current losses* are, in W/kg

$$p_{\mathrm{ed.cl}} = \frac{1}{24\rho_m}\sigma\omega^2 \mathrm{B}_{\mathrm{m}}^2 d^2 \tag{1.37}$$

where ρ_m is the mass density (kg/m^3).

According to Nippon Steel Corporation [1.41], the losses (in W/kg) resulting from the eddy-current "anomaly" are*

$$p_{\mathrm{eddy}} = p_{\mathrm{ed.cl}} + p_{\mathrm{ed.an}} = 1.628\frac{2\mathrm{L}}{\mathrm{d}}p_{\mathrm{ed.cl}}, \text{ where } \chi = \frac{p_{\mathrm{ed.cl}} + p_{\mathrm{ed.an}}}{p_{\mathrm{ed.cl}}} = 2 \text{ to } 3 \tag{1.38}$$

(for amorphous magnetics, $\chi = 10\text{–}300$), d is the thickness of sheet, $\omega = 2\pi f$, B_{m} is the maximum flux density, σ is that electric conductivity (S/m), ρ_{m} is the density of the metal (kg/m^3), 2L is the distance between the domain walls creating parallel strips of thickness d.† From Pry's et al. formula (1.38), it follows that the eddy-current losses can be reduced by the refinement of the magnetic domain structure. For this purpose, a *laser irradiation* of iron sheets crosswise to the direction of rolling has been used [1.41].

For instance, at $B_{\mathrm{m}} = 1.7$ T, $d = 0.30$ mm, and $f = 50$ Hz, before the lasing, the losses were equal to 1.23 W/kg, whereas after the lasing, they were equal to 0.97 W/kg [1.41]. The hysteresis losses in anisotropic iron sheets typically amount from 18% to 20% of P_{Fe}.

* Pfuetzer J. et al.: Nanocristaline materials ... Vacuumschmelze GmbH, Hanau, 1997.

†

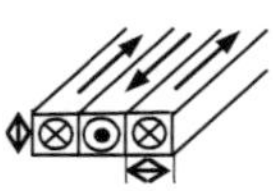

The *hysteresis losses* are equal to the area of hysteresis loop (1.30) multiplied by the frequency of remagnetization f. Currently, there is no sufficiently accurate method of theoretical calculation of the hysteresis loop (except the discrete Preisach mathematical model—Figure 1.20). This is why the hysteresis losses are calculated with the help of semiempirical formulae. One of the oldest and most popular formulae for the hysteresis losses is the Steinmetz formula, from 1891, which expresses the hysteresis losses (in W/kg), at the frequency f (in hertz):*

$$p_h = \eta f B_{\mathrm{m}}^x \tag{1.39}$$

where η is the semiempirical coefficient depending on the chemical constitution, thermal treatment, and mechanical working of steel (Table 1.6); B_{m} is the maximum in-time flux density (in T) at sinusoidal remagnetization; $x = 1.5–3.0$ is the semiempirical exponent depending on the type of steel, which as per E. Jezierski [1.5] is

- For isotropic, hot-rolled electrical steel, at $B_{\mathrm{m}} = 0.5–1.0$ T, $x = 1.6$
- For anisotropic, cold-rolled transformer steel, at $B_{\mathrm{m}} \leq 1.45$ T, $x \approx 2$;
- For anisotropic, cold-rolled transformer steel, at $1.5 \leq B_{\mathrm{m}} \leq 1.6$ T, $x \approx 2.25$
- For anisotropic, cold-rolled transformer steel, at $B_{\mathrm{m}} = 1.7$ T, $x \approx 2.60$

In the range of variation of the maximum flux density and in the case of electrical sheets used in electric machines and transformers, one also uses the simplified, semiempirical Richter formula [1.5]

$$P_h \approx \varepsilon \frac{f}{100} B_{\mathrm{m}}^2 M_{\mathrm{Fe}} \tag{1.40}$$

where P_h is the hysteresis power losses, in W; ε is a constant depending on the type of steel: for sheets without silicon admixture, $\varepsilon = 4.4–4.8$ m^4/(H kg); for generator sheets with medium siliconizing (about 2% Si), $\varepsilon = 3.8$ m^4/(H kg), and for transformer sheets with silicon content 4%Si, $\varepsilon = 1.2–2.0$ m^4/(H kg); f is the frequency (in Hz); B_{m} is the maximum in-time flux density (in T) at sinusoidal remagnetization; M_{Fe} is the iron mass (in kg). By adopting the power exponent $x = 2$, in simplified calculations, one can consider jointly the hysteresis losses and eddy-current losses, which are proportional to $B_{\mathrm{m}}^2 f^2$ (6.16).†

In more accurate calculations, however, the total iron losses P_{Fe} (1.35) should be calculated on the basis of the curves of the power-per-unit loss, in W/kg, $p_{\mathrm{Fe}} = f(B_{\mathrm{m}})$ of electrical sheets at a given frequency f and a given thickness d of the sheet (Figures 1.23 and 1.24a), where $p_{\mathrm{Fe}} = p_{\mathrm{eddy}} + p_h$ are the per-unit losses, in W/kg, measured and provided by a manufacturer of steel.

* Such synthetic formulae experience "second youth" in modern mechatronic environment, when one wish to reach competitiveness by "rapid design," "reduced models," and to minimize "time to market."

† Generally, $p_{\mathrm{Fe}} \approx cB_m^a f^b$, where for cold-rolled Japanese steel RG8H, $b = 1.63$ and for $0 \leq B_m \leq 1.2$ T, a = 1.8; for $B_m \geq 1.4$ T, $a = 1.9$.

The total power losses in iron of laminated core can then be calculated from the dependence

$$P_{Fe} = k_{ad}\, p_{Fe}\, M_{Fe} \tag{1.41}$$

where k_{ad} (equal from 1.05 to 1.15) is the *coefficient of additional losses* as a result of constructional and processing factors of building the core, and M_{Fe} is the mass of iron core, in kg.

1.2.3.7 Superposition of Remagnetization Fields

In present power electronics systems, there often appears an additional overmagnetization of the core by a DC field or by a field of another frequency.

At a superposition of slow-alternating and higher-frequency fields, the total losses for remagnetization are equal to the area of basic hysteresis loop plus a sum of all partial cycles BB_2 (Figure 1.20a) (Bozorth [1.2], p. 441).

Tellinen et al. [1.42] proposed a simple approximate formula for the per-unit hysteresis losses at superposition of AC field B_m and DC field B_a:

$$p_{h0} = p_h\left[1+\eta\left(\frac{B_a}{B_m}\right)^2\right] \tag{1.42}$$

where p_h is the per-unit losses at symmetric remagnetization with amplitude of flux density B_m; $\eta = 1$–0, for example, for cold-rolled sheet $\eta \approx 0.9$ at $B_m > 1.0$ T. A similar formula can be found in Ref. [7.2] (p. 23).

The total per-unit losses p_{Fe} in iron at the superposition of DC and AC fields, as per Ref. [1.42], can be expressed by this simple formula, with accuracy up to ±15%:

$$p_{Fe} = p_h\left[1+\eta\left(\frac{B_a}{B_m}\right)^2\right] + p_{add}\frac{\sum_{\nu=1}^{\infty}\nu^2 B_{\nu m}^2}{B_{1m}^2} \tag{1.43}$$

for B_{1m} from 1.0 to 1.6 T.

In a similar way like the power losses in Equation 1.41, one can calculate the apparent power P_{app} consumed for the excitation of the core. Using corresponding graphs for $p_{app} = f(B_m)$ (Figures 1.23 and 1.24b), we get

$$P_{app} = p_{app}\, M_{Fe} \tag{1.44}$$

1.2.3.8 Amorphous Strips

In the 1970s–1980s, Allied Corporation [1.36] discovered, and implemented to industrial production, a new magnetic material METGLASS of revolutionary low per-unit losses p_{Fe} and consumption of reactive power for magnetization (Figure 1.24). One of the early papers on METGLASS application was published in Poland by the author (J. Turowski) jointly with S. Partyga in 1982 [9.6] and very recently by Maria Dems et al. [1.50], [1.51], [1.52].

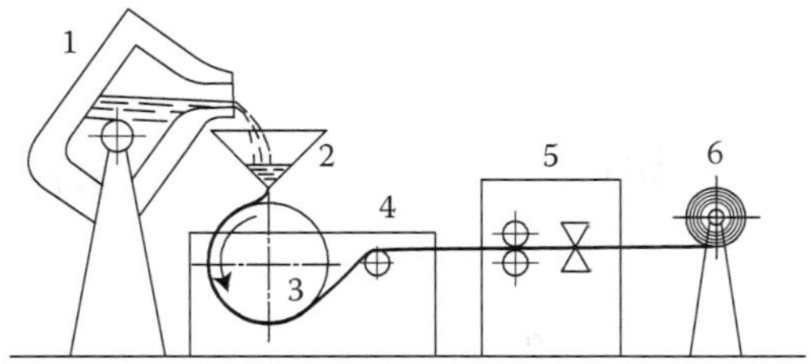

FIGURE 1.25 Schematic of manufacturing of the amorphous strips (Adapted from METGLAS Electromagnetic Alloys. What if? Allied Corporation, USA 1981.): 1—Induction furnace with liquid metal; 2—batcher of metal; 3—rotating drum; 4—solidification of metal within time of 1 ms; 5—measurement and control of dimensions of strip in a feedback system; 6—winding the strip on a roll.

Allied's discovery consisted in establishing that a thin layer of fluid metal of type FeBSi, FeBSiC, and others, when cooled at a high rate, about 1,000,000°C/s (Figure 1.25), freezes to a thin (30–50 μm) elastic strip of frozen fluid, with amorphous structure, similar to glass. From it comes its name "METGLASS" (metallic glass). The reason for the rapid solidification of fluid metal is that its atoms cannot follow up their order and create crystals. Amorphous strips demonstrate per-unit loss p_{Fe} and magnetizability q_{Fe} about 4 times lower than the best cold-rolled anisotropic iron sheets (M4).

This was the second (after cold-rolled sheets) revolution in the construction of power transformers, reactors, and other electromagnetic apparatus (Table 1.7), especially because the cost of this material dropped down from about 150 $/kg in 1972 to about 3.3 $/kg in 1987. It affected both power supply equipment and power electronics.

1.2.3.9 Rotational Hysteresis

At rotational remagnetization, which appears at rotational fields in electric machines, there can also occur the above-described nonreversible rotation of the vector of magnetic polarization, which for soft magnetic materials is the fourth process, beyond the cycle shown in Figure 1.20.

At the axial remagnetization (Figure 1.20a), the field energy, W_h, dissipated for hysteresis, in J/m^3, equals the area of the loop (1.30), but at the rotational remagnetization, the energy W_{h0} of rotational hysteresis is determined as the energy necessary for the rotation of the sample by 360°, that is

$$W_{h0} = \int_{0}^{2\pi} M(\Theta) \cdot \mathrm{d}\Theta \tag{1.45}$$

where $M(\Theta)$ is the curve of the torque, in J/m^3.

This phenomenon in rotating electric machines causes, depending on the maximal flux density B_{m}, about 1.8 times the increase of hysteresis losses, in comparison with the axial (AC) remagnetization. When the flux density B_m increases from zero, these losses first grow, reaching a maximum at about 1.5–1.6 T, and then they decrease to almost zero [1.27], [2.3]. The hysteresis angle $\angle(\boldsymbol{B}, \boldsymbol{H})$ is the biggest in section II of the magnetization curve (Figures 1.19c and 1.20a), and then decreases almost to zero in section III. Recently, a novel test fixture to measure rotational core losses in machine lamination was published [10.35].

1.2.3.10 Types of Magnetization Curves

Magnetization curves $B = f(H)$ of ferromagnetics are determined experimentally. The magnetization curve called *initial* or *primary* (curve 1 in Figure 1.20a) describes the magnetization process of a sample after its previous complete demagnetization. Demagnetization can be reached by a periodic remagnetization of the metal at successive reductions of the magnetic field intensity amplitude. The initial curve is situated near the growing (right) branch of the hysteresis loop. Their ordinates, with a pretty good accuracy, can be substituted by an *arithmetic mean* of the upper and lower ordinates of the maximal loop (curve 2 in Figure 1.20a). The initial magnetic characteristic (curve) practically overlaps with the curve, passing through vertexes of the family of hysteresis loops, but just slightly lower (Figure 1.21). The curve connecting vertexes of symmetric hysteresis loops is called the *vertex curve*, *commutational curve*, or *basic curve*.

Ideal magnetic characteristics (curve 3 in Figure 1.20a) can be obtained by simultaneous magnetization of a sample by a DC field and an AC field with amplitude being reduced progressively to zero. The measurement is carried out in the way that at every value of DC field, the AC field amplitude is being gradually reduced, from the value equivalent to full technical saturation down to zero. The *ideal* or *hysteresisless* magnetization curve, obtained in this way, is characterized by lack of inflexion point. The course of an ideal magnetization curve is very close to the line running through the centers of the horizontal chords of the hysteresis loop corresponding to saturation (Figure 1.21).

1.2.3.11 Curie Point

The magnetization process of ferromagnetic bodies described above undergoes changes while increasing to higher temperatures, while the values of μ, $\boldsymbol{J}_{\text{is}} \approx \boldsymbol{B}_{\text{sat}}$, B_r and H_C gradually decrease. Initially, these changes are very slow (Figure 1.26), and only after passing a certain temperature T_C, called Curie point, does the body lose its ability to keep the microzones of spontaneous magnetization (*domains*), and becomes *paramagnetic*, with the relative permeability μ_r close to one. This phenomenon is reversible. The Curie temperature is 770°C for iron and 690°C for silicon iron (4% Si).

The Curie temperature phenomenon can be utilized for a direct conversion of thermal energy into mechanical energy, by means of the so-called Curie Motor. In the Curie Motor, an iron disk rotating in the field of a permanent magnet is heated from one side of the magnet by a gas burner to the *Curie point* and, due to that, the magnet attracts the cold part of the disk, situated on the other side of the magnet.

In metrology apparatus, electricity meters, and electric tachogenerators with permanent magnets, the almost linear dependence of the flux density B on temperature [1.33]

$$B = B_0[1 + \beta(t - t_0)] \tag{1.46}$$

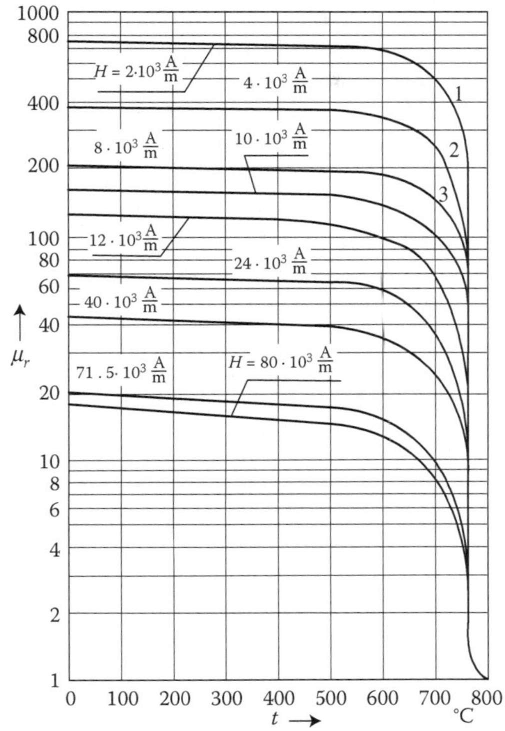

FIGURE 1.26 Dependence of the relative magnetic permeability, μ_r, on the steel temperature (t) at the magnetic field intensity H from 2×10^3 to 80×10^3 A/m. The *Curie* point for (1) $B = 750 \times 0.4\pi \times 10^{-6} \times 2 \times 10^3 = 1.885$ T; (2) 1.9 T; (3) 2T (After A.V. Donskoi. *Elektrichestvo* 5, 1951).

can be a reason of significant errors. The so-called *thermomagnetics*, with the Curie temperature from 65°C to 120°C, made as shunts of external circuit of magnet, enable thermocompensation of the decrease of useful flux density.

Permanent magnets with *rare earth* have lower Curie temperatures T_C than usual iron (e.g., Japanese Hicorex-Nd from Hitachi Corporation has $T_C = 300°C$) and higher sensitivity to temperature fluctuations, which requires greater effort for the elimination of this influence. This is an important disadvantage of these materials, and in REPM motors, it is often necessary to apply special magnetic *shunt thermo-compensators*. Thermomagnets used for thermocompensator should have the Curie temperature between 65° and 120°C, corresponding to the maximal rated (nominal) temperature t_N and the nominal flux density B_N of permanent magnet applied in electric machines.* An increase of *Curie* temperature can also be achieved, for instance, with the substitution of iron by cobalt.

* Kordecki A.: Thermocompensation of permanent magnet flux density changes in electric machines. *Sci. Archives of Inst. of Electromachine Systems*, Technical University of Wroclaw (in Polish). 39, 1987, 5–19.

Iron, depending on the temperature, exists in three *allotropic forms*. The space lattice of pure iron in normal temperature (*α-Fe, ferrite*) corresponds to the form of *body-centered lattice* (Figure 1.5a). This form possesses ferromagnetic characteristics, which it loses at the *Curie* temperature, transitioning into the *paramagnetic* form *β*-Fe without change of the lattice structure. At a temperature of 910°C, β-Fe passes into *γ*-Fe (*austenite*) *phase-centered lattice* (Figure 1.5b) and next, at 1400°C, it passes into the *δ*-Fe form with the same structure like α-Fe. These transformations of the metal structure are often presented in phase graphs as a function of temperature and contents of admixtures.

1.2.3.12 Nonmagnetic Steel

The structural transformations of iron enable obtaining nonmagnetic steel and cast iron, which have paramagnetic properties, but maintain at the normal (room) temperature the permanent *austenite structure γ*-Fe. The relative permeability of such materials does not exceed 1.05–1.5, and their electric conductivity does not exceed $(0.5\text{–}1.4) \times 10^6$ S/m.

The nonmagnetic austenite steels [1.13], [1.9] can be divided into

- Manganic, nickelous, nickelous-manganic, and stainless chromic-nickelous steel
- Nonmagnetic cast steel
- Nonmagnetic cast iron: nickelous, manganic, copperized, and nickelous-manganic (nomag) [1.9]

Manganic steels used for bonnets and shrouds of rotors of large electric machines contain, for instance, 0.25–0.35% C, 18–19% Mn, 1% Cr, and 3–5% Ni. The limit of tear resistance is at the same time about 100 daN/mm^2. Orientational composition of nonmagnetic *copperized cast iron* used for light section (thin walls) casting is 3.4–3.9% C, 2.6–3% Si, 6.8–7.5% Mn, 1.5–2% Cu, 0.4–0.6% Al, and 0.3–0.7% P [1.13] and the tear resistance limit is 12–16 daN/mm^2. Some steel types (with small contents of Ni and Mn) can lose the nonmagnetic properties after being heated up to 400–500°C, or at cold working while passing from the form *γ* to α.

Typical elements made of the *nonmagnetic cast iron* include [1.13] covers, containers, sleeves, and other parts of oil switch gears, consoles, yoke beams, and other parts of large power transformers, elements in current instrument transformer housings, flanges, tubes and other reinforcement parts of distribution equipment, magnetic core pressing disks of stators and rotors, internal fly wheels, winding consoles and bus-bars in electric machines, covers and parts of welding transformers, parts of crane electromagnets, and so on. The nonmagnetic alloys should preferably be easy to form (or mold), because they are often used for manufacturing of complicated details. From this viewpoint, however, these materials are inferior to the nonferromagnetic metals, like brass, bronze, and aluminum alloys. An advantage of nonmagnetic steels is their big resistivity, which is another factor helping to reduce power losses due to eddy currents, in addition to small permeability.

1.2.3.13 Influence of Various Factors on Properties of Magnetic Materials

Magnetic properties of various materials generally depend on the chemical constitution, way of manufacturing, and thermal working. They can be generally divided into groups of properties structurally sensitive and structurally nonsensitive (Bozorth [1.2]). The properties that are *structurally sensitive*, that is, sensitive to such factors as general chemical constitution, admixtures, mechanical stresses, temperature, crystallographic structure, and orientation of crystals, include the permeability μ_r, the coercive force H_C, and the hysteresis losses P_h. The properties that are *structurally nonsensitive* include the polarization of saturation J_{is}, the Curie point, the magnetostriction at saturation λ_y, and the constant of magnetic anisotropy K.

An advantageous phenomenon from the viewpoint of electromagnetic calculation of constructional elements made of steel is the fact that significant differences in magnetic properties appear only in the region of weak fields (Figure 1.27), and

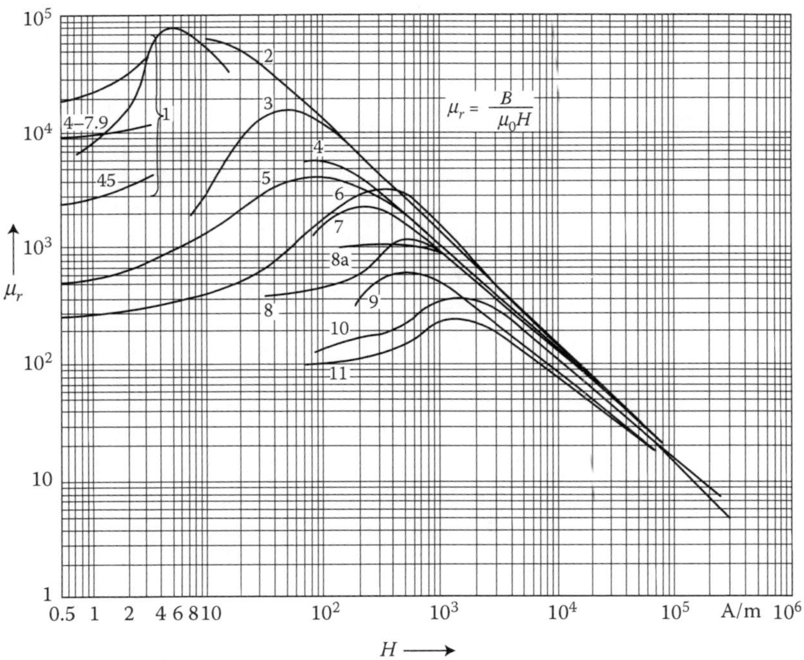

FIGURE 1.27 Relative magnetic permeability μ_r of iron and its alloys as a function of the magnetic field intensity H (After Turowski J.: (1) *Calculations of Electromagnetic Components of Electric Machinery and Equipment.* (in Polish). Warsaw: WNT, 1982; and (in Russian). Moscow: Energoatomizdat, 1986.): 1—permalloy; 2—cold-rolled, anisotropic transformer sheets in most advantageous direction [100] of magnetization; 3—electrolytic iron melted in vacuum; 4—cold-rolled transformer sheets in most disadvantageous direction [111] of magnetization; 5—hot-rolled transformer sheets 4% Si; 6—hot-rolled constructional carbon steel 0.3% C; 7—alloys with 0.23% C; 8 and 8a—cast steel; 9—annealed gray cast iron; 10—alloys with 1.78% C; 11—nonannealed gray cast iron. (After A. Donskoi. *Elektrichestvo 5/1951*, and other sources).

particularly in the growing part of the curve $\mu = f(H)$. With the growth of magnetic field intensity, in the dropping part of the curve $\mu = f(H)$, differences in magnetic properties of different alloys are gradually decreasing, and at $H = 10^4$ A/m all the curves practically join each other into one curve (Figures 1.27 and 1.29 [later]). More information can be found in [10.10].

The purer a metal is, the steeper is its magnetization curve and the bigger its maximum permeability; at higher field intensities, these differences become lesser. This property is important from the viewpoint of generalized results in the calculation of strongly saturated iron parts (e.g., calculation of impulse magnetization current at switching-on of a transformer or a turbine-generator). Adding alloy-forming components to iron causes the reduction of its flux density.

On the basis of Table 1.8, one can approximately evaluate the flux density, which at a given magnetic field intensity corresponds to the sort of constructional steel when the contents of admixtures do not exceed 2%. As it follows from the table, the biggest influence on the worsening of magnetic properties of an alloy is carbon C. This is why cast iron, which always contains more than 1.7% carbon (mainly in the form of graphite) and other admixtures, has the lowest magnetizability (magnetic permeability). For these reasons, among others, the yoke beams of the biggest power transformers are sometimes made of *spherical cast iron* of high strength, because small permeability is linked with small eddy-current losses, as it follows from the next considerations. The spherical cast iron has a fine-grained structure of graphite (contrary to lamellar microstructure of graphite in ordinary cast iron), which makes its strength properties closer to the cast iron (tear resistance 70 daN/mm² and bending strength 100 daN/mm²). The chemical composition of cast iron used for yoke beams of large power transformers (H.W. Kerr et al. Proc. IEE 4/64) is as follows:

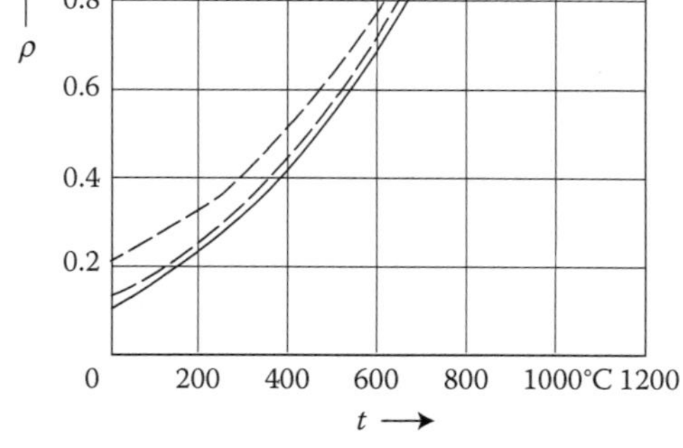

FIGURE 1.28 Changes of the resistivity of ferromagnetic metals, ρ, by heating to high temperatures: 1—electrolytic iron; 2—0.11% C; 3—1% C. (After *A. Donskoi*, 1951).

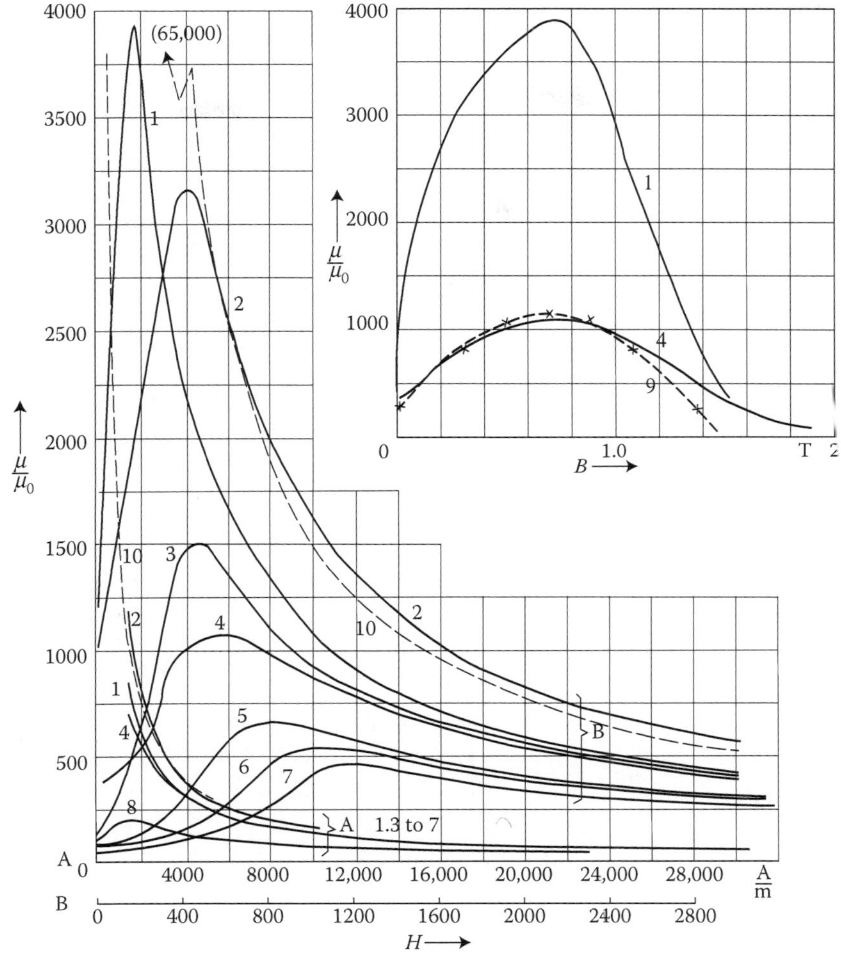

FIGURE 1.29 Curves of the relative permeability for various sorts of steel: 1—hot-rolled, isotropic transformer sheets 4% Si; 2—general-purpose constructional carbon steel, 0.3% C, hot-rolled; 3—constructional steel ST4s, $\sigma_{20C} = 6.20 \times 10^6$ S/m, of general purpose (Polish standard PN-61/H829 20); 4—cast iron; 5 through 7—constructional steels ST35, ST55, ST65 quality steel [6.2] (PN-66/H-84019); 8—gray cast iron; 9—analytical approximation of curve 4 with the function $\mu_r = 1100 \sin((\pi/1.6)B + 0.2)$; see Section 7.1; 10—anisotropic sheet ET4 0.3 mm, rolling direction [100] $\mu_{r\,\max} = 65\,000$ (Figure 1.32).

3.5% C, 2% Si, 0.2% Mn, 1.5% Ni, at properties: $\sigma = 1.82 \times 10^6$ S/m, $\mu_{r,\max} = 1400$, or an alloy of 3% C, 2.8% Si, 2.4% Mn, 24% Ni, and $\sigma = 0.98 \times 10^6$ S/m, $\mu_{r.\max} = 1.03$. The spherical cast iron is used, for example, for the manufacture of automotive parts, such as crankshafts and distribution shafts, cylinders, piston rings, and so on.

Typical changes of *iron resistivity* as a function of temperature are presented in Figures 1.6 and 1.28. The character of these curves is also approximately the same for most all iron alloys. Knowing the resistivity at a temperature of 20°C, one can

TABLE 1.8
Reduction of the Flux Density $B(H)$ in Low-Carbon Constructional Steel by Enriching Elements

H	B	Reduction of Flux Density by 1% Enriching Element						
(10^2 A/m)	(T)	C	Si	Mn	Cr	Mo	Cu	Al
25	1.65	−0.50	−0.055	−0.080	−0.090	−0.150	−0.035	−0.075
50	1.75	−0.38	−0.044	−0.040	−0.050	−0.095	−0.032	−0.070
100	1.90	−0.30	−0.031	−0.040	−0.030	−0.078	−0.015	−0.060
300	2.08	−0.28	−0.025	−0.030	−0.020	−0.060	−0.010	−0.065

Source: Adapted from *Handbook of Electrical Materials*. (in Russian) Vol. 2, Moscow: Gosenergoizdat, 1960.

approximately evaluate it for any temperature, for example—at hardening, annealing, or induction smelting.

The fact that admixtures reduce at the same time, the magnetic permeability μ and conductivity σ helps reduction of errors in calculating eddy-current power losses and resistivity of steel (Chapters 2 and 3) caused by material dispersion of these values, because in the mentioned formulae, there typically appears the ratio $\sqrt{\mu/\sigma}$. For instance, for the Polish constructional steel (Figure 1.29):

Type of Steel	$\mu_{r\,40}$ at $H = 40$ A/cm	$\sigma_{20°C} \times 10^6$ S/m	$\sqrt{\mu_{40}/\sigma_{20°C}} \times 10^{-6}\,\Omega\sqrt{s}$.
St 4s general purpose	290	6.2	7.67
St 35 quality steel	260	4.50	8.52
St 55 quality steel	255	4.64	8.31
St 65 quality steel	220	4.57	7.78

1.2.3.14 Types of Magnetic Permeability

The ratio of flux density to magnetic field intensity, $\mu = B/H$, in a case when the point (B, H) lies on the primary magnetization curve, is called *primary permeability* or simply, *permeability*. The limit that approaches the primary permeability (or the *amplitude permeability*, $\mu_a = B_m/\mu_0 H_m$ [1.37]), at H approaching zero, is called *initial permeability*, μ_i.

The highest value on the permeability curve (Figure 1.29) is called the *maximal permeability*, μ_{max}. It occurs at relatively weak fields.

The permeability at submagnetization, μ_Δ [1.37], also called the *permeability of partial cycle* characterizes the properties of a body in a direct (submagnetizing) field and in a superimposed field, which can be alternating. If on a sample, apart of the direct field of intensity H_b, there is an additional alternating magnetic field of intensity H_Δ that is causing the alternating flux density B_Δ (Figure 1.20a), then $\mu_\Delta = B_\Delta/H_\Delta$.

Reversible permeability, μ_{rev}, is the limit to which tends to μ_Δ, when H_Δ tends to zero. In a demagnetized material, at the zero submagnetization field, the reversible permeability μ_{rev} and initial permeability μ_{in} are coinciding with each other. The permeability of partial cycle μ_Δ and the reversible permeability μ_{rev} depend on the value of submagnetization field H_b (at increasing H_b, the permeability μ_Δ gets smaller) as well as they depend on the magnetic history of the sample. The permeability μ_Δ also depends on H_Δ. The permeability of partial cycle μ_Δ serves for the characterization of materials from which one makes cores of output transformers, magnetic amplifiers, and coils for *pupinizations* of telecommunication lines [1.2].

Return permeability is the notion analogical to the permeability of partial cycle μ_Δ, but with the difference that changes of sample magnetization are not due to the superposition of an alternating field, but by a partial demagnetization and remagnetization of a permanent magnet, at the opening and closing armature of its magnetic circuit. Opening the armature of a fresh magnetized magnet causes reduction of the flux density from value B_r to B_2 (Figure 1.20). All next cycles of opening and closing will run through the *recoil loop* B_1B_2 (Figure 1.20). The straight line B_1B_2 is called *recoil straight line*. It can be considered as an approximate parallel to the demagnetization curve in point B_r. The tangent of the angle of inclination of the recoil straight line B_1B_2 is called the *recoil permeability*. This concept is utilized for the calculation of permanent magnets and synchronous generators with permanent magnets.

Differential (dynamic) permeability (1.48) is utilized sometimes for the characterization of the steepness of the magnetization curve ($\mu_{dyn} = \mathrm{d}B/\mathrm{d}H$). It is the coefficient of inclination of a tangent to the magnetization curve. As the flux density increases from zero, the differential permeability quickly increases and then decreases. When the differential permeability drops to a value merely a few times bigger than the permeability of air, then the iron is considered saturated.

Ideal permeability is the ratio B/H found from the ideal magnetization curve. This permeability, as can be seen from curve 3 in Figure 1.20a, has no extreme value (no inflexion point). It has, instead, a very large initial value, at $H = 0$.

Nonlinear permeability is the value obtained from the differentiation of the right-handed side of the second Maxwell equation 2.2 with consideration of the functional dependence $\mu = f(H)$. In a case when investigating the media of varying permeability, the first Maxwell equation 2.1 remains with no change (in metals, at power frequencies, the displacement currents can be ignored), whereas in the second Maxwell equation 2.2, we must differentiate in terms of the time the composite function $\mu(\boldsymbol{H}) \cdot \boldsymbol{H}$:

$$\begin{aligned}\mathrm{curl}\boldsymbol{E} &= -\frac{\partial[\mu(H)\cdot\boldsymbol{H}]}{\partial t} = -\frac{\partial \boldsymbol{B}}{\partial \boldsymbol{H}}\cdot\frac{\partial \boldsymbol{H}}{\partial t} = -\left[\frac{\partial\mu(H)}{\partial H}\cdot\frac{\partial H}{\partial t}\right]\boldsymbol{H} - \mu(\boldsymbol{H}) \\ &\times \frac{\partial \boldsymbol{H}}{\partial t} = -\left[\mu(H) + H\frac{\partial\mu(H)}{\partial H}\right]\frac{\partial \boldsymbol{H}}{\partial t} = -\mu_n\frac{\partial \boldsymbol{H}}{\partial t}\end{aligned} \tag{1.47}$$

Generally, then

$$\mathrm{curl}\boldsymbol{E} = -\mu_n\frac{\partial \boldsymbol{H}}{\partial t}$$

where μ_n is called dynamic, nonlinear, or differential magnetic permeability, equal to

$$\mu_n(H) = \frac{\partial \boldsymbol{B}}{\partial \boldsymbol{H}} = \mu(H) + H\frac{\partial \mu(H)}{\partial H} \tag{1.48}$$

The form of Equation 1.48 presents a simple, though cumbersome, way of finding the curve $\mu_n(H)$ from the magnetization curve of a given type of steel (Figure 1.29).

1.2.3.15 Permeability at High Frequencies

At the power frequencies, the permeability does not depend on the frequency, but begins to decrease at radio frequencies (RFs). At the frequency of light, it reaches the value of one [1.2].

Another problem is with the dielectric permittivity $\varepsilon = \varepsilon_0\ \varepsilon_r$ and the displacement current density $(\mathrm{d}D/\mathrm{d}t) \approx \varepsilon(\mathrm{dE}/\mathrm{d}t)$. In electromagnetic literature, it is commonly accepted (with no discussion) that in normal metals, the current density $\boldsymbol{J}_{\mathrm{metal}} = \sigma\boldsymbol{E} \gg \varepsilon_{\mathrm{metal}}(\mathrm{dE}/\mathrm{d}t) \approx j\omega\varepsilon_{\mathrm{metal}}\boldsymbol{E}$. It means that it is commonly accepted that the polarization current of particles in metal, represented by $\varepsilon_{\mathrm{metal}}$, is much smaller and masked by "electron gas" of free electrons. This is one of the most interesting questions and still an open problem of solid-state physics, not sufficiently explained. There exist absolutely controversial opinions from "$\varepsilon_{\mathrm{metal}} \to \infty$ at $f = 0$," as per Ashcroft [1.1, p. 407], to even $\varepsilon_{\mathrm{metal}} = 0$. The authors' evaluation shows however that $\varepsilon_{\mathrm{metal}}$ is rather nearer to ≈ 1.

Complex permeability is the permeability taking into consideration a phase shift (in time) of the flux density B_{m} and the magnetic field intensity H_{m} in case both values are alternating sinusoidally (first harmonics).* Like any complex number, one can express it by module μ and argument ψ

$$\underline{\mu} = \frac{B_{\mathrm{m}}}{H_{\mathrm{m}}} = \mu \mathrm{e}^{-j\psi} \tag{1.49}$$

where μ is the permeability in the top, return point of hysteresis, called *amplitude permeability*, $\mu_a = (1/\mu_0)\,(B_{\mathrm{m}}/H_{\mathrm{m}}) = \mu\mathrm{e}^{-j\psi}$, at $B_{\mathrm{m}} = \mathrm{const}$ [1.37].

Remagnetization of an iron sample needs the delivery of both reactive and active power to the excitation circuit. Results from that apparent power equals the product of the effective (rms) value of flux density and conjugate effective (rms) value of magnetic field intensity, or vice versa. If we assume sinusoidal changes $\boldsymbol{H}_{\mathrm{m}}^* = \boldsymbol{H}_{\mathrm{m}}\,\mathrm{e}^{-j\omega t}$ and after (1.49) $\boldsymbol{B}_{\mathrm{m}} = \mu\boldsymbol{H}_{\mathrm{m}}\,\mathrm{e}^{-j\psi}\,\mathrm{e}^{j\omega t}$, then we obtain the apparent power per volume (VA/m³) of the sample at frequency f hertz (where $\boldsymbol{H}_{\mathrm{m}}$, $\boldsymbol{B}_{\mathrm{m}}$–modules)

$$\boldsymbol{S}_{1V} = f\oint(\boldsymbol{H}_{\mathrm{m}}^* \cdot \mathrm{d}\boldsymbol{B}_{\mathrm{m}}) = \frac{f}{2}\int_0^T j\omega\mu H_{\mathrm{m}}^2 \mathrm{e}^{-j\psi}\,\mathrm{d}t = \frac{\pi B_{\mathrm{m}}^2 f}{\mu}(\sin\psi + j\cos\psi) \tag{1.50}$$

* According to the Polish standard BN-85/3382-20 [1.30], for coils with magnetic cores of impedance Z, only one of these values must be sinusoidal, whereas $\mu = Z/Z_0$, where Z_0 – at $\mu_r = 1$.

On the basis of Equation 1.50, knowing the hysteresis losses P_V per volume, or the magnetization power Q_V of the sample, one can calculate the argument ψ of the complex permeability $\underline{\mu}$:

$$\sin\psi = \frac{\mu P_v}{\pi B_{\mathrm{m}}^2 f} \quad \text{or} \quad \cos\psi = \frac{\mu Q_v}{\pi B_{\mathrm{m}}^2 f} \tag{1.51}$$

The complex permeability (1.49) can also be represented as

$$\underline{\mu} = \mu' - j\,\mu'' = \mu\,\mathrm{e}^{-j\psi} = \mu\,(\cos\psi + j\sin\psi) \quad \text{or} \quad \underline{\mu} = \mu\,(1 - j\,\mathrm{tg}\,\delta) \tag{1.52}$$

Considering Equation 1.51, we shall obtain the so-called *conservative permeability*

$$\mu' = \frac{2Q_v}{\omega H_{\mathrm{m}}^2} \tag{1.53}$$

consumption permeability

$$\mu'' = \frac{2P_v}{\omega H_{\mathrm{m}}^2} \tag{1.54}$$

and *loss angle* (δ)

$$\mathrm{tg}\delta = \frac{P_v}{Q_v} = \frac{\mu''}{\mu'} \tag{1.55}$$

The above equations do not consider losses from eddy currents, which are also a reason of phase shifts. These phase shifts can be accounted for either by a corresponding increase of power in the above equations or by the introduction of separate arguments derived from the calculation of eddy currents (Section 3.2).

The assumption of sinusoidal alternation in time of H and B is equivalent to the substitution of the real hysteresis loop by a corresponding inclined ellipse with the same area as in the real hysteresis loop.

Equivalent (*effective*) *complex permeability* [2.8], [1.37] is an imaginary permeability of a package of sheets at alternating flux density, in which eddy currents, displacing the flux, cause an imaginary reduction of the effective permeability $\underline{\mu}_e$ and its shift in phase, whereas $\underline{\mu}_e = B_{m,\mathrm{average}}/H_{\mathrm{m}}$ (see Section 6.2.1). The $\underline{\mu}_e$ serves, among others, for the evaluation of the *quality factor* of coils with iron: $Q = \omega L/R$ (Bozorth [1.2], pp. 610–612). The $\underline{\mu}_e$ is also the permeability of a hypothetic uniform core equivalent to a core of the same dimensions, but containing nonmagnetic gaps and parts with various and/or nonuniform materials.

Tensor permeability appears at the analysis of anisotropic media. It will be discussed in Section 2.3 (Equation 2.82).

Imaginary permeability (fictitious, supposed), μ_{im}. This term is used (Bozorth [1.2], p. 670) for the permeability of a cylindrical iron sample in which, due to finite length, appears as the internal demagnetization field H_{demagn} (Figure 1.34 [later in the chapter]) taken into account by means of the demagnetization coefficient N (1.58)

$$\mu_{\text{im}} = \frac{B}{H_{\text{ext}}} \quad \text{and} \quad \frac{1}{\mu_{\text{im}}} \approx \frac{1}{\mu} + \frac{N}{\mu_0} (\mu \gg 1) \tag{1.56}$$

The value of the coefficient N depends on the ratio m of the length to the diameter of the sample; for example, at $m = 10$ and $\mu > 300$, $\mu_i = 60–70 \approx \text{const}$ ([1.2], p. 673).

Quasi-permeability (seeming permeability), μ_q. The author (J. Turowski) introduced this notion in 1982 (J. Turowski [1.16], p. 35) to denominate* an equivalent fictitious permeability of a solid iron, where a conductor with AC is running nearby. This name is useful in the application of the method of *mirror images* (Section 5.2) for alternating currents and fields, when the "mirror is not fully clean" due to the eddy-current reaction and iron saturation.

1.2.3.16 Magnetic Anisotropy

A ferromagnetic monocrystal, being a body with a uniform, accurate crystal structure, has different properties along various crystallographic axes. The main crystallographic directions in an elementary isometric lattice of crystal of iron are nominated by the axes [100], [110], [111] (Figure 1.30). The monocrystal of iron has the biggest permeability and smallest hysteresis losses at magnetization in the direction parallel to the [100] edge of cube, the smallest permeability and biggest hysteresis losses in the direction of the longest diagonal [111], and average values in the direction of the smaller diagonal [110].

The anisotropic properties of iron are utilized in the manufacturing of low-loss, *cold-rolled* electrical sheets of high magnetizability and low per-unit losses. The isotropic, *hot-rolled* transformer sheets (4% Si), formerly used extensively for power transformer cores, but now practically eliminated, as well as the motor and some generator sheets

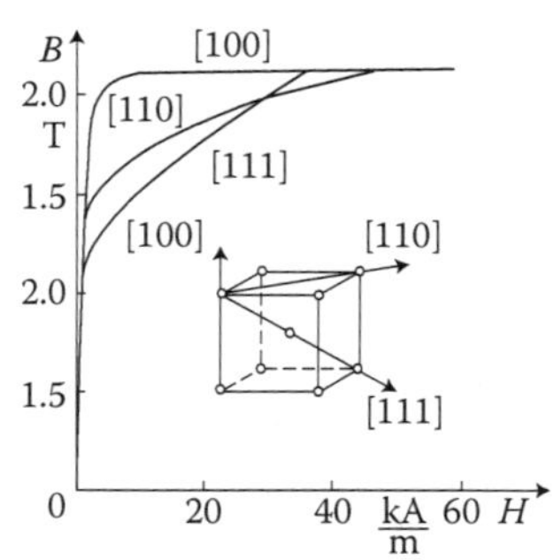

FIGURE 1.30 Magnetization characteristics of silicon iron along the axes of the most advantageous [100], medium [110], and the worst magnetization [111] and lowest permeability direction.

* The quasi-permeability *Paradigm* helps to accelerate modeling and calculation of a system with nonlinear conducting iron, with simulation of nonlinearity, hysteresis, and eddy current influence.

(about 2% Si), have *isotropic, polycrystalline* structures with the random orientation of crystals. Due to that, their magnetic properties in various directions are practically identical. Higher-quality types include *cold-rolled* sheets, which are divided into

1. *Generator isotropic sheets* (in Poland EP—Figure 1.23), with almost the same magnetic properties like hot-rolled sheets or those with minimal anisotropy; the cold-rolling here has mainly the purpose to obtain better smoothness and equalization, which gives better filling factor of iron core.
2. *Anisotropic transformer sheets*, also called *textured* or *grain-oriented* (in Poland ET—Figure 1.23), in which most of the crystals are oriented in the most advantageous direction [100], in accordance with the rolling direction (Bochnia 2006 [1.55]).

Specific features of the textured sheets manufacturing process include the use of iron with the Si content of 2–3%, cold rolling with interoperation annealing for recrystallization, and final thermal processing at a high temperature (1100°C) in hydrogen atmosphere, which causes a reduction of the contents of carbon in the iron to a value of about 0.005% [1.2]. Cold-rolled transformer sheets are typically delivered as double sided, isolated with a thin ceramic layer, called *carlit*. Generator sheets can also be covered by an electroinsulating organic coating, or not isolated, which helps to punch it.

The industrial manufacturing methods of textured sheets typically do not fully exploit all their capabilities. The texture obtained (crystal layout) and traditionally used gives only one advantageous direction of magnetization, [100] or [001] (Figure 1.31a). The scheme of the most advantageous texture, not yet achieved in industrial practice, is presented in Figure 1.31b. Such texture, achieved already in the laboratory [1.45], [1.46], would allow to totally eliminate the [111] direction with the most difficult magnetic properties, and to achieve the two most advantageous directions along the [100]

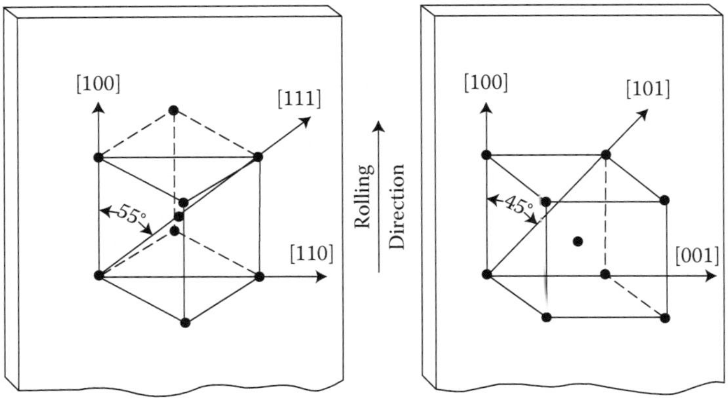

FIGURE 1.31 Texture of crystals in cold-rolled anisotropic sheets: (a) achieved so far and traditionally used, the so-called cube-on-edge Goss texture; (b) the most advantageous cube-on-face texture. (Adapted from Kolbinski K. and Słowikowski J.: *Electrical Materials.* (in Polish), Warsaw: WNT 1988.)

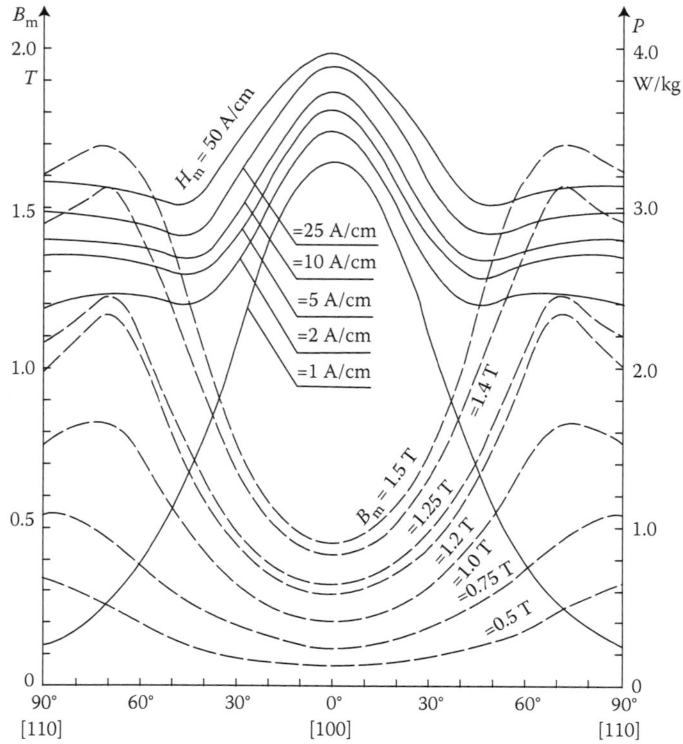

FIGURE 1.32 The flux density B_m (———) and the per-unit loss p (-------) of cold-rolled transformer sheets ET4, 0.30 mm, versus the angle of inclination from the rolling direction, at 50 Hz, after *ZPT Bochnia Works* (Poland). (Adapted from *Silicon Electrical Sheets*. Bochnia: Catalogue of Metallurgical Processing Plant, 1982.)

and crosswise [001] rolling. An industrial production of the latter sheets would open broad possibilities of their application in building of rotating electric machines.

Lately, there has been a trend in building iron cores in a manner to make the flux passing perpendicularly to the rolling direction as small as possible.

As one can see from Figure 1.32, the ratio of the per-unit losses in the most advantageous directions to the per-unit losses in the most difficult direction depends on the flux density, for example, in the range of B_m from 0.5 to 1.5 T the ratio is from 0.20 to 0.25. The relation of the magnetization energy in the direction of the most difficult magnetization to the energy in the direction of the easy magnetization is called *anisotropy constant*, K [1.22].

1.2.3.17 Magnetostriction

Magnetostriction describes the process of the change of crystals and dimensions of a body under the influence of magnetization (Figure 1.33). It is caused by the change of interatomic distances between the atoms of a lattice. Magnetostriction, which is characterized by the relative elongation, $\varepsilon = \Delta l/l$, can be positive or negative. It is one

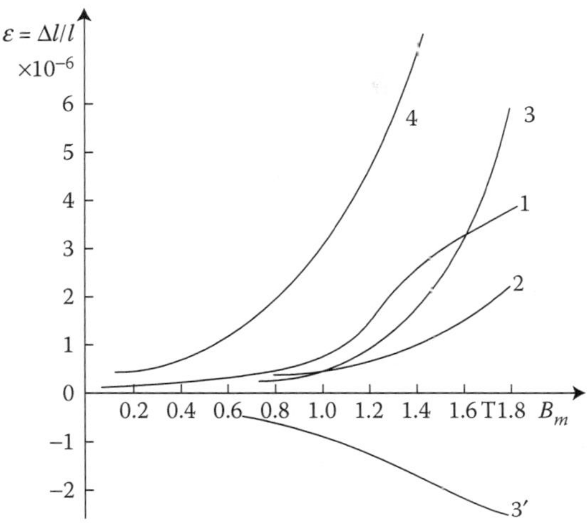

FIGURE 1.33 The magnetostriction curves, $\varepsilon = \Delta l/l$, of transformer sheets, after Narolski [1.5]: 1—hot-rolled sheet, per Knowlton; 2—cold-rolled sheet, after data of Armco; 3—cold-rolled sheet not annealed, per *BBC*; 3′—like 3, but annealed; 4—cold-rolled sheet not annealed.

of the sources of noise of electric machines and transformers, but it is also exploited in ultrasonic generators.

Ferrites [1.22]. Ferrites belong to the ceramic, magnetic materials obtained from magnetite Fe_2O_4 or from the double oxide $FeO \cdot Fe_2O_3$. In contrast to the normal ferromagnetics, ferrites have a very small electric conductivity and can be considered as insulators (Figure 1.35 [later in the chapter]).

Because of that, phenomena appear in ferrites that are hidden in other materials due to the screening effect of eddy currents. These phenomena are related to the so-called *ferromagnetic resonance*. Spins of electrons, considered as *spinning tops* with defined magnetic field, gyrating in an external magnetic field $\boldsymbol{H}$, perform a *precession* motion. This causes a tensor relation of the $\boldsymbol{B}$ and $\boldsymbol{H}$ vectors (Simonyi [1.11]). *Soft ferrites* (e.g., silicon ferrites), with a narrow hysteresis loop (Table 1.7), are used for high-frequency transformers, over 100 kHz. *Hard ferrites*, containing manganese and magnesium, can reach a hysteresis loop form close to square, which allows using them in computer memory storage. Other ferrites, for instance, $BaFe_{12}O_{19}$ (magnadur) are used as *permanent magnets* with very large coercive magnetic intensity (Figure 1.22)—for magnets with big air gaps. Ferrites can have an anisotropic structure. They are relatively cheap and are produced on the basis of powder metallurgy [1.22].

1.2.3.18 Demagnetization Coefficient

If an iron rod is placed in an external magnetic field, H_{ext}, on its ends will be created magnetic poles, which in turn produce in the rod its own field. This field is directed opposite to the external field and is called *demagnetization field* (Figure 1.34). The magnetic field intensity of the real (internal) magnetic field in any cross section of

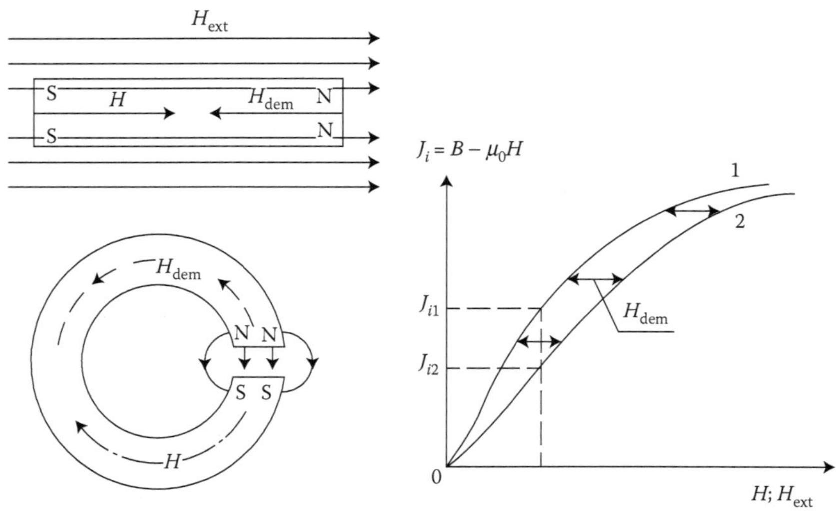

FIGURE 1.34 Demagnetization action of the ends of an iron rod and of an air gap; 1—the real curve: $B - \mu_0 H = f(H)$; 2—the lowered curve: $B - \mu_0 H = f(H_{ext})$; H_{dem}—the demagnetization field; H_{ext}—the external field.

the sample is the difference between the external field H_{ext} and the demagnetization field H_{dem}, that is

$$H = H_{ext} - H_{dem} \tag{1.57}$$

In a first-order approximation, the demagnetization field intensity is proportional to the magnetic polarization of the sample $\boldsymbol{J}_i$

$$H_{dem} \approx N\frac{\boldsymbol{J}_i}{H_0} = N\left(\frac{B}{\mu_0} - H\right) \tag{1.58}$$

The proportionality factor N is called *demagnetization factor*. It depends, first of all, on the shape of the sample. The demagnetization factor can be calculated exactly only for ellipsoid of revolution, placed in a uniform field. Inside of such ellipsoids, the field is always uniform and their demagnetization factor is expressed by the formula*

$$N = \frac{\frac{n}{\sqrt{n^2-1}}\ln\left(n+\sqrt{n^2-1}\right)-1}{n^2-1} = \frac{1-\frac{n}{\sqrt{1-n^2}}\arccos n}{1-n^2} \tag{1.59}$$

* Neiman LP., Kalantarov PL.: *Theoretical Fundamentals of Electrotechnics*. (in Russian) Moscow; GEI, 1948, p. 117.

in which n is the ratio of the ellipsoid revolution axis directed along the lines of the external field, to the axis perpendicular to it.* The first form of the expression (1.59) for N is more convenient when $n > 1$, and the second form when $n < 1$. By changing the ratio of axes, one can obtain approximate values of N for bodies having a shape different than ellipsoid. For instance, in the case of an infinitely extended plate placed across the field ($n = 0$), we can obtain the biggest possible value of $N = 1$.

In the case of a cylinder placed across the field, $N = 0.5$ [1.2], and a sphere ($n = 1$) we get $N = 0.33$, and finally, in the case of an infinite rod placed along the field ($n \to \infty$) we get $N = 0$. For cylindrical iron rods with the ratio of length to diameter $n' = 5$, the obtained N was ≈ 0.5, and at $n' = 100$, $N = 0.0045$ [1.9]. It follows from this, that iron rods placed inside a solenoid, of length some dozens times bigger than its diameter, can be considered as practically infinitely long.

However, at short samples, one has to consider the demagnetization effect. These conclusions are important from the viewpoint of model investigations with long rods and at the induction heating of short pins or steel charges.

The demagnetization effect reveals itself in machines and electric equipment with permanent magnets in the form of the so-called *recoil straight line* (Turowski [1.18]; [1.2], p. 285). It has a significant effect, among others, on the reduction of the resultant (apparent) permeability μ' of a sample of finite length (1.56).

1.2.4 Semiconductors and Dielectrics

The term *semiconductors* is used for a large group of bodies (elements, compounds, alloys, ceramic bodies, glassy, and liquid bodies), whose electric conductivity value is between the conductivities of metals and insulators (Figure 1.35).

Characteristic features of semiconductors include a strong dependence of their conductivity on temperature, impurities, light, radioactive radiation, and so on. At temperatures near the absolute zero, they become almost like ideal dielectrics,

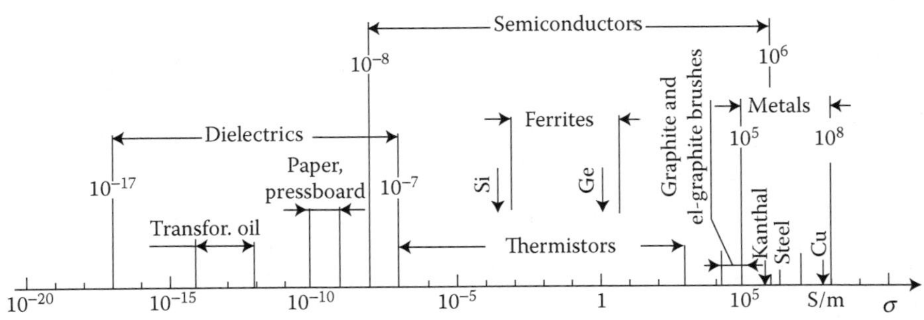

FIGURE 1.35 A comparative scale of conductivity of dielectrics, semiconductors, and metals at a temperature of 20°C.

* As per Ref. [10.2] (p. 28), the apparent permeability of such sample is $\mu' = 1/(1/\mu + N)$. So, at large values of N, the μ' does not depend on μ, but rather on N.

and at high temperatures, the conductivity increases inversely to metals. The electric conductivity of semiconductors increases under the influence of strong electric fields, and decreases in magnetic fields. The ability of semiconductors to change their properties under the influence of external factors is the basis of the operation of thermoresistive elements, that is, the so-called *thermistors (thermal resistors)*, nonlinear variable resistors (the so-called *varistors*), photoelements, and so on.

Thermistors, which have a significant negative thermal coefficient of resistivity, are used as simple and sensitive thermometers which allow it to register temperature changes of 0.0005°C, in systems for control and compensation of temperature, for measurement of flow and velocity of gases, pressure and humidity, mechanical stresses, and in dividers and stabilizers of voltage. The material for thermistors includes oxides of various metals, such as CuO, Mn_2O_4, UO_2, and others, in the form of pressured and sintered powders (Wyatt [1.22], p. 582).

Varistors (variable resistors) made of silicon carbide (SiC) are nonlinear resistor elements. The resistivity of varistors decreases with an increase of electric field intensity. In the high-voltage technology, varistors are employed in valve arresters, while in the low-voltage technology, they are used in voltage stabilizer systems, frequency multiplicators, in the control of rotational speed and reversible operation of motors, in computational technology, and other systems.

The specific properties of semiconductors result from their crystallographic structure, different from the structure of metals. In metals, adjacent atoms, whose external electrons interpenetrate, are held in the crystal lattice owing to the action of attraction forces (each nucleus attracts electrons of adjoining atoms) and repulsion (Pauli exclusion principle). In this situation, the external valence level is not fully filled. In semiconductors, however, the so-called *atomic bonds* (electrons pairwise) occur, in which atoms are so distributed that two or bigger number of atoms have joint electrons tending to attain the so-called *electron octet* (8 electrons). On a joint orbit, pairs of electrons with opposite spins bind at the same time, so that Pauli exclusion principle is not violated. In such a system, all valence electrons are then rigidly bound with atoms, which are tied together and there are no free electrons. Hence, conductivity does not exist. Such a state exists primarily at the temperature of absolute zero. At sufficiently high temperature of crystal, oscillations of atoms can increase so much that some bonds may be incidentally broken and the released electrons begin to behave as free electrons in metals, causing the conductivity of crystal. In the crystal, spots abandoned by such electrons emerge as the so-called *holes*, with a positive unit charge. Hence, the higher the conductivity of semiconductors, the higher their temperature.

In contrast to metals, in which the free allowed band of energy levels adheres directly to the band of levels filled with valence electrons, in semiconductors and dielectrics between such levels exists a *forbidden band* (*band gap*). If this band gap is very broad, the electrons cannot penetrate into the band of free levels and such a body will be a dielectric. In semiconductors, the band gap is smaller and corresponds to energy around 1–2 eV (Figure 1.36). By delivering thermal, photonic, or electric energy in semiconductors, a jump (*excitation*) occurs of an electron over the forbidden energy band to the *conduction band*. Such electrons become free and create *electron conductivity of semiconductor*. Into the vacancies (*holes*) left by electrons in

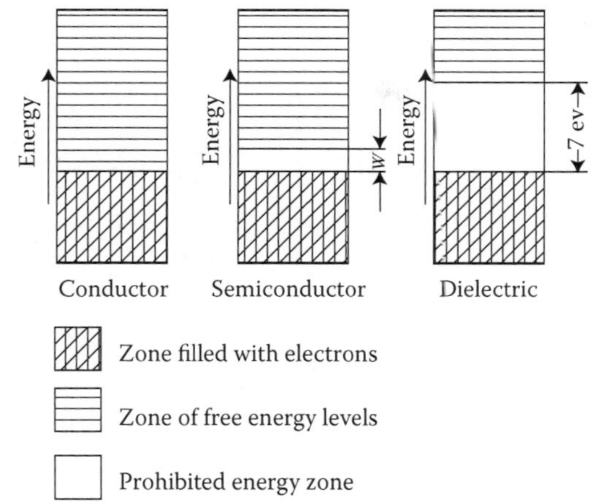

FIGURE 1.36 Differences in the distribution of permitted energy levels in conductors, semiconductors, and dielectrics. (Adapted from *Handbook of Electrical Materials.* (in Russian) Vol. 2, Moscow: Gosenergoizdat, 1960.)

the valence band, under the influence of electric field, electrons enter from neighboring bonds, leaving new holes, and so on. In effect, under the influence of an electric field, the holes of positive charge move in the direction opposite to the direction of electrons, and the resulting total current consists of an *electron* or *n* (negative) current, and of a *hole* or *p* (positive) current.

The described way of conductivity in ideal crystals with atomic bonds is called *intrinsic conductivity* and has the symbol *i* (*intrinsic*), and the corresponding materials are called *intrinsic* semiconductors (e.g., pure germanium, silicon). Since the motion of holes takes place with bigger inertia than motion of free electrons (the electron *mobility* is higher than the hole mobility), intrinsic semiconductors in principle have a character of electron conductivity.

A significant influence on semiconductor properties are admixtures and impurities (called dopants), even at relatively small contents (for instance, 10^{-5}–10^{-6}% of the total number of atoms). One type of impurity, or dopants, in semiconductors are called *acceptors* which capture (*accept*) electrons from the valence band and thereby leave holes. Such acceptor-doped semiconductors have conductivity of type *p*, due to the *positive* (+) charge of hole, and are called *p-type* semiconductors. The other type of impurities are called *donors*, which deliver (*donate*) free electrons to the conduction band of a semiconductor. Such a donor-doped semiconductor has conductivity of type *n*, due to *negative* (–) charge of the electron, and is called an *n-type* semiconductor. Such *doping-based conductivity* need less energy for exciting free carriers (*activation energy*, on the order of 0.01–0.1eV) and hence occurs at lower temperatures than the intrinsic conductivity (Figure 1.37).

By contacting the two types of semiconductors: the *electron-dominated n-type semiconductor* and the *hole-dominated p-type semiconductor*, an electrostatic potential barrier arises (*built-in potential* barrier layer) promoting easier current flow in one direction (from *p* to *n*, called forward bias) [2.40]. This asymmetry results in

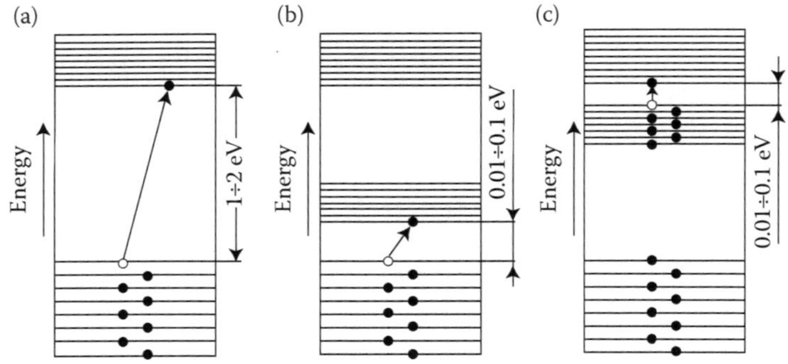

FIGURE 1.37 Influence of impurities (dopants) on the allowed energy levels in semiconductors: (a) intrinsic semiconductor, (b) semiconductor with "acceptor" doping of type p (positive), with hole conductivity, (c) semiconductor with "donor" doping of type n (negative), with electron conductivity. •—electron; o—hole. (Adapted from *Handbook of Electrical Materials.* (in Russian) Vol. 2, Moscow: Gosenergoizdat, 1960.)

the effect of rectification of an AC in *p–n junction* devices, such as silicon diodes. The application of two or more connected p–n junctions allows the build of more advanced, controlled semiconductor devices, such as transistors and thyristors [2.40].

1.2.4.1 Hall Effect in Semiconductors

The Hall effect, described earlier for metals (Figure 1.13), occurs even more distinctively in semiconductors, in which electrons and holes gather themselves in opposite ends of a plate, thus increasing the transverse electric field. Formulae (1.14) and (1.15) are also valid. The Hall constant for semiconductors of type p or n (without dopants of the opposite conductivity) equals to

$$R_H = \frac{3\pi}{8} - \frac{1}{N'e} \tag{1.60}$$

where N' is the density of electrons or holes, respectively, and e is the elementary charge (of electron or hole).

The Hall constant in the case of germanium plates is on the order of 10^{-3} Vm/(AT).

The Hall effect in semiconductors is utilized for measurements of magnetic field intensity (the so-called *hallotrons*).

Also, for the measurement of magnetic fields are utilized spirals made of bismuth wire, whose resistance changes under the influence of change of magnetic field intensity (the so-called *gaussotrons*).

Piezoelectricity, discovered in 1880 by Pierre and Paul Curie, consisting of mechanical deformation of some crystals lacking symmetry center (like quartz SiO_2, Seignette's salt $NaKC_4H_4O_6 \cdot 4H_2O$, or titanate barium $BaTiO_3$), under the influence of an external electric field, is utilized, among others, for building precise electromechanical elements of automatics and microdrives of power up to 50 W. Various types of "ratchet" mechanisms enable conversion of crystal's oscillation into unidirectional displacement of a rotor or a linear element. This effect is reversible.

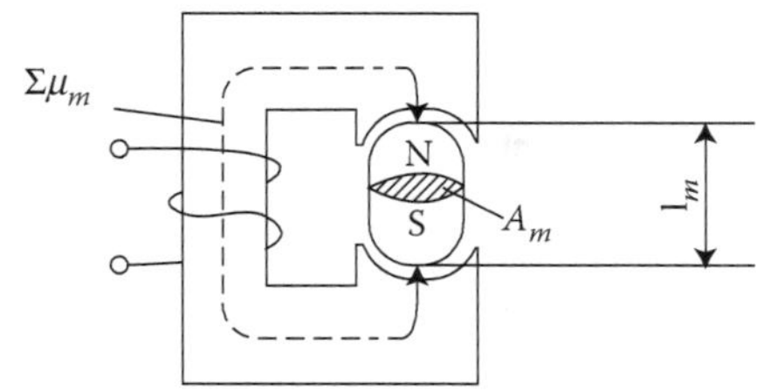

FIGURE 1.38 Main dimensions of magnetic circuit with permanent magnet.

At a mechanical deformation, for instance at strain, distances between ions increase, which causes an increase of dipole moments between pairs of ions. If a crystal does not have a center of symmetry, the dipoles of opposite signs are not self-balancing and some resultant dipole moment appears, which manifests itself by the emergence of opposing electric charges on the opposite surfaces of crystals, and eventually an electric field.

EXAMPLE

Determine the main dimensions of a DC machine with permanent magnets (J. Turowski [1.19]).

If it is assumed that the volume of a permanent magnet is $V_m = A_m l_m$, where A_m and l_m are the cross-sectional area and the length of the magnet, respectively (Figure 1.38), then the magnetic flux per pole is

$$\Theta_m = B_K A_m \tag{1.61}$$

where B_K is the flux density in the intersection point K of the recoil line 4 and the permeance line 3 (Figure 1.39) of external part of magnetic circuit. Then, the sum

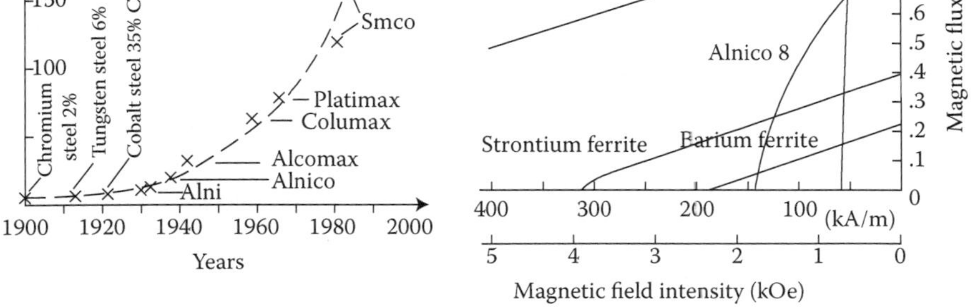

FIGURE 1.39 Progress in the development of permanent magnets: (a) the maximum energy product since 1900 and (b) typical demagnetization curves.

ΣF_m, of the mmf drops in the external circuit must be equal to the magnetomotive force $H_K l_m$ per pole:

$$\sum F_m = H_K l_m \tag{1.62}$$

where H_k is the field intensity of the magnet at the working point K. The volume (1.31) of the magnet is then:

$$V_m = \frac{\Phi \Sigma F_m}{B_K H_K} \tag{1.63}$$

Equation 1.63 shows that the volume and mass of the magnet, and therefore the entire motor, are lower; the higher is the specific magnet energy $w = BH/2$ per unit volume (J/m^3). The increase in magnet energy that is achieved with new materials (Figure 1.40) represents the major progress in the last few decades.

The excitation in large electric machines is produced by electromagnets whereas in small machines it is produced by permanent magnets. This is caused by the magnetic energy of permanent magnets, which is proportional to the cubic power (l^3) of the linear dimensions l and that of electromagnets to the product of flux $\Phi = Bs$ and magnetizing force $F_m = IN$, that is, to the fourth power (l^4) of the linear dimensions.

At the same time, if it is considered that the volume of the permanent magnet in an electric machine is inversely proportional (1.31) to its maximal energy $(BH)_{max}$, and that since the beginning of the twentieth century the specific energy of permanent magnets has increased more than 30 times (as shown in Figure 1.22), the impact on machine design and weight that has been achieved can be understood. During the same period, the per-unit power loss in laminated cores has been reduced almost 10 times, which again has improved the design parameters. It is expected that amorphous magnetic materials, high-temperature

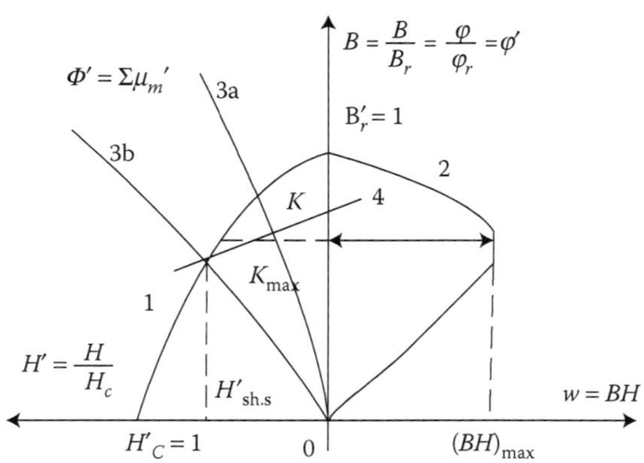

FIGURE 1.40 Evaluation of optimal working point K for minimum PM motor size: 1—demagnetization curve, 2—curve of magnet energy, 3—permeance line at no-load (a) and at short circuit (b) when stabilized by short circuit, 4—recoil line of magnet stabilization (compare loop B1–B2 in Figure 1.20).

superconductivity, and silicon micromechanics will have further significant impact on modern motors and their performance.

The mentioned low Curie point (Figure 1.26) sensitivity to changes of temperature and corrosion means that the thermal characteristics of REPM materials get worse at a lower Curie temperature t_C than usual iron. This is an important disadvantage of these materials and in REPM motors it is often necessary to apply special magnetic *shunt thermocompensators*. Thermomagnets used for thermocompensator should have the Curie temperature t_C between 65°C and 120°C, corresponding to the maximal rated (nominal) temperature t_n and the nominal flux density B_n of permanent magnet applied in electric machinesapplied in electric machine [1.33].

2 Fundamental Equations of Electromagnetic Field

2.1 PRIMARY LAWS AND EQUATIONS OF ELECTROMAGNETISM

From the moment when Maxwell, on the basis of the empirical laws discovered by Biot and Savart (1820) and Faraday (1831), formulated the general equations of electromagnetic field (1873) and introduced to these equations the hypothesis of displacement current, one could give up the traditional, historical way of teaching the electric and magnetic fields and, taking Maxwell's equations as a starting point, derive from them in a deductive way all the basic equations of the macroscopic electrodynamics.

The first and the second Maxwell's equations for bodies being in a relatively slow motion, with the velocity $v \ll c$, have the form:

$$\text{curl}\, \boldsymbol{H} = \boldsymbol{J}_{\text{total}} \tag{2.1}$$

$$\text{curl}\, \boldsymbol{E} = -\frac{\partial \boldsymbol{B}}{\partial t} \tag{2.2}$$

where the "total" current density

$$\boldsymbol{J}_{\text{total}} = \sigma \boldsymbol{E} + \frac{\partial \boldsymbol{D}}{\partial t} + \sigma\, (\boldsymbol{v} \times \boldsymbol{B}) + \rho \boldsymbol{v}_{\rho} + \text{curl}(\boldsymbol{D} \times \boldsymbol{v}) + \boldsymbol{J}_{\text{extern}} \tag{2.3}$$

The set of equations of electrodynamics, together with Equations 2.1 through 2.3, is completed by

- Two divergence (outflow) equations

$$\text{div}\, \boldsymbol{B} = 0 \tag{2.4}$$

$$\text{div}\, \boldsymbol{D} = \rho \tag{2.5}$$

- Two "*constitutive*" equations

$$\boldsymbol{D} = \varepsilon\, \mathbf{E}, \quad \boldsymbol{B} = \mu\, \boldsymbol{H} \tag{2.6}$$

- Lorentz force (*N*), with which the fields $\boldsymbol{E}$ and $\boldsymbol{B}$ act on a moving charge Q

$$\boldsymbol{F}_L = Q\, (\boldsymbol{E} + \boldsymbol{v} \times \boldsymbol{B}) \tag{2.7}$$

- The volume density (N/m^3) of Lorentz force

$$f_L = \rho\, \boldsymbol{E} + \boldsymbol{J} \times \boldsymbol{B} \tag{2.8}$$

- *Field energy* density (J/m^3)

$$w = \int_{D_o}^{D} \boldsymbol{E} \cdot \mathrm{d}\boldsymbol{D} + \int_{B_o}^{B} \boldsymbol{H} \cdot \mathrm{d}\boldsymbol{B} \tag{2.9}$$

- *Poynting* vector (W/m^2)

$$\boldsymbol{S} = \boldsymbol{E} \times \boldsymbol{H} \tag{2.10}$$

Sometimes, all the equations from Equations 2.1 through 2.6 are called Maxwell's *equations* (Hammond [2.8], p. 257).

Equation 2.3 is a different form of *Ohm's law*. Its particular components are: $\sigma\boldsymbol{E}$ is the *conduction* current density induced due to change of magnetic field in time (*transformation current*); $\partial\boldsymbol{D}/\partial t$ the *displacement* current density (primarily in dielectrics); $\sigma\,(\boldsymbol{v} \times \boldsymbol{B})$ the *conduction* current density induced by $\boldsymbol{B}$ in a conductor moving with a velocity $\boldsymbol{v}$; $\rho\boldsymbol{v}_\rho = \boldsymbol{v}_\rho\ \mathrm{div}\boldsymbol{D}$ the current density of charge density ρ_{free} convection with the velocity $\boldsymbol{v}_\rho$; curl $(\boldsymbol{D} \times \boldsymbol{v})$ the current density as a result of motion of electric field $\boldsymbol{E}$ or polarized dielectric or bound charges ρ_{bound}; $\boldsymbol{J}_{\mathrm{extern}} = \sigma\boldsymbol{E}_{\mathrm{extern}}$ the external current density supplied from external sources or caused by external electromotive forces of non-electromagnetic origin; for instance, due to a nonuniform charge concentration which causes diffusion currents, thermoelectric forces, etc. For example, at a varying charge density ρ, or temperature T:

$$\left.\begin{aligned} \sigma\boldsymbol{E}_{\mathrm{extern}} &= -d\,\mathrm{grad}\,\rho \\ \sigma\boldsymbol{E}_{\mathrm{extern}} &= -b\,\mathrm{grad}\,T \end{aligned}\right\} \tag{2.11}$$

where d is the diffusion coefficient, and b is the thermoelectric coefficient.

In fields of lower frequency (e.g., in power equipment), the displacement current density ($\partial\boldsymbol{D}/\partial t$) can often be neglected, in comparison to the conduction current density ($\boldsymbol{J} = \sigma\boldsymbol{E}$). Such fields are called *quasi-stationary*. In such fields, however, the varying in the time component $\partial\boldsymbol{B}/\partial t$ cannot be omitted.

The denominations of differential operators such as **curl** (circulation, rotation = rot), **div** (divergence, outflow), and **grad** (gradient = slope), suggest a physical meaning of the discussed relations. A more uniform viewpoint from the formal-mathematical stance, and faster, is the representation of these equations with the vector operator ∇, called "del" (or ∇), or *Hamilton's differential operator*. In the Cartesian (rectangular) coordinates, it has the form

$$\nabla = \mathbf{i}\frac{\partial}{\partial x} + \mathbf{j}\frac{\partial}{\partial y} + \mathbf{k}\frac{\partial}{\partial z} \tag{2.12}$$

Then, we can write

$$\text{curl}\,\boldsymbol{H} = \nabla \times \boldsymbol{H}, \quad \text{div}\,\boldsymbol{H} = \nabla \cdot \boldsymbol{H}, \quad \text{grad}\,V = \nabla\,V$$

and the Laplacian operator in Cartesian coordinates

$$\nabla \cdot \nabla = \nabla^2 = \frac{\partial^2}{\partial x^2} + \frac{\partial^2}{\partial y^2} + \frac{\partial^2}{\partial z^2} \tag{2.13}$$

The Laplacian operator is sometimes denoted as $\nabla^2 \equiv \Delta \equiv \text{Lap}$.

More generally,

$$\nabla^2 u = \sum_{i=1}^{n} \frac{\partial^2 u}{\partial x_i^2} \tag{2.13a}$$

From the vector Equations 2.1 through 2.9, immediately follow the scalar equations for the magnitudes of vector components, because the unit vectors **i**, **j**, **k** at both sides of the equations given below reduce themselves.

$$\frac{\partial H_z}{\partial y} - \frac{\partial H_y}{\partial z} = J_x + \frac{\partial D_x}{\partial t} + \sigma\,(v_y B_z - v_z B_y) + \rho v_x \tag{2.1a}$$

$$\frac{\partial H_x}{\partial z} - \frac{\partial H_z}{\partial x} = J_y + \frac{\partial D_y}{\partial t} + \sigma\,(v_z B_x - v_x B_z) + \rho v_y \tag{2.1b}$$

$$\frac{\partial H_y}{\partial x} - \frac{\partial H_x}{\partial y} = J_z + \frac{\partial D_z}{\partial t} + \sigma\,(v_x B_y - v_y B_x) + \rho v_z \tag{2.1c}$$

$$\frac{\partial E_z}{\partial y} - \frac{\partial E_y}{\partial z} = -\frac{\partial B_x}{\partial t} \tag{2.2a}$$

$$\frac{\partial E_x}{\partial z} - \frac{\partial E_z}{\partial x} = -\frac{\partial B_y}{\partial t} \tag{2.2b}$$

$$\frac{\partial E_y}{\partial x} - \frac{\partial E_x}{\partial y} = -\frac{\partial B_z}{\partial t} \tag{2.2c}$$

$$\text{div}\,\boldsymbol{B} = \frac{\partial B_x}{\partial x} + \frac{\partial B_y}{\partial y} + \frac{\partial B_z}{\partial z} = 0 \tag{2.4a}$$

$$\boldsymbol{D} = \boldsymbol{D}_x + \boldsymbol{D}_y + \boldsymbol{D}_z = \mathbf{i}\, D_x + \mathbf{j}\, D_y + \mathbf{k}\, D_z = \mathbf{i}\, \varepsilon E_x + \mathbf{j}\, \varepsilon E_y + \mathbf{k}\, \varepsilon\, E_z \tag{2.6a}$$

$$w = \frac{\varepsilon}{2}(E_x^2 + E_y^2 + E_z^2) + \frac{\mu}{2}(H_x^2 + H_y^2 + H_z^2) \tag{2.8a}$$

The differential Maxwell's equations are not fundamental laws of nature, but rather genial scientific hypotheses and natural law generalizations, verified in a large number of applications. They can be reduced to the integral laws of electrical engineering theory, with the help of *Stokes' theorem*, which states that the *linear integral of the field vector along a closed curve l equals to the flux of the* curl *of this vector over surface a bounded by this curve* (Figure 2.1).

Applying Stokes' theorem to the first Maxwell's equation (Figure 2.1)

$$\oint_l \boldsymbol{H} \cdot \mathrm{d}l = \iint_A \mathrm{curl}\, \boldsymbol{H} \cdot \mathrm{d}\boldsymbol{A} \tag{2.14}$$

we obtain the *law of flow*, known also as the Ampere's law equation:

$$\oint_l \boldsymbol{H} \cdot \mathrm{d}l = \iint_A \boldsymbol{J} \cdot \mathrm{d}\boldsymbol{A} = i \tag{2.15}$$

The scalar (dot) product $\boldsymbol{H} \cdot \mathrm{d}\boldsymbol{l} = H\, \mathrm{d}l \cos\alpha = \mathrm{d}U_m$ represents the magnetic voltage across the section of loop of length dl (Figure 2.1). At the same time, the integral $\iint_A \boldsymbol{J} \cdot \mathrm{d}\boldsymbol{A} = F$ is the *magnetomotive force* (MMF), also called *flow*.

The field $\boldsymbol{H}$ in Equation 2.15 is defined by the Biot–Savart law:

$$\boldsymbol{H} = \frac{i}{4\pi}\oint_l \frac{\mathrm{d}l \times \mathbf{1}_\mathrm{r}}{r^2} = \frac{i}{4\pi}\oiiint_V \frac{\vec{J} \times \vec{\mathbf{1}}_r}{r^2}\mathrm{d}V \quad \mathrm{d}V = \mathrm{d}l\mathrm{d}A, \quad i = \iint \boldsymbol{J} \cdot \mathrm{d}\boldsymbol{A} \tag{2.16}$$

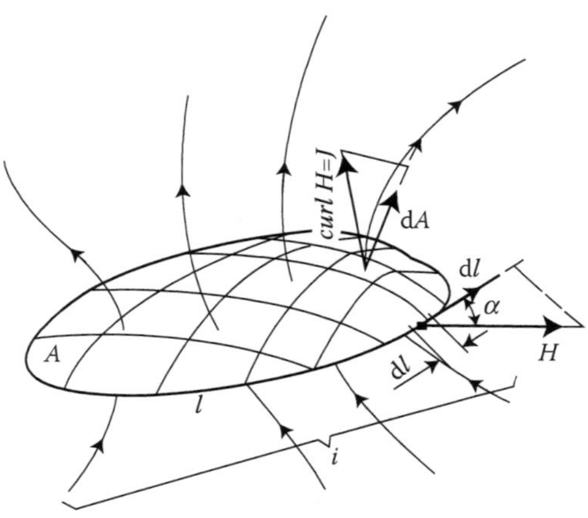

FIGURE 2.1 Illustration of the Stokes' theorem (curl ≡ rot).

which can also be represented as a vector component d$\boldsymbol{H}$ caused by a conductor element d$\boldsymbol{l}$

$$\mathrm{d}\boldsymbol{H} = \frac{i}{4\pi}\frac{\mathrm{d}\boldsymbol{l}\times\mathbf{1}_{\mathrm{r}}}{r^2};\quad \mathrm{d}H = \frac{i}{4\pi}\frac{\mathrm{d}l}{r^2}\sin\alpha \tag{2.16a}$$

where d$\boldsymbol{l}$ is the length of element of the conductor directed in accordance with the direction of the current i; $\mathbf{1}_{\mathrm{r}}$ is the unit vector pointing from a given element of conductor to the considered point; r the distance from this point to the element of conductor. The vectors, in the order d$\boldsymbol{l}$, $\mathbf{1}_{\mathrm{r}}$, d$\boldsymbol{H}$, form a dextrorotatory system (Figure 2.2).

If we apply Stokes' theorem to the second Maxwell's Equation 2.2

$$\iint_{A} \mathrm{curl}\,\boldsymbol{E}\cdot\mathrm{d}\boldsymbol{A} = \oint_{l} \boldsymbol{E}\cdot\mathrm{d}\boldsymbol{l} = -\iint_{A}\frac{\partial \boldsymbol{B}}{\partial t}\cdot\mathrm{d}\boldsymbol{A} \tag{2.17}$$

we obtain *Faraday's law of electromagnetic induction* in the mathematical Maxwell's formulation ([2.8], p. 234)

$$e = \oint_{l} \boldsymbol{E}\cdot\mathrm{d}\boldsymbol{l} = -N\frac{\mathrm{d}\Phi}{\mathrm{d}t} \tag{2.18}$$

and the experimental Faraday's formulation

$$\mathrm{d}Q = -\frac{\mathrm{d}\Phi}{R} \quad \text{or} \quad -N\frac{\mathrm{d}\Phi}{\mathrm{d}t} = R\,i \tag{2.18a}$$

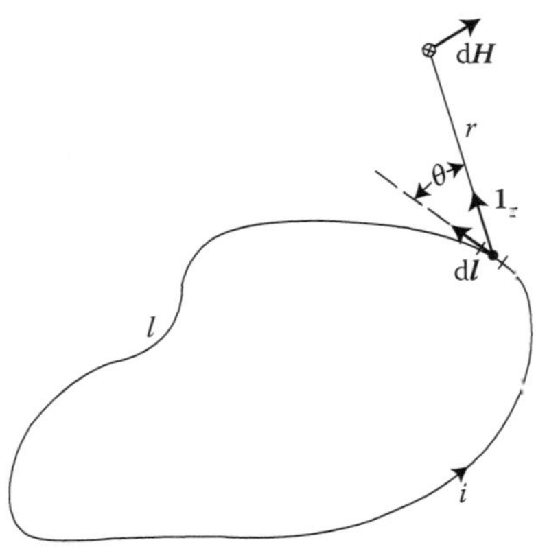

FIGURE 2.2 Illustration of the Biot–Savart Law.

In Equations 2.18, N is the number of turns enclosing the flux $\Phi = \iint_A \boldsymbol{B} \cdot \mathrm{d}\boldsymbol{A}$; $\mathrm{d}Q = i\,\mathrm{d}t$ is the electric charge. When the turns are so distributed that every of them encircles only a part of the total flux, then the electromotive force e is calculated from the equivalent number of turns $N \cdot k_w$, where $k_w < 1$ is the so-called *winding factor* (Turowski [1.18]). The minus sign means the well-known *electromagnetic inertia rule* of *Lenz* (1833), as follows:

In electric circuits and conducting bodies there exists a tendency to preserve in unchanged state the magnetic flux coupled with these circuits.

Any effort to change the flux in a circuit induces electromotive forces acting in direction opposite to these changes. This phenomenon appears most distinctively in superconductors [Meissner Effect (2.117), (2.118)].

Although Equations 2.18 and 2.18a are equivalent to each other, (2.18a) clearly explains that in a circuit with $R \approx 0$, the alternating flux $\mathrm{d}\Phi/\mathrm{d}t$ is completely displaced outside the circuit. Equation 2.18a concerns generally cutting conductor sections by the flux Φ. It is, therefore, more broad than Equation 2.18, which applies to closed circuits.

The flux encircled by a coil of N turns* can generally† undergo a change in time and space, and can be expressed by a composite function in the form

$$\Phi = f(t, x), \quad \text{where } x = x(t)$$

Derivative of the composite function, according to Equation 2.18, equals to

$$e = -N\frac{\mathrm{d}\Phi}{\mathrm{d}t} = -N\frac{\partial \Phi}{\partial t} - N\frac{\partial \Phi}{\partial x} \cdot \frac{\mathrm{d}x}{\mathrm{d}t} \tag{2.19}$$

The first component on the right-hand side of Equation 2.19 is called *transformation electromotive force* (it occurs in transformers and AC commutator machines), whereas the second component—*rotational electromotive force* (it occurs in rotational machines and other movable systems).

In the case when the investigated areas of field have character of circuits with uniform sections, it is more convenient to employ an integral form of the electromagnetic field equation. This is the reason for broad utilization of the (Faraday) *law of flow* (2.15) and the *law of electromagnetic induction* (2.18) in basic theory of electric machines and transformers. At more accurate investigations, when we wish to study the distribution of field in cross-section of these circuits, it is necessary to apply Equations 2.1 through 2.10, describing the field in any point of space, and not just resultant effect.

Equation 2.4, sometimes called the third Maxwell's equation [1.11], expresses *solenoidality*, that is, a *source-free* flux density (B) field, which means lack of free magnetic charges (poles). Equation 2.5 is a differential form of the Gauss' Law:

---

* In old Polish standards, the number of turns had the symbol z (*zwój*), and in German and Russian—w (Windung).

† Most generally, $\Phi = f(t, x, y, z)$, however, in electric machines $\Phi = f(t,x)$ is sufficient.

$$\oint_A \mathrm{d}\boldsymbol{D}\cdot \mathrm{d}\boldsymbol{A} = \begin{cases} \int_V \rho\,\mathrm{d}V & \text{when } \rho \text{ lies inside region } V \text{ of surface } A \\ 0 & \text{when } \rho \text{ lies outside surface } A \end{cases} \tag{2.20}$$

which means that: the flux of vector of electric flux density (electric displacement) $\boldsymbol{D}$ through a closed surface $\boldsymbol{A}$ equals to the electric charge enclosed inside this surface, $Q = \int_V \rho\,\mathrm{d}V$.

In mathematics, the Gauss's law is called the theorem or identity of Green and presented in the form:

$$\int_V \mathrm{div}\,\boldsymbol{D}\,\mathrm{d}V = \oint_A \boldsymbol{D}\cdot \mathrm{d}\boldsymbol{A} \tag{2.20a}$$

From Equation 2.20a, by substituting $\boldsymbol{D} = \Phi\nabla^2\Psi$ (Jackson [2.10], p. 59),* where Φ and Ψ are arbitrary scalar fields and $\partial/\partial n$ means derivative normal to the surface A, we obtain the *first Green's identity*, called also the *first formula of Green* [2.2], [2.10]

$$\int_V (\Phi\nabla^2\Psi + \nabla\Phi\cdot\nabla\Psi)\,\mathrm{d}V = \int_A \Phi\frac{\partial\Psi}{\partial n}\mathrm{d}A = \int_A \boldsymbol{D}\cdot\mathrm{d}\boldsymbol{A} \tag{2.21}$$

Reversing roles of Φ and Ψ in Equation 2.21, we get [2.10] the *second Green's identity*, called *"Green's Theorem"*:

$$\int_V (\Phi\nabla^2\Psi - \Psi\nabla^2\Phi)\,\mathrm{d}V = \oint_A \left(\Phi\frac{\partial\Psi}{\partial n} - \Psi\frac{\partial\Phi}{\partial n}\right)\mathrm{d}A \tag{2.22}$$

With the help of Equation 2.22 one can, for instance, transform the Poisson's differential equation $\nabla^2\Phi = -(\rho/\varepsilon)$ into an *integral equation* for points $\boldsymbol{x}$ located inside a volume V ([2.10], p. 59).

Electrostatics and magnetostatics are the particular and simplest cases of electrodynamics. Electrostatic and magnetostatic phenomenon are ruled by the same Maxwell's equations, which at the assumption of DC fields ($\partial\boldsymbol{D}/\partial t = \partial\boldsymbol{B}/\partial t = 0$) outside any sources ($\rho = 0$) and currents ($\boldsymbol{J} = 0$) in the region, for immovable media, take the form:

- For electrostatics:

$$\mathrm{curl}\,\boldsymbol{E} = 0, \quad \mathrm{div}\,\boldsymbol{D} = \rho, \quad \boldsymbol{D} = \varepsilon\boldsymbol{E} \tag{2.23}$$

- For magnetostatics:

$$\mathrm{curl}\,\boldsymbol{H} = 0, \quad \mathrm{div}\,\boldsymbol{B} = 0, \quad \boldsymbol{B} = \mu\boldsymbol{H} \tag{2.24}$$

* Here $\Delta \equiv \nabla^2$.

As one can see, these fields can exist and be investigated fully independent of each other. The equations curl $\boldsymbol{E} = 0$ and curl $\boldsymbol{H} = 0$ confirm that these fields are *irrotational*. Such fields are *potentional*, which means for their description one can introduce scalar functions of position $V(x, y, z)$ and $V_m(x, y, z)$ called *electric* and *magnetic potentials*, respectively. These potentials, per the principle of identity of vector calculus (curl grad $V \equiv 0$), fulfill the relations:

$$\boldsymbol{E} = -\text{grad } V \tag{2.23a}$$

$$\boldsymbol{H} = -\text{grad } V_m \tag{2.24a}$$

The next particular case of electrodynamics is *electroflow* field, in which we take into account the presence of currents steady in time. Assuming $(\partial \boldsymbol{D}/\partial t = \partial \boldsymbol{B}/\partial t = 0)$ and $\boldsymbol{J} \neq 0$, we obtain the fundamental equations of *electroflow* field (for immovable, conducting bodies):

$$\begin{gathered} \text{curl } \boldsymbol{H} = \boldsymbol{J}, \quad \text{curl } \boldsymbol{E} = 0 \\ \text{div } \boldsymbol{B} = 0, \quad \text{div } \boldsymbol{D} = 0 \\ \boldsymbol{J} = \sigma(\boldsymbol{E} + \boldsymbol{E}_{\text{extern}}), \quad \text{div } \boldsymbol{J} = 0 \end{gathered} \tag{2.25}$$

The equations div $\boldsymbol{J} = 0$ and curl $\boldsymbol{E} = 0$ are differential equivalents of the *1st* and *2nd Kirchhoff's Equations.*

2.2 FORMULATION AND METHODS OF SOLUTION OF FIELD DIFFERENTIAL EQUATIONS

The process of solution for electrodynamic problems comprises (Figure 1.1a) formulating differential equations for boundary value problems, finding a function which describes the field in adjacent areas, determining initial and boundary conditions, determining material properties and parameters, and at last—a solution of the task by determining variables on the basis of boundary conditions ("sewing together" of adjacent fields).

The best solution would be a simple *analytical* form, which is the shortest and most accurate description of field. Unfortunately, such solutions are possible only for very simple structures, and usually it becomes indispensable to apply *numerical methods* (Figure 1.1a). Their main advantage is the possibility of automated calculations for systems with complex structures.

2.2.1 Finding of Field Function

Many analytical problems of electrodynamics require solutions with different methods of *partial differential equations of second order* that describe phenomena that occur in electromagnetic fields. These equations are obtained from the fundamental equations of electrodynamics, (2.1) through (2.6), by elimination of one of the two variables, $\boldsymbol{E}$ or $\boldsymbol{H}$, by means of increasing the order of equation.

To accomplish this, we are utilizing the vector calculus identity

$$\text{curl curl } \boldsymbol{H} \equiv \text{grad div } \boldsymbol{H} - \nabla^2 \boldsymbol{H} \tag{2.26}$$

In *isotropic* and *linear* medium ($\mu_x = \mu_y = \mu_z = \mu = \text{const}$), from the law (2.4), it follows

$$\text{div } \boldsymbol{H} = 0 \tag{2.27}$$

At the same time, from Equations 2.1, 2.6, and 2.2, for immovable media ($v = 0$), we get

$$\text{curl curl } \boldsymbol{H} = \sigma \text{ curl } \boldsymbol{E} + \varepsilon \frac{\partial}{\partial t}(\text{curl } \boldsymbol{E}) = \left(\sigma + \varepsilon \frac{\partial}{\partial t}\right) \text{curl } \boldsymbol{E} = \left(\sigma + \varepsilon \frac{\partial}{\partial t}\right)\left(-\mu \frac{\partial \boldsymbol{H}}{\partial t}\right) \tag{2.28}$$

After substitution of Equations 2.27 and 2.28 through 2.26, we obtain the general differential equation for nonideal dielectrics:

$$\nabla^2 \boldsymbol{H} = \mu\sigma \frac{\partial \boldsymbol{H}}{\partial t} + \mu\varepsilon \frac{\partial^2 \boldsymbol{H}}{\partial t^2} \tag{2.29}$$

In an analogical way, one can develop the equation for the intensity of electric field, $\boldsymbol{E}$, in a medium with no free electric charges (div $\boldsymbol{D} = 0$)

$$\nabla^2 \boldsymbol{E} = \mu\sigma \frac{\partial \boldsymbol{E}}{\partial t} + \mu\varepsilon \frac{\partial^2 \boldsymbol{E}}{\partial t^2} \tag{2.30}$$

A similar equation for the conduction of current density, $\boldsymbol{J} = \sigma \boldsymbol{E}$, can be obtained by multiplying both sides of Equation 2.30 by the medium conductivity σ

$$\nabla^2 \boldsymbol{J} = \mu\sigma \frac{\partial \boldsymbol{J}}{\partial t} + \mu\varepsilon \frac{\partial^2 \boldsymbol{J}}{\partial t^2} \tag{2.30a}$$

2.2.2 Classification of Field Equations

Depending on the kind of field and properties of medium in Equations 2.29 and 2.30, we obtain a partial differential equation of function $u(x,y)$ in the form:

$$\mathrm{A}\frac{\partial^2 u}{\partial x^2} + 2\mathrm{B}\frac{\partial^2 u}{\partial x \partial y} + \mathrm{C}\frac{\partial^2 u}{\partial y^2} + F\left(x, y, u, \frac{\partial u}{\partial x}, \frac{\partial u}{\partial y}\right) = 0 \tag{2.31}$$

Then, depending on the sign of its discriminant, δ, where

$$\delta(x,y) = \mathrm{AC} - \mathrm{B}^2 \tag{2.32}$$

we distinguish (per du Bois–Reymond [2.12]) *elliptic* ($\delta > 0$), *hyperbolic* ($\delta < 0$), or *parabolic* ($\delta = 0$) equations. The most popular equations include

- Laplace's equation (elliptic)

$$\nabla^2 \boldsymbol{u} = 0 \tag{2.33}$$

  which is satisfied by potentials of electrostatic and magnetostatic fields. This equation is satisfied by any *holomorphic** function (i.e., a function of complex variable, which has derivatives in any point of a certain region).

  In a vector field, Laplace's equation has the form

$$\nabla^2 \boldsymbol{H} = \mathbf{i}\,\nabla^2 \boldsymbol{H}_x + \mathbf{j}\,\nabla^2 \boldsymbol{H}_y + \mathbf{k}\,\nabla^2 \boldsymbol{H}_z \tag{2.34}$$

- Poisson's equation (elliptic)

$$\nabla^2 \boldsymbol{u} = f(x_i) \quad (i = 1, 2, 3,\ldots) \tag{2.35}$$

  where $f(x_i)$ is a known function, defined in the investigated region (e.g., charge or current density distribution). It can also be the vector Equation 2.34 with $f(x_i) = \boldsymbol{F}(x_i)$.
- Helmholtz equation (elliptic)

$$\nabla^2 \boldsymbol{u} + k^2\,\boldsymbol{u} = f(x_i) \tag{2.36}$$

  which is applied in electrodynamics for description of variable sinusoidal fields (the so-called *harmonic*).

  Equation 2.36 can be *scalar* (at $\boldsymbol{u} = V$) or *vector* (when $\boldsymbol{u} = \boldsymbol{A}$, $f = \boldsymbol{F}$), *homogenous* (when $f = 0$), or *heterogeneous* (when $f \neq 0$). Equation 2.36 also describes *wave propagation in conducting media* (2.172).
- Conductivity equation (parabolic)

$$\nabla^2 \boldsymbol{u} = a^2 \frac{\partial u}{\partial t} + f(x_i) \tag{2.37}$$

  which is satisfied by the *temperature* $u(x, y, z, t)$ in a conducting medium. Equation 2.37 is satisfied also by components $\boldsymbol{E}$ and $\boldsymbol{H}$ of electromagnetic wave in conductors and field diffusion into conductors, as well as concentration in liquids (*diffusion equation*). The conductivity Equation 2.37 can also be homogenous ($f = 0$) and vectorial ($\boldsymbol{u} = \boldsymbol{A}$), scalar, and heterogeneous.

* *Holomorphic* function is also called *analytic function* and is used, among others, for description of sinusoidal alternating fields (the so-called *harmonic* functions).

- Wave equation in dielectrics (hyperbolic)

$$\nabla^2 \boldsymbol{u} = \frac{1}{v^2}\frac{\partial^2 u}{\partial t^2} + f \tag{2.38}$$

where v is a constant which expresses the wave velocity in dielectric.

The wave Equation 2.38 can be homogenous ($f = 0$) and vectorial ($\boldsymbol{u} = \boldsymbol{A}$), scalar, and heterogeneous.

All the equations mentioned above are linear and have a simple structure, but the theory of each of them is difficult, very broad, and varied, creating a different part of the mathematical analysis, which is in close connection with the theory of function of complex variable and differential geometry as well as boundary value problems. The difficulty of analytically finding a function that describes a field in a given region is mostly determined by the number of of field vectors and coordinates. This depends on the geometric structure of the field and the investigated part, and physical properties of the medium, that is, its nonlinearity or anisotropy. The simplest way for finding a solution is a system that can be represented in two dimensions, quasi three-dimensions (axisymmetric), or reduced to a problem of the so-called plane wave in a linear and isotropic medium.

In the case of *nonlinear media*, the field equations have a form even more complicated than the equations presented above, and often they can be resolved only with simplified approximated methods with application of analytical approximations, or by numerical (computer) methods.

Many practical problems of electrodynamics can be reduced to the *theory of potential fields or impedance multi-node networks*. Electric and magnetic potential fields with various configurations in many cases can be resolved accurately on the basis of general methods of theory of potential. Results of these calculations can be expressed with the help of rapid *special functions*. For solutions of two-dimensional potential fields one can use, presently less popular, the method of *conform mapping (isogonal transformation)* [2.3].

The potential fields include irrotational electrostatic and magnetostatic fields, as well as the permanent-in-time electric fields inside flat or solid conductors, as well as quasistatic alternating fields which fulfill Laplace's equation. Although Poisson's and Laplace's equations in principle describe static fields, they can also be applied to solution of alternating fields whose frequency is small enough that the effect of displacement currents are negligible (*quasi-stationary fields*). The structure of conducting elements is such that one can neglect the eddy-current effect (e.g., laminated iron cores, split and transposed conductors), or when conducting parts are outside the investigated region. Many fields fulfill these conditions. This is the reason for application as well as the practical significance of the method of potential fields.

2.2.3 Boundary and Initial Value Problems of Electrodynamics

Most of the calculation tasks of electrodynamics are reduced to solutions of the so-called *limit value problems* that consist of searching for a region of such a *harmonic function*,* which fulfills the limit (border) conditions. The *limit (border) conditions* are divided into *boundary* conditions (on a surface S or a boundary line L) and *initial* conditions (for time $t = t_0$).

The basis for determinating a full solution depend on conditions existing on the body surface and is the theorem on exclusivity of field, also called *uniqueness theorem*. According to the uniqueness theorem applied to solutions of Maxwell's equations, spatial and temporal boundary conditions univocally define the function describing the field in a linear and isotropic medium. This theorem states [1.11] that:

If for given time $t = t_0$ known are the electric and magnetic field intensities in an arbitrary point of area bounded by some surface, then one can, with Maxwell's equations, calculate all the electromagnetic values at a time instant t; while it is assumed that known are the tangential components of electric field $\mathbf{E}$ and magnetic field $\boldsymbol{H}$ in any point of the boundary surface from the initial time t_0 *to the time t.*

Another uniqueness theorem (Skilling, 1954) states that a vector field is defined univocally if there are given in the investigated region its following values:

$$\operatorname{div}\boldsymbol{H},\ \operatorname{curl}\boldsymbol{A} \quad \text{and} \quad \left(\frac{\partial A}{\partial n}\right)_S \quad \text{or} \quad (\boldsymbol{A})_{r\to\infty} \to 0, \quad \text{or} \quad \boldsymbol{A} \sim \frac{1}{r^2} \tag{2.39}$$

where S is the boundary surface of the region, $\partial/\partial n$ is the derivative along the direction normal to S.

There exist several uniqueness theorems of the solution of equations of electrodynamics, depending on what problem is stated. In this way, similar uniqueness theorems have been developed for the potential electrostatic and magnetostatic fields. In mathematics, it is called the Dirichlet problem.

The Dirichlet problem consists in determination of a harmonic function u, in a closed space D:

$$u(M) \quad \text{for } M \in D$$

which in each point P of the surface A enclosing the space D will be equal to the imposed boundary value $f(P)$

$$u(P) = f(P) \quad \text{for } P \in A \tag{2.40}$$

The Dirichlet problem has at most only one solution [2.12]. These are potentials of a single or dual layer and an external region of space charge. We distinguish an internal and external Dirichlet problem, depending on which region is investigated—inside or outside of the considered boundary surface. Condition (2.40) is called *boundary condition of first order* or *Dirichlet condition* of the existence of the described solution.

* A harmonic function in a certain space region is a function of three variables $u(x, y, z)$ of class C^2, which in each point of this region fulfils Laplace's equation.

The second major problem concerning the boundary values for harmonic function is the Neumann problem. It consists in determining a harmonic function u inside the area D, whose derivative along the normal du/dn in points P of the boundary surface A assumes the given (known) boundary values $g(P)$, that is,

$$\left(\frac{du}{dn}\right)_P = g(P) \quad \text{for } P \in A \tag{2.41}$$

Similarly as before, we recognize *internal and external Neumann problems.* The necessary and sufficient condition for existing solutions of the *internal Neumann problem* is when the function $g(P)$ satisfies the equation:

$$\int_A g(P)\, dA = 0 \tag{2.42}$$

The *external Neumann problem* does not require the above condition. A solution determined in this way is unique. The condition 2.41 is called a *boundary condition of second type* or a *Neumann condition.*

The third type of boundary problem, called Hankel problem, consists in determining an analogical harmonic function $u(M)$ for $M \in D$, which on the surface A bounding the area D fulfills the so-called *third-type (Hankel) condition*:

$$\left(\frac{du}{dn}\right)_P = h(P)\,[u(P) - v(P)] \quad \text{for } P \in A \tag{2.43}$$

where $h(P)$, $v(P)$ are the given functions, $u(P)$ the searched function.

In the problems of first, second, and third type [(2.40) through (2.43)] there can also exist functions of space and time $u(M, t)$, $u(P, t)$, $f(P, t)$, $h(P, t)$ and $v(P, t)$, respectively [2.12].

The existence of a single-harmonic function u, satisfying arbitrary partial differential Laplace's equation of second order, of wave or conductivity, is proven with the Cauchy's theorem.

Solution of the initial Cauchy problem for partial differential equation of second order needs, however, knowledge of the functions

$$\left.\begin{aligned} u(P, t_0) &= f(P) \\ \text{and} \qquad \left(\frac{du}{dt}\right)_{t=t_0} &= \varphi(P) \end{aligned}\right\} \tag{2.44}$$

in points P on the surface bounding the investigated region.

Just the finding of these bounding functions on a surface of an investigated region belongs to the most important and often the most difficult problems of technical electrodynamics. The most common method of finding the bounding functions consists in a solution of Maxwell's equations for the adjusted regions and then equating the obtained solutions to each other on the basis of general boundary conditions for fields (2.191) through (2.194). Nevertheless, in many cases a much simpler method is application of static *mirror images* (Chapter 5) or dynamic reflections of electromagnetic waves, that is, *wave method* (Section 4.3). These methods are primarily applicable to systems with rectilinear or circular boundary surfaces.

The uniqueness theorem is proved in principle for the cases when the material parameters, ε, μ, and σ, are constant and depend only on position, and do not depend on time (Tamm, 1949). That is why when applying these principles to a case of nonlinear and anisotropic media (Simonyi [1.11]) one must be careful and check each time the calculated results experimentally, or in another way.

2.2.4 Auxiliary Functions and Vectors

To simplify the solution of electromagnetic field equations, instead of the direct variables $\boldsymbol{E}$ and $\boldsymbol{H}$, there are introduced various auxiliary functions and vectors. With the help of them, it is easier to calculate field directly from spatial and temporal distributions of currents and charges constraining the field. Such auxiliary functions are presented below.

2.2.4.1 Electric and Magnetic Scalar Potentials

V_e and V_m of static and stationary fields (2.23) and (2.24):

$$\boldsymbol{E} = -\text{grad}\ V_{\text{e}} \quad \text{and} \quad \boldsymbol{H} = -\text{grad}\ V_{\text{m}} \tag{2.45}$$

The method of magnetic scalar potential can be used only outside of regions of eddy currents ($\boldsymbol{J} = 0$), to fields steady in time or changing so slowly that one can ignore the displacement currents ($\partial D/\partial t = 0$). Such fields include, in general, irrotational fields of power frequency (50 or 60 Hz). In such conditions, Equation 2.1 has the form: curl $\boldsymbol{H} = 0$, which per (2.24a) means that this field has the magnetic scalar potential V_m, while the magnetic field strength is expressed by the basic equation of this method (2.45).

The equation

$$V_{\text{m}}(x, y, z) = \text{const} \tag{2.46}$$

is the equation of magnetically *equipotential surface. Lines* of the magnetic *field* $\boldsymbol{H}$ are perpendicular to the *equipotential surfaces.* In order to determine an analytical equation of the field lines, it is enough to consider that the length element d$\boldsymbol{l}$ of field lines is parallel to the magnetic field strength $\boldsymbol{H}$ in this point. It means that its components, dx, dy, dz along the coordinate axes x, y, z, are proportional to the components $\boldsymbol{H}_x$, $\boldsymbol{H}_y$, $\boldsymbol{H}_z$ of the vector $\boldsymbol{H}$, that is

$$\frac{\mathrm{d}x}{H_x} = \frac{\mathrm{d}y}{H_y} = \frac{\mathrm{d}z}{H_z}, \quad \text{that is,}\ \frac{\mathrm{d}x}{\mathrm{d}y} = \frac{H_x}{H_y} \quad \text{and} \quad \frac{\mathrm{d}y}{\mathrm{d}z} = \frac{H_y}{H_z} \tag{2.47}$$

These are *force line equations*, not only of magnetic fields, but of any potential field, whose integrals have the form $f_1(x, y, z) = C_1$ and $f_2(x, y, z) = C_2$, where C_1 and C_2 are integration constants depending on coordinates of the considered point.

If d$\boldsymbol{l}$ is an increment in the $\boldsymbol{H}$ direction, then

$$|\text{grad}\, V_\text{m}| = \boldsymbol{H} = -\frac{\partial V_\text{m}}{\partial l}$$

and

$$H_x = -\frac{\partial V_\text{m}}{\partial x}; \quad H_y = -\frac{\partial V_\text{m}}{\partial y}; \quad H_z = -\frac{\partial V_\text{m}}{\partial z}.$$

Since for a permanent permeability, div $\boldsymbol{H} = 0$, the magnetic potential V_m fulfills the Laplace's equation

$$\text{div grad}\, V_\text{m} = \nabla^2 V_\text{m} = 0 \tag{2.48}$$

The scalar magnetic potential is generally a multivalued function. A linear integral of the vector $\boldsymbol{H}$ along an arbitrary closed circuit, not bounding any current, equals to zero: $\oint \boldsymbol{H} \cdot \text{d}\boldsymbol{l} = 0$. In the case, however, when the path of integration between two points A and B bounds certain current i k-times, or cuts certain circuit with current ki, the difference of magnetic potentials between the points A and B equals

$$V_\text{mA} - V_\text{mB} = \oint \boldsymbol{H} \cdot \text{d}\boldsymbol{l} - ki \tag{2.49}$$

Therefore, if we assume the potential of one of these points (e.g., B) equals zero, then the potential in point A, denoted V_mA, will be *non-univocal* and depending on integration path. Only in the case when this path does not cut a circuit with current

$$V_\text{mA} - V_\text{mB} = \oint \boldsymbol{H} \cdot \text{d}\boldsymbol{l}$$

2.2.4.2 Magnetic Vector Potential *A* and the Scalar Potential *V* of Electromagnetic Field (*Electrodynamic Potentials*)

From the principle of continuity of flux and flux density lines, div $\boldsymbol{B} = 0$ (2.4), using the vector identity, div curl $\boldsymbol{A} \equiv 0$, we define the *magnetic vector potential* $\boldsymbol{A}$ as

$$\boldsymbol{B} = \text{curl}\, \boldsymbol{A} \tag{2.50}$$

Substituting Equation 2.50 into Equation 2.2, and considering the vector identity curl grad $V \equiv 0$, yields

$$\boldsymbol{E} = -\frac{\partial \boldsymbol{A}}{\partial t} - \text{grad}\, V \tag{2.51}$$

The electrodynamic potentials $\boldsymbol{A}$ and V are functions not only of coordinates, but also of time ([2.14], p. 343/2).

After substitution of $\boldsymbol{H} = 1/\mu$ curl $\boldsymbol{A}$ (2.50) and (2.51) into the first Maxwell's equation curl $\boldsymbol{H} = \boldsymbol{J}_{\text{total}}$ (2.1), and utilizing the vector identity (2.26), we obtain for $\mu = \text{const}$, $\varepsilon = \text{const}$, $v_\rho \neq 0$, and $v = 0$

$$\text{grad div}\boldsymbol{A} - \nabla^2\boldsymbol{A} = -\mu\sigma\,\text{grad}V - \mu\sigma\frac{\partial \boldsymbol{A}}{\partial t} - \mu\rho\boldsymbol{v}_\rho - \mu\boldsymbol{J}_{\text{external}}$$

$$- \mu\varepsilon\frac{\partial^2 \boldsymbol{A}}{\partial t^2} + -\mu\varepsilon\,\text{grad}\frac{\partial V}{\partial t} \tag{2.52}$$

The magnetic vector potential $\boldsymbol{A}$ in Equation 2.50 is a multi-valued function and, according to Equation 2.39, one should choose for it the *divergence*, div $\boldsymbol{A}$, and determine boundary conditions.

Hence, we select the divergence in dependence on the task needs, in such a way so that we obtain proper equations in a simplest way.

For the electromagnetic fields, we adapt the Lorentz's condition [2.12]

$$\text{div}\,\boldsymbol{A} = -\mu\varepsilon\frac{\partial V}{\partial t} - \mu\sigma V \tag{2.53}$$

After substituting Equation 2.53 into Equation 2.52, we get the equation of *hyperbolic type*

$$\nabla^2\boldsymbol{A} - \mu\varepsilon\frac{\partial^2 \boldsymbol{A}}{\partial t^2} = \mu\sigma\frac{\partial \boldsymbol{A}}{\partial t} + \mu\rho\boldsymbol{v}_\rho - \mu\boldsymbol{J}_{\text{external}} \tag{2.54}$$

This is the *heterogeneous wave equation*, called *d'Alambert's equation*. For conducting media, it takes the form

$$\nabla^2\boldsymbol{A} = \mu\sigma\frac{\partial \boldsymbol{A}}{\partial t} \tag{2.54a}$$

Substituting Equation 2.51 into div $\boldsymbol{E} = \rho/\varepsilon$ (2.5), and considering Equation 2.53, as well as the identity div grad $V \equiv \nabla^2 V$, we obtain the *d'Alambert's scalar equation*

$$\nabla^2 V - \mu\varepsilon\frac{\partial^2 V}{\partial t^2} = \frac{\rho}{\varepsilon} + \mu\sigma\frac{\partial V}{\partial t} \tag{2.55}$$

In dielectrics ($\sigma = 0$), the last component disappears. In steady fields ($\partial/\partial t = 0$), Equation 2.55 becomes Poisson's equation (2.35) or Laplace's equation (2.33). In *quasi-stationary* fields ($\partial\boldsymbol{D}/\partial t = 0$), we obtain the *conductivity equation* (2.37). In the fields with no free electric charges ($\rho = 0$) and no conductivity currents ($\sigma = 0$), Equations 2.54 and 2.55 become the *wave equations* (2.38).

In *magnetostatic* and *slowly alternating* (power) fields, in which one can neglect the displacement currents ($\partial \boldsymbol{D}/\partial t = 0$), as well as in media where eddy currents can be ignored ($\sigma = 0$), the most convenient way is to adopt the condition

$$\operatorname{div} \boldsymbol{A} = 0 \tag{2.56}$$

Then, for the regions where only a given current density $\boldsymbol{J}$ exists, we obtain

$$\nabla^2 \boldsymbol{A} = -\mu \boldsymbol{J} \tag{2.57}$$

It means that the vector $\boldsymbol{A}$ components fulfill the *scalar Poisson's equation*

$$\nabla^2 A_x = -\mu J_x(x, y, z); \quad \nabla^2 A_y = -\mu J_y(x, y, z); \quad \nabla^2 A_z = -\mu J_z(x, y, z); \tag{2.57a}$$

which is identical with analogous equations for the scalar electric potential generated by electric charges (when one replaces $\boldsymbol{A}$ with V, and $\mu \boldsymbol{J}$ with ρ/ε). Due to this reason, the vector $\boldsymbol{A}$ is called the *vector potential of magnetic field of currents*, generated by the current of density $\boldsymbol{J}$.

In a region where there is no eddy-currents, that is, where the magnetic field is *irrotational* (curl $\boldsymbol{H} = 0$), the magnetic vector potential fulfills Laplace's equation

$$\nabla^2 \boldsymbol{A} = 0 \tag{2.58}$$

Having a given distribution of the current density $\boldsymbol{J}$, one can determine $\boldsymbol{A}$ by analogy with the known formula for electric potential

$$\boldsymbol{A} = \frac{\mu}{4\pi} \int_V \frac{\boldsymbol{J}\,\mathrm{d}V}{r} \tag{2.59}$$

It means that any component of $\boldsymbol{A}$ can be determined by integration (over a volume occupied by the current) of the corresponding component of the current density $\boldsymbol{J}$ divided by the distance r of the element of current density from the point in which we want to determine the components of $\boldsymbol{A}$, and by multiplying it by $\mu/4\pi$. In the case, for instance, when the current density of the excitation current has only one component J_x, the magnetic vector potential has also only one component A_x. In such a case, the field $\boldsymbol{A}$ of one current filament (Figure 2.3) of the current density $\boldsymbol{J}$, per Equation 2.59, is

$$\mathrm{d}\boldsymbol{A} = \frac{\mu_0}{4\pi} \boldsymbol{J} \frac{\mathrm{d}V}{r} = \frac{\mu_0}{4\pi} \boldsymbol{J}\,\mathrm{d}S \int_{-\infty}^{+\infty} \frac{\mathrm{d}l}{r} = -\frac{\mu_0}{2\pi} \boldsymbol{J}\,\mathrm{d}S \ln R \tag{2.59a}$$

where

$$r = \sqrt{R^2 + l^2}.$$

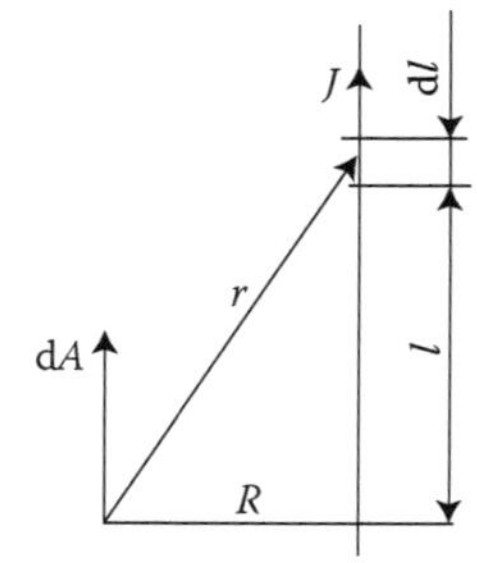

FIGURE 2.3 The magnetic vector potential $\boldsymbol{A}$.

In the case when the current $i = \int_S \boldsymbol{J} \cdot \mathrm{d}\boldsymbol{S}$ flows in a circuit made of a wire of transverse dimensions much smaller than the wire length and the distance r to the investigated point, formula (2.59) can be simplified to the form

$$\boldsymbol{A} = \frac{\mu}{4\pi} \oint_l \frac{\mathrm{i}\,\mathrm{d}\boldsymbol{l}}{r} \tag{2.60}$$

Using Stokes's theorem (2.14), one can determine, with the help of the magnetic vector potential $\boldsymbol{A}$, the magnetic flux

$$\Phi = \int_S \boldsymbol{B} \cdot \mathrm{d}\boldsymbol{S} = \int_S \operatorname{curl} \boldsymbol{A} \cdot \mathrm{d}\boldsymbol{S} = \oint_l \boldsymbol{A} \cdot \mathrm{d}\boldsymbol{l} \tag{2.61}$$

In this way, the difficult double surface integral has been substituted by a simpler linear integral.

The per-unit energy of electromagnetic field (in Ws/m^2), considering Equation 2.50 and the vector identity div $(\boldsymbol{H} \times \boldsymbol{A}) = \boldsymbol{A}$ curl $\boldsymbol{H} - \boldsymbol{H}$ curl $\boldsymbol{A}$, has the form

$$w_\mathrm{m} = \frac{\boldsymbol{HB}}{2} = \frac{1}{2\mu}\operatorname{curl}^2 \boldsymbol{A} = \frac{1}{2}\boldsymbol{H} \operatorname{curl} \boldsymbol{A} = \frac{1}{2}[\boldsymbol{A} \operatorname{curl} \boldsymbol{H} - \operatorname{div}\,(\boldsymbol{H} \times \boldsymbol{A})]$$

In the region of a volume V, the magnetic energy (in Ws), considering curl $\boldsymbol{H} = \boldsymbol{J}$ (2.61), equals

$$W_\mathrm{m} = \frac{1}{2}Li^2 = \frac{1}{2}\int_V \boldsymbol{H} \cdot \boldsymbol{B}\,\mathrm{d}V = \frac{1}{2}\int_V \boldsymbol{A} \cdot \boldsymbol{J}\,\mathrm{d}V = \frac{1}{2}\Phi i \tag{2.62}$$

where, per Equation 2.20a $\int_{V\to\infty} \operatorname{div}\,(\boldsymbol{H} \times \boldsymbol{A})\,\mathrm{d}V = \int_{S\to\infty} (\boldsymbol{H} \times \boldsymbol{A})\mathrm{d}S \to 0$, since $H \sim l/r^2; A \sim l/r$; $\mathrm{d}S \sim r^2$, hence $(\boldsymbol{H} \times \boldsymbol{A})\,\mathrm{d}S \sim l/\mathrm{r}$.

In *sinusoidal, quasi-stationary alternating fields* ($\partial D/\partial t = 0$), by applying the symbolic method, $\underline{A} = A\mathrm{e}^{\mathrm{j}\omega t}$, Equation 2.51 can be represented as

$$\boldsymbol{E} = -j\omega\boldsymbol{A} - \operatorname{grad} V \tag{2.51a}$$

and Equation 2.54 takes the form of *heterogeneous Helmholtz's equation* (2.36) in a region where the external current density $\boldsymbol{J}_{\text{extern}}$ exists

$$\nabla^2 \boldsymbol{A}_{\text{m}} - \alpha^2 \boldsymbol{A}_{\text{m}} = -\mu \boldsymbol{J}_{\text{extern}}. \tag{2.63}$$

$$\alpha = \sqrt{j\omega\mu\sigma} = (1 + j)k = \sqrt{2}\, ke^{j\frac{\pi}{4}}; \quad k = \frac{1}{\delta} = \sqrt{\frac{\omega\mu\sigma}{2}} \tag{2.64}$$

Solution of Equation 2.63, for a region with the external current density $\boldsymbol{J}_{\text{extern}}$ and outside of it ($\boldsymbol{J}_{\text{extern}} = 0$), has the form analogical to Equation 2.59

$$\boldsymbol{A} = \frac{\mu}{4\pi}\int\limits_V \frac{\boldsymbol{J}_{\text{extern}}\,\mathrm{e}^{-\alpha r}}{r}\,\mathrm{d}V \tag{2.65}$$

2.2.4.3 Electric Vector Potential *T*

From the first *differential Kirchhoff's law*, div $\boldsymbol{J} = 0$ (2.25), and from the vector identity div curl $\boldsymbol{T} \equiv 0$, and per analogy to $\boldsymbol{A}$ ($\boldsymbol{B}$ = curl $\boldsymbol{A}$), follows the notion

$$\boldsymbol{J} = \text{curl}\, \boldsymbol{T} \tag{2.66}$$

Since at the same time, curl $\boldsymbol{H} = \boldsymbol{J}$ (2.1), then curl ($\boldsymbol{H} - \boldsymbol{T}$) = 0. Considering the vector identity curl grad $V \equiv 0$, yields the relation between the potentials

$$\boldsymbol{H} - \boldsymbol{T} = -\text{grad}\, V_{\text{m}} \tag{2.67}$$

but, div $\boldsymbol{H}$ – div $\boldsymbol{T}$ = –div grad $V_{\text{m}} = \nabla^2 V_{\text{m}}$.

At μ = const, div $\boldsymbol{H} = 0$, hence

$$\nabla^2 V_{\text{m}} = \text{div}\, \boldsymbol{T}\ (\mu = \text{const}) \tag{2.68}$$

This method, introduced in 1977 by Carpenter, and known as "*T*-Ω" method [2.18], where $\Omega \equiv V_{\text{m}}$, is distinguished by a good accuracy in calculations of three-dimensional fields [2.25], the possibility of separation of the "electric" problem from the "magnetic" one, and by the liberty of selection of boundary conditions [2.18].

The problem of coupled fields, described as

$$\text{curl curl}\, \boldsymbol{T} = -j\omega\mu\sigma(\boldsymbol{T} - \text{grad}\, V_{\text{m}}) + \text{curl}\, \boldsymbol{J}_{\text{extern}}.$$

is usually resolved by the finite-element method (FEM) with the help of boundary integral circuits.

2.2.4.4 Hertz Vector *P*

Ref. [2.7] serves rather for simultaneous description of wave processes in electromagnetic fields of higher frequencies. It is expressed by the equation

$$\boldsymbol{P} = c^2 \int \boldsymbol{A} \cdot \mathrm{d}t \quad (c = 3 \times 10^8\,\text{m/s}) \tag{2.69}$$

After determination of the vector $\boldsymbol{P}$ with the help of the *wave equation*

$$\nabla^2 \boldsymbol{P} = \frac{1}{c^2}\frac{\partial^2 \boldsymbol{P}}{\partial t^2} \tag{2.70}$$

one can find all the remaining field quantities, like

$$\boldsymbol{A} = \frac{1}{c^2}\frac{\partial^2 \boldsymbol{P}}{\partial t^2} \tag{2.71}$$

$$\boldsymbol{B} = \operatorname{curl} \boldsymbol{A} = \frac{1}{c^2}\operatorname{curl}\frac{\partial^2 \boldsymbol{P}}{\partial t^2} \tag{2.72}$$

$$\boldsymbol{V} = -\operatorname{div} \boldsymbol{P} \tag{2.73}$$

$$\boldsymbol{E} = \operatorname{grad} V - \frac{\partial \boldsymbol{A}}{\partial t} = \operatorname{grad}\operatorname{div} \boldsymbol{P} - \frac{1}{c^2}\frac{\partial^2 \boldsymbol{P}}{\partial t^2} \tag{2.74}$$

2.2.4.5 Maxwell's Stress Tensor $\overleftrightarrow{T}$

J. Turowski [1.15, 1.17] helps to convert electromagnetic per-unit volumetric forces f_L (2.8) into the resultant surface force

$$\boldsymbol{F} = \int_V f_L \cdot \mathrm{d}V = \oint_S \boldsymbol{T}^n \cdot \mathrm{d}S \tag{2.75}$$

where $f_L = \rho \boldsymbol{E} + \boldsymbol{J} \times \boldsymbol{B} = \varepsilon\left(\boldsymbol{E}\operatorname{div}\boldsymbol{E} + \boldsymbol{B} \times \frac{\partial \boldsymbol{E}}{\partial t}\right) - \mu \boldsymbol{H} \times \operatorname{curl}\boldsymbol{H} = \operatorname{div}\boldsymbol{T}^n$—Lorentz force density [1.17]; $\boldsymbol{T}^n$ is the vector of per-unit surface force, which is a projection of the Maxwell's stress tensor $\overleftrightarrow{T}$ to the normal to body surface.

Generally, the Maxwell's stress tensor

$$\overleftrightarrow{T} = \varepsilon(\boldsymbol{n} \cdot \boldsymbol{E})\boldsymbol{E} + \mu(\boldsymbol{n} \cdot \boldsymbol{H})\boldsymbol{H} - \frac{1}{2}(\varepsilon E^2 + \mu H^2)\boldsymbol{n} \tag{2.76}$$

and its elements are

$$T_{ij} = \varepsilon \cdot E_i E_j + \mu \cdot H_i H_j - \frac{1}{2}(\varepsilon E^2 + \mu H^2) \qquad i, j = 1,2,3 \tag{2.77}$$

The component $\mathrm{d}F_i$ of the force $\mathrm{d}\boldsymbol{F}$ transferred by surface element $\mathrm{d}S_i$ of the vector $\mathrm{d}\boldsymbol{S}$ is

$$dF_i = \sum_{j=1}^{3} T_{ij}\, dS_j \tag{2.78}$$

From Equation 2.76 it follows that the modulus of surface force density of magnetic field is

$$\left|\vec{T}\right| = \frac{1}{2}\mu H^2 \tag{2.79}$$

The Maxwell's stress tensor is a very practical and useful tool for calculating forces on iron surfaces from the result of field distribution in the gap of electric machines and apparatus.

EXAMPLE

(J. Turowski: Coupled Fields. [1.17/2], pp. 252–253)

The force $\boldsymbol{F}$ acting on a surface $\boldsymbol{S}$ of the stator or rotor of electric machines

$$\mathbf{F} = \iint_S \boldsymbol{T}\, dS = \iint_S \frac{1}{\mu_0}\left[(\boldsymbol{B}\cdot \mathbf{n})\boldsymbol{B} - \frac{1}{2}\left|\boldsymbol{B}\right|^2 \boldsymbol{n}\right] dS$$

$$= \left(\mu_0 \iint_S \frac{H_n^2 - H_t^2}{2} dS\right)\boldsymbol{n} + \left(\mu_0 \iint_S \boldsymbol{H}_n \boldsymbol{H}_t\right) dSt = F_n n + F_t t; \tag{2.75a}$$

It can be seen that two forces, F_n and F_t, act on each metal element. Normally, motor cores are made of laminated iron and are not highly saturated. Therefore, one can assume $H_t = 0$. Hence, the force $F_t = 0$ and on the iron core act only the normal forces $F_n = 0.5\iint_S \boldsymbol{BH} dS$.

However, if the core is made of solid iron or a permanent magnet of low permeability, or if it is covered with a copper screen and the field is alternating, the equivalent permeability is low and the values $\boldsymbol{H}_n$ and $\boldsymbol{H}_t$ are comparable. In such a case, both forces F_n and F_t exist and should be considered, especially at impulse excitation of switch gears or fast relays.

From Equation 2.75a, it can be easily seen that the modulus of the surface force density of the magnetic field satisfies Equation 8.18 and the vector $\boldsymbol{H}$ divides the angle between $\boldsymbol{n}$ and $\boldsymbol{T}$ by half [1.17].

2.2.5 Methods of Solution of Field Equations

The main objective of technical electrodynamics is a scientifically based aid for *design and research work, qualification testing*, and ensuring proper *reliability* of power equipment, mainly for large power apparatus during operation. For this purpose, sufficiently accurate methods should be chosen, able to achieve the necessary goal rapidly, on the basis of mechatronics, automation, CAD, CAM, and CAE methods (Turowski [1.20]).

For different tasks, one can use different mathematical models and methods of calculation. The most popular methods, tested at the Technical University of Lodz and other research centers, can be divided into the following three groups [2.34]:

a. Analytical-numerical methods (ANM)
   1. Method of analytical prolongation and Fourier's series ([1.16], pp. 115 and 289), [1.15]
   2. Method of superposition of magnetic vector potential ([1.16], pp. 237 and 273)
   3. Method of mirror images (Chapter 5 and [1.16], pp. 34 and 244)
   4. Method of Fredholm's integral equations ([1.16], pp. 47 and 279)
   5. Method of magnetic charges [2.22]
b. Numerical, distributed mesh methods
   6. Finite element method (FEM) [2.17], [1.17], [7.1]
   7. Finite difference method (FDM) (Wiak et al. [2.37])
   8. Flow network method (FNM) and its relative reluctance network method (RNM) [2.34], [1.15], [1.20], [4.25]
   9. Boundary elements method (BEM) [1.17]
   10. Method of finite boxes (MFB)—as generalization of FDM [2.20], [2.35]
   11. Hierarchal FEM with Hermite elements of higher order [1.17], [2.23], [7.8]
c. Energy methods
   12. Hamilton's principle with its Euler–Lagrange's Equations and virtual works method ([1.16], p. 229; [1.20]), for transient processes; competitive to the expensive, commercial SABER software
   13. Tubes and slices method (Sykulski and Hammond [2.29])—computerization of the old (1909) Lehman's graphic method (Figure 5.35)
   14. Methods of solution of Inverse Problems (field synthesis), including ill-conditioned problems [1.17], [6.1]

Apart of the above 14 "pure" methods, there exist many variants of them, their modifications, and hybrid combinations. In spite of significant number of solution methods, only a few of them are broadly applied in practice. Besides the accurate but limited application of analytical methods, the most broadly used has been the FEM-2D. There is a growing interest in BEM and RNM-3D [7.23], [1.20] which is especially interesting for easy and rapid design of three-dimensional fields.

When selecting a method, one should consider its feasibility to satisfy the basic technical requirements, such as

- Easy to operate on a personal computer with standard capacity
- Easy introduction of input data, and getting of useful, integral design data
- Flexible in use and modifications at minimum cost and time
- Possible to analyze complex, three-dimensional geometrical forms, three-phase systems, thermal effects, nonlinear permeability, and so on
- Possibility of interactive operations, preferably with a simple, intelligent pre-processor (user interface), solver, and post-processor (Figure 4.22)

Although the FEM-2D is a good and popular tool, to satisfy all the requirements, especially for 3-D, interactive design, FEM-3D is often too heavy and too expensive. Much better now is the RNM-3D (see Chapters 4, 7, 10, and Figure 7.22 in [1.20]).

2.3 ANISOTROPIC MEDIA

Equations 2.3 and 2.6 can be expressed with the scalar values, ε, μ, σ, only in isotropic, uniform, and linear media. In general case, the media properties can be functions of coordinates and time

$$\varepsilon = \varepsilon(x, y, z, t), \quad \mu = \mu(x, y, z, t), \quad \sigma = \sigma(x, y, z, t)$$

In *anisotropic* media, but with *constant parameters* (i.e., linear media) each component occurring on the left side of Equations 2.3 and 2.6 is a general linear function of all components of the quantity existing at right-hand side. For instance, $B_x = B_x(H_x, H_y, H_z)$.

Let us consider the vector of magnetic field intensity $\boldsymbol{H}$ (Figure 2.4) in an anisotropic medium, with mutually perpendicular axes of anisotropy 1, 2, 3, to which correspond the different permeabilities μ_1, μ_2, μ_3. Let us initially assume that the anisotropy axes 1, 2, 3 do not align with the coordinate axes x, y, z.

To find the vector B_x we must make a projection of the vector $\boldsymbol{H}$ onto the axes of anisotropy 1, 2, 3, find components $B_1 = \mu_1$, H_1, B_2, B_3 along these axes, and then

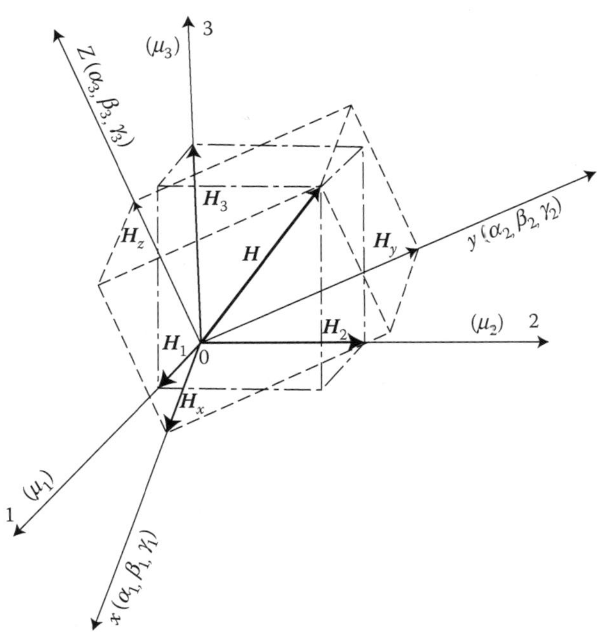

FIGURE 2.4 The vector $\boldsymbol{H}$ in a nonoverlapping coordinate systems of anisotropy 1, 2, 3 (μ_1, μ_2, μ_3) and the Cartesian coordinates x, y, z.

make again a projection back of these component onto the x-axis. Let us denote by α_1, β_1, γ_1 the *directional cosines* of the 0x axis of the Cartesian (rectangular) coordinate system with respect to the system of anisotropic axes 1, 2, 3, that is—the cosines of angles between the axis 0X and the axes 01, 02, 03.

Similarly, let us denote by α_2, β_2, γ_2 the *directional cosines* of the 0y axis, and by α_3, β_3, γ_3 the *directional cosines* of the 0z axis, according to the table given below

	1	2	3
X	α_1	β_1	γ_1
Y	α_2	β_2	γ_2
Z	α_3	β_3	γ_3

According to the theorem on projection of geometric sum of vectors, we can, on the basis of Figure 2.4, write a system of equations connecting corresponding components of the vector along different axes:

$$H_1 = \alpha_1 H_x + \alpha_2 H_y + \alpha_3 H_z$$

$$H_2 = \beta_1 H_x + \beta_2 H_y + \beta_3 H_z$$

$$H_3 = \gamma_1 H_x + \gamma_2 H_y + \gamma_3 H_z$$

By multiplying these values by corresponding permeabilities, we obtain the components of the flux density along anisotropy axes

$$B_1 = \mu_1 H_1, \quad B_2 = \mu_2 H_2, \quad B_3 = \mu_3 H_3 \tag{2.80}$$

After application of the same principle in reverse direction, we obtain components of flux density along the coordinates

$$\begin{aligned} B_x &= \alpha_1 B_1 + \beta_1 B_2 + \alpha_3 B_3 \\ B_y &= \alpha_2 B_1 + \beta_2 B_2 + \alpha_3 B_3 \\ B_z &= \alpha_3 B_1 + \beta_3 B_2 + \alpha_3 B_3 \end{aligned} \tag{2.81}$$

After substitution of Equation 2.80 into Equation 2.81, we have the general matrix equation for anisotropic media

$$\begin{pmatrix} B_x \\ B_y \\ B_z \end{pmatrix} = \begin{pmatrix} \mu_{11} & \mu_{12} & \mu_{13} \\ \mu_{21} & \mu_{22} & \mu_{23} \\ \mu_{31} & \mu_{32} & \mu_{33} \end{pmatrix} \begin{pmatrix} H_x \\ H_y \\ H_z \end{pmatrix} \tag{2.82}$$

where the matrix of permeability, composed of the elements

$$\begin{aligned}
\mu_{11} &= \mu_1\alpha_1^2 + \mu_2\beta_1^2 + \mu_3\gamma_1^2; \quad \mu_{12} = \mu_1\alpha_1\alpha_2 + \mu_2\beta_1\beta_2 + \mu_3\gamma_1\gamma_2;\\
\mu_{13} &= \mu_1\alpha_1\alpha_3 + \mu_2\beta_1\beta_3 + \mu_3\gamma_1\gamma_3\\
\mu_{21} &= \mu_1\alpha_1\alpha_2 + \mu_2\beta_1\beta_2 + \mu_3\gamma_1\gamma_2; \quad \mu_{22} = \mu_1\alpha_2^2 + \mu_2\beta_2^2 + \mu_3\gamma_2^2;\\
\mu_{23} &= \mu_1\alpha_2\alpha_3 = \mu_2\beta_2\beta_3 + \mu_3\gamma_2\gamma_3\\
\mu_{31} &= \mu_1\alpha_1\alpha_3 + \mu_2\beta_1\beta_3 + \mu_3\gamma_1\gamma_3; \quad \mu_{32} = \mu_1\alpha_2\alpha_3 + \mu_2\beta_2\beta_3 + \mu_3\gamma_2\gamma_3;
\end{aligned}$$

is the *symmetric tensor of permeability*, because, as can be seen: $\mu_{12} = \mu_{21}$, $\mu_{13} = \mu_{31}$ and $\mu_{23} = \mu_{32}$.

In an anisotropic medium, the vector of flux density will not have any more the same direction as the vector of magnetic flux intensity, and as well div $\boldsymbol{H} \neq 0$.

The coefficients μ_{ik} in Equation 2.82 depend on orientation of the coordinate axes x, y, z with respect to the axes of anisotropy 1, 2, 3. If the crystallography (anisotropy) axes do not align with the axes of the assumed coordinate system, then the permeability tensor contains six components and, for instance, the second Maxwell's Equations 2.2a,b,c creates the set of equations:

$$\begin{aligned}
\frac{\partial E_z}{\partial y} - \frac{\partial E_y}{\partial z} &= -\frac{\partial}{\partial t}(\mu_{11}H_x + \mu_{12}H_y + \mu_{13}H_z)\\
\frac{\partial E_x}{\partial z} - \frac{\partial E_z}{\partial x} &= -\frac{\partial}{\partial t}(\mu_{21}H_x + \mu_{22}H_y + \mu_{23}H_z)\\
\frac{\partial E_y}{\partial x} - \frac{\partial E_y}{\partial z} &= -\frac{\partial}{\partial t}(\mu_{31}H_x + \mu_{32}H_y + \mu_{33}H_z)
\end{aligned} \tag{2.83}$$

In the case when the coordinate system aligns with the anisotropy axes 1, 2, 3, we have

$$\alpha_1 = \cos 0 = 1; \quad \beta_1 = \cos\frac{\pi}{2} = 0; \quad \gamma_1 = \cos\frac{\pi}{2} = 0$$

$$\alpha_2 = \cos\frac{\pi}{2} = 0; \quad \beta_2 = \cos 0 = 1; \quad \gamma_2 = \cos\frac{\pi}{2} = 0$$

$$\alpha_3 = \cos\frac{\pi}{2} = 0; \quad \beta_3 = \cos\frac{\pi}{2} = 0; \quad \gamma_2 = \cos 0 = 1$$

and only the *directional cosines* $\alpha_1 = \beta_2 = \gamma_3 = 1$ remain, whereas all the other cosines are equal zero. Such a system is described to be of rectangular or orthotropic anisotropy (Kącki [9.1], p. 226). The tensor of magnetic permeability at the rectangular anisotropy has a form of a diagonal matrix:

$$\vec{\mu} = \begin{bmatrix} \mu_1 & 0 & 0 \\ 0 & \mu_2 & 0 \\ 0 & 0 & \mu_3 \end{bmatrix} \tag{2.83a}$$

However, since $\mu_1 = \mu_x$, $\mu_2 = \mu_y$, $\mu_3 = \mu_z$, we can finally write that in the case when the coordinate system aligns with the anisotropic axes

$$B_x = \mu_x H_z, \quad B_y = \mu_y H_y, \quad B_z = \mu_z H_z \tag{2.84}$$

where μ_x, μ_y, μ_z are the permeabilities in the directions of three mutually perpendicular anisotropy axes. In this case, Equation 2.83 are decoupled like (2.2a,b,c). A uniform medium in which the permeability tensor $\vec{\mu}$ can be expressed in the diagonal form (2.83a), in some orthogonal, Cartesian, or curvilinear coordinate system, is called *orthotropic* or with *rectangular anisotropy* [9.1].

An analogical consideration can be carried out in relation to the remaining two equations of system (2.3) and (2.6). Generally, we obtain three systems of equations ($i = x, y, z$) in relation to anisotropic media

$$D_i = \sum_k \varepsilon_{ik} E_{ik}; \quad B_i = \sum_k \mu_{ik} H_{ik}; \quad J_i = \sum_k \sigma_{ik} \left[E_k + E_{\text{extern},ik} + (v \times B)_k \right]; \tag{2.85}$$

where $\varepsilon_{ik} = \varepsilon_{ki}$, $\mu_{ik} = \mu_{ki}$, $\sigma_{ik} = \sigma_{ki}$. The previous comments and conclusions fully apply to these equations too.

Anisotropy has been becoming more and more important in practice, due to the broad introduction of textured magnetic materials, nonuniform electric and thermal conductivities of laminated iron cores, plasma anisotropy, Hall effect, and so on.

Electromagnetic calculations of nonlinear and anisotropic bodies belong to the most difficult ones. For instance, for fields with only one component of the current density $J_z = -\sigma \partial A_z / \partial t$ (2.51), from Equation (2.1) curl $\boldsymbol{H} = \boldsymbol{J}$, it follows: $J_z = \partial H_y / \partial x - \partial H_x / \partial y$. In turn, from Equation (2.50) $\boldsymbol{B}$ = curl $\boldsymbol{A}$, at $A_x = A_y = 0$, we get $H_x = v_x \partial A_z / \partial y$ and $H_y = -v_y \partial A_z / \partial x$, from which, after substitution, we obtain the equation for a medium with rectangular anisotropy:

$$\frac{\partial}{\partial x}\left(v_y \frac{\partial A_z}{\partial x} \right) + \frac{\partial}{\partial y}\left(v_x \frac{\partial A_z}{\partial y} \right) = \sigma \frac{\partial A_z}{\partial t} \tag{2.86}$$

which, among others, in the work by Pietruszka and Jablonski [2.26] was utilized for analysis of fields in transformer cores made of cold-rolled, anisotropic iron sheets (Figure 1.32), where $v_x = (1/\mu_x) = v_x(B_x)$, $v_y = 1/\mu_y = v_y(B_y)$ are the *directional reluctivities of medium*, which correspond to the components B_x, B_y depending on the flux density and on the angle α of the inclination of vector $\boldsymbol{B}$ from the rolling direction (Figure 1.32).

The mutual dependence $\mu_i = f(\mu_j, B_i)$, $i, j = x, y, z$, determined by shifting of boundaries and rotation of magnetic axes of domains (Turowski [2.31]), needs a separate investigation.

The density of electromagnetic forces in nonlinear anisotropic electric systems ($H = 0$) and magnetic systems ($E = 0$) can be calculated (Frolov [2.21]) with the help of the *Maxwell's stress tensor* (2.77), where T_{ij} is the projection of jth component of tensor onto the ith axis of the Cartesian coordinate system.

2.4 NONLINEAR MEDIA

Media whose physical properties, especially the resistivity $\rho = 1/\sigma$ (specific resistance), permeability μ, or permittivity ε, are not constant, but are changing under influence of change of field parameters or system energy, are called nonlinear. This variability significantly makes the field analysis difficult because it restricts the possibility of applying field superposition, which is very convenient in solutions of linear systems.

Equations 2.3 and 2.6 can, therefore, be nonlinear in certain cases.

2.4.1 Nonlinear Permittivity and Resistance

The permittivity ε (*electric permeability*) at high frequencies can depend on frequency and electric field intensity. Especially strong nonlinearity of the permittivity occurs in the so-called *seignetodielectrics* (the name originates from Seignette salt), called also *ferroelectrics*—it is because they comprise domains (regions) of spontaneous polarizations, hysteresis, Curie point, and so on [1.13].

Variations of resistivity should be taken into account, first of all, in nonlinear resistors, whose current–voltage characteristics remarkably deviate from a straight line.

Nonlinearity of conductors results primarily from the influence of temperature on their resistivity, in function of time and load.

Nonlinearity of semiconductor resistors (e.g., SiC) reveals a strong increase of their conductivity with increasing temperature (*thermistors*) or with increasing electric field (*varistors*)—see Section 1.2.4. Nonlinear resistances of semiconductors are characterized by the value of resistivity or conductivity at a given voltage, by means of *differential resistivity* and the *nonlinearity coefficient* [1.13]:

$$\beta = \frac{\mathrm{d}i}{\mathrm{d}u} \cdot \frac{u}{i} \tag{2.87}$$

The β coefficient is constant if the *current–voltage characteristic* of the resistor has the form

$$i = Bu^{\beta} \quad \text{or} \quad E = n\, J^{\alpha} \tag{2.88}$$

where $\alpha = 1/\beta$ and n are constants, depending on specifics of the material. This form is very convenient for analytical approximation of characteristics.

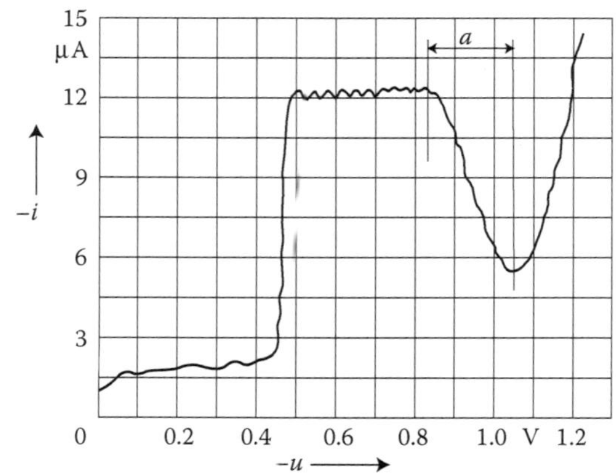

FIGURE 2.5 Current–voltage characteristic of a galvanic cell (polarogram), $i = f\ (u)$, for a liquid solution 1 mM $In_2(SO_4)_3$ and 1 M KSCN. (After Turowska M.: *Soc. Sci. Lodz, Acta Chim.*, 18, 1973, 73–84); *a*—region of the negative differential resistance (generating oscillations.)

Particularly high nonlinearity of resistance is shown by diffusion processes (Turowska Maria [2.30]) in *galvanic* cells, including negative differential resistance (Figure 2.5), similar to tunnel diodes.

"By appropriate adjustments of the value of negative resistance, by means of the applied voltage U, one can obtain conditions in which the total resistance of circuit equals zero and then such circuit can *oscillate with natural, resonance frequency, creating self-operated electrochemical oscillator*" [2.30].

In more general considerations concerning the influence of medium on electromagnetic field, instead of the equations $\boldsymbol{D} = \varepsilon\boldsymbol{E}$ and $\boldsymbol{B} = \mu\boldsymbol{H}$, the relation between the field vectors is expressed by means of Equation 1.25 and analogical to it equation $\boldsymbol{D} = \varepsilon_0\boldsymbol{E} + \boldsymbol{P}$, in which $\boldsymbol{J}_i$ and $\boldsymbol{P}$ are the vectors of *magnetic polarization* and *electric polarization*, respectively. They are equal to the magnetic and electric moments (1.24), respectively. These vectors do characterize the influence of medium on electromagnetic field. In *ferromagnetic* and *segnetoelectric* bodies, $\boldsymbol{P} = f(\boldsymbol{E})$ and $\boldsymbol{J}_i = \varphi(\boldsymbol{H})$ are complex functions, which in most cases can be represented only graphically.

From this point of view, the biggest difficulties are posed by the most important constructional material—steel—whose properties are distinctively nonlinear.

2.4.2 Nonlinear Magnetic Permeability

Most important tasks in this group include the problem of analytical representation of the nonlinear permeability of iron (Chapter 7). Among magnetics, only *paramagnetic* and *diamagnetic* materials have linear properties. However, on the basis of knowledge of physical properties of the body and geometry of its surface as well as field, one can in many cases obtain correct results at an assumption of constant magnetic permeability. Significant help in this case can be brought by auxiliary empirical investigations. Nonlinear properties of steel are so complicated that they are expressed only with the help of empiric magnetization curves. An analytical approximation of these curves often leads to functions too difficult for a more comprehensive mathematical analysis. Therefore, in computational design practice, a semi-empirical approach has become popular. In it basic Maxwell's equations are first resolved for a linear medium, with the help of Equations 2.36, 2.37, 2.54, and 2.57, and only later, in final formulae, the variable magnetic permeability is introduced. This convenient method is, theoretically, only partly justified (Chapter 7). Therefore, results obtained with this approach should be checked experimentally. The works aimed at justification and improvements of this method are worthwhile.

Full solution of Maxwell's equations with the variable permeability is possible only with numerical methods and, moreover, with significant limitations in economy and time.

Using the second Maxwell's equation for the nonlinear permeability (1.48), $\mu_n = \partial B/\partial H = \mu(H) + H\ \partial\mu(H)/\partial H$,

$$\operatorname{curl} \boldsymbol{E} = -\mu_n \frac{\partial \boldsymbol{H}}{\partial t} \tag{2.89}$$

and substituting into the identity (2.26) the Equation 2.4 div $\boldsymbol{B} = 0$ for isotropic media ($\mu_x = \mu_y = \mu_z = \mu$) at $\mathrm{d}\mu_i(x_i)/\mathrm{d}x_i = 0$, we obtain the equation

$$\nabla^2 \boldsymbol{H} = \mu_n(\boldsymbol{H})\sigma \frac{\partial \boldsymbol{H}}{\partial t} \tag{2.90}$$

At a nonuniform μ, using the identity of vector calculus div $(U\ \boldsymbol{A}) \equiv \boldsymbol{A}$ grad $U + U$ div $\boldsymbol{A}$, from Equations 2.4 and 2.26 we get

$$\operatorname{div} \boldsymbol{B} = \operatorname{div}(\mu \boldsymbol{H}) = \boldsymbol{H} \operatorname{grad} \mu + \mu \operatorname{div} \boldsymbol{H} \tag{2.91}$$

and from it

$$\operatorname{div}\boldsymbol{H} = -\frac{1}{\mu} \boldsymbol{H} \operatorname{grad} \mu \tag{2.92}$$

After substituting Equation 2.92 into Equation 2.26, and considering Equation 2.28, we obtain the equation for $\mu = \mu(H) \neq$ const :

$$\nabla^2 \boldsymbol{H} + \operatorname{grad}\left[\boldsymbol{H} \frac{\operatorname{grad}\mu}{\mu}\right] = \mu\sigma \frac{\partial \boldsymbol{H}}{\partial t} + \mu\varepsilon \frac{\partial^2 \boldsymbol{H}}{\partial t^2} \tag{2.93}$$

Using the magnetic vector potential, per Equation 2.50, in the form

$$\boldsymbol{H} = \nu \operatorname{curl} \boldsymbol{A} \tag{2.94}$$

where $\nu = \nu(B) = 1/\mu$ is the reluctivity of medium, from Equations 2.1 and 2.2 we get

$$\operatorname{curl} \boldsymbol{H} = \operatorname{curl}(\nu \operatorname{curl} \boldsymbol{A}) = \sigma \boldsymbol{E} + \sigma(\boldsymbol{v} \times \operatorname{curl} \boldsymbol{A}) + \boldsymbol{J}_{\text{external}} \tag{2.95}$$

Applying in turn the identity of vector calculus curl $U\boldsymbol{W} \equiv \text{grad}\, U \times \boldsymbol{W} + U\, \text{curl}\, \boldsymbol{W}$ to the left side of Equation 2.95, and considering the identities (2.26) and (2.51), we get

$$\text{grad}\,\nu \times \text{curl}\,\boldsymbol{A} + \nu\,\text{grad div}\,\boldsymbol{A} - \nu\nabla^2\boldsymbol{A} = -\sigma\frac{\partial \boldsymbol{A}}{\partial t} - \sigma\,\text{grad}\,V + \sigma(\boldsymbol{v} \times \text{curl}\,\boldsymbol{A}) + \boldsymbol{J}_{\text{external}} \tag{2.96}$$

Now, it is convenient to accept the condition for divergence (2.53) in the form

$$\text{div}\,\boldsymbol{A} = -\mu\sigma V \tag{2.97}$$

Thanks to it, Equation 2.96 takes the form

$$\text{grad}\,\nu \times \text{curl}\,\boldsymbol{A} - \nu\nabla^2\boldsymbol{A} = -\sigma\frac{\partial A}{\partial t} + \sigma(\boldsymbol{v} \times \text{curl}\,\boldsymbol{A}) + \boldsymbol{J}_{\text{external}} \tag{2.98}$$

The second equation for conductors ($\partial A/\partial t = 0$), per Equation 2.55, has the form

$$\nabla^2 V - \mu\sigma\frac{\partial V}{\partial t} = -\frac{\rho}{\varepsilon} \tag{2.99}$$

Equations 2.98 and 2.99 are resolved numerically (Komęza, Krusz [2.23]).

Influence of temperature on magnetic properties at temperatures 150–200°C is small, because the permeability of normal iron at an increase of body temperature until the Curie point (several hundred degrees Celsius) almost does not depend on temperature, but next, almost in a leap, iron loses its ferromagnetic properties (Figure 1.26). The Curie point, however, is highly significant in calculations of electromagnetic-thermal *coupled fields* occurring at induction heating, hardening, and melting of steel.

An important influence of temperature appears, on the other hand, in rare earth magnets used in small and special electric machines.

2.5 FUNDAMENTAL EQUATIONS OF MAGNETOHYDRODYNAMICS AND MAGNETOGASDYNAMICS

The subject of investigation of *magnetogasdynamics* is *plasma*, ionized gas being the fourth state of matter. At very high temperatures (over 10^4 K) gas becomes practically fully *ionized*, taking properties of conducting fluid. Partial ionization of gas can be obtained also at lower temperatures, about 2000–3000°C, if some special auxiliary means are applied, such as the so-called *potassium seeding*. In this way, one can obtain the so-called *cold plasma* not fully ionized.

Contrary to normal gases, an electromagnetic field has a strong dynamic effect on plasma, putting it in motion. Plasma motion causes, in turn, deformation of the magnetic field. At investigation of plasma motion, the complicated process of interaction of electrodynamic and gas dynamic forces and phenomena is considered. Properties of incompressible conductive fluids having features similar to plasma (e.g., mercury, Hg) but easier for laboratory investigations are subject of the science called *magnetohydrodynamics* (MHD).

Both these branches of science are significant, not only from the viewpoint of nuclear process control, astrophysics, geophysics, space science, and design of magnetohydrodynamic generators, but also from the viewpoint of purely industrial problems, such as transport, pumping, and electromagnetic mixing of melted metals in metallurgy, cooling of nuclear reactors with liquid sodium (Na), and so on.

The simplest theoretical analysis is limited to investigation of strong conducting properties, that is, electrically neutral (lack of separated charges and coulomb forces), nonmagnetic, not affected by gravitation forces fluid,* placed in a magnetic field. Such fluid, that is, plasma, is subordinated to the laws of electromagnetic fields, (2.1) through (2.10), considering that due to its high conductivity the displacement currents can be omitted.

Assuming the absence of external current sources in plasma ($E_{ext} = 0$), we can conclude that it moves under the influence of fields $\boldsymbol{E}$ and $\boldsymbol{B}$ according to the law

$$\boldsymbol{E} = \frac{\boldsymbol{J}}{\sigma} - (\boldsymbol{v} \times \boldsymbol{B}) = \frac{\text{curl}\,\boldsymbol{B}}{\mu\sigma} - (\boldsymbol{v} \times \boldsymbol{B}) \tag{2.100}$$

In some cases, one can assume that conductivity of plasma is infinitely high. Substituting $\sigma \to \infty$ into Equation 2.100, we get

$$\boldsymbol{E} + (\boldsymbol{v} \times \boldsymbol{B}) = 0 \tag{2.101}$$

As one can see, the behavior of plasma in the presence of magnetic field is significantly dependent on its conductivity. Effects of this are both electromagnetic and mechanical.

The result of substituting Equation 2.100 into the second Maxwell's Equation 2.2 we get

$$\frac{\partial B}{\partial t} = -\frac{1}{\mu\sigma}\text{curl}\,\text{curl}\,\boldsymbol{B} + \text{curl}(\boldsymbol{v} \times \boldsymbol{B}) = \text{curl}(\boldsymbol{v} \times \boldsymbol{B}) + \frac{1}{\mu\sigma}\nabla^2\boldsymbol{B} \tag{2.102}$$

since curl curl $\boldsymbol{B} \equiv$ grad div $\boldsymbol{B} - \nabla^2\boldsymbol{B}$, but div $\boldsymbol{B} = 0$. For plasma being in standstill ($v = 0$), the following equation is hence valid:

$$\frac{\partial B}{\partial t} = \frac{1}{\mu\sigma}\nabla^2\boldsymbol{B} \tag{2.103}$$

which, from the formal point of view, is identical to the equation of thermal conductivity (Fourier eqn.) or the equation describing the *diffusion* processes (2.37).

* "Fluid" means here gases or liquids.

From Equation 2.103, it follows that the initial field distribution of magnetic field will decay; in other words, the magnetic field will apparently diffuse from plasma of finite conductivity, with time constant equal to $1/\mu\sigma$ (Simonyi [1.11]), and the initial field distribution will disappear after the diffusion time

$$\tau_d = \mu\sigma L_e^2 \tag{2.104}$$

where L_e is the characteristic length of space variation of $\boldsymbol{B}$. For a copper sphere with a radius 1 cm, $\tau_d = 1$ s, for the liquid core of earth $\tau_d = 10{,}000$ years. When the conductivity of plasma is so high that one can ignore the second component in Equation 2.102, then we obtain

$$\frac{\partial \boldsymbol{B}}{\partial t} - \text{curl}(\boldsymbol{v} \times \boldsymbol{B}) = 0 \tag{2.105}$$

The left side of Equation 2.105 represents the full derivative of flux density for circuits moving in an alternating field, since

$$\frac{\mathrm{d}\boldsymbol{B}(r,t)}{\mathrm{d}t} = \frac{\partial \boldsymbol{B}(r,t)}{\partial t} + \frac{\partial \boldsymbol{B}(r,t)}{\partial r}\frac{\mathrm{d}\boldsymbol{r}}{\mathrm{d}t}$$

hence,

$$\frac{\mathrm{d}\boldsymbol{B}}{\mathrm{d}t} = \frac{\partial \boldsymbol{B}}{\partial t} + (\boldsymbol{v} \cdot \text{grad})\boldsymbol{B} = \frac{\partial \boldsymbol{B}}{\partial t} - \text{curl}(\boldsymbol{v} \times \boldsymbol{B}) \tag{2.106}$$

because, moreover, from the general theory of vector calculus it is known that $(\boldsymbol{v} \cdot \text{grad})\boldsymbol{B} = \boldsymbol{v} \cdot \text{div}\,\boldsymbol{B} - \text{curl}\,(\boldsymbol{v} \times \boldsymbol{B})$, and per Equation 2.4, $\text{div}\,\boldsymbol{B} = 0$.

Calculating the flux enclosed by an arbitrary selected closed circuit moving together with the plasma, we obtain therefore, on the basis of Equation 2.105, the conclusion that this flux does not increase in time because

$$\frac{\mathrm{d}}{\mathrm{d}t}\int_A \boldsymbol{B} \cdot \mathrm{d}\boldsymbol{A} = \int_A \left[\frac{\partial \boldsymbol{B}}{\partial t} + \text{curl}\left(\boldsymbol{B} \times \boldsymbol{v}\right)\right] \cdot \mathrm{d}\boldsymbol{A} = 0 \tag{2.107}$$

This means that the flux coupled with an arbitrary element of plasma is not changing, and moves together with plasma. Figuratively speaking, the magnetic field lines become "*frozen in*" or "*glued*" to the liquid substance, which due to this reason, never cuts the magnetic field lines. In the first approximation, one can assume that the flux density lines cannot go out from plasma or enter into it. This phenomenon creates the possibility of immediate conversion of thermal or nuclear energy into electric energy. A moving plasma of high energy, for example, due to thermonuclear reaction, is pulling with itself the lines of magnetic field, carrying out in this way the work of overcoming forces of field. The changes of magnetic field caused by moving flux density lines can induce in an external immovable circuit a useful electromotive force [1.11].

If we assume that the velocity $\boldsymbol{v}$ of the flux density lines is perpendicular to $\boldsymbol{B}$, from Equation 2.101, after vector multiplication of both sides by $\boldsymbol{B}$, we shall obtain the so-called "$\boldsymbol{E} \times \boldsymbol{B}$ – drift" of the conducting substance and force lines

$$\boldsymbol{v} = \frac{\boldsymbol{E} \times \boldsymbol{B}}{B^2} \tag{2.108}$$

The set of basic equations of plasma, together with Maxwell's equation, create the following *hydrodynamic equations:*

- Equation of plasma continuity and of mass conservation (ρ_m—plasma density)

$$\frac{\partial \rho_m}{\partial t} + \operatorname{div}(\rho_m v) = 0 \tag{2.109}$$

- *Equation of motion*, which without the last component is the Navier–Stokes equation for ordinary gasdynamics [1.11]

$$\rho_m \frac{\partial v}{\partial t} + \rho_m (v \cdot \operatorname{grad}) v = -\operatorname{grad} p + \rho_m \nu \nabla^2 v + \boldsymbol{J} \times \boldsymbol{B} \tag{2.110}$$

The left side of Equation 2.110 represents the product of plasma density $\rho_m(x, t)$ and acceleration ($d\boldsymbol{v}/dt$); the first component on the right-hand side of Equation 2.110—gradient of pressure $p(x, t)$; the second—force density of internal friction, the third—volume density of Lorentz's force (2.8) putting the conducting medium in motion with the velocity $\boldsymbol{v}$;

- Equation of fluid state

$$p = f(\rho_m, t) \tag{2.111}$$

Reynold's Magnetic Number, R_M—it is an often used dimensionless parameter which permits to distinguish the state in which diffusion of magnetic field from plasma appears, from the state in which magnetic field lines are frozen:

$$R_M = v\mu\, \sigma L = \frac{v\tau_d}{L} \tag{2.112}$$

where $\tau_d = \mu\, \sigma L_e^2$ (2.104).

The Reynold's Magnetic Number $\boldsymbol{R}_{\mathrm{M}}$ can be considered either as the ratio of linear dimension l of the flow field to the characteristic length

$$l_{\mathrm{e}} = \frac{1}{\mu \sigma v} \tag{2.113}$$

on which the conducting fluid is embraced by the magnetic field, or as the ratio of the flow velocity v to the characteristic velocity

$$v_{\mathrm{e}} = \frac{1}{\mu \sigma l} \tag{2.113a}$$

with which the magnetic field propagates inside the conducting fluid. If $R_{\mathrm{M}} \gg 1$ (i.e., $l \gg l_{\mathrm{e}}$ or $v \gg v_{\mathrm{e}}$), then the magnetic field will move together with plasma flux (the *frozen in* field) and motion of plasma will strongly influence the magnetic field. Whereas if $R_{\mathrm{M}} \ll 1$ (i.e., $l \ll l_{\mathrm{e}}$ or $v \ll v_{\mathrm{e}}$), then the motion of plasma will not remarkably influence the magnetic field. For liquids such as mercury or sodium in laboratory conditions $R_{\mathrm{M}} < 1$ (except for very high velocities). In geophysical and astrophysics conditions, the Reynold's magnetic number R_{M} can be very much higher than one.

2.5.1 MHD Generators

The phenomena described above are utilized in the so-called MHD generators (Figure 2.6), with which are connected hopes (rather illusory) of direct conversion of combustion energy into electric energy (Zygielbaum [2.38]). In a conducting gas, moving with velocity v_x, the electric current of density $\boldsymbol{J} = \sigma\,(\boldsymbol{B}\boldsymbol{v} - U/h)$ is generated along with the Lorentz's force of density $\boldsymbol{f}_L = \boldsymbol{J} \times \boldsymbol{B}$ impeding the motion of gas with the mechanical power $P_{m1} = J\,B\,v$ (W/m³).

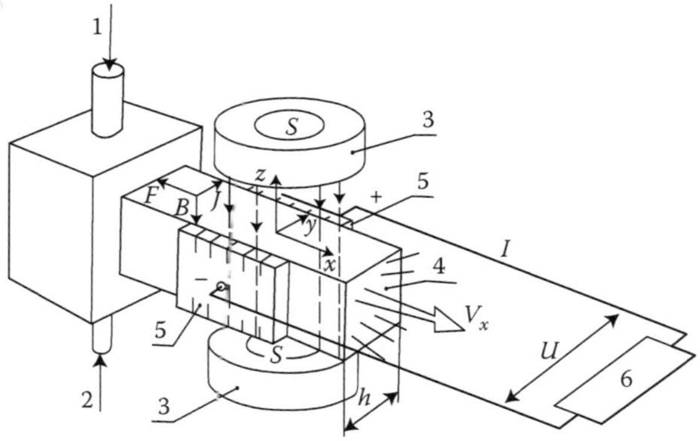

FIGURE 2.6 A scheme of magnetohydrodynamic (MHD) generator: 1—fuel with sodium or potassium seeding (up to 1%); 2—oxidizer; 3—superconducting electromagnet; 4—ionized gas (plasma); 5—electrodes; 6—receiver (load).

The conductivity σ of combustion products with 1% of potassium seeding, at 2500 K, is on the order of 10 S/m (while for copper, 58×10^6 S/m) and exhibits tensor properties, which is the reason of the Hall effect (Figure 1.13). To eliminate the harmful longitudinal Hall currents, the electrodes are divided into many mutually isolated segments, which work with independent electric circuits (Faraday's generator). In other MHD generators (of Hall type) the opposite electrodes (+) and (–) are short-circuited pairwise, and the axial Hall voltage is utilized as the working voltage. There have also been used mixed solutions (after Celinski, 1969).

2.5.2 Electric Machines and Apparatus

The effect of partial "freezing in" of field lines (Lentz rule) to a highly conductive substance occurs also in normal electric machines; for instance, as the field of armature winding at impulse short circuit of a synchronous machine (*subtransient* state), or as the so-called "*electromagnetic tail*" of an induction linear motor, and so on. Due to the limited conductivity of short-circuited conductors, this field however decays fast with growing distance of shortened circuit from the initial position, according to an exponential function.

For instance, in the most unfavorable moment of impulse short-circuit of a synchronous machine, the impulse field of armature reaction penetrates the solid rotor at $v = 0$, according to Equation 2.104 as the dependence (J. Turowski [1.16], p. 211):

$$H(x,t) = H_0\left[1 - \frac{4}{\pi}\sum_{n=1}^{\infty}\frac{(-1)^n}{2n-1}\cos\frac{(2n-1)\pi x}{2a}\right]\exp\left[-\frac{(2n-1)^2\pi^2}{4\mu\sigma a^2}t\right] \tag{2.114}$$

At the sinusoidal distribution:

$$H_0 = H_{\mathrm{m}}\cos\frac{\pi}{\tau}x$$

$$H(x,\ t) = H(1 - e^{-\frac{t}{T}})\cos\frac{\pi}{\tau}x \tag{2.115}$$

where the time constant of field penetration

$$T = \frac{\tau}{\pi}\mu\sigma \tag{2.116}$$

Considering the rotor motion $v = 2\tau n$, the full Equation 2.103 should be resolved. If, however, one omits the resistance of stator winding in comparison with its reactance, the flux

$$\Phi_m = \int_{-\tau/2}^{\tau/2} H_{\mathrm{m}}\cos\frac{\pi}{\tau}x\,\mathrm{d}x,$$

according to Equation 2.105, will be held constant and equal to Φ_m, causing changes in the short-circuit current of armature (Turowski [1.18]).

2.6 ELECTRODYNAMICS OF SUPERCONDUCTORS

Superconductivity, whose basic physical properties were described in Section 1.2.2, attracts a broad interest both from the viewpoint of technical applications and of the fundamental theoretical research.

Superconductors are metals whose resistivity at low temperatures (4 to about 100 K, depending on materials—Figure 1.15) falls down to an immeasurable low value. The described state is called *superconductivity.* If to an ideal superconductor with specific conductivity $\sigma \rightarrow \infty$ one applies Maxwell's Equations 2.1 through 2.6, then one obtains the system of equations

$$\boldsymbol{E} = \frac{\boldsymbol{J}}{\sigma} \rightarrow 0; \quad \frac{\partial \boldsymbol{B}}{\partial t} = \text{curl}\,\boldsymbol{E} = 0; \quad \text{and}\ \boldsymbol{B} = \text{const} \tag{2.117}$$

from which the last one suggests the possibility of freezing-in the magnetic field in a solid conductor. However, investigations of Meissner and Ochsenfeld (1933) have shown that the flux density in a *mono-coherent superconductor* always equals to zero, independently of method of its freezing (the Meissner effect). Hence, an *ideal* superconductor is at the same time an ideal diamagnetic satisfying the equations

$$\mu = 0 \quad \text{and} \quad B = \mu H = 0 \tag{2.118}$$

to which one can apply results of the theory of magnetic potential [1.12]. It primarily concerns all the so-called *soft superconductors* (Section 1.2.2). The magnetic flux is, however, displaced only from the superconductor itself. Therefore, if the superconductor creates a multiple-connected body (e.g., a ring), the flux inside the empty space can be either "frozen-in" if the flux existed there before cooling the superconductor, or displaced if the superconducting state existed there before appearing of the external field. Per this rule, the *superconductors of type II* (*hard*, Section 1.2.2), in the volume of which exist superconducting threads, can "catch" the field and delay overmagnetization.

The F. and G. London brothers showed in 1935 that the Meissner effect, like many other phenomenological properties of superconductors, can be described by means of Maxwell's equations, complemented by the additional Equations [1.24]

$$\text{curl}\,\Lambda \boldsymbol{J}_S = -\boldsymbol{B} \tag{2.119}$$

$$\frac{\partial(\Lambda \boldsymbol{J}_S)}{\partial t} = \boldsymbol{E} \tag{2.120}$$

where $\boldsymbol{J}_S$ is the current density of superconductivity (the so-called *supercurrent*); $\Lambda = m/(e^2\, n_s)$ is the constant which characterizes a superconductor, while m, e, and n_s are the mass, charge, and density of electrons of superconductivity, respectively.

The gradual increase of the critical magnetic field intensity H_C at decreasing temperature has been taken into account by the assumption that the number of superconducting electrons n_s increases from zero at $T = T_C$ to 100% at $T = 0$. The current density of these electrons, J_S, satisfies Equation 2.120, whereas the other electrons, in result of collisions with crystal lattice, produce the normal current density $J_n = \sigma E$ where σ is a function of temperature. This is the so-called *model of two electronic fluids* that obey different laws of electrodynamics. The resultant current density has two components: $\boldsymbol{J} = \boldsymbol{J}_S + \boldsymbol{J}_n$ of which the second one in steady state equals to zero, because in this case in a superconductor $\boldsymbol{E} = 0$.

From Equation 2.119 and the first Maxwell's equation one can obtain the Helmholtz equation

$$\nabla^2 \boldsymbol{H} = \frac{1}{\delta^2}\boldsymbol{H} \tag{2.121}$$

whose solution is analogical to Equation (2.173) in the case of a solid conductor. Therefore, in a superconducting half-space, the field distribution is exponential, that is

$$H = H_S e^{-\frac{z}{\delta}} \tag{2.122}$$

where (Wilkes [1.24]):

$$\delta = \sqrt{\frac{m}{\mu_0 n_s e^2}} \tag{2.123}$$

is the *equivalent depth of field penetration* into superconductor (2.181), that is, the distance from the surface at which the field is decreased $e = 2.7$ – times. This value is on the order of 10^{-6} cm (e.g., for tin (Sn): 5.2×10^{-6} cm, for lead (Pb): 3.9×10^{-6} cm (J. Turowski [1.15]). Although this thin layer can be neglected at macroscopic samples, for very thin layers and small masses, for instance for colloids, the effect of penetration becomes remarkable. Calculations of field distribution in other bodies, for instance in plates, is done in a similar way like the calculation of transient fields in analogical resistive metal bodies. The difference consists only in that there we consider a quasistatic process, while here—a static one.

Experiments have shown that the *theory of the Londons is correct only in an area of weak fields* ($H \ll H_C$), for $T \approx T_C$, and for low frequencies. Therefore, it was next extended by Gincburg and Landau, Pippard, and others [1.12].

The Londons' equations do not substitute Maxwell's equations. They only take into account "*responses*" of medium in the form of their microscopic transformations under the influence of external interaction on the medium by its freezing or destructive influence of magnetic field.

Recently, many strong *air-cored magnets* have already been produced, aimed for magnetohydrodynamic generators, plasma traps at investigation of controlled thermonuclear synthesis, fluid separators, etc. Production of strong fields practically with no power losses in windings creates a possibility of radical improvement of large power transformers, construction of large superconducting generators (e.g., 3000 and more MW), and electric motors without iron and power losses in excitation circuit (Sykulski [1.54]).

Already in 1983, for the first time in history there was set in motion in the USA (M. Superczynski) a ship with a superconducting motoring power of 3000 HP [2.28].

There were several other innovative examples, but the latest progress is not that impressive as it was expected in the beginning of the so-called high-temperature superconductors. Broad perspectives are still in the field of utilization of superconductivity in electronics, automation, and computer technology components, especially since the discovery in 1986 by J. Bednorz [1.31] the high-temperature superconductors (Figure 1.15).

2.7 ELECTRODYNAMICS OF HETEROGENEOUS MEDIA

In nonuniform media whose physical parameters are continuous functions of the coordinate vector $\boldsymbol{x}$, that is,

$$\sigma = \sigma(\boldsymbol{x}), \quad \varepsilon = \varepsilon(\boldsymbol{x}), \quad \mu = \mu(\boldsymbol{x}) \tag{2.124}$$

according to H. Rawa of Warsaw University of Technology [2.15], [2.27], the Maxwell's Equations 2.1 and 2.2, similar to the Londons' Equations 2.119 and 2.120, are correct only when we assume the condition div $\boldsymbol{D} = \rho$, where ρ is defined by the additional equation

$$\frac{\partial \rho}{\partial t} + \frac{\sigma(x)}{\varepsilon(x)}\rho = \sigma(x)E \cdot \text{grad}\, ln\frac{\varepsilon(x)}{\sigma(x)} \tag{2.125}$$

The solution is then determined by heterogeneous structure of the medium that is able to conduct electric current.

This hypothesis needs further verification research because, on the other hand, investigations of fields in nonlinear ferromagnetic media, based on the dependence $B(x) = \mu(x)\, H(x)$ (see Chapter 7) produce results satisfactory for technical purposes.

2.8 ELECTRODYNAMICS OF SEMICONDUCTOR DEVICES

Electrodynamic modeling and simulations may provide insight into physical phenomena in semiconductor devices and circuits, which is otherwise not possible just by experiments. Modern analysis, design, and optimization of semiconductor structures are based on physics-based Technology Computer-Aided Design (TCAD) software, which when coupled with a circuit simulator (e.g., *Spice*) is called a Mixed-Mode simulation [2.42], [4.35].

Static and dynamic behavior of charge carriers, that is, electrons (n) and holes (p), under influence of external fields is described by three systems of equations [2.19], [2.35]:

a. Maxwell's Equations 2.1 through 2.6;
b. Equations of current density for electrons and holes [2.40]

$$\left.\begin{aligned} J_n &= q\,\mu_n\, n\boldsymbol{E} + q\,D_n \text{ grad } n \\ J_p &= q\,\mu_n\, p\boldsymbol{E} - q\,D_p \text{ grad } p \end{aligned}\right\} \tag{2.126}$$

where μ_n, μ_p are the mobilities of electrons (n) and holes (p); D_n, D_p are the corresponding *diffusion coefficients*; n, p the concentration of electrons and holes, respectively; q the charge of electron ($q = 1.60219 \times 10^{-19}$ C);
c. Continuity equations for electrons and holes

$$\left.\begin{aligned} -\frac{1}{q}\text{div}\,\boldsymbol{J}_n - G + R + \frac{\partial n}{\partial t} &= 0 \\ \frac{1}{q}\text{div}\,\boldsymbol{J}_p - G + R + \frac{\partial p}{\partial t} &= 0 \end{aligned}\right\} \tag{2.127}$$

where G is the *generation* rate of carriers (1/cm³/s), for instance by thermal excitation or collision ionization or by an external radiation or light illumination (photoabsorption), R is the *recombination* rate.

After substituting Equation 2.126 into Equation 2.127 and considering div $\boldsymbol{D} = \rho$ and the identity div grad $V \equiv \nabla^2 V$, one obtains the corresponding Poisson's equation $\nabla^2 V = -\rho/\varepsilon$ Equation 2.35 for the electric potential V, which after numerical solution, together with Equation 2.127 allows to determine a distribution of the electric potential V and the density of electrons and holes (n, p) in a semiconductor device [2.19], [2.35], [2.36], [2.42], [4.35].

Electromagnetic wave propagation and radiative effects become more and more important for the integrated circuit domain, when today's submicron or nanometer-scale semiconductor devices are operated under high frequencies, often at $f > 1$ GHz. This is particularly important for the characterization of interconnected structures loaded at microwave frequencies [2.43]. Indeed, it is well known that electromagnetic compatibility (EMC) and signal integrity (SI) are strongly affected by the geometry of interconnects and by the possibly complex nonlinear/dynamic behavior of the electronic devices collocated at their terminations.

For example, in a recent paper [2.44], the semiconductor analysis is based on the time-domain *drift-diffusion model* [Equations 2.126 through 2.127] in conjunction with the electromagnetic model (Maxwell's equations), where the active semiconductor device model corresponds to the Poisson equation and the carrier transport equations obtained by splitting the Boltzmann transport equation (BTE) into its first two moments [2.35]. Numerical discretization and efficient solution of such a coupled system is a serious challenge and is still a subject of ongoing research [2.44], [2.45].

2.9 ELECTRODYNAMICS OF ELECTROCHEMICAL SYSTEMS

Fundamental tasks of *electrochemical analysis* [2.30] involve description of concentration of *the depolarizator* (carrier of electric charges in solution) in dependence on the distance from electrodes and on the time of electrolysis. Changes of concentration c (in mols per cm^3 of solution) of diffusing substance (depolarizator) in the investigated element can be described (Galus [2.4], p. 69) by the equation

$$\frac{\partial c}{\partial t} = -\mathrm{div}\, \boldsymbol{f}^{*}_{\mathrm{diff}} \tag{2.128}$$

where

$$\boldsymbol{f}_{\mathrm{diff}} = -D \,\mathrm{grad}\, c \tag{2.129}$$

is the flux of depolarizator [velocity of diffusion, in mol/(m^2s)]; D is the coefficient of its diffusion (m^2/s).

Equation 2.129, in the notation for one-dimensional diffusion ($\partial c/\partial y = \partial c/\partial z = 0$), is called *the first Fick's law of diffusion* [2.30] and has the form

$$f_x = -D\frac{\partial c}{\partial x} \tag{2.129a}$$

After substituting Equation 2.129 into Equation 2.128 and considering the vector identity div grad $c \equiv \nabla^2 c$, we obtain *the general equation of diffusion*

$$\nabla^2 c = \frac{1}{D}\frac{\partial c}{\partial t} \tag{2.130}$$

which, for the one-dimensional diffusion assumed here, bears the name of *the second Fick's law of diffusion* (1855) and has the form

$$\frac{\partial^2 c}{\partial x^2} = \frac{1}{D}\frac{\partial c}{\partial t} \tag{2.130a}$$

The mass of a substance released during electrolysis is defined by the *first Faraday's law*

$$m = k\, i\, t = k\, Q \tag{2.131}$$

and the second Faraday's law

$$k = \frac{M}{nF} \tag{2.132}$$

where m is the mass of the substance educed on electrode; M is the molecule mass (in mol) of the substance; n the number of electrons exchanged in the electrode process; $F = 96{,}500$ C—the Faraday's constant.

* Here $\boldsymbol{f}_{\mathrm{diff}} \equiv \boldsymbol{J}$, as per *J. Koryta*: *Electrochemistry* (in Polish), PWN, Warszawa, 1980, p. 91.

Equations 2.128 through 2.132 are used, among others, for description of substance diffusion into the electrode. The substance flow caused by convection is $\boldsymbol{J}_{\text{conv}} = c\,\boldsymbol{v}$, (after J. Koryta, PWN 1980, p. 93), that is, the flux density of substance.

In particular, in electrochemical oscillators (Figure 2.5) the concentration c, as a function of free current, has the character of an oscillating wave penetrating into the solution (Maria Turowska [2.30]) according to Equation 2.130a, in an identical way as electromagnetic wave of alternating electromagnetic field penetrating into a solid, conducting half-space (Figure 2.10 [later in the chapter]).

The density of heat flow in liquids (Fourier's law) is expressed by

$$q_t = -\lambda \frac{\partial T}{\partial x} \quad \text{or} \quad q_t = -\lambda \operatorname{grad} T.$$

2.10 GENERAL WAVE EQUATIONS

The solution of Maxwell's equations is relatively simple only in simple, linear, isotropic media. Let us accept this assumption and suppose that medium is stationary, free from moving electric charges and external fields. The calculated result of curl for both sides of Equation 2.1 and substitution of Equation 2.2 into it, per (2.28) we obtain Equation 2.29 which describes *wave processes in a nonideal dielectric* of $\mu = \text{const}$, $\varepsilon = \text{const}$:

$$\nabla^2 \boldsymbol{H} = \mu\sigma \frac{\partial \boldsymbol{H}}{\partial t} + \mu\varepsilon \frac{\partial^2 \boldsymbol{H}}{\partial t^2} \tag{2.133}$$

In the case of an ideal dielectric ($\sigma = 0$), on the right-hand side the first component disappears and Equation 2.133 transforms into the *wave equation of dielectric* (2.38)

$$\nabla^2 \boldsymbol{H} = \frac{1}{v^2} \frac{\partial^2 \boldsymbol{H}}{\partial t^2}; \qquad \left(v = \frac{1}{\sqrt{\mu\varepsilon}} \right) \tag{2.134}$$

In the case when the displacement current density ($\varepsilon(\partial E/\partial t)$) can be omitted in comparison with the conduction current density ($\sigma\boldsymbol{E}$), on the right-hand side of Equation 2.133 the second component disappears and Equation 2.133 transforms into the *wave equation in conductors*

$$\nabla^2 \boldsymbol{H} = \mu\sigma \frac{\partial \boldsymbol{H}}{\partial t} \tag{2.135}$$

which, if considered as scalar, has the same form as the equation of thermal conductivity 2.37 or the diffusion Equations 2.130 and 2.130a.

While solving Equation 2.133 we often limit it to analysis of only sinusoidal alternating processes, or to only one (first) harmonic of an arbitrary waveform in time (*monochromatic wave*):

$$\underline{\boldsymbol{E}}_m = \boldsymbol{E}_m e^{j\omega t} \quad \text{and} \quad \underline{\boldsymbol{H}}_m = \boldsymbol{H}_m e^{j\omega t} \tag{2.136}$$

Introducing the second expression from Equation 2.136 into Equation 2.133 we will see that the time factor $e^{j\omega t}$ is reduced on both sides, which finally gives the equation for complex amplitude of an alternating field

$$\nabla^2 \boldsymbol{H}_m = \Gamma^2 \boldsymbol{H}_m \tag{2.137}$$

where

$$\Gamma = \sqrt{j\omega\mu(\sigma + j\omega\varepsilon)} \tag{2.138}$$

2.10.1 Wave Equations in Metal

In the case of *conducting media*, the second component under root in Equation 2.138 is omitted and in Equation 2.137 instead of Γ will emerge the coefficient

$$\alpha = \sqrt{j\omega\mu\sigma} = (1 + j)k = \sqrt{2}ke^{j\frac{\pi}{4}} \tag{2.139}$$

called propagation constant of electromagnetic wave in space, where

$$k = \sqrt{\frac{\omega\mu\sigma}{2}} = \sqrt{\pi f \mu\sigma} \tag{2.140}$$

called *attenuation constant of wave in metal* (1/m).

In this case

$$\nabla^2 \boldsymbol{H}_m = \alpha^2 \boldsymbol{H}_m \tag{2.141}$$

Proceeding analogically, we obtain the equation for the electric field in conductors and in regions free from electric charges (div $\boldsymbol{D} = \rho = 0$)

$$\nabla^2 \boldsymbol{E}_m = \alpha^2 \boldsymbol{E}_m \tag{2.142}$$

The same equation can be obtained for current density if into Equation 2.142 we introduce $\boldsymbol{J}_m = \sigma \boldsymbol{E}_m$.

Since in formulae (2.133) and (2.141) on both sides occur the same vectors $\mathbf{i}\nabla^2 H_{mx} + \mathbf{j}\nabla^2 H_{my} + \mathbf{k}\nabla^2 H_{mz} = \alpha^2 (\mathbf{i}H_{mx} + \mathbf{j}H_{my} + \mathbf{k}H_{mz})$, after equalizing them to each other we obtain, in the rectangular (Cartesian) system of coordinates, three independent scalar differential equations

$$\nabla^2 H_{mx} = \alpha^2 H_{mx}; \quad \nabla^2 H_{my} = \alpha^2 H_{my}; \quad \nabla^2 H_{mz} = \alpha^2 H_{mz} \tag{2.141a,b,c}$$

Using the magnetic vector potential (2.50) $\boldsymbol{B} = \text{curl}\, \boldsymbol{A}$, at the assumption div $\boldsymbol{A} = 0$, both Equations 2.141 and 2.142 can be reduced to one (2.63) already discussed

$$\nabla^2 \boldsymbol{A} = \alpha^2 \boldsymbol{A} \tag{2.143}$$

One of the most convenient analytical methods for solving equations of type $\nabla^2\boldsymbol{A} = \alpha^2\boldsymbol{A}$ is Fourier's method.

2.11 FOURIER'S METHOD

The wave equation

$$\frac{\partial^2 H_{mx}}{\partial x^2} + \frac{\partial^2 H_{mx}}{\partial y^2} + \frac{\partial^2 H_{mx}}{\partial z^2} = \alpha^2 H_{mx} \tag{2.141a}$$

in the function of many variables is most conveniently solved with *Fourier's method of separation of variables*. This method consists in substituting a function of two or three variables $H_m(x, y, z)$ by a product or sum of products of functions dependent on only one variable. In the case of Equation 2.141a we write therefore

$$H_{mx}(x, y, z) = \sum_{n=1}^{\infty} C_n \cdot X_n(x) \cdot Y_n(y) \cdot Z_n(z) \tag{2.144}$$

Differentiating expression (2.144) and substituting it into Equation 2.141a, yields

$$X_n'' Y_n Z_n + X_n Y_n'' Z_n + X_n Y_n Z_n'' = \alpha^2 X_n Y_n Z_n$$

or after dividing by $X_n\, Y_n\, Z_n$

$$\frac{X_n''}{X_n} + \frac{Y_n''}{Y_n} + \frac{Z_n'' - \alpha^2 Z_n}{Z_n} = 0 \tag{2.145}$$

Since every component of Equation 2.145 is a function of a different and only one independent variable, all the components together will satisfy the relation (2.145) only then when they will be independent of x, y, z, which means—constant. These constants can be infinitely many

$$\frac{X_n''}{X_n} = A_n; \quad \frac{Y_n''}{Y_n} = B_n; \quad \frac{Z_n'' - \alpha^2 Z_n}{Z_n} = -(A_n + B_n); \qquad (n = 1, 2, 3, \ldots) \tag{2.146}$$

In this way, we have obtained three simplified differential equations, linear, of second order, with constant parameters

$$\begin{aligned} &X_n''(x) - A_n X_n(x) = 0 \\ &Y_n''(y) - B_n Y_n(x) = 0 \\ &Z_n''(z) - (a^2 - A_n - B_n) Z_n(z) = 0 \end{aligned} \tag{2.147}$$

If we assume that the arbitrary constants are negative, $A_n = -\beta_n{}^2$ and $B_n = -\eta_n{}^2$, then the general solution of Equations 2.147 takes the form

$$\begin{aligned} X_n(x) &= C_{1n}\cos\beta_n x + C_{2n}\sin\beta_n x = C_{3n}\sin(\beta_n x + \varphi_n) \\ Y_n(x) &= C_{4n}\cos\eta_n y + C_{5n}\sin\eta_n y = C_{6n}\sin(\eta_n y + \psi_n) \\ Z_n(z) &= C_{7n}\mathrm{e}^{-\sqrt{\alpha^2+\beta_n^2+\eta_n^2 z}} + C_{8n}\mathrm{e}^{\sqrt{\alpha^2+\beta_n^2+\eta_n^2 z}} \end{aligned} \tag{2.148}$$

If, instead, we assume that the constants are positive ($A_n = -\beta_n{}^2$ and $B_n = -\eta_n{}^2$), then

$$\left.\begin{aligned} X_n(x) &= D_{1n}\mathrm{e}^{-\beta_n x} + D_{2n}\mathrm{e}^{-\beta_n x} \\ Y_n(y) &= D_{3n}\mathrm{e}^{-\eta_n y} + D_{4n}\mathrm{e}^{-\eta_n y} \\ Z_n(z) &= D_{5n}\cos(\sqrt{\beta_n^2+\eta_n^2-\alpha^2 z}) + D_{6n}\sin(\sqrt{\beta_n^2+\eta_n^2 z-\alpha^2 z}) \\ &\qquad (\alpha^2 < \beta_n^2+\eta_n^2) \\ &\text{or} \\ Z_n(z) &= D_{5n}\mathrm{e}^{\sqrt{\alpha^2-\beta_n^2-\eta_n^2}} + E_{6n}\mathrm{e}^{\sqrt{\alpha^2-\beta_n^2-\eta_n^2 z}};\quad (\alpha^2 > \beta_n^2+\eta_n^2) \end{aligned}\right\} \tag{2.149}$$

From Equations 2.148 and 2.149, we choose of course this form which is the nearest to physical boundary conditions of the task at hand. For instance, for field in the gap of electric machine more convenient will be the system (2.148).

After substituting expression (2.148), or correspondingly Equation 2.149, into the initial function (2.144), and eliminating unnecessary constants at consideration of boundary conditions, we shall find the final solution of the equation. The solution in the case of a deformed curve of three variables has the form of a double trigonometric series. Since any field quantity cannot increase to infinity at an increase of distance, hence in solid bodies the constants C_{8n}, D_{2n}, and so forth, at growing exponential functions in formulae (2.148) and (2.149) practically always are equal zero (if calculation concerns a half-space). The other constants are determined from boundary conditions on the surface of the investigated body (see Chapter 6).

The above solution applies also to Laplace's equation $\nabla^2 V_m = 0$ for potential fields, if we substitute $\alpha = 0$. In this case, the general solution (2.144) can be represented by all possible combinations of products that satisfy the equation

$$V_\mu = \sum_n M_n \begin{Bmatrix}\sin\\\cos\end{Bmatrix}\beta_n x \begin{Bmatrix}\sin\\\cos\end{Bmatrix}\eta_n y \begin{Bmatrix}\mathrm{sh}\\\mathrm{ch}\end{Bmatrix} m_n z + \sum_n N_n \begin{Bmatrix}\mathrm{sh}\\\mathrm{ch}\end{Bmatrix}\beta_n x \begin{Bmatrix}\mathrm{sh}\\\mathrm{ch}\end{Bmatrix}\eta_n y \begin{Bmatrix}\sin\\\cos\end{Bmatrix} m_n z \tag{2.150}$$

where the symbol $\begin{Bmatrix}\sin\\\cos\end{Bmatrix}$ means "sin or cos."

Equations of type (2.141) can be resolved also by the application of substitutions different than (2.144) [2.12].

2.12 WAVE EQUATIONS IN CYLINDRICAL COORDINATES

Cylindrical coordinate systems are often used as the so-called quasi three-dimensional systems. The wave Equation 2.142 in cylindrical coordinates (Figure 2.7), in case of semiconducting media, has the form

$$\nabla^2 \boldsymbol{E}_m = \Gamma^2 \boldsymbol{E}_m$$

or

$$\mathbf{1_r}\left(\nabla^2 E_{mr} - \frac{2}{r^2}\frac{\partial E_{m\theta}}{\partial \theta} - \frac{E_{mr}}{r^2}\right) + \mathbf{1}_\theta\left(\nabla^2 E_{m\theta} - \frac{2}{r^2}\frac{\partial E_{mr}}{\partial \theta} - \frac{E_{m\theta}}{r^2}\right) + \mathbf{1_z}\nabla^2 E_{mz}$$
$$= \Gamma^2(\mathbf{1_r}E_{mr} + \mathbf{1}_\theta E_{m\theta} + \mathbf{1_z}E_{mz}), \qquad (\underline{\mathrm{E}}_m = E_m \mathrm{e}^{\mathrm{j}\omega t}) \qquad (2.151)$$

In cylindrical coordinates it is not possible, therefore, to separate particular scalar equations for particular components of field. This is possible only in rectangular (Cartesian) coordinate system.

Often, however, in specific cylindrical systems, one can assume that an axial symmetry of field vectors exists, that is $(\partial E_m / \partial \Theta) = 0$, and then we can obtain a system of three scalar equations, that is, *quasi three-dimensional*:

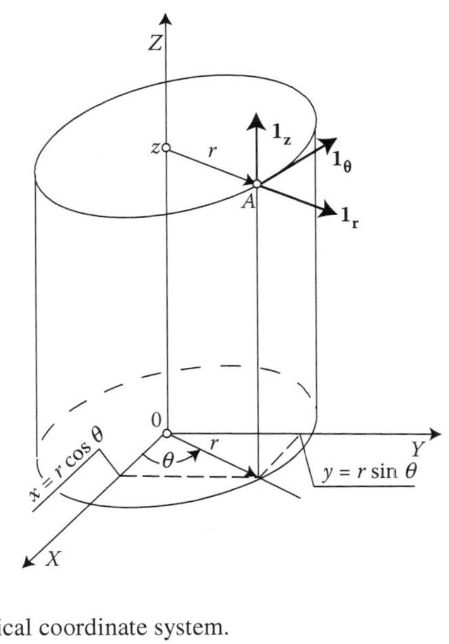

FIGURE 2.7 Cylindrical coordinate system.

$$\left.\begin{aligned}
&\frac{\partial^2 E_{mr}}{\partial r^2}+\frac{1}{r}\frac{\partial E_{mr}}{\partial r}-\frac{E_{mr}}{r^2}+\frac{\partial^2 E_{mr}}{\partial z^2}=\Gamma^2 E_{mr}\\
&\frac{\partial^2 E_{m\theta}}{\partial r^2}+\frac{1}{r}\frac{\partial E_{m\theta}}{\partial r}-\frac{E_{m\theta}}{r^2}+\frac{\partial^2 E_{m\theta}}{\partial z^2}=\Gamma^2 E_{m}\\
&\frac{\partial^2 E_{mz}}{\partial r^2}+\frac{1}{r}\frac{\partial E_{mz}}{\partial r}+\frac{\partial^2 E_{mz}}{\partial z^2}=\Gamma^2 E_{mz}
\end{aligned}\right\} \tag{2.152a,b,c}$$

These equations can also be resolved with the *method of separation of variables* by substituting $E_{mr}(r, z) = R(r) \cdot Z(z)$, and then proceeding like in the previous section.

Now we shall limit ourselves to a case when field depends only on one variable, r, which means $\partial E_m/\partial z = 0$. When we substitute into Equation 2.152 the new complex variable

$$p = r\sqrt{-\Gamma^2} \tag{2.153}$$

we shall obtain two types of Bessel's equations

- for E_{mr} and $E_{m\Theta}$:

$$\frac{\partial^2 E_{mr}}{\partial p^2}+\frac{1}{p}\frac{\partial E_{mr}}{\partial p}+\left(1-\frac{1}{p^2}\right)E_{mr}=0 \tag{2.154}$$

and

- for E_{mz}

$$\frac{\partial^2 E_{mz}}{\partial p^2}+\frac{1}{p}\frac{\partial E_{mz}}{\partial p}+E_{mz}=0 \tag{2.155}$$

General solution of Equation 2.154 can have the form

$$E_{mr}=C_1\cdot I_1(p)+C_2\cdot jN_1(p)\quad\text{or}\quad E_{mr}=C_3\cdot H_1^{(1)}(p)+C_4\cdot H_1^{(2)}(p) \tag{2.154a}$$

Solution of Equation 2.155 can have the form

$$E_{mz}=C_5\cdot I_0(p)+C_6\cdot jN_0(p)\quad\text{or}\quad E_{mz}=C_7H_0^{(1)}(p)+C_8\cdot H_0^{(2)}(p) \tag{2.155a}$$

where

$I_1(p)$ and $I_0(p)$ are the Bessel functions of the first kind, of first and zero order;
$N_1(p)$ and $N_0(p)$ are the Bessel functions of the second kind, of first and zero order, called Neumann functions;
$H_1^{(1)}(p)$ and $H_1^{(2)}(p)$ are mutually coupled Hankel functions of first order;
$H_0^{(1)}(p)$ and $H_0^{(2)}(p)$ are mutually coupled Hankel functions of zero order.

The Bessel function of nth order (where n is an integer) is the sum of series

$$I_n(p) = \sum_{\alpha=0}^{\infty} \frac{(-1)^\alpha}{\alpha!\,(n+\alpha)!}\left(\frac{p}{2}\right)^{n+2\alpha} \tag{2.156}$$

The Hankel functions, called also Bessel functions of the third kind, are related to the Bessel and Neumann functions by the dependences

$$\left.\begin{aligned} H_\nu^{(1)}(p) &= I_\nu(p) + jN_\nu(p) \\ H_\nu^{(2)}(p) &= I_\nu(p) - jN_\nu(p) \end{aligned}\right\} \tag{2.157}$$

and with the MacDonald function [2.1]

$$K_\nu(z) = \frac{1}{2}\pi H_\nu(z) \tag{2.158}$$

The significance of the Hankel functions $H_\nu(p)$ for practical applications consists primarily in the fact that among all the Bessel functions only $H_\nu(p)$ assume zero values when the complex argument tends to infinity. Particularly, $H^{(1)}$ is when imaginary part is positive, and $H^{(2)}$ is when it is negative (Jahnke [2.11]). In the opposite situation, both the functions grow to infinity.*

2.13 PLANE WAVE

The plane wave term is used for a transverse electromagnetic wave, that is, such a wave in which the field vectors **E** and **H** are perpendicular to each other and perpendicular to the direction of wave propagation. We shall also distinguish the plane wave which is polarized (linearly) or unpolarized.

The plane, polarized (linearly) *wave* is the simplest case of wave of alternating electromagnetic field. It corresponds to a field in which the vectors **E** and **H** are functions of only one variable. For example, it physically means that the vectors **E** and **H** in a given time instant lay in a plane perpendicular to the direction of wave propagation and their absolute values, directions, and senses are constant in this plane (Figure 2.8).

Such a wave is an abstract notion; however, in many cases very close to reality. For instance, when we consider a small section of the surface of a large sphere, like that created by fronts of radio waves or light emitted by a distant point source, or interior of a flat steel wall placed in a uniform magnetic field, then we can consider the investigated wave as a plane wave.

* Some scholars may doubt the need of the old methods of analysis at the current time. This is not right, because the old analytical solutions are often more accurate, simpler, and easier to use in practical, fast software tools.

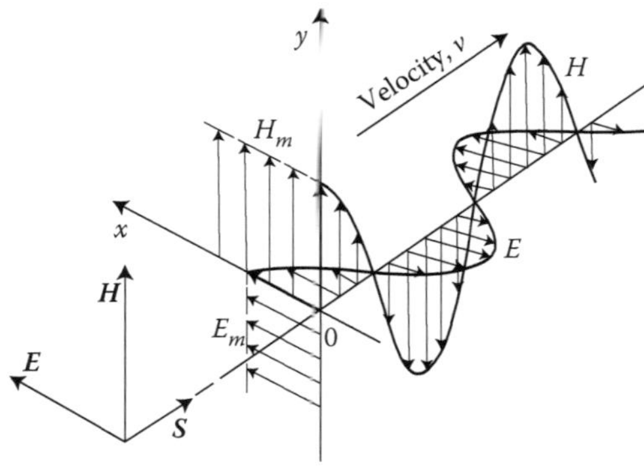

FIGURE 2.8 A scheme of a plane polarized wave; $\boldsymbol{S} = \boldsymbol{E} \times \boldsymbol{H}$, where $\boldsymbol{S}$ is the Poynting vector [VA/m^2].

The plane unpolarized wave or *natural wave* is such a wave in which the vectors $\boldsymbol{E}$ and $\boldsymbol{H}$, lying in the planes perpendicular to the OZ-direction of the wave propagation, have directions, senses, and absolute values dependent on a position in these planes (e.g., Figure 5.23). The plane unpolarized wave is more typical for elements of power equipment than the polarized wave, but is more complicated for calculations. Because of that, most calculations giving first approximation are based on the notion of the linearly polarized plane wave.

If the vectors $\boldsymbol{E}$ and $\boldsymbol{H}$ of electromagnetic wave are sinusoidal functions of time, such a wave is called *monochromatic* (the term borrowed from optics).

We will consider properties of plane waves in a rectangular (Cartesian) coordinate system. In accordance with the definition of plane wave, we shall assume that the vectors $\boldsymbol{E}$ and $\boldsymbol{H}$ lie in the XY plane perpendicular to the OZ direction of the wave propagation. It means that the components along the Z axis are equal to zero: $E_z = 0$ and $H_z = 0$. The Maxwell's Equations 2.1 through 2.4 of the unpolarized wave, at $\mu = \text{const}$, take the form:

$$\left.\begin{aligned} -\frac{\partial H_y}{\partial z} &= \sigma E_x + \varepsilon \frac{\partial E_x}{\partial t} \ \text{(a)} \qquad & \frac{\partial E_x}{\partial z} &= -\mu \frac{\partial H_y}{\partial t} \ \text{(b)} \\ \frac{\partial H_x}{\partial z} &= \sigma E_y + \varepsilon \frac{\partial E_y}{\partial t} \ \text{(c)} \qquad & \frac{\partial E_y}{\partial z} &= \mu \frac{\partial H_z}{\partial t} \ \text{(d)} \end{aligned}\right\} \tag{2.159}$$

$$\left.\begin{aligned} &\frac{\partial H_y}{\partial x} - \frac{\partial H_x}{\partial y} = 0 \quad \text{(a)} & &\frac{\partial E_y}{\partial x} - \frac{\partial E_x}{\partial y} = 0 \quad \text{(b)} \\ &\operatorname{div} \boldsymbol{H} = \frac{\partial H_x}{\partial x} + \frac{\partial H_y}{\partial y} = 0 \quad \text{(c)} & &\operatorname{div} \boldsymbol{D} = \frac{\partial E_x}{\partial x} + \frac{\partial E_y}{\partial y} = 0 \quad \text{(d)} \end{aligned}\right\} \tag{2.160}$$

The equation systems (2.159a,b) and (2.159c,d) connect pairs of the field components E_x with H_y, and E_y with H_x, respectively. These pairs will be independent from each other only in the case when their derivatives with respect to the X and

Y axes are equal to zero, it means—when they create two polarized plane waves in the direction of these axes or, what is equivalent—when they create one plane wave polarized along any mutually perpendicular axes. In all these cases, the remaining four Equations 2.160 will disappear.

The vectors ***E***, ***H*** of a plane wave create *orthogonal right-handed system*. Hence, they are mutually perpendicular. It should be stressed, however, that it concerns only plane waves and is not a general rule, though application of the orthogonal right-handed system is convenient at considering many practical cases (Simonyi [1.11]).

In the case of a linearly polarized plane wave, one can choose coordinate axes along the perpendicular vectors ***E*** and ***H*** and exploit only one pair from the four scalar Equation 2.159.

In the case of an unpolarized plane wave, in addition to Equation 2.159 the additional Equation 2.160 connecting both mentioned waves with each other, is also needed.

There exists a proof [1.11] that waves of cylindrical or spherical type can be represented as a superposition of series of polarized plane waves.

If in the same space one linearly polarized plane wave is superimposed by another polarized plane wave of the same frequency as the first one, but with vectors perpendicular to it in space and shifted in phase by an angle of φ, a result of this superposition would be a resultant wave polarized elliptically or circularly (of the motion in the form of corkscrew) [1.11].

In the following text, the term *plane wave* will be used in reference to the linearly polarized plane waves.

2.13.1 Plane Wave in a Dielectric

From Equation 2.159, at the conductivity $\sigma = 0$, after calculating again their derivatives with respect to z or t, we obtain two *wave equations* in a dielectric:

$$\frac{\partial^2 E_y}{\partial z^2} = \frac{1}{v^2}\frac{\partial^2 E_y}{\partial t^2} \quad \text{and} \quad \frac{\partial^2 E_x}{\partial z^2} = \frac{1}{v^2}\frac{\partial^2 E_x}{\partial t^2} \tag{2.161}$$

where

$$v = \frac{1}{\sqrt{\mu\varepsilon}} \tag{2.162}$$

is the velocity of wave propagation in dielectrics. In vacuum, in Equations 2.161 we should replace v with the velocity of light:

$$c = \frac{1}{\sqrt{\mu_0 \varepsilon_0}} \approx 3 \times 10^8\,\text{m/s} \quad \left(\mu_0 = 0.4\pi \times 10^{-6}\,\frac{\text{H}}{\text{m}};\quad \varepsilon_0 = \frac{1}{4\pi \cdot 9}10^{-9}\,\frac{\text{F}}{\text{m}} \right)$$

The wave Equations 2.161 satisfy any function of class C_2 (double differentiable) depending only on argument $(z \mp vt)$, which can be easily checked by the method of substitution. Hence, the general solution has the form

$$E_y = g_y(z - vt) + f_y(z + vt) \quad \text{and} \quad E_x = g_x(z - vt) + f_x(z + vt) \tag{2.163}$$

where g and f are arbitrary functions depending on the type of disturbance which generated the wave. It can be, for instance, the function $(2\pi/T)(z - vt)$.

The g functions represent the waves propagating with the velocity v in the direction of the Z axis without a change of their form. The f functions do not depend on the function g and represent a wave traveling in the direction opposite to sense of the $0Z$ axis. After substitution of the obtained solutions to Equation 2.159c,a, respectively, we get, for instance $(u_{1,2} = z \mp vt)$

$$\frac{dH_x}{dz} = \varepsilon \frac{d(g_y + f_y)}{dt} = \varepsilon\left(\frac{\partial g_y}{\partial u_1}\frac{du_1}{dt} + \frac{\partial f_y}{\partial u_2}\frac{du_2}{dt}\right) =$$

$$= -\varepsilon \upsilon\left(\frac{\partial g_y}{\partial u_1} \frac{\partial f_y}{\partial u_2}\right) = -\sqrt{\frac{\varepsilon}{\mu}}\frac{d}{dz}(g_y - f_y)$$

Since

$$\frac{df}{dz} = \frac{\partial f}{\partial u} \cdot \frac{du}{dz} = \frac{\partial f}{\partial z}$$

After integration with respect to z, we get

$$H_x = -\sqrt{\frac{\varepsilon}{\mu}}[g_y(z - vt) - f_y(z + vt)] \quad \text{and} \quad H_y = -\sqrt{\frac{\varepsilon}{\mu}}[g_x(z - vt) - f_x(z + vt)] \tag{2.164}$$

For all the wave components (E_y, H_x) and (E_x, H_y) we have, therefore, a resultant traveling wave consisting of two opposite *co-running* and *counter-running* waves (incident E_{inc}, H_{inc} and reflected E_{refl}, H_{refl})

$$\left.\begin{aligned} E &= F_1(z - vt) + F_2(z - vt) = E_{\text{inc}} + E_{\text{refl}} \\ H &= \sqrt{\frac{\varepsilon}{\mu}}\left[F_1(z - vt) - F_2(z + vt)\right] = H_{\text{inc}} + H_{\text{refl}} \end{aligned}\right\} \tag{2.165}$$

The ratio

$$\frac{E_{\text{inc}}}{H_{\text{inc}}} = -\frac{E_{refl}}{H_{refl}} = \sqrt{\frac{\mu}{\varepsilon}} = Z \tag{2.166}$$

has the dimension of electric impedance and is called the *wave impedance of dielectric*.

The wave impedance of vacuum and air is real and equals:

$$Z_0 = \sqrt{\frac{\mu_0}{\varepsilon_0}} = \sqrt{\frac{4\pi \times 10^{-7}\,\{\mathrm{H/m}\}}{\{1/4\pi \cdot 9\}10^{-9}\,\{\mathrm{F/m}\}}} = 377\,\Omega \tag{2.166a}$$

and, for instance, for the transformer oil $Z_{\text{oil}} = 243\ \Omega$.

From formula (2.166), it follows

$$\frac{\mu H^2}{2} = \frac{\varepsilon E^2}{2} \tag{2.167}$$

which means that the energy of electric and magnetic wave incident on a conducting surface and reflected from it are equal to each other. The above formulae apply to polarized plane waves.

As it follows from the assumptions for Equation 2.161, the propagation of electromagnetic wave in the way described by Equation 2.165 is dependent on the existence of displacement currents, $\varepsilon(\partial E/\partial t)$. Omission of these currents is equivalent to omission of radiation of electromagnetic energy by a source by means of "dielectric" waves of type (2.165).

As it follows from simple considerations, such radiation in the domain of electric machines and apparatus can be completely omitted. Let us assume that a source generates a harmonic field alternating in time ($\underline{E}_m = E_m e^{j\omega t}$). Then, at the distance r from the source, according to Equation 2.165, the field will have the value

$$\underline{E}_m \sim e^{j\omega\left(t-\frac{r}{v}\right)} = e^{j\omega t}\left(1 - \frac{j\omega r}{v} + \cdots\right)$$

where r/v is the time delay after which the wave travels the distance r. The effect of phase delay (time-lag), which is a result of displacement currents and a condition of energy radiation, can be neglected and omitted when

$$\frac{\omega r}{v} = \frac{2\pi r}{\lambda} \ll 1$$

(at $f = 50$ Hz, the wavelength $\lambda = c/f = 3 \times 10^8/50 = 6000$ km).

Since the wavelength λ at the power equipment frequencies is much larger than linear dimensions r, the above condition is always satisfied and we can consider that all electromagnetic quantities are in phase with each other.

Specific calculations of the electromagnetic field of a single oscillating electric dipole (Neiman [2.14], p. 350/2, Turowski [2.31]) with charge $Q = I_m\,(1/\omega)\cos\omega(t - r/v)$ or an element of alternating current $i = \mathrm{d}Q/\mathrm{d}t = I_m \cos\omega(t - r/v)$ of length $\mathrm{d}l$ (Figure 2.9) lead to the following formulae for the field components:

$$\mathrm{d}B_\varphi = \frac{\mu\,\mathrm{d}l}{4\pi r v}\left(\frac{\partial i}{\partial t} + \frac{\omega\lambda i}{2\pi r}\right)\sin\Theta \qquad \left(v = \frac{\lambda}{2\pi r}\omega r\right) \tag{2.168}$$

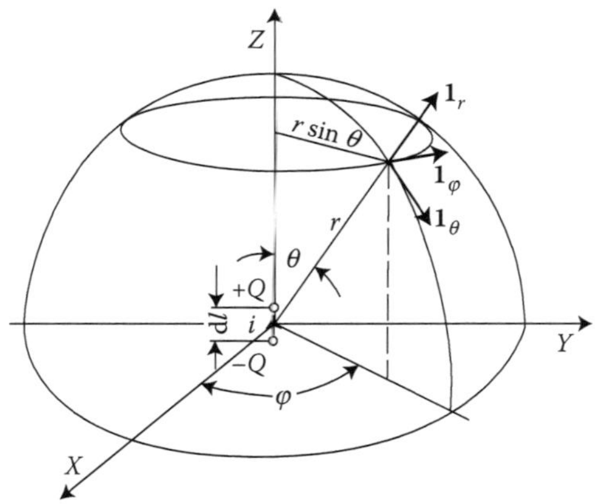

FIGURE 2.9 Spherical coordinate system with a dipole in center.

$$dE_r = \frac{2\mu\,dl}{4\pi r}\left(\frac{\omega\lambda i}{2\pi r} + \frac{\omega^2\lambda^2}{4\pi^2 r^2}Q\right)\cos\Theta \qquad \left(\lambda^2 = \frac{4\pi}{f\mu\sigma}\right) \tag{2.169}$$

$$dE_\Theta = \frac{\mu\,dl}{4\pi r}\left(\frac{\partial i}{\partial t} + \frac{\omega\lambda i}{2\pi r} + \frac{\omega^2\lambda^2}{4\pi^2 r^2}Q\right)\sin\Theta \tag{2.170}$$

In the above formulae, the amplitude of every following component differs from the amplitude of previous component by the multiplier $\lambda/2\pi r$. Thanks to it, one can practically investigate:

1. Near field, taking place in electric machines and other power equipment, where $\lambda \gg r$, and
2. Remote field, which occurs at radiation of radio waves, light, and so on, where $\lambda \ll r$.

At $\lambda \gg r$, in Equations 2.168–2.170 one can retain only the last components. Thanks to it, in electric machines and other power equipment the following equations are applicable

$$dB_\varphi = \frac{\mu i\,dl}{4\pi r^2}\sin\theta = \frac{\mu I_m e^{j\omega t}dl}{4\pi r^2}\sin\theta \tag{2.168a}$$

$$d\underline{E}_r = \frac{2dlQ}{4\pi\varepsilon r^3}\cos\theta = \frac{2I_m e^{j\left(\frac{\pi}{2}-\omega t\right)}dl}{4\pi\varepsilon r^2\omega}\cos\theta \tag{2.169b}$$

$$dE_\Theta = \frac{d\,lQ}{4\pi\varepsilon r^3}\sin\theta = \frac{I_m e^{j\left(\frac{\pi}{2}-\omega t\right)}dl}{4\pi\varepsilon r^3\omega}\sin\theta \tag{2.170a}$$

where $Q = I_m/\omega\, e^{j\omega t}$; $i = dQ/dt$.

Equation 2.168a is the same as the Biot–Savart law (2.16a). The two remaining equations have a form similar to the known equations for static dipole at $dl \ll r$.

The components (2.168a) through (2.170a) of the vectors $\boldsymbol{B}$ and $\boldsymbol{E}$ represent the electromagnetic field of dipole in a distance, small in comparison with the wavelength ($r \ll \lambda$). These vectors determine solely the reactive power of field ($\boldsymbol{E} \times \boldsymbol{H}$), since their instantaneous values are shifted in phase by $\pi/2$. In this way, a single dipole of low-frequency practically does not radiate any active power into near ambient space. This conclusion is correct until the wave leaving the dipole is not superimposed by a returning wave reflected from a dissipative (with power loss) conducting surface situated nearby. In this case, the dipole begins to radiate (induce) active power also in a small distance. This power equals to the power loss in metal surface [2.31].

In distances farther from the dipole, exceeding the wavelength ($r \gg \lambda$), in Equations 2.168a through 2.170a one can omit all components except the first ones. The quantity E_r can also be omitted completely, because both its components are very small in comparison with the first component of the quantity E_Θ. We have then

$$dB_\varphi = \frac{\mu_m dl}{2r\lambda} \sin\theta \cos\left(\omega t - \frac{2\pi r}{\lambda}\right) \tag{2.168b}$$

$$dE_\Theta \frac{\mu \omega I_m dl}{4\pi r} \sin\theta \cos\left(\omega t - \frac{2\pi r}{\lambda}\right) = \frac{dB\varphi}{\sqrt{\mu\varepsilon}} \tag{2.170b}$$

A wave with the components (2.168b) and (2.170b) is called *spherical*. It satisfies also the relation (2.167). The components E and H at $r \gg \lambda$ are mutually in phase and now represent the active power. The Poynting vector (2.10) $\boldsymbol{S}_r = \boldsymbol{E}_\Theta \times \boldsymbol{H}_\varphi$ is now directed, at any time and any point of space, along the r axis (Figure 2.9) and energy is transmitted outwards, in the radial direction. This energy does not return to the source anymore and is radiated by the *dipole operating as antenna*.

2.13.2 Plane Wave in a Conducting Half-Space

Let us consider a plane wave satisfying Equations 2.159 and 2.160, which penetrates to a metal half-space (Figure 2.10) and propagates in it perpendicularly to the surface.

In metals, at frequencies occurring in electric machines and power equipment (the first and higher harmonics) one can omit the displacement currents ($\varepsilon\, \partial E/\partial t$) in comparison with the conducting currents (σE). It follows from the ratio of densities of these currents in monochromatic fields ($E_m = E_m\, e^{j\omega t}$):

$$\frac{J_{\text{displ.}}}{J_{\text{conduct}}} = \frac{\varepsilon(\partial E/\partial t)}{\sigma E} = j\frac{\omega\varepsilon}{\sigma} \tag{2.170c}$$

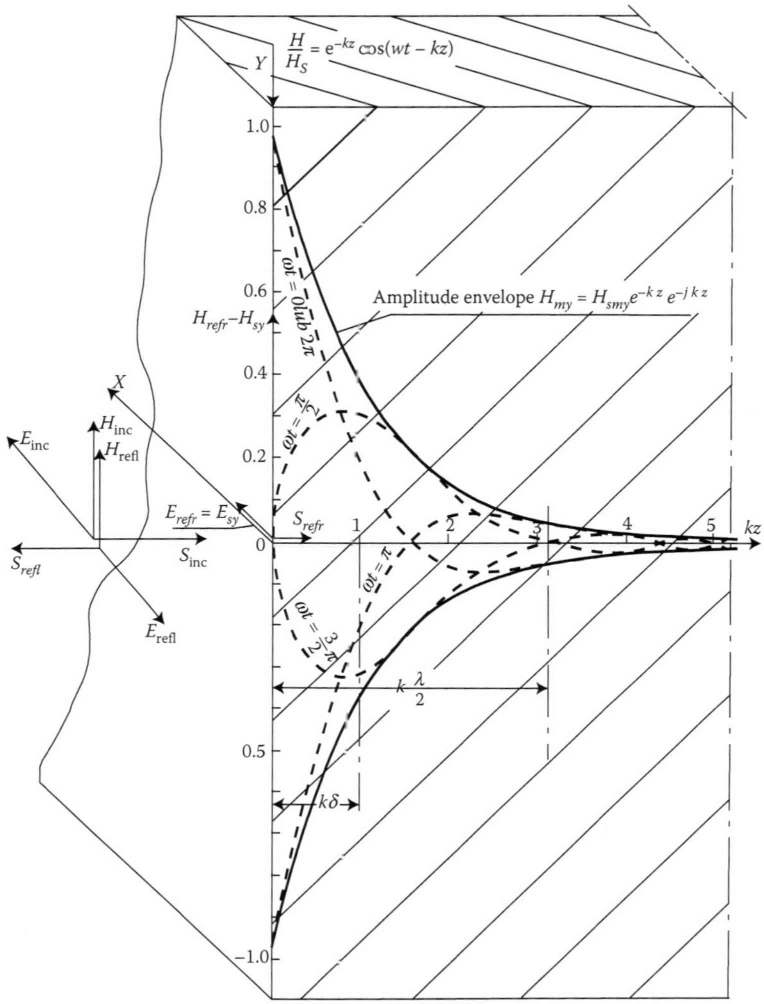

FIGURE 2.10 Penetration of electromagnetic plane wave into a metal half-space: envelopes of amplitudes and instantaneous field distributions, for several arbitrary time instants: t_1, t_2,...

For copper, for instance, $\sigma = 58 \cdot 10^6$ S/m; assuming $\varepsilon = \varepsilon_0^*$ and $f = 1080$ Hz, for this relation we get

* The permittivity ε and polarization of metal are not satisfactorily investigated, because these effects are shadowed by the strong electronic conductivity σ. In elementary approach to electrostatics there exists opinion ([1.1], p. 407), that at $f = 0$, in uniform field E, the permittivity of metal $\varepsilon \to \infty$ (which seems nonsense), "since supposed charges having free motion give in effect infinitely large polarizability." At increase of frequency, the permittivity has finite value depending on frequency (*Pashicki E.: Osnovy svierhprovodimosti*, Kiev, 1985, p. 45). Here compared is the current of polarization $\partial D/\partial t$ with ignoring the electron conductivity. A more accurate consideration rather suggests that in metal ε_r is nearer to a value between 1 and 6.

$$\frac{2\pi \cdot 1080\,(1/\text{s})}{58 \cdot 10^6\,(\text{S/m})} \cdot \frac{10^{-9}}{4\pi 9} \frac{\text{s}}{\Omega \cdot \text{m}} = 10^{-15}$$

As one can see, even at much higher permittivity ε and higher frequency, the density of hypothetic *displacement currents in conductor can be completely ignored in comparison with the conductivity currents.* Proceeding analogically in the case of currents in dielectric, from Equation 2.159 at $\varepsilon(\partial E/\partial t) = 0$, we obtain the plane wave equations in conductors (conductivity equations)

$$\frac{\partial^2 H_x}{\partial z^2} = \mu\sigma \frac{\partial H_x}{\partial t} \quad \text{and} \quad \frac{\partial^2 H_y}{\partial z^2} = \mu\sigma \frac{\partial H_y}{\partial t} \tag{2.171}$$

and in the case of monochromatic wave (2.136), satisfying Equation 2.141, we get

$$\frac{\partial^2 H_{mx}}{\partial z^2} = \alpha^2 H_{mx} \quad \text{and} \quad \frac{\partial^2 H_{my}}{\partial z^2} = \alpha^2 H_{my} \tag{2.172}$$

where $\alpha = (1 + j)k$ is defined by formula (2.139).

A general solution of Equations 2.172 has the form

$$H_{mx} = A_{1x}\text{e}^{-\alpha z} + A_{2x}\text{e}^{\alpha z}$$

Since the real part of the coefficient α, which decides on the wave attenuation, is positive, and the magnetic field intensity cannot increase to infinity, for the wave traveling along the OZ axis (Figure 2.10) we must accept $A_{2x} = 0$, from which

$$H_{mx} = A_{1x}\text{e}^{-\alpha z}$$

After substituting $z = 0$, we conclude that $A_{1x} = H^*_{msx} = |H_{msx}|\text{e}^{j\Psi}$ from which

$$H_{mx} = H_{msx}\text{e}^{-\alpha z} = H_{msx}\text{e}^{-kz}\text{e}^{-jkz} \tag{2.173}$$

This is the equation of envelopes of the wave amplitudes inside metal (Figure 2.10). An instantaneous value, in symbolic form, can be obtained by multiplying the last Equation 2.173 by the operator $\text{e}^{j\omega t}$

$$\underline{H}_{mx} = \left|H_{msx}\right|\text{e}^{-kz}\text{e}^{j(\omega t + \Psi - kz)} \tag{2.174}$$

* Index s according to standards PN-64/E-01 100 means surface value.

The instantaneous value of the magnetic field intensity (Figure 2.10) is therefore expressed by the formula

$$H_z = |H_{msx}|\, e^{-kz}\cos(\omega t + \Psi - kz) \tag{2.174a}$$

The electric field intensity, E, can be determined by applying to Equation 2.173 the first Maxwell's Equation 2.159c, from which

$$E_{my} = -\frac{\alpha}{\sigma}H_{ms}e^{-\alpha z} = -\sqrt{\frac{\omega\mu}{\sigma}}\left|H_{msx}\right|e^{-kz}e^{j(\omega t+\Psi-kz+\frac{\pi}{4})} \tag{2.175}$$

where

$$\frac{\alpha}{\sigma} = \frac{(1+j)k}{\sigma} = \frac{\sqrt{2}\,k}{\sigma}e^{j\frac{\pi}{4}}$$

As one can see, the electric field intensity $\underline{E}_m$ and the current density σE in metal with constant parameters have a phase shift of 45° ($\pi/4$) versus the magnetic field intensity, $\underline{H}_m$. The amplitudes $\underline{E}_m$ and $\underline{H}_m$ both decrease during penetration of the wave inside metal according to the exponential function e^{-kz}. Oscillation of these fields during penetration of wave inside the metal has an increasingly growing phase lag, by kz radians, versus the oscillation on the surface of the body. Solutions (2.174) and (2.175) concern a polarized plane wave, which has only the components E_y and H_x. In the case of a nonpolarized wave, there exists the second pair of components (E_x, H_y), which can be found analogically from Equations 2.172 and 2.159a

$$H_{my} = H_{msy}\,e^{-\alpha z}$$

$$E_{mx} = \frac{\alpha}{\sigma}H_{msy}\,e^{-az}$$

Moving now to the resultant values $H_m = \sqrt{H_{mx}^2 + H_{my}^2}$ and $H_{ms} = \sqrt{H_{msx}^2 + H_{msy}^2}$ we obtain the resultant field

$$H_m = H_{ms}e^{-\alpha z}, \qquad E_m = \frac{\alpha}{\sigma}H_{ms}e^{-\alpha z}, \qquad J_m = \alpha H_{ms}e^{-\alpha z}. \tag{2.176}$$

The wave impedance of metal is here a complex value:

$$Z_{\text{met}} = \frac{E_m}{H_m} = \frac{\alpha}{\sigma} = (1+j)\sqrt{\frac{\omega\mu}{2\sigma}} = \sqrt{\frac{\omega\,\mu}{\sigma}}\,e^{j\frac{\pi}{4}} \tag{2.177}$$

At a frequency of 50 Hz, the modulus of wave impedance is

$$\left.\begin{array}{l} \text{for copper} \quad |Z_{\text{Cu}}| = 2.7\times10^{-6}\,\Omega \\ \text{for steel} \quad |Z_{\text{St}}| = 2.4\times10^{-4}\,\Omega, \quad \text{at} \quad \mu_r = 1000 \end{array}\right\} \tag{2.177a}$$

As it can be seen, $Z_{diel} \gg |Z_{st}| \gg |Z_{Cu}|$, which have decisive consequences for design of screens (see Section 4.3).

The wavelength in metal, λ, is the depth z at which the phase of field components changes by 2π. Hence, from formula (2.174a) follows the condition $k\lambda = 2\pi$, from which

$$\lambda = 2\pi\sqrt{\frac{2}{\omega\mu\sigma}} = 2\sqrt{\frac{\pi}{f\mu\sigma}} = 2\pi\delta \qquad \lambda^2 = \frac{4\pi}{f\mu\sigma} \tag{2.178}$$

The wave velocity in metal:

$$v = \lambda f = 2\sqrt{\frac{\pi f}{\mu\sigma}} \tag{2.179}$$

At a frequency of 50 Hz and in the case of iron with the relative permeability $\mu_r = 800$ and the conductivity (at 20°C) $\sigma = 7 \times 10^6$ S/m, the ratio of velocity v and wavelength λ in the air and steel equals to

$$\frac{v_{air}}{v_{steel}} = \frac{\lambda_{air}}{\lambda_{steel}} = \frac{3 \times 10^8 \text{ m/s}}{0.3 \text{ m/s}} = 10^9 = \frac{\sin\varphi_1}{\sin\varphi_2} \tag{2.180}$$

φ_1, φ_2—see Figure 2.13 [later in the chapter] and Equation 2.202.

The ratio of amplitudes of field intensities at the distance $z = \lambda$ from the surface of solid conductor to their values on the surface, equals

$$e^{-k\lambda} = e^{-2\pi} = 0.00185$$

which means that at this depth the wave is practically extinguished (remains only 0.185%). This phenomenon is called *skin-effect*. In Table 2.1 are given lengths of waves in different conducting and semi-conducting materials at various frequencies.

2.13.3 Equivalent Depth of Wave Penetration and Impedance of Solid Conductors

An inverse of the attenuation coefficient k in formulae (2.140) and (2.174) is called *equivalent depth of penetration* of electromagnetic wave into metal half-space

$$\delta = \frac{1}{k} = \sqrt{\frac{2}{\omega\mu\sigma}} = \frac{1}{\sqrt{\pi f\mu\sigma}} = \frac{\lambda}{2\pi} \tag{2.181}$$

TABLE 2.1
Wavelength $\lambda = 2\pi\delta$ and the Equivalent Depth of Penetration δ in Various Constructional Materials and Media

Material	Slightly Saturated Steel	Saturated Steel	Copper	Aluminum	Steel at 800°C	Metal-Graphite Brushes	Plasma (Combustion Gases) +1% K, 2500°C	Dry Soil
σ in S/m =	8×10^6	8×10^6	58×10^6	34×10^6	$\times 10^6$	0.15×10^6	10	10^{-2}
μ_r	1000	300	1	1	1	1	1	1
				$\lambda = 2\pi\delta$				
f	mm		cm			m		km
5 Hz	17.8	32.5	18.7	24.42	143	—	—	—
50 Hz	5.6	10.2	5.9	7 7	45	1.16	141	4500
5 kHz	0.56	1.0	0.59	0.77	4.5	0.116	14.1	450
500 kHz	0.056	0.1	0.059	0.077	0.45	0.0116	1.41	45
				δ				
5 Hz	2.84	5.17	2.98	3.89	22.7	—	—	—
50 Hz	0.89	1.62	0.94	1.23	7.16	0.185	22	716
5 kHz	0.089	0.162	0.094	0.123	0.716	0.0185	2.2	71.6
500 kHz	0.0089	0.0162	0.0094	0.00123	0.0716	0.00185	0.22	7.16

This quantity has a linear dimension, and corresponds to e (= 2.718)–*fold reduction of the amplitude* of field with respect to the amplitude on the solid metal surface (Figure 2.10).*

For solid steel conductors: $\delta_{steel} \approx (1/a_p)\delta$, $a_p \approx 1.4$ (J. Turowski [1.16]). The definition (2.181) of the *equivalent depth of penetration* comes from the following considerations.

If the maximum value $\boldsymbol{J}_m$ of the current flux density (2.176) in the X direction (Figure 2.11)

$$J_m = \sigma E_m = J_{ms}\, e^{-\alpha z} \tag{2.182}$$

is integrated along the entire *Z* axis, we shall obtain the maximum value of resultant current I_{mx} flowing through the conducting layer of half-space in the width *b* in the *X* direction:

$$I_m = J_{ms} b \int_0^\infty e^{-\alpha z} dz = \frac{J_{ms}}{\alpha} b = \frac{J_{ms}}{\sqrt{2}} \delta b \cdot e^{-j\frac{\pi}{4}},$$

* It should be stressed here again, that "equivalent" value of δ is not the real depth of wave penetration. The real value of approximate wave extinction (to 0.18%) is about 3δ.

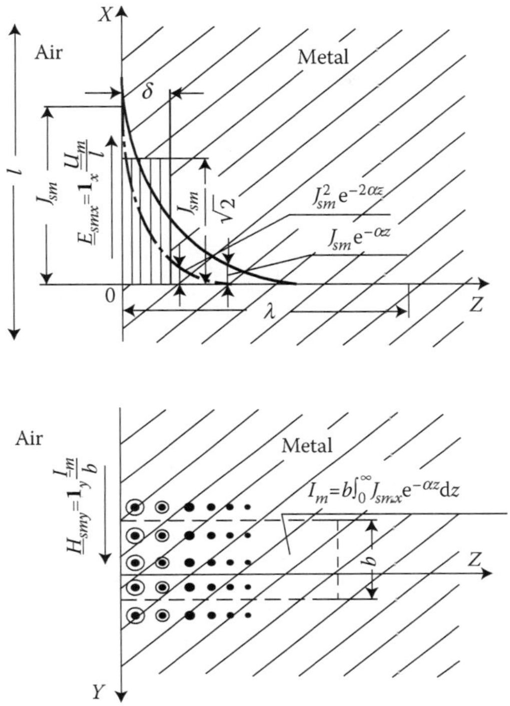

FIGURE 2.11 Illustration of the equivalent depth of wave penetration, δ, and the electrical resistance R of a solid metal half-space.

whose rms* value is

$$I = \frac{|I_m|}{\sqrt{2}} = \frac{|J_{ms}|}{2}\delta \cdot b = \frac{J_{\text{rms,s}}}{\sqrt{2}}\delta \cdot b \tag{2.183}$$

In this way, δ can be considered as a *thickness of such equivalent layer* with uniform current density $J_{\text{rms}} = (J_{\text{rms,s}}/\sqrt{2})$, in which flows the current equal to the real rms current I flowing in a solid metal half-space.

The magnetic field intensity on the surface of a solid metal half-space (solid conductor) equals in turn

$$H_{ms} = \int_0^\infty J_{mx}\text{d}z = A_{lm} = J_{m,\text{ surf.}} \tag{2.184}$$

where $A_{lm} = (I_m/b)$ is the *specific electric loading* of conductor, or the surface current density $J_{m,\text{surf.}}$ per unit length in the direction perpendicular to the direction of current flow.

* rms = root mean square.

From Figure 2.10 it follows that for a polarized plane wave incident on a solid metal half-space, causing a current flow of density J_x, the per-unit (1 × 1 m) impedance equals

$$r_1 + jx_1 = \frac{U_m/1}{I_m/b} = \frac{\alpha}{\sigma} = (1+j)\sqrt{\frac{\omega\mu}{2\sigma}} = \frac{1+j}{\sigma\delta} \tag{2.185}$$

and it is equal to the wave impedance (2.177) of metal.

If we assume that the wavelength λ in a cylindrical metal conductor is much smaller than the conductor radius ($D/2$), that is, $\lambda \ll D/2$, then the *resistance and reactance of this conductor at alternating current* is

$$R = X = \frac{1}{\pi D}\sqrt{\frac{\omega\mu}{2\sigma}} = \frac{1}{\pi D\sigma\delta} \tag{2.185a}$$

$$\text{tg}\varphi = \frac{X}{R} = 1$$

where l is the length of the conductor, X the internal reactance of the conductor. Thus, these are the internal resistance and reactance of a solid cylindrical conductor modeled by an equivalent tubular conductor of the wall thickness δ with uniform distribution of the rms value of current. After the investigation by Rosenberg, extended by Neiman [2.14], the variable permeability and hysteresis losses cause that, for ferromagnetic metals the above values of resistance and reactance should be multiplied by a more-or-less constant linearization coefficients (Section 7.2), which for steel are: $a_p \approx 1.4$ and $a_q \approx 0.85$, respectively:

$$r_{1,St} \approx 1.4\sqrt{\frac{\omega\mu_s}{2\sigma}} \quad \text{and} \quad x_{1,St} \approx 0.6 r_{1,St} = 0.85\sqrt{\frac{\omega\mu_s}{2\sigma}} \tag{2.185b}$$

where μ_S is the permeability on the metal surface, determined from the magnetization curve $H = H_{rms} = I/\pi D$

EXAMPLE

Resistance (in ohms) of a steel cylindrical wire, of the diameter $D > \lambda$ and the length l, equals to the resistance of tube of the same external diameter and thickness of walls equal δ, hence

$$Z = (a_p + ja_q)R = (1.4 + j0.85)\frac{l}{\pi D}\sqrt{\frac{\omega\mu_S}{2\sigma}}$$

where μ_S is from Figure 1.27 or Figure 1.29 for $H_{ms} = (\sqrt{2}\,I/\pi D)$; I is current in the wire.

If we carry out an analogical consideration, but replace J_m and I_m by the field H_m (2.176) and the flux $\Phi_{m1} = \int_0^\infty \mu H_m dz = (\mu/\alpha)H_{ms}$, then we can see that we shall obtain the same flux Φ_{m1} and the same *equivalent per-unit reluctance* (magnetic resistance) per 1 m of length along the x and y axes:

$$\underline{R}_{m1} = \frac{H_{ms}}{\Phi_{m1}} = R_{m1} + jR_{m1q} = (1+j)\frac{1}{\mu\delta}; \quad R_{m1} = R_{m1q} = \sqrt{\frac{\omega\sigma}{2\mu}}; \tag{2.186}$$

which exist in a solid metal half-space.

In the case of steel, we introduce also the linearization coefficients

$$R_{m1,\mathrm{St}} \approx 0.37\sqrt{2}R_{m1} = 0.52R_{m1}; \quad R_{m1q,\mathrm{St}} \approx 0.61\sqrt{2}\,R_{m1} = 0.86R_{m1q} \tag{2.186a}$$

In this way, in simplified, but sufficient calculations, a solid steel half-space can be substituted by an equivalent layer of thickness δ (2.181) and $\mu = \mu_S$, free of eddy-currents and skin-effect.* In the case of steel, the corrected equivalent value of the depth of wave penetration is: $\delta_{\mathrm{St}} \approx 0.71\delta$ (see Equation 7.30).

2.13.4 Field Diffusion into a Conductor

So far, we considered a monochromatic plane wave (2.136) satisfying Equation 2.141. At investigations of transient processes of a wave of any form, one can exploit the analogy between diffusion of gases and phenomenon of field diffusion in conducting media, since both phenomena are described with similar equations of type 2.171. Let us assume that at a certain initial time instant, on the surface of a half-space $z < 0$ (Figure 2.12) appeared a DC field with only one component H_0. Its diffusion to the area $z > 0$ will be described by the scalar Equation 2.171. Let us introduce a characteristic element Δt of time, during which we wish to look into the process of field penetration into a metal, and let us define the depth of diffusion (penetration) of magnetic field, δ_0, as the distance at which $H = 0.5 \cdot H_0$. Applying next the relative units $H_r = H/H_0$, $z_r = z/\delta_0$, $t_r = t/\Delta t$, let us transform Equation 2.171 into the form

$$\frac{\partial H_r}{\partial t_r} = \frac{\Delta t}{\mu\sigma\delta_0^2}\frac{\partial^2 H_r}{\partial z_r^2} \tag{2.187}$$

The value $(1/\mu\sigma)$, by analogy to the diffusion of gases, can be called *diffusion coefficient* of magnetic field into conductor.

Solution of Equation 2.187 can be expressed by the *error function* (also called the *Gauss error function*), Equation 7.6, for which convenient tables exist (Jahnke [2.11])—see Table 2.2.

$$H_r = \frac{2}{\sqrt{\pi}}\int_u^{\infty} e^{-u^2}\,\mathrm{d}u = F(u) \tag{2.188}$$

* It permits to substitute, also as a asimplified model in the numerical methods such as FDM, FEM, and RNM, a solid steel wall by an equivalent package of laminated iron, made of thin sheets with included counter magnetomotive (MMF) source corresponding to the eddy-current reaction [reactive reluctance $R_{m1q,\mathrm{St}}$ (2.186a)].

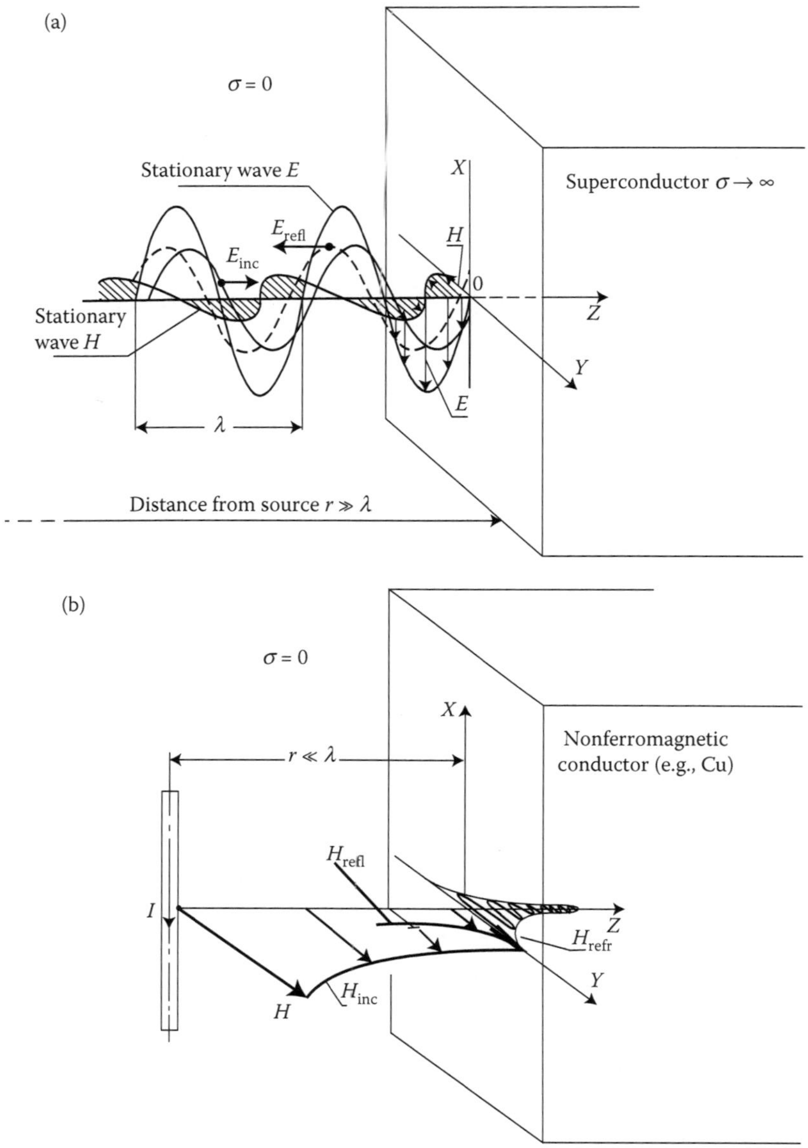

FIGURE 2.12 Reflection of plane waves from a conducting surface: (a) far wave radiation; (b) induction field near the source, stationary (only the H component is shown); inc—waves incident on the metal surfaces, refl—reflected, refr—refracted inside solid metal.

where

$$u = \sqrt{\frac{\mu \sigma \delta_0^2}{4 \Delta t}} \cdot \frac{z_r}{\sqrt{t_r}} \tag{2.189}$$

TABLE 2.2
The *F*(*u*) Function for Formula 2.188

u	*F*(*u*)	*u*	*F*(*u*)
0	1.0	0.564	0.425
0.0885	0.900	0.8	0.258
0.1	0.887	1.0	0.1577
0.3	0.671	1.163	0.100
0.477	0.500	1.386	0.05
0.5	0.480	1.822	0.01

Source: After Govorkov W.A.: *Electric and Magnetic Field.* (in Polish). Warsaw: WNT 1962 (Moscow, 1968).

and

$$\frac{2}{\sqrt{\pi}}\int_x^{\infty} e^{-t^2}\,dt = \mathrm{erfc}(x) = 1 - \Phi(x)$$

$$\Phi(x) = \mathrm{erf}(x) = \frac{2}{\sqrt{\pi}}\int_0^{x} e^{-t^2}\,dt \tag{2.189a}$$

For instance, the value of the Gauss integral $H_r = F(u) = 0.5$ corresponds to the argument $u = 0.477$ (from Table 2.2). At $z_r = t_r = 1$, the depth of diffusion of magnetic field into metal, according to the dependence (2.189), will be, with accuracy to 5%:

$$\delta_0 \approx \sqrt{\frac{\Delta t}{\mu\sigma}} \tag{2.190}$$

This notion is used in magneto-hydrodynamics, in investigations of magnetic missiles in contact with steel armor, at transient induction heating, and other rapid transient processes.

Another example of an analytical and numerical solution of the problem of field diffusion into metal is given in the work (Turowski [1.16], pp. 210–215).

2.14 REFLECTION AND REFRACTION OF PLANE WAVE

2.14.1 Boundary Conditions

The basis of field analysis on a boundary of two media, 1 and 2, with different constants, ε_1, μ_1, σ_1 and ε_2, μ_2, σ_2, are the boundary conditions:

1. Equality of the normal components of flux densities

$$B_{2n} = B_{1n} \tag{2.191}$$

following from the law (2.4), div $\boldsymbol{B} = 0$. Using the so-called *surface divergence*, this condition can be presented as

$$\text{Div}\,\boldsymbol{B} = B_{2n} - B_{1n} = 0 \tag{2.191a}$$

2. Equality of the tangential components of electric field intensity

$$E_{2t} = E_{1t} \tag{2.192}$$

This condition follows from Equation 2.17. Using the so-called *surface curl*, it can be presented as

$$\text{Curl}\,\boldsymbol{E} = \boldsymbol{n}(\text{E}_{2t} - E_{1t}) = 0 \tag{2.192a}$$

3. Equality of the tangential components of magnetic field intensity

$$H_{2t} = H_{1t} \tag{2.193}$$

which follows from Equation 2.1 and is valid in the case of continuous current distribution nearby boundary surface, that is, for a finite current density ($J \neq \infty$).

Only in the case of an ideal theoretical superconductor we have

$$H_{2t} - H_{1t} = J_{\text{surf}} \tag{2.193a}$$

where J_{surf} is the *linear density of the surface current* (A/m) flowing in a dimensionless boundary layer. Applying the surface curl, we get

$$\text{Curl}\,\boldsymbol{H} = \boldsymbol{n}(H_{2\text{t}} - H_{1\text{t}}) = J_{\text{surf}}$$

4. Condition for the normal components of electric field intensity

$$\varepsilon_2 E_{2\text{n}} - \varepsilon_1 E_{1\text{n}} = \rho_{\text{surf}} \tag{2.194}$$

corresponding to the surface divergence: Div $\boldsymbol{D} = \rho_{\text{surf}}$, where ρ_{surf} is the surface density of electric charge (C/m^2) placed in a dimensionless boundary layer. At an absence of such charge, there exists the equality

$$\frac{E_{2n}}{E_{1n}} = \frac{\varepsilon_1}{\varepsilon_2} \tag{2.194a}$$

Equations 2.191a, 2.192a, and 2.193 create the so-called system of "*Maxwell's surface equations.*"

Boundary conditions expressed by the magnetic vector potential $\boldsymbol{A}$, according to its definition 2.50 $\boldsymbol{B} = \text{curl}\,\boldsymbol{A}$, for a plane wave ($A_y = A_z = 0$), take the form

$$B_{2n} = B_{1n} \quad \text{which corresponds to} \quad \frac{\partial A_2}{\partial t} = \frac{\partial A_1}{\partial t} \tag{2.191b}$$

$$H_{2t} = \mathrm{H}_{1t} \quad \text{which corresponds to} \quad \frac{1}{\mu_2}\frac{\partial A_2}{\partial n} = \frac{1}{\mu_1}\frac{\partial A_1}{\partial n} \tag{2.193b}$$

where $\partial/\partial t$ and $\partial/\partial n$ denote derivatives along the tangential axis t and the normal n to the boundary surface, respectively.

2.14.2 Reflection and Refraction of a Perpendicular Plane Wave

If a plane electromagnetic wave, moving in medium 1 with parameters ε_1, μ_1, σ_1, meets the surface of medium 2 with parameters ε_2, μ_2, σ_2, then the wave (E_{inc}, H_{inc}) incident on this surface will partly penetrate (refracted) E_{refr}, H_{refr} into the second medium, and part will be reflected (E_{refl}, H_{refl}) – Figure 2.10.

The relationship between these components can be calculated using the earlier defined boundary conditions

$$E_{t\ \text{inc}} + E_{t\ \text{refl}} = E_{t\ \text{refr}}$$

$$H_{t\ \text{inc}} + H_{t\ \text{refl}} = H_{t\ \text{refr}} \tag{2.195}$$

If we define the *wave impedance of medium* as the ratio of the tangential components of electric and magnetic field (complex values)

$$\frac{E_{t\ \text{inc}}}{H_{t\ \text{inc}}} = Z_{\text{inc}} \quad \text{and} \quad \frac{E_{t\ \text{refr}}}{H_{t\ \text{refr}}} = Z_{\text{refr}} \tag{2.196}$$

and using formula 2.166

$$Z_{\text{inc}} H_{t\ \text{refl}} = -E_{t\ \text{refl}}$$

we obtain the important relations

$$E_{t\ \text{refl}} = M\,E_{t\ \text{inc}} \quad \text{and} \quad H_{t\ \text{refl}} = -M\,H_{t\ \text{inc}} \tag{2.197}$$

$$E_{t\ \text{refr}} = (1 + M)\,E_{t\ \text{inc}} \quad \text{and} \quad E_{t\ \text{inc}} = (1 - M)\,H_{t\ \text{inc}} \tag{2.198}$$

where

$$M = \frac{Z_{\text{refr}} - Z_{\text{inc}}}{Z_{\text{refr}} + Z_{\text{inc}}} \tag{2.199}$$

is called the *coefficient of wave reflection*.

If the plane wave impinges perpendicularly onto the boundary surface, the index t can be removed.

Comparing the dependences (2.166a) and (2.177a), we can see that the wave impedance of conductors is much, much smaller than the wave impedance of dielectrics, and often can be omitted ($Z_{air} \mp Z_{met} \approx Z_{air}$). For instance, if a wave moving in a dielectric impinges perpendicularly onto a conducting surface (Figure 2.10), from Equation 2.199 it follows that $M = -1$, and from Equation 2.197

$$E_{t\,\mathrm{refl}} \approx -E_{t\,\mathrm{inc}} \quad \text{and} \quad H_{t\,\mathrm{refl}} \approx H_{t\,\mathrm{inc}}$$

It means that the tangential component of resultant electric field intensity $E_{t\mathrm{refr}}$ on the conductor surface is very small (in a superconductor $E_{t\,\mathrm{refr}} = 0$), and the corresponding magnetic component is two times bigger than the incident value (Figure 2.10). The reflected wave, due to the change of sign of E, moves in the opposite direction against the incident wave, what is in accord with Equation 2.165. As a result of superposition of both the waves, a stationary wave emerges (Figure 2.12a) with node of wave E and amplitude of wave H on the surface of metal ($z = 0$). For instance, at $E_{x\,\mathrm{inc}} = -E_{x\,\mathrm{refl}}$

$$\underline{E}_{mx} = E_{mx\,\mathrm{inc}} \mathrm{e}^{j\omega(t-\frac{z}{v})} + E_{mx\,\mathrm{refl}} \mathrm{e}^{j\omega(t+\frac{z}{v})}$$

$$\approx E_{mx\,\mathrm{inc}} \mathrm{e}^{j\omega t} (\mathrm{e}^{-j\omega\frac{z}{v}} - \mathrm{e}^{j\omega\frac{z}{v}}) = -2jE_{mx\,\mathrm{inc}} \sin\frac{\omega z}{v} \mathrm{e}^{j\omega t}.$$

2.14.3 Near and Far Fields

From the theory of oscillating dipole, with the alternating electric moment Ql (2.168) through (2.170), it follows that electromagnetic field consists of two components: the near *induction field*, defined by formulae (2.168a), (2.160a), (2.170a), and the far *field of radiation* defined by formulae (2.168b) and (2.170b). The amplitude of the induction field is different from the amplitude of the radiation field, by the coefficient $\lambda/2\pi r$), where $\lambda = (v/f)$ is the wavelength (at frequency 50 Hz in vacuum $\lambda = 6000$ km), r is the distance from the source. It means that in power equipment, in which $r \ll \lambda$, the radiation field can be completely ignored. Only at very high frequencies and large distances from the source ($r \gg \lambda$) the induction field can be neglected in comparison with the radiation field. The far radiation field is a fundamental subject in the research and theory of radio-communication and tele-technology. Components of the induction fields—formulae 2.168a through 2.170a, contrary to the radiation field—Equations 2.168b through 2.170b, do not change their phases in dielectrics in space between the source and surface of reflection (Figure 2.12b). However, they are subordinated to the same rules of reflection on the boundary surface of different media.

Both fields, after penetration into a metal behave in the same way, that is, they have a character of a moving attenuated wave with significantly reduced wavelength—see formulae 2.173 through 2.179. As a result of this similarity in properties of the induction field and the wave radiation field, a plane induction field at power frequencies from the dielectric side is also often called a plane wave, though it is stationary in space (only alternating in time) and does not have the characteristic for all waves periodic distribution along the 0Z axis.

2.14.4 Oblique Reflection, Refraction, and Guiding of a Wave

A Reflection and refraction of electromagnetic wave incident at a certain angle to the boundary of two media represents a somewhat more difficult problem. In the case when one of the vectors, $\boldsymbol{E}$ or $\boldsymbol{H}$, of the plane wave incident obliquely on a boundary surface, is parallel to this surface, then the second vector can be decomposed into components—parallel and perpendicular to this plane. The parallel component, together with the first (parallel) vector, creates the considered before perpendicular wave. The normal component, instead, with the same parallel vector creates a plane wave moving in parallel to the plane (Figure 3.1). In the case when the boundary plane is a well conducting surface, and the wave incidents from the dielectric side, the perpendicular wave together with its reflection creates a *stationary wave* (Figure 2.12a). If we neglect the decaying wave penetrating the metal (Figure 2.12b), which in the case of ideal superconductor does not exist, then the only wave spreading further is the wave *moving along the conductor.* (*Note:* In the case of a nonideal conductor, on the stationary wave is superimposed a small wave moving towards the conductor surface and delivering energy of power losses.)

The conduction surface plays a *guiding role*. The electromagnetic wave incident on its surface changes its moving direction and moves along the metal surface. Per this principle, the electromagnetic energy of field can encircle metal obstacles (e.g., transformer covers Figure 3.2) and can be guided by means of transmission line conductors, the so-called *waveguides*,* and so on. The surface, in any case, should be evidently satisfied by the boundary conditions 2.195.

If we consider a "far" monochromatic plane wave of the velocity along the direction of propagation defined by unit vectors $\boldsymbol{s}_1$, $\boldsymbol{s}_2$, $\boldsymbol{s}_3$ (Figure 2.13), then for both the dielectric 2.165 and for the conductor 2.176 the wave can be represented in the form

$$\underline{\boldsymbol{E}}_m = \boldsymbol{E}_m e^{j\omega\left(t-\frac{rs}{v}\right)} \quad \text{and} \quad \boldsymbol{H}_m = \boldsymbol{H}_m e^{j\omega\left(t-\frac{rs}{v}\right)}$$

where $\boldsymbol{r} = \mathbf{i}x + \mathbf{j}y + \mathbf{k}z$ is the radius vector of the considered point of reflection (distance from the origin of system of coordinates).

From the condition of equality of tangent components 2.195 on the boundary surface XY (Figure 2.13), we get, for instance, for E_x or E_y

$$E_{m1i} e^{j\omega_1\left(t-\frac{rs_1}{v_1}\right)} + E_{m3i}\, e^{j\omega_3\left(t-\frac{rs_3}{v_1}\right)} = E_{m2i} e^{j\omega_2\left(t-\frac{rs_2}{v_2}\right)} \tag{2.200}$$

where $i = x$ or y.

Since this equation must be satisfied for any arbitrary time t and for any points in the XY plane, the corresponding moduli and arguments of Equation 2.200 must be equal to each other:

$$E_{m1t} + E_{m3t} = E_{m2t}$$

* Guiding of electromagnetic wave inside waveguides is possible only at high frequencies and under the condition of generating the longitudinal field component either in vector $\boldsymbol{E}$ or $\boldsymbol{H}$.

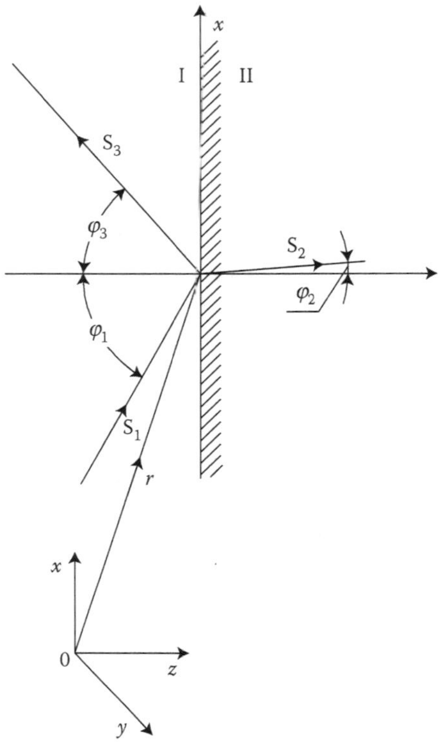

FIGURE 2.13 Reflection and refraction of a plane wave on the boundary of two media: I (ε_1, μ_1, σ_1, v_1, ω_1) and II (ε_2, μ_2, σ_2, v_2, ω_2).

$$\omega_1 = \omega_2 = \omega_3 \quad (i = x, y) \tag{2.201}$$

$$\frac{s_{1i}}{v_1} = \frac{s_{2i}}{v_2} = \frac{s_{3i}}{v_1}$$

The components s_z are absent here, since Equation 2.200 concerns tangent components only.

If the vector $\mathbf{s}_1$ lies in the XZ plane, then $s_{1y} = 0$, hence $s_{3y} = s_{2y} = 0$. This means that the directional vectors of all three waves lie in one plane, XZ, called *plane of incidence*. Since $s_{jx} = s_j \sin \varphi_j$, where $j = 1, 2, 3$, and $s_j = 1$, hence

$$\frac{\sin\varphi_1}{v_1} = \frac{\sin\varphi_2}{v_2} = \frac{\sin\varphi_3}{v_1} \quad \text{or} \quad \frac{\sin\varphi_1}{\sin\varphi_2} = \frac{v_1}{v_2} \tag{2.202}$$

If we compare the last formula 2.202 with formula 2.180, we can see that the electromagnetic *wave inside a metal always propagates perpendicularly to the metal surface*, practically independently on the glancing angle onto the surface from the dielectric side (except of exact $\varphi_1=\pi/2$). This conclusion, together with the previously mentioned identity of interaction inside metal of the near and far fields, allows to consider all solid metal parts located in an alternating field as bodies with incident *plane wave*, if the radius of curvature of the surface of these parts is much larger than $\lambda/2$ (2.178).

3 Transfer and Conversion of Field Power

3.1 POYNTING'S THEOREM: POYNTING VECTOR*

The basis for investigating energy motion in electromagnetic fields is Poynting's theorem [3.3] and the Poynting vector (Turowski [1.15], [3.4]). Poynting's theorem says that:

The electromagnetic power P_s flowing into (or flowing out of) a closed volume, across an enclosing surface A, equals to the surface integral of the normal component of the Poynting vector $\boldsymbol{S}_n$ over the entire enclosed surface A:

$$P_s = \oint\!\!\oint_A \boldsymbol{S} \cdot \mathrm{d}A = \oint\!\!\oint_A \boldsymbol{S}_n \, \mathrm{d}A \tag{3.1}$$

where

$$\boldsymbol{S} = \boldsymbol{E} \times \boldsymbol{H} \tag{3.2}$$

is the Poynting vector, which determines the power and direction of the electromagnetic power flux passing through a surface unit perpendicular to the energy flow direction. Of course,

$$\boldsymbol{S}_n = \boldsymbol{E}_t \times \boldsymbol{H}_t \tag{3.2a}$$

Poynting's theorem can be applied to calculation of active, reactive, and apparent power flowing into the investigated space.

Both the power P_s and the vector $\boldsymbol{S}$ can therefore be active, reactive, or apparent, depending on the character and phase of the field components $\boldsymbol{E}$ and $\boldsymbol{H}$. At sinusoidal variations of these components, we calculate a complex Poynting vector similar to the apparent electric power, that is,

$$\boldsymbol{S}_s = \boldsymbol{S}_p + j\boldsymbol{S}_q = \frac{1}{2}(\boldsymbol{E}_m \times \boldsymbol{H}_m^*) = \frac{1}{2}(\boldsymbol{E}_m^* \times \boldsymbol{H}_m)^* \tag{3.3}$$

* This vector in Russian literature is sometimes also called Umov–Poynting vector. Umov, however, investigated motion of mechanical energy in elastic media and the Umov vector $\boldsymbol{U} = (\mathrm{d}W/\mathrm{d}V)V$ cannot be identified with the Poynting vector $\boldsymbol{E} \times \boldsymbol{H}$.

In this sense, more correct would be the formulation in 1980 by the author ([1.16], pp. 20–21 and 232–233) the *generalized vector of power density* $\boldsymbol{T} = \boldsymbol{S} + \boldsymbol{U} + q_F + q_N + q_{SB} + \boldsymbol{M} + \boldsymbol{E}_e + \boldsymbol{I} + \boldsymbol{S}_{ch} + \cdots$ and the corresponding generalized theorem which better matches the energy conservation law.

In this case, we obtain the complex Poynting vector, which consists of active and reactive power flowing across the surface unit.

Poynting's theorem has been proven for any linear and nonlinear medium having hysteresis as well as for nonuniform and anisotropic media. The application and experimental verification of the $\boldsymbol{S}$ vector in numerous works of the author (*JT*) [2.31, 2.33, 2.34, 4.13, 4.21, 6.15, 7.17, 7.18] and of others have shown its great practical usefulness.

After scalar multiplication of the first Maxwell's equation (2.1) by $\boldsymbol{E}$, and of the second one (2.2) by $\boldsymbol{H}$, and adding them together, we get, for a uniform, isotropic medium:

$$\boldsymbol{E}\ \text{curl}\ \boldsymbol{H} - \boldsymbol{H}\ \text{curl}\ \boldsymbol{E} = \frac{\partial}{\partial t}\left(\frac{\varepsilon E^2}{2} + \frac{\mu H^2}{2}\right) + \boldsymbol{E}\cdot\boldsymbol{J} + \rho\boldsymbol{E}\cdot\boldsymbol{v}_\rho + \sigma\boldsymbol{E}\left(\boldsymbol{E}_{\text{ext}} + \boldsymbol{v}\times\boldsymbol{B}\right) \tag{3.4}$$

Considering the vector identity

$$\boldsymbol{E}\ \text{curl}\ \boldsymbol{H} - \boldsymbol{H}\ \text{curl}\ \boldsymbol{E} = -\text{div}(\boldsymbol{E}\times\boldsymbol{H})$$

and introducing the Poynting vector (3.2), we can transform Equation 3.4 into the form

$$-\text{div}\left(\boldsymbol{E}\times\boldsymbol{H}\right) = \frac{\partial}{\partial t}\left(\frac{\varepsilon E^2}{2} + \frac{\mu H^2}{2}\right) + sE^2 + r\boldsymbol{E}\cdot\boldsymbol{v}_\rho + \sigma\boldsymbol{E}\left(\boldsymbol{E}_{\text{ext}} + \boldsymbol{v}\times\boldsymbol{B}\right)$$

After integration of both sides over the volume V of the investigated space, and after application of Green's theorem $\oint\!\!\oint_A \boldsymbol{D}\cdot \mathrm{d}\boldsymbol{A} = \iiint_V \rho\,\mathrm{d}V$ Equation 20.20a to the left side, we obtain finally

$$-\oint_A \boldsymbol{S}\cdot\mathrm{d}\boldsymbol{A} = \frac{\partial W_{\text{em}}}{\partial t} + \int_V \sigma E^2\,\mathrm{d}V + \int_V \rho\boldsymbol{E}\cdot\boldsymbol{v}_\rho\,\mathrm{d}V + \int_V \sigma\boldsymbol{E}(\boldsymbol{E}_{\text{ext}} + \boldsymbol{v}\times\boldsymbol{B})\mathrm{d}V \tag{3.5}$$

where

$$\frac{\partial W_{\text{em}}}{\partial t} = \int_V \frac{\partial}{\partial t}\left(\frac{\varepsilon E^2}{2} + \frac{\mu H^2}{2}\right)\mathrm{d}V \tag{3.6}$$

is the power delivered for increasing the electrical and magnetic energy accumulated in a electromagnetic field in the volume V. The second component on the right-hand side of Equation 3.5 is the power loss due to a current flowing in the resistive volume V. The two last components are related to the conversion of electromagnetic energy of the field into electromechanical energy and to the increased velocity of electric charges ρ, or with braking of a conducting medium which moves with the velocity $\boldsymbol{v}$. The left side of Equation 3.5 represents the total power flux flowing from outside into the volume V across the external surface area A. This energy is consumed, dissipated, or accumulated in the volume V. Equation 3.5, therefore, expresses the fundamental Poynting's theorem formulated in the beginning.

Since Poynting's theorem is a specific case of the general *law of conservation of energy*, there is no reason to think it is not also valid for media of another character, for instance, media containing internal sources or receivers of electric energy of a type different than mentioned above (e.g., thermoelectric, electrochemical, etc.). In the latter case, this additional energy should be added or subtracted on the right-hand side of Equation 3.5.

On this basis, in 1980 the author formulated (J. Turowski [1.16, 1.20, 3.4]) the notion of the complex *Generalized Power Density Vector* $\boldsymbol{T}$ and, by analogy to Equation 3.1—the *Generalized Theorem on Power Density Vector* for coupled fields:

$$\boldsymbol{T} = \boldsymbol{S} + \boldsymbol{U} + \boldsymbol{q}_{\mathrm{F}} + \boldsymbol{q}_{\mathrm{N}} + \boldsymbol{q}_{\mathrm{SB}} + \boldsymbol{M} + \boldsymbol{E}_{\mathrm{e}} + \boldsymbol{I} + \boldsymbol{S}_{\mathrm{ch}} + \cdots \quad \text{and} \quad \boldsymbol{P}_{\mathrm{s}} = \oint\!\!\oint_A \boldsymbol{T} \cdot \mathrm{d}\boldsymbol{A} \tag{3.5a}$$

where $\boldsymbol{S} = \boldsymbol{E} \times \boldsymbol{H}$ is the complex Poynting vector, $\boldsymbol{U} = \mathrm{v}\,(\mathrm{d}W_{\mathrm{mech}}/\mathrm{d}V)$ is the Umov vector, $\boldsymbol{q}_{\mathrm{F}} = \lambda\, grad\, \Theta$ the thermal power conduction flux density (Fourier's law), $\boldsymbol{q}_{\mathrm{N}}$ the thermal power convection flux density (Newton's law), $\boldsymbol{q}_{\mathrm{SB}}$ the thermal power radiation flux density (Stefan–Boltzmann law), $\boldsymbol{M}$ the light emittance power, $\boldsymbol{E}_{\mathrm{e}}$ irradiance power incident on the surface, $\boldsymbol{I}$ the acoustic sound intensity, $\boldsymbol{S}_{\mathrm{ch}}$ the electrochemical power flux density, and so on.

This generalized vector $\boldsymbol{T}$ and its theorem (3.5a) are much more useful for electromechanical engineering, energy conversion analysis, and for the new, modern discipline of *Mechatronics** [1.20].

The Poynting vector (3.2) designates the direction and density of the field energy flow in a given point of space. We can, therefore, with its help, calculate the power and its distribution in any spot of the investigated system. In the case of AC fields, the vector moduli (not time dependent) of particular components of the Poynting vector are the scalar components of the vector product $\frac{1}{2}(\boldsymbol{E}_{\mathrm{m}} \times \boldsymbol{H}_{\mathrm{m}}^{*})$ (3.3) of the coupled fields:

$$\left.\begin{aligned} S_{\mathrm{x}} &= \frac{1}{2}(E_{\mathrm{my}}H_{\mathrm{mz}}^{*} - E_{\mathrm{mz}}H_{\mathrm{my}}^{*}) \\ S_{\mathrm{y}} &= \frac{1}{2}(E_{\mathrm{mz}}H_{\mathrm{mx}}^{*} - E_{\mathrm{mx}}H_{\mathrm{mz}}^{*}) \\ S_{\mathrm{z}} &= \frac{1}{2}(E_{\mathrm{mx}}H_{\mathrm{my}}^{*} - E_{\mathrm{my}}H_{\mathrm{mx}}^{*}) \end{aligned}\right\} \tag{3.7}$$

* Technical committee on Mechatronics of the International Federation for the Theory of Machines and Mechanisms adapted in Prague the definition: "*Mechatronics is the synergistic combination of precision mechanical engineering, electronic control and system thinking in the design of products and manufacturing processes*" (It should be completed by *Engineering Electromagnetics*).

It should be stressed, however, that the product ($\boldsymbol{E} \times \boldsymbol{H}$) is the Poynting vector only in the case when both fields $\boldsymbol{E}$ and $\boldsymbol{H}$ are generated from the same source and mutually related by Maxwell's equations, at $\boldsymbol{J} \neq 0$. Poynting's theorem concerns also fields of DC currents, but the product $\boldsymbol{E} \times \boldsymbol{H}$ will not be the measure of energy of superimposed electrostatic and magnetostatic fields, because in static conditions ($\partial/\partial t = 0$, $\boldsymbol{J} = 0$) the whole derivation of Poynting's theorem loses its sense, because it is based on the assumption of $\boldsymbol{J} \neq 0$.

A more accurate analysis (A. A. Vlasov, Moscow, 1955) showed also groundlessness of the supposition that the Poynting vector (3.2) is an ambiguous quantity, what supposedly would follow from the identity $\oint_A \operatorname{curl} \boldsymbol{P} \cdot \mathrm{d}A \equiv 0$ and therefore $\oint_A \boldsymbol{S} \cdot \mathrm{d}A = \oint_A (\boldsymbol{S} + \operatorname{curl} \boldsymbol{P}) \cdot \mathrm{d}A$, where $\boldsymbol{P}$ is any vector.

3.2 PENETRATION OF THE FIELD POWER INTO A SOLID CONDUCTING HALF-SPACE

Let us consider a typical case of conversion of electromagnetic field energy into a thermal energy in a conducting medium. In the case of a plane wave, the field components in the direction of field motion equal zero. If we investigate the motion of a plane wave in metal, then as it follows from Equations 2.180 and 2.202, it can be accepted with high accuracy that such a wave always moves in the direction perpendicular to the metal surface. If we select such a coordinate system that the XY plane overlaps with the surface of metal half-space, then in formulae (3.7) one should substitute $E_z = 0$ and $H_z = 0$, from which it follows: $S_x = S_y = 0$. Then, there remains only the component $S_z = S$.

Using the field components (2.173) and (2.175), we obtain

$$\frac{1}{2} E_{mx} H^*_{my} = \frac{1}{2}\sqrt{\frac{\omega\mu}{\sigma}} \left|H_{msy}\right|^2 \mathrm{e}^{-2kz} \mathrm{e}^{j\frac{\pi}{4}}$$

$$\frac{1}{2} E_{my} H^*_{mx} = -\frac{1}{2}\sqrt{\frac{\omega\mu}{\sigma}} \left|H_{msx}\right|^2 \mathrm{e}^{-2kz} \mathrm{e}^{j\frac{\pi}{4}}$$

from where the field power density, in VA/m², at the distance z from the metal surface

$$S(z) = \frac{1}{2}\sqrt{\frac{\omega\mu}{\sigma}} \left|H_{ms}\right|^2 \mathrm{e}^{-2kz} \left(\cos\frac{\pi}{4} + j\sin\frac{\pi}{4}\right) \tag{3.8}$$

where $H^2_{ms} = H^2_{msx} + H^2_{msy}$.

Modulus of the active component of the Poynting vector, corresponding to the active power (in W/m²) flowing through a surface unit, equals to

$$S_F(z) = \sqrt{\frac{\omega\mu}{2\sigma}} \frac{\left|H_{ms}\right|^2}{2} \tag{3.9}$$

Modulus of the reactive component of the Poynting vector, corresponding to the reactive power (in var/m^2) flowing through a surface unit, has the same value, but different units

$$S_q = S_p$$

As it follows from formula (3.9), the power of an electromagnetic wave entering into a solid metal half-space is attenuated in the same way as a field, that is, according to exponential functions, but much faster (2.173, 2.175). In Figure 2.11, we presented a graph of attenuation of the current density (field attenuation) and of square of attenuation (the power attenuation). Beyond the depth $z = \lambda/2$ from the surface ($k\lambda = 2\pi$), only $e^{-2k\pi} \times 100\% = 0.185\%$ of the total energy is left to be consumed by the conducting medium. This is why one can assume that an electromagnetic wave is damped at the depth 2× shorter than the wavelength, given in Table 2.1. If the thickness of a metal part is larger than half of a wavelength (in the case of steel, at 50 Hz and $\mu_r = 300–1000$, larger than 5–3 mm, and in the case of copper larger than 30 mm), then at one-sided wave penetration such a part can be considered infinitely thick (the so-called *half-space*) because an electromagnetic wave is almost completely extinguished before reaching the opposite surface. The electromagnetic wave which entered such a half-space moves only in one direction until its complete extinction and has no possibility to get outside.

This is why one can recognize that the Poynting vector **S** at any point z inside a solid metal half-space (Figure 2.11) is the measure of the total per-unit power consumed in the space on the right-hand side from this point. Per the same principle, one can assume that the power consumed by a solid metal half-space equals the Poynting vector **S** value on the surface of this space (at $z = 0$) multiplied by the area of this surface. If the Poynting vector has unequal values in different points on this surface, the power loss should be calculated by integration of $\mathbf{S}(x, y, z = 0)$ on the whole surface of the body.

According to Equation 3.9, the Poynting vector in the complex form on the surface of a metal half-space is expressed by the formula:

$$S = (1 + \mathrm{j})\sqrt{\frac{\omega\mu}{2\sigma}}\frac{|H_{ms}|^2}{2} \tag{3.10}$$

and its active component equals to the loss density on the surface, P_1 (in W/m^2):

$$S = P_1 = a_p\sqrt{\frac{\omega\mu_s}{2\sigma}}\frac{|H_{ms}|^2}{2} = \frac{\omega}{a_p}\sqrt{\frac{\omega\sigma}{2\mu_s}}\frac{|\Phi_{m1}|^2}{2} \tag{3.10a}$$

where $a_p = 1$ for $\mu = \text{const}$, and $a_p \approx 1.4$ for steel (Chapter 7) [3.4].

According to Equation 2.176, the per-unit flux Φ_{my1} on the surface of a metal half-space per 1 m of width of its track equals, in Wb/m

$$\Phi_{m1} = \mu\int_0^{\infty} H_{ms}e^{-\alpha z}\,\mathrm{d}z = \frac{\mu}{\alpha}H_{ms} = (1 - \mathrm{j})\sqrt{\frac{\mu}{2\omega\sigma}}H_{ms} = \mu H_{ms}\frac{\delta}{1 + \mathrm{j}} \tag{3.11}$$

and according to Turowski ([1.16], p. 30), its modulus

$$|\Phi_{ml}| = a_p \sqrt{\frac{\mu_s}{\omega\sigma}} H_{ms} \tag{3.11a}$$

The reactive component of the Poynting vector equals the per-unit consumption of reactive power by a steel half-space, in var/m²

$$S_q = P_{q1} = a_q \sqrt{\frac{\omega\mu_s}{2\sigma}} \frac{|H_{ms}|^2}{2} = \frac{a_q}{a_p} \omega \sqrt{\frac{\omega\sigma}{2\mu_s}} \frac{|\Phi_{m1}|^2}{2} \tag{3.12}$$

where $a_q = a_p = 1$ for $\mu = \text{const}$ and $a_q \approx 0.85$ and $a_p \approx 1.4$ for steel.

EXAMPLE

Calculate the per-unit power P_1 and the maximum magnetic flux Φ_{m1} for a surface unit of a steel element with the wall thickness $d > \lambda/2$, on which one side impinges a plane electromagnetic wave of 50 Hz, creating on its surface the resultant field $H_{ms} = 40$ A/cm.

SOLUTION

Assuming the average value of steel conductivity $\sigma_{St,20°C} = 7 \times 10^6$ S/m (Table 1.6) and from Figure 1.29, curve 2, the permeability on the surface $\mu_{rs} \approx 320$ for $H_{ms} = 4000$ A/m, per Equation 3.10a, we calculate the per-unit power

$$P_1 = a_p \sqrt{\frac{2\pi\, 50 \cdot 320 \cdot 0.4\pi \cdot 10^{-6}}{2 \cdot 7 \cdot 10^6}} \cdot \frac{4000^2}{2} = a_p \cdot 760 \frac{\text{W}}{\text{m}^2} = 1064 \text{ W/m}^2$$

and per Equation 3.11a, the per-unit flux in the steel wall

$$|\Phi_{m1}| = a_p \sqrt{\frac{320 \cdot 0.4\pi \cdot 10^{-6}}{2\pi \cdot 50 \cdot 7 \cdot 10^6}} \cdot 4000 = a_p \cdot 1.71 \cdot 10^{-3} = 2.4 \cdot 10^{-3} \text{ Wb/m}$$

These numbers can be considered as initial values at assessments of permitted electromagnetic loads of steel constructional elements from the viewpoint of local overheating hazards (Chapter 9).

The power in a metal half-space per volume unit, in W/m³

$$w(z) = \frac{1}{\sigma} J^2_{rms}(z) = \frac{1}{2} \sigma E^2_{ms}\, e^{-2kz} = \frac{\omega\mu}{2} H^2_{ms}\, e^{-2kz} = \sqrt{2\omega\mu\sigma}\, P_1 e^{-2kz} \tag{3.13}$$

where P_1 is the active power per surface unit (W/m²) in the metal half-space.

3.3 POWER FLUX AT CONDUCTORS PASSING THROUGH A STEEL WALL

The power of electric current flowing through electric line conductors is transported by an electromagnetic field surrounding these conductors. Physically it is explained that although electrons inside the conductors move along the electric field with velocity barely ca. 0.2 mm/s, the impulse and electromagnetic field move with velocity near 300,000 km/s. With the same speed is transmitted electric energy. Mathematically, the whole flux of power flowing along conductors can be calculated by the integration of the power flux density, that is, the Poynting vector, through an infinite plane perpendicular to the conductor axes.

Let us consider the power flow in the case when two current-carrying conductors pass through a metal wall (Figure 3.1), which may be a cover of a power transformer, or a metal wall with a thickness $d > \lambda$. From the total power flux flowing in surroundings of the conductors and meeting the barrier in the form of impenetrable transformer cover, one can distinguish the following components of Poynting vector in particular points of space (Figure 3.1):

- The main vector $\boldsymbol{S}_g$ corresponding to the power density flowing into the transformer. It is directed along the conductors. This vector can be split into

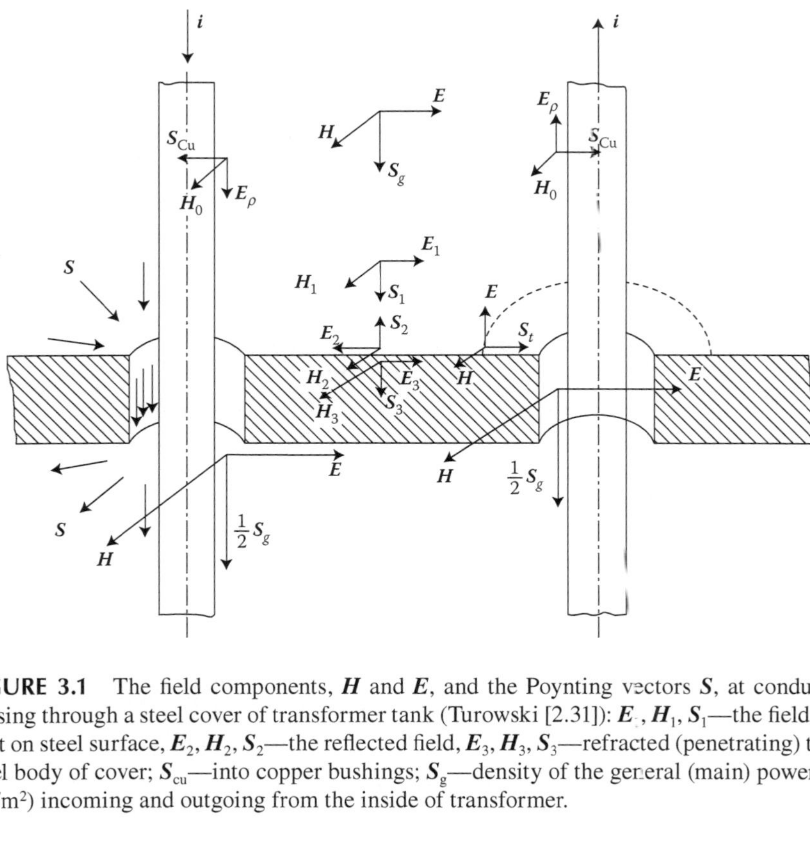

FIGURE 3.1 The field components, $\boldsymbol{H}$ and $\boldsymbol{E}$, and the Poynting vectors $\boldsymbol{S}$, at conductors passing through a steel cover of transformer tank (Turowski [2.31]): $\boldsymbol{E}_1$, $\boldsymbol{H}_1$, $\boldsymbol{S}_1$—the field incident on steel surface, $\boldsymbol{E}_2$, $\boldsymbol{H}_2$, $\boldsymbol{S}_2$—the reflected field, $\boldsymbol{E}_3$, $\boldsymbol{H}_3$, $\boldsymbol{S}_3$—refracted (penetrating) to the steel body of cover; $\boldsymbol{S}_{cu}$—into copper bushings; $\boldsymbol{S}_g$—density of the general (main) power flux (W/m^2) incoming and outgoing from the inside of transformer.

two components: the component of power which is the incident on the cover surface, $\boldsymbol{S}_1$, and the component of reflected power, $\boldsymbol{S}_2$.

- The vector of power penetrating the solid steel of cover, $\boldsymbol{S}_3$. It represents the power losses in the cover.
- The vector of power penetrating into the current-carrying conductors, $\boldsymbol{S}_{\mathrm{Cu}}$. It is directed perpendicularly to the surface of the conductors and its active component equals to the per-unit losses in the conductors.

The above considerations follow from the following simple equations

$$E_{\mathrm{p}} = \frac{Ri}{l}, \quad H_0 = \frac{i}{2\pi r}, \quad S = E_{\mathrm{p}} H_0 = \frac{Ri^2}{2\pi rl}$$

which, after considering Poynting's theorem (3.1), gives the total losses in conductors:

$$\Delta P_{\mathrm{cu}} = \oint_A S \cdot \mathrm{d}A = \frac{Ri^2}{2\pi rl} 2\pi rl = Ri^2$$

In an analogical way, the reactive component of the $\boldsymbol{S}_{\mathrm{Cu}}$ vector determines the internal reactance of conductors.

An investigation of the electromagnetic field and power losses in the cover alone, the component $\boldsymbol{S}_{\mathrm{Cu}}$ can be skipped, assuming that the bushings are made of a very good conductor.

- The refracted vector $\boldsymbol{S}_3$ of the power-penetrating metal of the cover through all its surfaces; it is directed perpendicularly to the cover surfaces. This vector will be used in the calculation of losses in covers and other constructional metal parts of electric machines and apparatus.

The main flux of power $\boldsymbol{S}_{\mathrm{main}} \equiv \boldsymbol{S}_{\mathrm{g}}$ (Figure 3.1), carrying the power flowing into or out of the transformer, squeezes almost totally through the isolation (porcelain) gap around the bushing conductor if the thickness of the cover is sufficiently big ($d > \lambda$) and has no other holes filled with dielectric in which an electromagnetic field would exist. The power flux flowing into the transformer does not depend on the thickness of the mentioned isolation gap, because the thinner is the gap the bigger will be value of the vector $\boldsymbol{E}$, and therefore vector $\boldsymbol{S}$. In the case of a metallic connection, when the isolation gap does not exist, a short-circuit occurs and then no power flux or current can penetrate through the cover.

The total power flow into a transformer exclusively through holes filled by dielectric can be easily checked quantitatively. In a single-phase transformer, both on the primary and the secondary side, only two bushings are present. The electric field intensity in the central plane of the hole in the steel sheet amounts to

$$E = \frac{u}{r \ln (R/\mathrm{r}_1)}$$

where u is half of the line voltage, that is, $2u = u_{\text{line}}$; R the radius of the hole for bushing; r_1 the radius of the conductor in bushing. Usually, the distance between bushings is so large in comparison with the radius of hole that one can assume that in the hole exists the magnetic field intensity $H = i/(2\pi r)$.

Since both fields are perpendicular to each other, the modulus of the vector of power density in the hole per surface unit has the value

$$S = EH = \frac{ui}{2\pi \ln(R/r)} \cdot \frac{1}{r^3} \tag{3.14}$$

and an element of the considered surface is $dA = 2\pi r \cdot dr$. The instant power flowing into the transformer tank by two holes, according to Poynting's theorem, therefore, equals

$$p = 2\iint_A S\,dA = \frac{2ui}{2\pi \ln(R/r_1)} \int_{r_1}^{R} \frac{2\pi r \cdot dr}{r^2} = 2ui = u_p\, i$$

which proves that the entire instantaneous power passes only by the isolation of the bushing holes (Figure 3.2).

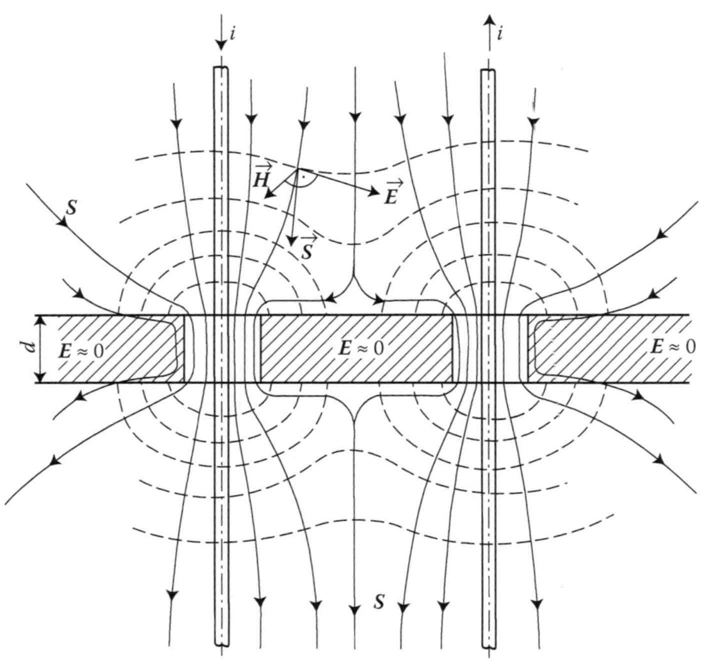

FIGURE 3.2 A schematic picture of the power flow through the cover of a transformer tank, the Poynting vector (**S**) lines, with ignoring the eddy current losses. (Adapted from Turowski J.: Electromagnetic field and losses in the transformer housing. (in Polish). *"Elektryka" Science Papers*, Technical University of Lodz., No. 3, 1957, pp. 73–63.)

In Figure 3.2, the continuous lines represent the power flux lines (lines of the Poynting vector $\boldsymbol{S}$) and the dashed lines represent the lines of the force of an electric field $\boldsymbol{E}$. Lines of vector $\boldsymbol{H}$ lie in planes parallel to the surface of the tank cover. Figure 3.2 does not present quantitative interdependences. Such plots are created on the basis of perpendicularity of the lines of vector $\boldsymbol{S}$ to the planes of vectors $\boldsymbol{E}$ and $\boldsymbol{H}$.

In case when the thickness d of a cover or screen is smaller than the wavelength ($d < \lambda$) in metal (Table 2.1), a certain fraction of the field power penetrates into the tank directly through the cover, which in this case becomes transparent to the electromagnetic field (Section 4.3).

3.4 POWER FLUX IN A CONCENTRIC CABLE AND SCREENED BAR

Reasoning as above and assuming that in formula (3.14) u means the voltage between conductors of the cable, we can see that the total power ui transmitted by the cable is transported by the electromagnetic field moving in parallel to the cable axis in dielectric enclosed between internal and external conductors of cable. This power flows, however, not as a uniform flux but is more concentrated near the surface of the internal conductor and decays according to Equation 3.14 inversely proportionally to square of the distance r from the cable axis (Figure 3.3).

A similar picture of the power flux distribution occurs in screened bars used in power generators and transformer systems of power stations. In such a case, each of three bars of a threephase system is enclosed in a cylindrical screen. The screens are either grounded or connected together on both ends, directly or through reactors. In such systems, the magnetic field of bar gets out indeed outside of the screen, but the electric field exists practically only in the insulation space between bar and screen. Therefore, the power, similarly as in a concentric cable, is transported in this case also only through the enclosed space.

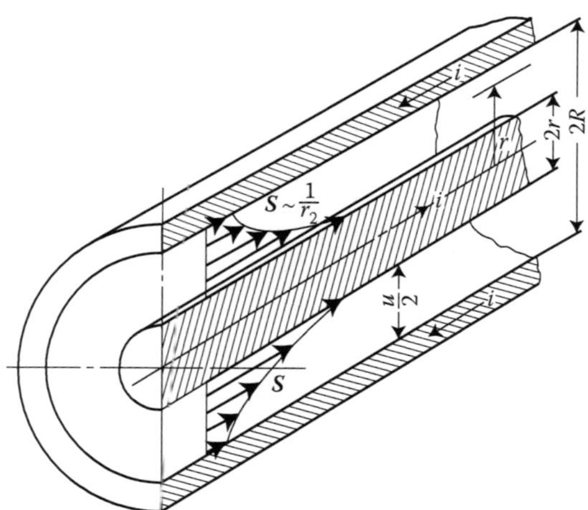

FIGURE 3.3 Distribution of the Poynting vector (power density) in cross-section of a concentric cable.

3.4.1 Factors of Utilization of Constructional Space

One of the most important factors of technology advancement is reduction of space (limiting outlines) occupied by electromagnetic equipment. Especially it is seen in switching stations and devices with SF_6, occupying several times smaller space than conventional constructions.

The main reserves are in a nonuniform distribution of the Poynting vector (power density), measure of which can be the *factor of utilization of constructional space*

$$\eta_S = S_{min}/S_{max} \tag{3.15}$$

The factor (3.15) can assume values within the range $0 \le \eta_S \le 1$. For instance, for a concentric cable, $\eta_s = r_1^2/R^2$, which means that a small change of one of the diameters causes a significant change of η_S. For the interwinding gap of a transformer, however, $\eta_S \approx 1$.

Another factor could be the ratio $\eta_P = P_{var}/P_{unif}$ of the total powers $P = \iint_A S\, dA$ at a nonuniform P_{var} distribution to the flux P_{unif} at a uniform distribution of the Poynting vector (power density).

Another factor here is the highest possible value S_{max}, which in turn is limited by the electric field strength of space. As per Equation 3.14, for cables with a constant utilization of conductors (u = const, i = const, r = const)

$$\eta_S = \exp\left(-\frac{ui}{\pi r_1^2 S_{max}}\right) \tag{3.16}$$

Assuming, for example, for a concentric cable: $R/r_1^2 = e = 2.718$, we get $\eta_S = 1/e^2 = 0.136$. The maximum value of the Poynting vector for such a cable, with parameters 35 kV, 400 A, cross-section 185 mm^2, $r_1 = 7.5$ mm, $R = er_1 = 2.7 \cdot 7.5 = 20.2$ mm, would be

$$S_{max} = \frac{35 \cdot 400}{2\pi \cdot 0.75^2} = 4000\,\text{kW/cm}^2.$$

It is not difficult to estimate how the transmission power of the cable could be increased if the factor η_S was made higher. It is possible by means of multilayer structures. These conclusions should of course take into account material and processing limitations.

3.5 POWER FLUX IN A CAPACITOR, COIL, AND TRANSFORMER

A capacitor and a cylindrical coil can be considered as the simplest constructional elements. They can, at the same time, serve as models of more complex systems, for instance—transformers. In a *parallel-plate capacitor* (Figure 3.4) connected to an alternating voltage u, in the part in which the field is uniform, we have

$$\boldsymbol{E} = \mathbf{1}_z \frac{u}{a} \quad \text{and} \quad \boldsymbol{H} = \mathbf{1}_\Theta \frac{i}{2\pi r} \tag{3.17}$$

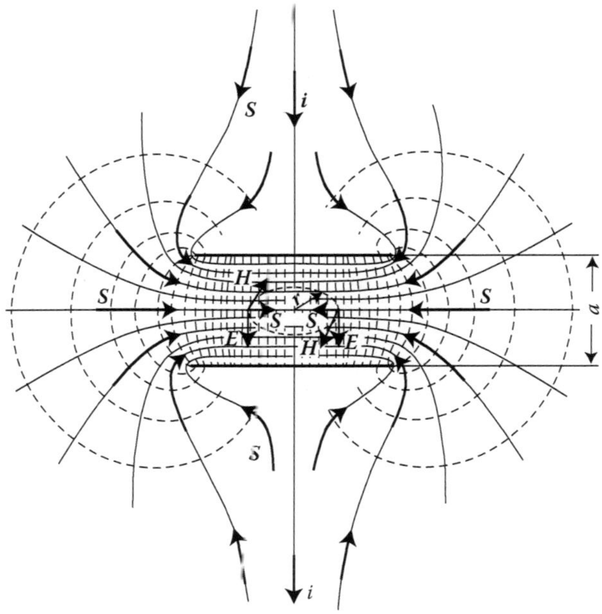

FIGURE 3.4 Distribution of power flux in a parallel-plate capacitor; ***S***—Poynting vector = power flux density (VA/m^2).

Ignoring the edge deformations of the field (the so-called *fringing fields*) and taking into account the sense of vectors (3.17) we can see that the Poynting vector in any considered moment is directed toward the capacitor center axis, and equals

$$\boldsymbol{S} = \boldsymbol{E} \times \boldsymbol{H} = -\mathbf{1}_r \frac{u \cdot i}{2\pi r a}$$

After multiplying this value by the lateral surface $2\pi r a$, we can see that the entire power flux entering to capacitor field by its lateral surface equals to $u \cdot i$ delivered to capacitor, as expected. The power flow into or out of the capacitor occurs along the equipotential lines, which in Figure 3.4 is shown by the arrows, at a moment of increasing voltage. At decreasing voltage, the sense of the instantaneous vectors ***S*** changes to the opposite one. The power flux in a capacitor connected to an alternating voltage oscillates with frequency $2f$.

In a *resistance-less cylindrical coil* (Figure 3.5), supplied by an alternating voltage, the magnetic field ***H*** (dashed lines) inside the coil in a considered moment is directed toward the top, and outside—toward the bottom. The electric field ***E*** is tangential to the rings created by turns and directed according to the electromotive force (EMF) induced by the flux of the coil. As a result, the vector ***S*** (continuous lines) goes out from the coil surface to the ambient surroundings in both directions.

Similarly as before, it is easy to show (Turowski [1.15]) that the entire flux of an electromagnetic field delivered to the coil leaves (or enters) by its lateral surfaces—external and internal. The direction of power flow changes with frequency $2f$. During

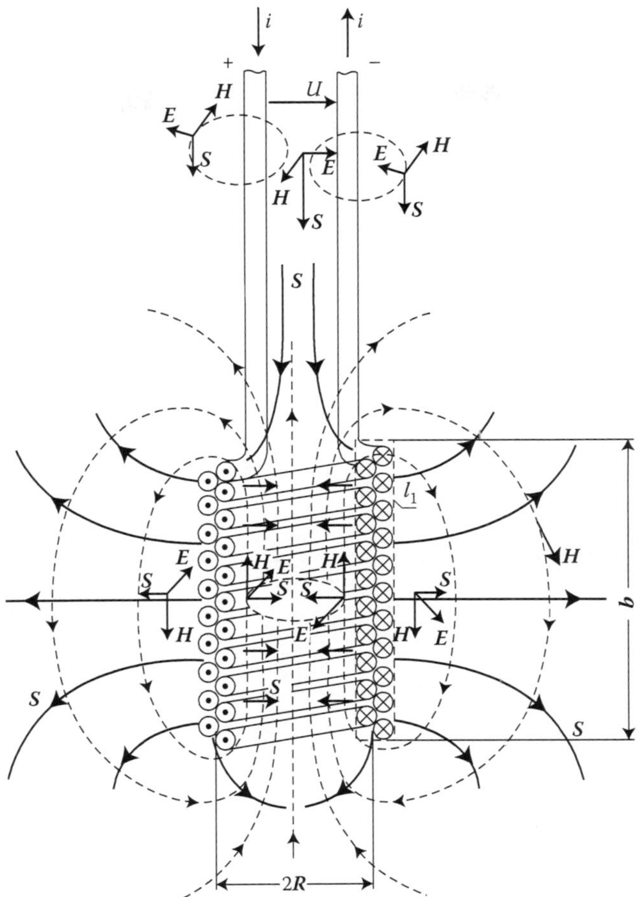

FIGURE 3.5 Distribution of power flux in a cylindrical coil.

one quarter period of the network frequency the magnetic energy is accumulated in the coil field, and then, during the next quarter period, it is given back to the network. This is the so-called reactive power.

The active power flux (power losses) $P_1 = S_p$ (in W/m^2) has always the same sense—toward the receiver. In Figure 3.5, the lines (continuous) are shown of the Poynting vector $\boldsymbol{S}$ at the moment when the reactive power is delivered to the coil from the network side. In a similar way, a distribution of power in a coil with iron core or in a transformer at no load will occur.

The described distribution of electromagnetic power concerns a smooth coil as a whole. It could be, of course, subdivided into particular twists around different discrete elements, such as turns of coil, leads, interconnections, and so on, as it was done in the work of Leites [3.2]. However, it is not necessary in practice.

In the *short-circuit condition* of a power transformer almost all the magnetic flux is displaced into the interwinding gap. There, as a leakage field, it induces in both windings electric fields of inverse senses. Therefore, approximately in the center of the gap there exists a cylindrical surface on which $\boldsymbol{E} = 0$, and hence $\boldsymbol{S} = 0$. Thus, the gap is a barrier through which at short-circuit condition no power flux can pass from the primary to the secondary winding, except the power loss in secondary winding.

In a *loaded power transformer* the senses of $\boldsymbol{E}$ in both windings are the same. Thanks to it the power flux $\boldsymbol{S}$ transfers from the primary winding to the secondary winding and to the magnetic core without obstacles. In the author's book [1.15] the flow and distribution of power flux in a loaded transformer and in conductors in motion are discussed in more detail.

As we can see from the above figures, such a graphic illustration of the power flow $\boldsymbol{S}(x, y)$ is quite simple for any electromechanical system. It is only necessary to determine lines of fields $\boldsymbol{H}$ and $\boldsymbol{E}$. It can be sometimes useful in practice, like for instance that case in Figure 3.2. In the next section, let us consider such power flow in rotating machines.

3.6 POWER FLUXES AND THEIR CONVERSION IN ROTATING MACHINES

3.6.1 Power Flux of Electromagnetic Field in Gap of Induction Machine

As shown by Bron [3.1] and other authors, the *Poynting* vector lines can be determined for any kind of machine operation. The angular speed of rotor $\omega = 2\pi n = \omega_s(1 - s)$, where n is the rotational speed (in rev/s), $\omega_s = 2\pi f$ is the synchronous speed (at the assumption that the number of pairs of poles $p = 1$), and $s = (n_s - n)/n_s$ is the slip.

At the synchronous speed, $\omega = \omega_s$, no current flows in the rotor. A rotating field of amplitude $\boldsymbol{B}_{\delta m}$* (Figure 3.6a) induces in the gap δ the electric field $\boldsymbol{E}_{\delta m} = -\boldsymbol{v}_s \times \boldsymbol{B}_{\delta m}$, directed toward the observer, where $v_s = \pi D n_s$ is the linear speed, $n_s = f/p$ is the rotational speed (in 1/s), and D is the internal diameter of stator. In this case, the vector $\boldsymbol{S}$ is directed along the stator circumference. The right-hand picture in Figure 3.6a shows lines of the vector $\boldsymbol{S}_q = \boldsymbol{E}_{\delta m} \times \boldsymbol{H}_{\delta m}$. In this case, no exchange of energy between the stator and rotor exists.

At no-load, $\omega \approx \omega_s$, the picture is similar, but at a relatively big no-load current.

At load (motor operation, Figure 3.6b), the rotor is delayed in relation to the rotating field ($\omega < \omega_s$). Due to the currents induced in rotor conductors, in the gap appears the flux density $\boldsymbol{B}_{mr}$ directed along the circumference. It gives the resultant field in the gap, $\boldsymbol{B}_m = \boldsymbol{B}_{\delta m} + \boldsymbol{B}_{mr}$, and a new active component S_p directed toward the rotor. Integration of this active component, according to Poynting's theorem (3.1), equals to the power output P on the motor shaft and losses in rotor. The losses in rotor are comparatively small.

At rotor speed $\omega \geq \omega_s$ (generator operation) the pictures are the reverse of Figure 3.6a.

* Old standard symbols were $\boldsymbol{B}_{mp}$, n_1, $\omega_1 = 2\pi f$.

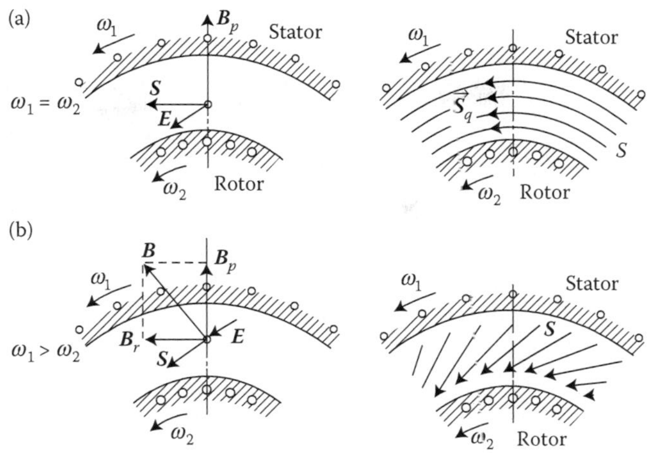

FIGURE 3.6 Power transfer in the air gap of an induction machine: (a) no-load operation, (b) motor operation $B_p \equiv B_\delta$. (Modified from Bron, O.B.: *Electromagnetic Field as a Form of Matter.* (in Russian). Moscow: Gosenergoizdat, 1962.)

At braked rotor ($\omega = 0$) the state is similar as with a short-circuited transformer; the rotor consumes only loss.

3.6.2 Power Flux of Electromagnetic Field in Air Gap of Synchronous Machine

In a smooth air gap of synchronous machine, similar to an induction machine, a sinusoidal rotating field of amplitude $\boldsymbol{B}_{\delta m}$ equals (Turowski J. [1.18])

$$\boldsymbol{B} = \boldsymbol{B}_{\delta m} \cos(\omega t - \alpha), \quad \alpha = \frac{\pi}{\tau} x, \quad \tau = \frac{\pi D}{2p} \tag{3.18}$$

At no-load, and at synchronous rotation with the angular velocity $\omega_s = 2\pi f = 2\pi p n_s$, in the air gap exists only the flux density $\boldsymbol{B}_0$ directed radially (Figure 3.7). This field induces in the gap the electric field

$$\boldsymbol{E}_0 = v \boldsymbol{B}_0 = \boldsymbol{B}_{\delta m} \frac{\omega D}{2} \cos(\omega t - \alpha) \tag{3.19}$$

The vector $\boldsymbol{E}_0$ is proportional to $\boldsymbol{B}_0$ and directed along the machine axis toward the observer in Figure 3.7a. Both these vectors give the instantaneous Poynting vector $\boldsymbol{S}_0$, whose modulus is

$$\boldsymbol{S}_0 = \boldsymbol{E}_0 \boldsymbol{H}_0 = \frac{\boldsymbol{B}_{\delta m}^2}{\mu_0} \cdot \frac{\omega D}{2} \cos^2(\omega t - \alpha) \tag{3.20}$$

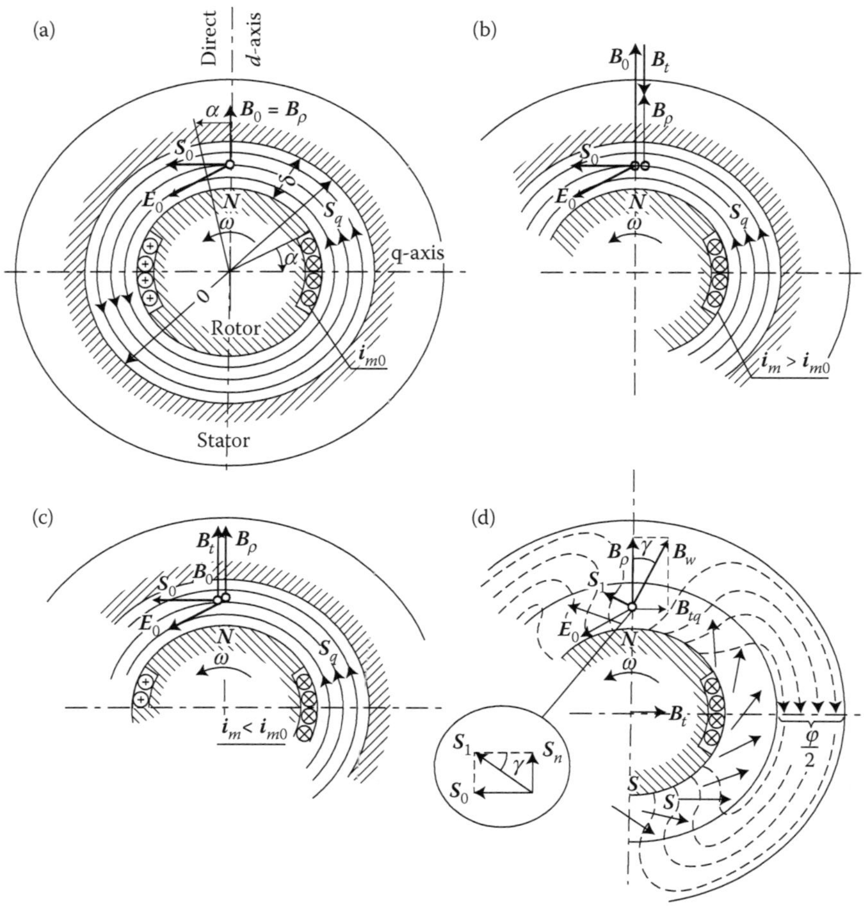

FIGURE 3.7 Motion of the electromagnetic power flux in a synchronous machine: (a) no-load operation; (b) inductive load (overexcitement); (c) capacitive load (underexcitement); (d) active load of generator $B_t \equiv B_a$. (After Bron, O.B.: *Electromagnetic Field as a Form of Matter*. (in Russian). Moscow: Gosenergoizdat, 1962.)

The vector $\boldsymbol{S}_0$ is tangential to the circumference of the machine and its lines in the gap are parallel to the stator surface. From Poynting's theorem (3.1), we get the instantaneous power:

$$p = \boldsymbol{S}_0\, l\delta = \frac{\boldsymbol{B}_{\delta\mathrm{m}}}{\mu_0}\omega \cdot \frac{Dl\delta}{2}\cos^2(\omega t - \alpha) \tag{3.21}$$

where l is the active length of the armature (stator).

Let us now consider the typical cases of machine operation.

At an ideal inductive load (Figure 3.7b) or *capacitive load* (Figure 3.7c) the flux of an armature reaction $B_t \equiv B_a \equiv B_d$ of the maximal value $\boldsymbol{B}_{\mathrm{dm}}$ is directed along the magnetic axis of poles. The $\boldsymbol{B}_{\mathrm{dm}}$ is in the opposite direction with respect to the pole flux when the load is inductive, and in accord at a capacitive load. In this way, by controlling the excitation current I_m so that to keep $\boldsymbol{B}_{dm}$ constant (to reduce the voltage fluctuations) we obtain the same picture of the vector $\boldsymbol{S}$ lines as in Figure 3.7a. The regulation of the excitation current I_m does not therefore affect the movement of active power.

At an ideal active load (Figure 3.7d), the field lines start to "pull" rotor (like springs) after stator field—in motor, or to "pull" stator field after rotor—in generator. As a result, the vector $\boldsymbol{S}$ obtains an additional radial component toward rotor, or vice versa, respectively.

By loading a generator at constant excitation $\boldsymbol{B}_{m0} = \boldsymbol{B}_{\delta m} = \text{const}$, we cause a resultant rotation of the effective flux density vector $\boldsymbol{B}_{me}$ by the angle γ with respect to the pole axis and increase of flux density in the gap according to dependence

$$\boldsymbol{B}_{me} = \frac{\boldsymbol{B}_{\delta m}}{\cos\gamma} \tag{3.22}$$

Ignoring details, we can note that the angle γ is not much smaller than the *angle of load θ of synchronous machine* [1.18]. The biggest difference at active load can be determined from the ratio $\sin\gamma/\sin\theta \approx x_a/x_s$.* The Poynting vector $\boldsymbol{S}_1$ will be turned by the same angle, since the vector $\boldsymbol{E}_0$ will not change its position nor value.

Modulus of the normal component of the Poynting vector $\boldsymbol{S}_1$ toward armature surface (Figure 3.7d) is

$$\boldsymbol{S}_n = S_0\, tg\,\gamma = \frac{B_{mp}^2}{\mu_0}\cdot\frac{\omega D}{2} tg\,\gamma \cos^2(\omega t - \alpha) \tag{3.23}$$

and is the measure of the power transferred from the air gap to armature (stator) per its surface unit.

The total power flux entering the stator at a constant excitation is

$$P = \int_0^{2\pi} S_n l \frac{D}{2}\,d\alpha = \frac{B_{mp}^2}{\mu_0}\omega\frac{D^2 l}{4}\,tg\gamma\int_0^{2\pi}\cos^2(\omega t - \alpha)\,d\alpha = \frac{B_{mp}^2}{\mu_0}\pi\omega\frac{D^2 l}{4}\,tg\gamma \tag{3.24}$$

Formula (3.24) offers the possibility of *rapid* determination of main dimensions of the machine, or the power which can be obtained from the object with given dimensions.

EXAMPLE

Examine what active power can be obtained from a turbine-generator with a rotation speed of 50 rev/s, 50 Hz, and main dimensions at: $l = 4.7$ m, $D = 1.31$ m,

---

* x_a/x_s—the ratio of reactances of armature reaction and synchronous one, respectively. In turbogenerators of 120–300 MW it is around 0.9 [1.18].

$\delta_{air} = 80$ mm, working at a constant excitation corresponding to no-load flux density in the gap (δ) $\boldsymbol{B}_{\delta m} \equiv \boldsymbol{B}_{mp} = 1.06$ T, at the load angle in the range $\theta = 10$–$13°$.

Solution

In the first approach, we can adapt tg $\gamma \approx tg\ \theta = 0.176$–$0.231$. From formula (3.24), we obtain

$$P = \frac{B_{\delta m}^2}{\mu_0}\pi\omega\frac{D^2 l}{4} tg\,\gamma = \frac{1.06^2}{4\pi \cdot 10^{-7}}\pi \cdot 314\frac{1.31^2 \cdot 4.7}{4}(0.176 \text{ to } 0.231)$$
$$= 313 \text{ to } 410.6\,\text{MW}$$

Evidently, windings and magnetic cores should be designed for this power.* The resulting flux density in the gap would be varying in this situation within the limits (3.22)

$$\boldsymbol{B}_{me} = \frac{B_{\delta m}}{\cos\gamma} = \frac{1.06}{0.985 \div 0.974} = 1.076 \text{ to } 1.088\,\text{T}.$$

The power rotating together with the rotating field in the air gap of a generator at no-load, according to dependence (3.21), is

$$P = \frac{B_{\delta m}^2}{\mu_0}\omega\frac{Dl\delta}{2} = \frac{1.06^2}{4\pi \times 10^{-7}}314\frac{1.31 \cdot 4.7 \cdot 0.08}{2} = 69 \times 10^6 \text{ var} = 69\,\text{M var}$$

Using a similar method, one can carry out an analogical calculation for constant resultant flux density in the gap, determine electromagnetic torque, and so on.

A similar analysis can be carried out for other types of machines, for instance electrostatic generators (see [1.15]).

Dynamics of turbogenerator operation can be investigated on the basis of the equation relating the electrodynamic and electromechanical energy conversion

$$\oint_A \boldsymbol{S} \cdot dA = \frac{dW_{acc}}{dt} + M\frac{d\varphi}{dt} \tag{3.25}$$

where $\int_V \boldsymbol{S} \cdot dA = \pi D L \cdot E_0 H_{tq}$, in which $H_{tq}(t) = 1/\mu_0 \boldsymbol{B}_{tq}(t)$ is the quadrature component of the field in the gap, W_{acc}—the energy accumulated in electromechanical system, $M = M(t)$—the shaft output torque, φ—the angle of shaft rotation.

* It is now a routine job for the specialists [1.3], [1.18], [1.16], [2.3], [2.5], [4.11], [4.32], [5.7], [5.12], [6.3], [6.5], [8.14], [8.15], [8.23], [10.3].

4 Screening of Constructional Parts

4.1 TYPES AND GOALS OF SCREENING AND SHUNTING

One of the most effective methods to tackle harmful thermal and power loss effects of big concentrations of leakage fields, affecting constructional elements of electric machines and power equipment, is *electromagnetic* and *magnetic screening* (shunting).

As *screens*, in a broad understanding of this word, we shall call metallic barriers in the form of walls, metal meshes, short-circuited turns, screening magnetic shunts, and so on, whose purpose is protecting regions against external electric or magnetic field, as well as protecting the surrounding environment from the field escaping outside a certain region.

From the point of view of applications, the screens can be divided into *electrostatic, magnetic* (including laminated shunts), and *electromagnetic*. The first ones (*electrostatic screens*) have a broad application in high-voltage technology and high frequency. They will not be, therefore, discussed in this book due to lack of space.

Magnetic screens are usually applied as housings or shunts made of laminated steel at AC, or of solid magnetic material at DC. Their objective is to carry away the magnetic flux from a protected region or to redirect the flux into a defined path.

Electromagnetic screens are formed by metal housing, walls, or short-circuited turns, which push away the alternating magnetic flux on the principle of eddy-current reaction. A special case of such screens are eddy-current concentrators of the field, based on the idea illustrated in Figure 8.4a (the author's logo for the ISEF Conference), applied in Japan for electric railway by Professor K. Bessho [1.29].

One can distinguish *lengthwise screens* (along the flux lines) and *crosswise screens*—perpendicular to the penetrating flux.

From the point of view of screening effectiveness, most screens can be divided into *closed*, *open*, or *half-closed* for magnetic or electromagnetic field.

The *closed screens* are such that magnetic or electromagnetic field can reach the opposite side of screen exclusively by penetrating through its uniform wall.

The *open screens* are where the main part of a magnetic or electromagnetic field has a possibility to reach the opposite side of the screen mainly by another way, for example, by encircling its wall.

The half-closed screens are those in which a magnetic or electromagnetic field reaches the opposite surface of screen by both ways discussed above.

This division has a fundamental significance when considering the design of screening circuits. Designers should, first of all, make sure that their screening system will belong to the closed screens, because in another case, it can have a small screening effectiveness. A screen closed in the magnetic or electromagnetic sense does not necessarily have to be closed mechanically. It depends mainly on the configuration of the external field on the surface of screened area and on the position of the screen not mechanically closed. Sometimes, mistakenly located screens or magnetic shunts may not weaken the threat, but actually they may significantly increase it (Figure 10.5).

Figure 4.1 shows a cylindrical magnetic screen laminated with iron sheets of $\mu \gg \mu_0$, placed in an AC magnetic field, in two different positions. In the case (a) it is a closed screen, where there is no component of magnetic field $\boldsymbol{H}$ tangential to the circle in gap, due to which the Poynting vector (2.10) in the gap $\boldsymbol{S} = 0$ and the field in the screened area is very low. The flux of power, S_P, can enter the screened area only in the case (b)—*half-closed* screen.

A typical example of a different action (as closed or open) of the same cylindrical electromagnetic screens are high-current bars of large turbine generators, in situations when these screens are isolated (Figure 4.33 [later in the chapter]) and not connected electrically at the ends with housings of a generator and generator transformer.

For the external field from the other bars, the considered tubular screen is electromagnetically closed (Figure 4.34a [later in the chapter]). For the field of its own bar, however, this tube is only an open, apparent screen, since independently on its thickness, on the external surface of the screen will flow practically the same eddy currents (Figure 4.34b [later in the chapter]) and the same field will appear as on the internal surface of the tube. This phenomenon explains, among others, the possibility of the existence of external fields of adjacent bars in spite of the presence of impenetrable screens.

The closed, that is, correct, electromagnetic screens can then be divided into *penetrable* ("transparent") and *impenetrable* ones.

The *penetrable* screens are electromagnetic screens where an electromagnetic field incident on the screen from one side, after penetration to the opposite wall, has still a large enough value (Figure 4.7 [later in the chapter]) that it cannot be ignored from the technical point of view. From this immediately follows the definition of

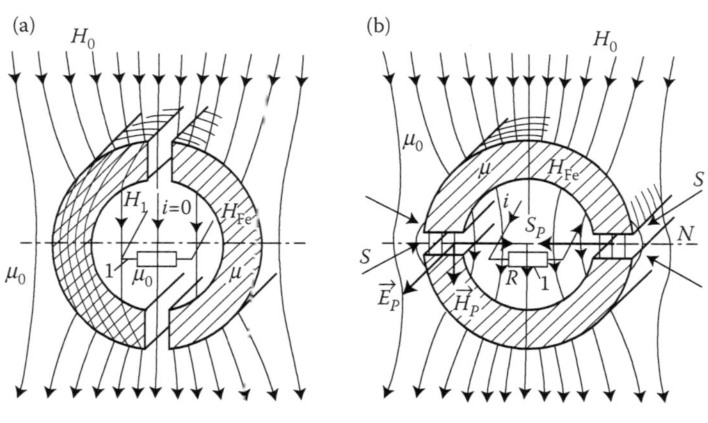

FIGURE 4.1 A magnetic screen in external magnetic field H_0—the screen type depends on its position with regard to the field: (a) closed, (b) half-closed; 1—the power received from the AC field by the circuit closed through the resistance R.

impenetrable screens, which in theoretical considerations can be substituted by an infinite half-space (Figure 2.11).

Electromagnetic screens, in a general sense, also include metal walls where electromagnetic waves impinge from both sides (Figure 4.11 [later in the chapter]). This type of incorrect screen is, for instance, a transformer cover for an electromagnetic field of bushings (Figure 3.2) or a tube where a circuit flows inside (Figure 4.34b [later in the chapter]). Instead, tank walls of power transformers are correct for use as screens.

Problems of screening have been broadly discussed in the telecommunication literature [4.1], [4.2], [4.37]. However, there exist some differences between the problems in the high-frequency technologies and weak fields, and similar problems in the design and construction of large electric machines and power transformers with strong fields. In the former case, such problems like power losses, heating, iron saturation, and so on are ignored, as less important; in the latter case, they have the most important significance. Different also is the role of electromagnetic forces and geometric dimensions due to large differences in the field strength, wavelengths, and their frequencies.

For these reasons, the theory and practice of screening in some cases requires renewed elaboration for the need of building and testing of large power apparatus.

4.2 MAGNETIC SCREENS

4.2.1 Spherical and Crosswise Cylindrical Screens

Measurement apparatus and other small devices can be screened from external magnetic fields, and contrariwise, with the help of closed housings made of magnetic material. In such a case, lines of the external magnetic fields tend to pass by the way of the lowest magnetic resistance (reluctance) and are concentrated inside the screen walls, almost not penetrating into the internal region. Hence, the principle of building such screens is to avoid air gaps situated across the path of flux in the screen. One can apply, though, gaps along the path of flux. In the case of AC magnetic fields, magnetic screen should be laminated in the direction of lines of the magnetic field (Figure 10.5). It is because if the screen was made of solid iron, the effective thickness of the screen would be reduced only to a thin layer of depth of penetration of electromagnetic field into the iron (Table 2.1). It could also be assisted by an additional electromagnetic screening effect from eddy currents, but it would be accompanied by an increase of power losses in the screen.

The calculation of a closed cylindrical screen in a steady-state (DC) field is done similar to the calculation of a closed spherical screen.

Such an *empty iron sphere*, placed in a uniform, permanent magnetic field (Figure 4.2), can be notionally considered as two spheres placed inside each other—the bigger one with permeability μ and radius R_2 and the smaller one with μ_0 and radius R_1. As it is known, a sphere made of a material different from the environment, placed in a uniform magnetic field, undergoes magnetic polarization. Due to that, apparently two displaced spheres, of positive and negative space "magnetic charges," are created (Figure 4.2a). Such spheres in external space produce the same field as two heteropolar "magnetic charges" placed in the center and creating a magnetic dipole. The displacement d in Figure 4.2a has a dimension of the order of one particle.

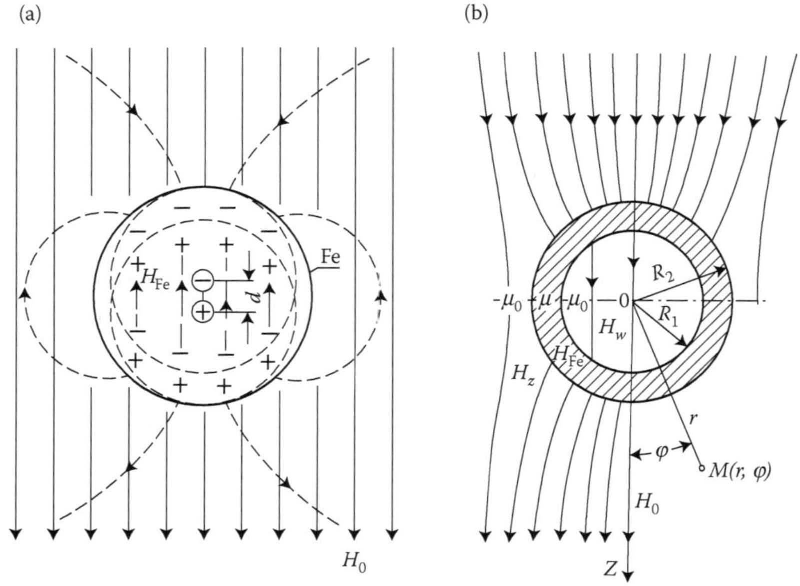

FIGURE 4.2 Principle of calculation of a spherical magnetic screen: (a) polarization of sphere in a uniform field; (b) cross section of the screen.

Hence, the field outside the screen is a result of superposition of the field of a certain equivalent dipole placed in the center of the screen, and the external uniform field. Sequentially, the field inside a uniform sphere placed in a uniform field is also a uniform field (Figure 4.2a). Therefore, the field in the screen wall will also consist of the uniform field and the field of the equivalent dipole placed in common center of both spheres.

And, finally, the field in the internal region of screen, H_w, is also a uniform field resulting from the superposition of internal fields of polarization of both spheres on the external field. Utilizing these principles and the magnetic scalar potential (2.24a), a formula has been developed (J. Turowski [1.15/1]) for the magnetic field strength H_w inside a *spherical* magnetic screen placed in the permanent uniform field H_0:

$$H_w = \frac{H_0}{1 + \frac{2}{9}\left(1 - \frac{R_1^3}{R_2^3}\right)\left(\frac{\mu_0}{\mu} + \frac{\mu}{\mu_0} - 2\right)} \approx \frac{H_0}{1 + \frac{2}{9}\left(1 - \frac{R_1^3}{R_2^3}\right)\mu_r} \tag{4.1}$$

At calculation of the magnetic field strength inside a crosswise cylindrical screen, we consider the circles in Figure 4.2b as traces of a crosscut of two coaxial cylinders, and then the magnetic field strength H_w inside a crosswise *cylindrical* iron screen placed in the permanent uniform field H_0

$$H_w = \frac{H_0}{1 + \frac{1}{4}\left(1 - \frac{R_1^2}{R_2^2}\right)\left(\frac{\mu_0}{\mu} + \frac{\mu}{\mu_0} - 2\right)} \approx \frac{H_0}{1 + \frac{1}{4}\left(1 - \frac{R_1^2}{R_2^2}\right)\mu_r} \tag{4.2}$$

The simplified form in formulae (4.1) and (4.2) concerns the practical cases when $\mu \gg \mu_0$. As it follows from comparison of these formulae, a spherical screen of the same ratio R_2/R_1 is better than a cylindrical screen.

As a result of the transformation of formula (4.2), we can find the needed thickness d (Figure 4.3) of cylindrical screen, in dependence on the internal radius $R_1 = R_w$ and the required percentage of reduction of the field ($h_w = H_w/H_0 \times 100\%$):

$$\frac{d}{R_1} = \frac{1}{\sqrt{1 - \frac{4}{\mu_r} \cdot \frac{100 - h_w}{h_w}}} \tag{4.2a}$$

As one can see from Figure 4.3, for the reduction of the internal field strength to a few percent, at the screen diameter D = 5 through 10 cm and $\mu_r = 1000$, it is enough to apply a screen of a few millimeters thick. However, next reduction, below 2–3% of the external field, is at a cost of significant and fast increase of the screen thickness.

The bigger the diameter of the tube, the bigger the thickness necessary to obtain the same field reduction. To reduce the external field to about 5% in the tube of diameter 1 m, the thickness of the tube wall should be about 2 cm, and to 2%—about 6 cm. These examples show how difficult and expensive is the screening of space by means of magnetic screens placed crosswise the field. However, the same does not apply to screens placed along the field lines.

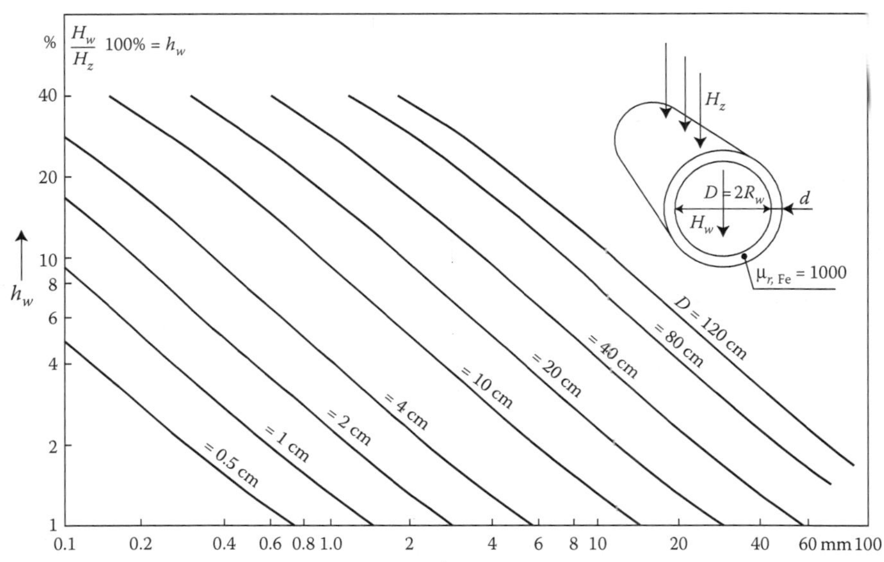

FIGURE 4.3 Effectiveness of magnetic screening as a function of the wall thickness d and the diameter D of crosswise cylinder. (After Turowski J.: *Calculations of Electromagnetic Components of Electric Machinery and Equipment.* (in Polish). Warsaw: WNT, 1982; and (in Russian). Moscow: Energoatomizdat, p. 94, 1986 [1.16/1].)

The influence of the permeability μ_r of a crosswise screen (4.2a) explains how the sort of steel ($\mu_r = 2000–1000$) used for the screen has a small significance, but its strong saturation ($\mu_r = 100$) is significant.

Crosswise magnetic screens made of laminated shunts are often placed under pressing yoke beams of large power transformers (Figure 4.18 [later in the chapter]) and in the end region of turbine generators (Figure 4.27 [later in the chapter]). However, at the same time, they play a role of parallel screens, turning the leakage flux.

4.2.1.1 Magnetic Screening of a Double-Conductor Line

An inverse job to Figure 4.2, that is, magnetic shielding of some environment from the field of a double-conductor line (Figure 4.4), can be resolved utilizing the magnetic vector potential (2.60) for two parallel conductors:

$$A_z = \frac{\mu I}{4\pi}\int_{-\infty}^{+\infty}\left(\frac{1}{r_1}+\frac{1}{r_2}\right)dz = \frac{\mu I}{2\pi}\ln\frac{r_2}{r_1} \tag{4.3}$$

In case of potential fields, in Equation 2.151, the right-hand side disappears and the equations on the left-hand side are resolved for $A(r, \theta)$ by substitution of the so-called "cylindrical harmonics" $A(r,\theta) = R(r)\cdot\Theta(\theta)$ and application of the method of separation of variables ([4.3], p. 73), similarly as in Equation 2.144.

After transformations and taking into account boundary conditions, in Ref. [4.3] (p. 287), there were developed solutions for regions 1, 2, 3, shown in Figure 4.4.

For the external field it was obtained, among others

$$A_{3z} = \frac{4\mu I}{\pi}\sum_{n=0}^{\infty}\frac{1}{(2n+1)}\left(\frac{c}{Q}\right)^{2n+1}\frac{\cos(2n+1)}{(\mu_r+1)^2-(\mu_r-1)^2(a/b)^{4n+2}} \tag{4.4}$$

from where

$$B_\rho = \frac{1}{\rho}\frac{\partial A_z}{\partial\Theta} \quad \text{and} \quad B_\Theta = -\frac{\partial A_z}{\partial\rho} \tag{4.5}$$

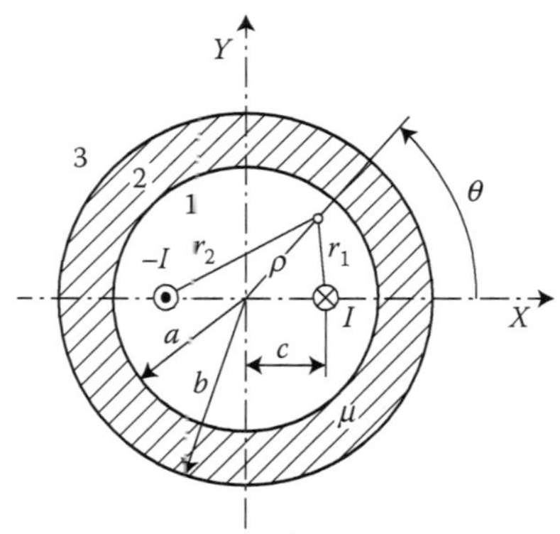

FIGURE 4.4 Magnetic screen of a double-conductor line.

4.2.2 Longitudinal Magnetic Screens

In power transformers, more often than crosswise screens, longitudinal magnetic screens are used along the lines of the magnetic field. Such screens serve the weakening of the field in the region adjacent to the screen, for instance, in a tank wall (Figure 4.18 [later in the chapter]). Calculation of such a field can be conveniently carried out with the help of the well-known method of magnetic circuits and *equivalent reluctance network*.

Let us consider a simple case, when an infinitely long solid steel cylinder of diameter D, significantly bigger than the depth of AC electromagnetic field penetration into steel (D > λ), is screened by layers of transformer sheets of thickness 0.35 mm (Figure 4.5a). These sheets in the package have a break in their circumference in order to cut the path for circular eddy currents, which could needlessly heat the screen. Let us investigate the effect of such magnetic screening in a case when, on the cylinder is wound an excitation winding with flow (ampere-turns) equal to $F_{m1} = \sqrt{2}IN_1$ per unit length ($\underline{I}_m = I_m e^{j\omega t}$). We can model it by the equivalent network presented in Figure 4.5b, from which

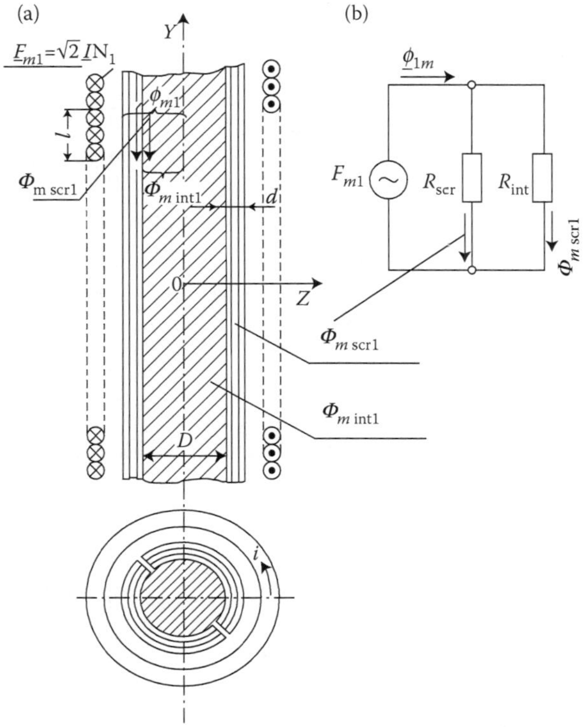

FIGURE 4.5 Longitudinal magnetic screen (shunt) of a solid steel (internal) part; $\underline{F}_{m1} = \sqrt{2}\underline{I}N_1$—maximum in time magnetomotive force (ampere-turns) per unit length (A/m); I—rms current; N_1—number of turns per meter; $\underline{\Phi}_{m1} = \underline{\Phi}_{m,scr1} + \underline{\Phi}_{m,int1}$—total maximum in time magnetic flux (Wb); $\underline{\Phi}_{m,scr1}, \underline{\Phi}_{m,int1}$—components of total flux in screen (shunt) and internal steel part, respectively; R_{scr1}—per-unit reluctance ($1/H$) of magnetic screen (laminated shunt); R_{int1}—per-unit reluctance ($1/H$) of the internal part made of solid steel.

$$\underline{\Phi}_{m\,\text{int}1} = \frac{F_{m1}}{\underline{R}_{\text{int}}};\quad \underline{\Phi}_{m\text{scr}1} = \frac{F_{m1}}{R_{\text{scr}}};\quad \frac{\underline{\Phi}_{m\,\text{int}1}}{\underline{\Phi}_{m1}} = \frac{R_{\text{scr}}}{\underline{R}_{\text{int}} + R_{\text{scr}}} \tag{4.6}$$

$$\underline{\Phi}_{m1} = \underline{\Phi}_{m,\text{scr}1} + \underline{\Phi}_{m\,\text{int}1};\quad R_{\text{scr}} = \frac{1}{\mu_{\text{scr}} d} \tag{4.7}$$

where R_{scr} is the reluctance (magnetic resistance) of the magnetic screen (shunt), in 1/H, per unit of length and width of the path of the flux $\Phi_{m\,\text{scr}1}$.

At diameters much bigger than the wavelength λ in metal (Table 2.1), the radius of curvature does not play a significant role, and the reluctance of steel cylinder can be calculated like for a solid metal half-space (Figure 2.10). The complex *magnetic flux* in a *solid metal half-space*, per 1 m along and across the flux path, according to Equation 2.176 is

$$\underline{\Phi}_{m\,\text{int}1} = \mu_{\text{int}} H_{ms\,\text{int}1} \int_0^{\infty} \mathrm{e}^{-\alpha z}\, \mathrm{d}z = \frac{\mu_{\text{int}} H_{ms\,\text{int}}}{\alpha} = \frac{H_{ms\,\text{int}}}{(1+j)(k/\mu_{\text{int}})} \tag{4.8}$$

and its modulus $\Phi_{m\,\text{int}1} = \sqrt{\mu/\omega\sigma}\ |H_{ms}|$.

Since at sufficiently long solenoid, $H_{ms\,\text{int}} = \sqrt{2} I N_1 = F_{m1}$, considering Equations 4.6 and 2.139, we obtain the *per-unit reluctance of solid conductor*:

$$R_{\text{internal}} = \frac{H_{ms\,\text{int}}}{\underline{\Phi}_{m\,\text{int}1}} = (1+\mathrm{j})\sqrt{\frac{\omega\sigma}{2\mu_{\text{int}}}} \tag{4.9}$$

The nonlinear permeability of steel can be taken into account as in Equation 2.186a. The reluctance of a solid conductor is a complex value, because it contains in itself the reactive magnetic reluctance R_{m1r} caused by the oppositely directed magnetomotive force F_{eddy} (reaction) of eddy currents. It is similar to the electric reactance X in electric circuits, which is caused by the electromotive force of self-inductance E_S.

The ratio of fluxes is equal to the inverse of the ratio of absolute values of both determined reluctances, that is

$$\frac{\underline{\Phi}_{m\,\text{scr}1}}{\underline{\Phi}_{m\,\text{int}1}} = \frac{|R_{\text{int}}|}{R_{\text{scr}}} = \mu_{\text{scr}}\sqrt{\frac{\omega\sigma}{\mu_{\text{int}}}}\, d = cd;\quad \left(c = \mu_{\text{scr}}\sqrt{\frac{\omega\sigma}{\mu_{\text{int}}}}\right) \tag{4.10}$$

where d is the thickness of the magnetic screen laminated from n electric steel sheets.

The reluctance of a solid steel wall is on the order of the reluctance of one transformer sheet if it is not saturated over the value corresponding to the relative permeability of about $\mu_r = 1000$.

The power losses in a solid steel element, according to Equation 3.10a, are determined by the magnetic field strength $\boldsymbol{H}_{ms\,\text{int}}$ on its surface (of internal solid part) or the magnetic flux Φ_{int} existing in a small surface layer, proportional to $\boldsymbol{H}_{ms\,\text{int}}$. Since inside a laminated screen the eddy currents are very weak, the magnetic field strength $\boldsymbol{H}_{m\,\text{scr}}$ is the same in the entire cross section of the screen and the same as on its surface ($\boldsymbol{H}_{ms\,\text{int}} = \boldsymbol{H}_{m\,\text{scr}} = \boldsymbol{H}_{ms\,\text{scr}} = \boldsymbol{F}_{m1}$). Considering formulae (4.6), we can hence formulate the following principles for magnetic screens:

1. If the flow F_{m1} acting along the investigated element is *constant*, the losses in the solid steel element after magnetic screening (shunting) *will be the same* as they were before screening. The magnetic screen therefore does not give any effect, except increasing the resultant flux $\boldsymbol{\Phi}_{m1}$ (Figure 4.5).
2. If the total flux $\boldsymbol{\Phi}_{m1}$ in the investigated core *is the same* after magnetic screening as before the screening, the losses in the screened solid steel wall will decrease in relation determined by the *power coefficient of magnetic screening* (of internal solid part, i.e., on the inner wall of the transformer tank):

$$p_m \approx \left|\frac{\underline{\Phi}_{\text{int}1}}{\underline{\Phi}_1}\right|^2 = \left|\frac{R_{\text{scr}}}{R_{\text{scr}} + \underline{R}_{\text{int}}}\right|^2 = \left|\frac{1}{1 + (1 + \mathrm{j})\mu_{\text{scr}} d\sqrt{\frac{\omega\sigma}{2\mu_{\text{int}}}}}\right|^2 \tag{4.10a}$$

It is the ratio of the losses in the magnetically screened (shunted) element to the losses in the not screened solid steel element placed in the magnetic field parallel to its surface.

After substitution to the above formula the symbols from Equation 4.10 and the determination of the modulus of the complex fraction, we can find *the power coefficient of magnetic screening* (Figure 4.6):

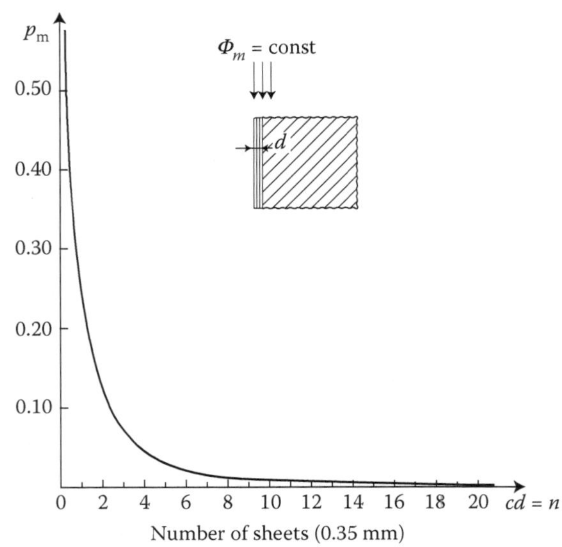

FIGURE 4.6 The magnetic screening (shunting) power coefficient, p_m, versus the number n of transformer iron sheets (0.35 mm thick) in a longitudinal magnetic screen (shunt).

$$p_m = \frac{2}{\left(\sqrt{2}+cd\right)^2 + c^2d^2} = \frac{2}{\left(\sqrt{2}+n\right)^2 + n^2} \approx \frac{1}{n^2} \tag{4.11}$$

where d is the thickness of screen (shunt), and cd is the dimensionless coefficient (4.10) approximately equal to the number n of flat sheets in the screen (shunt) of transformer steel of thickness 0.35 mm (Figure 10.5/3).

Formula (4.11) can be approximately applied both to a laminated screen that consists of sheets packaged of n layers of transformer sheets (Figure 10.5/3), parallel to the screened surface, as well as to a screen created from strips of thickness d positioned perpendicularly (Figure 10.5/1) to the screened wall. This second screen (shunt) is evidently better.

For technical purposes, in the first approach, we assume $\mu = \text{const}$ and $\sigma = \text{const}$.

Assuming next, as an example, the values

$$\sigma_{\text{int}} = 7 \times 10^6 \text{ S/m};\ \mu_{\text{scr}} = 1200 \times 4\pi \times 10^{-7} \text{ H/m};$$

$$\mu_{\text{int}} = 500 \times 4\pi \times 10^{-7} \text{ H/m}$$

we obtain $c = \mu_{\text{int}}\sqrt{\omega\sigma/\mu_{\text{int}}} = 2.85 \times 10^3$ 1/m $\approx$ 0.3 mm.

Hence, even at a screen consisting of one transformer sheet of thickness 0.35 mm ($d = 0.35 \times 10^{-3}$ m), both reluctances (magnetic resistances) are similar to each other ($cd \approx 1$), and in weaker fields, a noticeable screening effect should already appear. At one sheet ($cd \approx 1$) and $\mu_{\text{scr},r} = 1200$, the losses after such screening will be

$$p_{\text{m\%}} = \frac{2}{2.41^2 + 1^2} 100\% = 30\%$$

From Figure 4.6, it follows that for reduction of losses in a solid steel element to about 3%, at permanent value of maximum flux Φ_{m1}, it is enough to use 5–6 transformer sheets, providing they will not be affected by excessive saturation, that is, if their $\mu_{\text{scr},r} \geq 1000$. However, one has to remember that formula (4.11) corresponds to ideal screening, when the magnetic flux is parallel to the surface of screen (shunt). In real systems, the AC leakage flux (e.g., in transformers) or its parts often penetrate the screen also perpendicularly to its surface, which in case of sheets placed parallel to the surface causes a significant increase of p_m due to the induced eddy-current reaction—see Section 4.5.

4.3 ELECTROMAGNETIC SCREENS: WAVE METHOD OF MODELING AND CALCULATION

4.3.1 Practical Convenience at Application of Wave Method of Calculation

Electromagnetic screens can be calculated with different methods. It follows from various interpretations of the fundamental Equation 2.133. A well-known and popular method is based on the application of the formal analogy of wave impedances and wave propagation in long transmission lines and electromagnetic screens (Turowski [1.16], p. 84). The advantage of this method is the possibility of easy modeling of processes in metals with the help of electric circuits, and the disadvantage is mathematical formality with no clear physical representation. Also, the direct method of solution of the differential Equation 2.135 and substitution of boundary conditions (see Section 4.10) has a formal mathematical character.

The *wave method* of modeling and the calculation of screens, proposed by J. Turowski in 1957 [4.21], [2.34], which loses nothing from mathematical accuracy of direct methods, and delivers a clear physical picture of the electromagnetic processes taking place inside metals and on their surface. Owing to it, even a simple quantitative analysis of wave processes allows the designer to draw practical conclusions for designer, without the need of cumbersome modeling and calculation. The wave method enables the determination of the field behind the screen on the basis of only wave impedances of the media, without the need to look for the boundary conditions behind the screen.

4.3.2 Penetrable (Translucent) Screen with a One-Sided Incident Wave

We shall resolve this problem with the help of the wave method of modeling and the calculation of electromagnetic wave processes in metal [2.32].

Let us consider an infinitely vast, penetrable, flat screen with a complex wave impedance of its material (2.177), Z_2, conterminous with two different media with complex wave impedances Z_1 and Z_3 (Figure 4.7). Let us assume that a *plane, polarized*, and *harmonic wave* impinges on such a screen perpendicularly from one side.

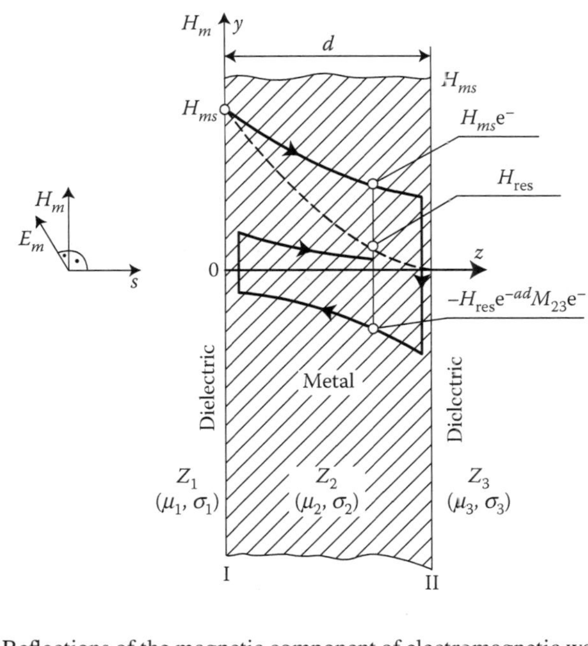

FIGURE 4.7 Reflections of the magnetic component of electromagnetic wave that impinges and penetrates a thin screen from one side.

This case is representative for such constructional parts like a transformer tank wall and closed screens made of nonferrous metals, overlaid on steel parts for loss reduction caused in them by eddy currents. The electromagnetic wave, after penetration into a conducting wall, decays inside under the influence of the eddy-current reaction, according to the dependence (2.176—see Figure 2.10)

$$H_m = H_{ms}e^{-\alpha z} = H_{ms}e^{-kz} \cdot e^{jkz} \tag{4.12}$$

$$E_m = \frac{\alpha}{\sigma} H_{ms}e^{-\alpha z}$$

In a screen of limited thickness, $d < \lambda/2$, the electromagnetic wave (of incident impedance Z_{inc}) after meeting another medium (of refraction impedance Z_{rfr}) reflects Z_{rfl} from it with the coefficient

$$M_{\text{inc-rfr}} = \frac{Z_{\text{rfr}} - Z_{\text{inc}}}{Z_{\text{rfr}} + Z_{\text{inc}}} = -\frac{H_{m\ \text{rfl}}}{H_{m\ \text{inc}}} = \frac{E_{m\ \text{rfl}}}{E_{m\ \text{inc}}} \tag{4.13}$$

In particular, from the surface II (Figure 4.7) with the coefficient

$$M_2 = M_{23} = \frac{Z_3 - Z_2}{Z_3 + Z_2} = -\frac{H_{m\ \text{rfl II}}}{H_{m\ \text{incII}}} = \frac{E_{m\ \text{rfl II}}}{E_{m\ \text{incII}}} \tag{4.13a}$$

and from the surface I, during return of the wave

$$M_1 = M_{21} = \frac{Z_1 - Z_2}{Z_1 + Z_2} = -\frac{H_{m\ \text{rfl I}}}{H_{m\ \text{incI}}} = \frac{E_{m\ \text{rfl I}}}{E_{m\ \text{incI}}} \tag{4.13b}$$

where the wave impedance of metal is expressed by the formula (2.177)

$$Z_{\text{met}} = \frac{E_m}{H_m} = \frac{\alpha}{\sigma} = (1 + j)\sqrt{\frac{\omega\mu}{2\sigma}}$$

and for dielectrics—by the formula (2.166)

$$Z_{\text{diel}} = \sqrt{\frac{\mu_0}{\varepsilon_0}} \gg |Z_{\text{met}}|$$

Since the permeability μ and the permittivity ε always appear in this case under the square root, in a first approach, one can assume that the wave impedance of iron Z_{Fe} is constant. These waves during reflections do not change the axes of vectors, but at most their sign, and therefore instead of the corresponding vectors $\boldsymbol{E}_{\text{m}}$ and $\boldsymbol{H}_{\text{m}}$ we can consider only their vector moduli.

Figure 4.7 shows schematic* envelopes of amplitudes of the real component of magnetic component of wave $H_m(z)$ penetrating through the plane $z = 0$. As it follows from formulae (2.166a) and (2.177a), if the media 1 and 3 are dielectrics, then their impedances are real and $|Z_2| \ll Z_3$ and the magnetic component reflects itself with the opposite sign ($M_1 = M_2 = +1$),† as in Figure 4.7.

Generally, the resultant magnetic field strength $H_{ms}(z)$ in the distance z from the surface at $z = 0$ is a result of superposition of the initial field $H_{ms}e^{-\alpha z}$ and the reflected fields:

$$\begin{aligned} H_m(z) = {} & H_{ms1}e^{-\alpha z} - (H_{ms1}e^{-\alpha d}M_2)e^{-\alpha(d-z)} + \\ & + (H_{ms1}e^{-\alpha d}M_2e^{-\alpha d}M_1)e^{-\alpha z} + \\ & - (H_{ms1}e^{-2\alpha d}M_1M_2e^{-\alpha d}M_2)e^{-\alpha(d-z)} + \\ & + (H_{ms1}e^{-3\alpha d}M_1M_2^2e^{-\alpha d}M_1)e^{-\alpha z} + \\ & - (H_{ms1}e^{-4\alpha d}M_1^2M_2^2e^{-\alpha d}M_2)e^{-\alpha(d-z)} + \\ & + \cdots \end{aligned} \tag{4.14}$$

Each component of this sum corresponds to a consecutive wave reflection (Figure 4.7). After grouping components of the series (4.14), as shown below, we can find its solution:

$$\begin{aligned} H_m(z) = {} & H_{ms1}[e^{-\alpha z}(1 + M_1M_2e^{-2\alpha d} + M_1^2M_2^2e^{-4\alpha d} + \\ & + M_1^3M_2^3e^{-6\alpha d} + \cdots) - M_2e^{-\alpha(2d-z)}(1 + M_1M_2e^{-2\alpha d}) + \\ & + M_1^2M_2^2e^{-4\alpha d} + M_1^3M_2^3e^{-6\alpha d} + \cdots)] = \\ = {} & (H_{ms1}[e^{-\alpha z} - M_2e^{-\alpha(2d-z)}]\frac{1 - M_1^nM_2^ne^{-2\alpha dn}}{1 - M_1M_2e^{-2\alpha dn}} \end{aligned} \tag{4.15}$$

Since the quotient of this geometrical progression $|M_1M_2\, e^{-2\alpha d}| < 1$, and the number of reflections is infinite ($n \to \infty$), the numerator of the last fraction is equal to 1. Multiplying and dividing the obtained result by $e^{-\alpha d}$, we obtain (Figure 4.8)

$$H_m(z) = H_{m1}\frac{e^{\alpha(d-z)} - M_2e^{-\alpha(d-z)}}{e^{\alpha d} - M_2M_2e^{-\alpha d}} \tag{4.16}$$

At $Z_1 \gg Z_2$ and $Z_3 \gg Z_2$, $M_2 = M_1 = +1$, from which

$$H_m(z) = H_{ms1}\frac{\text{sh}\alpha(d - z)}{\text{sh}\alpha d} \tag{4.16a}$$

---

* The correct height of ordinate is $H_{ms1}\, e^{-kz}$.

† Compare coefficient Equations 4.13a and 4.13b.

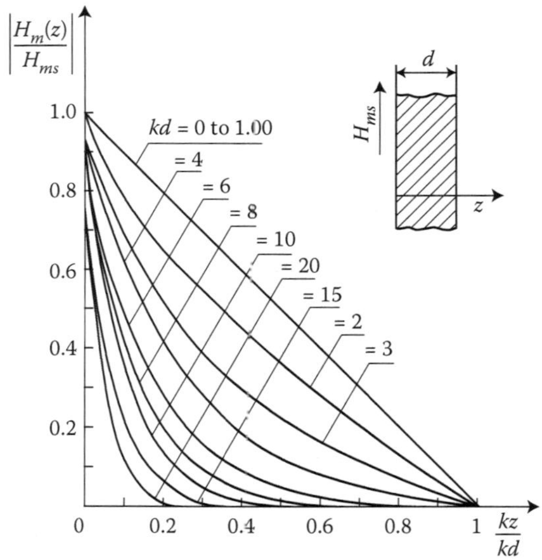

FIGURE 4.8 Magnetic field distribution, as per Equation 4.16, inside a metal wall surrounded by dielectrics, at one-sided excitation. (Adapted from Turowski J.: *Calculations of Electromagnetic Components of Electric Machinery and Equipment*. (in Polish). Warsaw: WNT, 1982; and (in Russian). Moscow: Energoatomizdat, 1986 [1.16/1].)

After substituting $z = d$ to Equation 4.16, we obtain the magnetic field strength on the opposite surface of the screen:

$$H_{ms2} = H_{ms1} \frac{1 - M_2}{e^{\alpha d} - M_2 M_2 e^{-\alpha d}} = \frac{H_{ms1}}{e^{\alpha d} - M_2 M_2 e^{-\alpha d}} \frac{2Z_2}{Z_2 + Z_3} \quad (4.17)$$

In these equations, H_{ms1} is the magnetic field strength that would exist on the surface of screen if its thickness was infinity large (practically, more than 3 equivalent depths of penetration, i.e., 3δ). In a rough approximation, H_{ms1} equals the resultant magnetic field strength on the surface of a screen with finite dimensions.

In a similar way, one can determine the resultant electric field $E_m(z)$, whose envelopes are presented schematically in Figure 4.9. If the screen is surrounded by a dielectric, the electric component according to the dependence (4.13a,b) reflects with the positive sign, as it is shown in Figure 4.9.

In this case, one can also directly utilize Maxwell's equations, by introducing expression (4.16) into Equation 2.159a and after neglecting the displacement current $\partial D/\partial t$. Then, we shall obtain

$$E_{mx} = -\frac{1}{\sigma} \frac{\partial H_{my}}{\partial z}$$

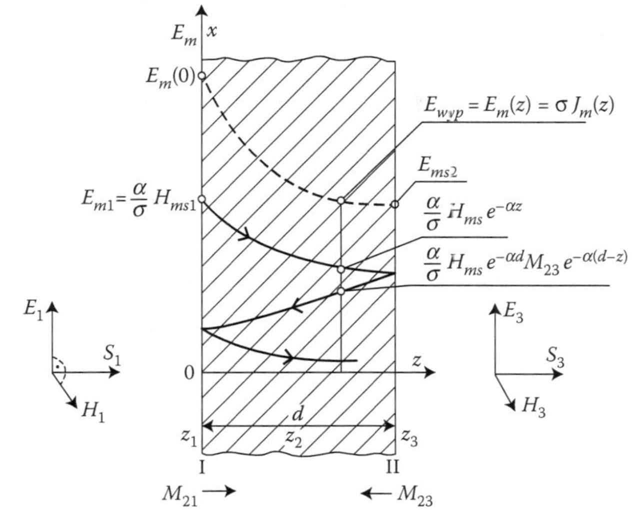

FIGURE 4.9 Reflection (interference) of the electric component of electromagnetic wave that impinges and penetrates a thin screen from one side.

from which (Figure 4.10)

$$E_m(z) = \frac{\alpha}{\sigma} H_{ms1} \frac{e^{\alpha(d-z)} + M_2 e^{-\alpha(d-z)}}{e^{\alpha d} - M_1 M_2 e^{-\alpha d}} = \frac{1}{\sigma} J_m(z) \tag{4.18}$$

Substituting $z = d$, we obtain the electric field strength on the opposite surface of the screen:

$$E_{ms2} = \frac{\alpha}{\sigma} H_{ms1} \frac{1 + M_2}{e^{\alpha d} - M_1 M_2 e^{-\alpha d}} = \frac{\alpha}{\sigma} \frac{H_{ms1}}{e^{\alpha d} - M_1 M_2 e^{-\alpha d}} \cdot \frac{2Z_3}{Z_2 + Z_3} \tag{4.19}$$

Let us consider some cases of particular screening.

At $Z_1 \gg Z_2$ and $Z_3 \gg Z_2$, $M_2 = M_1 = +1$, from which as per Equation 4.18

$$E_m(z) = \frac{\alpha}{\sigma} H_{ms1} \frac{\text{ch}\alpha(d - z)}{\text{sh}\alpha d} = \frac{1}{\sigma} J_m(z) \tag{4.18a}$$

At $d > \lambda/2$, what corresponds approximately to $d \to \infty$, formulae (4.16) through (4.18a) transition into Equation 2.176, which means that $H_m(z) = \sigma/\alpha\, E_m = H_{ms}\, e^{-\alpha z}$.

4.3.2.1 Thin Screens

Thin screens include walls where one can adopt $|\alpha| \approx \sqrt{2}kd \ll 1$, which allows to substitute exponential functions by two first components of the series

$$e^{-\alpha d} \approx 1 \pm \alpha d \tag{4.20}$$

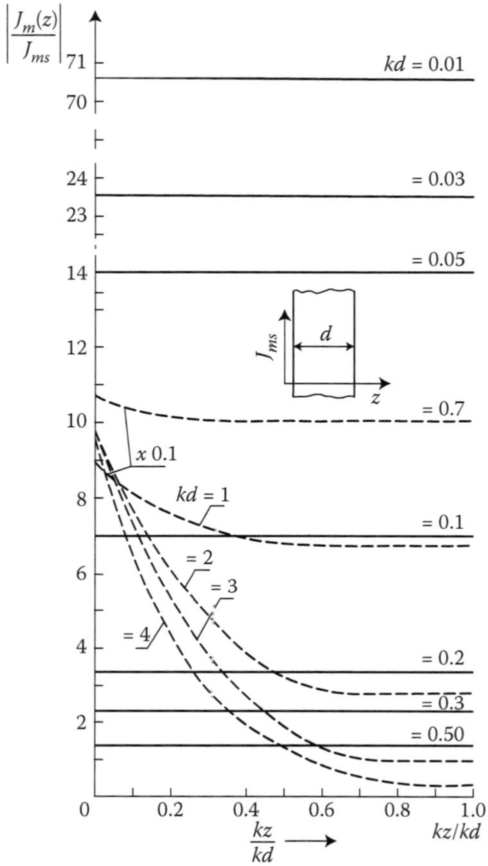

FIGURE 4.10 Distribution of the current density (4.18) inside a metal wall surrounded by a dielectric, at one-sided excitation. (Adapted from Turowski J.: *Calculations of Electromagnetic Components of Electric Machinery and Equipment.* (in Polish). Warsaw: WNT, 1982; and (in Russian). Moscow: Energoatomizdat, 1986 [1.16/1].)

For orientation, we can define typical values of the coefficient k for the cases:

- Of copper $k_{Cu} = \sqrt{\dfrac{\omega\mu_0\sigma}{2}} = \sqrt{\dfrac{2\pi 50 \times 0.4 \times 10^{-6} \times 54 \times 10^{6}}{2}} =$

$$= 104\ \text{m}^{-1} \approx 1\ \text{cm}^{-1}$$

- Of aluminum $k_{Al} = \sqrt{\dfrac{\omega\mu_0\sigma}{2}} = 82\ \text{m}^{-1} \approx 0.82\ \text{cm}^{-1}$

- Of steel $k_{St} = \sqrt{\dfrac{\omega\mu\sigma}{2}} = \sqrt{\dfrac{2\pi 50(500 \text{ to } 1000) \times 0.4 \times 10^{-6} \times 7 \times 10^{6}}{2}} =$

$$= (830 \text{ to } 1180)\ \text{m}^{-1} \approx (8.3 \text{ to } 11.8)\ \text{cm}^{-1}$$

As one can see, in the case of copper, the dimensionless coefficient *kd* is numerically equal to the thickness of copper plate in centimeters. As a thin screen, we can then consider a nonmagnetic screen of thickness of the order of millimeters or an iron screen on the order of tenth parts of millimeter.

After substituting the dependence (4.20) into Equations 4.17 and 4.19, we obtain the simplified formulae

$$H_{ms2} \approx H_{ms1} \frac{Z_2(Z_1 + Z_2)}{Z_2(Z_1 + Z_3) + \alpha d(Z_2^2 + Z_1 Z_3)} \tag{4.21}$$

and

$$E_{ms2} \approx \frac{\alpha}{\sigma} H_{ms1} \frac{Z_3(Z_1 + Z_2)}{Z_2(Z_1 + Z_3) + \alpha d(Z_2^2 + Z_1 Z_3)} \tag{4.22}$$

4.3.2.2 Thin Screens in Dielectric

If a screen is located in dielectric, then from formulae (2.166a) and (2.177a), it follows that $Z_1 = Z_3 = 377\,\Omega \gg |Z_2| = \sqrt{2}k/\sigma$.

Next, substituting the left equality to Equations 4.21 and 4.22, and neglecting Z_2 in comparison with the remaining impedances, we obtain

$$H_{ms2} \approx \frac{H_{ms1}}{2 + (377\,\Omega)\sigma_2 d} \approx \frac{H_{ms1}}{(377\,\Omega)\sigma_2 d} \tag{4.23}$$

$$E_{ms2} \approx \frac{(377\,\Omega)H_{ms1}}{2 + (377\,\Omega)\sigma_2 d} \approx \frac{H_{ms1}}{\sigma_2 d} \tag{4.24}$$

For instance, for a steel sheet of thickness 0.5 mm and the average conductivity $\sigma = 7 \times 10^6$ S/m, we obtain

$$H_{ms2} = \frac{H_{ms1}}{2 + 377 \times 7 \times 10^6 \times 0.5 \times 10^{-3}} = 0.76 \times 10^{-6} H_{ms1}$$

For the same screen, but made of copper ($\sigma_{Cu} = 54 \times 10^6$ S/m), we obtain $H_{ms2} = 10^{-7} H_{ms1}$.

The effectiveness of screening of the plane wave is therefore very high, thanks to the internal reflections of the wave. However, it must be once more emphasized that these considerations concern only plane waves.

4.3.2.3 Thin Screens on the Surface of Iron

In the case when a thin copper screen adheres by its one side to a steel plate, then from Equations 2.166a and 2.177a we conclude that $Z_1 = 377\,\Omega \gg |Z_3| = 2.4 \times 10^{-4}\,\Omega$ and $|Z_2| = 2.7 \times 10^{-6}\,\Omega$. After removing from formula (4.21) small values of low order, we obtain the approximate formula

$$H_{ms2} \approx \frac{H_{ms1}}{1 + \alpha d\sqrt{\mu_3\sigma_2/\mu_2\sigma_3}} \approx \frac{H_{ms1}}{\sigma_2 d}\sqrt{\frac{\sigma_3}{\omega\mu_3}}\, e^{-j\frac{\pi}{4}} \tag{4.25}$$

For instance, for $\sigma_2 = \sigma_{Cu} = 54 \times 10^6$ S/m; $\sigma_3 = \sigma_{Steel} = 7 \times 10^6$ S/m; $\mu_{3r} = 500$ to 1000, and $d = 5$ mm:

$$|H_{ms2}| \approx H_{ms1} \frac{1}{54 \cdot 10^6 \cdot 0.5 \cdot 10^{-3}} \sqrt{\frac{7 \cdot 10^6}{2\pi 50(500 \text{ to } 1000)0.4\pi \cdot 10^{-6}}} = \\ = (0.156 \text{ to } 0.22)H_{ms1}.$$

In this system, as one can see, the screening effectiveness is significantly smaller than in the previous system.

From the two last examples, it might be implied that the screening effectiveness can be increased by making a clearance between the screen and the steel surface. Such treatment was even made at some constructions of foreign transformers. However, a more accurate analysis (Turowski et al. [4.22]) showed that a visible influence of such an insulation gap between copper screens and steel surface is evident only at very large dimensions of gap or at high frequencies (see Section 4.10).

4.3.2.4 Thick Screens

Thick screens, that is, such screens in which $|\alpha d| \gg 1$, are rarely seen at power frequencies as proper screens of copper or aluminum. However, one can consider as such the steel plates of thickness bigger than the wavelength in iron (2.178), that is, over 4 to 5 mm.

In this sense, one can often see a thick, nonmagnetic screen in high-frequency technologies, where the wavelength λ of an electromagnetic wave in metal can be in the order of fractions of millimeters (Table 2.1). If we assume $|\alpha d| \gg 1$, we can obtain on the basis of Equation 4.17, known from literature (e.g., H. F. Schwenkhagen, 1959), the formula

$$H_{ms2} = H_{ms1} \frac{2|Z_2|}{|Z_2| + |Z_3|} e^{-kd} = H_{ms1} \frac{2\sqrt{\mu_2 \sigma_3 / \mu_3 \sigma_2}}{1 + \sqrt{\mu_2 \sigma_3 / \mu_3 \sigma_2}} e^{-kd} \tag{4.26}$$

4.3.3 Penetrable Screen at Wave Incident from Both Sides

The penetrable double-sided screens include electrostatic screens in power transformers placed in interwinding region, sheets in transformer core, thin transformer covers, flat copper, or aluminum conductors, and other constructional elements placed in an alternating electromagnetic field, where the values of this field are equal to each other on both surfaces of the body (Turowski [2.31], [2.41], [4.18]).

We can now assume that on a screen placed in a dielectric, two identical electromagnetic waves are incidents from both sides. Figure 4.11a demonstrates a graph of envelopes of amplitudes of the magnetic component H_m, and Figure 4.11b demonstrates the electric component E_m of the electromagnetic waves penetrating through the planes $z = 0$ and $z = d$.

According to formula (2.173), and the assumption that the coefficient of reflection of electromagnetic wave M (Equation 2.199) from the surface of dielectric equals +1,

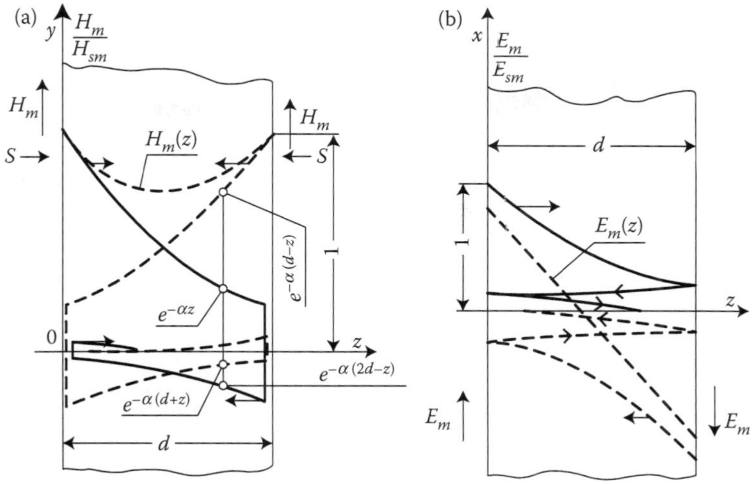

FIGURE 4.11 Internal reflections of electromagnetic waves in a thin sheet [2.31] at a double-sided symmetric excitation: (a) reflections of the H_m component; (b) reflections of the E_m component.

we conclude (Turowski [1.15/1]) that the resultant magnetic field inside the screen is a result of the superposition of a series of reflected waves (Figure 4.11).

We can take advantage of formula (4.16) and, according to the superposition principle, add to the quantity $H_m(z)$ the quantity $H_m(d-z)$. Then, at $Z_1 = Z_3 = Z_{\text{surface}}$ $(M_1 = M_2 = M)$

$$H_m(z) = \frac{H_{ms1}}{e^{\alpha d} - M^2 e^{\alpha d}} \left[e^{\alpha(d-z)} - Me^{-\alpha(d-z)} + e^{\alpha z} - Me^{-\alpha z} \right] \tag{4.27}$$

The resultant electric field strength E_m equals, instead, the difference of $E_m(z)$ (formula (4.18)) and $E_m(d-z)$, from which

$$E_m(z) = \frac{\alpha}{\sigma} H_{ms1} \frac{e^{\alpha(d-z)} + Me^{-\alpha(d-z)} - e^{\alpha z} - Me^{-\alpha z}}{e^{\alpha d} - M^2 e^{-\alpha d}} \tag{4.28}$$

If we assume, as before, that $M = +1$, then after a few transformations we obtain from the above formulae the following values (Figure 4.12):

$$H_m(z) = H_{ms1} \frac{\mathrm{ch}\,\alpha\left(\frac{d}{2} - z\right)}{\mathrm{ch}\,\alpha\frac{d}{2}} \quad \text{and} \quad E_m(z) = \frac{\alpha}{\sigma} H_{ms1} \frac{\mathrm{sh}\,\alpha\left(\frac{d}{2} - z\right)}{\mathrm{ch}\,\alpha\frac{\alpha d}{2}} \tag{4.29}$$

For a case of antisymmetric excitation, see (Turowski [1.16/1], p. 78) and Figure 4.15 (later in the chapter).

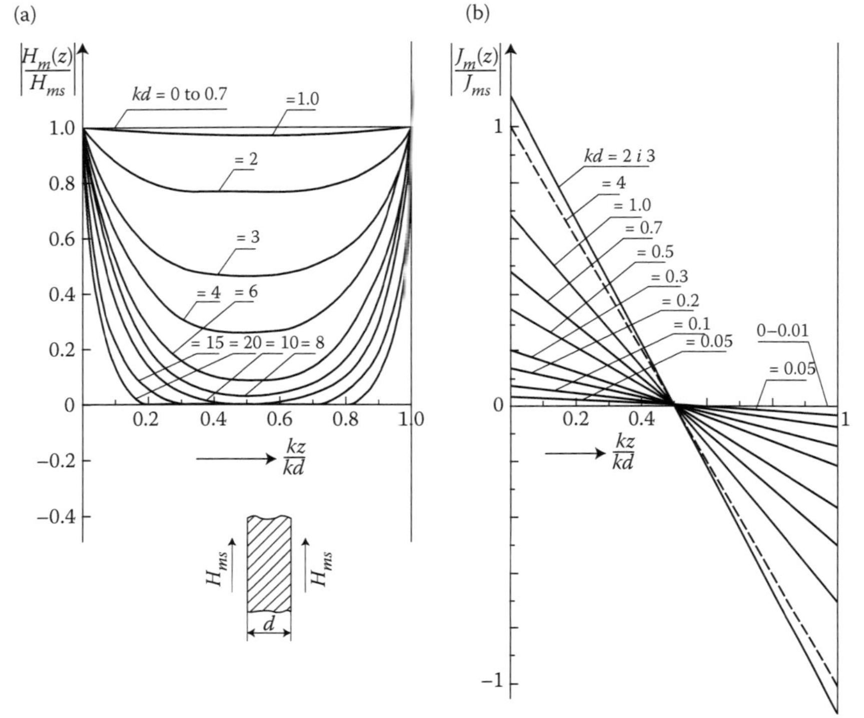

FIGURE 4.12 Distribution of (a) the field $H_m(z)$ and (b) the current density $J_m(z)$, in a screen excited symmetrically by the field H_{ms}, for instance, in a sheet in laminated transformer core. (Adapted from Turowski J.: *Calculations of Electromagnetic Components of Electric Machinery and Equipment.* (in Polish). Warsaw: WNT, 1982; and (in Russian). Moscow: Energoatomizdat, 1986 [1.16/1].)

4.4 POWER LOSSES IN SCREENS

4.4.1 Poynting Vector and Power Losses in a One-Sided Screen

Active and reactive losses in a screen are calculated on the basis of Poynting's theorem (Chapter 3) (Turowski [2.41]).

The complex Poynting vector, at the depth z from the screen surface, is equal to the vector product of the complex rms* value $\boldsymbol{E}_\mathrm{m}$ of the electric field strength (4.18) and the conjugate rms value $\boldsymbol{H}^*_\mathrm{m}$ of the magnetic field strength (4.16):

$$\boldsymbol{S}_\mathrm{s} = \boldsymbol{S}_\mathrm{p} + \mathrm{j}\boldsymbol{S}_\mathrm{q} = \frac{1}{2}(\boldsymbol{E}_\mathrm{m} \times H^*_\mathrm{m}) =$$

$$= \mathbf{1}_\mathrm{n} \frac{\alpha}{2\sigma}\left|H_{ms1}\right|^2 \frac{\mathrm{e}^{\alpha(d-z)} + M_2\mathrm{e}^{-\alpha(d-z)}}{\mathrm{e}^{\alpha d} - M_1 M_2\,\mathrm{e}^{-\alpha d}} \times \frac{\mathrm{e}^{\alpha^*(d-z)} + M^*_2\mathrm{e}^{-\alpha^*(d-z)}}{\mathrm{e}^{\alpha^* d} - M^*_1 M^*_2\mathrm{e}^{-\alpha^* d}} \tag{4.30}$$

where, according to the dependence (2.149), $\alpha = (1 + \mathrm{j})k$ and $\alpha^* = (1 - \mathrm{j})k$.

* rms = root mean square.

If the coefficients M_1 and M_2 (4.13) contain wave impedances of dielectrics, which are real, and wave impedances of metals, which are complex, the latter can be omitted in comparison with the former. If, instead, in formulae (4.13) exist only the wave impedances of metals, which all have the same arguments, then the multiplier $e^{j\pi/4}$ is eliminated and only the moduli remain. Thus, in practice, the reflection coefficients can be treated as real and instead of the complex impedances Z_1, Z_2, Z_3, we can take for calculations only their moduli.

Considering the facts mentioned above and removing irrationality in the denominator of Equation 4.30, we obtain the Poynting vector

$$S_s = S_p + jS_q = \frac{|H_{ms1}|^2}{2}\sqrt{\frac{\omega\mu}{2\sigma}}\,\frac{\left[e^{2k(d-z)} - M_2^2 e^{-2k(d-z)} + 2M_2 \sin 2k(d-z)\right] + \cdots}{} \\ \frac{\cdots + j\left[e^{2k(d-z)} - M_2 e^{-2k(d-z)} - 2M_2 \sin 2k(d-z)\right]}{e^{2kd} + M_1^2 M_2^2 e^{-2kd} - 2M_1M_2 \cos 2kd} \tag{4.31}$$

This is, at the same time, the apparent power entering the screen through a unit surface at $z = 0$. If the screen is placed in a dielectric, then practically all this power remains in the screen, in the form of power loss, or goes back as a reactive power, because, as follows from the examples to formulae (4.23) and (4.24), the reflection of the magnetic component of electromagnetic wave from the opposite surface is almost ideal and the resultant Poynting vector exiting through the opposite surface is negligibly small. It cannot be, however, considered as absolutely accurate in the case when a nonmagnetic screen adheres to the iron surface. Generally, the active power dissipated in a screen per unit surface, in the form of heat losses, can be found from the difference of the per-unit active power flux S_p entering through the plane $z = 0$ and the flux leaving through the surface $z = d$.

$$P_1 = \kappa\sqrt{\frac{\omega\mu_2}{2\sigma}}\cdot\frac{|H_{ms1}|^2}{2} \tag{4.32}$$

where

$$\kappa = \frac{e^{2kd} - M_2^2 e^{-2kd} + 2M_2 \sin 2kd - 1 + M_2^2}{e^{2kd} - M_1^2 M_2^2 e^{-2kd} - 2M_1M_2 \cos 2kd} \tag{4.33}$$

and usually $\mu_2 = \mu_0$ (nonmagnetic screens).

In an analogical way, we can determine the reactive power consumed by the screen per unit surface (var/m^2):

$$Q_1 = \xi\sqrt{\frac{\omega\mu_2}{2\sigma}}\cdot\frac{|H_{ms1}|^2}{2} \tag{4.34}$$

where

$$\xi = \frac{e^{2kd} - M_2^2 e^{-2kd} - 2M_2 \sin 2kd - 1 + M_2^2}{e^{2kd} + M_1^2 M_2^2 e^{-2kd} - 2M_1 M_2 \cos 2kd} \tag{4.35}$$

Formulae (4.32) and (4.34) differ from the basic formula (3.10a) by the coefficients κ and ξ.

In practice, the most often occurring types of screening systems are (1) a screen surrounded by dielectric, and (2) a screen adhering to iron. Let us consider them separately.

4.4.1.1 Screen Surrounded by Dielectric

In the case when a screen is surrounded from both sides by dielectric (or air), in the formulae for the screening coefficient κ, we can utilize the following properties of the system:

$$Z_1 = Z_3 \gg |Z_2| \quad \text{and} \quad M_1 = M_2 = +1.$$

and then we obtain [1.15/1]:

1. For *thin* screens ($kd \ll 1$)

$$\kappa = \frac{kd}{2\left(\frac{|Z_2|}{|Z_1|}\right)^2 + 2\frac{|Z_2|}{|Z_1|}kd + (kd)^2} \approx \frac{1}{kd} \tag{4.36}$$

   As we can see, the losses in this part of the curve $\kappa(kd)$ increase with the decrease of kd, according to hyperbolic dependence. However, a more accurate formula shows that down from a certain, very small limit value ($kd_{\text{limit}} \approx \sqrt{2}|Z_2|/|Z_1|$) they again decrease to zero.

2. For *thick* screens ($kd \gg 1$)

$$\kappa = \frac{e^{2kd} + 2\sin 2kd}{e^{2kd} - 2\cos 2kd} \tag{4.37}$$

   at $kd \to \infty$, $\kappa \to 1$.

3. For *intermediate* screen thicknesses, after substituting to Equation 4.33, the values $M_1 = M_2 = +1$, we obtain the general formula comprising also two previous cases:

$$\kappa = \frac{\operatorname{sh} 2kd + \sin 2kd}{\operatorname{ch} 2kd - \cos 2kd} \tag{4.38}$$

Figure 4.13 shows a plot of the coefficient $\kappa(kd)$ for the case of a screen placed in dielectric. This graph is correct with high accuracy also for a screen adhering to iron.

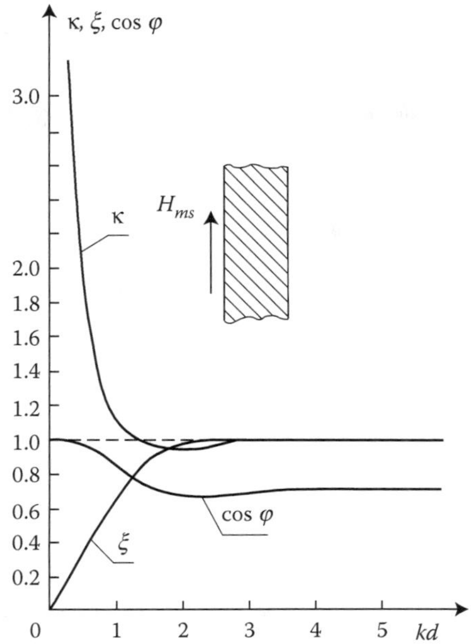

FIGURE 4.13 The screening coefficient κ of active power (4.33) and ξ of reactive power (4.35), and $\cos\varphi = \kappa/\sqrt{\kappa^2 + \xi^2}$ in a one-sided screen surrounded by dielectric. (Adapted from Turowski J.: *Calculations of Electromagnetic Components of Electric Machinery and Equipment.* (in Polish). Warsaw: WNT, 1982; and (in Russian). Moscow: Energoatomizdat, 1986 [1.16/1].)

4.4.1.2 Screen Adhering to Iron

When a screen made of copper or aluminum (Figure 4.9) protects a solid steel surface, there are valid dependences

$$|Z_1| \gg |Z_3| > |Z_2|, \quad M_1 \approx +1, \quad M_2 = \frac{|Z_3| - |Z_2|}{|Z_3| + |Z_2|}$$

According to formula (2.177a), Z_{Steel} is on the order of 100 Z_{Cu}. Hence, one can assume with a high accuracy that $M_2 = +0.98 \approx +1$.

Therefore, for *thin* screens ($kd \ll 1$) (Turowski [1.15/1])

$$\kappa = \frac{2kd + |Z_2|/|Z_3|\cdot\left((2kd)^2/2\right)}{\left(|Z_2|/|Z_3|\right)^2 + |Z_2|/|Z_3|2kd + \left((2kd)^2/2\right)} \approx \frac{1+\left(|Z_2|/|Z_3|\right)kd}{\left(|Z_2|/|Z_3|\right)+kd} \tag{4.36a}$$

When $kd \to 0$, then $\kappa \to 0$.

Assuming $|Z_2|/|Z_3| = |Z_{\text{Cu}}|/Z_{\text{Steel}} = 1/100 \ll 1$, we get $\kappa = 1/0.01 + kd \approx 1/kd$ and therefore, for not too small values of *kd*, it is a hyperbolic function similar to Equation 4.36 although with lower accuracy. For remaining screens, the same formulae (4.37) and (4.38) are valid accordingly, but with slightly lower accuracy (Turowski [1.15/1]).

Consequently, the graphs in Figure 4.13 can be accepted as valid with a sufficient accuracy also in this case.

In this system, to the losses in screen one should also add the losses in the screened steel element, which can be calculated with the help of the basic formula (3.10a) and from the field strength (4.25). Thus, in total, the losses in screen and steel per surface unit, at $kd < 1$, equal

$$\begin{aligned} P_{1\text{scr+St}} &= \kappa\sqrt{\frac{\omega\mu_2}{2\sigma_2}}\cdot\frac{|H_{ms1}|^2}{2}+\sqrt{\frac{\omega\mu_3}{2\sigma_3}}\cdot\frac{|H_{ms1}|^2}{2\sigma_2^2 d^2}\cdot\frac{\sigma^3}{\omega\mu_3} \\ &\approx \frac{1}{kd}\sqrt{\frac{\omega\mu_2}{2\sigma_2}}\cdot\frac{|H_{ms1}|^2}{2}\left(1+\frac{1}{d\sigma_2}\sqrt{\frac{\sigma_3}{2\omega\mu_3}}\right)\approx\frac{1}{2d_{\text{scr}}\sigma_{\text{scr}}}|H_{ms1}|^2 \end{aligned} \tag{4.39}$$

In the case of copper screens, assuming $\sigma_3 = \sigma_{\text{Steel}} = 7\times10^6$ S/m, $\sigma_2 = \sigma_{\text{Cu}} = 54\times10^6$ S/m, $\mu_3 = 1000\times4\pi\times10^{-7}$ H/m, we obtain the losses per surface unit:

$$P_{1\text{Cu+St}} \approx \kappa\sqrt{\frac{\omega\mu_0}{2\sigma_2}}\frac{|H_{ms1}|^2}{2}\left(1+\frac{a_d}{d}\right) \tag{4.40}$$

(Here, $a_d = 5.5\times10^{-5}$ m).

For example, at $d = 1$ mm $= 10^{-3}$ m, the second component in formula (4.40) is equal to $a_d/d = 0.055$, therefore, less than 6% of the losses in screen itself. Hence, even at small thicknesses of copper screens, *almost all losses pass from the protected steel part to the screen* where they have a low value thanks to the high conductivity of copper. Losses in the total system (screen+ steel) are just a few percent higher than the losses in the screen alone.

Screening coefficient of power The ratio of losses in the whole screened system (screen + steel) to the losses in nonscreened steel is called the *coefficient of electromagnetic screening* of power. It is equal to, as per Equation 4.32

$$p_e = \frac{\kappa\sqrt{\frac{\omega\mu_2}{2\sigma_2}}\frac{|H_{ms1}|^2}{2}\left(1+\frac{a_d}{d}\right)}{1.4\sqrt{\frac{\omega\mu_3}{2\sigma_3}}\frac{|H_{ms1}|^2}{2}}k_H = \frac{\kappa}{\sqrt{2}}\sqrt{\frac{\mu_2\sigma_3}{\mu_3\sigma_2}}\left(1+\frac{a_d}{d}\right)k_H \tag{4.41}$$

(Here, $a_d = 5.5\times10^{-5}$ m).

In formula (4.41), the magnetic permeability μ_3 should be determined from the surface value H_{ms1}, because it concerns the steel in nonscreened state. At $kd < 1$ (i.e., in the case of a copper sheet of thickness $d < 1$ cm)

$$p_e = \frac{1}{\sqrt{2}k_d}\sqrt{\frac{\mu_2\sigma_3}{\mu_3\sigma_2}}\left(1+\frac{a_d}{d}\right)k_H = \frac{1}{\sigma_2 d}\sqrt{\frac{\sigma_3}{\omega\mu_3}}\left(1+\frac{a_d}{d}\right)k_H \tag{4.42}$$

At $kd > 1$, one can accept $\kappa = 1$ and $a_d/d = 0$, from which, as per Equation 4.41

$$p_e = \sqrt{\frac{\mu_2\sigma_3}{2\mu_3\sigma_2}}k_H \tag{4.43}$$

If medium 3 is not ferromagnetic, then in formula (4.43), the coefficient 2 in the denominator under the root does not appear.

The coefficient $k_H = (H_{ms1scr}/H_{ms1})^2 \gg 1$ takes into account the change of the field configuration and a change in the H_{ms} value on the surface after overlaying the screens on the steel wall (4.68), (4.69), [4.18].

The distribution of the per-unit power losses (in W/m²) inside a screen that is excited from one side (Figure 4.7), as per Equation 4.18, is expressed by the formula

$$w(z) = \frac{1}{2\sigma}|J_m(z)|^2 = \frac{\varphi_1}{d}\sqrt{\frac{\omega\mu}{2\sigma}}\frac{|H_{ms1}|^2}{2}$$

$$\varphi_1 = 2kd\frac{\operatorname{ch}2k(d-z)+\cos 2k(d-z)}{\operatorname{ch}2kd+\cos 2kd} \tag{4.44}$$

At $d > \lambda/2$, that is, $d \to \infty$, formula (4.44) transitions into Equation 3.13.

4.4.2 Poynting Vector and Power Loss at Double-Sided Symmetric Incidence of Waves

For the determination of the apparent value of the Poynting vector in the screens described in Section 4.3.3, it is necessary to multiply the complex rms value of the electric field intensity E_{rms} (4.29) by the conjugate rms value of magnetic field intensity H_{rms} (4.29), or vice versa, accordingly to Equation 3.3 [2.31]:

$$\begin{aligned} \boldsymbol{S}_{\mathrm{S}} &= \boldsymbol{S}_{\mathrm{P}} + \mathrm{j}\boldsymbol{S}_{\mathrm{Q}} = \frac{1}{2}(\boldsymbol{E}_{\mathrm{m}} \times \boldsymbol{H}^*_{\mathrm{m}}) \\ &= \mathbf{1}_{\mathrm{n}}\frac{1}{2}\frac{\alpha}{\sigma}|H_{ms}|^2\frac{\operatorname{sh}[k((d/2)-z)+\mathrm{j}k((d/2)-z)]\cdot\operatorname{ch}[k((d/2)-z)-\mathrm{j}k((d/2)-z)]}{\operatorname{ch}(1+\mathrm{j})(kd/2)\cdot\operatorname{ch}(1-\mathrm{j})(kd/2)} \end{aligned}$$

from where

$$S_{\mathrm{S}} = \frac{1}{2}\frac{\alpha}{\sigma}|H_{ms}|^2\frac{\operatorname{sh}(kx+\mathrm{j}kx)\cdot\operatorname{ch}(kx-\mathrm{j}kx)}{\operatorname{ch}(1+\mathrm{j})(kd/2)\cdot\operatorname{ch}(1-\mathrm{j})(kd/2)} \tag{4.45}$$

where $x = \dfrac{d}{2} - z$

$$\operatorname{sh}(kx+\mathrm{j}kx)\cdot\operatorname{ch}(kx-\mathrm{j}kx) = \frac{\operatorname{sh}2kx+\mathrm{j}\sin 2kx}{2}$$

$$\alpha(\operatorname{sh}2kx+\mathrm{j}\sin 2kx) = k\,[\operatorname{sh}2kx-\sin 2kx+\mathrm{j}(\operatorname{sh}2kx+\sin 2kx)]$$

and

$$\text{ch}\left(\frac{kd}{2} + \text{j}\frac{kd}{2}\right)\text{ch}\left(\frac{kd}{2} - \text{j}\frac{kd}{2}\right) = \frac{\text{ch}\,kd + \cos kd}{2}$$

After substituting the last dependences into formula (4.45), we get

$$\boldsymbol{S}_\text{S} = \boldsymbol{S}_\text{P} + \text{j}\boldsymbol{S}_\text{Q} = \frac{\text{k}}{2\sigma}H_{ms}\Big|^2 \frac{(\text{sh}\,2kx - \sin 2kx) + \text{j}(\text{sh}\,2kx + \sin 2kx)}{\text{ch}\,kd + \cos kd} \tag{4.46}$$

In the center of the sheet, that is, at $z = d/2$ (Figure 4.11), $\boldsymbol{S}_s = 0$, hence, the whole flux penetrating into the sheet through both its surfaces decays before reaching the sheet center. Substituting $z = 0$ into Equation 4.46 and after doubling up the flux of power $\boldsymbol{S}_s$ entering through this surface, we obtain, consequently, the reactive and active power losses in the whole sheet per unit of the surface (Figure 4.14):

- The active power, in W/m²

$$P_1 = \boldsymbol{S}_\text{p} = \zeta\sqrt{\frac{\omega\mu}{2\sigma}}|H_{ms}|^2 \tag{4.47}$$

- The reactive power

$$Q_1 = \boldsymbol{S}_\text{q} = \psi\sqrt{\frac{\omega\mu}{2\sigma}}|H_{ms}|^2 \tag{4.48}$$

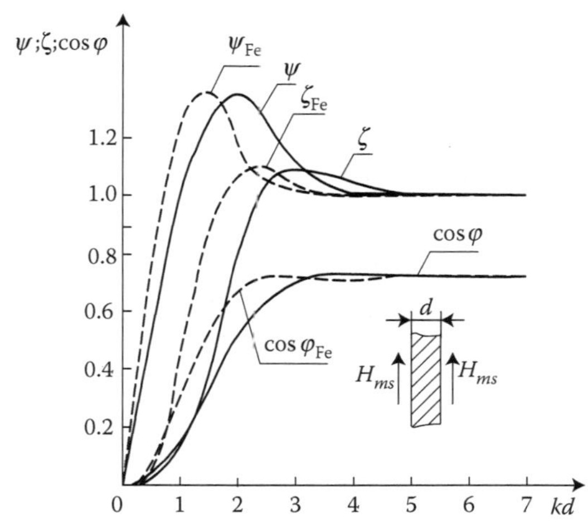

FIGURE 4.14 The coefficients: ζ—of the active power in screen (4.50), ψ—of the reactive power (4.50), and cos φ in a screen excited symmetrically at $\mu = \text{const}$ (solid lines) and in the case of a nonlinearity $\mu = \mu(H)$ (broken lines). (Adapted from Turowski J.: Calculation of losses and temperatures in the rail pipe screen. (in Polish). *Archiwum Elektrotechniki*, 1, 1963, 61–79 [2.33].)

- The electric power factor

$$\cos\varphi = \frac{S_p}{|S_s|} = \frac{\zeta}{\sqrt{\zeta + \psi^2}} \tag{4.49}$$

where

$$\zeta = \frac{\operatorname{sh} kd - \sin kd}{\operatorname{ch} kd + \cos kd} \text{ and } \psi = \frac{\operatorname{sh} kd + \sin kd}{\operatorname{ch} kd + \cos kd} \tag{4.50}$$

Simple computer programs, based on the above-described approach, allowed the development of original and rapid, hybrid, three-dimensional (3D), analytical–numerical, and network methods (ANM-3D, RNM-3D), which are tools to calculate (Turowski [1.16/1], [1.17]) not only coefficients (4.50) for symmetric excitation (Figure 4.14), but also for an antisymmetric excitation (Figure 4.15). Antisymmetric

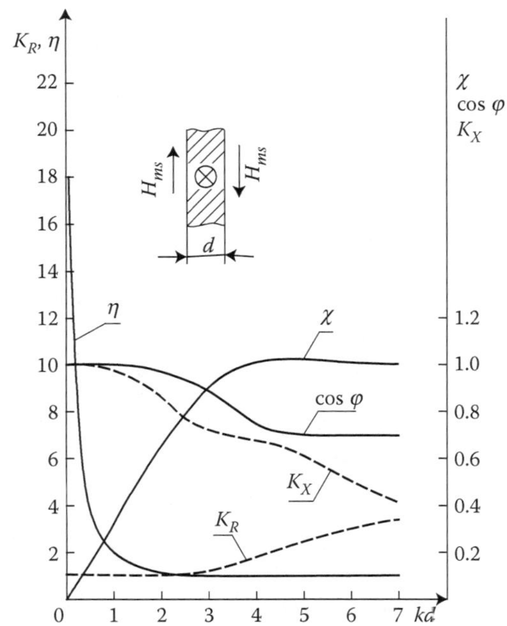

FIGURE 4.15 The coefficients: η—of the active power in screen, χ—of the reactive power, and cos φ (4.51), in a screen excited antisymmetrically at μ = const (solid lines); $K_R(kd)$, $K_X(kd)$—the coefficients of increase of resistance R and decrease of reactance X of a rectangular bar at alternating current (AC). (Adapted from Turowski J.: *Calculations of Electromagnetic Components of Electric Machinery and Equipment.* (in Polish). Warsaw: WNT, 1982; and (in Russian). Moscow: Energoatomizdat, 1986 [1.16/1].)

excitation occurs, for example, in bars conducting current, in deep-slot induction motors with the slot depth of $d/2$, and so on.

$$P_1 = \eta\sqrt{\frac{\omega\mu}{2\sigma}}|H_{ms}|^2, \quad Q_1 = \chi\sqrt{\frac{\omega\mu}{2\sigma}}|H_{ms}|^2, \quad \cos\varphi = \frac{\eta}{\sqrt{\eta^2+\chi^2}} \tag{4.51}$$

where

$$\eta = \frac{\operatorname{sh} kd + \sin kd}{\operatorname{ch} kd - \cos kd} = \frac{2}{kd}K_R \quad \text{and} \quad \chi = \frac{\operatorname{sh} kd - \sin kd}{\operatorname{ch} kd - \cos kd} = \frac{kd}{3}K_X \tag{4.52}$$

$K_R = R_{AC}/R_{DC}$ and $K_X = X/X_0$ are the coefficients of increased resistance R and decreased reactance X of a rectangular bar at alternating current (AC) and $f > 0$; R_{DC}, X_0 are the values at a uniform distribution of the current density in the bar.

Formulae 4.50 concern, however, the metals with a constant magnetic permeability, $\mu = \text{const}$.

More detailed investigations by the author [2.33], based on the ideas of E. Rosenberg and L. Neiman (Chapter 7), of the behavior of the electromagnetic field in a ferromagnetic screen of varying magnetic permeability, $\mu = \mu(H)$, showed that at the magnetic field intensity on the sheet surface (H_{ms}) is bigger than the value corresponding to the maximum value of permeability, the electromagnetic wave decays faster than it is described by the exponential dependence (2.176) obtained for an idealized iron with a constant permeability $\mu = \text{const}$ (such as on the surface). The nonlinearity of magnetic permeability in iron as a function of magnetic field intensity, $\mu = \mu(H)$, reduces the depth of wave penetration into solid iron by about 1.4 times. Experimental research by Turowski and Janowski [4.24] on steel tube screens confirmed this conclusion and showed that the analogical reduction of penetration depth into iron also appears in the case of internal superpositions of waves entering a steel sheet through opposite surfaces. It means that the nonuniform distribution of flux in a sheet (e.g., electrical sheets) appears at thicknesses 1.4 times smaller, or at frequencies 2 times smaller, than it is determined by the calculations using classical formulae (e.g., using the Thomson's equations) with the assumption of $\mu = \text{const}$. On the basis of the above considerations, one can introduce the correction for the equivalent depth of field penetration:

$$\frac{1}{\mathrm{k}_{\mu(H)}} = \sqrt{\frac{2}{\omega\,\mu(H)\sigma}} = \frac{1}{1.4\,\mathrm{k}_{\mu=\text{const}}}$$

It means that $\mathrm{k}_{\mu(H)} \cdot d = 1.4 \cdot k_{\mu=\text{const}} \cdot d = k(1.4d)$. Thus, at strong fields, the coefficients ζ and ψ should be determined not for the real sheet thickness d, but for the substitute value of $1.4d$. In Figure 4.14, the broken lines show the course of such corrected curves ζ_{Fe}, ψ_{Fe}, and $\cos\varphi_{Fe}$ for solid iron.

In the case when on the iron surface exists a magnetic field with the intensity lower than that corresponding to the maximal permeability, that is, in the case of weak fields, less interesting from the point of view of power equipment, the equivalent depth of wave penetration into solid iron will be only slightly bigger than that calculated at $\mu = \text{const}$ [7.11]. In such a case, one can utilize formulae (4.50) and the curves ζ and ψ in Figure 4.14 (continuous lines).

The total *losses in a plate* made of metal of $\mu = \text{const}$ equal, therefore

$$P = \frac{\zeta}{2}\sqrt{\frac{\omega\mu}{2\sigma}}\oint_A |H_{ms}|^2 \mathrm{d}A \tag{4.53}$$

where A is the total, double-sided surface of the plate. Approximately, formula (4.53) can also be used for thin iron sheets if we introduce, if needed, the permeability μ under the integral. As it follows from Figure 4.14, at $kd > 3\ldots4$ the coefficient ζ is practically equal to 1.

For the steel wall thickness bigger than the wavelength, $d > \lambda$ (2.178), and practically at $kd > 2.5$ (Figure 4.14), the coefficient $\zeta_{\text{Fe}} \approx 1$, and one can use the approximate formula

$$P_{\text{steel}} \approx \frac{\mathrm{a_p}}{2}\sqrt{\frac{\omega\mu_0}{2\sigma}}\oint_A \sqrt{\mu_r}\cdot|H_{ms}|^2 \mathrm{d}A \tag{4.54}$$

where A is the total, double-sided external surface of the investigated plate, and $H_{ms} = H_{ms}(x, y)$ is the maximal value of the magnetic field strength (time module) on the X,Y surface of the body; $\mathrm{a_p} \approx 1.4$—the linearization coefficient of steel (Chapter 7). In case of thin steel plates ($kd < 2.5$) and a symmetric excitation (Figure 4.14), the active power (in watts) is

$$P_{\text{steel}} = \zeta_{\text{Fe}}\frac{\mathrm{a_p}}{2}\sqrt{\frac{\omega\mu_0}{2\sigma}}\oint_A \sqrt{\mu_{rs}(x,y)}\cdot|H_{ms}(x,y)|^2 \mathrm{d}x\mathrm{d}y \tag{4.55}$$

The distribution of power density, in W/m^2, inside a screen excited double-sided, symmetrically (for instance, Figure 4.14), according to Equation 4.29

$$w(z) = \frac{1}{2\sigma}|J_m(z)|^2 = \frac{\varphi_2}{d}\sqrt{\frac{\omega\mu}{2\sigma}}\frac{|H_{ms}|^2}{2};\quad \varphi_2 = 2\mathrm{k}d\frac{\operatorname{ch}k(d-2z)-\cos(d-2z)}{\operatorname{ch}kd-\cos kd} \tag{4.56}$$

At $d > \lambda/2$, that is, $d \to \infty$, Equation 4.56 takes the form of Equation 3.13.

The distribution of volumetric power density, in W/m^3, inside a screen excited antisymmetrically (flat conductor) from both sides of the metal sheet (Figure 4.15), has the form

$$w(z) = \frac{\varphi_3}{d}\sqrt{\frac{\omega\mu}{2\sigma}}\frac{|H_{ms}|^2}{2};\quad \varphi_3 = 2kd\frac{\operatorname{ch}k(d-2z)+\cos k(d-2z)}{\operatorname{ch}kd-\cos kd} \tag{4.57}$$

4.5 SCREENING OF TRANSFORMER TANKS

Strong leakage fields, which generate losses and local excessive heating in inactive constructional parts, have resulted in practically all large transformers (over 70–100 MVA) being built with magnetic (shunts) or electromagnetic screening. Computer analysis [4.23], with the author's formulae (Turowski [7.18], [1.16/1]) for the power losses in nonscreened steel walls of tanks of large transformers, showed (Figure 4.16) a dramatic increase of these losses in transformers of power rated higher than 50–60 MVA—for single-phase transformers, and 75–90 MVA—for three-phase transformers. This value can be higher or lower, depending on the relative distance of the tank wall (Figure 4.18 [later in the chapter]) from windings, a_k/a_r.

Such screening is also used in smaller transformers with stronger leakage fields (e.g., furnace transformers, autotransformers, with large short-circuit voltage).

Lack of screens in large transformers, and especially in autotransformers, could cause additional huge losses and local overheating of tanks and other constructional parts [1.15/1], [1.16]. When choosing the type of screen, we should consider, among others, their influence on the change of leakage field distribution in the windings region, and especially the radial component of flux density (Figure 4.17).

Magnetic screens (shunts) can protect tanks more reliably than electromagnetic screens [4.23], [4.28], [10.19]. However, they cause an increase of the radial component of the stray field in the external winding. It creates an increase of additional losses and short-circuit forces from the radial component in the external winding (Figure 4.17a).

With electromagnetic screens, the losses in tank are a little bigger than with magnetic screens (shunts), but they reduce the radial component of the stray field and remarkably reduce additional losses in external winding from this component (Figure 4.17b). However, on the edges of the Cu (or Al) screens, a dangerous concentration of field emerge (Figure 4.21a [later in the chapter]).

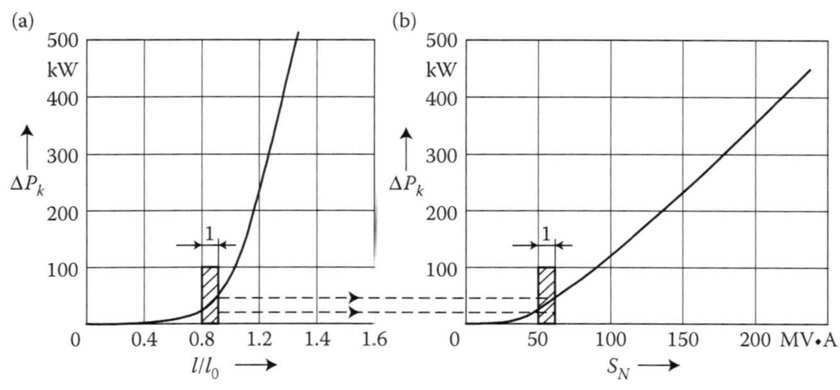

FIGURE 4.16 Power losses in a transformer tank (7.95): (a) versus linear dimensions, (b) versus rated power; 1—region above which screening is unavoidable. (Adapted from Turowski J.: *Calculations of Electromagnetic Components of Electric Machinery and Equipment*. (in Polish). Warsaw: WNT, 1982; and (in Russian). Moscow: Energoatomizdat, 1986; Turowski J. and Byczkowska L.: Computer calculation of power losses in tanks of transformers. (in Polish). *Ibid*, June 1970, No. 2, pp. 102–111 [1.16/1], [4.23].)

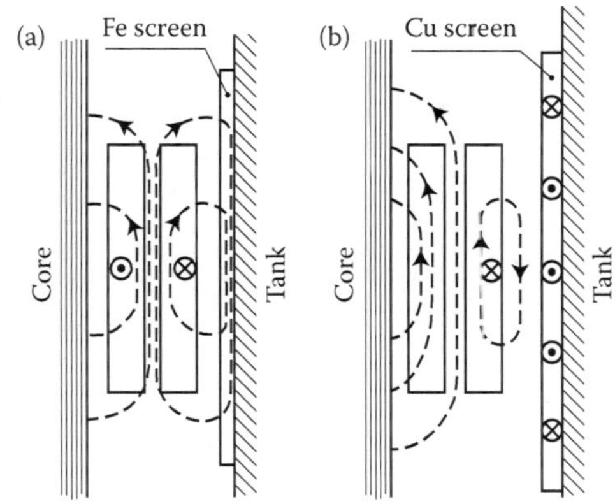

FIGURE 4.17 The principle of operation of screens of transformer tanks: (a) magnetic Fe screens (shunts), (b) electromagnetic Cu or Al screens. (Adapted from Turowski J.: *Calculations of Electromagnetic Components of Electric Machinery and Equipment.* (in Polish). Warsaw: WNT, 1982; and (in Russian). Moscow: Energoatomizdat, 1986 [1.16].)

The influence of screens on the internal winding is weaker and opposite to that described above.

4.5.1 Magnetic Screening (Shunting) of Tanks

Magnetic screens are used for the protection of tanks up to the biggest rated powers of transformers. Often they are preferred over the electromagnetic screens. They are fastened to the internal surface of a tank wall in the form of packages of transformer sheets, where they play the role of magnetic shunts, on the principle described in Section 4.2.2 (Figure 4.22 [later in the chapter]).

In large transformers (Figure 4.18), the value of the resultant flux Φ_δ is determined practically by the reluctance of interwinding gap R_δ, which is much higher than the other reluctances connected in series. They can be calculated by transforming the known formula for the short-circuit voltage of the transformer [1.5]:

$$R_\delta = R_\delta \frac{h_R}{\mu_0 \delta'}; \quad \delta' = \delta + \frac{a_1 + a_2}{2}; \quad h_R = \frac{h}{K_R} \tag{4.58}$$

where $K_R \approx 1 - ((a_1 + a_2 + \delta)/\pi h)$—the Rogowski's coefficient [1.5]; a_1, a_2, a_c, a_t, δ—the radial dimensions defined in Figure 4.18.

The distribution of the stray flux Φ_δ between the core Φ_c and tank Φ_t is determined mainly by the reluctances of oil regions between these surfaces and the interwinding gap. In the first approach, one can assume (in the case of a transformer with $h_R \gg a_c + a_t$)

$$\frac{\Phi_c}{\Phi_t} = \frac{R_t}{R_c} \approx \frac{a_t}{a_c} \tag{4.59}$$

The flux Φ_t that enters the tank is practically independent of the magnetic material of which the tank wall is built (except Cu or Al screening). This is why in the case

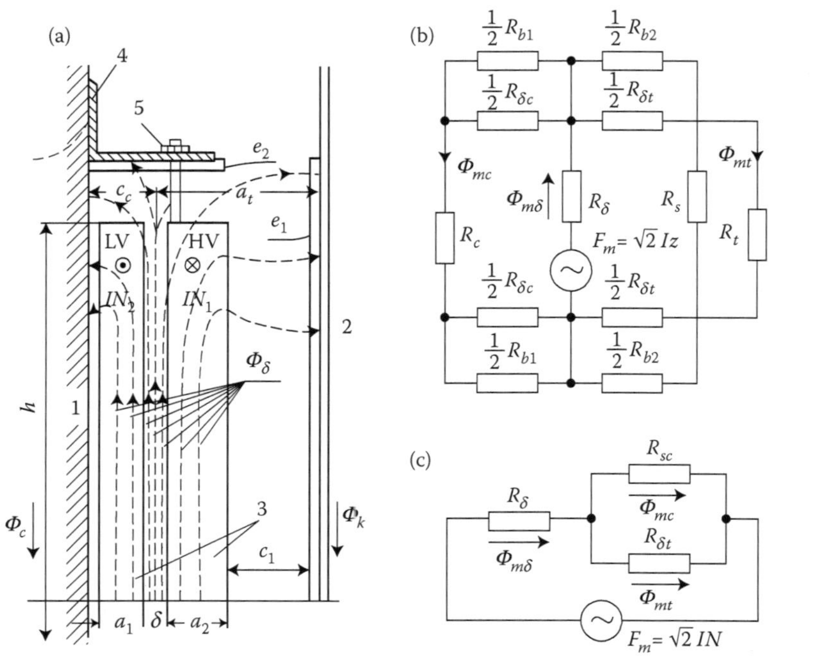

FIGURE 4.18 Scheme of distribution of leakage fluxes in a transformer of 240 MVA or more: (a) cross section, in a relative scale of 1:10, 1—core, 2—tank, 3—windings, 4—yoke beam, 5—yoke bolt, e_1—longitudinal screen, e_2—crosswise screen (flux collector); (b) a simplified equivalent circuit; $F_m = \sqrt{2}IN$—maximum mmf of one winding; $\Phi_{m\delta}$, Φ_{mc}, Φ_{mt}—the leakage fluxes in gap (δ), core (c), and tank (t); R_δ—the reluctance of equivalent gap δ', $R_{\delta c}$—between gap and core, R_c—of core with adjusted oil layer, R_{b1}, R_{b2}—between gap and yoke beams, $R_{\delta t}$—between gap and tank, R_t—of tank with adjusted oil layer, R_s—of yoke bolt. (Adapted from Rizzo M., Savini A., and Turowski J.: *IEEE Transactions on Magnetics*, 36(4), July 2000, 1915–1918; Turowski J. et al. Stray losses control in core- and shell type transformers. Part I: Upgrading of energy-saving and reliability of large transformers. Chapter in the book: *International Advanced Research Workshop on Transformers—ARWtr2007* X.M. Baiona, Spain, October 2007, pp. 56–68 [1.15], [4.12], [4.31].)

of a tank screened by magnetic screen, conclusion 2 from Section 4.2.2 (Equation 4.10a) is valid.

The thickness d of a magnetic screen (shunt) should be chosen on the principle that the gap flux Φ_δ should not saturate the shunt for μ_r values less than, for example, $\mu_r \approx 1000$. From Equations 4.58 and 4.59, it follows that

$$\Phi_\delta \approx \Phi_t + \Phi_c \approx \frac{\sqrt{2}\, IN \mu_0\, \delta'}{h_R} \quad \text{and} \quad \Phi_t \approx \Phi_\delta \frac{a_c}{a_t + a_c} \tag{4.60}$$

At $a_t/a_c \leq 3$–3.5 formulae (4.60) have sufficient accuracy for this purpose. At higher values of a_t/a_C (transformers with highest voltages), a part of the Φ_t flux closes itself in the oil layer between the tank and the external winding. It can be taken into account by introducing into the right part of the second Equation 4.60 the reduction coefficient

$$k_d \approx \frac{\mu_r d}{\mu_r d + c_1} \tag{4.61}$$

Neglecting the correction (4.61) gives a result with some reserve.

One can, therefore, be certain that the screen permeability will not be smaller than $\mu_r = 1000$, if the thickness d satisfies the condition $d \geq \Phi/B_{1000}$, which means that, considering Equation 4.60, the *minimal thickness of magnetic screen (shunt)*, from the point of view of its saturation, can be estimated from the formula

$$d_{\min} \geq \frac{a_c}{a_t + a_c} \cdot \frac{\sqrt{2}\, IN\, \mu_0\, \delta'}{B_{1000}\, h} \tag{4.62}$$

where B_{1000} is the maximum flux density related to the magnetic permeability in screen $\mu_r \approx 1000$ (about 1.7 T for anisotropic, cold-rolled steel, and 1.4 T for isotropic hot-rolled steel). The neglected reluctance of solid steel wall (2.186) gives an additional reserve at calculating $d_{\min}$.

The surface area of screens (shunts) should be as big as possible, but they do not have to be obligatorily continuous along the tank circumference. Interruptions of the shunts along the tank circumference should be positioned opposite of the windows of the magnetic core. The influence of the shunt interruptions along the tank circumference l_{tank} on the power losses in the tank can be evaluated by comparison of the total effective reluctance of shunts R_{sh} with the total reluctance R_{steel} of the uncovered tank wall. The coefficient p'_m of loss reduction in the tank due to magnetic screening (shunting) in this case is

$$p'_m \approx \left| \frac{(R_{\text{sh}}/l_{\text{sh}})}{(R_{\text{sh}}/l_{\text{sh}}) + (R_{\text{steel}}/l_{\text{tank}})} \right|^2 \tag{4.63}$$

where l_{sh} and l_{tank} are both the total resultant length of shunts and the length of tank along its circumference.

In spots of high field concentrations, as for instance on the surface of core leg, external packages are cut into parallel strips (Figure 10.5/3) of widths from one to several centimeters. This enables the reduction of surface eddy currents and heating of shunts [10.18].

In the case of interruptions in shunts along the tank circumference, we can use the same graph in Figure 4.6, but with introducing into the abscissa axis instead of the number of sheets $n = cd$ the value

$$c'd' = cd \frac{l_{\text{sh}}}{l_{\text{tank}}}$$

Magnetic screens (shunts) can cause some increase of the zero reactance X_0 of transformer. At the same time, the zero resistance R_0 of the transformer will decrease.

4.5.1.1 Influence of Eddy Currents and Saturation on Magnetic Screens (Shunts)

Formulae (4.11) and (4.63) and Figure 4.6 concern ideal screening, that is, such a case when the magnetic flux Φ is parallel to the sheets surface (runs along the sheets of shunt).

In a real transformer, if the tank is covered by shunts made as a package of flat transformer sheets, the leakage flux will penetrate through the sheets planes, inducing simultaneously in them the eddy currents. Due to large resistivity of transformer iron and small thickness of the sheets, these currents are not very big. However, at some configurations, they can even increase the losses in the screened system and push out the penetrating flux (see Ref. [4.29] and Section 10.5). The effect of this push-out of flux is an apparent, and sometimes remarkable, reduction of effective permeability of iron ("quasi-permeability μ_q"; Section 5.1.2). On the basis of experiments performed on a big model (Figure 10.3) with central wall dimensions of 2 m × 0.8 m × 10 mm, it was established that such *quasi-permeability* μ_q can be even 3 times smaller than the real one ($\mu_q \approx \mu_r/3$). After substitution of the correspondingly reduced value $c_q d = cd/3 \approx n/3$ into formula (4.11), we shall obtain the new formula for magnetic *screening (shunting) coefficient*, which in this case is closer to reality:

$$p_m = K \frac{18}{(4.2+n)^2 + n^2} k_{\Phi} \tag{4.64}$$

where n is the number of transformer sheets in the flat magnetic screen (shunt), K (= 1.3–1.5) is the reserve coefficient taking into account the nonlinearity, three-dimensionality of shunt, and other effects; k_{Φ} is the coefficient of the way of field excitation; for zero-sequence flux: $k_{\Phi} > 1$, according to the conclusion from Section 4.2.2.

The lower limit of applicability of formula (4.64) and of the number of sheets n in the package of shunt is the *minimal permissible thickness of screen* $d_{\min}$ (4.62). The *upper limit* is the economy of material for shunts and the required insulation distances between windings, conducting leads and tank walls.

A graph of the coefficient (4.64) can be obtained from the curve presented in Figure 4.6 if we shift each of the ordinates by multiplying them with 3.

It should be emphasized here that the changes in iron saturation cause at measurements a remarkable dispersion of these coefficients. However, most of the points lie below the curve determined from Equation 4.64. Finally, we select the number of transformer sheets on magnetic screen (shunt) from a compromise between Equations 4.62 through 4.64, taking the bigger from these two values:

$$n = \frac{a_c}{d_0(a_t + a_c)} \frac{\sqrt{2} I N \mu_0 \delta'}{B_{1000} h} \quad \text{or } n = \frac{l_k}{l_e}\sqrt{\frac{9K}{p_m} - 4.4} - 2.1 \tag{4.65}$$

where $K = 1.3$–1.5; $d_0 = 0.35 \times 10^{-3}$ m—the thickness of the transformer sheet.

EXAMPLE

Task: For a single-phase transformer of 250/3 MVA, $242/\sqrt{3}/13.5$ kV, with the data (Figure 4.18a): $a_c = 0.114$ m, $a_t = 0.573$ m, $IN = 226.000$ A, $\delta' = 0.129$ m (4.58), h = 1.79 m, select the thickness of screen and number of 0.35-mm sheets in a magnetic screen (shunt) covering half of the tank circumference.

SOLUTION

For an isotropic transformer sheet $B_{1000} = 1.4$ T, as per Equation 4.65, we get

$$d_{min} \geq \frac{0.114}{0.573+0.114} \frac{\sqrt{2}\times 226\times 10^{3}\times 4\pi\times 10^{-7}\times 0.129}{1.4\times 1\times 79}$$
$$= 3.43\times 10^{-3}\,\text{m, from which}$$

$$n = \frac{3.43\text{ mm}}{0.35\text{ mm}} \approx 10\text{ sheets}$$

Assuming now the coefficient of power loss reduction in the tank (magnetic screening coefficient) $p_m = 12\%$ and, on average, the reserve coefficient $K = 1.4$, we get the second value

$$n = 2\sqrt{\frac{9\cdot 1.4}{0.12}} - 4.4 - 2.1 \approx 18\text{ sheets}$$

As a result, we obtain a package consisting of 18 sheets, with a total thickness $d = 18 \times 0.35 = 6.3$ mm.

We should also add a filling factor, which will finally give the sum ≈8 mm.

One can also resolve a reverse problem. These are, of course, approximate calculations that require experimental and technological verification. The width of sheets placed flatly is also verified from the heating point of view (Figure 10.5).

Magnetic screens made of sheets positioned *perpendicularly* to the tank surface (4 in Figure 10.5) act as ideal screens and can be calculated from formulae (4.11) and (4.65), and Figure 4.6, by substitution of d as the screen thickness. However, in this case, one should expect an increase of the screening coefficient p_m, as a result of the action of the field component directed along the tank circumference.

In the biggest transformers and autotransformers, the leakage flux of gap Φ_δ is, from the point of view of its value, comparable with the main flux of slightly smaller but also large transformers. That is why the paths of the leakage flux should be considered with almost the same care as of the main flux in the core. It should not be allowed to let the leakage flux have random paths in the region between core and tank.

The analysis presented above is simplified, but very helpful in a rapid constructional decision taking. A more complete analysis should be based on a 3D model (Figure 4.20 [later in the chapter] and Figure 7.22), which in turn also has its own restrictions. They can be explained with the help of the formulae presented in this chapter.

4.5.2 Electromagnetic Screening of a Transformer Tank

Electromagnetic screens are made of *copper* or *aluminum* sheets overlaid on the internal surface of tank walls. For a screen to fulfill its role, it must be, as it was mentioned in Section 4.1, closed for the leakage field. Since in power transformers, the induced eddy currents flowing on the internal surface of the tank also close themselves across the tank circumference, it is also possible to make some splits in the screen, for instance, in order to omit a pocket for tap changer on load. However, in such a case, the ends of the screen should be electrically well connected with the tank iron and with each other, in order to close the screen circuit.

The general theory of electromagnetic screens was given in Sections 4.3 and 4.4. However, it concerns a plane, a polarized electromagnetic wave, whereas in a transformer tank, there also exists a field perpendicular to the tank and eddy-current-creating loops. Nevertheless, in the first approach, those general formulae can be considered as satisfactory, and then one should introduce some corrections. The power losses per unit of a screened surface can be approximately calculated with the help of formula (4.40). At thicknesses of screen bigger than 10 mm, they practically do not depend on the screen thickness. Below this limit, on the other hand, the losses grow with the reduction of thickness d (4.39). This means that the selection of the screen thickness d should be a result of compromise between the reduction of loss and economy of copper. Most often applied are copper screens of thickness 4–5 mm, and aluminum screens 7–8 mm thick [4.21]. Too small a thickness of electromagnetic screen can cause excessive local heating of screens.

From the point of view of recalculatingmeasured short circuit losses to the rated current and the standard temperature of 75°C, interesting is the determination of the *dependence of losses in screened tank walls on the current and temperature*. From formula (4.39), it follows that the losses are proportional to $|H_{ms1}|^2$, therefore, practically to the square of the current. At the same time, we can find the dependence on temperature from formula (4.39) at the assumption that the binominal in parentheses is constant:

$$P_{1\text{Cu-Steel}} = \frac{1}{kd}\sqrt{\frac{\omega\mu_2}{2\sigma_2}}\frac{|H_{ms1}|^2}{2}\left(1+\frac{a_d}{d}\right) = \frac{|H_{ms1}|^2}{2\sigma_2 d}\left(1+\frac{a_d}{d}\right) \tag{4.66}$$

As it follows from formula (4.66), these losses are proportional to the resistivity of the screen material ($1/\sigma_2$), which means they depend on the temperature like the fundamental losses I^2R in windings, and not like the additional losses in windings, which are inversely proportional to the resistivity ($\sigma_2 = 1/\rho_2$). These two opposite dependences can sometimes compensate each other. In effect, there are transformers in which additional load losses do not depend on the temperature.

Since the thickness of copper screens is always smaller than 1 cm ($kd < 1$), the coefficient of loss reduction in tank at ideal (total) screening should be calculated from formula (4.42). In practice, one can assume the average material data

$$\sigma_2 = \sigma_{\text{Cu}} = 54 \times 10^6 \text{ S/m} \quad \text{or } \sigma_2 = \sigma_{\text{Al}} = 34 \times 10^6 \text{ S/m}$$

$\sigma_3 = \sigma_{\text{Steel}} = 7 \times 10^6$ S/m, $\mu_3 = (150$ to $200) \times 0.4\pi \times 10^{-6}$ H/m, $\omega = 2\pi \times 50$ 1/s and $a_d/d \approx 0$, from which, after substitution into formula (4.42), we obtain the electromagnetic screening coefficient

$$p_e \approx \frac{d_1}{d} k_H; \qquad (d_1 = 2 \times 10^{-4} \text{ m}) \tag{4.67}$$

In the above formula, a relatively small value of the permeability μ_3 was used for the nonscreened tank wall, assuming that the magnetic field intensity appearing at that time on the steel surface (H_{ms1}) is on the order of 10^4 A/m. The coefficient $k_H \approx (H_{ms,\text{scr}}/H_{ms\text{Steel}})^2 \geq 1$ takes into account an increase of the field H_{ms} and the power loss due to the eddy-current reaction after the introduction of electromagnetic (Al) screen. At a three-phase short circuit or load (Figure 4.17a) one can accept [4.18] approximately

$$k_{H3} \approx 49 \frac{(c_t + b)^2 (a_c + a_t)^2}{h^2}; \qquad (b = 0.4 \text{ m}) \tag{4.68}$$

This pushing out of flux from the tank wall gives $k_{H3} = 1$ to 3.

Equiphase flux in a three-phase transformer is the zero-sequence flux, which appears at the asymmetric load, or the third harmonic of flux, which appears at the lack of a third harmonic in the magnetization current, that is, at the lack of a neutral conductor in the Yy connection. The zero-sequence flux $\Phi_{m0i}(I_{0i})$ can be calculated from the theory of symmetric components 1, 2, 0, where i = U, V, W, and phase currents are

$$I_U = I_{U1} + I_{U2} + I_{U0}; \quad I_V = I_{V1} + I_{V2} + I_{V0}; \quad I_W = I_{W1} + I_{W2} + I_{W0}$$

After solving these equations, and assumption $a = e^{j\frac{2\pi}{3}}$, we have

$$I_{U1} = \frac{1}{3}(I_U + aI_V + a^2 I_W); \quad I_{U2} = \frac{1}{3}(I_U + a^2 I_V + aI_W); \quad I_{U0} = \frac{1}{3}(I_U + I_V + I_W)$$

The equiphase fluxes $\Phi_{m0i}(I_{0i})$, instead of symmetric fluxes at electromagnetic screens, are almost completely pushed into the narrow gap c_k (Figure 4.19b), which causes a big increase of H_{mse} and produces an effect of the order of $k_{H0} = 15$ to 20. One then can adopt

$$k_{H0} \approx \left(1 + \frac{c_1}{c_k}\right); \quad (c_1 = 0.37\,\text{m}) \tag{4.69}$$

In the last case, the losses (4.69) in the tank wall after screening are not reduced, but they only pass from the steel wall to the screen.

Since at $kd \geq 1$, the losses practically do not depend on the screen thickness d, whereas at $kd < 1$ we have $\kappa \approx 1/kd$, the selection of screen thickness should be a compromise between the tank loss reduction and the economy of copper or aluminum.

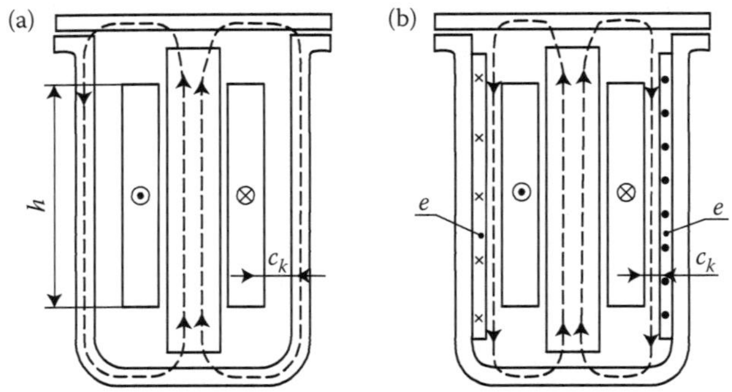

FIGURE 4.19 Increase of the magnetic field intensity H_{ms} on tank surface, at equiphase flux of a three-phase transformer, after coating the walls with an electromagnetic screen: (a) a steel tank without screen (small magnetic field intensity); (b) a tank with an electromagnetic (Cu) screen (*e*).

The remarkable changes of the coefficient κ (Figure 4.13) in the dependence on frequency and resistivity of the screen show a significant influence of these parameters on the losses. In the case of copper, as it was stated in Section 4.3.1, $k_{Cu} \approx 100$ 1/m, and in the case of aluminum, $k_{Al} \approx 100\sqrt{35/54} = 83$ 1/m. From Figure 4.13, it follows that the optimal thickness of a copper screen is 10 mm ($kd \approx 1$) and of an aluminum screen—about 12.5 mm. Usually, a transformer tank is screened with copper sheets 4–5 mm thick, which corresponds in Figure 4.13 to the coefficient $\kappa \approx 2$. In order to obtain the same losses with an aluminum screen, its thickness should be increased in such a way that the reduction of the κ coefficient should compensate the reduction of the screen conductivity existing in the denominator in formula (4.66), that is

$$\kappa_{Al} \approx \kappa_{Cu}\sqrt{\frac{\sigma_{Al}}{\sigma_{Cu}}} = 2\sqrt{\frac{35}{54}} = 1.55 \tag{4.70}$$

From this number, as per Figure 4.13, the corresponding $kd \approx 1/\kappa = 0.65$, from which the required thickness of aluminum screen is $d_{Al} = 0.65/83 \times 10^3 \approx 8$ mm, which means it is nearly 2 times larger.

Investigations in Siemens Works (W. Dietrich, 1967) confirmed this conclusion, by determining the number around 8 mm as the optimal thickness for an aluminum screen.

After inserting the removable part of the transformer into a tank, one more electromagnetic screen is added on top, to the oblique part of the cover, connected electrically with the lower aluminum screen. In this way, one ensures effective screening of the cover, frame, screw bolts,* and so on.

Using formulae (4.18) and (4.24) for a screen placed in dielectric, valid also for a nonscreened tank wall, it is worthwhile noting that eddy currents are not distributed symmetrically on both surfaces of the tank.

* Presently, more popular is welding of both parts of tank, and thus the problem of bolts overheating hazard is not so common.

On the external tank surface ($J_{s2} = \sigma E_{s2}$), the losses decrease inverse-proportionally to the tank thickness d. Since the thickness d of the tank walls of large transformers is typically from 10 to 15 mm ($d > \lambda/2$), we can conclude that the loops of eddy currents are closed in the internal surface of the tank.

An additional analysis showed that the three-phase leakage field creates in a tank a type of running field around its circumference. It is like in a linear induction motor. As a result, there exists a type of frontal wave swelling and a long tail wave (Figure 4.20d). The direction of such a field and loss concentration is determined by the supply sequence *U–V–W*, or vice versa.

4.5.3 Three-Dimensional Computer Analysis and Interactive Design of Screens

Among the numerical methods of field calculation, mentioned in Section 2.2, the most popular include FEM-2D (*finite element method, two-dimensional*) and RNM-3D (*reluctance network method, three-dimensional*), which is one of the most effective and easiest. Nowadays, 2D methods have little use in a scientifically based design.

All the methods are absolutely equivalent in the scientific sense, but absolutely not equivalent in their effectiveness, time of calculation (e.g., days to weeks with FEM-3D vs. less than 1 s with RNM-3D), *user friendliness, and cost-effectiveness.* The RNM-3D computer program, designed by the author (Janusz Turowski) and his young (at that time) students Marek Turowski and Miroslaw Kopec [2.32], [4.19], [4.20], [4.27], [4.28], [7.21], [7.23], [7.31], has been received and implemented with enthusiasm by more than 45 transformer works and research institutions all over the world, including the United States, Canada, Brazil, Mexico, Europe, Asia, and Australia.

Results of rapid calculations with the RNM-3D software are presented in the form of 3D color graphs of field and power loss density (Figure 4.20d). They show total loss and hot-spots in tanks, forces in windings, and so on.

RNM-3D Program User's Input Questionnaire:

QUESTIONNAIRE (cf. Figure 4.20a, e)
For three-dimensional (*RNM-3D*) computation of a leakage field, eddy-current losses and hot-spots in a tank wall and selection of electromagnetic and/or magnetic screens in three-phase power transformers, on a popular PC, within *less than 1 second of CPU time for each structural variant.*

Electrical data:

Rated power S_N .... MVA

Frequency f .... Hz

Rated phase currents HV/LV .... / .... A

Number of series turns per phase .... /..................

Rated phase voltage HV/LV .... / .... kV

Dimensions in mm:

$h_{u\text{HV}} =$ .... $h_{u\text{LV}} =$ ....

$a_1 =$ .... $a_2 =$ .... $\delta =$ ....

$a_C =$ .... $a_T =$ .... $\frac{1}{2}A_k =$ ....

$h_k =$ .... $h_p =$ .... $h_d =$ ....

$b_1 =$ .... $b_2 =$ .... $b_1' =$ ....

$\frac{1}{\pi}l_{1m} =$ .... $\frac{1}{\pi}l_{2m} =$ .... $b_3 =$ ....

$c' =$ .... $c'' =$ .... $c''' =$ .... $D =$ ....

$c^{\text{IV}} =$ .... $M =$ .... $c^{\text{v}} =$ ....

Material of other proposed elements (reluctances):

Copper screened, XYZ..

Aluminum screened, XYZ..................................

Screened with laminated iron (shunts), XYZ...

Made of solid steel, XYZ.....................................

Made of nonmagnetic solid steel, XYZ...

Other features, XYZ..

Insulation or air gap, XYZ....................................

Other components to be investigated, XYZ..... (see Figure 4.20a)

....

.................., Date................ Signature.............

Professor Janusz Turowski PhD, DSc, El. Eng.

PhD, DSc, Elec. Engineer, Dr hc. University of Pavia, Italy;

past Member of CIGRE. Sr. Member IEEE

The RNM-3D program consists of (Figure 4.20c): (1) *preprocessor* calculating elementary reluctances and mmf's in particular branches of the equivalent 3D network (Figure 4.20a), on the basis of the dimensions and parameters of a transformer [4.19], [4.20] delivered by the user on a special questionnaire (previous page); (2) *solver*—the authors' program for the solution of arbitrary, large electric networks of over 1000 or more nodes, using *modified nodal approach* [4.27] and *sparse matrix*

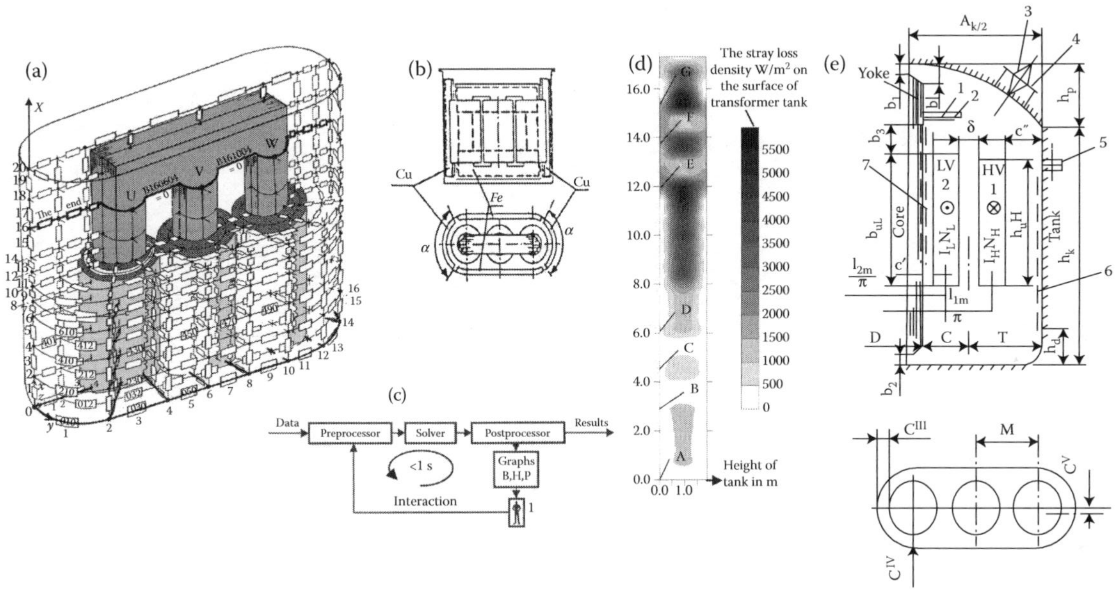

FIGURE 4.20 Interactive computer design, with the RNM-3D code, of Cu–Fe screens in tank of a three-phase transformer: (a) three-dimensional RNM-3D model. (Adapted from Turowski J.: Reluctance model of stray field in transformer. (in Polish). *Rozprawy Elektrotechniczne*, (4), 1984, 1121–1144; Turowski J. and Turowski M.: A network approach to the solution of stray field problems in large transformers. *Rozprawy Elektrotechniczne*, PAN. (2), 1985, 405–422; Turowski J., Kopeć M., and Turowski M.: *COMPEL—The International Journal for Computation and Mathematics in Electrical Engineering*. London: James & James, 1990, Vol. 9, Supplement A, pp. 167–170 and pp. 302–304 [4.19], [4.20], [7.21].); (b) screens configuration. (Adapted from Turowski J.: *Proc.of Jubilee Conference "The Transformer Production is 100 Years Old"*. Budapest, June 18–21, pp. 470–477 [4.26].); (c) flowchart of the interactive RNM-3D design. (Adapted from Turowski J. and Byczkowska L.: Computer calculation of power losses in tanks of transformers. (in Polish). *Ibid*, June 1970, No. 2, pp. 102–111 [4.23].), 1—designer; (d) losses in the tank of an autotransformer (100 MVA, 220/132/11 kV), calculated by the *RNM–3Dasm* program, with no screens in this case. (Adapted from Zwoliński G.: Stray fields and losses in asymmetric transformer structures. PhD thesis. Technical University of Lodz, 1999 (supervisor J. Turowski) [7.30].)—one can see the forehead and the tail of magnetic field.

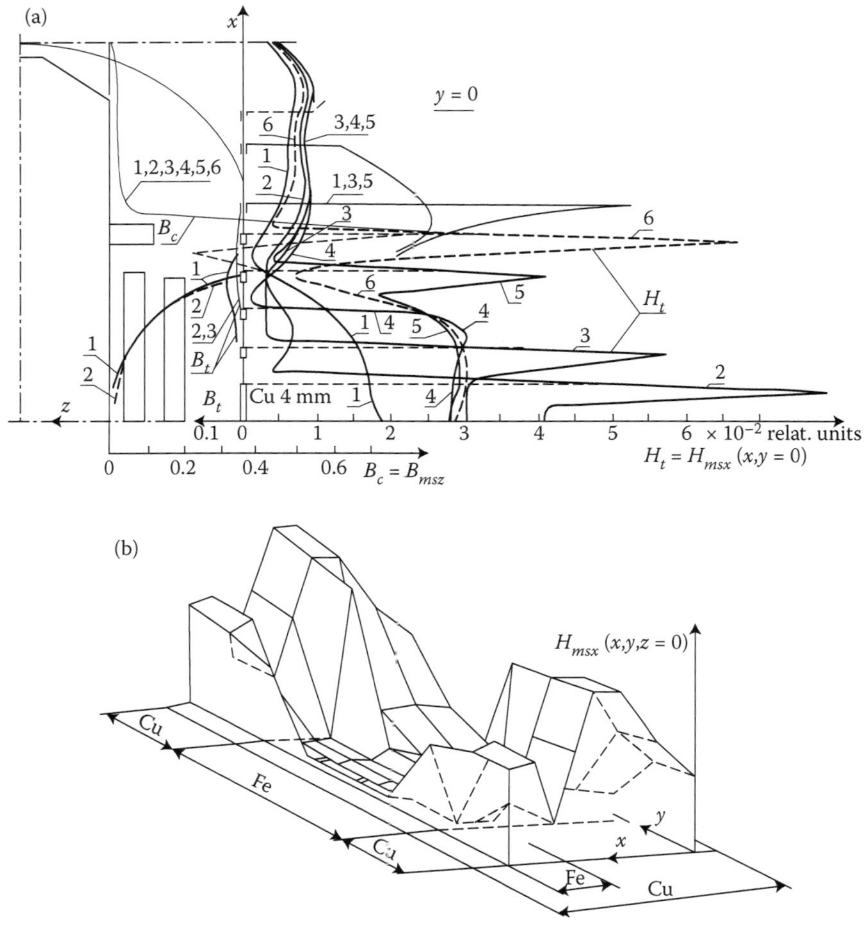

FIGURE 4.21 A fast, interactive selection of screens with the help of the RNM-3D program (Figure 4.20): (a) two-dimensional analysis of the influence of height of tank screen (Cu, 4 mm) on the H_t and B_t fields on the internal surface of the tank and core ($B_c = B_{msz}$, Turowski [4.26]); (b) three-dimensional distribution of the H_{msx} field at a combined Cu–Fe screen (Figure 4.20b)—as per author's patent. (Adapted from Turowski J.: *Calculations of Electromagnetic Components of Electric Machinery and Equipment.* (in Polish). Warsaw: WNT, 1982; and (in Russian). Moscow: Energoatomizdat, 1986; Turowski J., Kopeć M., and Turowski M.: *ISEF'89. Electromagnetic Fields in Electrical Engineering. COMPEL.* London: James & James, 1990, pp. 113–116; and *ITMA'91.* New Delhi, India [1.16], [4.28].)

methods [4.20], [4.28],* and (3) *postprocessor,* for recalculation of results from the solver and displaying them as 2D graphs (Figure 4.21a) or spatial 3D distributions (Figure 4.21b) of $H_m(x, y, z)$—for tank loss and hot-spots, $B_m(x, y,z)$—for the

* Here we used the MSR-1100 program in Fortran ([4.20], [4.27] of J. Turowski, M. Turowski, and M. Kopeć). The author (J.T.), using diploma students, applied also C++ version, but with worse results, and then Java version—with very good effects [4.31]. A precondition for the "rapid design approach" success, however, is a good Knowledge Base for "stapling" all the 3 components.

mechanical stresses in windings, and $P_1(x, y, z)$—for the screening and total loss, according to the user's preferences.

The computation results, after the introduction of the data, are obtained in less than 1 s, in the form of graphs that can be rotated and analyzed by changing the corresponding reluctances.

By changing values of reluctances in the network (Figure 4.20a), for example, in the first approach $R_m = \infty$ for Cu screen (in comparison with Fe) or $R_m = 0$ for Fe screen (shunt), designer 1 (Figure 4.20c) can investigate rapidly the effects of different screens and dimensions (Figures 4.21 and 4.22) and select the best configuration for subsequent detailed investigations. The RNM-3D graphs are expressed in relative units, in relation to the values in the interwinding gap (δ).

The tangential component of the field (H_{ms}) on the tank surface (3.10a) is responsible for local loss and possible excessive heating. However, although H_{ms} is much bigger on the surface of Cu screen than on the steel surface, the total losses in a screened tank are much smaller than those without any screens, according to the author's basic formula (in W):

$$P = \frac{1}{2}\sqrt{\frac{\omega\mu_o}{2\sigma_{St}}}\left[p_e \iint_{A_e} \sqrt{\mu_{\text{rse}}}\,|H_{ms}|^2\, dA_e + p_m \iint_{A_m} \sqrt{\mu_{\text{rsm}}}\,|H_{ms}|^2\, dA_m + \text{a}_\text{p} \iint_{A_{St}} \sqrt{\mu_{\text{rs}}}\,|H_{ms}|^{2x}\, \text{d}A_{St}\right]$$
$$\approx \text{I}^2(a + bI) \tag{4.71}$$

where $p_e = \sqrt{\dfrac{\mu_2\sigma_3}{2\mu_3\sigma_2}}$; $p_e \ll 1$, $p_m \ll 1$ are the screening coefficients of tank walls—electromagnetic (4.67) and magnetic (4.64), respectively, A_e, A_m, A_{St} are the surface

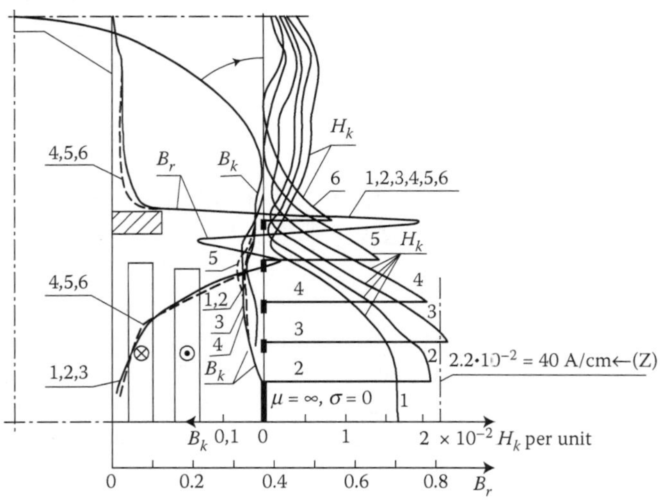

FIGURE 4.22 Two-dimensional analysis of the influence of height 1,2,3,... of magnetic screen (shunt) of tank on the field $H_{\text{ms}} = H_{\text{t}}$ on the surface of transformer tank walls; B_t—normal to tank, B_c—radial in windings. (Adapted from Turowski J.: *Proc. of Jubilee Conference "The Transformer Production is 100 Years Old."* Budapest, June 18–21, pp. 470–477 [4.26].)

areas of the corresponding electromagnetic (*e*) and magnetic (*m*) screens and of the uncovered steel (St), respectively; $\mu_s = \mu_{rs} \cdot \mu_0$, $\mu_s = \mu_s(H_{ms})^*$ is the absolute magnetic permeability on the steel surface, μ_{rs} is the relative magnetic permeability, $\mu_0 = 0.4\pi \times 10^{-6}$ H/m is the magnetic constant; and $x = 1.1$ to 1.14 is the specific constructional coefficient for various transformers.

The reluctance network method and the RNM-3D program have also been used for modeling and calculation of crosswise shunts [4.6], [4.20], [4.31] (Figure 4.18, e_2), design of large Mvar reactors, HV phenomena, currents, heat, oil and liquid flow, and so on.

Screw bolts connecting the tank with cover are known to be a serious overheating hazard of large transformers (Figure 9.3) if they are not covered with corresponding screens and connections.

Screw bolts bracing yoke beams along core legs can also be a source of loss and overheating hazard if they are not screened (usually by Cu tubes). At assumption of a uniform field, the power losses per unit length can be obtained by multiplying the Poynting vector (3.10a) by the screw bolt circumference πD and by the screening coefficient $p_{scr} \ll 1$ (4.42) if this is a steel screw screened by a Cu or Al tube. The power losses, in watt per meter of tube length, are

$$P_1 = a_p p_{scr} \pi D \sqrt{\frac{\omega\mu_0}{2\sigma}} \frac{|H_{ms}|^2}{2} \tag{4.72}$$

In order to reduce the screw losses, *bracing screws* are also made of nonmagnetic steel, which however is more expensive. In this case, the depth of electromagnetic wave penetration (2.181) significantly exceeds the diameter of the screw. Then, internal reflections of wave occur, which require using the formulae for cylindrical conductors (J. Turowski [1.16], p. 287). It should be stressed that the total power losses in well-isolated screws, even with a large number of them, are relatively small, although an overheating hazard exists.

4.6 INDUCTION MOTORS WITH A SCREENED AND MULTILAYER ROTOR

4.6.1 General Characteristics

In the chemical and petroleum industry, and in physical and other equipment, hermetic electric machines with a screened rotor are applied, for example, for driving pumps and mixers operating at high pressures and temperatures (up to some hundreds atm and 500°C) or at deep vacuum [1.15/1], [4.4]. Liquids existing in gaps of such motors are often conducting. Thermal isolation of the motor from the chemical apparatus operating at a temperature of 500°C is achieved by placing between such apparatus a tube made of *austenitic steel*. The thermal conductivity of these steels

* Nonlinear permeability $\mu(H)$ can be a cause of steel loss proportionality to H^n at $n < 2$ or $n > 2$, see Section 7.3.

is barely 13–29 W/(m K), that is, about 30 times smaller than the thermal conductivity of copper and about 4 times smaller than the thermal conductivity of iron. Such motors are produced in the power range from fractions of kW to several dozens of kW. They are usually supplied at an increased frequency (for instance, 200 Hz) for acquiring higher rotational speed, or at a frequency of 50 Hz.

For protection against chemical environments, the rotors are often coated with a conducting layer of aluminum, acid-resistant steel, and so on. The rotor is closed in a special screening case made of material with magnetic permeability near to one, of high electric resistivity and high mechanical strength. For such a purpose, the best are high-grade steels with austenitic structure (see Chapter 1). Thickness of such a screening case, depending on the pressure in apparatus and constructional clampings, varies from 1 to 2–3 to even 6 mm. The introduction of a screening case to the gap of the induction motor causes a corresponding increase in the magnetization current and slip, as well as a decrease of reliability.

For instance, the motor adapted from a normal short-circuited induction motor of 7 kW, 50 Hz, with a number of pole pairs $p = 1$, with a screening case with walls 1.28 mm thick, attained the highest efficiency of 57.5%, at the slip of 10% and effective power of 6.6 kW [4.4]. Such motors require special bearings (graphite, teflon, etc.) resistant to aggressive environments.

The discussed motors, from the theory point of view, belong to the whole family of induction machines with a stratified stator and rotor, such as drag-cup motors, linear motors, with solid steel rotors, and so on, and can be analyzed jointly.

4.6.2 Fundamental Equations of Stratified Induction Motors

Not entering into details of constructional calculations and the design of the motor itself, which are given in specialist literature [4.4], let us consider the methods of electromagnetic calculations [4.10] of multilayer motor with rotor coated by a nonmagnetic conducting layer and closed in motionless cylinder made of paramagnetic steel. This cylinder is placed in the gap between stator and rotor.

The considered model (Figure 4.23) consists of six layers, from which the top and the bottom layers are ideal, nonconductive ferromagnetics ($\sigma = 0$, $\mu = \infty$). The actual three-phase stator winding of an induction motor is simulated here by an infinitely thin current layer placed on the internal surface of the stator, that is, between the layers (1) and (2). So represented specific electric loading of the finite length of stator L^S rotates along the circumference—the z axis. The rotor and screen also have the finite length, L^R. The laminated iron cores of stator and rotor are, instead, infinitely wide (in the direction of the x axis).

With a sufficiently large diameter of rotor with respect to the thicknesses of layers d_i ($i = 1, 2, 3, 4, 5$), one can employ the rectangular (Cartesian) coordinate system XYZ, where the specific electric loading has the form

$$A_x = \sum_k \sum_\nu A_{mxk\nu} e^{j\left(\omega t - \frac{\pi}{\tau_z} z\right)} \cdot \cos\left(k \frac{\pi}{L^R} x\right) \tag{4.73}$$

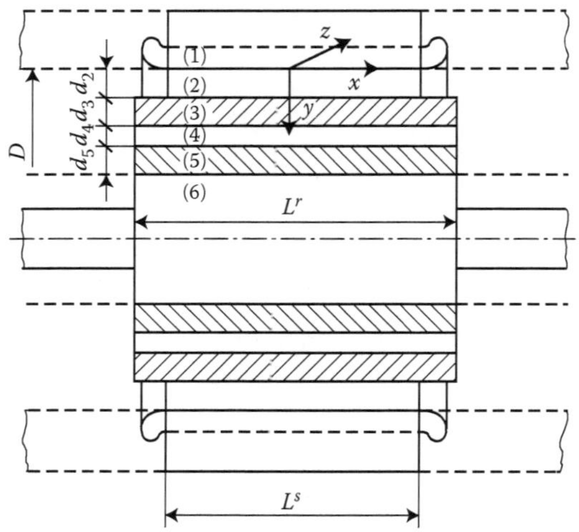

FIGURE 4.23 The rotation motor with a double-layer screened rotor. (Adapted from Mendrela E.A., Mendrela E.M., and Turowski J.: Flux density field, currents, forces and power losses and integral parameters in induction motors with rotary and linear multi-layer shielded rotor. (in Polish). *Report of IMET Tech. Univ. of Lodz*, No. 2687 CPBR 5.7, Phase I, October 1987 [4.10].)

where $\tau_z = \pi D/2p$ is the pole pitch; $\nu = 1, -5, 7, \ldots$ is the harmonics of armature (stator) winding phases, while the amplitude of the specific electric loading is

$$A_{mxk\nu} = \frac{4}{\pi k}\sin\left(k\frac{\pi L^S}{2L^R}\right)A_{mx\nu} \tag{4.74}$$

$k = 1, 3, 5, 7, \ldots$—the harmonics of analytic continuation (with Fourier series) [1.16] in the direction of the x axis.

The function (4.73) can also be presented in the form [4.10]

$$A_x = \sum_k \sum_k \frac{1}{2} A_{mxk\nu}\left\{\exp\left[j\left(\omega t - k\frac{\pi}{L^r}x - \nu\frac{\pi}{\tau_z}z\right)\right] + \exp\left[j\left(\omega t + k\frac{\pi}{L^r}x - \nu\frac{\pi}{\tau_z}z\right)\right]\right\} \tag{4.75}$$

in which $L^r \equiv L^R$ is the length of rotor core (Figure 4.23) and the k, νth harmonic is the sum of the two waves moving in the direction between the x and z axes, with the same linear speed v_x, where

$$v_{z\nu} = 2f\tau_{z\nu}, \quad v_{xk} = \pm 2f\tau_{xk}, \quad \tau_{z\nu} = \frac{\tau_z}{\nu}, \quad \tau_{sk} = \frac{L^R}{k} \tag{4.76}$$

Since one assumes the possibility of motion of the ith layer with an arbitrary speed (v_{iz}), the slip of the ith layer with regard to k, νth harmonic is expressed by the dependence

$$s_{iv} = 1 - \frac{v_{iz}}{v_{zv}} \tag{4.77}$$

The specific electric loading (4.75) in a shorter notation has the form

$$A_x = \sum_r \sum_k \sum_v \frac{1}{2} A_{mxrkv} \exp\left[j\left(\omega t - k\frac{\pi}{\tau_{xrk}} x - \frac{\pi}{\tau_{zv}} z \right)\right] \tag{4.78}$$

where $\tau_{xrk} = L^R/rk$; $r = 1, -1$; $k = 1, 3, 5, 7, \ldots$; $v = 1, -5, 7, -11, 13, \ldots$

4.6.2.1 Field Equations

Equations 4.73 and 4.75 present the enforcing field of the system, that is, boundary conditions for the field in particular layers.

The electromagnetic field in the ith layer of the system is described by Equation 2.142:

$$\Delta E_{mi} = \alpha_{mi}^2 E_{mi}, \quad \alpha_i = \sqrt{j\omega \mu_i \sigma_i s_i} \tag{4.79}$$

Assuming that in particular layers of the motor currents flow only in the directions parallel to the surface ($J_y = 0$, $E_y = 0$), from Equation 4.79, we have only two scalar equations:

$$\frac{\partial^2 E_{mj}}{\partial x^2} + \frac{\partial^2 E_{mj}}{\partial y^2} + \frac{\partial^2 E_{mj}}{\partial z^2} = \alpha_{irkv}^2 E_{mj} \tag{4.80}$$

$j = irkvx; irkvy.$

Applying the Fourier method of separation of variables (2.144) and the boundary condition (4.78), E. A. Mendrela, E. M. Mendrela, and J. Turowski [4.10] obtained the general solution for the ith layer and k,vth harmonic:

$$E_{mxi}(y) = C_i' \exp(\beta_i y) + C_i'' \exp(-\beta_i y) \tag{4.81}$$

where

$$\beta_{irkv}^2 = \left(\frac{\pi}{\tau_{xrk}}\right)^2 + \left(\frac{\pi}{\tau_{zv}}\right)^2 + \alpha_{irkv}$$

C_i', C_i'', is the integration constants that should be determined from the boundary conditions $B_{yi} = B_{y(i+1)}$ and $H_{zi} = H_{z(i+1)}$ on the boundary of adjacent layers. For the layers (1) and (6), it is additionally assumed that the field disappears at $y = \pm\infty$.

From Equation 2.5, div $\boldsymbol{D} = 0$, and from Equation 4.81, one can find

$$E_{miz} = -\int \frac{\partial E_{mix}}{\partial x} dz,$$

and from Equation 2.2, curl $\boldsymbol{E} = -j\omega \boldsymbol{B}$, one can find the components B_{mxi}, B_{myi}, B_{mzi}.

The influence of rotor slotting on the field in the gap is taken into account with the help of the Carter coefficient (k_C): $\delta_{\text{ideal}} = \delta \cdot k_C$ ([1.16], p. 102), where $k_C = t_1/(t_1 + \gamma\delta)$; $\gamma = ((b_4/\delta)^2/5 + (b_4/\delta))$; $t_1 = \pi D/Q$ is the slot pitch; Q is the number of slots on the rotor circumference, and b_4 is the opening of the slot (J. Turowski [1.16], Table 4.1).

After determining the vector components $\boldsymbol{E}_{mi} = 1/\sigma_1 \boldsymbol{J}_{mi}$ and $\boldsymbol{B}_{mi}$, one can calculate the density of forces in particular layers:

$$\left.\begin{aligned} f_{ix}(y_i, x) &= \frac{1}{2}\mathbf{Re}\left\{-J_{iz}(y_i, x)\boldsymbol{B}^*_{iy}(y_i, x)\right\} \\ f_{iz}(y_i, x) &= \frac{1}{2}\mathbf{Re}\left\{J_{ix}(y_i, x)\boldsymbol{B}^*_{iy}(y_i, x)\right\} \end{aligned}\right\} \tag{4.82}$$

and the density of power losses:

$$\Delta p_i(y_i, x) = \frac{1}{2}\frac{1}{\sigma_1}\mathbf{Re}\{J_{ix}(y_i, x)J^*_{ix}(y_i, x) + J_{iz}(y_i, x)J^*_{iz}(y_i, x) \tag{4.83}$$

The total active power of $rk\nu$th harmonics, penetrating through the entire surface till the ith layer, can be calculated with the help of the Poynting vector (3.7)

$$S_y = \frac{1}{2} \times \frac{1}{\mu_i}\left(E_{mz}B^*_{mx} - E_{mx}B^*_{mz}\right)$$

and *Poynting*'s theorem (3.1)

$$P_{pirk\nu} = \frac{1}{2}\frac{1}{\mu_i}\int_0^{L^{sr}}\int_{-L^r/2}^{L^r/2} \text{Re}(S_y)\,dz\,dy =$$

$$= \frac{1}{2\mu_i}L^{sr}L^r\text{Re}\left\{E_{mzrk\nu i}(y_i)B^*_{mxrk\nu i}(y_i) - E_{mxrk\nu i}(y_i)B^*_{mxrk\nu i}(y_i)\right\}_{\substack{x=0\\z=0}} \tag{4.84}$$

The active power dissipated (i.e., loss) in the ith layer equals the difference between the powers (4.84) flowing in and out of this layer $P_i = P_{pi} - P_{p(i+1)}$. At the same time, the dissipated active power equals the sum of the mechanical power and the power loss dissipated in the ith layer. The total active power of the ith layer equals to

$$P_i = \sum_r\sum_k\sum_\nu P_{irk\nu} \tag{4.85}$$

The total power loss in the ith layer is

$$\Delta P_i = \sum_r\sum_k\sum_\nu s_i P_{irk\nu}; \quad s_i = \frac{v_{si} - v_i}{v_{si}} \tag{4.86}$$

The mechanical power of the *i*th layer is

$$P_{mi} = \sum_r \sum_k \sum_\nu P_{irk\nu}(1 - s_i) \tag{4.87}$$

and the driving force acting on the *i*th layer

$$F_{iz} = \sum_r \sum_k \sum_\nu \frac{1}{v_{z\nu}} P_{irk\nu} \tag{4.88}$$

where $v_{z\nu}$ is the speed of the νth harmonic of the field.

Correspondingly, the driving torque of the rotating motor is $T_i = \frac{1}{2} F_{iz} D_i$.*

The *field attenuation coefficient* (or screening coefficient) of the *i*th layer is

$$K_{ti} = \left\{ \frac{|B_{y(i+)}(y_{i+1})|}{|B_{y(i)}(y_i)|} \right\}_{\substack{x=0\\y=0}} = f(s) \tag{4.89}$$

where $B_{yi}(y_i) = \sum_r \sum_k \sum_\nu B_{myrk\nu i}(y_i, x = 0, z = 0)$.

The computation programs On the basis of the developed equations, Mendrela E. A., Mendrela E. M., and Turowski J. [4.10] developed two numerical programs for calculations of a six-layer motor:

1. POLE (*field* in Polish)—program for the calculation of magnetic field distributions, currents, forces, and losses along the *x* axis, at any speed of any layer of the motor
2. CH-K1—program for determining the characteristics of the driving force, rotational torque, power loss, and attenuation (screening) coefficient, as a function of the slip *s*, at a permanent value of the stator current and supply frequency

The programs save results on the computer disk, ready for printing or plotting of characteristics. With the help of these codes, one can calculate linear or rotational motors of various types, with a finite length of stator and rotor.

Figures 4.24 through 4.26 present characteristics calculated for the rotational motor with a double-layer screen of rotor, using the data (*cf.* Figure 4.23)

$\sigma_2 = 0$, $\sigma_3 = 46.8 \times 10^6$ S/m; $\sigma_4 = 0$, $\sigma_5 = 46.85 \times 10^6$ S/m;
$d_2 = 0$, $d_3 = 1$ mm; $v_3 = 0$ (motionless screen); $d_4 = 0.5$ mm,
$d_5 = 1$ mm, $L^S = 0.107$ m, $L^R = 0.130$ m; the rotor diameter
$D_2 = 8.5$ cm, and so on

* The former symbol of the rotation torque was *M*.

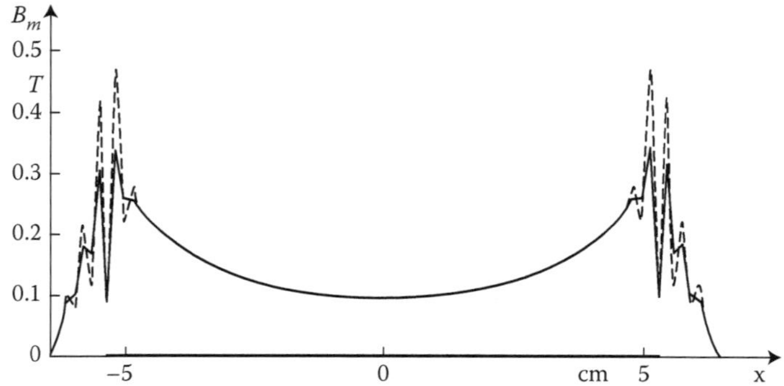

FIGURE 4.24 The radial flux density (B_{my}) of the induction motor with a screened rotor: continuous line—the 5th layer (rotor winding), dashed line—the 3rd layer (the motionless screen of stator). Computed with the POLE (*Field*) program. (Adapted from Mendrela E.A., Mendrela E.M., and Turowski J.: Flux density field, currents, forces and power losses and integral parameters in induction motors with rotary and linear multi-layer shielded rotor. (in Polish). *Report of IMET Tech. Univ. of Lodz*, No. 2687 CPBR 5.7, Phase I, October 1987 [4.10].)

In the screened motors dedicated to high temperatures, screens are made of austenitic steel (70% Fe, 18% Cr, 9% Ni, 2% Mn, 1% Ti; $\sigma = 1.24 \times 10^6$ S/m; instantaneous strength 75 dN/mm²), and the conducting coating of steel rotors is made of copper (Wiszniewski et al. [4.4]).

With the increase of the screen wall thickness (d_3), the ratio of the power delivered to the rotor with respect to the power consumed by stator decreases rapidly. An increase of the thickness of the conducting layer (d_5) on the rotor surface compensates significantly the disadvantages of screen thickening. However, an increase of the thickness of conducting layer on the rotor surface leads to a significant increase of the magnetization current. That is why, in practice, there are applied conventional squirrel-cage rotors with slots.

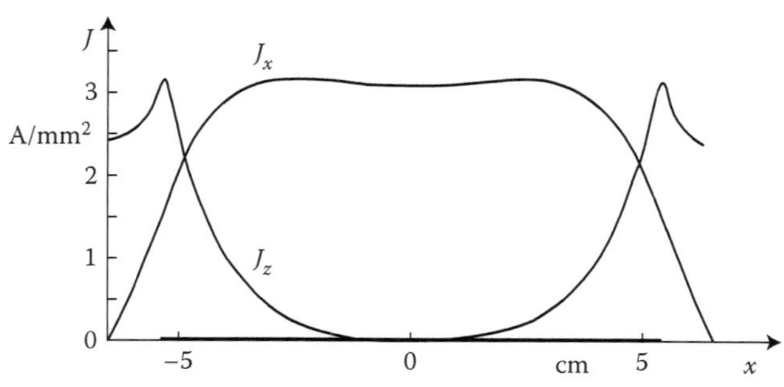

FIGURE 4.25 The current density in the induction motor with a screened rotor (Figure 4.23). Results of the POLE (*Field*) program. (Adapted from Mendrela E.A., Mendrela E.M., and Turowski J.: Flux density field, currents, forces and power losses and integral parameters in induction motors with rotary and linear multi-layer shielded rotor. (in Polish). *Report of IMET Tech. Univ. of Lodz*, No. 2687 CPBR 5.7, Phase I, October 1987 [4.10].)

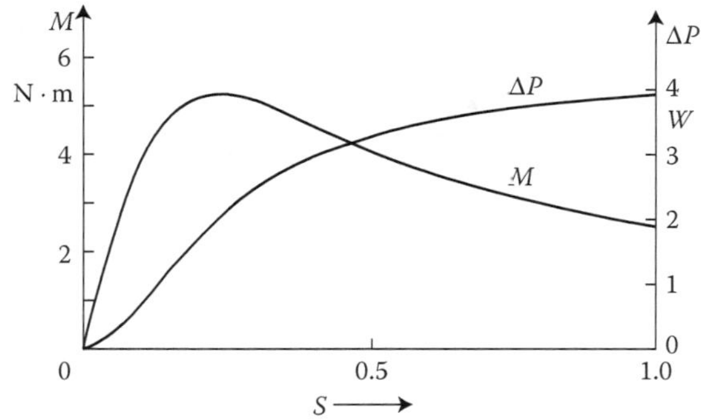

FIGURE 4.26 The rotation torque (M) and the power losses (ΔP) in the induction motor with a screened rotor (Figure 4.23). Computed with the CH-K1 program. (Adapted from Mendrela E.A., Mendrela E.M., and Turowski J.: Flux density field, currents, forces and power losses and integral parameters in induction motors with rotary and linear multi-layer shielded rotor. (in Polish). *Report of IMET Tech. Univ. of Lodz*, No. 2687 CPBR 5.7, Phase I, October 1987 [4.10].)

In order to attain the highest possible efficiency, the screening tube made of austenitic steel should be as thin as possible. In laboratory equipment of low power (up to 1 kW) one can use a tube with a wall thickness of the order of several millimeters. In industrial equipment, this is not tolerable. That is why at sufficiently high pressures, one should apply a design in which the stress on a thin screening tube is equilibrated by a correspondingly reinforced stator construction. At sufficiently thin screens (below 2 mm), the calculation of a screened electric motor can be carried out with the help of methods used in the calculation of normal motors, but in order to reduce the power losses in screens and magnetization current, some reduced flux densities should be applied, and in order to reduce heating, the current density in stator and rotor windings should be decreased [4.4].

In the case when the screened rotor is to be applied for operation with a conducting liquid in the gap (for instance, liquid metals, electrolytes, etc.), the layer of this liquid can be considered as an infinite number of empty (cup) rotors, whose slip is changing in a continuous manner, from $s = 1$ (close to the surface of stator) to $s = s_R$ (close to the surface of rotor). The magnetic field of such a machine is a result of the superposition of the fields of stator and of all "rotors."

4.7 SCREENING IN LARGE GENERATORS

4.7.1 Magnetic Screening and Shaping the Field

The most important problems in building large generators, similar to transformers, include additional losses in constructional elements, and especially in clamping plates or stator core. Harmful edge phenomena in solid rotor, especially on its ends, can be caused by two principal reasons: (1) displacing the field toward the endwindings due to eddy-current reaction (J. Turowski [1.16], J. Lasocinski [6.11]), and (2) influence of the stray field of endwindings of stator and rotor [1.16].

Application in these places of bronze or nonmagnetic steel frequently does not bring the effect of reduction of local excessive heating, due to considerable thickness of these parts. Further reduction of stray losses in these parts can be achieved by the application of magnetic screens (packages of electric iron sheets) or electromagnetic screens (copper or aluminum).

4.7.1.1 Magnetic Screening in Large Generators

One of the best solutions for screening of the end parts of a turbogenerator is the application of packaged magnetic screens (Figure 4.27), especially simultaneously both (a) and (c) in Figure 4.27. They are shunting the stray flux by attracting it from solid metal parts and ensuring a more beneficial distribution of leakage field, both from the point of view of additional losses as well as electrodynamic forces (Turowski [4.20]).

Theoretical calculation of the magnetic field in such a complex system as the space occupied by endwindings is troublesome (J. Turowski [1.16], pp. 191–202) and always needs an application of rough simplifications, which can be a source of significant errors. Especially dangerous for this area is the operation of a turbine generator at underexitation ([1.16], pp. 193–194).

For a general analysis of the endwinding region of a turbogenerator, the FEM in connection with the magnetic vector potential T (2.66) may be a good analysis tool. If we assume that only the axial component T_z exists, then $J_z = 0$, but the radial and circumferential components still exist. From Equations 2.2, 2.66, and 2.67, we obtain the equation (Jack et al. [4.9])

$$\frac{1}{r}\frac{\partial}{\partial r}\left(\frac{r}{\sigma_\theta}\frac{\partial T_z}{\partial r}\right)+\frac{1}{r}\frac{\partial}{\partial \theta}\left(\frac{r}{r\sigma_r}\frac{\partial T_z}{\partial \theta}\right)=\mu_z\frac{\partial}{\partial t}\left(T_z-\frac{\partial V_\mu}{\partial z}\right) \tag{4.90}$$

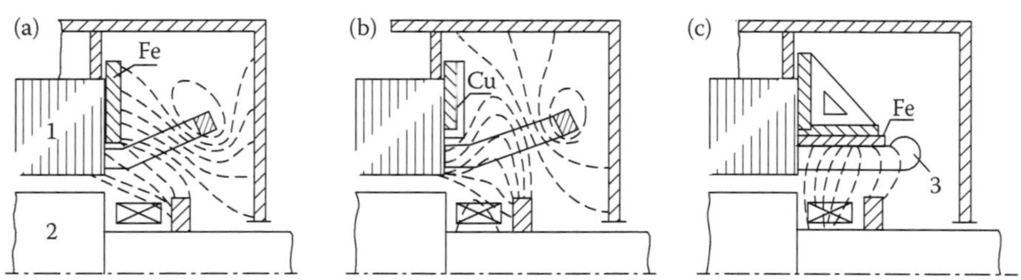

FIGURE 4.27 Configuration of leakage field of endwindings and reduction of additional losses in large turbogenerators with the help of cylindrical screen: 1—stator, 2—rotor, 3—endwinding; (a) Fe—a magnetic screen, packaged in the axial and radial direction, attracting the magnetic field; (b) Cu—an electromagnetic, conducting screen, displacing the magnetic field, but concentrating the edge field; (c) Fe—a magnetic prolongation of the active iron, with the purpose to improve the compensation of ampere-turns of endwindings of the stator and rotor. (Adapted from Turowski J.: *Calculations of Electromagnetic Components of Electric Machinery and Equipment.* (in Polish). Warsaw: WNT, 1982; and (in Russian). Moscow: Energoatomizdat, p. 193, 1986 [4.7].)

which can be resolved with the help of the finite element method (FEM) [4.9]. For conducting regions, Equation 4.90 should be resolved together with the equation

$$\mathrm{div}(\mu T) - \mathrm{div}(\mu \text{ grad } V_{\mu}) = 0 \tag{4.91}$$

which follows from the substitution (2.67) $\boldsymbol{H} = \boldsymbol{T} - \text{grad } V_m$ into (2.4) div $\boldsymbol{B} = 0$.

4.7.1.2 Screening of Windings and Conductors

The teeth of the magnetic core in an electric machine are also a type of magnetic screen for slots in which the winding is located, except the bottom at turbogenerator short circuit (Section 8.4 and [8.17]). Due to that, almost all the flux density lines are concentrated in the teeth, and in the slot the flux is rarefied, practically the whole circumferential force, acting on the armature of a loaded electric machine, is carried by the teeth. Bars of winding, instead, carry moderate forces of attraction to the iron surface.

It is an exceptionally positive phenomenon due to the limited mechanical strength of winding insulation. At a large saturation, however, in the places of narrowing, or at machine short-circuit, the lines of flux density are expulsed from iron to the slot, and the mechanical loading of the winding bars and its insulation grow.

The latter effect is especially dangerous in the case of an impulse short-circuit of large turbogenerators, when the attraction forces can cause erosion of the bottom insulation in the slot (J. Turowski [1.16], pp. 202–206 and [8.17]; Savini, Turowski [8.15], [8.16]). Reports of CIGRE'84 SC-11 (Kranz R.D. Special Report 11-00), manufacturers* and users† confirmed that the problem of the force in slots of large turbogenerators has recently become one of the most dangerous ones from the point of view of safety and reliability of these machines. The decisive factors are iron saturation and slot proportions [8.17].

In a similar way, if a conducting wire is placed in an iron tube (Figure 4.4), it can be released to an arbitrary degree from the action of mechanical forces caused by the external field (see Section 8.3). Then the whole load is taken over by the steel screen (tube). Such a screen in an alternating field should be packaged in the plane perpendicular to the conductor axis (Figure 4.1).

Such a solution would be irrational and, hence, in alternating currents one applies *electromagnetic screens* (see Section 4.9), which also reduces the external field close to conductors, and are cheaper and simpler. It should be emphasized that the magnetic screening of electric machine windings does not reduce electromotive force induced in the armature bars. The flux density in the region occupied by the bar is, in fact, weakened, but the flux density lines "jumping" from tooth to tooth cross the slot with correspondingly higher velocity.

4.7.2 Electromagnetic Screening in Generators

Screening plates of turbogenerators are made of a well-conducting, nonmagnetic metal. They have a form of a ring (Figure 4.27b) and are placed between the

* Petel M.R. Crashing stresses in hollow conductors in large water-cooled generator stator slot. *IEEE Trans.* PAS 1985, No 5.

† Wei C., Nai S. Analysis forces in stator slots. *ICEM'88*, Pisa, Vol. I.

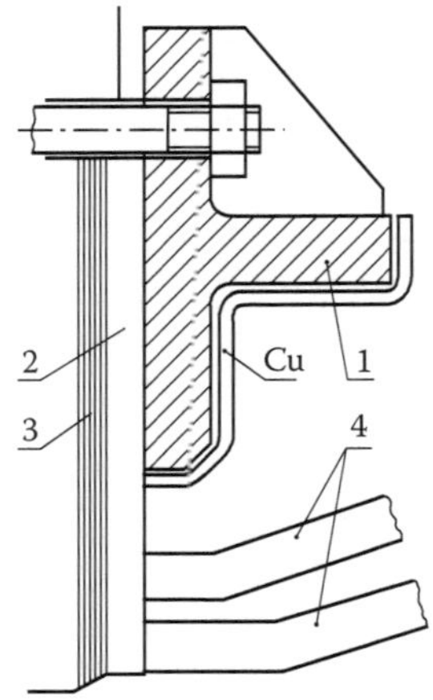

FIGURE 4.28 A copper screen of the periphery flange (1) of an 800 MVA turbogenerator, (Adapted from Doi S., Ito K., and Nonaka S.: *IEEE Transactions*, PAS-104(7), 1985, 1856–1862), (2) distancing plate, (3) package of stator core, (4) endwinding of the stator windings.

endwinding and peripheral packages of stator core (Figure 4.28). The magnetic field in the turbogenerator endwinding region can be calculated with the help of the reluctance network method (RNM) [4.5], [4.19], or FEM [4.7], [4.9]. After the determination of the leakage field, the power losses in screen, P_0, as per Ref. [4.7], can be calculated from the formulae

$$P_0 = \frac{\omega^2 B^2 d^2 \sigma}{24}\left[\frac{3}{kd}\left(\frac{\text{sh}\, kd - \sin kd}{\text{ch}\, kd - \cos kd}\right)\right] \quad \text{for } d < 2\sqrt{2}\delta \tag{4.92}$$

and

$$\frac{\omega^2 B^2 \sigma}{24}\left(3\delta \frac{dl}{d+l}\right) \quad \text{for } d > 2\sqrt{2}\delta \tag{4.93}$$

where B is the flux density of the flux penetrating the screen; δ is given by Equation 2.181.

The work [4.7] also contains results of thermal investigations.

4.8 SCREENING AT INDUCTION HEATING

The aim of induction surface hardening of machine elements, for instance gears, is to provide a high hardness to only working sections of the teeth, and to maintain plastic properties of internal or adjacent portions of the material, that is, surface hardening (Figure 6.17).

At thermal processing of big gearwheels (with modulus higher than 6) one often applies a successive processing of every tooth. In that case, for protection of neighboring teeth against dehardening due to repeated heating, protection screens made of copper sheets of thickness about 2 mm are superimposed (Figure 4.29a).

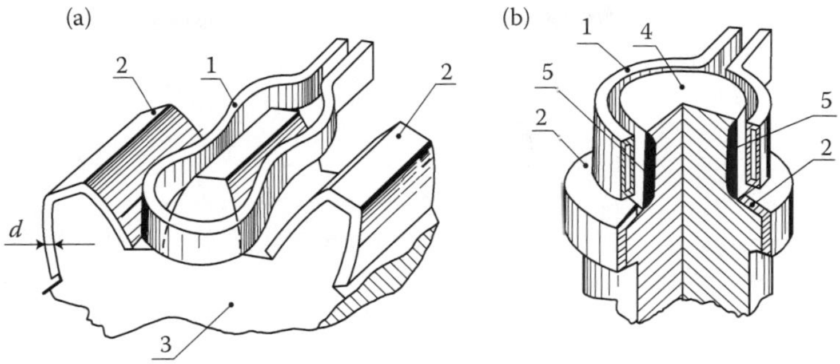

FIGURE 4.29 Application of protecting screens at (a) successive induction hardening of teeth of gear wheel; (b) at thermal processing of shaft neck; 1—inductor of high frequency, 2—copper screens, 3—gear wheel, 4—shaft, 5—heated region. (Adapted from Lozinskiy M. G. Surface hardening and induction heating of steel (in Russian), Moscow 1949. "Mashgiz" [2] in [1.15/1]; courtesy of "Mashgiz".)

Figure 4.29b presents, instead, the application of electromagnetic screens for the protection of shaft collar, which we do not want to heat at surface hardening of the shaft neck. Such screens, which at the same time operate as thermal screens, are made of electrolytic copper.

At the thickness of a hardened layer of the order of two equivalent depths of the AC field penetration $\delta = 1/a_p\sqrt{2/\omega\mu_s\sigma}$ into steel (Table 7.4/1), the thickness of screen should not be smaller than 4 times the equivalent depth of the AC field penetration into copper. According to dependence (2.181), the equivalent depth of penetration of electromagnetic wave into copper, at temperature 20°C, equals $\delta_{Cu} \approx 66/\sqrt{f}$ mm. Therefore, the thickness of copper screen at 200 kHz should be 0.6 mm, and at frequency $f = 10$ kHz, it should be 2.5 mm.

Since the power dissipated in the metal is proportional to $\sqrt{\mu/\sigma}$ (see Equation 3.10a), copper will consume 30–50 times less power than steel.

Sometimes, a deep superheating of spots neighboring with parts protected against the action of the electromagnetic field is necessary. In such a case, it is recommended to apply screens in the form of welded construction empty inside. Through the space between the walls of such a screen (of dimensions 1–1.5 mm) a cooling water is passed. The thickness of each wall of the screen should not be smaller than the depth of penetration of eddy currents into copper (3δ) at the applied frequency.

4.9 SCREENING OF BARS AND CONDUCTOR WIRES

4.9.1 Cylindrical Screen of a Single Conductor Wire

The aim of screening of bars and conductors is to reduce the external field in the direct neighborhood of conductor, for reduction of electrodynamic forces acting on the conductor as well as for the safety of the equipment (protection against arc propagation at a multiphase short-circuit), safety of staff (earthed housing of bars), model investigations, and so on (J. Turowski [2.33], [2.41]).

Magnetic screening of conductors (Figures 4.1 and 4.4), although possible in principle, has so far not found a broader application in the domain of AC currents. Electromagnetic screening has appeared much more beneficial.

4.9.1.1 Electromagnetic Screens

Let us consider the electromagnetic field and power losses in a metal cylindrical screen surrounding a single conductor with AC current (Figure 4.30). Let us assume at the same time that the second conductor is in such a distance that its field does not have a significant influence on the phenomena occurring in the screen. The electromagnetic field of the conducting wire has inside the screen wall only the axial component of electric field intensity: $\boldsymbol{E}_z = \boldsymbol{E}$. The magnetic field, instead, has only the component $\boldsymbol{H}_\theta = \boldsymbol{H}$ with the surface value

$$H_{m0} = \frac{\sqrt{2}I}{2\pi R_0} \tag{4.94}$$

Since both the field vectors $\boldsymbol{E}_\mathrm{m}$ and $\boldsymbol{H}_\mathrm{m}$ have only one component each, on further considerations, we can, in accordance with Equations 2.151 and 2.152c, limit our investigations to their vector modulus E_m and H_m and the scalar equation

$$\frac{\partial^2 E_m}{\partial r^2} + \frac{1}{r}\frac{\partial E_m}{\partial r} = \alpha^2 E_\mathrm{m} \tag{4.95}$$

Applying now the substitution (2.153)

$$p = p = r\sqrt{-\alpha^2} = \mathrm{j}\alpha r = x\mathrm{j}\sqrt{\mathrm{j}}, \quad \text{where } x = r\sqrt{\omega\mu\sigma} \tag{4.96}$$

that is, $p = p = \mathrm{j}(1+\mathrm{j})r\sqrt{\omega\mu\sigma/2} = -(1-\mathrm{j})rk = -xe^{j\frac{\pi}{4}}$

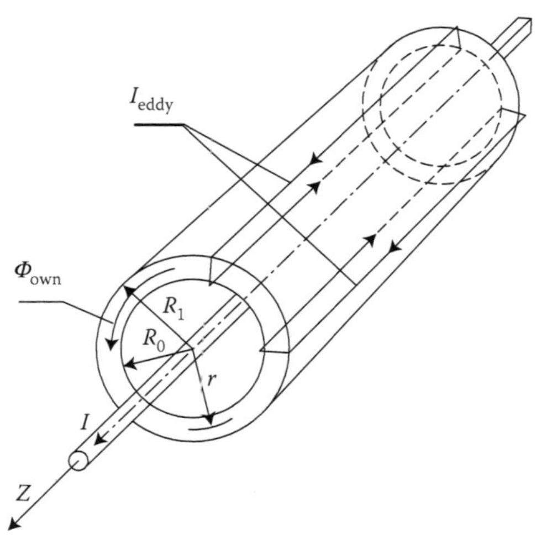

FIGURE 4.30 Eddy currents (I_{eddy}) in the screen of a single conducting wire; Φ_{own}—the own flux inside the screen wall.

we obtain the equation of type (2.155)

$$\frac{\partial^2 E_m}{\partial p^2} + \frac{1}{p}\frac{\partial E_m}{\partial p} + E_m = 0 \tag{4.97}$$

solutions of which can have the form (2.155a) with the complex argument $x\mathrm{j}\sqrt{\mathrm{j}}$..

The solution of Bessel's equation with the complex argument $x\mathrm{j}\sqrt{\mathrm{j}}$ can be expressed with the help of Thomson (Kelvin) function of the *first*, *second*, and *third* type, for which there exist tables [2.1] as

$$\left.\begin{aligned} \operatorname{ber} x + \mathrm{j}\operatorname{bei} x &= I_0(x\mathrm{j}\sqrt{\mathrm{j}}) \\ \ker x + \mathrm{j}\operatorname{kei} x &= K_0(x\sqrt{\mathrm{j}}) \\ \ker x - \mathrm{j}\operatorname{kei} x &= K_0(x\mathrm{j}\sqrt{\mathrm{j}}) \\ \operatorname{her} x + \mathrm{j}\operatorname{hei} x &= H_0^{(1)}(x\mathrm{j}\sqrt{\mathrm{j}}) \end{aligned}\right\} \tag{4.98}$$

where

$$\operatorname{her} x = \frac{2}{\pi}\operatorname{kei} x \quad \text{and} \operatorname{hei} x = -\frac{2}{\pi}\ker x \tag{4.99}$$

Since the argument (4.96) exists only in the first, third, and fourth of Equations 4.98, more convenient here is to utilize the second form of solution (2.155a) containing

$$E_m(p) = \mathrm{C}_1 H_0^{(1)}(p) + \mathrm{C}_2 H_0^{(2)}(p) \tag{4.100}$$

At the infinite value of the complex argument, the Hankel's functions acquire zero values, namely, $H^{(1)} = 0$ when the imaginary part of argument is positive, and $H^{(2)} = 0$ when it is negative [2.11]. In the opposite case, both functions grow to infinity. Since here $\operatorname{Im}\{x\mathrm{j}\sqrt{\mathrm{j}}\} > 0$, and the increase of electric field intensity to infinity is not possible, the function $H_0^{(2)}$ in formula (4.100) must disappear, which means the constant $\mathrm{C}_2 = 0$.*

Hence

$$E_m(p) = \mathrm{C}_1 H_0^{(1)}(x\mathrm{j}\sqrt{\mathrm{j}}) = \mathrm{C}_1(\operatorname{her} x + \mathrm{j}\operatorname{hei} x) \tag{4.101}$$

After substitution of Equation 4.99 into formula (4.101), we obtain

$$E_m = \frac{2\mathrm{C}_1}{\pi}(\operatorname{kei} x - \mathrm{j}\ker x) \tag{4.102}$$

* It concerns thick walls ($d > \lambda$), because in opposite case the case shown in Figure 4.11b occurs, and $\mathrm{C}_2 \neq 0$. In such a situation Equation 4.28 is valid.

From Equations 4.102 and 2.2, we get the magnetic field intensity (J. Turowski [2.33])

$$E_m = -\frac{\mathrm{j}}{\omega\mu}\frac{\mathrm{d}E_m}{\mathrm{d}r} = -\mathrm{j}\sqrt{\frac{\sigma}{\omega\mu}}\frac{2\mathrm{C}_1}{\pi}(\mathrm{kei}'x - \mathrm{j}\,\mathrm{ker}'x) \tag{4.103}$$

On the internal surface of screen $x = x_0 = R_0\sqrt{\omega\mu\sigma}$

$$H_{m0} = -\mathrm{j}\sqrt{\frac{\sigma}{\omega\mu}}\frac{2\mathrm{C}_1}{\pi}(\mathrm{kei}'x_0 - \mathrm{j}\,\mathrm{ker}'x_0)$$

from which

$$\frac{2\mathrm{C}_1}{\pi} = \mathrm{j}\sqrt{\frac{\omega\mu}{\sigma}}\frac{H_{m0}}{\mathrm{kei}'\,x_0 - \mathrm{j}\,\mathrm{ker}'\,x_0}$$

As a result of substituting $2\mathrm{C}_1/\pi$ into formulae (4.102) and (4.103), we obtain

$$E_m = \mathrm{j}\sqrt{\frac{\omega\mu}{\sigma}}H_{m0}\frac{\mathrm{ker}\,x + \mathrm{j}\,\mathrm{kei}\,x}{\mathrm{ker}'\,x_0 + \mathrm{j}\,\mathrm{kei}'\,x_0} \tag{4.104}$$

and

$$H_m = H_{m0}\frac{\mathrm{ker}'\,x + \mathrm{j}\,\mathrm{kei}'\,x}{\mathrm{ker}'\,x_0 + \mathrm{j}\,\mathrm{kei}'\,x_0} \tag{4.105}$$

The wave impedance of the screen material

$$Z = \frac{E_m}{H_m} = \mathrm{j}\sqrt{\frac{\omega\mu}{\sigma}}\frac{\mathrm{ker}\,x + \mathrm{j}\,\mathrm{kei}\,x}{\mathrm{ker}'\,x + \mathrm{j}\,\mathrm{kei}'\,x} \tag{4.106}$$

The Thomson's functions ker x and kei x and their derivatives vary in dependence of x as it is shown in Figure 4.31 [2.1]. From Figure 4.31, it follows that the amplitudes of the electric field (E_m) and that of the magnetic field intensity (H_m) decrease very fast with the increase of the distance from the conducting wire axis, and somewhat slower with the increase of frequency, permeability, and conductivity.

Moving away from the axis of the conducting wire, the phase of the vectors $\boldsymbol{E}_\mathrm{m}$ and $\boldsymbol{H}_\mathrm{m}$ changes. A theoretical increase of the field to infinity at $r \to 0$ (and $R_0 \to 0$) is quite natural at the assumption of an infinitely thin (dimensionless) filament of current.

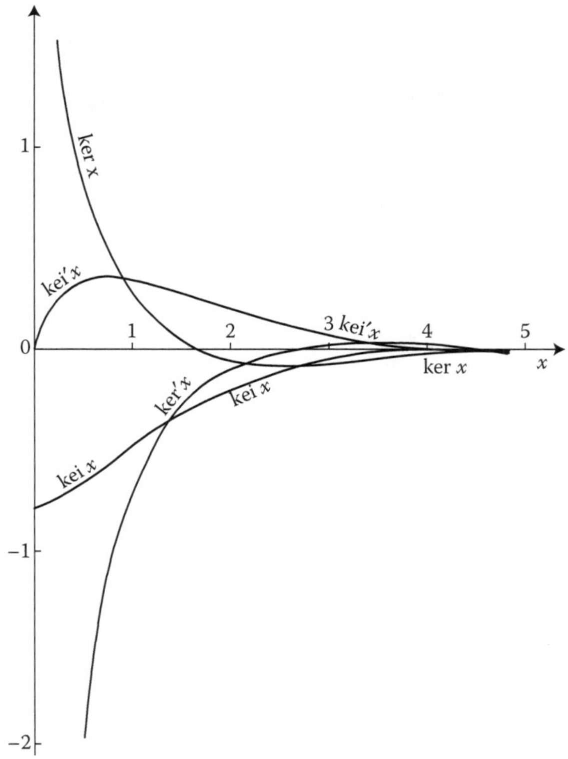

FIGURE 4.31 Plots of the Thomson's functions and their derivatives.

In order to estimate the order of magnitude of the argument x, let us assume for steel $\mu = (500 \text{ to } 1000) \times 0.4 \times \pi \times 10^{-6}$ H/m; $\sigma_{\text{Steel}} = 7 \times 10^6$ S/m (at 20°C), and for aluminum $\sigma_{\text{Al}} = 35 \times 10^6$ S/m; $f = 50$ Hz. Then, we shall obtain

- For steel

$$\frac{x}{r} = \sqrt{\omega\mu\sigma} = 11.9 \text{ to } 16.7 \text{ cm}^{-1} \tag{4.107}$$

- For aluminum

$$\frac{x}{r} = \sqrt{\omega\mu\sigma} = 1.2 \text{ cm}^{-1}$$

At a modeling frequency of 100 Hz, these numbers are, respectively, 17.9–24 cm^{-1} and 1.7 cm^{-1}.

As previous simple verification calculations have shown, even at values $x = 1$, one can utilize approximately (with accuracy of a few percent) asymptotic formulae for $x > 100$. At increasing x over 1, this error decreases rapidly. Since, usually in practice, $x \gg 1$ (r $\gg$ 1 cm), the error resulting from using the asymptotic formulae is negligibly small. Thanks to it, after performing several transformations, we obtained (J. Turowski [2.33]) convenient asymptotic formulae that can be very helpful in developing modern rapid design programs based on fast and simple calculations:

$$\ker x + \mathrm{j}\,\mathrm{kei}\, x = \sqrt{\frac{\pi}{2x}}\mathrm{e}^{-(1+\mathrm{j})\frac{x}{\sqrt{2}}}\mathrm{e}^{-\mathrm{j}\frac{\pi}{8}}\left(1 - \frac{1-\mathrm{j}}{8\sqrt{2x}}\right) \tag{4.108}$$

$$\ker x - \mathrm{j}\,\mathrm{kei}\, x = \sqrt{\frac{\pi}{2x}}\mathrm{e}^{-(1-\mathrm{j})\frac{x}{\sqrt{2}}}\mathrm{e}^{-\mathrm{j}\frac{\pi}{8}}\left(1 - \frac{1+\mathrm{j}}{8\sqrt{2x}}\right) \tag{4.109}$$

$$\ker' x + \mathrm{j}\,\mathrm{kei}' x = -\sqrt{\frac{\pi}{2x}}\mathrm{e}^{-(1+\mathrm{j})\frac{x}{\sqrt{2}}}\mathrm{e}^{-\mathrm{j}\frac{\pi}{8}}\left[1 + \frac{3(1-\mathrm{j})}{8\sqrt{2x}}\right] \tag{4.110}$$

$$\ker' x - \mathrm{j}\,\mathrm{kei}' x = -\sqrt{\frac{\pi}{2x}}\mathrm{e}^{-(1-\mathrm{j})\frac{x}{\sqrt{2}}}\mathrm{e}^{-\mathrm{j}\frac{\pi}{8}}\left[1 + \frac{3(1+\mathrm{j})}{8\sqrt{2x}}\right] \tag{4.111}$$

Substitution of the latter formulae into formulae (4.104) through (4.106) yields

$$\begin{aligned} E_m &= -\sqrt{\frac{\omega\mu}{\sigma}}H_{m0}\sqrt{\frac{R_0}{r}}\mathrm{e}^{-(1+\mathrm{j})k(r-R_0)}\mathrm{e}^{\mathrm{j}\frac{\pi}{4}}\frac{(8\sqrt{2}x - 1 + \mathrm{j})R_0}{(8\sqrt{2}x_0 + 3 - 3\mathrm{j})r} \\ &\approx -\sqrt{\frac{\omega\mu}{\sigma}}H_{m0}\sqrt{\frac{R_0}{r}}\mathrm{e}^{-\alpha(r-R_0)}\mathrm{e}^{\mathrm{j}\frac{\pi}{4}} \end{aligned} \tag{4.112}$$

$$H_m = H_{m0}\sqrt{\frac{R_0}{r}}\mathrm{e}^{-(1+\mathrm{j})k(r-R_0)}\mathrm{e}^{\mathrm{j}\frac{\pi}{4}}\frac{(8\sqrt{2}x + 3 + 3\mathrm{j})r_0}{(8\sqrt{2}x_0 + 3 - 3\mathrm{j})r} \approx H_{m0}\sqrt{\frac{R_0}{r}}\mathrm{e}^{-\alpha(r-R_0)} \tag{4.113}$$

$$Z = -(1+\mathrm{j})\sqrt{\frac{\omega\mu}{2\sigma}}\frac{(8\sqrt{2}x - 1 + \mathrm{j})R_0}{(8\sqrt{2}x + 3 - 3\mathrm{j})r} \approx -\frac{\alpha}{\sigma} \tag{4.114}$$

On the internal surface of the screen, $r = R_0 (x = x_0)$, hence

$$E_{m0} = -\sqrt{\frac{\omega\mu}{\sigma}}H_{m0}\mathrm{e}^{\mathrm{j}\frac{\pi}{4}}\frac{8\sqrt{2}x - 1 + \mathrm{j}}{8\sqrt{2}x_0 + 3 - 3\mathrm{j}} \approx -\sqrt{\frac{\omega\mu}{\sigma}}H_{m0}\mathrm{e}^{\mathrm{j}\frac{\pi}{4}} \tag{4.115}$$

$H_m = H_{m0}$ and $Z_0 = -\alpha/\sigma$.

For most cases appearing in practice, approximate formulae are not only satisfactory, but they clearly demonstrate parametric dependences. This is very convenient for qualitative evaluation of processes and, what is even more important, can be very useful in he development of modern simple and rapid calculation programs. More accurate formulae should be applied only at small diameters of screens (<2 cm), screens from nonferrous (color) metals, and at frequencies lower than 50 Hz.

4.9.1.2 Power Losses in Screen

The *apparent Poynting vector* can be obtained by multiplying the electric field intensity, as per formula (4.112), by conjugate value of the magnetic field intensity, as per formula (4.113):

$$\begin{aligned} S_S &= S_p + jS_q = \frac{1}{2}(\boldsymbol{E}_m \times \boldsymbol{H}_m{}^*) \\ &= \mathbf{1}_n \sqrt{\frac{\omega\mu}{\sigma}} \frac{H_{m0}^2}{2} \sqrt{\frac{R_0}{r}} e^{-2k(r-R_0)} e^{j\frac{\pi}{4}} \frac{(8\sqrt{2}x - 1 + j)(8\sqrt{2}x + 3 + 3j)R_0^2}{(8\sqrt{2}x_0 + 3 - 3j)(8\sqrt{2}x_0 + 3 + 3j)r^2} \\ &\approx -\mathbf{1}_n \sqrt{\frac{\omega\mu}{\sigma}} \frac{H_{m0}^2}{2} \sqrt{\frac{R_0}{r}} e^{-2k(r-R_0)} e^{j\frac{\pi}{4}} \end{aligned} \tag{4.116}$$

On the internal surface of screen, $r = R_0(x = x_0)$, and hence the apparent power (in VA/m^2) penetrating the screen is

$$S_S = S_p + jS_q = -\mathbf{1}_n \sqrt{\frac{\omega\mu}{\sigma}} \frac{H_{m0}^2}{2} e^{j\frac{\pi}{4}} \frac{8\sqrt{2}x_0 - 1 + j}{8\sqrt{2}x_0 + 3 - 3j} \approx -\mathbf{1}_n(1 + j)\sqrt{\frac{\omega\mu}{2\sigma}} \frac{H_{m0}^2}{2} \tag{4.117}$$

The minus sign here means that the energy flow is directed from the conducting wire outwards.

Let us consider the influence of various factors on power losses:

1. *Influence of screen curvature on losses:* If we assume that the thickness of the screen is not large in comparison with its diameter, then one can skip the screen curvature and calculate the power by multiplying the *Poynting vector* (4.117) by the doubled surface area of the screen. In the opposite case, the *Poynting vector* (4.117) should be multiplied by the internal surface area of the screen and by the coefficient

$$\left(1 + \frac{D_0}{D_1}\right) \tag{4.118}$$

2. *Influence of variability of the steel magnetic permeability, hysteresis losses, and loss from eddy-current anomaly:* If the screen is made of steel, the ferromagnetic properties of the material should be considered. Owing to uniformity of the system along the conductor axis, the permeability (in case of steel screen) on the surface, necessary for calculation, can be determined from the curve of permeability $\mu(H)$ for constructional steel (Figures 1.27 and 1.29). Variability of the permeability inside steel and the relatively small hysteresis loss can be taken into account by multiplying the active

component of the *Poynting vector* (4.117) by the constant linearization coefficient a_p (see Section 7.2 and Table 7.4).

As has been shown by results of the author's investigations (J. Turowski [2.41]) and others, the losses from the "eddy-current anomaly" (1.38) in constructional steel do not have such significance as in high-quality electrotechnical sheets. They are contained completely in the adopted semiempirical coefficient a_p, potentially shifting it at most toward the higher limit of its range (1.3–1.5), while the most often used is its average value of $a_p \approx 1.4$ (7.26). This suggestion is also supported by opinions of other researchers who particularly investigated the problem of losses from "anomaly" of eddy currents in various steel sorts.*

3. *Influence of thickness of screen walls:* At thicknesses of screen walls larger than the length of the electromagnetic wave in metal (2.178), there is no need to take into account the influence of the thickness (4.120), because in this range, the eddy-current losses are exactly the same as in infinitely thick (half-space) sheets (except dependence (4.118)). However, such circumstances can appear only in steel screens of thickness of several millimeters, or in screens made of nonferrous metal of thickness of several centimeters, or at a higher frequency—on the order of hundreds of hertz. At normal wall thicknesses and power frequencies, one should take into account the internal reflection of electromagnetic wave from screen walls, because a wave incident on the screen surface penetrates to the opposite wall, not being able to be extinguished before reaching the opposite wall. A detailed analysis of this phenomenon with the assumption that identical plane waves are incident on the wall from both sides symmetrically (see Section 4.4.2) led to the conclusion that the active power dissipated in the wall should be multiplied by the coefficient ζ, and the reactive power by the coefficient ψ according to formula (4.50). Graphs of both these coefficients for metal with a constant permeability (ζ, ψ) and for steel (ζ_{Fe}, ψ_{Fe}) are presented in Figure 4.14. Bearing in mind the above remarks, we can finally write the relatively simple and rapid parametric formulae for the losses.

The active power losses

$$P = \pi D_0 l\left(1 + \frac{D_0}{D}\right) a_p \zeta \sqrt{\frac{\omega\mu_s}{2\sigma}}\frac{H_{m0}^2}{2} = a_p \zeta P_0 \tag{4.119}$$

where the power losses in thick screens with $\mu_S = \text{const}$ and $H_{m0} = \dfrac{1}{\sqrt{2}\pi R}$ are

$$P_0 = \pi D_0 l\left(1 + \frac{D_0}{D}\right)\sqrt{\frac{\omega\mu_s}{2\sigma}}\frac{H_{m0}^2}{2} \tag{4.120}$$

* Breilsford, JIEE No. 43/1948; Steward, Proc. IEE 1950; Drushinin, Electricestvo No. 8/1956, and others.

The consumption of reactive power

$$Q = \mathrm{a_q}\psi P_0 \tag{4.121}$$

and the consumption of apparent power

$$S = \sqrt{P^2 + Q^2} = \sqrt{\mathrm{a_p^2}\zeta^2 + \mathrm{a_q^2}\psi^2}\, P_0 \tag{4.122}$$

The power coefficient of the screen as a power receiver

$$\cos\varphi = \frac{P}{S} = \frac{\mathrm{a_p}\zeta}{\sqrt{\mathrm{a_p^2}\zeta^2 + \mathrm{a_q^2}\psi^2}} \tag{4.123}$$

In the case of nonmagnetic screens, we introduce to formulae (4.119) through (4.123) the semiempirical linearization coefficients $a_p = 1$, and in the case of steel screens $\mathrm{a_p} \approx 1.4$ and $\mathrm{a_q} \approx 0.85$ (7.26).

In the case of screen shapes different than a cylindrical tube, one can also utilize approximately the above given formulae, after determination of the magnetic field intensity H_m for a mean radius corresponding to the profile of the screen cross section and multiplying the Poynting vector by the surface area of the actual screen.

As we can see from formula (4.123), the cos φ of the screen considered as a power receiver depends solely on the parameter kd of screen. It has, starting from $kd \approx 3$ (Figure 4.14), practically a constant value, approaching to about 0.8, which corresponds to tg $\varphi = 0.7$ and $\varphi \approx 35.5°$.

EXAMPLE

Calculate power losses and the power factor for a steel screen in the form of cylindrical tube of length 2.94 m, diameters 2.5/2.88 cm, $\sigma_{Fe} = 7 \times 10^6$ S/m, at the bar current $I = 150$ A, and frequency $f = 50$ Hz.

SOLUTION

From formula (4.94): $H_m = \dfrac{\sqrt{2}I}{2\pi R_0} = \dfrac{\sqrt{2}\cdot 150}{2\pi(2.5\cdot 10^{-2}/2)} = 2700$ A/m.

From Figure 1.29 (curve 4): $\mu_r = 420$.

According to formula (2.140)

$$kd = \sqrt{\frac{\omega\mu\sigma}{2}}d = \sqrt{\frac{2\pi 50\times 420\times 0.4\pi\times 10^{-6}\times 7\times 10^{6}}{2}}\cdot\frac{0.38\times 10^{-2}}{2} = 1.44$$

From Figure 4.14, we find $\zeta_{Fe} = 0.8$; $\psi_{Fe} = 1.35$; cos $\varphi_{Fe} = 0.55$; then, as per formula (4.120)

$$P = \pi\times 2.5\times 10^{-2}\times 2.94\left(1+\frac{2.5}{2.88}\right)\sqrt{\frac{2\pi 50\times 420\times 0.4\pi\times 10^{-6}}{2\cdot 7\times 10^{6}}}\times\frac{2700^2}{2} = 182\text{ W}$$

and as per formula (4.119): $P = 1.4 \times 0.8 \times 182 = 203$ W; from measurements (J. Turowski et al. [4.24]), it was obtained $P = 184$ W. As per formula (4.121): $Q = 0.85 \times 1.35 \times 182 = 210$ var; from measurement [4.21], it was obtained $Q = 250$ var.

As per formula (4.122): $S = \sqrt{1.4^2 \times 0.8^2 + 0.85^2 \times 1.35^2} \times 182 = 294$ VA; from measurement [4.24], it was obtained $S = 310$ VA. $\cos\varphi = (203/294) = 0.69$; from measurement [4.24], it was obtained $\cos\varphi = 0.60$.

EXAMPLE

Calculate the power losses in an aluminum cylindrical tube screen with an internal diameter of 0.85 m and a wall thickness of 5 mm, if the current of the bar placed in the center is 10,000 A ($\sigma_{Al} = 34.8 \times 10^6$ S/m).

SOLUTION

From formula (4.94): $H_{m0} = \sqrt{2} \times 10{,}000/\pi \times 0.85 = 53 \times 10^2$ A/m, and from formula (4.120), per 1 m along the axis of impenetrable screen, the losses are

$$\frac{P_0}{l} = \pi \times 0.85(1+1)\sqrt{\frac{2\pi \times 50 \times 0.4\pi \times 10^{-6}}{2 \times 34.8 \times 10^6}} \times \frac{(53 \times 10^2)^2}{2} = 180 \text{ W/m}$$

$$k = \sqrt{\frac{\omega\mu\sigma}{2}} = \sqrt{\frac{2\pi \times 50 \times 0.4\pi \times 10^{-6} \times 34.8 \times 10^6}{2}} = 82.8 \text{ m}^{-1}$$

$$kd = 82.8 \times 5 \times 10^{-3} = 0.414$$

The coefficient of loss reduction in the screen wall due to its "transparency" (penetrability), according to Figure 4.14, equals $\zeta = 0.02$. Losses in the wall of penetrable screen, per 1 m of length along the conductor axis, according to Equation 4.119, therefore equal

$$P_1 = ap\,\zeta\,P_0 = 1 \times 0.02 \times 180 \text{ W/m} = 3.8 \text{ W/m}$$

4.9.2 Cylindrical Screen in a Transverse, Uniform Field

The magnetic field in the aerial space outside (H_0) and inside (H_{int}) an infinitely long, cylindrical, metal screen (Figure 4.32) can be described with the help of Laplace's equation (2.33), in the cylindrical coordinates, resulting from the definition (2.24)

$$\frac{\partial^2 V_m}{\partial r^2} + \frac{1}{r}\frac{\partial V_m}{\partial r} + \frac{1}{r^2}\frac{\partial^2 V_m}{\partial \vartheta^2} = 0 \tag{4.124}$$

Applying the method of separation of variables ([4.3], p. 73), similarly as in Equation 2.144

$$V_m = R(r) \cdot \Theta(\vartheta) = R \cdot \Theta$$

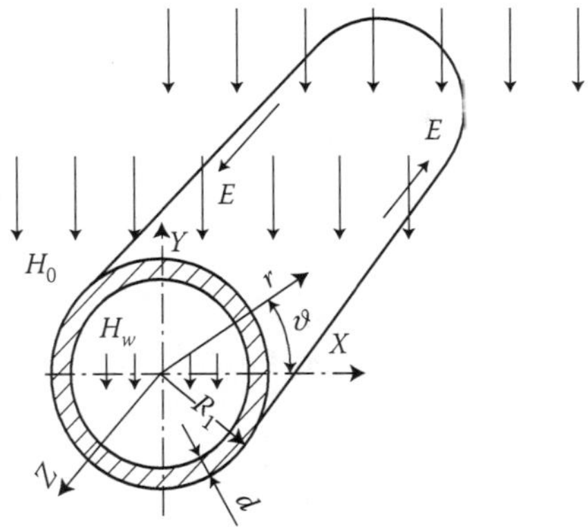

FIGURE 4.32 Cylindrical screen in a uniform transverse field.

we obtain two differential equations:

$$\frac{\partial^2 R}{\partial r^2} + \frac{1}{r}\frac{\partial R}{\partial r} - \beta^2 R = 0 \quad \text{and} \quad \frac{\partial^2 \Theta}{\partial \vartheta^2} + \beta^2 \Theta = 0 \tag{4.125}$$

The general solutions of Equations 4.125 have the form

$$R = \mathrm{C}_1 r + \frac{\mathrm{C}_2}{r} \quad \text{and} \quad \Theta = \mathrm{C}_3 \sin \beta\vartheta + \mathrm{C}_4 \cos \beta\vartheta$$

Since the magnetic potential cannot contain functions that are multiples of the angle ϑ, the constant β should equal unity. From the symmetry condition, instead, it follows that during the transition from the angle ϑ to $(\pi/2 + \vartheta)$, the function Θ should not change its sign. Hence, $\mathrm{C}_4 = 0$. From $\boldsymbol{H} = -\text{grad } V_m$, it follows, therefore, that the resultant field is

$$\boldsymbol{H} = \mathbf{1_r} \boldsymbol{H_r} + \mathbf{1}_\vartheta \boldsymbol{H}_\vartheta = \mathbf{1_r}\left(A - \frac{B}{r^2}\right)\sin \vartheta + \mathbf{1}_\vartheta \left(A + \frac{B}{r^2}\right)\cos \vartheta \tag{4.126}$$

Since at $r \to 0$, the magnetic field intensity cannot grow to infinity, and at $\vartheta = \pi/2 H_{\text{rint}} = H_{\text{int}}$, then inside the screen, $B = 0$ and $A = H_{\text{int}}$. Hence, the field components inside the screen ($r \le R_1 - d$):

$$H_{r,\text{int}} = H_{\text{int}} \sin \vartheta = pH_0 \sin \vartheta \qquad \text{and} \qquad H_{\vartheta,\text{int}} = H_{\text{int}} \cos \vartheta = pH_0 \cos \vartheta \tag{4.127}$$

where $p = H_{\text{int}}/H_0$ is the coefficient of screening.

Since, in turn, the field in infinity is not deformed by the influence of the screen, when $r \to \infty$ and $\vartheta = \pi/2$, $H = H_0$, from which $A = H_0$, and then for the field outside the screen ($r \geq R_1$), we find the components

$$H_{r0} = H_0\left(1 - \frac{R_1^2}{r^2}W\right)\sin\vartheta \qquad \text{and} \qquad H_{\vartheta 0} = H_0\left(1 + \frac{R_1^2}{r^2}W\right)\cos\vartheta \tag{4.128}$$

where $W = B/H_0R_1^2$ is a complex value, called the *coefficient of reflexive interaction* or *of screen reaction* [4.2], because the second component in formulae (4.128) is the measure of deformation of the external field caused by the presence of the screen. It can be used, for instance, for the detection of a metal mass; for example, a ship, in the region of the uniform magnetic field of the earth.

Inside the mass of screen walls, Equation 2.142 is valid. Since the field is crosswise, the magnetic field intensity will have only one component, along the axis ($E_z = E$), whereas $E_r = 0$. We can therefore utilize the scalar equation in cylindrical coordinates:

$$\nabla^2 E_m = \frac{\partial^2 E_m}{\partial r^2} + \frac{1}{r}\frac{\partial E_m}{\partial r} + \frac{1}{r^2}\frac{\partial^2 E_m}{\partial \vartheta^2} + \frac{\partial^2 E_m}{\partial z^2} = \alpha^2 E_m \tag{4.129}$$

Assumption of an *infinitely long tube* for the analysis eliminates the last component ($\partial E_m/\partial z = 0$). Assuming that the screen has thin walls ($d \ll R$) made of a nonmagnetic metal, one can neglect variation of r within the thickness of the metal wall. Kaden [4.2] reduced the problem to the solution of the equation:

$$\frac{\partial^2 E_m}{\partial r^2} + \frac{1}{R^2}\frac{\partial^2 E_m}{\partial \vartheta^2} = \alpha^2 E_m \tag{4.130}$$

which, after solving with the method of separation of variables (2.144), allowed [4.3] to finally obtain the formula for the *screening coefficient*

$$P = \frac{H_{\text{int.}}}{H_0} = \frac{1}{\text{ch}\,\alpha_1 d + \frac{1}{2}\left(K_1 + \frac{1}{K_1}\right)\text{sh}\,\alpha_1 d} \tag{4.131}$$

and for the *coefficient of reflexive interaction* (*screen reaction*)

$$W = \frac{\frac{1}{2}\left(K_1 - \frac{1}{K_1}\right)\text{sh}\,\alpha_1 d}{\text{ch}\,\alpha_1 d + \frac{1}{2}\left(K_1 + \frac{1}{K_1}\right)\text{sh}\,\alpha_1 d} = \frac{1}{2}p\left(K_1 - \frac{1}{K_1}\right)\text{sh}\,\alpha_1 d \tag{4.132}$$

where $K_1 = (\mu_0/\mu)\alpha_1 R_1$ and $\alpha_1 = \sqrt{\alpha^2 + 1/R_1^2}$.

Usually, $|\alpha_1| \cdot R_1 \gg 1$ and one can assume $\alpha_1 \approx \alpha = (1 + j)k$.

Analysis of formulae (4.131) and (4.132) leads to the following conclusions:

1. The *internal field*, similar to the external field, is uniform and has the same direction, but is shifted in phase due to the screen reaction. It is taken into account by the complex screening coefficient p (4.131).
2. At the frequency limit value of $f = 0$, the screening effect of iron screens ($\mu_{Fe} \gg \mu_0$) does not reduce itself to zero like in nonmagnetic screens, but we obtain magnetostatic screening caused by shunting of magnetic flux by such screen. The screening coefficient will be in this case

$$p = \frac{1}{1 + (1/2)(\mu/\mu_0) \cdot (d/R)} \quad (\text{at } \alpha \to 0 \text{ and } \mu \gg \mu_0) \tag{4.133}$$

The formula (4.133) agrees, with high accuracy ($d^2/R^2 \approx 0$), with formula (4.2) developed on the basis of the magnetostatic law. This indirectly confirms the adoption of the simplifications employed in Equation 4.130.

When the frequency increases, the iron screen begins to act as an electromagnetic one (see the curve A in Figure 10.30), for which $|K > 1|$.

The screening effect is often described with the help of *screen damping*:

$$b_S = \ln\left|\frac{H_0}{H_{int}}\right| = \ln\frac{1}{p} \tag{4.134}$$

A plot of the *screen damping* for a steel cylindrical screen in a transverse field, as a function of the coefficient $kd = d\sqrt{\omega\mu\sigma/2}$, is presented, among others, in the works of Kaden [4.2] and (J. Turowski [1.15/1], p. 177).

3. *Screen reflexive reaction* on an external field is equivalent to the action of a dipole placed in the cylinder axis (Figure 4.2). The electric moment of this dipole is proportional to the coefficient W defined by formula (4.132). When the frequency is increases, W changes asymptotically from zero to some limit value. This moment is also proportional to the square of the screen radius R. From Equations 4.128, it follows that the magnetic field intensity of the reflexive reaction decreases proportionally to square of the distance (r^2) from the screen.

The limit value of the coefficient W of screen reflexive reaction, at magnetostatic screening, equals –1, whereas at increasing frequency, this coefficient approaches +1.

4.9.3 Screens of Busbars in Generator Unit Systems of Power Stations

Metal screens (usually made of aluminum alloys) employed for the protection of busbars that connect large generators with generator transformers (Hologa [4.8]) operate in two types of field: (i) in its *own field* of the bar

$$H_{own} = \frac{\sqrt{2}I}{2\pi R} \tag{4.135}$$

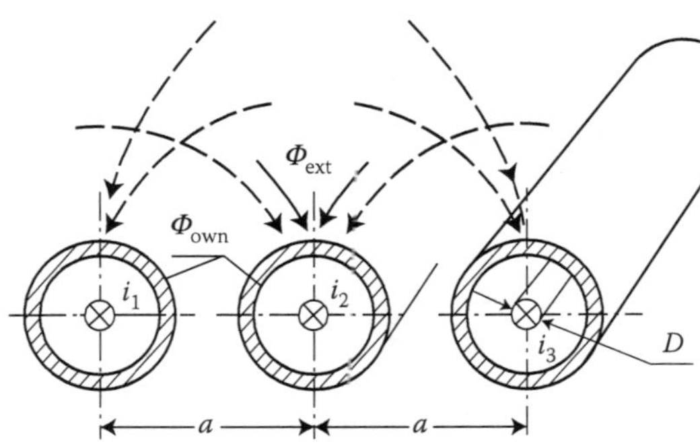

FIGURE 4.33 The own magnetic field (Φ_{own}) and the external magnetic field (Φ_{ext}) in a three-phase system of screened busbars.

and (ii) in an *external field* H_{ext} from the other bars (Figure 4.33) as long as these bars are not completely closed on their ends by the generator's frame or transformer tank.

For the external field, whose value depends on the distances and distribution of the bars, a cylindrical screen creates a closed electromagnetic protection operating very effectively. However, for the own field of the current-carrying bar, a cylindrical tube creates an incorrect, open screen because, independently of its thickness, on the external surface of the screen will occur practically the same induced currents and the same field as on the internal surface of the tube. The last effect explains, among others, the possibility of existence of the external field on the surfaces of neighboring bars, in spite of the existence of their screens (Figure 4.34) (J. Turowski [2.33]).

The losses from the own field of bar can be calculated with the help of formulae (4.119) through (4.123). The losses from the external field and the screening coefficient, at a sufficiently large distance of screens, can be approximately calculated on the basis of the equations given in Section 4.9.2.

In a case when screens have big dimensions in comparison to their distances, one should take into account the nonuniformity of the external field distribution on their surface and influence of the *proximity effect* (J. Turowski [1.16], pp. 286–287) on the distribution of eddy currents in the screens. This problem can be resolved in a

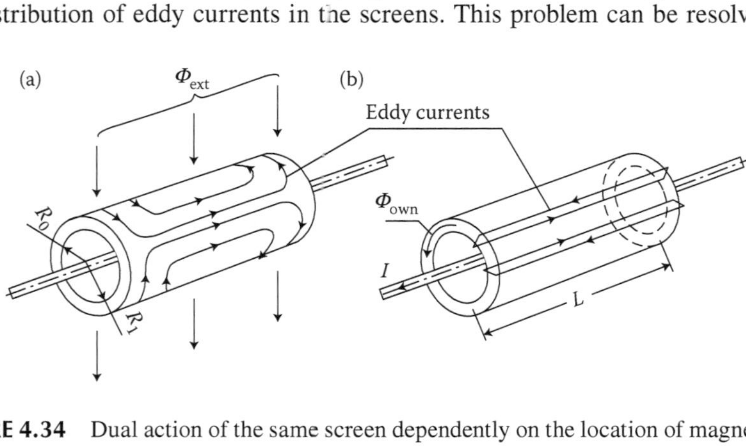

FIGURE 4.34 Dual action of the same screen dependently on the location of magnetic field source: (a) closed screen; (b) open screen, incorrect.

simplified way if we assume that the investigated screen is made of an *ideal* (superconducting) *conductor.*

In the *method of conformal mapping*, developed by W. Zarebski [4.33], the currents induced in cylindrical screens by neighboring bars are substituted by fictitious currents, placed in the plane of three busbars, in such a way that they produce on the surface of screen, together with excitation currents, the same magnetic field that exists there in reality (the mapping method).

4.9.3.1 Isolated Screens

Let us consider, first, the case when a current I flows only in one bar of a screened, infinitely long, three-phase busbar system (R, S, T), for instance, in the bar R in screen L_1 (Figure 4.35).* The current flowing in the bar R of screen L_1 produces in point P of the neighboring screen L_2 the magnetic field intensity $H_m = \sqrt{2}I/2\pi c$. In the first approach, we assume that eddy currents in screen are so large that they dislodge completely the electromagnetic field from the region embraced by the screen. This assumption is approximately right if a screen has sufficiently large conductivity and thickness of some millimeters (e.g., 5 mm), which is often made in practice. According to this assumption, the normal component of magnetic field H'_{mn} produced by the eddy currents flowing in the screen L_2 is equal and oppositely directed to the component H_{mn} (Figure 4.35) produced by the current flowing in the screen L_1. The same compensation of the normal components of magnetic field intensity will be obtained if the investigated screen will be substituted by two equal fictitious currents $I'' = I$ and $I' = -I$. The current I'' will be placed in the axis of the screen, so that it produces on the surface of the screen only the tangential component of the field. The second current I' will be placed in the distance b from the screen axis (Figure 4.35).

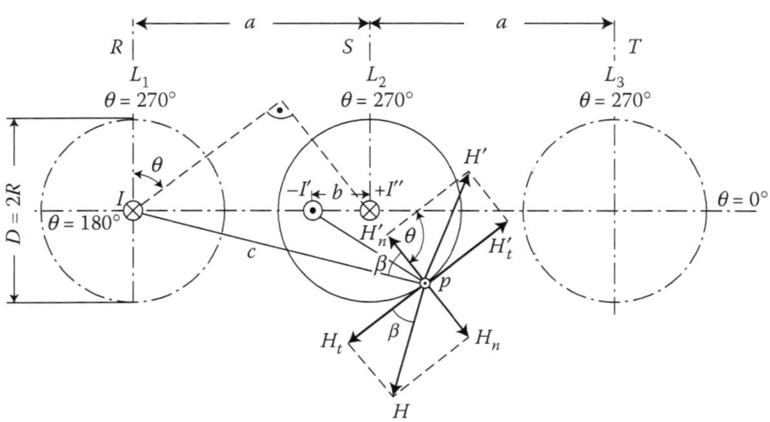

FIGURE 4.35 Mapping of the magnetic field intensity in the point P of the screen L_2 of phase S, produced by the currents in phase R and in the screens, with the help of fictitious currents $+I$ and $-I$. (Adapted from Zarębski W.: Calculation of eddy currents induced in the bus-bar screens. (in Polish). *Przegląd Elektrotechniczny*, XXXVIII(8), 1962, 334–339 [4.33], courtesy of Dr. W. Zarębski.)

* For simplification, in Figure 4.35 indexes m are skipped.

From the geometric dependences (Figure 4.35), we can find the normal (n) and tangential (t) components:

$$
\begin{aligned}
H_{mn} &= \frac{1}{\sqrt{2}\pi c}\sin\beta = \frac{1}{\sqrt{2}\pi}\frac{a\sin\theta}{a^2+R^2+2aR\cos\theta} \\
H_{mt} &= \frac{1}{\sqrt{2}\pi c}\cos\beta = \frac{1}{\sqrt{2}\pi}\frac{R+a\sin\theta}{a^2+R^2+2aR\cos\theta}
\end{aligned}
\tag{4.136}
$$

Moduli of the components H'_{mn} and H'_{mt} can be expressed by the very same equations as Equation 4.136, but instead of the parameter a, we should replace it by the parameter b. The distance b of the substituting current I' from the axis can be found from the condition $H_{mn} = H'_{mn}$, from which $b = R^2/a$.

Now, on the surface of the screen remains the resulting tangential field, H_{result}, which is the sum of three components

$$
H_{result} = H_{mt} - H'_{mt} - H''_{m} = \frac{\sqrt{2}I}{\pi}\frac{R+a\cos\theta}{a^2+R^2+2aR\cos\theta}
\tag{4.137}
$$

where $H''_{mt} = H''_{m} = I/\sqrt{2}\pi R$.

Having determined the distribution of the field H_{result} (4.137) on the surface, and assuming that the internal field of the screen $H_{int} = 0$, from the law of flow (Ampere's law: $\sqrt{2}J_1 \mathrm{d}l = H_{result}\mathrm{d}l + H_{n1}d + H_{int}\mathrm{d}l + H_{n1}d$), we can now determine the rms* value (J_1) of the linear current density along the screen circumference in a single screen in the vicinity of which a conductor with the current I is located:

$$
J_1 = \frac{H_{result}}{\sqrt{2}} = \frac{I}{\pi}\frac{R+a\cos\theta}{a^2+R^2+2aR\cos\theta}
\tag{4.138}
$$

Utilizing the method presented above, Zarebski [4.33] calculated eddy currents induced inside all three screens, when the current $+I$ flows in the bar of phase L_1 and the current $-I$ flows in the bar of phase L_2. A three-phase current can therefore be simulated by two such pairs of currents, shifted in time by 120°.

The resultant current densities for the screens L_1, L_2, and L_3 of phases R, S, T, obtained by Zarebski [4.33] by means of such geometrical summation, are presented in Figures 4.36 and 4.37, for various geometrical proportions occurring in industrial practice.

The power losses in the screens are

$$
P = \int_{C}^{\pi D} (J_1\,\mathrm{d}x)^2 \frac{1}{\sigma d \cdot \mathrm{d}x}
\tag{4.139}
$$

* rms = root mean square.

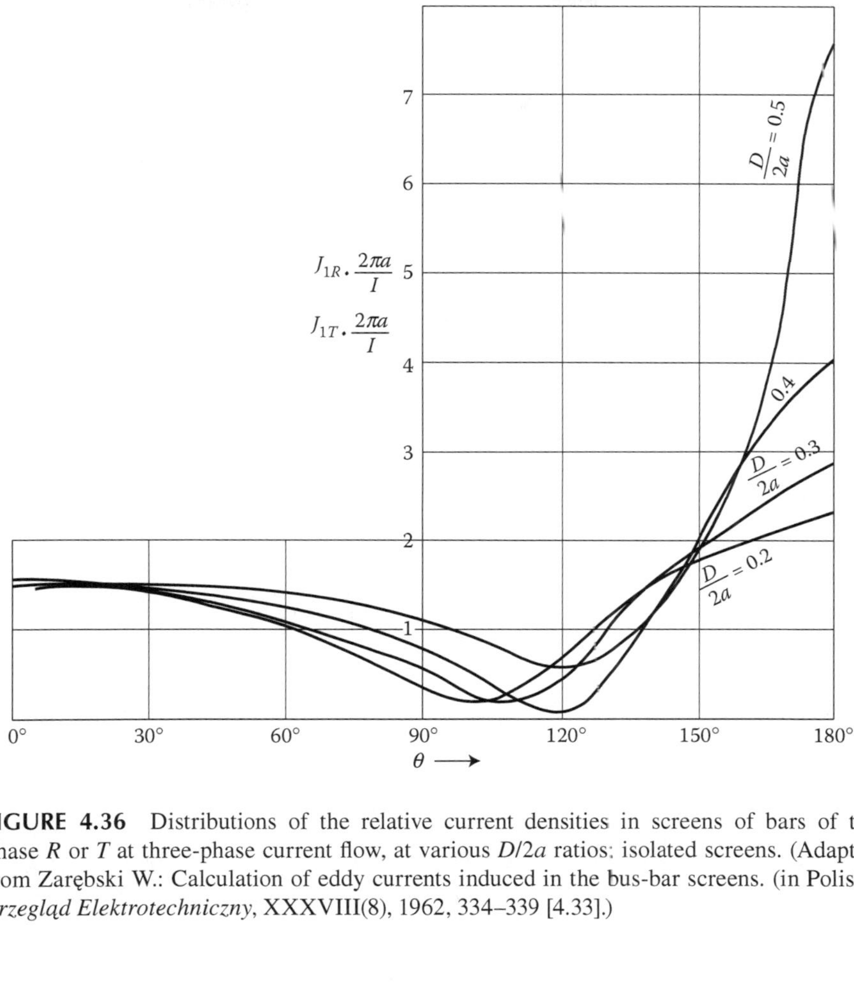

FIGURE 4.36 Distributions of the relative current densities in screens of bars of the phase R or T at three-phase current flow, at various $D/2a$ ratios; isolated screens. (Adapted from Zarębski W.: Calculation of eddy currents induced in the bus-bar screens. (in Polish). *Przegląd Elektrotechniczny*, XXXVIII(8), 1962, 334–339 [4.33].)

Formula (4.139) gives, as per Ref. [4.33], accurate results in the range of dimensions and forms of screened busbars occurring in practice, that is, the bars with aluminum tube screens of diameters (D) of 0.7–1 m and of thicknesses (d) 5–8 mm, at the ratios of diameter of screen to the distance (a) between bus axes D/a of 0.6–0.9. At smaller D/a ratios, one should also take into account the losses in screen from the own field of bar (formula 4.119). The presented results were verified experimentally by Dr. W. Zarębski in a regular power station.

4.9.3.2 Connected Screens

Busbar connections of large generators with generator (unit) transformers are both equipped with the so-called *isolated* screens (Figure 4.38a) and *connected* screens, that is, connected either directly (Figure 4.38b) or by means of corresponding reactors (Figure 4.38c), which reduce currents in screens. In the work [4.33], an approximate method of calculation of power losses in connected screens was also proposed, on the basis of the theory presented above, of isolated screens.

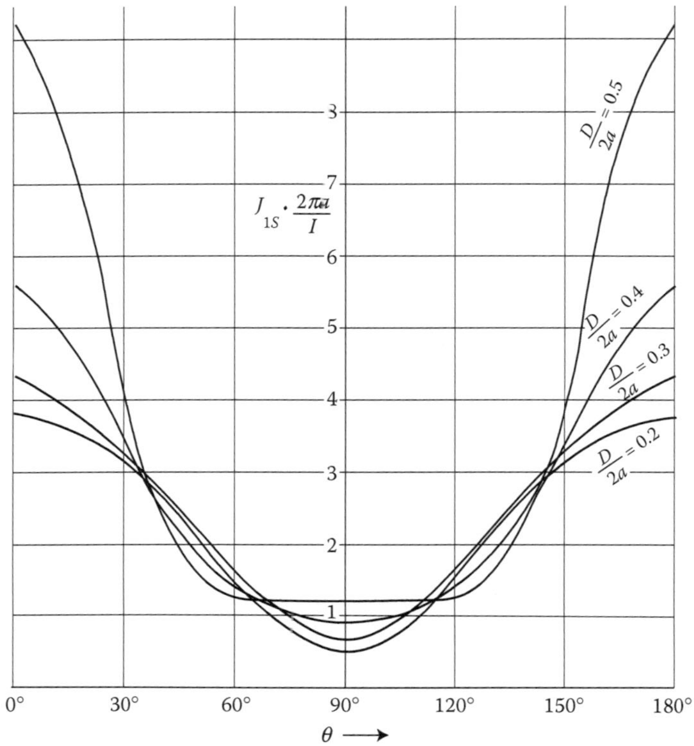

FIGURE 4.37 Distributions of the relative current density in the screen of bar of the phase S, at three-phase R, S, T current flow, at various $D/2a$ ratios; isolated screens. (Adapted from Zarębski W.: Calculation of eddy currents induced in the bus-bar screens. (in Polish). *Przegląd Elektrotechniczny*, XXXVIII(8), 1962, 334–339 [4.33].)

Another type of screened busbars consist of triple-wire screened cables, the construction of which was presented in catalogues of "Elektrobudowa"-Katowice company [4.8], and their electromagnetic and thermal calculations in the works by G. Szymanski and A. Patecki [4.15].

An extended theory and design of screens, shunts, and metal housings were given in another J. Turowski's book ([1.16], pp. 60–98). Special multilayer systems with screens of extraordinary importance and their calculation methods are used in superconducting generators (J. Turowski [1.16], pp.188–191, 216–221).

4.10 ELECTROMAGNETIC FIELD IN MULTILAYER SCREENS

4.10.1 Two-Layer Conductor

Let us assume that a conducting half-space consists of two uniform conducting metal layers of different parameters (Figure 4.39). The first layer has the thickness d and the second layer occupies the rest of the half-space. For the first and the second conductors, Equations 2.142 and 2.159 are valid:

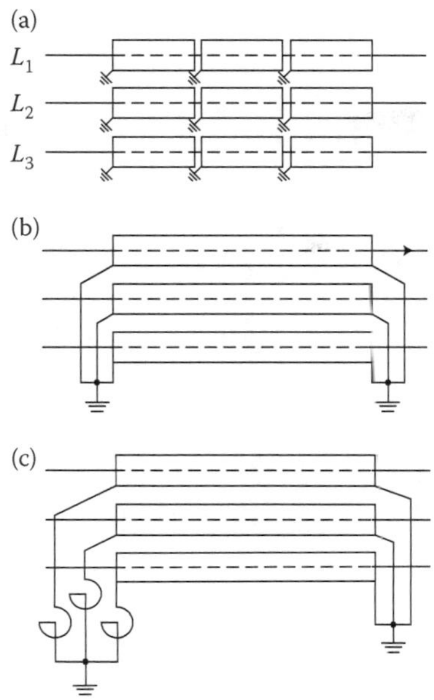

FIGURE 4.38 Types of three-phase busbar screen connections: (a) isolated screens, grounded in one point; (b) directly connected; (c) connected by reactors. (Adapted from Zarębski W.: Calculation of eddy currents induced in the bus-bar screens. (in Polish). *Przegląd Elektrotechniczny*, XXXVIII(8), 1962, 334–339 [4.33].)

$$\nabla^2 E_m = \alpha^2 E_m \quad \text{and} \quad \frac{\partial E_x}{\partial z} = -\mu \frac{\partial H_y}{\partial t}$$

Assuming that onto the surface of a body impinges a polarized, monochromatic, plane electromagnetic wave, we can represent the above equations, for both layers, in the form

$$\frac{\mathrm{d}^2 E_{m1}}{\mathrm{d}z^2} = \alpha_1^2 E_{m1} \quad \frac{\mathrm{d}E_{m1}}{\mathrm{d}z} = -j\omega\mu_1 H_{m1} \tag{4.140}$$

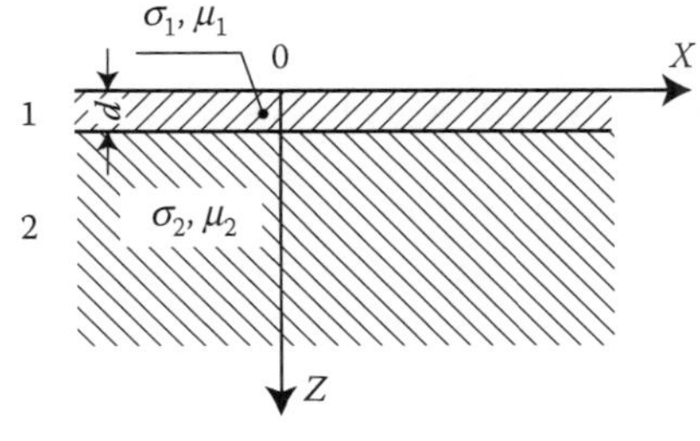

FIGURE 4.39 A double-layer conductor.

and

$$\frac{d^2E_{m2}}{dz^2} = \alpha_2^2 E_{m2} \quad \frac{dE_{m2}}{dz} = -j\omega\mu_2 H_{m2} \tag{4.141}$$

From these equations, carrying on similarly like with Equation 2.172, we shall obtain solutions for the electric field intensity:

$$E_{m1} = A_2 e^{\alpha_1 z} - B_1 e^{-\alpha_1 z} \quad \text{for } 0 < z < d \tag{4.142}$$

$$E_{m2} = A_2 e^{\alpha_2 z} - B_1 e^{-\alpha_2 z} \quad \text{for } z > d \tag{4.143}$$

where

$$\alpha_1 = (1+j)\,k_1; \quad \alpha_2 = (1+j)\,k_2; \quad k_1 = \sqrt{\frac{\omega\mu_1\sigma_1}{2}}; \quad k_2 = \sqrt{\frac{\omega\mu_2\sigma_2}{2}}$$

and for the magnetic field intensity

$$H_{m1} = -\frac{\alpha_1}{j\omega\mu_1}(A_1 e^{\alpha_1 z} - B_1 e^{-\alpha_1 z}) \; przy \quad 0 < z < d \tag{4.144}$$

$$H_{m2} = -\frac{\alpha_2}{j\omega\mu_2}(A_1 e^{\alpha_2 z} - B_1 e^{-\alpha_2 z}) \; przy \quad z > d \tag{4.145}$$

Some constants can be determined from the condition of equality to zero of the magnetic field intensity in infinity ($z \to \infty$, for the solid conductor 2, Figure 4.39). Therefore

$$A_2 = 0, \quad E_{m2} = B_2 e^{-\alpha_2 z} \tag{4.146}$$

$$H_{m2} = B_2 \frac{\alpha_2}{j\omega\mu_2} e^{-\alpha_2 z} \tag{4.147}$$

On the boundary surfaces, the principle of equality of the tangential components in both media is valid:

$$E_{m1} = E_{m2} \quad \text{and} \quad H_{m1} = H_{m2} \quad \text{for } z = d$$

and

$$E_{m1} = E_{mS}, \quad \text{for } z = 0$$

Substituting the above boundary conditions for $z = d$ into Equations 4.142, 4.144, 4.146, and 4.147 yields

$$\left.\begin{aligned} &A_1 e^{\alpha_1 d} - B_1 e^{\alpha_1 d} = B_2 e^{-\alpha_2 d} && \text{(148a)} \\ &-\frac{\alpha_1}{j\omega\mu_1}(A_1 e^{\alpha_1 d} - B_1 e^{-\alpha_1 d}) = B_2 \frac{\alpha_2}{j\omega\mu_2} e^{-\alpha_2 d} && \text{(148b)} \end{aligned}\right\} \tag{4.148}$$

After double-sided multiplication of the first of Equations 4.148 by $\alpha_2/j\omega\mu_2$, and subtraction of the second Equation 4.148, we obtain

$$\frac{B_1}{A_1} = \frac{\mu_1 k_2 + \mu_2 k_1}{\mu_2 k_1 - \mu_1 k_2} e^{2\alpha_1 d} \tag{4.149}$$

Substituting the boundary condition for $z = 0$ into Equation 4.142 yields

$$A_1 + B_1 = E_{mS} \tag{4.150}$$

From Equations 4.148 through 4.150, one can find all the necessary constants: A_1, B_1, and B_2, and also the magnetic and electric field intensity and the current density in any point of the conducting bodies. This method is correct in cases when the *curvature radius* is much bigger than the thickness of the first conductor.

Impedance of a double-layer conductor, per unit length in the X direction, is

$$Z = \frac{\int E_m \cdot d1}{I_{m1}} = \frac{E_{ms}}{I_{m1}} = r_1 + j\omega L_1 \tag{4.151}$$

The relationship between the current and the magnetic field intensity on the conductor surface is expressed by Equation 2.184 and Figure 2.11 ($A = I_1$). This dependence is also right for a double-layer conductor that occupies a half-space. Hence, the impedance (see Equations 4.142 and 4.144)

$$Z = \frac{E_{ms}}{H_{ms}} = \frac{A_1 + B_1}{B_1 - A_1}\frac{j\omega\mu_1}{\alpha_1} \tag{4.152}$$

The above Z is at the same time the *wave impedance* of a screened solid conductor.

In result of the substitution of Equation 4.149 into Equation 4.152, we obtain, as per Simonyi [1.11] and [1.15/1]

$$Z = \frac{\alpha_1}{\sigma_1}\frac{\text{sh}\,\alpha_1 d + (\mu_2 k_1/\mu_1 k_2)\text{ch}\,\alpha_1 d}{\text{ch}\,\alpha_1 d + (\mu_2 k_1/\mu_1 k_2)\text{sh}\,\alpha_1 d} \tag{4.153}$$

It is easy to notice that when both the layers are made of the same material ($\mu_1 = \mu_2$; $\sigma_1 = \sigma_2$), or when the screen thickness $d = 0$, expression (4.153) equals the wave impedance of the second solid conductor (α_2/σ_2); see formula (2.177).

In the case when the thickness of layer (d) of the first conductor, or its magnetic permeability, or frequency of the field alternation is so large that the equivalent depth of field penetration $\delta_1 = \sqrt{2/\omega\mu_1\sigma_1}$ (formula (2.181)) is smaller than the thickness d, the influence of the second conductor becomes negligibly small and in extreme case ($d \to \infty$), formula (4.153) will be simplified to the form $Z = \alpha_1/\sigma_1$.

The ratio of the wave impedance of a screened conductor (double layer) to the wave impedance of a conducting space with parameters σ_1, μ_1, according to formulae (4.153) and (2.185), is

$$\frac{Z}{Z_1} = \frac{\text{sh}\,\alpha_1 d + (r_2/r_1)\text{ch}\,\alpha_1 d}{\text{ch}\,\alpha_1 d + (r_2/r_1)\text{sh}\,\alpha_1 d} \tag{4.154}$$

where

$$r_i = \frac{1}{\sigma_i\delta_i} \quad \text{and} \quad \delta_i = \sqrt{\frac{2}{\omega\mu_i\sigma_i}} \quad (i = 1 \text{ or } 2)$$

4.10.2 Influence of Insulation under the Screen

In some screened systems, a layer of dielectric or air is recommended and applied between the screen and the screened steel wall (e.g., transformer tank). Such a three-layer system (Figure 4.40), for sinusoidal fields $E = E_m e^{j\omega t}$, in the ith layer is described by the equation of type (2.137)

$$\nabla^2 E_{mi} = \Gamma_i^2 E_{mi}, \quad \Gamma_i = \sqrt{j\omega\mu_i(\sigma_i + j\omega\varepsilon_i)} \tag{4.155}$$

where for particular layers

$$\Gamma_i = \alpha_I = \sqrt{j\omega\mu_i\sigma_i} = (1 + j)k_i, \quad k_i = \sqrt{\frac{\omega\mu_i\sigma_i}{2}} \quad (i = 2, 4)$$

$$\Gamma_i = \sqrt{-\omega^2\mu_i\varepsilon_i} \quad (i = 1, 3)$$

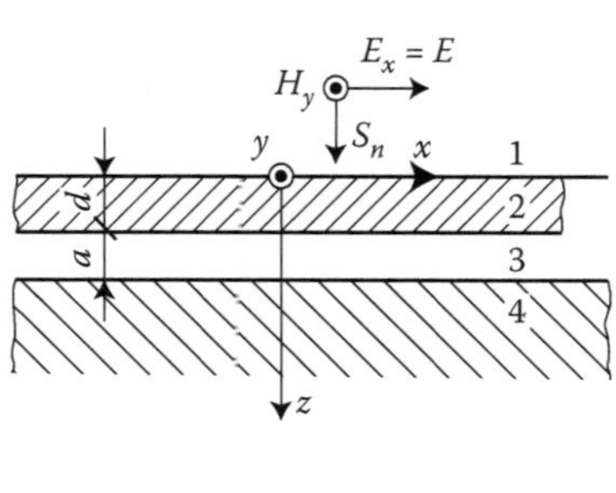

FIGURE 4.40 A double-layer screen with a distancing insulation (a); 2—metal screen; 1, 3—dielectrics; 4—solid steel.

A solution of the system of Equations 4.155, taking into account the boundary conditions (2.191) through (2.193), can be found similarly to the previous general solution ([1.16], p. 67), which leads to the conclusion that at power frequencies (50, 60 Hz)

> the insulation gap ("a" in Figure 4.40) under the screen does not play any role and its application is redundant and in power transformers and machines absolutely needless [4.22].

However, at high frequencies (of MHz order), a remarkable influence of the dimension (*a*) of this gap on the effectiveness of screening can appear. It is utilized in radiolocation protection of ships, vehicles, aircrafts, and similar equipments, and even battleships* (J. Turowski [1.16], pp. 83–84).

A convenient method of simulation of multilayer screens can be the analogy with the theory of long transmission line (J. Turowski [1.16], pp. 84–88).

Screening of a single conductor or a group of conductors can be investigated with the help of the method of mirror images (J. Turowski [1.16], pp. 95–98), the method of superposition of vector potentials (J. Turowski [1.16], pp. 236–240), or the presently more popular numerical methods.

At the end of this chapter, it is worth mentioning that the heritage of the "pre-computer era" has left plenty of ingenious analytical solutions that deliver clever, parametric, and more clear descriptions of phenomena, with the same or better accuracy than numerical methods, and are often much faster than complicated numerical modeling and simulations.

* After the World War II, in the Baltic See there were left numerous drifting magnetic mines, sensitive to any changes of the earth magnetic lines, caused by metal mass of vessels. They were especially dangerous for fishing vessels which travelled out of cleaned fairways So, those ships needed special screens or active coils.

5 Magnetic Fields Near Iron Surfaces

5.1 METHOD OF MIRROR IMAGES

5.1.1 Singular Images of Direct Current

The method of images is based on the analogy with optics. The analogy of electrostatic $\boldsymbol{E} = -\text{grad } V$ and magnetostatic $\boldsymbol{H} = -\text{grad } V_m$ fields gives us the possibility to consider them in the same way. Let us start with the magnetostatics, where $\boldsymbol{B} = \mu\boldsymbol{H}$. The method of mirror images, in the case of constant magnetic permeability μ_i of two adjacent media, consists of substitution of two media I and II (Figure 5.1a) by only one medium I (Figure 5.1b) or II (Figure 5.1c) and an auxiliary fictitious, image current i_1. The value of this fictitious current i_1 is determined by the rule of *uniqueness of the field*, specified by the boundary conditions (2.191) and (2.193):

$$\mathrm{B}_{\mathrm{II}n} = \mathrm{B}_{\mathrm{I}n} \quad \text{and} \quad \mathrm{H}_{\mathrm{II}n} = \mathrm{H}_{\mathrm{I}n}$$

If a conductor, carrying current *i,* is placed in parallel to the boundary plane of two media of different finite permeabilities μ_I and μ_II (Figure 5.1a), then by proper selection of fictitious additional currents i_1 and i_2 we can substitute

1. The field in region I of the real system (Figure 5.1a) by a field of the currents i_1 and i_2 placed in the same medium I (Figure 5.1b)
2. The field in region II of the real system (Figure 5.1a) by a field of the current $i - i_1$ placed only in the medium II (Figure 5.1c)

The fictitious additional currents i_1 and i_2 are determined on the basis of the boundary conditions (2.191) and (2.193) and from Ampere's law of flow (2.15): $H = i/2\pi r$. For the tangential H_t and normal μH_n components in point A (Figure 5.1b,c) we have the corresponding equations

$$\frac{i}{2\pi r}\sin\alpha - \frac{i_2}{2\pi r}\sin\alpha = \frac{i - i_1}{2\pi r}\sin\alpha$$

$$\mu_\mathrm{I}\frac{i}{2\pi r}\cos\alpha + \mu_\mathrm{I}\frac{i_2}{2\pi r}\cos\alpha = \mu_\mathrm{II}\frac{i - i_1}{2\pi r}\cos\alpha$$

which after solution give the relations

$$i_2 = i_1 = \mathrm{M}i \quad \text{and} \quad i - i_1 = \mathrm{m}i \tag{5.1}$$

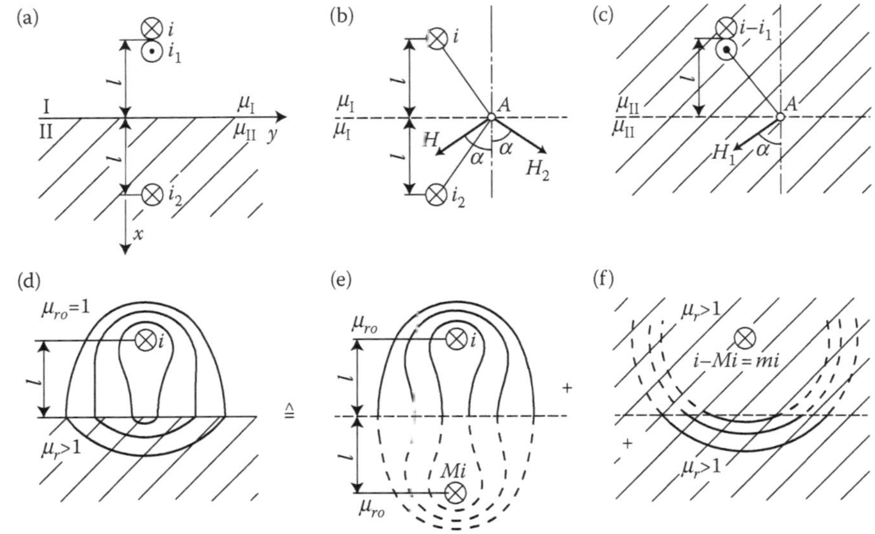

FIGURE 5.1 Method of mirror images: (a), (b), (c)—determination of equivalent currents $i_1 = i_2 = \mathrm{M}i$; (d), (e), (f)—drawing of field-line diagram of the real current i nearby a steel surface.

where

$$\mathrm{M} = \frac{\mu_{\mathrm{II}} - \mu_{\mathrm{I}}}{\mu_{\mathrm{II}} + \mu_{\mathrm{I}}} \quad \text{and} \quad \mathrm{m} = 1 - \mathrm{M} = \frac{2\mu_{\mathrm{I}}}{\mu_{\mathrm{II}} + \mu_{\mathrm{I}}} \tag{5.2}$$

are the mirror image coefficients of the current i.

On the basis of Equations 5.1 and 5.2, we can now easily build a picture of the field of an isolated current placed outside the iron surface (Figures 5.1d and 5.2a) or diamagnetic (Figure 5.2b) nearby the boundary surface.

If in the magnetostatic setup (Figure 5.1) the medium I is air ($\mu_1 = \mu_0$) and the material II is iron, then

$$\mathrm{M} = \frac{\mu_r - 1}{\mu_r + 1} \approx 1 \quad \text{and} \quad \mathrm{m} = \frac{2}{\mu_r + 1} \approx 0 \tag{5.2a}$$

Using this method, one can calculate the magnetic field of currents in nonuniform systems of media, with the help of a simple principle of superposition with the application of the law of flow (Ampere's law) (2.15) or Biot–Savart law (2.16).

Smythe ([4.3], p. 123) developed this method for the scalar potential of field $V_{\mathrm{I,II}} = f(x, y, z)$, and Štafl ([6.7], p. 76)—for the magnetic vector potential (2.50) $\boldsymbol{A}(x, y)$ of the exciting current i.

In the latter case, it is assumed that the solution of Laplace's equation $\nabla^2 \boldsymbol{A} = 0$ for region I (Figure 5.1a) is the superposition $\boldsymbol{A}^{\mathrm{I}} = \boldsymbol{A}_0 + \boldsymbol{A}_1$ of the primary *forcing* (excitation) field

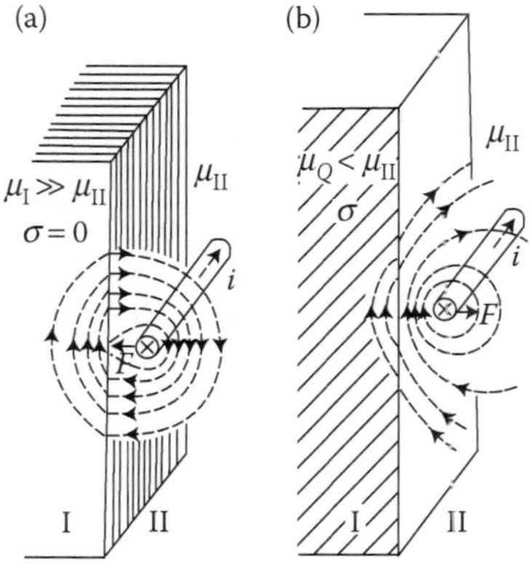

FIGURE 5.2 Mirror "image" of AC current in a metal wall: (a) laminated iron; (b) solid good conducting. (Adapted from Turowski J.: Losses and local overheating caused by stray fields. (in Polish). *Scientific Archives of the Technical University of Lodz "Elektryka*, No. 11, 1963, pp. 89–179 (*Habilitation (DSc) dissertation; 1st Ministry Award* [2.41].)

$$A_0 = -\frac{\mu_I i}{2\pi} \ln \sqrt{(x+1)^2 + y^2} \tag{5.3}$$

superimposed with a response of the iron surface, that is, with the *interfering field* $A_I = f(x) \cos \lambda y$, created as an effect of placing iron in proximity of a filament with current i.

After substituting Equation 5.3 into Laplace's equation, one obtains a standard differential equation:

$$f''(x) - \lambda^2 f(x) = 0 \quad \text{from where } f(x) = \mathrm{C}e^{+\lambda x} + \mathrm{D}e^{-\lambda x} \tag{5.4}$$

Since $A^I(x \to \infty) \neq \infty$, hence D = 0, from which, at any value of λ

$$A^I = A_0 + \mathrm{C}e^{-\lambda x} \cos \lambda y = -\frac{\mu_I i}{2\pi} \ln \sqrt{(x+l)^2 + y^2} + \int_0^\infty \mathrm{C}(\lambda) e^{\lambda x} \cos \lambda y \, d\lambda \tag{5.5}$$

and analogically, $A^{II}(x \to \infty) \neq \infty$, hence C = 0, therefore

$$A^{II} = -\frac{\mu_2 i}{2\pi} \ln \sqrt{(x+l)^2 + y^2} + \int_0^\infty \mathrm{C}(\lambda) e^{-\lambda x} \cos \lambda y \, d\lambda \tag{5.6}$$

The constants C(λ) and D(λ) are determined from the boundary conditions (2.191) and (2.193) [6.7] approaching finally to the same conclusions as in Figure 5.1d.

5.1.1.1 Analogy in Electrostatic Fields

By analogy, we can analyze in the same way the *method of images* in electric fields of a charge q. A linear electric charge of density $+q$ (Figure 5.3a) produces electric field lines entering perpendicularly into a conducting metal plate P.

Due to the high conductivity of the plate, its potential is constant or zero at all its points. If it is zero, then the induced charges produce in the plate P a potential equal and opposite to that of the inducing charge. But, the same effect would be caused by an equivalent line charge but of opposite sign, $q_1 = -q$, placed in the same distance on the opposite side of the plate P.

We can use this "image" charge to calculate the field anywhere near the plate. This is clearly a great simplification. This image theory was developed by P. Hammond* ([2.8], p. 94) to even more complicated geometry systems.

More generally, when the plate P separates two regions of different dielectric constants, ε_1 and ε_2, with $q_1 = k_1 q$, then the field in the region of ε_2 can be calculated as if instead of the real charge q was put $q_2 = k_2 q$. From the equality on the boundary surface of potentials $V_1 = V_2$ and field boundary conditions $E_{1t} = E_{2t}$ and $\varepsilon_1 E_{1n} = \varepsilon_2 E_{2n}$, we can analogically determine the image coefficients for charge

$$\mathrm{M_e} = \frac{\varepsilon_1 - \varepsilon_2}{\varepsilon_1 + \varepsilon_2} \quad \text{and} \quad \mathrm{m_e} = \frac{2\varepsilon_2}{\varepsilon_1 + \varepsilon_2} \tag{5.6a}$$

We can remove now from the computational model all conducting surfaces and introduce only corresponding real and fictitious charges and then apply superposition.

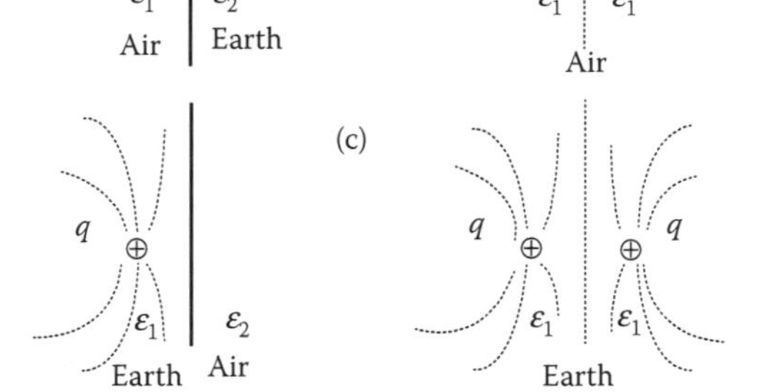

FIGURE 5.3 Method of electric charge images: (a) real configuration of charge lines in a conducting plate P ($\varepsilon_1 < \varepsilon_2$); (b) equivalent image model; (c) at $\varepsilon_1 > \varepsilon_2$.

* Professor *Percy Hammond* of Southampton University, UK, is Doctor Honoris Causa of the Technical University of Lodz, 1993. Professor J. Turowski was his supervisor in this procedure.

(In earth $\varepsilon_r = 2–15$, whereas $\varepsilon_0 = 1/\mu_0 c^2 \approx 8.85 \times 10^{-12}$ F/m.) Such geometric model is very convenient for simple computer modeling of complex electric and magnetic structures.

5.1.2 Application of the DC Field Theory to AC Fields

The theory of an electrostatic field limited by iron boundaries is often also applied to the solution of constructional systems with an AC excitation. Such approximation is possible everywhere where the effect of a large value of magnetic permeability μ_{Fe} significantly predominates over the effect of reaction of eddy currents induced by AC field in a reflecting surface of solid iron. The saturation of the iron is usually of less hindrance. For instance, when a steel core, where an AC current flows nearby, is laminated in such a way that eddy currents can not freely develop in it (Figure 5.2a), the magnetic flux lines diagram is the same as for a DC current flowing nearby the iron surface (Figure 5.1). If, on the other hand, the AC current flows nearby the surface of solid metal wall or core, laminated parallel to its surface, the field picture depends strongly on the metal conductivity σ and the current frequency. Eddy currents induced in the surface of metal generate an apparent reduction of the iron permeability ("quasi-permeability" μ_Q of material, e.g., in formulae (5.2a)).

In the case of solid steel bodies ($\mu_r \gg 1$), a noticeable reduction of their quasi-permeability, due to the eddy-current reaction, occurs only at very high frequencies of the field. At very high frequencies, at quick-varying transient states or at high conductivities σ, especially in nonferrous metals ($\mu_r = 1$), metal can act like a *diamagnetic* ($\mu_Q < 1$), driving out the AC magnetic field from the region occupied by metal (Figure 5.2b). In an extreme case, in Formula (5.2a) the quasi-permeability takes the value of $\mu_Q \approx 0$, which gives a fully negative reflected current and causes pushing out of the conductor wire from the metal surface. The magnetostatic image method is then strictly valid only for DC currents, or for AC when an iron wall is laminated in the direction of field lines (Figure 5.2a). If we wish to use this method to AC currents nearby a solid metal wall, we have to consider the deformation of image by the eddy-current reaction and saturation of iron. In such a case, in the first approach, we can apply the mirror image method very approximately:

1. For a solid steel wall, using semiempirical image coefficients $M_{QSt} = 0.4$ to 0.6 (instead of +1)—only for the normal component $H_{ms,n}$ from the air side
2. For a solid copper wall, using $M_{QCu} = -0.9$ to -0.8 (instead of -1)—only for the tangential component $H_{ms,t}$ from the air side

Following the view of magnetic field lines obtained experimentally with the help of iron filings, these lines yet at frequencies of hundreds Hz enter the solid iron surface practically perpendicularly, as shown in Figure 5.2a. Other measurements of magnetic field intensity (J. Turowski [2.41], [5.15]) show (Figure 5.4) that at frequency of 50 Hz the field on the surface of solid steel is somewhat weaker than at full positive mirror image, as if the current Mi imaged in the iron (5.1) were about 0.4–0.6 of the real current [5.15]. In this way, in the proximity of a solid steel surface, the field at an AC current can be estimated approximately

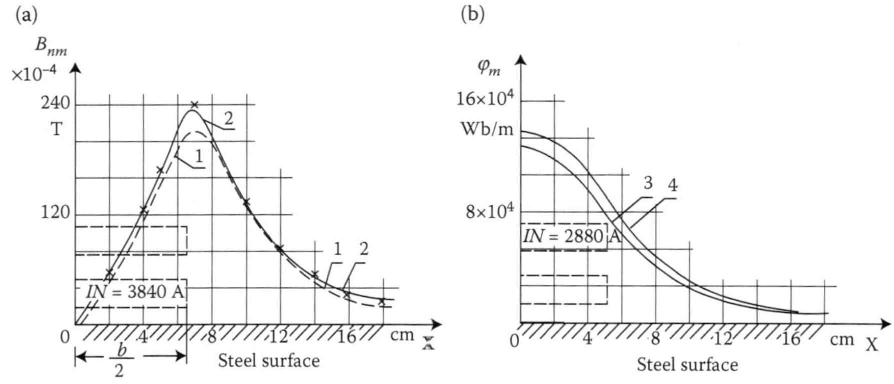

FIGURE 5.4 An experimental verification of the mirror image of an alternating current of 50 Hz in a solid steel plate: (a) model at ampere-turns $IN = 3840$ A. (Adapted from Turowski J.: Losses and local overheating caused by stray fields. (in Polish). *Scientific Archives of the Technical University of Lodz "Elektryka*, No. 11, 1963, pp. 89–179 (*Habilitation (DSc) dissertation; 1st Ministry Award* [2.41].) (b) Experimental verification of phase shifting of flux density at $IN = 2880$ A; 1— the flux density on the surface, measured in the presence of steel; 2—the doubled normal component on the X axis, in the presence of dielectric; ××—points calculated at full one-sided mirror image; 3—the measured flux Φ_{mx}. 4—The flux calculated as $\Phi_{mx} = \int_0^\infty B_{m,n}\,dx$ at $B_{m,n}$ per the curve 1. (Adapted from Turowski J.: Losses and local overheating caused by stray fields. (in Polish). *Scientific Archives of the Technical University of Lodz "Elektryka*, No. 11, 1963, pp. 89–179 (*Habilitation (DSc) dissertation; 1st Ministry Award)* [2.41]; Turowski J.: Power losses in transformer covers at various methods of their reduction. (in Polish) *Archiwum Elektrotechniki*, (4), 1959, 529–556 [5.15].)

like at a DC current, by corresponding successive reduction of the fictitious imaged current Mi.

The mirror image of DC current and permanent magnets in a steel surface will be called here *static images*, and the substitutional image of AC current and AC electromagnets, taking also into account the eddy-current reaction—*dynamic images.*

In Section 6.5 (Figure 6.12) is presented the method of images for *transient currents* in a thin conductor layer as well as the reflection coefficient M (6.46) which, however, in this case is a function of time.

5.1.2.1 Mirror Image Coefficients of Alternating Currents in Metal Surfaces

Utilizing the notion of equivalent quasi-permeability, μ_Q, also for an AC current, one can evaluate the approximate coefficients (5.2) of image in flat conducting metal surfaces, as

$$\mathrm{M}_Q \approx \frac{\mu_{Qr} - 1}{\mu_{Qr} + 1} \tag{5.7}$$

Since the field on the metal surface is calculated as a *superposition* of the field of the real current i and its image $i_2 = \mathrm{M}i$ (Figure 5.1b), hence, instead of M_Q, it is more convenient to make use of the *mean coefficient of mirror image*

$$\eta_Q = \frac{i + M_Q i}{2i} = \frac{1 + M_Q}{2} \quad (5.8)$$

which is laden with a smaller evaluation error.

Values of M_Q, η_Q, and μ_Q are univocal for the whole surface and have a simple form only when the wall in the vicinity of which the AC current flows (Figure 5.1) and when the current itself fulfills the following conditions:

1. *Superconductors* ($\sigma_{II} \to \infty$) or high frequency ($f \to \infty$)—then the body behaves as an ideal diamagnetic with the properties:

$$M_Q = -1, \quad \eta_Q = 0, \quad \mu_{Qr} = \frac{1 + M_Q}{1 - M_Q} \approx 0 \quad (5.9)$$

   which applies to a frequency $f \geq 5$ kHz for steel, and $f \geq 50$ kHz for Cu and Al (J. Turowski [5.15]).
2. A *nonmagnetic body* ($\mu_r = 1$) without eddy currents ($\sigma_{II} = 0$)—then

$$M_Q = 0, \quad \eta_Q = 0.5, \quad \mu_{Qr} = 1 \quad (5.10)$$

3. A *ferromagnetic without eddy currents* ($\mu_r \gg 1$, $\sigma_{II} = 0$), for example, a powdered core or a core laminated with iron sheets, like in Figure 5.2a—then

$$M_Q = M = \frac{\mu_r - 1}{\mu_r + 1}, \quad \eta_Q = \frac{1 + M}{2}, \quad \mu_{Qr} = \mu_r \quad (5.11)$$

4. An ideal ferromagnetic ($\mu_r \to \infty$):

$$M_Q = 1, \quad \eta_Q = 1, \quad \mu_{Qr} \to \infty \quad (5.12)$$

In all intermediate cases (with $f > 0$), the reaction of eddy currents should be taken into account by decreasing the coefficient $M_Q = 1$ for solid steel to the value of the order of $M_{QSt} = 0.4–0.6$ ($\eta_Q = 0.7–0.8$), and increasing $M_Q = -1$ for solid copper to the value of the order of $M_{QCu} = -0.9$ to -0.8 ($\eta_Q = 0.05$ to 0.1).

The first definition and evaluation of the equivalent coefficients for the theory of mirror images in solid iron was applied effectively and verified experimentally at an industrial design by the author in 1962 (J. Turowski [2.32], [2.41]), where he stated that the coefficients can be applied, first of all, for determination of the *normal component* $H_{ms,n}$ of magnetic field intensity on the steel surface on the dielectric side (region I) and of the *tangential component* $H_{ms,t}$ at the surface of a good conductor (Cu, Al), because only in these cases the determined field components, in region I, demonstrate relatively stable values and small dependence on steel saturation (Figure 5.5).

Application of M_Q to other field components, or of $m_Q = 1 - M_Q$ for evaluation of the field inside metal (region II), is not recommended due to significant scattering of these values. Inside a metal, one should apply the Maxwell's equations taking

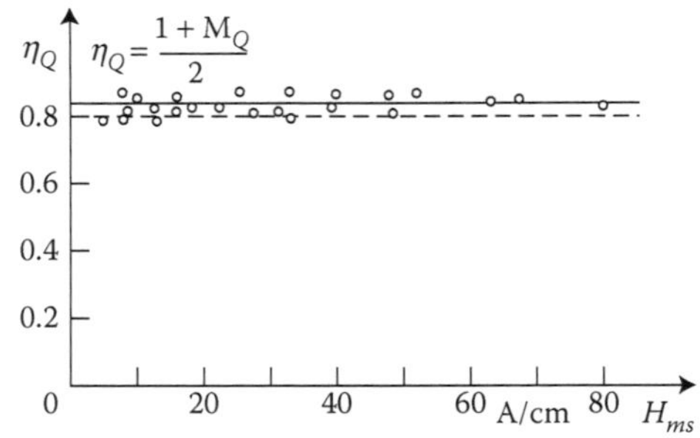

FIGURE 5.5 The coefficients of a not-fully-reflected mirror image of AC current in a solid steel wall, as a function of the field H_{ms} on the steel surface; – o – o – measurements [2.32], [10.16]; - - - calculated by formula (5.13).

into account the boundary conditions from the dielectric side (region I). In 1970, the above formulae were confirmed experimentally by Janowski and J. Turowski [10.16] (Figure 5.5) and extended by Kazmierski [5.9] toward insulation zone and transformer windings (Figure 5.6).

In 1977, K. Zakrzewski and J. Sykulski [5.19] confirmed the above conclusions theoretically, by development of a general formula (5.13) for the normal ($M_n = M_Q$) component of field, $H_{m,n}$:

$$M_n = \frac{4\mu_{IIr}\left|\int_{-L/2}^{L/2}\left\{\int_0^{\infty}\frac{\lambda(e^{-c_1\lambda}-e^{-c_2\lambda})}{\mu_{IIr}\lambda+\sqrt{\lambda^2+\alpha^2}}\sin\left[\lambda(y+l)d\lambda\right]\right\}dl\right|}{\left|\ln\left[\frac{c_2^2+\left(y-\frac{L}{2}\right)^2}{c_1^2+\left(y-\frac{L}{2}\right)^2}\cdot\frac{c_2^2+\left(y+\frac{L}{2}\right)^2}{c_1^2+\left(y+\frac{L}{2}\right)^2}\right]\right|}-1 \tag{5.13}$$

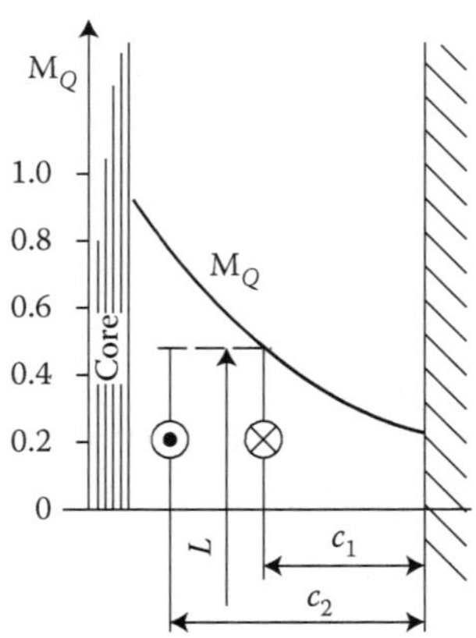

FIGURE 5.6 Variability of the equivalent coefficient M_Q of transformer winding mirror image at the surface of a solid steel wall versus the spot of measurement in a flat system. (Adapted from Kaźmierski M.: An estimation of approximate methods for the analysis of electromagnetic fields in transformers. *Rozprawy Elektrotechniczne*, 16(1), 1970, 3–26 [5.9].)

5.1.3 Magnetic Images of Current in an Iron Cylinder

Similarly as in Section 5.1.1, the magnetic field of a single conductor inside or outside of a metal cylinder can be determined on the basis of the boundary conditions (2.191) and (2.193). In result of the analysis presented in Refs. [5.1] and [6.7] one can conclude that *if a current i is placed inside a solid iron cylinder in point A* (Figure 5.7a), then

1. The field in the iron of real system can be substituted by the field of the current i in point A and of the current $\left(-\frac{\mu_r - 1}{\mu_r + 1}i\right)$ in point B placed in a uniform iron medium. The points A and B are at the same time inverse points, that is, fulfilling for any point P on the circle the dependence

$$\mathrm{OA} \cdot \mathrm{OB} = \mathrm{OP}^2$$

2. The field in the air of a real system can be substituted by the field of the current $\left(\frac{2\mu_r}{\mu_r + 1}i\right)$ in point A and of the current $\left(-\frac{\mu_r - 1}{\mu_r + 1}i\right)$ in point O—both of them placed in a uniform air medium.

Figure 5.7b shows the field of flux density (B) of the current i described above, at the assumption of $\mu_r = 0$ and OA:OC:OB = 8:10:12.5. As one can see, almost all the field is concentrated inside the iron cylinder. At $\mu \to \infty$, the flux density in the air is very small in comparison with the flux density in the iron.

In an opposite case, when the current i is placed inside a *hollow iron tunnel* (then in Figure 5.7a inside the circle occurs the air of magnetic permeability equal 1, and outside—the iron of permeability μ_r), the so created field in the air is produced by the real current i in point A (Figure 5.7a) and its mirror image $((\mu_r - 1)/(\mu_r + 1)i)$ in point B, whereas the field in the external iron is produced by the equivalent current $(2/(\mu_r + 1)i)$ in point A and the current $((\mu_r - 1)/(\mu_r + 1)i)$ in point O.

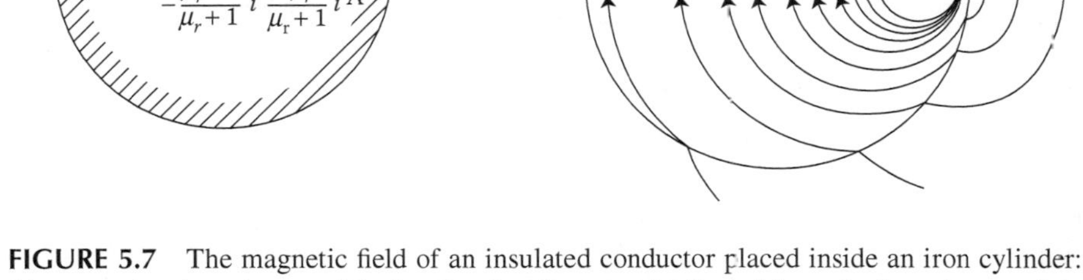

FIGURE 5.7 The magnetic field of an insulated conductor placed inside an iron cylinder: (a) magnetic mirror image of the insulated real current i reflected from the boundary surface (*note*: the currents whose values are given above the axis produce field in the iron, and those below the axis—in the air); (b) the resultant field [5.1], [6.7].

In such a system, almost all the lines in the air are closed by the surrounding iron, whereas in the previous case these lines were closed mainly within the body of iron cylinder, practically not going out to the outside air.

Both these cases can serve as a basis for the method of calculation of reactance in closed slots of electric machines, magnetic screens protecting ambient space from interference of strong fields produced for example, by busbars carrying DC currents or by electric machines (J. Turowski [1.15/1], p. 195),* and so on.

Calculation of field with the method of magnetic images in cases when the conductor is placed *outside the cylinder* is carried out analogically and is presented in Hague [5.1] and other books.

5.1.4 Multiple Mirror Images

Although the method of mirror images was introduced already in the mid-nineteenth century, it is still being developed, especially for multiple images of AC currents in surfaces of complex structures. This method is convenient for computer calculations, when we are interested in the *rapid design* and reduced models allowing for fast solutions.

5.1.4.1 Mirror Images of Current in Crossing Flat Iron Surfaces

If a conductor with current is placed between two boundary iron surfaces that create an angle of π/n, where n is an integer, then a mirror image of the current can be found on a circle passing through the real current. The center of this circle is in the crossing point of the boundary surfaces. In Figure 5.8 are shown the images for angles defined by $n = 1, 2, 3, 4, 5, 6$. The arrows on chords show the sequence of creating the images. The hatched region represents iron, and the not-hatched—air. When the conductor is placed in the air, the images are unidirectional (have equal sign). When the conductor is placed in the iron, the images have opposite sign. A magnetic field in the region where the real current is placed can be established as a *superposition* of the field from the real current and all its images, placed in the same uniform medium, in which the real current was located.

With the help of this method one can, for instance, quickly develop analytical formulae for *impedance of trapezoidal slots* (J. Turowski [5.15]) of induction motors (see Section 5.5.4), even with the consideration of saturation of teeth iron (Figures 5.9 and 5.30 [later in the chapter]). After mapping of these slots with the help of solid copper or aluminum tubes (Figure 5.31 [later in the chapter]), subsequent calculations are carried out for tubular screens (see Section 4.9.1).

5.1.4.2 Conductor Placed between Two Iron Surfaces

Shown in Figure 5.10b and c are consecutive phases of creating images of a real current placed between two iron surfaces of finite permeability (Figure 5.10a) (J. Turowski [5.15]). The real current i first is imaged in the upper metal surface. According to Equations 5.1 and 5.2a, after the first image reflection, the field in the metal is generated by the current $i - \mathrm{M}i = \mathrm{m}i$ (Figure 5.10c) and the field in the air area—by the pair

* For example, protection of military or fishing ships from postwar magnetic mines.

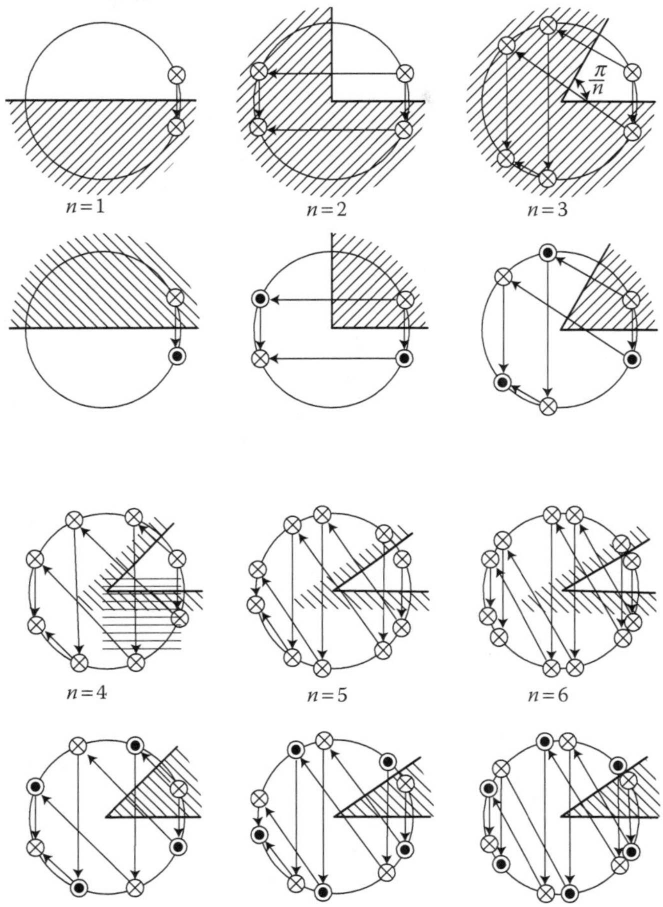

FIGURE 5.8 Mirror images of current in two crossing under angle π/n boundary surfaces of air and iron of infinitely large permeability (hatched area) by B. Hague [5.1].

of currents i and Mi (Figure 5.10b). Now, the two last currents for the lower iron surface substitute the whole system located over this surface. This is why we can consider the second image reflection as an image of only the currents i, Mi in the lower iron surface. At this time, we obtain for the field in the air four fictitious currents: Mi, i, Mi, M^2i, flowing in the identical direction, and for the field in the lower iron area—two currents: mi and mMi. At subsequent, third image reflection, the whole lower system, for the upper iron area can be substituted by air at which this time four currents are placed: two of them were already considered at the first image reflection, and therefore now imaged are only the two last currents which arrived after the second image reflection. In this way, at every next image reflection, to the series of currents creating the field in air added are from one end two currents reduced in the described way. They have the direction like the real current. At the same time, to the series creating the field in iron, added are also the currents from the opposite end, reduced in the same way.

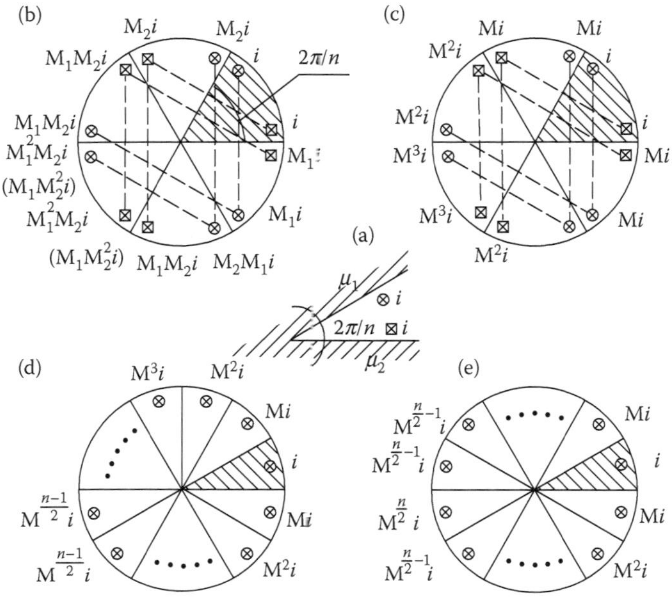

FIGURE 5.9 Mapping (simulation) of a real system: (a) (center) of currents placed between iron walls, by means of a field of fictitious currents placed in the area of $\mu_r = 1$; (b) at different saturation values μ_1, μ_2 of the neighboring iron walls $M_1 = (\mu_{r1} - 1)/(\mu_{r1} + 1)$, $M_2 = (\mu_{r2} - 1)/(\mu_{r2} + 1)$; (c) at identical saturation $M_1 = M_2 = M$; (d) at an odd n, (e) at an even n. (Adapted from Turowski J., Andryszczak W., and Pelikant A.: Application of the method of mirror images to calculation of the slot impedances, taking into account the saturation of the teeth. (in Polish). *"Elektryka" Scientific Papers of Techn. University of Lodz*, (74), 1983, 187–198 [5.17].)

Since the number of such images is infinitely large, hence all three series (Figure 5.11) of the fictive currents extend to infinity, being decreased according to a geometric series of a common ratio M < 1 (5.2a).

5.1.4.3 Conductor Encircled by Steel from Three Sides

According to the principle given above, one can believe that the whole system located on the right-hand side from the side surface of steel (Figure 5.12) can be, for this part of steel region, substituted by air in which a single series of currents is placed (Figure 5.10b). The reflected image of this series in lateral surface gives, for air, a double series (Figure 5.12b). In the lateral region of steel appears, instead, a very weak field produced by a single series of small currents m*i* (Figure 5.12c). In the upper and lower steel regions, the field is produced by a double series of the fictive currents (Figure 5.12d and e).

The method of simulation (mapping) of the *magnetic field created by a conductor placed nearby a steel plate* was given in the author's book ([1.16], pp. 95–98).

5.1.5 Mirror Images of Magnets and Circuits with Direct Currents

Per an analogy with electrostatics, one can conclude that if near a boundary surface, in the region with permeability μ_{II} is placed a magnetized core having magnetic

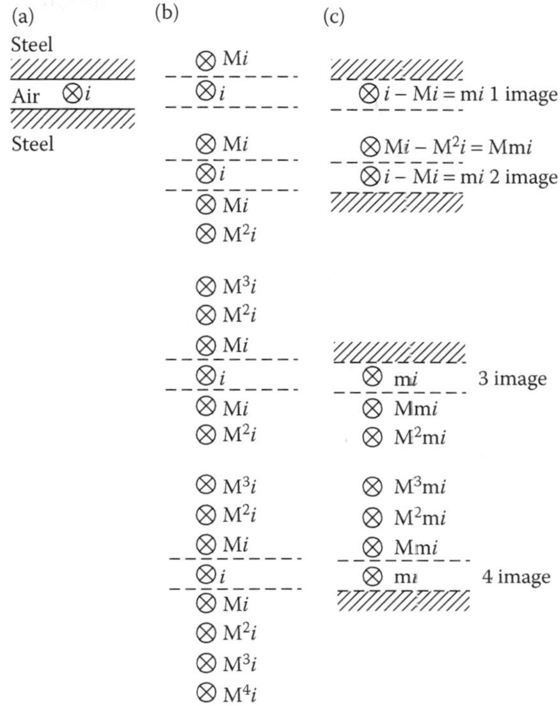

FIGURE 5.10 Consecutive phases of multiple mirror images of a current placed between two iron surfaces of μ = const: (a) real system; (b) series of fictive currents which produce the field in the air; (c) series of fictive currents which produce the field in the upper and lower iron region, respectively. (Adapted from Turowski J.: Power losses in transformer covers at various methods of their reduction. (in Polish) *Archiwum Elektrotechniki*, (4), 1959, 529–556 [5.15].)

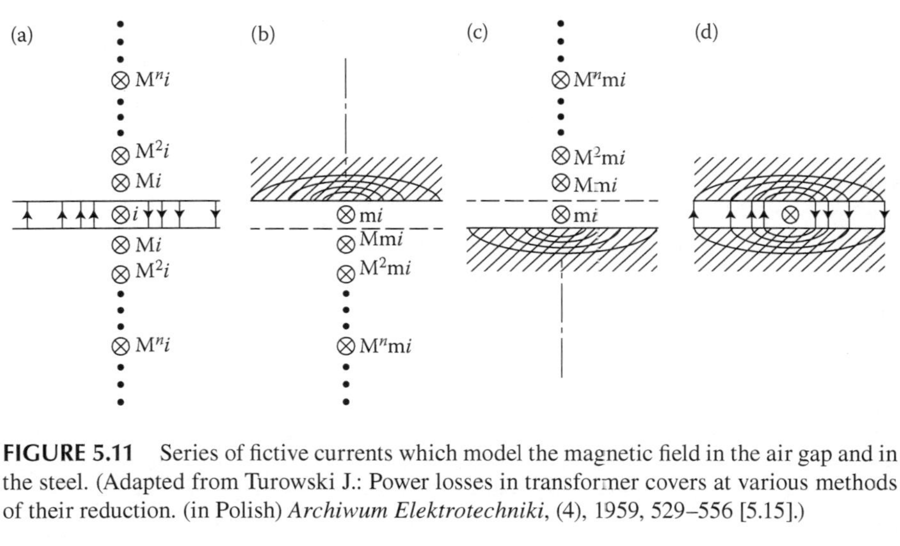

FIGURE 5.11 Series of fictive currents which model the magnetic field in the air gap and in the steel. (Adapted from Turowski J.: Power losses in transformer covers at various methods of their reduction. (in Polish) *Archiwum Elektrotechniki*, (4), 1959, 529–556 [5.15].)

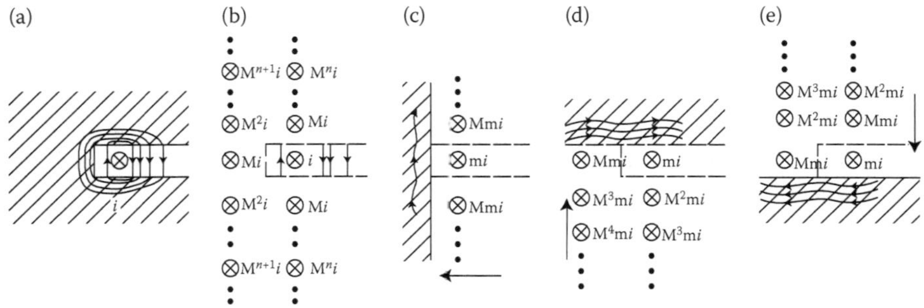

FIGURE 5.12 Series of fictive currents which model the real magnetic field in the slot and in the surrounding steel. (Adapted from Turowski J.: Power losses in transformer covers at various methods of their reduction. (in Polish) *Archiwum Elektrotechniki*, (4), 1959, 529–556 [5.15].)

momentum **K** (Figure 5.13a), the field in medium II equals the field of the real core of magnetic momentum **K** and its mirror image with the magnetic momentum—M**K** [5.8], [5.26], where M is the coefficient (5.2)

$$\mathrm{M} = \frac{\mu_{\mathrm{II}} - \mu_{\mathrm{I}}}{\mu_{\mathrm{II}} + \mu_{\mathrm{I}}} = \frac{\mu_r - 1}{\mu_r + 1}$$

In a similar way are imaged *loops of direct currents* (Figure 5.13b), which also have the magnetic momentum. Both types of sources can be investigated as dipoles of magnetostatic charges (Hammond [5.8]). In a general case, the mirror image method can be applied for combination of fields produced by steel cores and coils with current.

From a simplified analysis and electrostatic analogies follow the initial recommendations for analysis of mirror images of circuits with DC current [5.8]:

1. Currents i in sections of circuits placed in air (Figure 5.14a), parallel to the steel surface, *considered like circuit elements*, are imaged in the steel

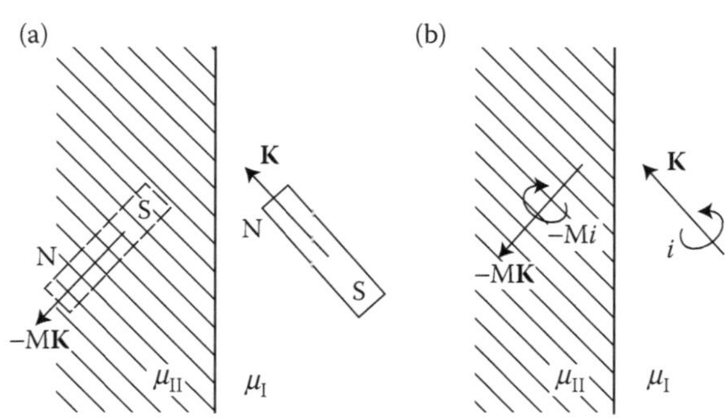

FIGURE 5.13 Mirror images of: (a) magnets; (b) equivalent current loops M = $(\mu_r - 1)/(\mu_r + 1)$. (Adapted from Nietushil A.B. and Polivanov K.M.: *Bases of Electrical Engineering. Part 3. Theory of Electromagnetic Field.* (in Russian). Moscow: Gosenergoizdat, 1956 [5.26]; Hammond P.: Electric and magnetic images. *Proc. IEE—Part C*, 107(12), Sep. 1960, 306–313 [5.8].)

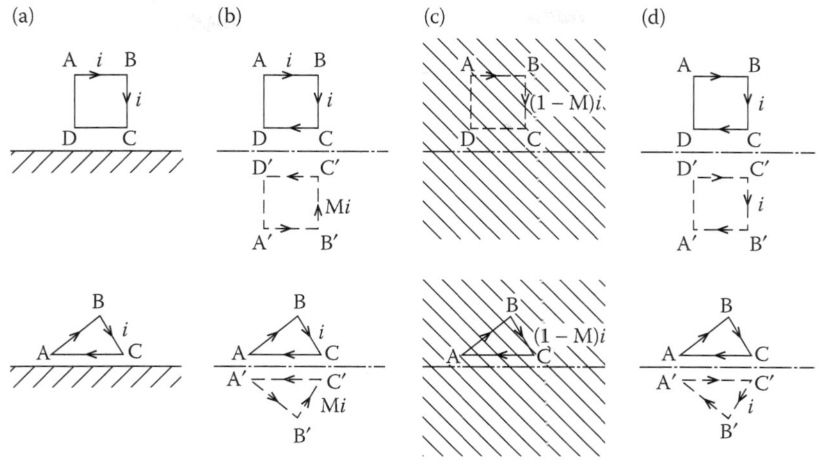

FIGURE 5.14 Mirror images of current loops in a steel surface (a, b) and in a superconductor (d): (a) real currents nearby the steel surface; (b) real current i and fictitious (imaged) current Mi creating jointly the field in the air; (c) fictitious (imaged) current $(1 - \mathrm{M})i$ creating the field in the steel; $\mathrm{M} = (\mu_r - 1)/(\mu_r + 1)$; $(1 - \mathrm{M}) = 2/(\mu_r + 1)$. (After Carpenter C.J.: *Proc. IEE—Part A: Power Engineering*, 107(35), October 1960, 487–500 [5.7]; Hammond P.: *Proc. IEE—Part C*, 107(12), Sep. 1960, 306–313 [5.8].)

surface with the same direction, as +Mi (Figure 5.14b), and produce jointly with i the field in the air.

2. Currents i in circuit sections placed in the air (Figure 5.14a) perpendicularly to the steel surface are imaged with opposite direction (Figure 5.14b) as –Mi and create jointly with i the field in the air.
3. Currents i located in the air (Figure 5.14a) create in the steel such a field like the current $(1 - \mathrm{M})i$ located in the same place like i but in the steel medium (Figure 5.14c).
4. Currents i in sections of circuits placed in steel (Figure 5.15a), parallel to the dielectric surface, are imaged with opposite directions (Figure 5.15b) as –Mi and create jointly with i the field in the steel.
5. Currents i in sections of circuits placed in steel (Figure 5.15a) perpendicular to the dielectric surface are imaged in the steel surface with the same direction, as +Mi (Figure 5.15b), and produce jointly with i the field in the steel.
6. Currents i located in the steel (Figure 5.15a and d) create in the air, a field like the current $(1 + \mathrm{M})i$ located in the same place like i, but in the air medium (Figure 5.15c and f).

At inclined sides of the circuit, the directions of currents and their images are determined on the basis of their components—parallel and perpendicular to the surface.

These conclusions have a character of auxiliary hypothesis, which however can be verified on the basis of the principle of uniqueness of field (the uniqueness theorem [2.8]) and boundary conditions.

The presented indications concern all kinds of circuits and their sections, as well as all combinations of media. One can also apply them in the first approach to AC

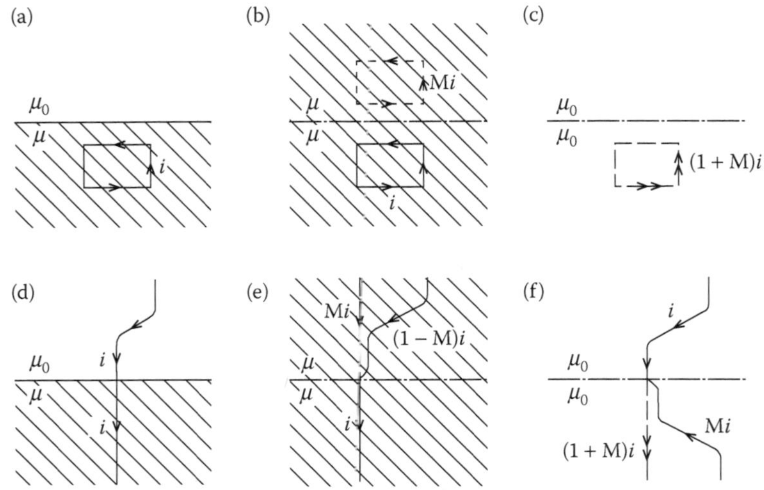

FIGURE 5.15 Mirror image (in dielectric) of a DC current circuit located inside steel (a, b, c), and of a conductor crossing both media (d, e, f): (a), (d)—real system of currents; (b), (e)—field in the steel; (c), (f)—field in the dielectric. (After Hammond P.:. *Proc. IEE—Part C*, 107(12), Sep. 1960, 306–313 [5.8].)

currents, considering the conditions given in Section 5.1.2, with utilization of formula (5.7) for quasi-permeability of the analyzed media

$$M_Q = \frac{\mu_{Q\mathrm{II}} - \mu_{Q\mathrm{I}}}{\mu_{Q\mathrm{II}} + \mu_{QI}} \tag{5.14}$$

The mirror image of an AC current in solid metal, with consideration of eddy-current reaction, is easy to resolve only in cases when the eddy-current reaction is so big that the metal can be considered as an ideal diamagnetic (e.g., in superconductors). Then, the current in the imaged loop has the same twist as the real current (Figure 5.14d). In intermediate cases, the reaction of eddy-current losses can be considered as some apparent reduction of the permeability μ_r in relation to its real value (quasi-permeability μ_{Qr}).

One should remember, however, that the method of mirror images of current conductors can be applied only to *closed* circuits. And, although the imaged currents can be analyzed, in part being images of corresponding parts of the real circuit, these parts should be added in such a way that there does not exist anywhere a break in the current flow as well as in the nodes on the media boundary; the principle where real current continuity and all its images is held (Figure 5.15e and f).

5.2 FIELD OF ENDWINDINGS IN ELECTRIC MACHINES

For finding the field of endwindings in electric machines, in static and transient conditions, at magnetic and nonmagnetic pressing endrings of stator, screens, and so on, one can apply one of the simplest and evident methods, that is, the method of *mirror images*.

5.2.1 Mirror Image in Solid Steel Wall

In 1956, J. Turowski [2.31] proposed considering AC current-carrying conductors as a series of consecutive *electric dipoles* which are imaged by reflection in nearby located metal surfaces. In 1960, P. Hammond [5.8] used this method for determining mirror images of endwindings. Equivalent circuits of the endwindings, determined in this way, are shown in Figures 5.16 and 5.17, in a case when the core surface is screened by an ideal conductor (Figure 5.16) and when the core with the relative permeability μ_r is not screened (Figure 5.17). The left side of the figures shows a real system, and the right side—the equivalent system, which produces an equivalent field in the air (b) and in the iron (c).

The field of medium I is superimposed on the field of medium II, and vice versa, according to the principles (1) through (6) given in Section 5.1.5.

The reaction of eddy currents flowing in the end-face surfaces of core, and the limited screen conductivity, are cause for the images to be a little "fogged" and "blurred." In any case, the picture from Figure 5.17 is closer to reality than that from Figure 5.16. Figure 5.17 is especially applicable for analysis of the field of endwindings of rotors of induction machines, in which during operation the current frequency is very small.

5.2.2 Influence of Air Gap

Carpenter [5.7] gave the method of considering of the air gap at analysis of mirror images of endwindings. While at the end-face surfaces of stator and rotor cores of

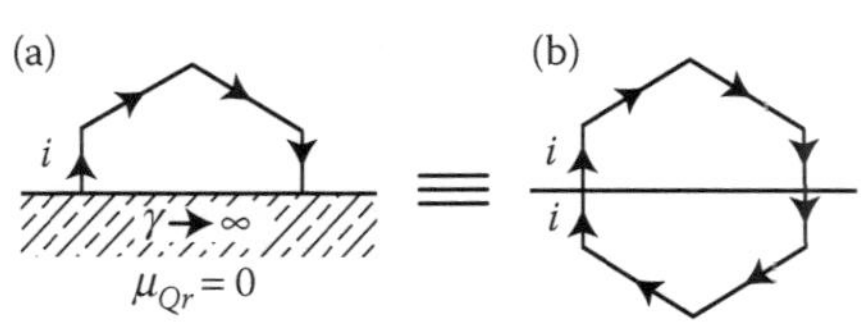

FIGURE 5.16 Mirror image of endwinding of coil [5.8] in an ideal electromagnetic screen.

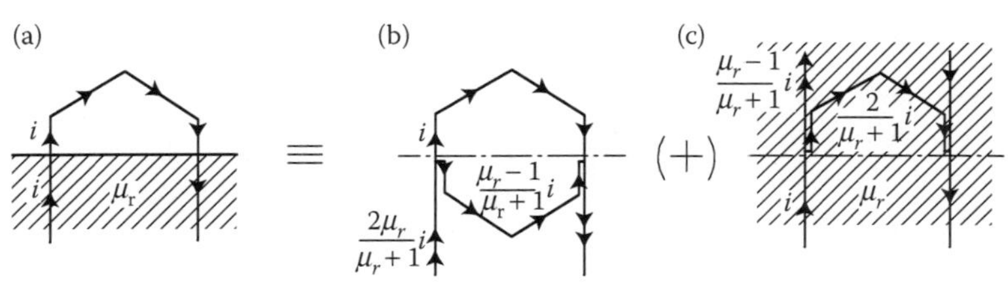

FIGURE 5.17 Determination of magnetic field of endwinding in iron core with permeability μ_r with the method of mirror image, (a) real system, (b) field in air, (c) field in iron (Adapted from Hammond P.: *Proc. IEE—Part C*, 107(12), Sep. 1960, 306–313 [5.8].)

infinite permeability, the tangential component of magnetic field intensity equals zero, but in the air gap the tangential component is the main component of field, determined by the magnetic voltage of the gap. For the parts of endwindings which are parallel to the iron surface, the air gap, as a small part of the magnetic reluctance on the way of magnetic lines surrounding such conductors, has no significant influence on their image, and the end-face surfaces of core of machine can be for these parts considered as a uniform steel block. In the case of conductors entering perpendicularly to the gap, the induction lines run as shown in Figure 5.18a, so as if they were created by branched conductors running along the gap and conducting half the current (Figure 5.18b). The equivalent circuit, determined in this way, is in turn imaged in the iron surface, according to Figure 5.14b, and creates an equivalent system of conductors located in the air (Figure 5.18c), which creates in the air area the same field as in the real system.

Taking into account the principles described above, and the fact that the picture of the field presented in Figure 5.18a takes place on the whole circumference of machine, the pictures presented in Figures 5.16 and 5.17 can be made more accurate by considering the air gap. In this way, for the field in the air we obtain a system of equivalent currents presented in Figure 5.19.

The limited volume of this book does not allow presentation of detailed methods and formulae for calculating inductances of endwindings. They can be found in the works by the author (J. Turowski [1.16]) and others [5.4], [5.6], [5.7], and so on. Another investigation of the influence of the air gap closed at the ends is discussed in Section 6.4 of this book.

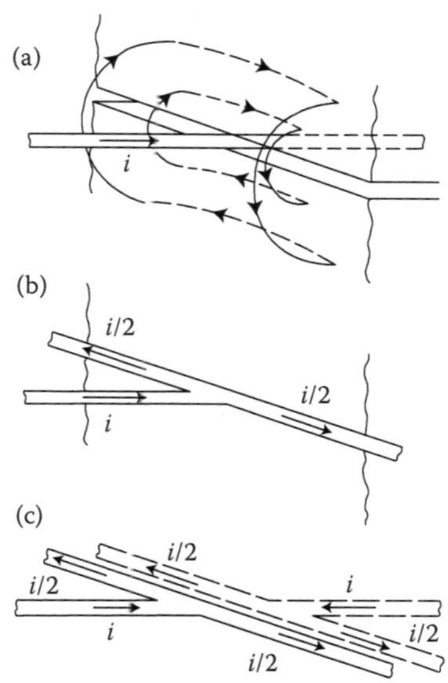

FIGURE 5.18 Mirror image of a conductor wire entering perpendicularly to air gap in an iron block of infinite permeability: (a) real field, (b) circuit substituting the action of air gap; (c) circuit substituting the action of entire iron block with the air gap. (Adapted from Carpenter C.J.:. *Proc. IEE—Part A: Power Engineering*, 107(35), October 1960, 487–500 [5.7].)

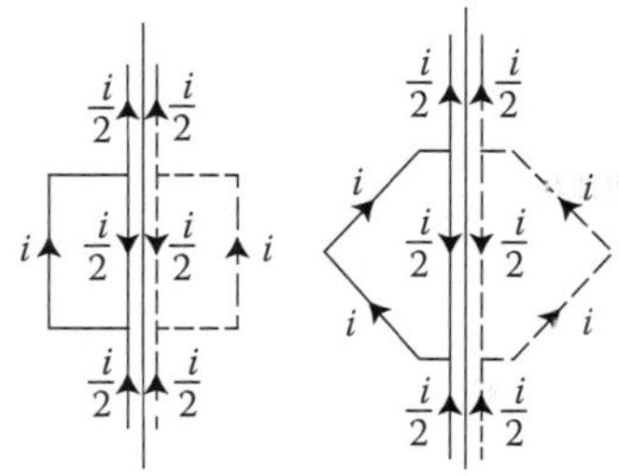

FIGURE 5.19 Mirror image of endwindings in the infinitely permeable wall with an air gap: (a) single-layer winding, (b) double-layer winding. (Adapted from Carpenter C.J.:. *Proc. IEE—Part A: Power Engineering*, 107(35), October 1960, 487–500 [5.7].)

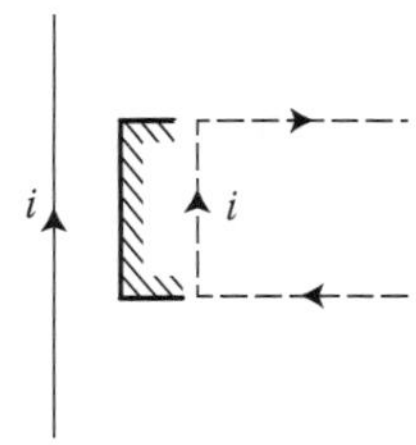

FIGURE 5.20 Mirror image of current in the steel beam. (Adapted from Carpenter C.J.:. *Proc. IEE—Part A: Power Engineering*, 107(35), October 1960, 487–500 [5.7].)

5.2.3 Influence of Constructional Elements

Different steel clamping parts of electric machines and transformers, electric apparatus, switching stations, and so on, have the form of relatively narrow metal stripes. In such a case, the field nearby a current carrying conductor can be determined with a substitution of the steel strip by a half-closed, infinite circuit presented in Figure 5.20 with broken line.

In a case when the iron permeability is finite, one can use formulae (5.1) and (5.2). The general problem of eddy-current reactions at AC is still open, from the theoretical point of view, and can be evaluated in various, more or less accurate ways, depending on the adopted model.

5.3 FIELD OF BUSHINGS

5.3.1 Dynamic Mirror Image of Currents

In Figures 3.1 and 3.2, we presented the flow of electric power through a transformer cover. In order to find the dynamic images of AC current in solid metal, we can take an element dl of a conductor crossing through the cover (Figure 5.21) as an *electric dipole* acting in a space of dimensions negligibly small in comparison with the length of electromagnetic wave ($r \gg \lambda$). The electromagnetic field of such dipole is described by formulae (2.168a), (2.169a), and (2.170a).

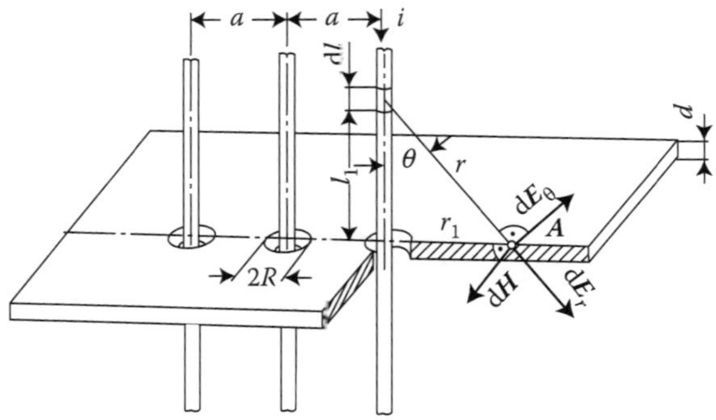

FIGURE 5.21 Components of field on the cover surface. (After Turowski J.: Losses and local overheating caused by stray fields. (in Polish). *Scientific Archives of the Technical University of Lodz "Elektryka*, No. 11, 1963, pp. 89–179 (*Habilitation (DSc) dissertation; 1st Ministry Award)* [2.41]; Turowski J.: Losses in single and three phase transformers covers. (in Polish). *Rozprawy Elektrotechniczne*, (1), 1959, 87–119 [4.16].)

If the steel plate through which the bushing passes is sufficiently thick (over 6–7 mm), the electromagnetic field practically does not penetrate through it. The magnetic field intensity in point A of transformer cover (Figure 5.21) can be therefore calculated by integrating, in the known way, the expression (2.168a):

$$\mathrm{d}H = \frac{i\,\mathrm{d}l}{4\pi r^2}\sin\Theta$$

with regard to dl, from zero to infinity. As a result, we conclude that onto the surface of the iron plate impinges an electromagnetic wave with the magnetic field intensity $H_{\text{inc}} = i/4\pi r_1$.

This wave, under the influence of the eddy-current reaction, is reflected from the surface of metal of the wave impedance $|Z_2| \ll |Z_1| = Z_{\text{air}}$ (see formulae 2.166a and 2.177a), according to the dependence (2.197), which produces on the surface the resultant value $H_S = 2H_{\text{inc}} = i/2\pi r$, which is in agreement with the (Ampere's) *law of flow* (2.15).

Therefore, the dynamic mirror image of an AC current perpendicular to a solid steel surface has the same direction as the real current. Also, the static image of a single conductor (Figure 5.15c) must have the same direction, because in opposite case the *law of flow* $2\pi rH = i$ would not be satisfied. Hence, both methods—dynamic and static—give the same result.

5.3.2 Field on the Cover Surface

The resultant magnetic field intensity on the cover surface can be calculated from the *law of flow*, with application of the *superposition* principle. As experience shows (J. Turowski [4.16]), for practical applications one can omit some heterogeneity of the medium, caused by the variable steel permeability, the eddy-current reaction, and demagnetizing action of the holes for bushings.

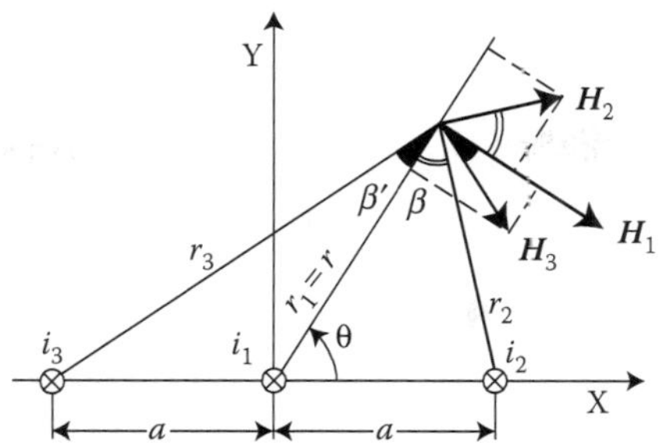

FIGURE 5.22 Mirror image of endwindings in the infinitely permeable wall with an air gap: (a) single-layer winding, (b) double-layer winding. (Adapted from Turowski J.: Losses in covers of three phase transformers caused by the electromagnetic field of bushings (in Polish). *Science Reports "Elektryka"*, No. 4, Tech. University of Lodz, 1958, pp. 79–101 [7.16].)

The components H_Θ and H_r of the instantaneous value of magnetic field intensity, in accordance with Figure 5.22, equal

$$H_\Theta = \frac{i_1}{2\pi r_1} + \frac{i_2 \cos\beta}{2\pi r_2} + \frac{i_3 \cos\beta'}{2\pi r_3}; \quad H_r = \frac{i_2 \sin\beta}{2\pi r_2} - \frac{i_3 \sin\beta'}{2\pi r_3}$$

Taking into account the phase shifts of three-phase currents i_1, i_2, i_3 and after solving the integral

$$\frac{H_m^2}{2} = \int_0^T (H_\Theta^2 + H_r^2)\,dt$$

we find (J. Turowski [7.16]) the maximum value of magnetic field intensity in any point on the surface cover of a three-phase transformer, in the cylindrical coordinates (Figure 5.22)

$$H_{ms} = \frac{Ia}{\sqrt{2}\pi r}\sqrt{\frac{3r^2 + a^2}{r^4 - 2a^2r^2\cos 2\Theta + a^2}} \tag{5.15}$$

and in the rectangular coordinates

$$H_m = \frac{Ia}{\sqrt{2}\pi r}\sqrt{\frac{3x^2 + 3y^2 + a^2}{(x^2 + y^2)(x^4 + y^4 + 2x^2y^2 - 2a^2x^2 + 2a^2y^2 + a^4)}} \tag{5.16}$$

In a similar way, one can determine (J. Turowski [4.16]) the maximum value of magnetic field intensity on the surface cover of a single-phase transformer (Figure 5.23), in the cylindrical coordinates

$$H_m = \frac{Ia}{\sqrt{2}\pi}\frac{1}{\sqrt{r^4 - (a^2r^2/2)\cos 2\theta + (a^4/16)}} \tag{5.17}$$

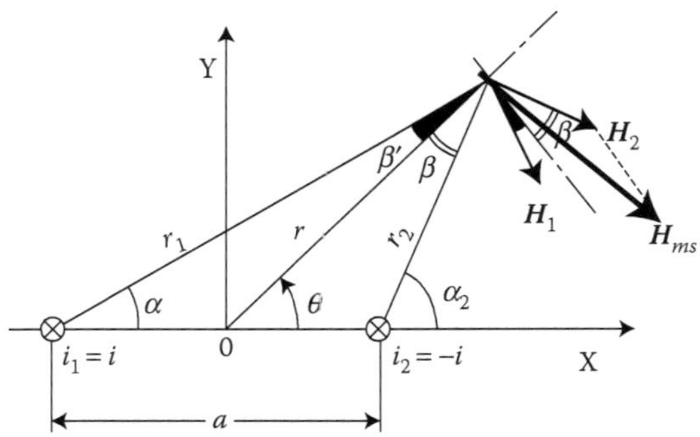

FIGURE 5.23 Determination of the resultant field on the surface of cover of a single-phase transformer. (Adapted from Turowski J.:. Losses in single and three phase transformers covers (in Polish). *Rozprawy Elektrotechniczne*, (1), 1959, 87–119.)

5.4 FIELD OF BARS NEARBY A STEEL SURFACE

The components of magnetic field intensity of a rectangular bar with no skin effect, in point (y, z) of ambient space (Figure 5.24), can be determined from the classical formulae

$$\left.\begin{aligned} H_y &= -\frac{i}{2\pi ab}\left[\begin{aligned}&\left(z+\frac{a}{2}\right)(\alpha_3-\alpha_1)-\left(z-\frac{a}{2}\right)(\alpha_4-\alpha_2)+\\ &+\left(y+\frac{b}{2}\right)\ln\frac{r_3}{r_4}-\left(y-\frac{b}{2}\right)\ln\frac{r_1}{r_2}\end{aligned}\right]\\ H_z &= -\frac{i}{2\pi ab}\left[\begin{aligned}&\left(y+\frac{b}{2}\right)(\alpha_4-\alpha_3)-\left(y-\frac{b}{2}\right)(\alpha_2-\alpha_1)+\\ &-\left(z+\frac{a}{2}\right)\ln\frac{r_3}{r_1}-\left(z-\frac{a}{2}\right)\ln\frac{r_4}{r_2}\end{aligned}\right]\end{aligned}\right\} \tag{5.18}$$

Technical calculations of the field outside the bar, the bar cross section can often be substituted (e.g., J. Turowski et al. [4.28]) by a *current flow (ampere-turns) concentrated* in one axis (line AB in Figure 5.24). Then, the angles $\alpha_1 = \alpha_2$, $\alpha_3 = \alpha_4$, $\alpha_3 - \alpha_1 = \alpha_4 - \alpha_2 = \beta$, and the distances $r_1 = r_2 = r$ and $r_3 = r_4 = R$. Hence, formulae (5.18) can be simplified to the form

$$H_y = \frac{-i}{2\pi b}\beta \quad \text{and} \quad H_z = \frac{i}{2\pi b}\ln\frac{R}{r} \tag{5.19}$$

In the case of determinating the need of these components in many points, the values R, r, and β can be taken directly from a winding drawing in the scale. In case of many bars, one can apply the *principle of superposition*. In this way the graph in Figure 5.25 was drawn, for the upper half of windings of large transformer, substituted by a system of flat bars imaged in the steel surface.

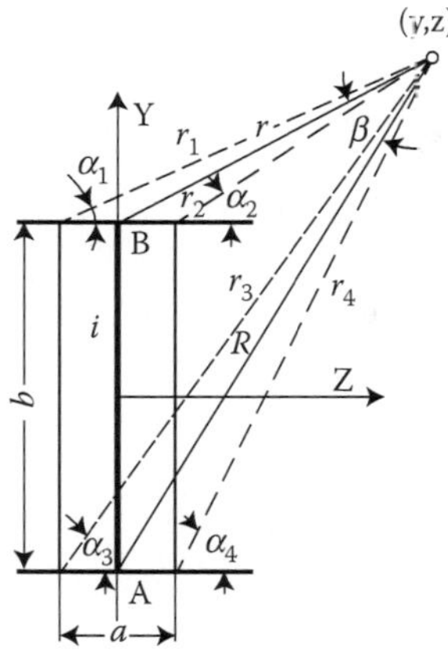

FIGURE 5.24 Determination of the magnetic field of a rectangular bar.

In the case when a single bar with DC current is located nearby a steel wall (1 in Figure 5.26), the magnetic field on the steel surface from the air side can be determined as a superposition of the field of this bar and its univocal image 1′. Analytically, the normal component H_{z0} on the steel surface from the air (or oil) side is expressed in this case by the formula (5.14):

$$H_{z0}(g) = \frac{iN}{2\pi a_1 b}\left\{ \begin{array}{l} (g+a_1)\ln\dfrac{\left(y+\dfrac{b}{2}\right)^2+(g+a_1)^2}{\left(y-\dfrac{b}{2}\right)^2+(g+a_1)^2} - g\ln\dfrac{\left(y+\dfrac{b}{2}\right)^2+g^2}{\left(y-\dfrac{b}{2}\right)^2+g^2}+ \\ +2\left(y+\dfrac{b}{2}\right)\left[\operatorname{arctg}\dfrac{y+\dfrac{b}{2}}{g}-\operatorname{arctg}\dfrac{y+\dfrac{b}{2}}{g+a_1}\right]+ \\ +2\left(y-\dfrac{b}{2}\right)\left[\operatorname{arctg}\dfrac{y-\dfrac{b}{2}}{g}-\operatorname{arctg}\dfrac{y-\dfrac{b}{2}}{g+a_1}\right] \end{array} \right\} \quad (5.20)$$

The resultant field on the steel surface from the air (or oil) side is of two bars 1 and 2 (Figure 5.26) can be found as a superposition

$$\mathrm{H}_{z0} = \mathrm{H}_{z0}(g) - \mathrm{H}_{z0}(g+\delta+a_1) \quad (5.21)$$

If we neglect the winding thickness, we can obtain a simplified formula for both bars (Figure 5.26)

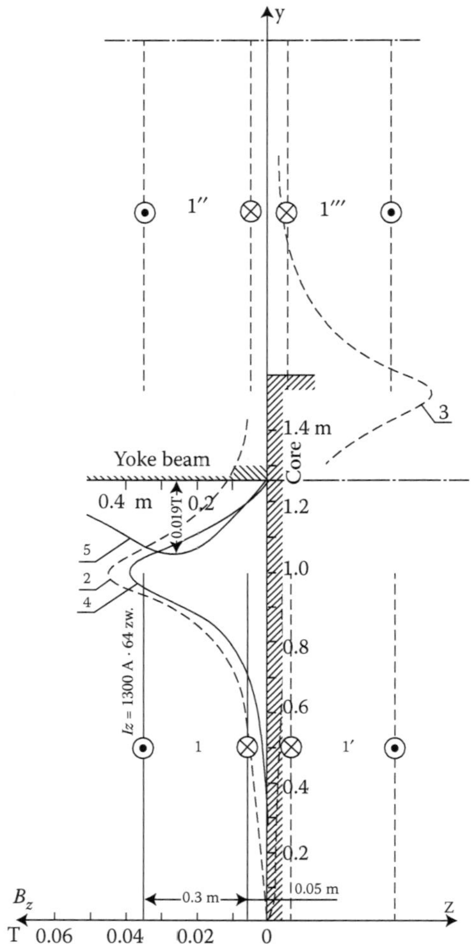

FIGURE 5.25 Determination of the flux density distribution on the surface of transformer core with the method of winding substitution by parallel bars: 1—real ampere-turns linearly concentrated; 1′, 1‴, 1‴—imaged ampere-turns; 2—distribution of B_z component without a yoke beams; 3—flux density B_z from ampere-turns 1″, 1‴ imaged in iron beam; 4—the resultant distribution of B_z component when yoke beams exist; 5—the distribution of B_y component on the surface of yoke beam. (Adapted from Turowski J.: Losses and local overheating caused by stray fields. (in Polish). *Scientific Archives of the Technical University of Lodz "Elektryka*, No. 11, 1963, pp. 89–179 (*Habilitation (DSc) dissertation; 1st Ministry Award)* [2.41].)

$$H_{z0} = \eta_Q \frac{iN}{2\pi b}\left[\ln\frac{c_1^2+\left(y+\dfrac{b}{2}\right)^2}{c_1^2+\left(y-\dfrac{b}{2}\right)^2} - \ln\frac{c_2^2+\left(y+\dfrac{b}{2}\right)^2}{c_2^2+\left(y-\dfrac{b}{2}\right)^2}\right]\Bigg\} \tag{5.21a}$$

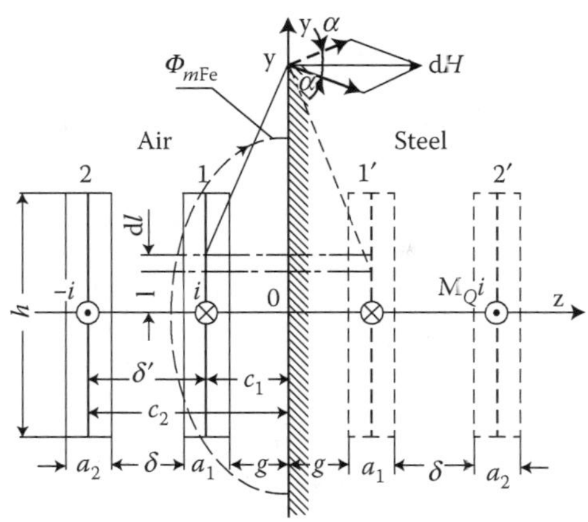

FIGURE 5.26 Determination of magnetic field of two bars on the steel surface. 1, 2—real ampere-turns; 1′, 2′—imaged ampere-turns. (Adapted from Turowski J.: Losses and local overheating caused by stray fields. (in Polish). *Scientific Archives of the Technical University of Lodz "Elektryka*, No. 11, 1963, pp. 89–179 (*Habilitation (DSc) dissertation; 1st Ministry Award)* [2.41].)

where $\eta_Q = 1 + (M_Q/2) \leq 1$ is the semiempirical coefficient (5.8) considering the possible not full reflection of the image of current in the steel surface, due to the eddy-current reaction (at AC currents) and steel saturation.

5.5 LEAKAGE FIELD IN TRANSFORMERS AND IN SLOTS OF ELECTRIC MACHINES

5.5.1 Application of the Method of Multiple Mirror Images

In a transformer window, or in a slot of an electric machine, multiple reflections of mirror images of windings take place that create a series of currents, shown in Figures 5.9 through 5.12, for the same permeability μ = const of surrounding surfaces. This method can also be applied when imaging surfaces have significantly different permeabilities (Figure 5.27), for instance, in a transformer when equivalent (quasi-) permeabilities μ_Q (taking into consideration a screening action of eddy currents), are different for tank walls, and different for the cover and the bottom. The mirror image coefficients M and M_1 are determined by formulae (5.2) and (5.7), but their actual values depend on the influence of eddy currents on the *quasi-permeability* μ_Q. Depending on the condition of the surface, current frequency, and steel saturation, one can distinguish the following typical values of the image coefficients:

$M = M_1 = 0$ for $\mu_r = 1$ – lack of tank
$M = 1$ for $\mu_r \approx \infty$ – iron screen (shunt) packet of sheets
$M_Q \approx -1$ for $\mu_{Qr} \approx 0$ – superconductor, high frequency, or impulse excitation
$M_Q = -0.9$ to -0.8 – copper screen of steel wall
$M_Q = 0.4$ to 0.6 – not screened steel wall of tank

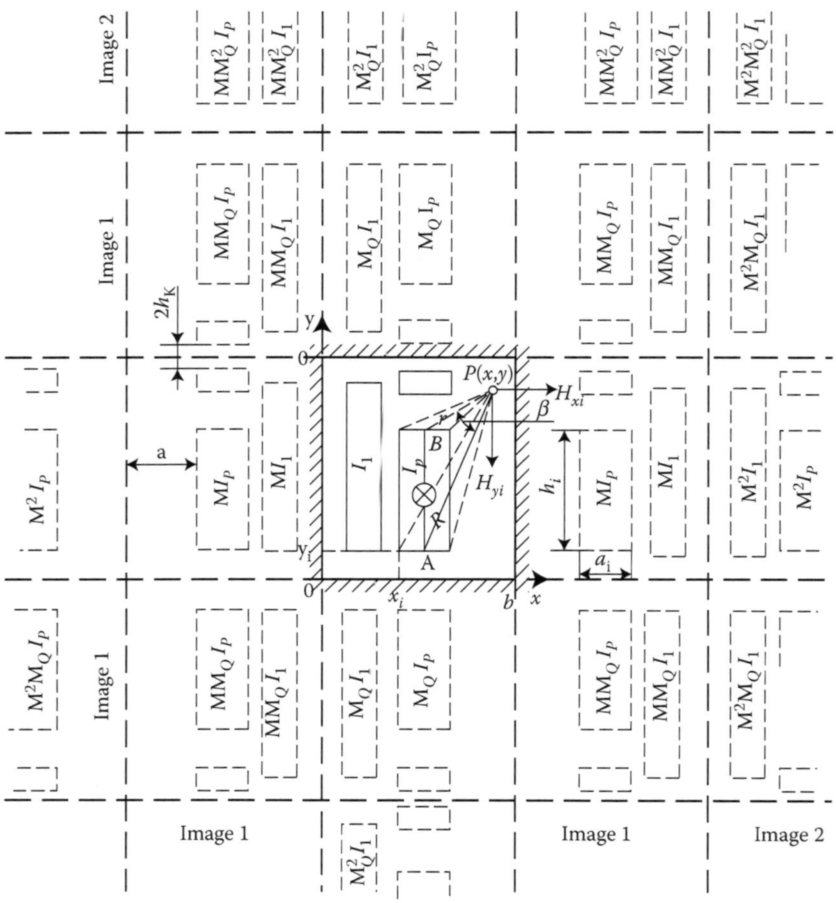

FIGURE 5.27 Field of mirror images of winding in a transformer window, with different (due to screens or shunts) image coefficients: M—in side walls, M_Q—in cover and bottom. (Adapted from Turowski J.: (1) *Technical Electrodynamics*. (in Polish). Warsaw: WNT, 1969, 1st edition [1.15].)*

At such calculations, one should remember that this is an approximate method and only in the first approach one can assume that the AC images in a nonideal steel have the same features (shape, geometrical distributions, harmonics, etc.) as the real currents. The equivalent image quasi-coefficient $M_Q = (\mu_{Qr} - 1)/(\mu_{Qr} + 1)$ can be applied only for determination of the normal component H_n on the steel surface or only for determination of the tangential component H_t on the surface of copper or aluminum. One should also remember that the value of M_Q depends also on the distance of the current from the tank wall (Figure 5.6).

* This method was most broadly applied in the PhD thesis of Dr. P. Jezierski, under J. Turowski's supervision, 1959, Award of Ministry of Education and Science.

Other solved cases of mirror images in saturated steel walls are presented in Chapter 7—for transformer covers, and in Chapter 8—for forces in slots of electric machines. Examples of applications and methods of calculations are given in two other books by the author (J. Turowski [1.15/1], p. 212; [1.16], pp. 244–247 and 256).

5.5.2 Method of Approximate Solution of a Field in a Transformer Window with the Help of Fourier Series

5.5.2.1 Method of Analytical Prolongation

This method is based on the *multiple mirror images approach* (Figure 5.27). In a case when the magnetic window is not closed, for instance, over the edge of transformer windings, the distance h_k is chosen as small as possible to obtain a rapid convergence of the series, but sufficiently large enough to make sure that the field of imaged fictitious currents will not distort the real field by more than a permitted error of ε relative units.

According to the author's experience (J. Turowski et al. [5.14]), the recommended distance of a fictitious mirror should be in the order of

$$h_k \approx \frac{2}{\pi} h_i \left(\frac{1}{\varepsilon} - 1 \right) \tag{5.22}$$

which, for example, for the assumed permitted error $\varepsilon = 0.05$ gives $h_k = 12h_i$. According to North [5.21], this condition is $2h_k \gg h_i$, which, at excessively large values of the distances h_k to the tank walls, may lead to an excessively big number of harmonics that have to be considered. Other authors of work [5.14] gave the values (5.22) for bodies in motion, for instance, for linear induction motors (J. Turowski [1.16/1]).

Using the magnetic vector potential $\boldsymbol{A}$, the field of real and imaged currents (Figure 5.27) can be described with the help of Poisson's (2.57) and Laplace's (2.28) equations

$$\nabla^2 \vec{A} = \begin{cases} -\mu_0 \vec{J} & \text{in the winding cross sections} \\ 0 & \text{in spaces with no current } (\boldsymbol{J} = 0) \end{cases} \tag{5.23}$$

From the general solution

$$\boldsymbol{A} = \sum_m \sum_n \sum_p C_{mnp} \begin{Bmatrix} \cos \\ \sin \end{Bmatrix} mx \begin{Bmatrix} \cos \\ \sin \end{Bmatrix} ny \begin{Bmatrix} \cos \\ \sin \end{Bmatrix} pz \tag{5.24}$$

where the symbol in brackets means "cos or sin." For given boundary conditions, connected with the flux density $\boldsymbol{B} = \text{curl}\ \boldsymbol{A}$ on the surface, and given constraints of the current densities $\boldsymbol{J}$, per Equation 2.59 $\boldsymbol{A} = \mu/4\pi \int_V \boldsymbol{J} \cdot \mathrm{d}V/r$, the field in the investigated region can be calculated for a selected number of components of Fourier's series.

As one can see, this method allows for a relatively simple investigation of three-dimensional (3D) fields XYZ, which was successfully utilized in several works by E. Mendrela et al. [4.10], [5.14].

Calculation of a leakage field in a *shell type transformer* with the help of magnetic vector potential $\boldsymbol{A}$ was carried out, among others, by Smythe in his work [4.3], p. 292.

The Rogowski's method (1909), described in detail in the first 1968 edition of this book ([1.15/1], p. 209) was based on an example of a two-dimensional (2D) field of a sandwich-winding transformer. Calculations were carried out with the help of Poisson's equation. A solution, based on Fourier series can be considered now as a particular case for solution (5.24), in which Figure 5.27 was taken into account only one row of the real and imaged coils in the direction of Y axis, at $M_Q = 1$ and $M = 0$ with only one component of the current density $J = J_z = \sum_{n=1}^{\infty} a_n \sin npy$. The final solution has the form of a separate series $A_i = \sum_{n=1}^{\infty} A_{ni}$ for particular regions $i = \mathrm{I}, \mathrm{II}, \ldots$.

The Roth method (1927) is also described in detail in the book (J. Turowski [1.15/1]). It is a solution (5.24) for the whole 2D field of real and imaged currents in Figure 5.27, at ideal mirror image at $M = M_Q = 1$, $\mu_{Fe} = \infty$ and only one component of the current density $J = J_z$.

The solution has the form of double Fourier's series (5.24):

$A = A_z = \sum_{h=1}^{\infty} \sum_{k=1}^{\infty} A_{hk}$, from which we find the components of flux density

$$B_x = \frac{\partial A_z}{\partial y}, \quad B_y = -\frac{\partial A_z}{\partial x}, \quad B_z = 0 \tag{5.25}$$

Next, we obtain equations of field lines, which follow from the condition that an element dl of the line in any point is parallel to the vector $\boldsymbol{B}$, which means

$$\frac{\mathrm{d}y}{\mathrm{d}x} = \frac{B_y}{B_x} \quad \text{and} \quad \frac{\mathrm{d}y}{\mathrm{d}z} = \frac{B_y}{B_z} \tag{5.26}$$

After substituting Equation 5.25 into Equation 5.26, we obtain the equation

$$\frac{\partial A_z}{\partial y}\mathrm{d}y + \frac{\partial A_z}{\partial x}\mathrm{d}x = 0 \tag{5.27}$$

which, as we can see, is the total differential of function $A(x, y)$. It means that the lines of force of the magnetic field $\boldsymbol{B}$ and $\boldsymbol{H}$ are at the same time the *lines of constant magnetic vector potential*

$$A_z(x, y) = \text{const} \tag{5.28}$$

Using this condition, one can find the course of lines of flux density $\boldsymbol{B}$ (J. Turowski [1.15/1], p. 216), similar to Figure 7.16, but containing less useful information.

In the author's works [5.15], [4.21], [5.17], [8.17], and others, a solution was given, with the help of the series, of fields for particular cases: a transformer cover

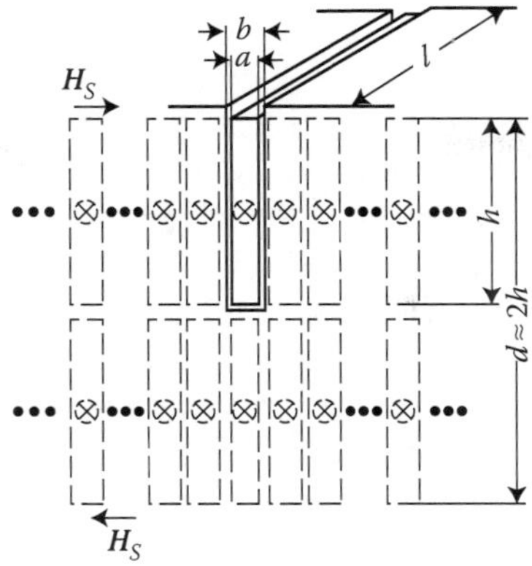

FIGURE 5.28 Multiple mirror images of a conductor in a deep-slot of an electric machine. (Adapted from Turowski J.: Wave method of calculating the screens and elements of electric machines. (in Russian). *Elektricziestwo*, (12), 1965, 53–58.)

with nonmagnetic inserts (6.35), slots for synchronous machines (Figure 8.8), and deep-slot induction motors (Figure 5.28) with rectangular and trapezoidal slots (Figure 5.30 [later in the chapter]).

In the work [4.10] (by Mendrela, J. Turowski et al.), Fourier series were used for the computer modeling of a 3D field of induction motors with various kinds of solid rotors. In Figure 5.7, we present a classical method of determining magnetic field in a circular slot, in Figure 5.9—in a trapezoidal slot, and in Figures 5.11 and 5.12—in a rectangular slot. These methods are applicable primarily for simple systems, because in more complicated structures, the field of imaged currents becomes too complicated for analytical methods. For instance, the imaged currents near circular surfaces have a nonuniform density. However, these simple systems already allow resolution to problems like electromagnetic processes in slots of deep-slot electric machines (Section 5.5.4), forces in slotting part of electric machines (Chapter 8), influence of slots width and teeth saturation on the field and crashing forces in slots of electric machines (J. Turowski [1.16], p. 96), and so on.

5.5.3 Numerical Methods: Mesh Methods of Solution of Leakage Magnetic Field in Power Transformers

Presently, there exist at least 15 basic methods of field calculations and much more hybrid or mixed methods. *All of them are theoretically absolute equivalent, but not equivalent in practical effectiveness.* Differences in time of 3D computation can be from many days to less than a second. The most rapid and easy seems to be RNM-3D (J. Turowski et al. [7.23]).

5.5.3.1 Reluctance Network Method

The main advantage of RNM is its special, competitive possibility for rapid (in time less than 1 s) solution of 3D stray fields of a transformer (e.g., by the authors' hybrid, partly "intelligent" package of "RNM-3D" programs (J. Turowski [1.17], Section 4), [7.21], [7.23], [4.28], [8.21]), whereas other, complex commercial programs typically need from several hours to many days for large problems (Table 5.1).

The secret of this success is that the RNM-3D program is dedicated to a specific class, although quite large, of industrial power transformers. It is based on an expert system, where knowledge base, interactive approach (Figure 4.20c), and machine intelligence are hidden inside the source code [1.20]. Thanks to the built-in knowledge and components, the RNM-3D user operates from outside with a simple tool in the form of Ohm's and Kirchhoff's laws. This is similar to the way in which a sophisticated, intelligent, modern automobile is controlled from outside by a user with the simple, traditional steering wheel.

The RNM-3D method is presented briefly in Chapter 4 (Figures 4.20 through 4.22) and then in Chapter 7 (Figures 7.20 and 7.21). It is an integral method, which does not generate local numerical errors, as is the case of differential methods, such as FDM, FEM, and so on. Similarly, RNM, like Fourier's method, enables simple and rapid, automatic calculation (Figure 4.20c) on a simple, basic computer. For creating a correct physical model and satisfactory accuracy, it is necessary to have a subject matter expert who knows the structure of the investigated object well (J. Turowski [2.34], J. Turowski et al. [4.31], [8.13]). At 2D calculations, RNM-2D is also rapid and simple, but not so much better than the commercial FEM tools as shown in Table 5.1 as in the case of 3D problems.

RNM-3D is presented here as an example of a method for successful construction of rapid tools to solve difficult industrial problems. If one wants to solve a

TABLE 5.1

Influence of the Chosen Method, from the Point of View of Time, Difficulty, and Cost of Computer Modeling and Simulation of 3D Stray Fields and Losses in Large Power Transformers

STRAY FIELDS AND LOSSES IN LARGE POWER TRANSFORMERS

RELUCTANCE NETWORK METHOD THREE-DIMENSIONAL **RNM-3D**		FINITE ELEMENT METHOD THREE-DIMENSIONAL **FEM-3D**
Low, secondary school level	←User qualification→	High, university level
Regular PC	← Hardware →	High quality
Simple Ohm's and Kirchhoff's laws	← Theory →	Calculus of variations
Small, smoothed by integral method	←Errors generation→	Large due to differential method
Low price	← Software →	Expensive and sophisticated
Half an hour	Time of elaboration ← of model →	Days to weeks
Simple analytical approximation	← **Nonlinear medium**→	Cumbersome iterations
Analytical thermal hazard signal	←**Artificial intelligence**→	Separate cumber some design
Less than 1 second	CPU time of computation of one design variant ←→	**Hours to many days**

different class of objects using RNM, such as linear motors [4.10], [8.20], [8.21], [8.26], separators [8.4], shell-type transformers [7.33], and so on, then one should build another, dedicated preprocessor.

FDM (Figures 8.14 and 8.15) and FEM (Figures 5.34 [later in the chapter], 7.15, 7.20, and 8.11) are described shortly in [1.16], ([1.17/2], pp. 148–150) and broader in (Zienkiewicz [2.17], Chari and Silvester [7.1]). Due to troublesome analysis of errors in stability and convergence, generated due to the differential approach used in these methods, the FDM-2D and FEM-2D are less dependent on professional experience of designer and on geometry of investigated system (Figure 5.34 [later in the chapter]). Owing to it, they are more general and very popular in education and research. However they are much more difficult and computationally more demanding and expensive in their 3D versions (Figure 7.15), especially in comparison with RNM-3D (Table 5.1).

5.5.4 Slot in a Deep-Slot Induction Machine

5.5.4.1 Rectangular Slot

Formulae for reactance and resistance of bars of deep-slot induction machines are usually developed from Maxwell's equations. It is a cumbersome job. This can be achieved much easier, however, starting from the *wave method* of calculation of screens (Turowski [4.21]) presented in Section 4.3 and from the *method of multiple mirror images* (Figure 5.12).

Using the method of *mirror images*, we can substitute the real current, together with its images, by a solid copper sheet of thickness $d = 2h$, located in the air (Figure 5.28). On both surfaces of this sheet occur equal maximum magnetic field intensities, H_{ms}, of opposite directions:

$$H_{ms} = \pm \frac{\sqrt{2} I}{h} \tag{5.29}$$

The electric field intensity on the surface was obtained by adding to each other the two expressions (4.18) for $z = 0$ and $z = d$, and assuming the image coefficients $M_1 = M_2 \approx -1$. Hence

$$\begin{aligned} E_{ms} &= E_m(0) + E_m(d) = \frac{\alpha}{\sigma} H_{ms1} \left(\frac{e^{\alpha d} + e^{-\alpha d}}{e^{\alpha d} - e^{-\alpha d}} + \frac{1+1}{e^{\alpha d} - e^{-\alpha d}} \right) = \\ &= \frac{\alpha}{\sigma} H_{ms1} \frac{e^{\alpha h}(e^{\alpha h} + e^{-\alpha h}) + e^{-\alpha h}(e^{-\alpha h} + e^{\alpha h})}{e^{\alpha h}(e^{\alpha h} + e^{-\alpha h}) - e^{-\alpha h}(e^{-\alpha h} + e^{\alpha h})} = \frac{\alpha}{\sigma} H_{ms} \frac{\text{ch}\alpha h}{\text{sh}\alpha h} = \\ &= \frac{k}{\sigma} H_{ms} \frac{(\text{sh}\alpha kh + \sin 2kh) + \text{j}(\text{sh}2kh - \sin 2kh)}{\text{ch}2kh - \cos 2kh} \end{aligned} \tag{5.30}$$

where $\alpha d = (1 + \text{j})\, kd$, after (2.139).

The apparent power P_S lost in the conductor equals the power of electromagnetic wave penetrating into an equivalent copper plate by its upper surface (3.1), that is, equals to the Poynting vector multiplied by the area of the upper surface of the conductor ($l \cdot b$, Figure 5.28):

$$P_S = P + jQ = \frac{1}{2}\mathbf{1}_A(\boldsymbol{E}_{ms} \times \boldsymbol{H}^{\bullet}_{ms})l \cdot b$$
$$= H^2_{ms}\frac{klb}{2\sigma}\frac{(\mathrm{sh}2kh + \sin 2kh) + \mathrm{j}(\mathrm{sh}2kh - \sin 2kh)}{\mathrm{ch}2kh - \cos 2kh} \quad (5.31)$$

The coefficient K_R of increase of the resistance (active) at AC current (R_{AC}), with respect to the resistance R_{DC} at DC current (Figure 5.29), can be immediately obtained from Equations 5.29 and 5.31

$$K_R = \frac{R_{\mathrm{AC}}}{R_{\mathrm{DC}}} = \frac{P}{I^2}\sigma\frac{bh}{l} = kh\frac{\mathrm{sh}2kh + \sin 2kh}{\mathrm{ch}2kh - \cos 2kh} \quad (5.32)$$

The inductive reactance of a conductor, produced by the slot leakage flux passing through the slot within the height h (Figure 5.28), at a uniform current distribution (e.g., with a bundle of thin series conductors) in the slot, has the form

$$\mathrm{X}_{h,\mathrm{uniform}} = \omega L = \omega\mu\frac{h}{3b}l = \frac{2k^2}{\sigma}\frac{h}{3b}l \quad (5.33)$$

The coefficient K_L *of reduction of the conductor reactance* due to the skin effect follows also from Equations 5.29 and 5.31:

$$K_L = \frac{X}{X_{h,\mathrm{uniform.}}}\frac{Q}{I^2}\frac{\sigma 3b}{2k^2hl}\frac{3}{2kh}\frac{\mathrm{sh}\,2kh - \sin 2kh}{\mathrm{ch}\,2kh - \cos 2kh} \quad (5.34)$$

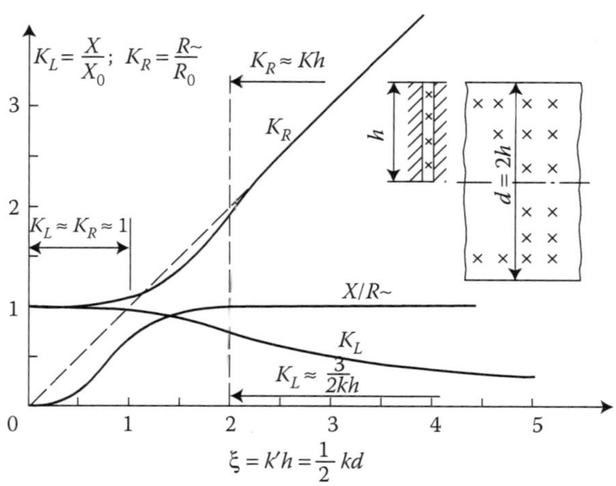

FIGURE 5.29 Relative electric resistances and reactances of deep-slot, cup and double-sided linear rotors; h—height of the slot, d—thickness of cup wall or of rotor of double sided linear motor. (Adapted from Turowski J.: *Calculations of Electromagnetic Components of Electric Machinery and Equipment.* (in Polish). Warsaw: WNT, 1982; and (in Russian). Moscow: Energoatomizdat, 1986 [1.16].)

In the case of significant differences in the dimensions, $b > a$ and $d > 2h$ (Figure 5.28), in the above formulae instead of k should be introduced the coefficient

$$k' = k\sqrt{\frac{a}{b} \cdot \frac{2h}{d}}, \quad k = \sqrt{\frac{\omega\mu\sigma}{2}},$$

in which we have considered the calculated increase of conductor conductivity due to neglecting the insulation layer between conductor and slot surface at the calculation.

5.5.4.2 Trapezoidal and Bulb Slots

Also for these slots it is convenient to utilize the method of *multiple mirror images*, but this time—of circular images (Figures 5.8 and 5.9). A trapezoidal bar made of Cu or Al is simulated as a section of a solid copper (Cu) tube (Figure 5.30) with an excitation current $I_0 = nI$ in the tube center (Figure 5.31a), where I is the current of slot, and $n = 2\pi R_0/b_2$ is the number of fictitious slots imaged together with the real one. A further solution for a trapezoidal slot of normal shape (Figure 5.31a) follows (per Turowski et al. [5.17]) as for Figure 4.30.

In this way, we obtain the distribution of current density and field in a trapezoidal slot in the form

$$\left.\begin{aligned} J_m &= \sigma E_m = \frac{nI}{\sqrt{2}\,\pi R_0}\frac{\alpha}{\sigma}\frac{I_1(\alpha R_1)\cdot K_0(\alpha r) + K_1(\alpha R_1)\cdot I_0(\alpha r)}{I_1(\alpha R_1)\cdot K_1(\alpha R_0) - K_1(\alpha R_1)\cdot I_0(\alpha R_0)} \\ H_m &= \frac{1}{\mathrm{j}\omega\mu}\frac{\mathrm{d}E_m}{\mathrm{d}r}\frac{nI}{\sqrt{2}\,\pi R_0}\frac{I_1(\alpha R_1)\cdot K_1(\alpha r) - K_1(\alpha R_1)\cdot I_1(\alpha r)}{I_1(\alpha R_1)\cdot K_1(\alpha R_0) - K_1(\alpha R_1)\cdot I_1(\alpha R_0)} \end{aligned}\right\} \quad (5.35)$$

where $I_0(\alpha r)$ is the modified *Bessel* function of first order with the index $\nu = 0$; $K_0(\alpha r)$ is as above, but of second order (*McDonald's* function), and $I_1(p) = \mathrm{d}I_0/\mathrm{d}p$; and $K_1(p) = -\mathrm{d}K_0/\mathrm{d}p$ $(p = \alpha r)$ [2.1].

For a rectangular slot (Figure 5.30a), from Equations 5.35 one obtains in the limit of $r \to \infty$ the known dependences (Turowski [1.16])

$$J_m = \frac{\sqrt{2}I}{b}\alpha\frac{\mathrm{ch}\,\alpha z}{\mathrm{ch}\,\alpha h} \quad \text{and} \quad H_m = \frac{\sqrt{2}I}{b}\frac{\mathrm{sh}\,\alpha z}{\mathrm{sh}\,\alpha h} \qquad (5.36)$$

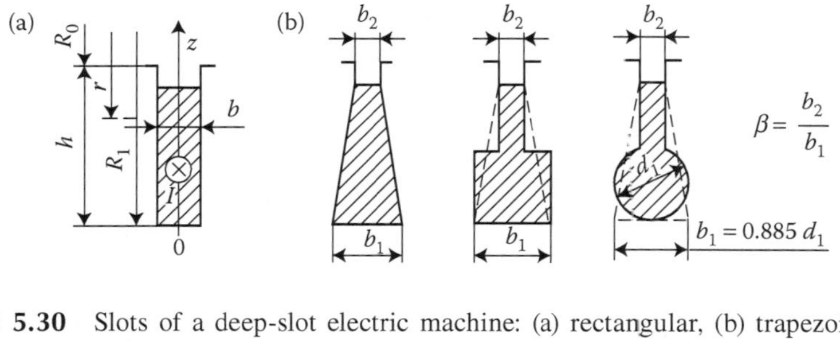

FIGURE 5.30 Slots of a deep-slot electric machine: (a) rectangular, (b) trapezoidal and bulb.

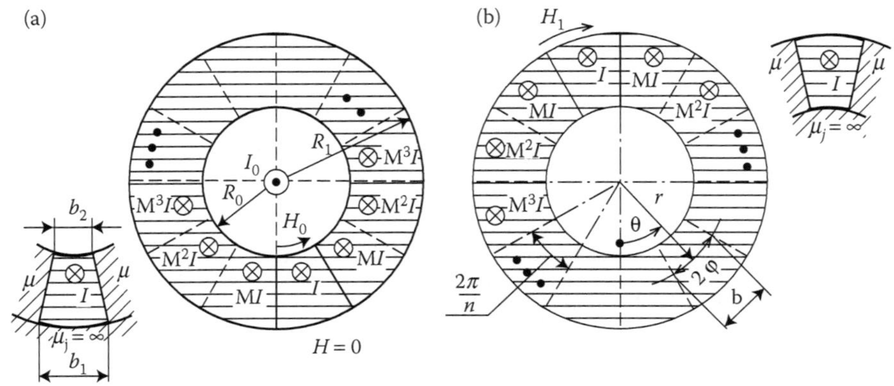

FIGURE 5.31 Simulation of trapezoidal slots with the help of solid copper tubes with the current I_0 in the tube axis: (a) fictitious internal excitation $I_0 = nI$; (b) external excitation. (Adapted from Turowski J., Andryszczak W., and Pelikant A.: Application of the method of mirror images to calculation of the slot impedances, taking into account the saturation of the teeth. (in Polish). *"Elektryka" Scientific Papers of Techn. University of Lodz*, (74), 1983, 187–198 [5.17].)

The power consumption and impedances of the trapezoidal slot can be found with the help of Poynting's theorem (3.1). When neglecting the yoke saturation ($\mu \to \infty$), at the slot bottom we get $H_m = 0$. Hence, the whole power of field ($\boldsymbol{S} = \boldsymbol{E} \times \boldsymbol{H}$) entering the slot through the gap b_2 (Figure 5.30b) remains in the slot as active losses and reactive magnetization power:

$$I^2R + \mathrm{j}I^2X = \int_0^{\mathrm{L}} \int_{-\phi}^{\phi} \frac{1}{2}\left(\boldsymbol{E}_m \times \boldsymbol{H}_m^{\bullet}\right)_{r=R_0} \cdot R_0 \,\mathrm{d}\theta \,\mathrm{d}L \tag{5.37}$$

After many transformations and simplifications, in Ref. [5.17] were obtained the coefficients of resistance increase K_{Rt} and reduction of reactance K_{Lt} for trapezoidal slots (Figure 5.32):

$$K_{Rt} \approx \frac{1+\beta}{2\beta} K_R \qquad (0.4 < \beta \le 1) \tag{5.38}$$

$$K_{Lt} \approx \frac{1+\beta}{2\beta} K_L \qquad (0.8 < \beta \le 1) \tag{5.39}$$

where K_R and K_L are the coefficients (5.32) and (5.34) for a rectangular deep-slot, and $\beta = b_2/b_1$ represents the proportions of trapezoid.

Formulae (5.38) and (5.39) in many cases are sufficient for initial calculation of rotors with trapezoidal slots. For more accurate calculations, especially at $\beta < 0.8$ it is recommended to use the more accurate formulae (5.35) and to consider the saturation of teeth ($\mathrm{M} = (\mu_r - 1)/(\mu_r + 1)$), by introducing, according to Ref. [5.17], the excitation current (Figure 5.31a)

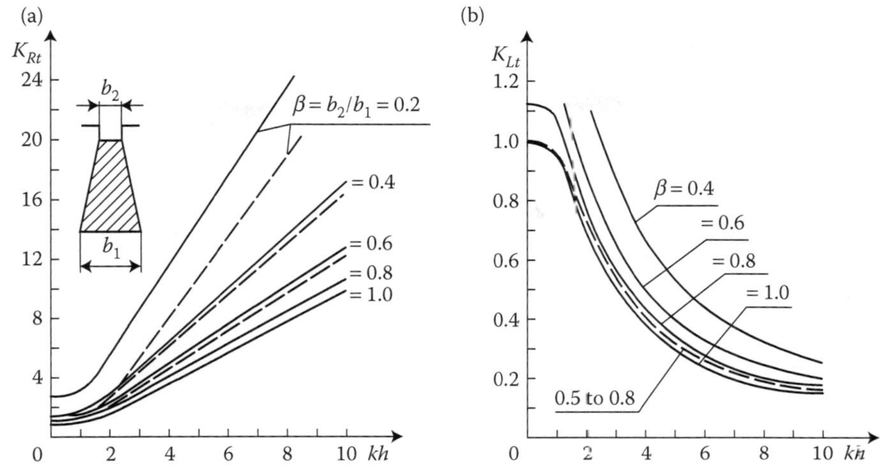

FIGURE 5.32 The coefficients of increase of resistance K_{Rt} and decrease of reactance K_{Lt} after simplified formulae (5.38), (5.39) for trapezoidal slots (continuous lines) and more accurate formulae (broken lines)—(see Turowski [1.16], p. 171).

$$\sqrt{2}\, I_0 = 2\int_0^{\pi} H_{m0}(\theta) R_0 \,\mathrm{d}\theta = \frac{4\varphi R_0 H_{m0}}{\ln \mathrm{M}} (\mathrm{M}^{n/2} - 1) \tag{5.40}$$

whereas $J(\Theta) = H_{m0}\ \mathrm{M}^{\Theta/2\varphi}$ is the averaged continuous current density distribution which substitutes the discrete distribution (Figure 5.31), from which we get

$$H_{m0}(\Theta) = H_{m0}\mathrm{M}^{\Theta/2\varphi}, \quad H_{m0} = \frac{\sqrt{2}\, I}{\mathrm{b}_2}, \quad \varphi = \mathrm{arc\ tg}\frac{b_2}{2R_0}$$

Then, the integral (5.37) will be multiplied by the coefficient

$$\frac{\mathrm{M} - 1}{\ln \mathrm{M}} \tag{5.41}$$

and by the same coefficient should be multiplied the coefficients (5.32), (5.34), (5.38), and (5.39).

A *distribution of temperature* in the slot can be in turn calculated as for foil windings, after (Turowski [1.16], Figure 5.39 [later in the chapter]).

5.5.5 Field in the Gap of Electric Machine

5.5.5.1 Analytical Methods

The known and popular Carter's coefficient k_C ([1.16], [1.18], [5.22]) for calculating an equivalent ideal gap $\delta_i = k_C \cdot \delta$ (Figure 5.33) is being continuously improved by various authors, for example, [5.23], [5.24], [5.25].

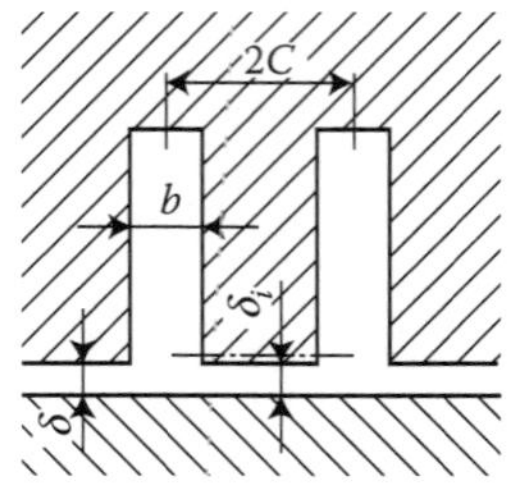

FIGURE 5.33 The real air gap δ and the ideal one $\delta_i = k_C \cdot \delta$.

Smythe ([4.3], p. 296) developed his own formula for an equivalent smooth gap at the slotted surface:

$$\delta_i = \frac{\pi\left((b/2) + C\right)\delta}{\pi C + b\,\mathrm{arctg}(2\delta/b) + \delta \ln[1 + (b/2\delta)^2]} \tag{5.42}$$

On pages 382–407 of the Smythe's book [4.3] there are given methods of calculating eddy currents in various conducting elements.

In the work ([1.16], p. 130), the author developed simple analytical formulae for amplitudes of *slot's pulsation* of the flux density in the gap:

$$\frac{B_{m\max}}{B_{m\,\mathrm{gap}}} = 2 - \frac{B_{m\min}}{B_{m\,\mathrm{gap}}} = \frac{2}{1 + B_{m\min}/B_{m\max}} \tag{5.43}$$

where $B_{m\min}/B_{m\max} = 2\,\delta_i(b + 2\,\delta_i)$.

Formulae (5.43) are in good agreement with calculations of a field by the FEM (Figure 5.34), thus validating and completing this method.

5.5.5.2 Meshed Numerical Methods

Application of the FEM (Rizzo, Savini, Turowski [5.12]) together with an automatic analysis of higher field harmonics, with the help of *rapid discrete Fourier transforms* and application of the Cooley–Turkey algorithm, allowed the Polish–Italian team to carry out effective harmonics analysis [5.13] in the gap (Figure 5.34). The developed method permits faster and more accurate calculation of the pulsation losses in pole shoes than with former analytical methods.

5.5.5.3 Graphical-Numerical Methods

The distribution of flux density under the pole can be determined graphically* by drawing reciprocally perpendicular tubes and equipotential surfaces (Figure 5.35).

If they are drawn in such a way that proportions of all mesh cells are equal (Turowski [1.15/1], p. 162), that is, $\delta_1/a_1 = \delta_2/a_2 = \cdots = \delta_x/a_x$, which is easy to achieve after a few corrections, the flux density in the center of the section a_x will be equal to

* T. Lehmann: Méthode graphique ..., *Lumière Electr.* 1909, p. 103, 137 and *Rev. Gen. Electr.* 1933, p. 351.

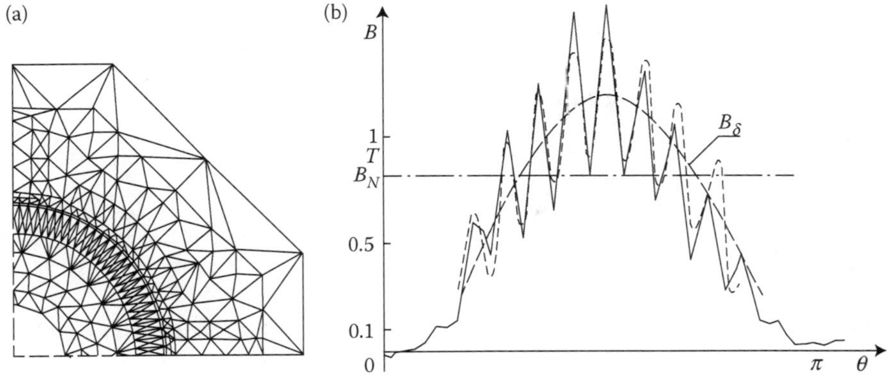

FIGURE 5.34 Calculation of flux density in the air gap of a DC motor, with the FEM: (a) the FEM mesh for the motor with compensation winding, (b) the calculated slot pulsation ------ FEM, –·–·– the author's formulae (5.43)—for motor without compensation windings. (Adapted from Rizzo M., Savini A., and Turowski J.: *Internat. Conference on Electric Machines. ICEM'88*. Pisa, Italy. September 12–13, 1988, Vol. 1, pp. 85–88 [5.12]; Rizzo M., Savini A., and Turowski J.: *Proc. of the Internat. Conference on Electric Drives ICED'88*. Polana Brasov, Romania, September 20–22, 1988, Paper 8.1–10, pp. 1–9 [5.13].)

$B_x = \Delta\Phi/(a_x l) = \Phi/(m\, a_x l)$, and the flux $\Phi = \Sigma\Delta\Phi = mB_0\delta_x/n$, where $\Delta\Phi$ is the flux of a single tube; Φ is the total flux of the pole; m is the number of tubes; n is the number of series meshes (slices) of the tube.

Actually, it is even simpler to assume all the mesh cells with equal dimensions $\delta_x = a_x$.

Some difficulty in the drawing of tubes will emerge when in the investigated region some distributed ampere-turns appear. In such a case, it is more convenient to apply the RNM (Figure 5.36), with an equivalent fictive magnetomotive force (mmf), corresponding to the popular electromotive force (emf).

5.5.5.4 Determination of the Magnetomotive Force (mmf) of Sources

Owing to the slenderness of a typical winding of large, core-type transformers, in many cases, especially when the field outside the winding is investigated, the

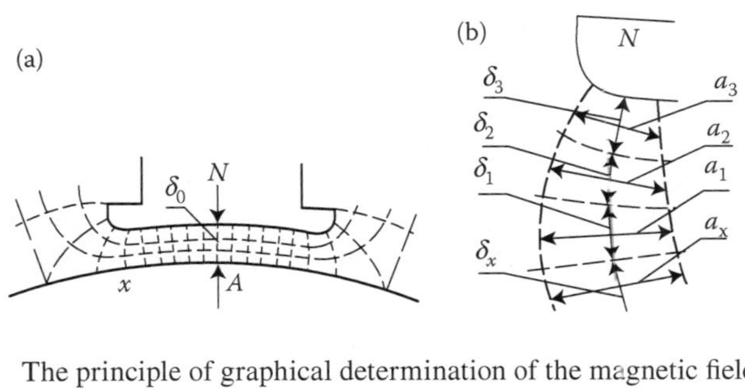

FIGURE 5.35 The principle of graphical determination of the magnetic field in air gap.

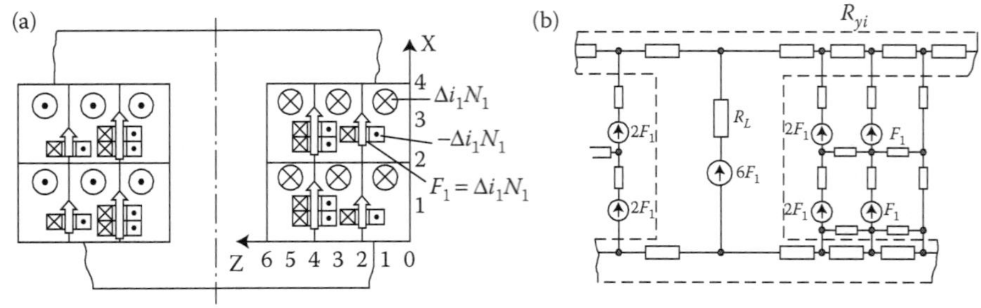

FIGURE 5.36 Modeling of magnetic circuits of a reactor with remarkable coils thickness, using RNM: (a) the cross section of the wound central leg; (b) the equivalent network: R_L—reluctance of central magnetic leg, R_{yi}—reluctance of the *i*th section of the yoke; ⊙, ⊗ —actual currents; ⊡, ⊠—fictive currents; ⥣—mmf compensating the fictive currents; F_i—elementary, equivalent, residual mmf ⊕. (Adapted from Turowski J.: Reluctance model of stray field in transformer. (in Polish). *Rozprawy Elektrotechniczne*, (4), 1984, 1121–1144 [4.19].)

magnetomotive force (mmf) of the winding can be evaluated by assuming that the entire current flow of the high voltage (HV) and the low voltage (LV) winding $F = \sqrt{2}\, I_{HV} N_{HV} = \sqrt{2}\, I_{LV} N_{LV}$ is concentrated in a filament layer (current sheet) in the axes of the HV or LV windings.

If we wish to account for finite dimensions of winding cross sections, we can use a well-known principle by which the field of a current flow, subdivided into elements $\Delta i_1 N_1$ (Figure 5.36), is replaced by a network of circuits.

The procedure consists of shifting the elementary current $\Delta i_1 N_1$ from one mesh cell, for example, ($x = 3, z = 1$), across the border between the nodes 22–24, to another mesh cell, say ($x = 3, z = 3$). This produces in the border a new fictive *mmf* source with the value of the shifted flow $F = \Delta i_1 N_1$. At the same time the flow in the mesh cell ($x = 3, z = 3$) is doubled. As a result, the first mesh ($x = 3, z = 1$) becomes empty. This can be explained by postulating that in the axis of the border between the adjacent mesh cells we can insert a fictive coil (rectangles in Figure 5.36), with identical but opposite flow ($-\Delta i_1 N_1$) from the first mesh, and simultaneously compensate the action of this coil by means of a fictive *mmf* of the same but oppositely directed flow. This implies an additional rule:

> *Every shifting of ampere-turns ΔilNl across any mesh border creates an mmf (flow) $F = \Delta i_l N_l$ in the axis of this border and doubles the flow in the adjacent mesh cell.*

The next shifting of ampere-turns follows the same rule, but this time with the entire enlarged mesh current flow—that is, its own and added.

By repeating this operation, we can replace a real coil of a distributed flow (Figure 5.36a) with a circuit network of reluctances and lumped magnetomotive forces (Figure 5.36b). In an analogous way, one can shift the current flow to the yoke axis.

In the case of height asymmetry of HV and LV windings (Figure 5.37a), an additional transverse flow F_z (Figure 5.37b) should be introduced. This radial flux can be found

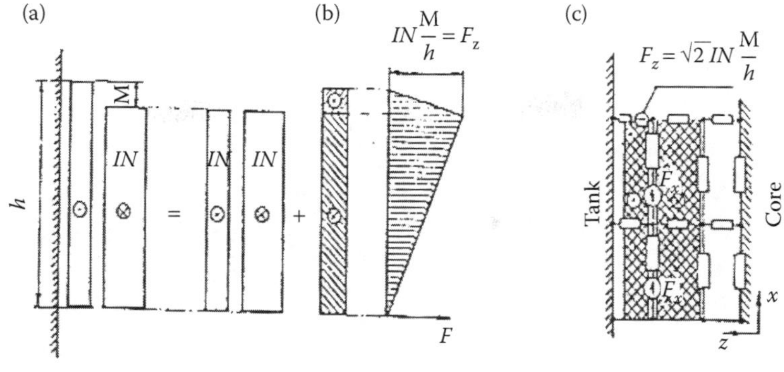

FIGURE 5.37 The influence of winding asymmetry on the RNM model [1.17]: (a) decomposition of ampere-turns into a symmetrical component $F = \sqrt{2}IN$; (b) radial component with maximum flow $F_z = \sqrt{2}IN \cdot M/h$; (c) equivalent reluctance network with addition source F_z in the radial branch, where F_z is an element of the main $F = \sqrt{2}IN$ winding flow.

using a well-known ([1.5], pp. 119–138, [1.56]) rule of ampere-turn decomposition into axial and radial components (Figure 7.17). In this way, any asymmetry can be considered, with the *accuracy depending on the number of discretization nodes introduced.*

A scientific development of the graphic Lehmann's* method, popular among designers in the precomputer era, was the Sykulski's and Hammond's computer method of "Tubes and Slices" [2.29]. The "TaB" method is a direct confirmation of the author's RNM method.

One can also use some less frequently used, physical models of the flow field, with the help of electrolytic tank or a semiconducting paper. However, in 2D field systems, the most convenient are FEM-2D (Figures 5.34 and 8.11) and the finite difference method FDM-2D (Figures 8.14 and 8.15).

Although in recent decades FEM-2D has become the most popular, it is becoming less and less useful in practice. Modern calculations, design, and research require 3D modeling of physical fields. In such cases, FEM-3D is too expensive, labor intensive, and less comprehensible. In particular, for large power transformers, the FEM-3D volume mesh technique is usually prohibitive even on machines with a large amount of RAM memory [9.13].

For the purpose of 3D field computation, a much better, easier, faster, and more cost-effective method is the authors' RNM-3D (see Table 5.1) [1.17], [2.34], [4.20], [4.27]–[4.29], and so on, or hybrid approaches utilizing a combination of analytical methods presented in this book, and successfully demonstrated recently in the works [1.48], [9.14].

5.6 FIELD OF CONDUCTORS NEARBY A STEEL WALL

Calculations of a pair of conductors placed nearby a magnetically screened *(shunted)* steel wall (Figure 5.38) can be started [5.20], [6.7] by considering a single conductor

* T. Lehmann: Méthode graphique …. *Lumiére Electr.* 1909, p. 103, 137 and *Rev. Gen. Electr.* 1933, p. 351.

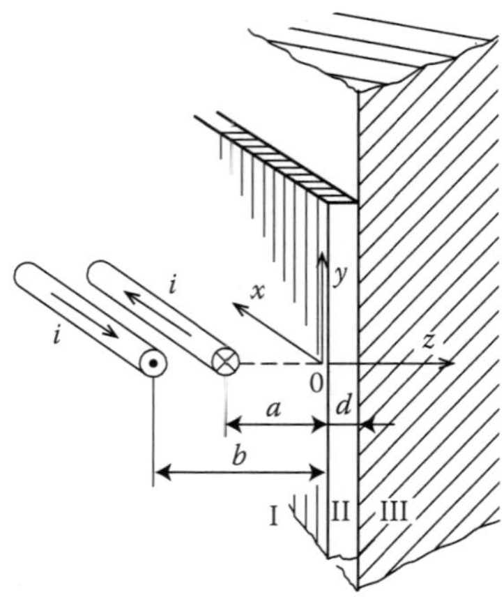

FIGURE 5.38 A pair of conductors nearby a screened steel surface [5.19].

with a current $i = I_m \cos \omega t$. The particular regions are assigned as (Figure 5.38): I—dielectric, II—screen (shunt) packaged from strips of electric iron sheets, III—solid steel half-space with a constant permeability μ.

Hence, $\sigma_{\mathrm{I}} = \sigma_{\mathrm{II}} = 0$.

The magnetic vector potential $\boldsymbol{A}$, defined by the equation $\boldsymbol{B} = \text{curl}\, \boldsymbol{A}$, satisfies the equations

- In regions I and II:

$$\nabla^2 \boldsymbol{A} = 0 \tag{5.44}$$

- In region III:

$$\nabla^2 \boldsymbol{A} = \alpha^2 \boldsymbol{A} \tag{5.45}$$

According to Figure 5.38, we assume $A_y = A_z = 0$, $A_x = A$, and $\partial A/\partial x = 0$, from which we obtain the scalar equations

- For region I:

$$\frac{\partial^2 A_{\mathrm{I}}}{\partial^2 y^2} + \frac{\partial^2 A_{\mathrm{I}}}{\partial^2 z^2} = 0 \tag{5.46}$$

- For region II:

$$\frac{\partial^2 A_{\mathrm{II}}}{\partial^2 y^2} + \frac{\partial^2 A_{\mathrm{II}}}{\partial^2 z^2} = 0 \tag{5.47}$$

- For region III:

$$\frac{\partial^2 A_{\text{III}}}{\partial^2 y^2} + \frac{\partial^2 A_{\text{III}}}{\partial^2 z^2} = \alpha^2 A_{\text{III}} \tag{5.48}$$

where A_{I}, A_{II}, and A_{III} are the magnetic vector potentials from one of the conductors.

$$\alpha = \sqrt{j\omega\mu_{\text{III}}\sigma_{\text{III}}} = (1 + j)\,k_{\text{III}}, \quad k_{\text{III}} = \sqrt{\omega\mu_{\text{III}}\sigma_{\text{III}}/2} \tag{5.49}$$

Next, we assume that the magnetic potential A_{I} in the dielectric consists of two parts

$$A_{\text{I}} = A_{\text{Io}} + A_{\text{Iw}} \tag{5.50}$$

where A_{Io} originates from the current $\boldsymbol{I}_{\text{m}} = I_{\text{m}}\, e^{j\omega t}$ and from its full, positive mirror image in an ideal ($\mu \to \infty$, $\sigma = 0$) superconducting ferromagnetic (Figure 5.39); and A_{Iw} is an additional component representing the deformation reaction of eddy currents and steel saturation effects on the field in dielectric.

The magnetic vector potential of a single conductor in the distance a (Figure 5.39), of an infinitely thin filament, according to Equation 2.60 and Figure 5.39, is described as

$$A = \frac{\mu}{4\pi}\int\limits_{l}\frac{i\,dl}{r} = \frac{\mu}{4\pi}\int\limits_{-\infty}^{+\infty}\frac{\mathrm{d}x}{\sqrt{x^2 + y^2 + (a + z)^2}} =$$

$$= -\frac{\mu}{4\pi}\ln\sqrt{x^2 + y^2 + (a + z)^2} + C \tag{5.51}$$

from which the maximum value of the complex magnetic potential A_{Io} for both conductors (the real one and its mirror image—Figure 5.39) is

$$A_{\text{Io}} = -\frac{\mu_0 I_{\text{m}}}{4\pi}\ln\{[y^2 + (a + z)^2][y^2 + (a - z)^2]\} \tag{5.52}$$

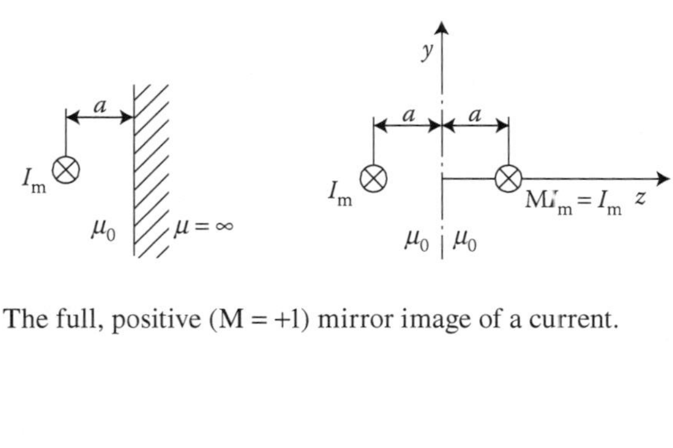

FIGURE 5.39 The full, positive (M = +1) mirror image of a current.

Expression (5.52) fulfills the Laplace Equation 5.44.

Owing to the symmetry of the system with respect to the OZ axis (Figure 5.40), the interfering field A_{Iw} should be an even function along the Y axis. Hence, for any nth harmonic of the space distribution of the interfering field A, we foresee a solution in the form of function

$$A_{Iw,n} = f_n(z) \cos \lambda_n y \quad (n = 0, 1, 2, 3) \tag{5.53}$$

Substituting Equation 5.53 into Equation 5.46 yields the equation

$$f_n''(z) - \lambda_n f_n(z) = 0 \tag{5.54}$$

whose solution has the form

$$f_n(z) = C_n e^{\lambda_n z} + Dn e^{-\lambda_n z} \tag{5.55}$$

Going back to Equation 5.53, we can write

$$A_{Iw,n} = (C_n e^{\lambda_n z} + Dn e^{-\lambda_n z}) \cos \lambda_n y \tag{5.56}$$

A general solution of Equation 5.46 for the interfering field A_{Iw} is a sum of the particular solutions (5.56) for particular harmonics

$$A_{Iw} = \sum_{n=0}^{\infty} A_{Iw,n} \tag{5.57}$$

Since the constant λ can have any value, a more general solution for linear systems can be obtained (Zakrzewski, Sykulski [5.19]) by considering the full spectrum of periodic distribution in the form of integral

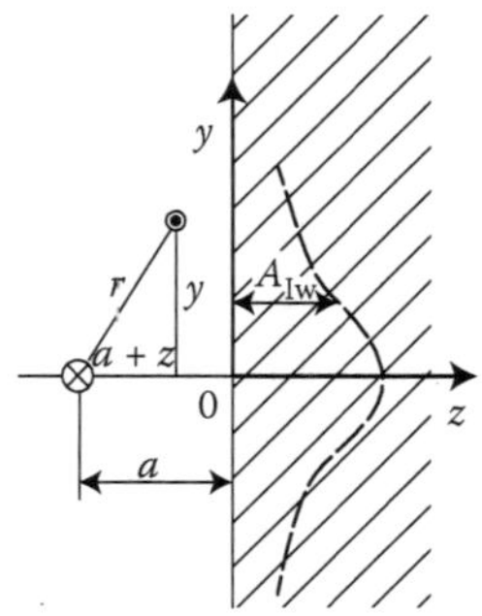

FIGURE 5.40 Determination of the vector potential A_{Iw} of the interfering eddy currents.

$$A_{Iw} = \int_0^\infty [C(\lambda)e^{\lambda z} + D(\lambda)e^{-\lambda z}] \cos \lambda y \cdot d\lambda \tag{5.58}$$

In the region of dielectric, the z coordinate is limited within the range $-\infty \le z \le 0$. Since at $z \to -\infty$, $A_{Iw} \ne \infty$; hence, $D(\lambda) = 0$, from which we get

$$A_{Iw} = \int_0^\infty C(\lambda)e^{\lambda z} \cos \lambda y \, d\lambda \tag{5.59}$$

Finally, according to Equation 5.50, the vector potential in dielectric

$$A_I = -\frac{\mu_0 I_m}{4\pi} \ln\{[y^2 + (a+z)^2][y^2 + (a-z)^2]\} + \int_0^\infty C(\lambda)e^{\lambda z} \cos \lambda y \, d\lambda \tag{5.60}$$

By an analogical procedure, for an electrically nonconducting, magnetic screen one can write

$$A_{II} = \int_0^\infty [E(\lambda) + F(\lambda)] \cos \lambda y \cdot d\lambda \tag{5.61}$$

In the region of solid steel, instead

$$A_{III} = \int_0^\infty G(\lambda)e^{-\sqrt{\lambda^2+\alpha^2}\, z} \cos \lambda y \cdot d\lambda \tag{5.62}$$

The components of flux density for the particular regions I, II, and III are determined from $\boldsymbol{B} = \text{curl}\, \boldsymbol{A}$, as

$$B_z = \frac{\partial A}{\partial y} \quad \text{and} \quad B_y = -\frac{\partial A}{\partial z} \tag{5.63}$$

The constants $C(\lambda)$, $E(\lambda)$, $F(\lambda)$, and $G(\lambda)$ are determined from the boundary conditions for particular media

$$\mu_0 H_{Iz} = \mu_{II} H_{IIz}, \quad H_{Iy} = H_{IIy} \text{—for } z = 0$$

$$\mu_{II} H_{IIz} = \mu_{III} H_{IIIz}, \quad H_{IIy} = H_{IIIy} \text{—for } z = d$$

After determining the constants in the above way, and substituting into Equations 5.60 through 5.62, we obtain

- For the dielectric I (cf. Figure 5.38):

$$A_{\mathrm{I}} = -\frac{\mu_0 I_{\mathrm{m}}}{4\pi} \ln \frac{y^2 + (a+z)^2}{y^2 + (a-z)^2} + \\ + \frac{\mu_{\mathrm{II}} I_{\mathrm{m}}}{2\pi}(n-1)\int_0^\infty \frac{\mathrm{e}^{-(a-z)\lambda}}{\lambda} \frac{1 + r\,\mathrm{e}^{-2d\lambda}}{n + r\,\mathrm{e}^{-2d\lambda}} \cos \lambda y \,\mathrm{d}\lambda \tag{5.64}$$

- For the magnetic screen (shunt) II:

$$A_{\mathrm{II}} = \frac{\mu_{\mathrm{II}} I_{\mathrm{m}}}{2\pi}(n-1)\left[\int_0^\infty \frac{r\,\mathrm{e}^{-(a+2d-z)}}{\lambda(n + r\,\mathrm{e}^{-2d\lambda})} \cos \lambda y \,\mathrm{d}\lambda + \int_0^\infty \frac{\mathrm{e}^{-(a+z)\lambda}}{\lambda(n + r\,\mathrm{e}^{-2d\lambda})} \cos \lambda y \,\mathrm{d}\lambda\right] \tag{5.65}$$

- For the region of solid steel (half-space) III:

$$A_{\mathrm{III}} = \frac{\mu_{\mathrm{II}} I_{\mathrm{m}}}{2\pi}(n-1)\int_0^\infty \frac{r+1}{n + r\,\mathrm{e}^{-2d\lambda}} \cdot \frac{\mathrm{e}^{-(a+d)\lambda + \sqrt{\lambda^2+\alpha^2}\,(d-z)}}{\lambda} \cos \lambda y \,\mathrm{d}\lambda \tag{5.66}$$

Analogically, the field of the second conductor carrying the current $-I_{\mathrm{m}}$ in the distance b can be obtained from formulae (5.64) through (5.66) after changing $+I_{\mathrm{m}}$ into $-I_{\mathrm{m}}$ and a into b.

The magnetic vector potential of two conductors (Figure 5.38) can be determined by the application of the principle of superposition (Zakrzewski, Sykulski [5.19]):

- For the region of dielectric I:

$$A_{1,2} = -\frac{\mu_0 I_{\mathrm{m}}}{4\pi} \ln \frac{y^2 + (a+z)^2}{y^2 + (b-z)^2}\,\frac{y^2 + (b-z)^2}{y^2 + (a-z)^2} + \\ + \frac{\mu_{\mathrm{II}} I_{\mathrm{m}}}{2\pi}(n-1)\int_0^\infty \frac{1}{\lambda}\left[\mathrm{e}^{-(a-z)\lambda} - \mathrm{e}^{-(b-z)\lambda}\right] \frac{1 + r\,\mathrm{e}^{-2d\lambda}}{n + r\,\mathrm{e}^{-2d\lambda}} \cos \lambda y \,\mathrm{d}\lambda \tag{5.67}$$

- For the region of magnetic screen (*shunt*) II:

$$A_{II,2} = \frac{\mu_{\mathrm{II}} I_m}{2\pi}(n-1)\left[\begin{array}{l} \displaystyle\int_0^\infty \frac{1}{n + r\,\mathrm{e}^{-2d\lambda}} \cdot \frac{\mathrm{e}^{-(a+2d-z)\lambda} - \mathrm{e}^{-(b+2d-z)\lambda}}{\lambda} \cos \lambda y \,\mathrm{d}\lambda + \\ \displaystyle + \int_0^\infty \frac{\mathrm{e}^{-(a+z)\lambda} - \mathrm{e}^{-(b+z)\lambda}}{\lambda\left(n + r\,\mathrm{e}^{-2d\lambda}\right)} \cos \lambda y \,\mathrm{d}\lambda \end{array}\right. \tag{5.68}$$

$$A_{\mathrm{III},2} = \frac{\mu_{\mathrm{II}} I_{\mathrm{m}}}{2\pi}(n-1)\int_0^\infty \frac{r+1}{n+r\,\mathrm{e}^{-2d\lambda}} \times$$
$$\times \frac{1}{\lambda}\left[\mathrm{e}^{-(a+d)\lambda+(d-z)\sqrt{\lambda^2+\alpha^2}} - \mathrm{e}^{-(b+d)\lambda+(d-z)\sqrt{\lambda^2+\alpha^2}\,(d-z)}\right] \tag{5.69}$$

where

$$\frac{\mu_{\mathrm{III}}\lambda - \mu_{\mathrm{II}}\sqrt{\lambda^2+\alpha^2}}{\mu_{\mathrm{III}}\lambda + \mu_{\mathrm{II}}\sqrt{\lambda^2+\alpha^2}}, \quad \frac{\mu_{\mathrm{II}}+\mu_0}{\mu_{\mathrm{II}}-\mu_0} \tag{5.70}$$

The corresponding components of flux density, B_x and B_y, can now be determined from Equation 5.63. Due to the complicated form of the formulae, it is necessary to use a computer for the calculations. At simplified calculations, however, it is easier to utilize the method of multiple mirror images (see Section 5.1.4).

5.7 ADDITIONAL LOSSES IN FOIL WINDINGS OF TRANSFORMER

Additional losses in foil windings caused by a strong skin effect at the ends of windings are an important factor in the design of small and special transformers. The problem is difficult since only the boundary condition: $\oint_{\Sigma l_i} \boldsymbol{H}\cdot \mathrm{d}\boldsymbol{l} = IN$ leads to the Fredholm equation:

$$y(x) = g(x) + \lambda\int_a^b K(x,s)\cdot y(s)\cdot \mathrm{d}s \tag{5.71}$$

of second order, whose solution (N. Mullineux [5.11]) is too complicated for engineering design use. Practically the same results, but in a much easier way, were obtained many years ago by the author [5.18] using the well-known deep-slot theory (Figures 5.28 and 5.41).

The active $J_a(y)$ and the reactive $J_r(y)$ eddy-current density distribution and the coefficient k_{ad} of additional losses in foil windings of transformers can be calculated very fast [5.18] and, as it is expected by industry [7.6], "essential results are clearly and simply presented":

$$\begin{aligned} J_a &= J_0\left\{1 + \frac{\sqrt{2}\,ka}{M} e^{-kh\left(\frac{1}{2}-\frac{y}{h}\right)} \cos\left[kh\left(\frac{1}{2}-\frac{y}{h}\right) - \frac{\pi}{4}\right]\right\} \\ J_r &= -J_0\left\{1 + \frac{\sqrt{2}\,ka}{M} e^{-kh\left(\frac{1}{2}-\frac{y}{h}\right)} \sin\left[kh\left(\frac{1}{2}-\frac{y}{h}\right) - \frac{\pi}{4}\right]\right\} \end{aligned} \tag{5.72}$$

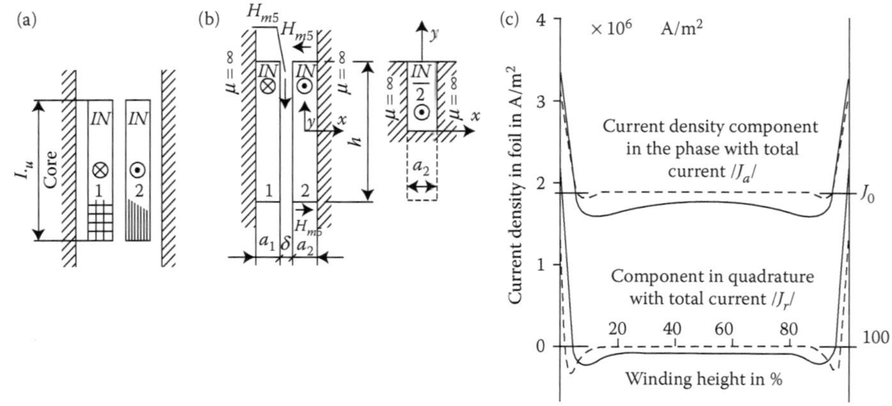

FIGURE 5.41 A transformer with foil winding (2): (a) real, (b) deep-slot model (5.27) [5.18], (c) current density distribution J_a, J_r along the height of the foil winding, calculated with the methods: - - - - author's formulae (14) from [5.18]; —— Mullineaux et al. [5.11].

From these considerations follows the coefficient k_{ad} of additional losses in foil winding:

$$k_{\text{ad}} = \frac{P}{J_0^2(ahl/\sigma)} \approx 1 + \frac{4a}{\text{M}h}\left(1 + \frac{ka}{2\text{M}}\right) \tag{5.73}$$

where

$$J_0 \approx \frac{NI}{ah}; \quad \text{M} \approx \pi\frac{2a_2}{a_1 + a_2 + \delta}; \quad k = \sqrt{\frac{\omega\mu\sigma}{2}} \tag{5.74}$$

$a = a_2 - a_i$, where a_i is the total thickness of the foil insulation.

Formulae 5.72 were next used (V. Preiningerova, "Transformers," BEZ-Bratislava, No. 2, 1981) for thermal calculation of heating of the foil windings in distribution transformers (J. Turowski [1.16], p. 282).

Here again, although such analytical transformations need special attention and discipline, the designer obtains a valuable reward in the form of a parametric answer and rapid calculation, because analytical formulae, correctly programmed, give results in microseconds. This approach satisfies one of the main imperatives of the modern mechatronics—a rapid, interactive design. Whereas the solution by Mullineaux et al. [5.11] is very difficult and time-consuming, and could be hardly repeated by anyone, not to mention university students.

6 Electromagnetic Phenomena in Metals with Constant Permeability

6.1 APPLICATION OF MULTIPLE REFLECTIONS OF ELECTROMAGNETIC WAVE

In Section 4.3, we presented fundamentals of calculation of electromagnetic field inside metal walls using the method of multiple reflections of the wave inside metal sheets. Using that method, in Section 4.4 the active and reactive power consumed in these walls and electromagnetic screens were calculated. In Section 4.9.1, the coefficients of power loss in cylindrical screen were developed, and in Section 5.5.1—impedances of rotor bars of a deep-slot induction motor. Other applications of this method will be given in the next sections.

6.2 ELECTRICAL STEEL

One of the most important tasks in the area of electric machines and power transformers is the selection of proper thickness of electric iron sheets [5.15], depending on their properties, the frequency used, and so on. The sheet thickness reduction is favorable for reduction of losses caused by eddy currents and elimination of nonuniformity of the flux density distribution inside the sheet cross section, which gives a better utilization of the material. Opposing the reduction of sheet thickness is the increase of the labor-consuming effort of packaging as well as overlapping of cores and reduction of the coefficient of core cross section (insulation coefficient) a_i.

6.2.1 INSULATION COEFFICIENT, a_i

The measure of filling a core by iron can be expressed by the coefficient of insulation:

$$a_i = \frac{d - d'}{d} = \frac{kd - kd'}{kd} \tag{6.1}$$

where d is the sheet thickness and d' is the thickness of between-sheets insulation.

The coefficient k (2.140) can be assessed for flux densities used in power transformers:

- For a hot-rolled (isotropic) steel, at flux densities 1.45–1.5 T, the relative permeability $\mu_r = 680–530$ (Figure 1.29), the conductivity $\sigma = 1.7 \times 10^6$ S/m, and $\mu_0 = 0.4\pi \times 10^{-6}$ H/m, it can be calculated as

$$k = \sqrt{\frac{\omega\mu\sigma}{2}} = \sqrt{\frac{2\pi \times 50(680\ldots530) \times 0.4\,\pi \times 10^{-6} \times 1.7 \times 10^6}{2}}$$
$$= (478\ldots422)\text{m}^{-1}$$

- For a cold-rolled (anisotropic) steel (Hipersil), at flux densities 1.65–1.7 T, the relative $\mu_r = 1640–840$ [5.2], and the conductivity $\sigma = 2 \times 10^6$ S/m, the coefficient $k = (805\ldots577)$ m^{-1}

Since $d' \ll d$, we will not make a significant error if we assume for the k component in formula (6.1) an average value of $k \approx 600$ m^{-1} = const, independent of material properties and saturation. Moreover, assuming a constant thickness of the insulation layer, $d' = 0.035$ mm (e.g., insulating varnish of iron sheets of thickness $d = 0.35–0.5$ mm), we get

$$a_i = \frac{kd - 0.021}{kd} \tag{6.2}$$

6.2.2 Coefficient of Flux Expulsion, a_S

Investigations of various authors, for example, [6.12], have shown that the distribution of magnetic field intensity in a sheet of variable permeability is not much different from the distribution obtained with the classic method of the assumption of an equivalent μ: $\mu_{\text{equiv.}}$ = const (Turowski [2.41]). Therefore, one can use the formula (4.29)

$$H_m(z) = H_m \frac{\text{ch}\alpha\left(\frac{d}{2} - z\right)}{\text{ch}(\alpha d/2)}$$

which represents the resultant distribution of magnetic field intensity inside a sheet, along the OZ axis (Figure 6.1).

In Figure 6.1 and in these formulae, only the moduli of vectors $\boldsymbol{H}_m$ and $\boldsymbol{B}_m$ are represented, because these vectors have practically only one component in the core. After integration of expression (4.29), from zero to d, we obtain the *complex maximum flux* in a sheet per unit of its width in the Y direction (Figure 6.1)

$$\Theta_{m1} = \int_0^d B_m(z)\text{d}z = \frac{\mu H_{ms}}{\text{ch}(\alpha d/2)} \int_0^d \text{ch}\alpha\left(\frac{d}{2} - z\right)\text{d}z = \frac{2\mu H_{ms}}{\alpha}\frac{\text{sh}(\alpha d/2)}{\text{ch}(\alpha d/2)} \tag{6.3}$$

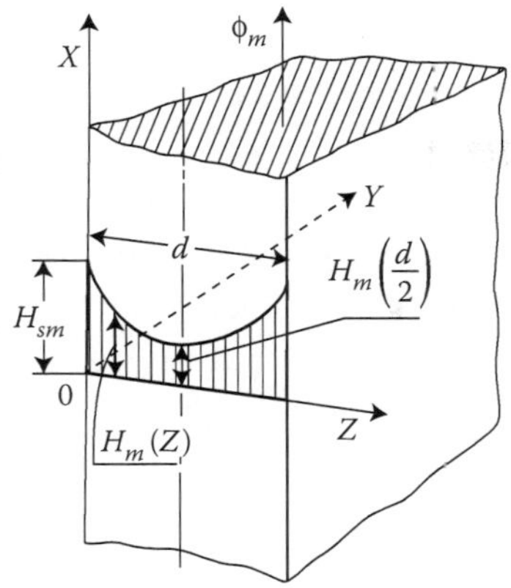

FIGURE 6.1 Distribution of magnetic field intensity inside an electric sheet. (Adapted from Turowski J.: Selection of optimal thickness of electrotechnical sheet from the standpoint of electrical properties of the core. (in Polish). *Przegląd Elektrotechniczny*, 8, 1964, 361–368 [6.15].)

Considering, after formula (2.139), $\alpha = (1 + \mathrm{j})k$, and after some transformations, we obtain the time-dependent modulus of magnetic flux

$$|\Phi_{m1}| = \frac{\sqrt{2}\mu}{k} H_{ms} \sqrt{\frac{\operatorname{ch} kd - \cos kd}{\operatorname{ch} kd + \cos kd}}$$

The same flux at uniform field distribution would be expressed by the formula

$$|\Phi_{m01}| = \mu H_{ms} d$$

The coefficient of flux expulsion in the sheet cross section, due to skin effect, therefore has the form (Figure 6.2)

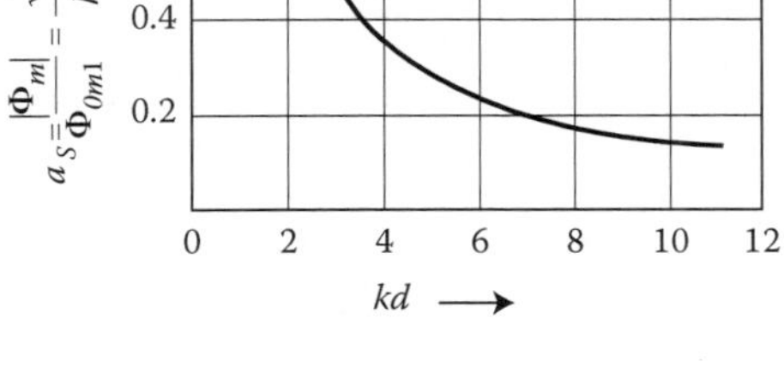

FIGURE 6.2 The coefficient of expulsion, a_S, of an alternating flux inside the sheet. (Adapted from Turowski J.: Selection of optimal thickness of electrotechnical sheet from the standpoint of electrical properties of the core. (in Polish). *Przegląd Elektrotechniczny*, 8, 1964, 361–368 [6.15].)

$$a_s = \frac{|\Phi_{m1}|}{\Phi_{m01}} = \frac{\sqrt{2}}{kd}\sqrt{\frac{\operatorname{ch} kd - \cos kd}{\operatorname{ch} kd + \cos kd}} \tag{6.4}$$

If the hyperbolic and trigonometric functions in Equation 6.4 are expanded into a series, and if we assume that $kd \ll 1$ (electrotechnical steel), and moreover only the two first components of the series are considered, we shall obtain $a_s = 1$. At $kd > 3$, the coefficient a_s tends to $\sqrt{2}/kd$ (hyperbole).

The *resultant coefficient of core utilization* a_w is a product of the coefficients a_i and a_s:

$$a_w = \sqrt{2}\,\frac{kd - 0.021}{(kd)^2}\sqrt{\frac{\operatorname{ch} kd - \cos kd}{\operatorname{ch} kd + \cos kd}} \tag{6.5}$$

A graph of the core utilization coefficient a_w (Figure 6.3) has its maximum at $kd = 0.75$, reaching the value of $a_w = 0.97$. According to the author's research [2.34], [6.15], taking into consideration the variation of permeability shifts (Figure 4.14) the a_w graphs to the left (curve 2 in Figure 6.3). The remarkable sharp extreme of the a_w curve clearly defines the region of optimal thickness (d_{opt}) of the sheet from the viewpoint of the cross-section utilization. From Figure 6.3, for $a_w \geq 0.9$ we obtain

$$0.15 < kd_{\text{opt}} < 1.1 \tag{6.6}$$

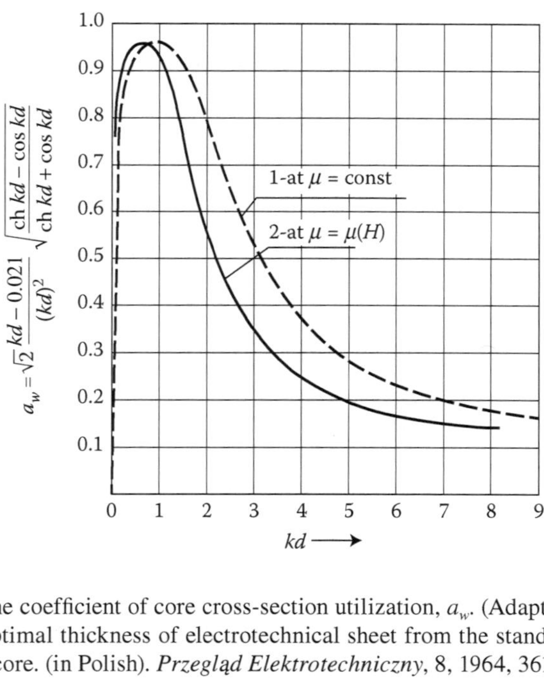

FIGURE 6.3 The coefficient of core cross-section utilization, a_w. (Adapted from Turowski J.: Selection of optimal thickness of electrotechnical sheet from the standpoint of electrical properties of the core. (in Polish). *Przegląd Elektrotechniczny*, 8, 1964, 361–368 [6.15].)

which corresponds to the optimal thickness of hot-rolled sheet ($k \approx 440$ 1/m):

$$0.34 \text{ mm} < d_{\text{opt}} < 2.5 \text{ mm} \tag{6.7}$$

and for cold-rolled (anisotropic) sheets ($k \approx 700$ 1/m):

$$0.25 \text{ mm} < d_{\text{opt}} < 1.58 \text{ mm} \tag{6.8}$$

As it follows from the given formulae, the coefficient of core cross-section utilization, a_w, is not an obstacle for even a remarkable increase of sheet thickness. The optimal thickness—from the point of view of magnetic conductivity of core—is about 1 mm for hot-rolled (isotropic) sheets, and about 0.7 mm for cold-rolled (anisotropic) sheets (Figure 6.3).

6.2.3 Hysteresis Losses

Using the Richter's formula (1.40) for hysteresis losses, we assume that the losses per volume unit of sheet (in W/m³) equal

$$P_h = \frac{\varepsilon'}{d} f \int_0^d \left|B_m\right|^2 \mathrm{d}z \tag{6.9}$$

where ε' is a constant depending on the sort of steel (for transformer sheets, $\varepsilon' \approx 166$ m/H).

Then, by substituting to this formula the maximum value of magnetic field intensity, taken from formula (4.29), we obtain

$$P_h = \varepsilon' f \frac{\mu^2 \left|H_{ms}\right|^2}{d} \int_0^d \left|\frac{\operatorname{ch} \alpha((d/2) - z)}{\operatorname{ch}(\alpha d/2)}\right|^2 \mathrm{d}z$$

and after integration and transformations, we obtain the losses (in W/m³):

$$P_h = \varepsilon' f |B_{ms}|^2 \frac{1}{kd} \frac{\operatorname{sh} kd + \sin kd}{\operatorname{ch} kd + \cos kd} \tag{6.10}$$

In the formulae for the hysteresis losses, the average (in space) value of maximum (in time) flux density occurs. This is why here we should make a corresponding substitution. From formula (4.29) we obtain the average, complex flux density

$$B_{m,\text{aver}} = \frac{1}{d} \int_0^d B_m(z)\,\mathrm{d}z = \frac{B_{ms}}{d} \int_0^d \frac{\operatorname{ch}\alpha((d/2) - z)}{\operatorname{ch}(\alpha d/2)} \mathrm{d}z = B_{ms} \frac{\operatorname{sh}(\alpha d/2)}{(\alpha d/2)\operatorname{ch}(\alpha d/2)}$$

and its modulus

$$\left|B_{m,\text{aver}}\right| = \left|B_{ms}\frac{\operatorname{sh}\dfrac{\alpha d}{2}}{\dfrac{\alpha d}{2}\operatorname{ch}\dfrac{\alpha d}{2}}\right| = \frac{\sqrt{2}\left|B_{ms}\right|}{kd}\sqrt{\frac{\operatorname{ch} kd - \cos kd}{\operatorname{ch} kd + \cos kd}} = |B_{ms}|\,a_s \tag{6.11}$$

After substituting Equation 6.11 into Equation 6.10, we obtain

$$P_h = P_h = \varepsilon' f \left|B_{m,\text{aver}}\right|^2 \frac{kd}{2}\frac{\operatorname{sh} kd + \sin kd}{\operatorname{ch} kd - \cos kd}\cdots \tag{6.12}$$

The hysteresis losses per volume unit of sheet (in W/m³), at the assumption of a uniform distribution of flux density in sheet cross section, equal

$$P_{h0} = P_{h0} = \varepsilon' f\,|B_{m,\text{aver}}|^2 \tag{6.13}$$

Hence, the coefficient a_h of increase of hysteresis losses due to the condensation of flux caused by skin effect (Figure 6.4)

$$a_h = \frac{kd}{2}\frac{\operatorname{sh} kd + \sin kd}{\operatorname{ch} kd - \cos kd} \tag{6.14}$$

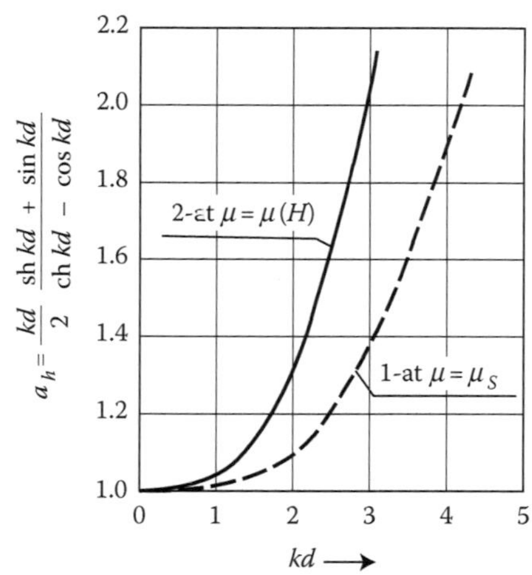

FIGURE 6.4 The coefficient a_h of increase of hysteresis losses due to the condensation of flux caused by skin effect. (Adapted from Turowski J.: Selection of optimal thickness of electrotechnical sheet from the standpoint of electrical properties of the core. (in Polish). *Przegląd Elektrotechniczny*, 8, 1964, 361–368 [6.15].)

6.2.4 Losses Caused by Eddy Currents (Eddy-Current Losses)

The active power losses from eddy currents per volume unit of sheet (in W/m^3) in the function of average flux density can be found by substituting dependence (6.11) into Equation 4.47 and dividing the result by the thickness d:

$$P_{\text{eddy}} = \frac{\zeta}{d}\sqrt{\frac{\omega\mu}{2\sigma}}\frac{\left|B_{m,\text{aver}}\right|^2}{\mu^2 a_s^2} = \frac{k^3 d}{2\mu^2}|B_{m,\text{aver}}|^2\frac{\text{sh}\,kd - \sin kd}{\text{ch}\,kd - \cos kd} \tag{6.15}$$

Analogically, assuming a uniform distribution of flux density in the sheet cross section, the losses can be obtained from formula (6.15) if assuming $kd \ll 1$ we substitute the hyperbolic and trigonometric functions with the two first components of their expansion into a series. Then, we obtain the popular simplified formula (Jezierski et al. [5.2])

$$P_{\text{eddy0}} = \frac{1}{24}\sigma\omega^2|B_{m,\text{aver}}|^2 d^2 \tag{6.16}$$

In formula (6.16) the thickness of sheet (d) occurs. Therefore, we cannot take Equation 6.16 as a reference for investigation of influence of sheet thickness with the help of the accurate formula (6.15). We shall express these losses in units of value $(\omega/4\mu)|B_{m,\text{aver}}|^2$ which have the unit of power. In this case, we shall obtain the following expression for the relative eddy-current losses per volume unit (Figure 6.5):

$$p_{\text{eddy}} = kd\frac{\text{sh}\,kd - \sin kd}{\text{ch}\,kd - \cos kd}\cdots \tag{6.17}$$

At $kd > 3$, practically $p_{\text{eddy}} = kd$, and then we obtain another known, simplified formula for the losses (in W/m^3)

$$p_{\text{eddy}} = \frac{1}{24}\sigma\omega^2|B_{m,\text{aver}}|^2 d^2\frac{3}{kd} \tag{6.18}$$

From Figure 6.5, it follows that the eddy-current losses are the factor that is decisive in the selection of optimal sheet thickness, because they are growing rapidly with the thickness (proportional to d^2, and then to d).

Assuming, therefore, as a criterion for optimal sheet thickness the lower limit of expression (6.6), *we obtained optimal thickness for a hot-rolled (isotropic) steel the thickness of about 0.35 mm (formula 6.7) and for a cold-rolled (anisotropic) transformer steel—about 0.25 mm (formula 6.8).* The former of these thicknesses is, as we know, widely used for hot-rolled transformer steel. The latter thickness shows rationality for reduction of thickness of the popular cold-rolled (anisotropic) transformer steel. These recommendations, proved by the author in 1964 [6.15], have been broadly recognized and used in practice (e.g., Steel Works Catalogue [1.30]).

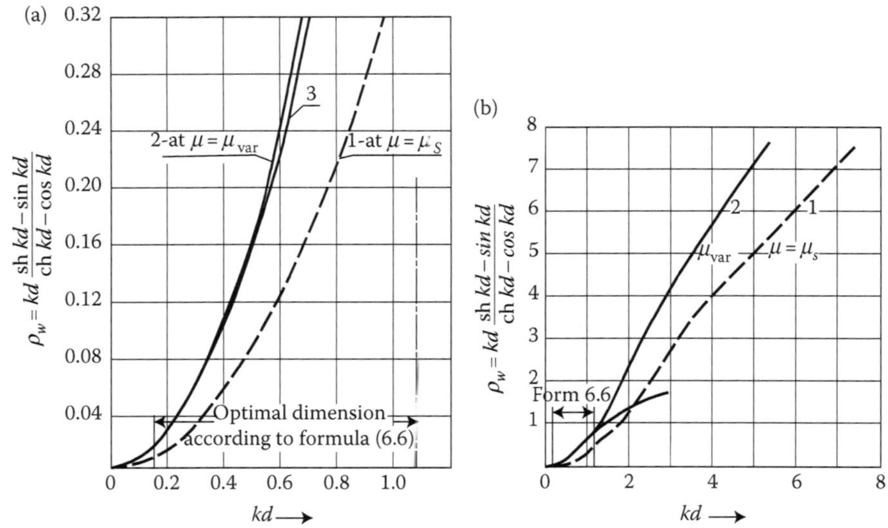

FIGURE 6.5 The eddy-current losses dependence on the coefficient *kd*: 1—curves from the classical formulae; 2—curves taking into account the reduced depth of field penetration into iron; 3—curves taking into account the reduction of the resultant iron sheet magnetic permeability due to the nonuniform distribution of flux in sheet cross section. (a) Small thickness and (b) big thickness. (Adapted from Turowski J.: Selection of optimal thickness of electrotechnical sheet from the standpoint of electrical properties of the core. (in Polish). *Przegląd Elektrotechniczny*, 8, 1964, 361–368 [6.15].)

The developed formulae allow also to calculate an increase of eddy-current losses in iron of another sheet thickness or at another frequency [6.15].

6.2.5 Reactive Power Consumption

The reactive power consumption by an iron sheet, per volume unit (in var/m^3), can be calculated with a method analogical to the active power consumption, after formulae (4.48) and (6.11):

$$Q_1 = \frac{\psi}{d}\sqrt{\frac{\omega\mu}{2\sigma}}\frac{|B_{m,\text{aver}}|^2}{\mu^2 a_S^2} = \frac{k^3 d}{2\sigma\mu^2}|B_{m,\text{aver}}|^2 \frac{\text{sh } kd + \sin kd}{\text{ch } kd - \cos kd} \tag{6.19}$$

At electrotechnical steel, $kd \ll 1$. We can therefore substitute in formula (6.19) the hyperbolic and trigonometric functions by two first components of their expansion into series. Then, we obtain a new simplified formula (Turowski [6.15]):

$$Q_1 = \frac{k^2}{\sigma\mu^2}|B_{m,\text{aver}}|^2 = \frac{\omega}{2\mu}|B_{m,\text{aver}}|^2 \tag{6.20}$$

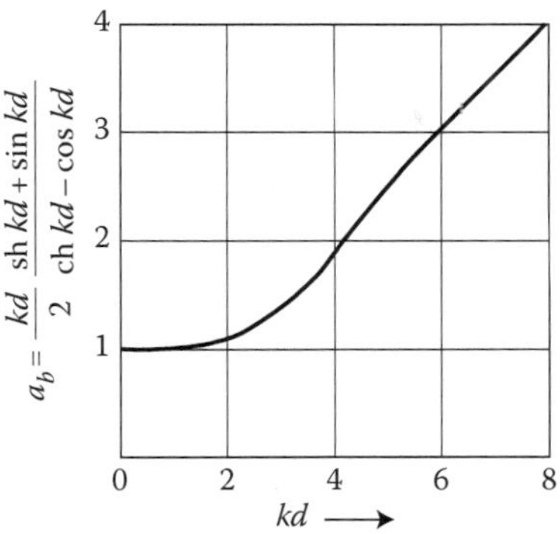

FIGURE 6.6 The coefficient a_b of the increase of reactive power consumption due to the flux skin effect in a sheet cross section. (Adapted from Turowski J.: Selection of optimal thickness of electrotechnical sheet from the standpoint of electrical properties of the core. (in Polish). *Przegląd Elektrotechniczny*, 8, 1964, 361–368 [6.15].)

As we can see from (6.20), the reactive power consumption by an iron sheet does not depend on the thickness of the electrical iron sheet. The coefficient a_b of the increase of reactive power consumption due to the flux skin effect in sheet cross section

$$a_b = \frac{Q_1}{Q_{01}} = \frac{kd}{2}\frac{\text{sh}kd + \sin kd}{\text{ch}kd - \cos kd} \tag{6.21}$$

shows (Figure 6.6) that the reactive power consumption by iron sheet does not depend on the electrical iron sheet thickness until the thickness reaches around 3 mm, and then it increases linearly, proportionally to $kd/2$.

Using the graphs of per-unit reactive power consumption by a transformer sheet, given by Jezierski et al. [5.2], based on laboratory-measured magnetization characteristics, the author [6.15] stated (Table 6.1), that in practice one can use the author's formula (6.20), multiplying it by a constant reduction coefficient of 0.67. This fact was confirmed, in principle, by the investigations of Neiman (7.28) and Agarwal (7.18). Thanks to it, J. Turowski [6.15] developed a convenient parametric formula for the *consumption of reactive power by the core* (in var/m^3):

$$Q_{01} \approx 2.1\frac{f}{\mu}\left|B_{m,\text{aver}}\right|^2 \tag{6.22}$$

A broader discussion of the laminated core properties can be found, for example, in the books [1.5], [5.2], [7.2], and in manufacturers' catalogues.

TABLE 6.1
Comparison of Formula (6.22) with Experimentally Verified Graphs of Reactive Power Consumption by Transformer Sheets

$\lvert B_{m,\text{aver}} \rvert$	1.3	1.4	1.5	1.6	1.7	$\frac{\text{Wb}}{\text{m}^2} = T$
μ	1500	1050	670	400	200	$10^{-6}\frac{\text{Wb}}{\text{Am}}$
Q_0 [1.5]	18.0	28.5	47	93	170	$\frac{\text{var}}{\text{kg}}$
Q_0—Equation 6.22	16	28.2	47	90	184[a]	$\frac{\text{var}}{\text{kg}}$

Source: After Jezierski E.: *Transformers. Theory*. Warsaw: WNT, 1975 [1.5]; Turowski J.: Selection of optimal thickness of electrotechnical sheet from the standpoint of electrical properties of the core. (in Polish). *Przegląd Elektrotechniczny*, 8, 1964, 361–368 [6.15].

[a] The specific mass of iron sheet was assumed to be 7.55×10^3 kg/m³.

6.3 POWER LOSSES AT CURRENT INTERSECTIONS THROUGH A SCREEN

6.3.1 Single-Phase Bushing System

According to the formulae (4.53) and (5.17), the active power losses in a nonmagnetic screen (J. Turowski [4.16]) intersected perpendicularly by two bars carrying current (Figure 5.23) amount to

$$P = \zeta \frac{I^2 a^2}{2\pi^2}\sqrt{\frac{\omega\mu}{2\sigma}}\iint_A \frac{r\mathrm{d}r\mathrm{d}\theta}{r^4 - (a^2r^2/2)\cos 2\theta + (a^4/16)} \tag{6.23}$$

Due to the existence of discontinuity points of the integrand function on the axes of bushings, the entire surface of the screen (transformer cover) was divided into three integration subareas (Figure 6.7)

$$\iint_A f(r,\Theta)\,\mathrm{d}A = \iint_{A_1} f(r,\Theta)\,\mathrm{d}A + \iint_{A_2} f(r,\Theta)\,\mathrm{d}A + \iint_{A_3+A_3'} f(r,\Theta)\,\mathrm{d}A \tag{6.24}$$

where $f(r, \Theta)$ is an integrand of Equation 6.23. Separation of the subarea A_3' is needed only at calculations related to steel plates (Chapter 7). Integrals for the subareas A_1 and A_3 are resolved similarly (Rizhik [6.6])

$$\iint_{A_1,A3} f(r,\theta)\mathrm{d}A = \int_{0,k_{2a}}^{k_{1,3a}} r\mathrm{d}r \cdot 2\int_0^{x} \frac{\mathrm{d}\theta}{r^4 - (a^2r^2/2)\cos 2\theta + (a^4/16)} = 2\int_{0,k_{2a}}^{k_{1,3a}} \frac{\pi r\mathrm{d}r}{\sqrt{p^2 - q^2}} \tag{6.25}$$

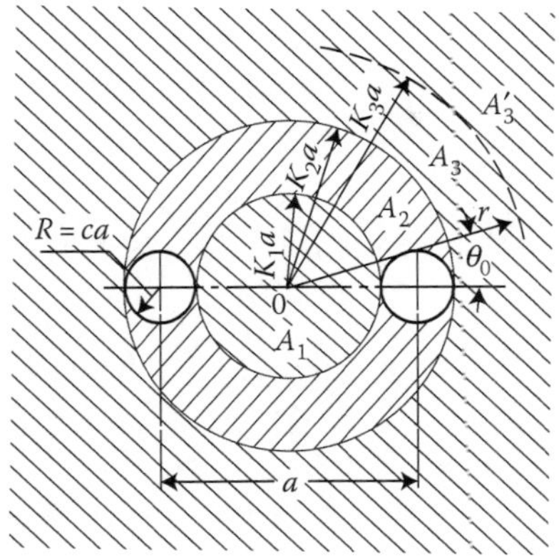

FIGURE 6.7 The limits of integration of formula (6.23). (Adapted from Turowski J.: Losses in single and three phase transformers covers. (in Polish). *Rozprawy Elektrotechniczne*, (1), 1959, 87–119 [4.16].)

where for

$$r \lessgtr \frac{a}{2}, \sqrt{p^2 - q^2} = \pm\left(\frac{a^4}{16} - r^4\right)$$

After solution of both integrals (6.25) in terms of r, at the assumption of an infinitely vast screen ($k_3 \to \infty$), we conclude that the sum of integrals (J. Turowski [4.16]) amounts to

$$\iint_{A_1} f(r,\theta)\,\mathrm{d}A + \iint_{A_3 + A_3'} f(r,\theta)\,\mathrm{d}A = \frac{2\pi}{a^2}\ln\frac{1+4c^4}{4c^2(1-c^2)} \tag{6.26}$$

where $c = R/a$. The mutual dependence between the constants R, a, c, k_1, k_2, k_3 is illustrated in Figure 6.7.

The power losses in the subarea A_2, containing discontinuity, can be calculated by substitution of the approximately circular hole by a hole in the form of a section of ring, and then we obtain

$$\iint_{A_2} f(r,\theta)\,\mathrm{d}A = \int_{k_1 a}^{k_2 a} r \cdot 2 \int_{\theta_o}^{\pi-\theta_o} \frac{\mathrm{d}\theta\mathrm{d}r}{r^4 - (a^2r^2/2)\cos 2\theta + (a^4/16)} = \frac{4}{a^2}\int_{T_1}^{T_2} \frac{\operatorname{arc} tg\, t}{t}\mathrm{d}t \tag{6.27}$$

where

$$t = \frac{r^2 + a^2/4}{r^2 - a^2/4}\, tg\,\theta_0 \quad \text{and} \quad T_{1,2} = \frac{(1 \mp c)^2 - 2c^2}{c \mp 1} \approx c \mp 1$$

Since usually $c^2 = (R/a)^2 \ll 1$ (e.g., around 0.03 or less), the result of integration in the limits $c \mp 1$ will differ very little from the result of integration within the limits ∓ 1. Therefore

$$\int_{T_1}^{T_2} \frac{\operatorname{arc} tg\, t}{t}\, \mathrm{d}t \approx 2\int_{1}^{2} \frac{\operatorname{arc} tg\, t}{t}\, \mathrm{d}t = 2G \tag{6.28}$$

where $G = 1.832\ldots$ is the Catalan constant (Rizhik [6.6]).

The final formula for the active power losses was obtained by the author as a result of substitution of the dependence (6.28) into Equation 6.27, and the latter formula together with Equation 6.26 into Equation 6.24, and into Equation 6.23

$$P = \frac{\operatorname{sh} kd - \sin kd}{\operatorname{ch} kd + \cos kd} \frac{I^2}{\pi} \sqrt{\frac{\omega\mu}{2\sigma}} \left[\ln \frac{1 + 4c^4}{4c^2(1 - c^2)} + 1.17 \right] \tag{6.29}$$

At small values of c, one can omit their second and fourth powers with respect to 1.

By moving the second component to under the logarithm, one can replace the function in square brackets by $1.93 \ln 1/c$, with an approximation error smaller than 1.7% (J. Turowski [4.16]). The coefficient ζ depending on kd (Equation 4.50 and Figure 4.14) can be, for $kd < 2.3$, replaced approximately by the parabola $0.2(kd)^2$; but for $kd \geqslant 2.3$, $\zeta = 1$. In this way, we can obtain the following simplified formulae for power losses in a single-phase system (Figure 6.7):

$$\text{For } kd < 2.3\text{: } P \approx 4.4 \times 10^{-2}\, I^2\, d^2 \sqrt{\omega^3 \mu^3 \sigma^3}\, \ln \frac{a}{R} \tag{6.29a}$$

$$\text{For } kd \geqslant 2.3\text{: } P \approx 0.44\, I^2 \sqrt{\frac{\omega\mu}{\sigma}}\, \ln \frac{a}{R} \tag{6.29b}$$

In formulae (6.29a) and (6.29b) all the values are in the SI units (m, kg, s, A).

For an average-heated copper $k \approx 100$ 1/m, hence, the value of the dimensionless coefficient kd equals approximately to the thickness of the copper screen, in cm.

6.3.2 Three-Phase Bushing System

Analogically as before, using the dependences (4.53) and (5.15), in the work [7.16], the author developed (J. Turowski [6.17], [7.16]) the formula for power losses in three-phase bushing system passing through a nonmagnetic screen (Figure 5.21)

$$P = \zeta \frac{I^2}{2\pi} \sqrt{\frac{\omega\mu}{2\sigma}} \left(\ln \frac{1 - 4.5c^2}{4c^6} + \frac{2.12 - 7.9c^2}{1 - 5.35c^2} \right) \tag{6.30}$$

After removing the components with c^2 in brackets of formula (6.30)—as very small, and applying the previous approximation of ζ, we obtain the simplified, but acceptable formulae

$$\text{For } kd < 2.3: \quad P \approx 1.13 \times 10^{-2} I^2 d^2 \sqrt{\omega^3 \mu^3 \sigma} \left(0.74 + 6 \ln \frac{a}{R} \right) \tag{6.30a}$$

$$\text{For } kd \geqslant 2.3: \quad P \approx 0.113\, I^2 \sqrt{\frac{\omega \mu}{\sigma}} \left(0.74 + 6 \ln \frac{a}{R} \right) \tag{6.30b}$$

Comparing formulae (6.30a) and (6.30b) with formulae (6.29a) and (6.29b) we can see that the losses in a three-phase system, at all other conditions the same, are about 1.5 times bigger than the losses in a single-phase system.

In Figure 6.8 is shown is a space distribution of square of magnetic field intensity on the surface of transformer cover in relation to its value on the $x = 0$ axis, that is, $H_m^2 / H_{m,x=0}^2$, which represents approximate relative loss density on the surface of the cover with three-phase bushing system. The graph was produced on the basis of formula (5.16). The iron saturation can change a little this loss distribution, particularly—reduce the peaks of field. The losses in iron cover are given by formulae (7.43) and (7.44).

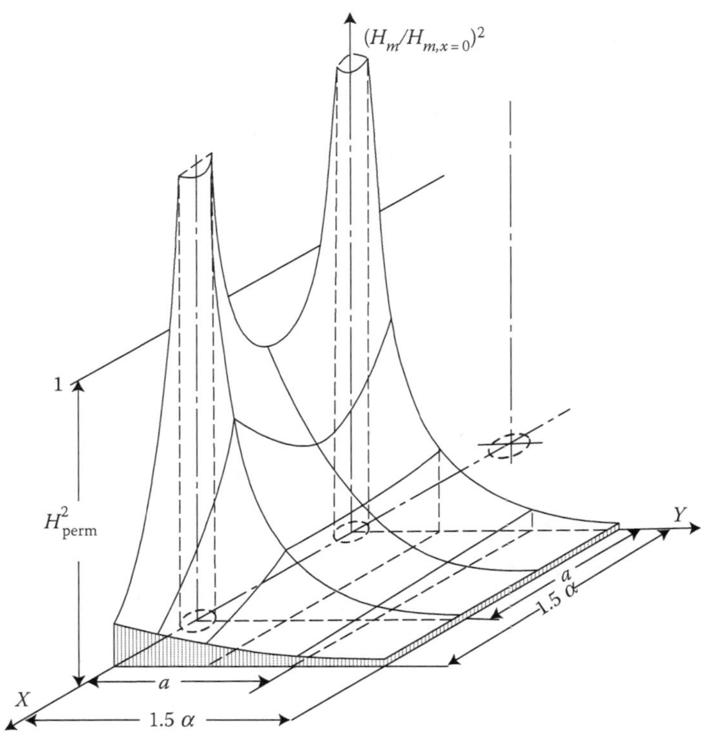

FIGURE 6.8 Distribution of active power losses on the cover of a three-phase power transformer at μ = const; 1—excessive heating line, 2—the distance H^2_{perm}—the relative value permitted from the point of view of local excessive heating hazard, see Equations 9.10, 9.28 and Figure 9.2 [6.17], [7.16].

6.4 POWER LOSSES IN STEEL COVERS WITH GAPS AND NONMAGNETIC INSERTS BETWEEN BUSHING HOLES

In order to reduce power losses in cover plates, nonmagnetic gaps are produced between bushing holes (J. Turowski [5.15], [6.17]), to easily cut the magnetic flux of bushings (Figure 6.9b). It causes such a weakening of the field in the cover that one can consider it as unsaturated, that is, of $\mu = \text{const}$. On the basis of Figures 5.10 and 5.11, the field on the surface layer of such a cover can be determined using mapping in the form of an infinite series of mirror-imaged currents (Figure 6.9). Simultaneously, for simplification, we can ignore the influence of the circular form of the holes on boundary phenomena at the gap and we can assume that for the dimensionless surface cover-layer the mirror images behave like at DC currents.

The magnetic field intensity in the steel, in the axes of the series of fictitious currents, amounts to (Figure 6.9a)

$$H_y = \frac{mi}{2\pi y} + \frac{Mmi}{2\pi(y+e)} + \frac{M^2 mi}{2\pi(y+2e)} + \cdots + \frac{M^n mi}{2\pi(y+ne)} + \cdots =$$

$$= \frac{mi}{2\pi e}\sum_{n=0}^{\infty}\frac{M^n}{\dfrac{y}{e}+n} = \frac{mi}{2\pi e}\frac{1}{M^{y/e}}\sum_{n=0}^{\infty}\frac{M^{n+y/e}}{n+y/e}$$

Since (per Rizhik [6.6]) $\sum_{k=1}^{\infty} M^k/k = \ln 1/(1-M) = \ln 1/m$, hence at $y/e = N = \text{integer} > 1$, we can write

$$H_y = \frac{i}{4\pi e}\frac{m}{M^N}\left(\sum_{k=1}^{\infty}\frac{M^k}{k} - \sum_{k=1}^{N-1}\frac{M^k}{k}\right) = \frac{i}{4\pi e}\frac{m}{M^N}\left(\ln\frac{1}{m}\sum_{k=1}^{N-1}\frac{M^k}{k}\right)$$

where $M = 1 - m$, after (5.2).

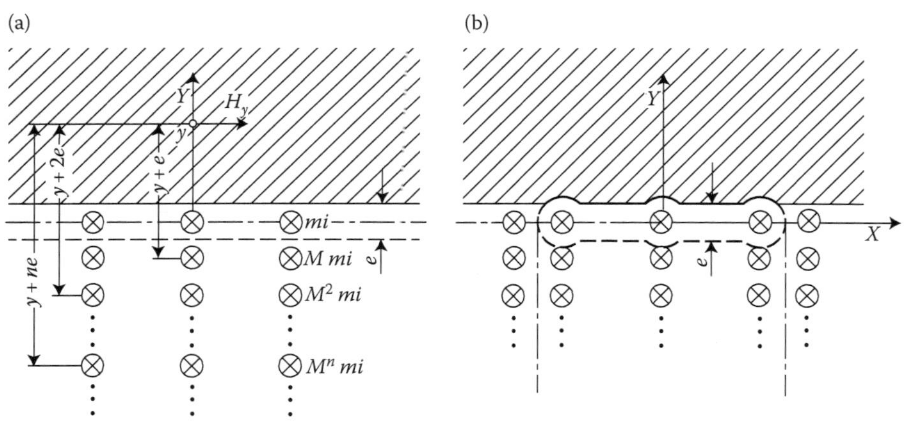

FIGURE 6.9 Fictitious currents series that create the real magnetic field in the upper part of transformer cover: (a) open gap, (b) closed gap.

Now we can utilize the already known formulae for power losses from Section 6.3. For this purpose, we are looking for such a single equivalent current i' which on the axis of the series of fictive currents would give a magnetic field intensity close to the real field, that is

$$\frac{i}{4\pi e}\frac{m}{M^N}\left(\ln\frac{1}{m}-\sum_{k=1}^{N-1}\frac{M^k}{k}\right)=\frac{i'}{2\pi y}$$

from which

$$\left(\frac{i'}{i}\right)^2=\frac{N^2}{M^{2N}}\left(m\ln\frac{1}{m}-m\sum_{k=1}^{N-1}\frac{M^k}{k}\right)^2 \tag{6.31}$$

The function M^k, with an accuracy sufficient for practical applications, at $k < 100$ can be approximated by $M^k \approx 1 - 0.0043$ k (J. Turowski [5.15]), and then

$$\sum_{k=1}^{M-1}\frac{M^k}{k}\approx\sum_{k=1}^{N-1}\left(\frac{1}{k}-0.0043\right)=\left(\sum_{k=1}^{N-1}\frac{1}{k}\right)-0.0043\,(N-1)$$

Whereas a sum of the harmonic series at sufficiently large values of N [6.4] can be expressed as

$$\sum_{k=1}^{N-1}\frac{1}{k}\approx\ln(N-1+1)+0.577=\ln N+0.577$$

After substituting these formulae into Equation 6.31, we obtain, at the assumption of the average relative permeability of iron $\mu_r = 400$, the approximate formula

$$\left(\frac{i'}{i}\right)^2\approx 25\times10^{-6}\,\frac{N^2}{M^{2N}}[4.73-\ln N+0.0043\,(N-1)]^2$$

Consecutively, with an accuracy sufficient for practical applications, the above formula can be approximated by a straight line (J. Turowski [5.15]) described by the equation

$$\left(\frac{i'}{i}\right)^2\approx 2.1\times10^{-3}\,\frac{y}{e} \tag{6.32}$$

This is an approximate course of the active power losses as a function of the width of nonmagnetic gap, e (Figure 6.9b). From Figure 6.10, plotted for different distances y = 5, 8, 10, 12, 15, and 25 cm, it follows that the power losses dramatically

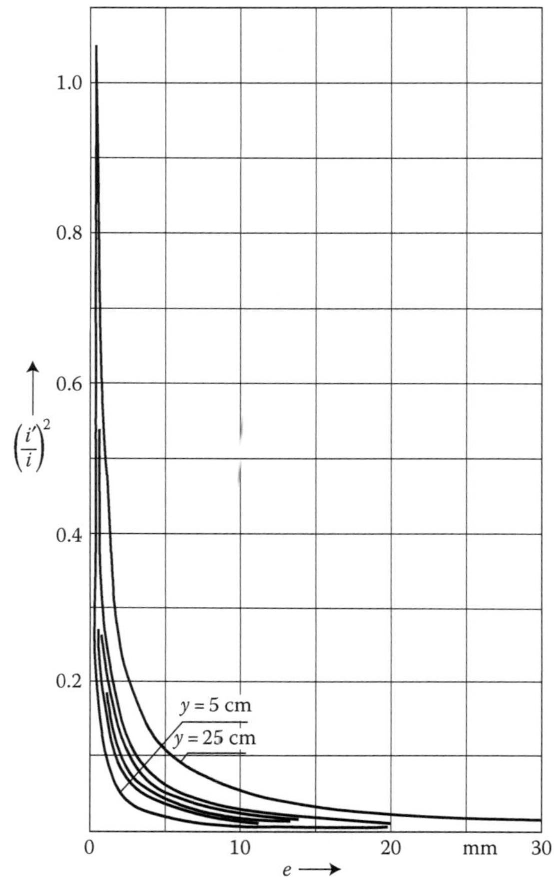

FIGURE 6.10 Theoretical dependence of the losses in transformer cover on the width e of nonmagnetic gap or insert, for different distances $y = 5$, 8, 10, 12, 15, and 25 cm. (Adapted from Turowski J.: Power losses in transformer covers at various methods of their reduction. (in Polish) *Archiwum Elektrotechniki*, (4), 1959, 529–556 [5.15].)

decrease with increase of the gap from zero to 3–5 mm. At broadening the gap *e over* 10–15 mm, the course of the curve of losses is so smooth that further broadening *does not deliver any remarkable benefits.* In reality, broadening of the gap causes "swelling" of the flux in the gap, and due to that, above a certain width the reluctance of gap stabilizes, and so do the losses—at a level corresponding to a few millimeters of the gap width. On the basis of Figure 6.10, we can roughly accept that $(i'/i)^2 = 0.1$.

The losses in the system with an open gap (Figure 6.9a) can be calculated by multiplying Equation 6.30b by this coefficient

$$P \approx 1.13 \times 10^{-2} \times I^2 \sqrt{\frac{\omega\mu}{\sigma}} \left(0.74 + 6 \ln \frac{a}{R}\right) \tag{6.33}$$

In the case of a gap closed on both ends, the field operating in the y direction is created by five series of currents (Figure 6.9b) instead of three in the open gap case (Figure 6.9a). This is why the previous losses (6.33) should be increased by the ratio $(5/3)^2$. Then, we shall obtain the losses in the transformer cover with a nonmagnetic gap or insert (J. Turowski [5.15]).

$$P \approx 3.15 \times 10^{-2} I^2 \sqrt{\frac{\omega\mu}{\sigma}} \left(0.74 + 6 \ln \frac{a}{R} \right) \tag{6.34}$$

Losses in the cover of a single-phase transformer can be determined [5.15] in the same way, by multiplying Equation 6.29b by the coefficient of 0.08. This time this coefficient takes into account also the multiple mirror images along the x axis

$$P \approx 3.5 \times 10^{-2} \times I^2 \sqrt{\frac{\omega\mu}{\sigma}} \ln \frac{a}{R} \tag{6.35}$$

The developed formulae were verified experimentally by the author in his work [5.15] and by other specialists, including Jovanovisč in Macedonia, 1966; P. Pelz in Czech Academy of Science, 1962, M. Kazmierski in the Power Institute in Lodz, 1969, as well as by computer calculations with BEM—Yang Junyou et al. ICEF'92, China, p. 97; plus several transformer works, and others.

6.5 TRANSIENT-INDUCED PROCESSES

6.5.1 Eddy Currents

Let us consider a course and distribution of eddy currents induced in an infinitely vast metal sheet with a negligibly small thickness. This induction is caused by a sudden change of the external magnetic field.

If $\boldsymbol{A}_1$ is the magnetic vector potential of the external excitation field, $\boldsymbol{A}$ is the magnetic vector potential of eddy currents, which is always parallel to the XY surface of sheet, and $\boldsymbol{J}_l$ and σ_l are the current density and the sheet's conductivity, respectively (cf. formula 2.184), then according to Equations 2.3, 2.23, and 2.51, the electric field intensity inside the sheet in point P (Figure 6.11) equals to

$$\boldsymbol{E} = \frac{\boldsymbol{J}_l}{\sigma_l} = -\frac{\partial(\boldsymbol{A}_{1t} + \boldsymbol{A})}{\partial t} - \text{grad } V \tag{6.36}$$

and according to Equation 2.50, $\boldsymbol{B} = \text{curl } \boldsymbol{A}$, the flux density inside the sheet ($A_z = 0$) has the form

$$B_x = -\frac{\partial A_y}{\partial z}; \qquad B_y = -\frac{\partial A_x}{\partial z} \tag{6.37}$$

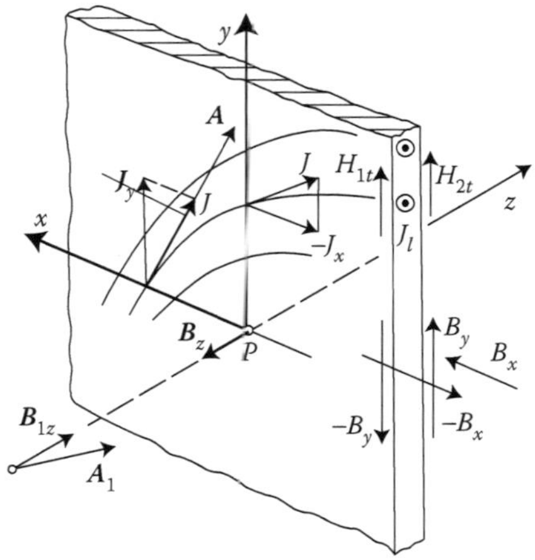

FIGURE 6.11 Induction of instantaneous eddy currents in a metal sheet, by an external field $\boldsymbol{B}_1 = \text{curl}\, \boldsymbol{A}_1$. (Adapted from Hańka L.: Induced transient phenomena in planar bodies. (in Czech). *Elektrotechn.*, Obzor 8, 1962 (51), 378–382 [6.9].)

If i is the current between point P and edge of the sheet or infinity (Figure 6.11), then from the surface curl (2.193a) and dependence (6.37) we have

$$J_x = -2H_y = -\frac{2}{\mu}\frac{\partial A_x}{\partial z};\; J_y = -2H_x = -\frac{2}{\mu}\frac{\partial A_y}{\partial z};\; \boldsymbol{J}_l = -\frac{2}{\mu}\frac{\partial \boldsymbol{A}}{\partial z} \tag{6.38}$$

After substituting dependences (6.38) into Equation 6.36, we obtain (Hanka [6.9])

$$\nu\frac{\partial \boldsymbol{A}}{\partial z} = \frac{\partial(\boldsymbol{A}_{1t} + \boldsymbol{A})}{\partial t} - \text{grad}\, V \tag{6.39}$$

where $\nu = 2/\mu\sigma$. After application of $\text{curl}_z\, A = B_z$ to both sides of Equation 6.39, we can obtain the equation for the normal component of flux density:

$$\nu\frac{\partial B_z}{\partial z} = \frac{\partial(B_{1z} + A_z)}{\partial t} \tag{6.40}$$

A sudden approach of an external magnetic field (e.g., a magnetic bullet) B_{1z} to the sheet will induce in the sheet eddy currents which, according to Lenz rule, at the instant $t = 0$ will generate their own field counteracting the flux density B_{1z} in this region of the sheet (Figure 6.12). Due to this assumption, we can formulate (after Hanka [6.9]) the initial conditions at the instant $t = 0$ for the flux density over the sheet (Figure 6.12c):

$$(B_z)_{t=0} = B_{1z}(x, y, -z) \tag{6.41}$$

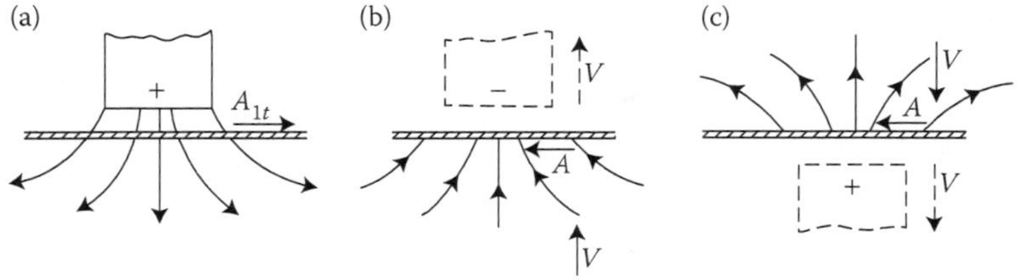

FIGURE 6.12 The field of transient (instantaneous) eddy currents: (a) the primary field excited in a stepwise manner; (b) mirror image simulating the field of eddy currents under the metal sheet; (c) mirror image simulating the field of eddy currents over the metal sheet. The real field is a superposition of the field of eddy currents and of the primary field. (Adapted from Haňka L.: Induced transient phenomena in planar bodies. (in Czech). *Elektrotechn.*, Obzor 8, 1962 (51), 378–382 [6.9].)

Further evolution is described by Equation 6.40, but already without the external field B_{1z} which henceforth remains constant:

$$v = \frac{\partial B_z}{\partial z} = \frac{\partial B_z}{\partial t} \tag{6.42}$$

A solution to the above equation can be any differentiable function of argument $\pm(z + vt)$

$$B_z = B_{1z}(x, y, -z - vt) \tag{6.43}$$

Hence, the flux density B_z changes in such a way as if the field generated by eddy currents over the sheet was penetrating into the sheet with the velocity v, that is, as if fictitious sources of field (mirror images), upper and lower, were moving away from the sheet with the speed v.

In the case when the primary field is excited by a current-carrying conductor (Figure 6.13), the calculations can be most conveniently carried out with the help of the magnetic vector potential Equations 6.38 and 6.39, with the assumption of $v = 0$. This procedure is completely analogical like with the flux density B_z. The magnetic vector potential A_1 of the primary field can be found from Equation 2.60:

$$A_1(x, z) = \frac{\mu_o i}{4\pi} \int_{-\infty}^{+\infty} \frac{\mathrm{d}y}{r'} = -\frac{\mu_o i}{4\pi} \ln[x^2 + (z - h)^2] \tag{6.44}$$

Whereas correspondingly to Equation 6.43

$$A = A_x = \frac{\mu_0 i}{4\pi} \ln[x^2 + (z + vt + h)^2]$$

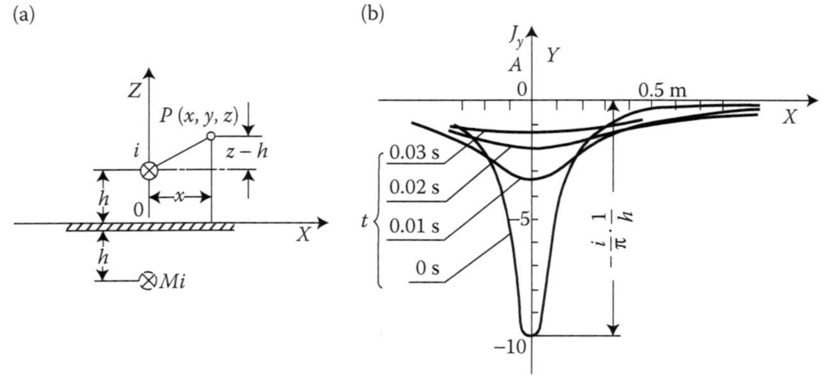

FIGURE 6.13 Distribution of eddy currents on the surface of a copper sheet in function of time, induced by a sudden switching on (at $t = 0$) of nearby flowing current. (a) Image model and (b) induced current distribution. (Adapted from Haňka L.: Induced transient phenomena in planar bodies. (in Czech). *Elektrotechn.*, Obzor 8, 1962 (51), 378–382 [6.9].)

The constant surface current density in the sheet (6.38)

$$J_y = \left(-\frac{2}{\mu_0}\frac{\partial A}{\partial z}\right)_{z=0} = -\frac{i}{\pi}\frac{vt + h}{x^2 + (vt + h)^2} \tag{6.45}$$

Figure 6.13b presents calculated in this way (by L. Haňka [6.9]) distribution of current density in a copper sheet of thickness 1.35 mm, for different time instants after the current switching, at $i = 3.14$ A and $h = 0.1$ m. For a sheet of this thickness, $v = 20$ m/s. In a thinner sheet, eddy currents are extinguished faster than in a thick one. In the work [6.9], also a field of magnetic dipole was considered.

6.5.2 Mirror Image Coefficients

On the basis of formula (6.45), one can evaluate the fictitious image coefficient M_Q (5.7) of a single current-carrying conductor in a thin metal sheet (Figure 6.13a):

$$M_Q = -\frac{1}{\pi}\frac{vt + h}{x^2 + (vt + h)^2} \tag{6.46}$$

As we can see, it depends on the time elapsed from the instant of switching-on the current.

6.6 SOLID ROTOR OF INDUCTION MOTOR

Let us investigate the solid rotor of an m-phase induction motor, made of steel with a constant permeability μ (theoretical assumption) placed in a periodic rotational field of a stator (Figure 6.14). The field of three-phase stator flux density $B = B_m \cos(\pi/\tau x)$ rotates, as one knows, along the stator internal surface with the synchronous rotational

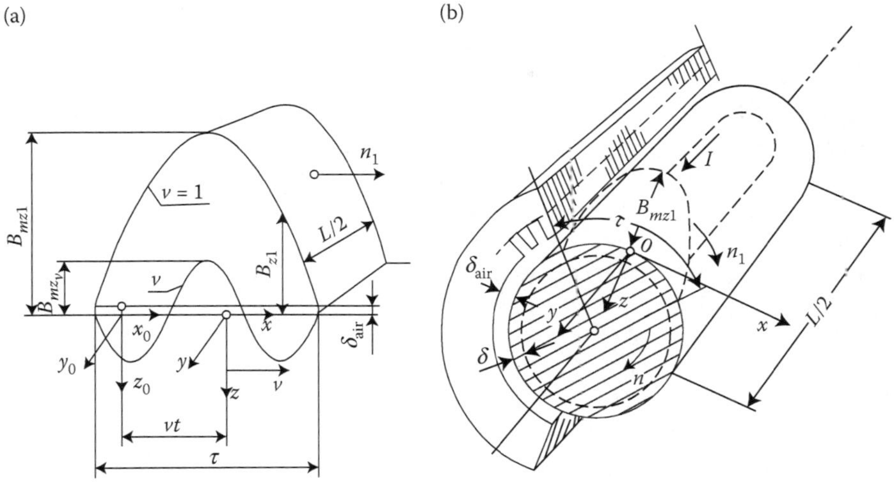

FIGURE 6.14 The solid rotor in a rotating field of stator: (a) x_0, y_0, z_0—the coordinate system connected with the stator; (b) x, y, z—the coordinate system connected with the rotor.

speed n_s* and on the rotor surface with the (relative) slip speed $n_s - n = \text{sn}$, which induces currents of frequency $f_2 = pn_s = sf$ in the squirrel cage or mass of the rotor, where $f = pn_s$ is the supply voltage frequency and $s = (n_s - n)/n_s$ is the slip.

Depending on the frequency $f_2 = sf$ and on the current value, the current and the field B penetrate shallower or deeper into the rotor mass. In the starting moment, the slip $s = 1$ ($n = 0$, $f_2 = f_1 = 50$ Hz), and the equivalent depth of field penetration (2.181) into the rotor mass $\delta = \sqrt{2/\omega_2 \mu \sigma_2}$ amount approximately to $\delta \approx 1$–2 mm, depending on saturation (μ) of the rotor steel. During the rotor run, with the normal slip s of 3–5%, the depth is $\delta \approx 5$–10 mm.

If the stator current is sinusoidal in time, the electromagnetic field in the air gap (δ_{air}) and in the steel of rotor is described, according to the Maxwell's theory (Section 2.10), with the help of Laplace equation $\nabla^2 \boldsymbol{F}_{air} = 0$ for the air gap, and with the Helmholtz equation $\nabla^2 \boldsymbol{F}_{steel} = \alpha^2 \boldsymbol{F}_{steel}$—for the rotor steel, where in place of $\boldsymbol{F}_{air}$ and $\boldsymbol{F}_{steel}$ one can substitute the vectors $\boldsymbol{H}_{m,air}$ and $\boldsymbol{E}_{m,air}$—in the air, and $\boldsymbol{H}_m$, $\boldsymbol{E}_m$, $\boldsymbol{J}_m$—for the rotor steel.

If the equivalent depth (δ) of field penetration (2.181) into the rotor mass is much smaller, as usual, than the rotor diameter $D = 2R$, the calculation can be carried out in rectangular (Cartesian) coordinates, rotating together with the field of speed n_s (Figure 6.14), according to the equations of type (2.141). In the opposite case, one should apply the more difficult equations of type (2.151) for a cylindrical system.

As applicable criteria for the rectangular coordinate system one can adopt the condition

$$R \geq \frac{\lambda}{2} = \pi\delta, \text{ that is, } \quad \frac{R}{\delta_s} = R\sqrt{\frac{s\omega\mu\sigma_2}{2}} \geq \pi, \quad \omega = 2\pi f \tag{6.47}$$

* Old symbol of n_s was n_1 (*primary field*).

As it follows from formulae (4.112) through (4.116), the substitution of the cylindrical coordinate system by the rectangular system at frequency 50 Hz causes satisfactorily small error even at small rotor diameters, for instance, for D over 5–6 cm (Turowski [1.16]). At small slips (small rotor frequencies) this error is naturally higher that at bigger slips.

An analysis of the field in the air gap of a motor with a solid ferromagnetic rotor in a rectangular coordinate system ($D \gg \delta$) was carried out, among others, by Dr. J. Lasocinski [6.11], the author's PhD student.

Calculation of the magnetic circuit of induction motors with a solid rotor is carried out in the same way as for squirrel-cage motors, using the equivalent circuit and the RNM (Turowski [1.18]). The basic difference consists in calculation of the rotor parameters $X_2' = X_2'(s)$ and $R_2' = R_2'(s)$ and the rotor current $I_2' = I_2'(s)$, all of them reduced to the stator parameters. These values, similarly like in deep-slot motors (Figure 5.29) strongly depend on the value of slip s and on the steel saturation $\mu = \mu(H)$. This saturation can be taken into account with the help of the linearization coefficients a_p and a_q (7.26).

In the stationary coordinate system x_0, y_0, z_0 connected with the stator (Figure 6.14a), the flux density of νth, nth space harmonics of the rotating field, in an arbitrary position x_0, changes (Turowski [1.18]) according to the dependence

$$B_{mz\nu n} = B_{mz\nu} \cos n\frac{\pi}{L} y \cdot \mathrm{e}^{\mathrm{j}(\omega t - \nu(\pi/\tau)x_o)} \tag{6.48}$$

where*

$$B_{mz_\nu} = \frac{\sigma_\pm}{\nu}\frac{\zeta_\nu}{\zeta_1} B_{mz1}; \qquad B_{mz1} \approx \frac{\mu_0}{2\delta_{\text{air}}} F_{m0} \tag{6.49}$$

where $\underline{F}_{m0} = \underline{F}_{ms} + \underline{F}_{m2}$; $\underline{F}_{m0} = 2\frac{\sqrt{2}}{\pi} m\, z_1 \frac{\zeta_1}{p} I_0$.

$$\sigma_\pm = \sin\nu\frac{\pi}{2} = \pm 1; \quad \nu = +1, -5, +7, -11, +13, \ldots, -(6\mathrm{a}-1), (6\mathrm{a}+1); \quad a = 1, 2, 3, \ldots$$

Calculations of the field in the rotor are most conveniently carried out in the coordinate system x, y, z connected with the rotor of the speed $v = v_s(1 - s)$, where $v_s = \omega t/\pi$.

Moving over to the new coordinate system (Figure 6.14), we substitute into Equation 6.48

$$x_0 = x + vt = x + (1 - s) - (6a - 1)\frac{\omega\tau}{\pi} t, \quad y_0 = y; \quad z_0 = z$$

* At the assumption for the stator yoke: $\mu_{Fe} \gg \mu_0$.

from where, in the new coordinate system

$$\underline{B}_{mz_\nu n} = \underline{B}_{mz_\nu} \cos n\frac{\pi}{\mathrm{L}}y \cdot \mathrm{e}^{\mathrm{j}(1-\nu+s\nu)\omega t} \cdot \mathrm{e}^{-\mathrm{j}\nu(\pi/\tau)x} \tag{6.50}$$

Assuming that the rotor diameter D ≫ δ, one can utilize Equation 2.141c which describes the field in an arbitrary point of rotor, immovable relatively to the system x, y, z and being overmagnetized with the frequency $f_2 = sf$. This equation, for the amplitude of the component $H_{\mathrm{mz}_\nu n} = B_{\mathrm{mz}_\nu n}/\mu$ inside the steel, at $\mu = \mathrm{const}$, has the form

$$\frac{\partial^2 H_{\mathrm{mz}\nu n}}{\partial x^2} + \frac{\partial^2 H_{\mathrm{mz}\nu n}}{\partial y^2} + \frac{\partial^2 H_{\mathrm{mz}\nu n}}{\partial z^2} = \alpha_\nu^2 H_{\mathrm{mz}\nu n} \tag{6.51}$$

where

$$\alpha_\nu = \sqrt{\mathrm{j}(1-\nu+s\nu)\omega\mu\sigma} = \sqrt{1-\nu+s\nu}\,\alpha \tag{6.52}$$

$$\alpha = (1+\mathrm{j})\,k = \sqrt{2}k\,\mathrm{e}^{\mathrm{j}\pi/4}, \quad k = \frac{1}{\delta} = \sqrt{\frac{\omega\mu\sigma}{2}}$$

The solution of Equation 6.51 with the help of Fourier method (Section 2.11) at periodic field distribution conducts to a result in the form Equation 2.148, where $H_{ym} = X(x) \cdot Y(y) \cdot Z(z)$. From the above conditions it follows (Figure 6.14) that $C_{2n} = C_{5n} = 0$ and $\cos \beta_n \tau/2 \cos \eta_n y = 0$ for all y and $t = 0$ and $\cos \beta_n x \cos \eta_n(L/2) = 0$ for all x, from which

$$\beta_n = \nu\frac{\pi}{\tau}; \quad \eta_n = (2n-1)\frac{\pi}{L} \tag{6.53}$$

where the constants $n = 1, 2, 3, \ldots, m$ determine the number and order of harmonics of the H_z field distribution on the surface of rotor ($z = 0$) along the machine axis. Also, $C_{8n} = 0$, because the field cannot grow to infinity at $z \to \infty$.

The general solution of Equation 6.51 is a sum of particular solutions. Usually, it is satisfactory to accept $\nu = 1$. Then

$$H_{mzn} = \cos\frac{\pi}{\tau}x\sum_{n-1}^{\infty} C_n e^{-\alpha_n z} \cdot \cos(2n-1)\frac{\pi y}{L} \tag{6.54}$$

where as

$$\alpha_n = \sqrt{s\alpha^2 + \left(\frac{\pi}{\tau}\right)^2 + (2n-1)\frac{\pi^2}{L^2}}$$

At $x = y = z = 0$ and $t = 0$, $H_{z\nu n} = \dfrac{1}{\mu} B_{z\nu n}$, hence after Equation 6.49

$$C_n = H_{z\nu n} = \frac{1}{\mu} B_{mz1} \frac{\sigma_\pm}{\nu} \frac{\zeta_\nu}{\zeta_1} \quad (B_{mz1} = B_{m,\mathrm{air}}), \quad \text{where } \sigma_\pm = \sin \nu \frac{\pi}{2} = \pm 1;$$

$$\nu = +1, -5, +7, -11, +13, \ldots, -(6a-1), (6a+1); \; a = 1, 2, 3, \ldots$$

For steel of parameters $\mu_r = 500$–1000 and $\sigma = 7 \times 10^6$ S/m, $|\alpha^2| = \omega\mu\sigma = (140$ to $280) \times 10^4$ 1/m. This is why at a low number of field harmonics (e.g., $n \le 7$) and at a larger slip $s = (n_1 - n)/n_1$, one can accept that

$$\mathrm{Re}(s\alpha^2) \gg \left(\frac{\pi}{\tau}\right)^2 + (2n-1)^2 \frac{\pi^2}{L^2}, \text{ that is, } \alpha_n \approx \sqrt{s}\,\alpha \tag{6.55}$$

Then, the component $e^{-\alpha z}$ in formula (6.54) can be brought before the sum, which then appears to be nothing different than a Fourier distribution of the curve of field distribution on the surface of rotor $H_{mzs} = f(y)$. Remembering the maximum flux density in the gap $B_{m,\mathrm{air}} = \mu H_{mzs}$, we finally obtain

$$H_{mz} \approx \frac{1}{\mu} B_{m,\mathrm{air}} e^{j(s\omega t - (\pi/t)x)} e^{-\sqrt{s}\alpha z} \sum_{n=1}^{N} \cos(2n-1)\frac{\pi}{L} y$$

where N is the number of field harmonics along the machine length taken into account.

The amplitude of the radial component of field is

$$H_{mz} \approx \frac{1}{\mu} B_{m,\mathrm{air}}(y) e^{-j(\pi/\tau)x} \cdot e^{-\sqrt{s}\alpha z} \tag{6.56}$$

Analogical equations like Equation 6.56 can be obtained for all the field components $\boldsymbol{H}_m$, $\boldsymbol{E}_m$, and $\boldsymbol{J}_m$.

If the machine is sufficiently long ($L \gg \tau$) then in order to investigate the field near its length center, its length can be assumed as $L \to \infty$. In such a case, $E_x = E_z = 0$, $H_y = 0$, $\partial H_z/\partial y = 0$,

$$\cos(2n-1)\frac{\pi}{\mathrm{L}} y = 1,$$

and

$$\alpha_n = \sqrt{s\alpha^2 + \left(\frac{\pi}{\tau}\right)^2} = \sqrt{s}\alpha$$

At such assumptions, from formula (6.56) and the Maxwell's equation $\partial E_y/\partial x = -\partial H_z/\partial t$, it follows

$$E_{my} = \frac{J_{my}}{\sigma} = -js\omega\mu\int H_{ms}\mathrm{d}x + C = \frac{s\omega\tau}{\pi}B_{m,\mathrm{air}}\mathrm{e}^{-\sqrt{s\alpha}z}\cdot\mathrm{e}^{-\mathrm{j}(\pi/\tau)x} \tag{6.57}$$

where for $x = 0$, $E_y = 0$; and hence from it: $C = 0$. In the spot at $x = \tau/2$ and $z = 0$ appears the maximum value of current on the surface of rotor:

$$J_{my} = J_{ms}\mathrm{e}^{-\sqrt{s}kz}\mathrm{e}^{-\mathrm{j}\left(\sqrt{s}kz+\pi x/\tau\right)};\quad J_{ms} = \frac{s\omega\sigma\tau}{\pi}B_{m,\mathrm{air}} \tag{6.58}$$

Substituting Equation 6.57 to the Maxwell's equation $\partial E_y/\partial z = -\partial H_x/\partial t$, yields

$$H_{mx} = \frac{1}{\mathrm{j}s\omega\mu}\frac{\partial E_{my}}{\partial z} = \frac{\mathrm{j}\sqrt{s\alpha\tau}}{\pi\mu}B_{m,\mathrm{air}}\mathrm{e}^{-\sqrt{s\alpha}z}\cdot\mathrm{e}^{-\mathrm{j}\pi x/\tau} \tag{6.59}$$

Dividing Equation 6.59 by Equation 6.56, we can find the ratio of the tangential component H_{mx} and the normal component H_{mz} of the field in rotor

$$\left|\frac{B_{mx}}{B_{m,\mathrm{air}}}\right| = \left|\frac{H_{mx}}{H_{mz}}\right| = \sqrt{2s}k\tau/\pi;\quad \left(k = \sqrt{\frac{\omega\mu\sigma}{2}}\right) \tag{6.60}$$

which means that the tangential field on the rotor surface

$$|H_{mx}| = \frac{\tau}{\pi}\sqrt{\frac{s\omega\sigma}{\mu}}B_{m,\mathrm{air}} \tag{6.60a}$$

The active and reactive power entering into the rotor, per unit surface, according to formula (3.3), equals (in VA/m^2)

$$\begin{aligned} S_z(z=0) &= S_{pz} + jS_{qz} = -\frac{1}{2}(E_{my}\cdot H^*_{mx}) \\ &= \frac{(1+\mathrm{j})\mathrm{k}}{2\sqrt{s\omega\mu\sigma^2}}J^2_{ms} = \frac{(1+\mathrm{j})\tau^2}{2\sqrt{2}\pi^2}\sqrt{\frac{\sigma}{\mu}}s\omega\sqrt{s\omega}B^2_{m,\mathrm{air}} \end{aligned} \tag{6.61}$$

The total power losses in the rotor can be found by integrating $S_{pz}(z = 0)$, according to Poynting's theorem 3.1:

$$m_2 J_2'^2 R_2' = \left[\int\limits_{D}^{\pi D} \int\limits_{-L/2}^{L/2} S_{pz} \mathrm{d}x\mathrm{d}y \right]_{z=0} = a_p \frac{DLk}{2\sqrt{s}\omega\mu\sigma^2} J_{ms}^2 \tag{6.62}$$

The maximum flow (ampere-turns) of rotor, per one pole pair

$$\frac{2\sqrt{2}}{\pi} m_2 \frac{N_1\zeta_1}{p} I_2' \int\limits_0^{\infty} \int\limits_0^{\tau} J_{mx} \mathrm{d}z\mathrm{d}x = -\frac{(1+\mathrm{j})\tau}{\pi\sqrt{sk}} J_{ms} \tag{6.63}$$

By comparison of Equation 6.62 with Equation 6.63, we define the rotor current reduced to the stator winding

$$I_2' = \frac{\tau \mathrm{p}}{2\sqrt{s}km_1N_1\zeta_1} J_{ms} = \frac{\sqrt{s}\omega\sigma \mathrm{p}\tau^2}{2\pi k m_1 N_1 \zeta_1} B_{mp} \tag{6.64}$$

as well as the resistance and reactance of the rotor winding, reduced to the stator winding (omitting the *reactive power* from the main flux in rotor)

$$R_2' \approx a_p \frac{2m_1(N_1\zeta_1)^2}{p} \cdot \frac{\sqrt{s}kL}{\sigma\tau} k_{end}; \; X_{2s}' \approx \frac{a_q}{a_p} R_2' = sX_2' \tag{6.65}$$

where $a_p \approx 1.4$ and $a_q \approx 0.85$ are the linearization coefficients (7.26) of nonlinear permeability; $k_{\text{end}} \approx 1 + 2\tau/\pi L$ is the Gibbs's correction coefficient of endwinding considering the finite length of machine [1.16]; $k = \sqrt{\omega\mu\sigma/2}$; $\tau = \pi D/2p$ is the pole pitch, D is the internal diameter of stator, L is the length of the machine, and p is the number of pole pairs.

In the secondary circuit of equivalent circuit of induction motor (Figure 6.15), one should introduce (Turowski [1.18]) R_2'/s and X_2'. It means that these parameters strongly depend on the ratio $\sqrt{\mu/s}$ or correspondingly $\mu = \mu(I_2')$.

It can be also done using the approximate method by Postnikov [6.5], or by a broader field analysis [7.2], [6.3].

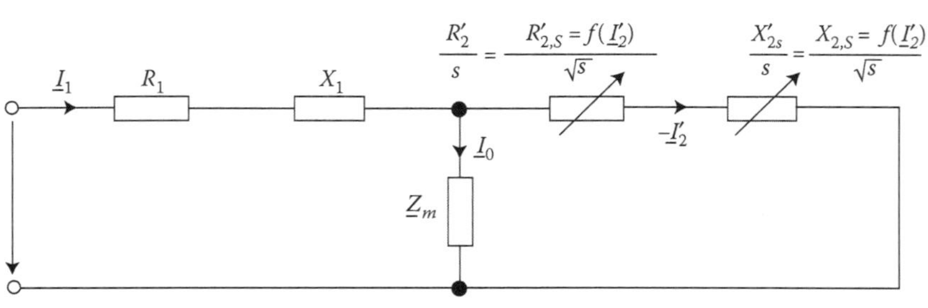

FIGURE 6.15 Equivalent circuit of induction motor with a solid steel rotor.

The magnetic voltage in the rotor is calculated as the integral $V_{m2} = \int_0^{\tau} H_{mx}\big|_{z=0} \mathrm{d}x$.

A method of computer calculation, similar to the one given above, is given in Section 4.6 and in Ref. [1.16]. The programs presented there comprise this type of motor as well as the induction motor with a cup rotor, discussed below.

6.7 CUP-TYPE ROTOR

The cup-type two-phase induction servomotors, so-called Ferraris motors (Figure 6.16), have immovable, laminated steel cores 1 and 3 and a rotor in the form of aluminum cup 2. The magnetic field in the gaps of these machines is varying during operation, from *oscillating* through *elliptic* to *circular* (Turowski [1.18], [6.8]).

Let us consider, as before, the field in the center of machine, in the expanded XZ plane. The stator winding is substituted by an infinitely thin current layer of surface density J_l directed along the Y axis. The slots are considered approximately by widening of the air gap δ_1 with the help of the Carter's coefficient (Turowski [1.16]). Let us assume first that the distribution of the linear load of stator J_i (A/m) along its circumference creates an immovable field oscillating in time. In symbolic calculus it is expressed as

$$\underline{J}_{ml} = J_{ml0} e^{j\omega t} \cos \nu x \qquad \left(\nu = \frac{\pi}{\tau} \right) \tag{6.66}$$

In accordance with dependence (2.60), the magnetic vector potential is also a periodic function in time and space:

$$\underline{A}_m = Z\, e^{j\omega t} \cos \nu x \tag{6.67}$$

where $Z = f(z)$ is the searched function, $A_m = A_{my}$, $A_x = A_z = 0$.

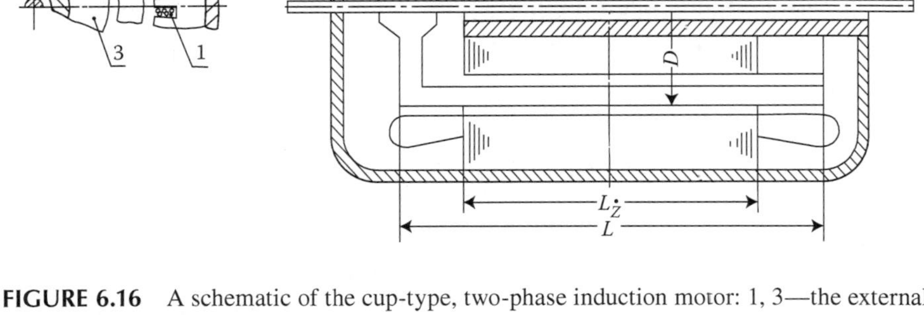

FIGURE 6.16 A schematic of the cup-type, two-phase induction motor: 1, 3—the external and internal laminated steel stators, 2—the empty aluminum (cup) rotor.

After substituting Equation 6.67 into Equation 2.58, we obtain, for the air gaps δ_1 and δ_2 (regions I and III in Figure 6.16)

$$\frac{\partial^2 Z}{\partial z^2} - \nu^2 Z = 0 \tag{6.68}$$

and after substituting Equation 6.67 into Equations 2.57a and 2.51, and by considering $J = \sigma E$, we obtain for the conducting material of rotor (region II)

$$\frac{\partial^2 Z}{\partial z^2} - \nu^2 Z = \alpha^2 Z \tag{6.69}$$

Solution of these equations leads to the following expressions of the complex magnetic vector potential for regions I, II, and III (Figure 6.16):

$$\begin{aligned} \underline{A}_{mI} &= (C_1\,\mathrm{e}^{\nu z} + C_2\,\mathrm{e}^{-\nu z})\,\mathrm{e}^{\mathrm{j}\omega t}\cos \nu x \\ \underline{A}_{mII} &= (C_3\mathrm{e}^{\alpha_1 z} + C_4\mathrm{e}^{-\alpha_1 z})\mathrm{e}^{\mathrm{j}\omega t}\cos \nu x \\ \underline{A}_{mIII} &= (C_5\,\mathrm{e}^{\nu z} + C_6\,\mathrm{e}^{-\nu z})\,\mathrm{e}^{\mathrm{j}\omega t}\cos \nu x \end{aligned} \tag{6.70}$$

where $\alpha_1 = \sqrt{\alpha^2 + \nu^2} = \varphi_1 + \mathrm{j}\psi_1$.

From Equation 2.50, $\boldsymbol{B} = \mathrm{curl}\,\boldsymbol{A}$, we find the components

$$B_{mx} = -\frac{\partial A_m}{\partial z} \quad \text{and} \quad B_{mz} = \frac{\partial A_m}{\partial x} \tag{6.71}$$

and then we determine the constants C_1, C_2, C_3, C_4, C_5, C_6 from the boundary conditions:

a. On the surface $z = 0$, according to formula (2.193a), the tangential component H_{mx} changes stepwise from 0 at $z = -0$ to $H_{mx} = J_{ml}$ (6.66) at $z = +0$.
b. On the surfaces $z = \delta_1$ and $\delta_2 = \delta_1 + d$ the boundary conditions (2.191) and (2.193) should be satisfied.
c. The tangential component of magnetic field intensity on the surface of internal stator (at $z = \delta_1 + d + \delta_2$) should be equal to zero.

The current density in the rotor wall is determined from formula (2.51), and the active power losses in the rotor—from the Poynting's vector (3.3), by subtracting the values for $z = \delta_1 + d$ from the value of this vector at $z = \delta_1$.

Since at the circular rotating field the maximum value of field appears consecutively in all points of rotor circumference, whereas at the immovable oscillating field it is only in two points (in the area of two poles), the *power losses in a rotor at rotating field are 2 times greater than at oscillating field* of the same amplitude of the tangential value of the flux density B_{mx0} on the stator surface.

This ratio, for the period $T = 2\pi/\omega$ on the double pole pitch 2τ ($\tau = \pi D/2p$), equals

$$\frac{P_{\text{rotat}}}{P_{\text{osc}}} = \frac{\frac{1}{T}\int_0^T \frac{1}{2\tau}\int_0^{2\tau} B_0^2 \cos^2\left(\omega t - \frac{\pi}{\tau}x\right) \mathrm{d}t\, \mathrm{d}x}{\frac{1}{T}\int_0^T \frac{1}{2\tau}\int_0^{2\tau} B_0^2 \cos^2 \omega t \sin^2 \frac{\pi}{\tau} x \mathrm{d}t\, \mathrm{d}x} = 2 \tag{6.72}$$

It means that at calculations of current density, power losses, and flux density of a machine with rotating field one can utilize the formulae obtained for an immovable wave (oscillating field), but with multiplying the power losses in rotor by 2 and the rms value of current and the flux density components B_{mx} and B_{mz}—by $\sqrt{2}$. From a detailed analysis of solutions, performed in the work of *Lopukhina* [Elektrichestvo 5/60] (in Russian), it follows that for the rotor wall thickness d ≤ 1 mm the normal component of flux density in the gap B_z practically does not change when passing through the rotor wall, and the skin effect in the wall is practically negligible, even at frequency 500 Hz.

The presented method is satisfactorily accurate for cases when the rotor is longer than the stator (as shown in Figure 6.16). In the case of the same length, three-dimensional field should be analyzed, considering the component J_x of current density in the rotor and variations of the flux density along the machine axis, like in Ref. [1.16].

In the case of an *elliptic rotating field*, one should similarly consider changes of the field amplitude along the circumference. In result of that, the coefficient (6.72) will be smaller than 2. Other methods of calculation of thin-walled rotors are given in Section 4.6 and in the works by Turowski [1.16] and Mendrela EA & EM and Turowski [4.10].

6.8 PRINCIPLES OF INDUCTION HEATING

As described in previous chapters, the effects of displacement and concentration of eddy currents, called also *skin effect* or *proximity effect*, are fundamentals of the metal induction heating.

If d is the thickness of the metal surface layer (Figure 4.9) heated to the temperature higher of Curie point (Figure 1.26), and δ_2 is the equivalent depth of penetration of electromagnetic field into solid metal (formula 2.181) with the parameters of heated metal (σ_2, μ_2), at induction heating we can distinguish three basic electromagnetic states (Figure 6.17):

I. Initial state ($d = 0$)—the metal is uniform and cold, with the parameters $\sigma_3 = c'\sigma_2$ and $\mu_3 = c''\mu_2$.
II. Transient state ($d < \delta_2$), when due to short duration of the process there exists only a thin layer ($0 \le z \le d$, Figure 6.17) of hot metal, and a deeper region ($z \ge d$) of cold metal. Such state occurs during the *surface hardening*.
III. Stationary state ($d > \delta_2$)—a uniform hot metal with the parameters σ_2, μ_2. This state applies to heating for melting, forging, through-hardening, and drawback treatment.

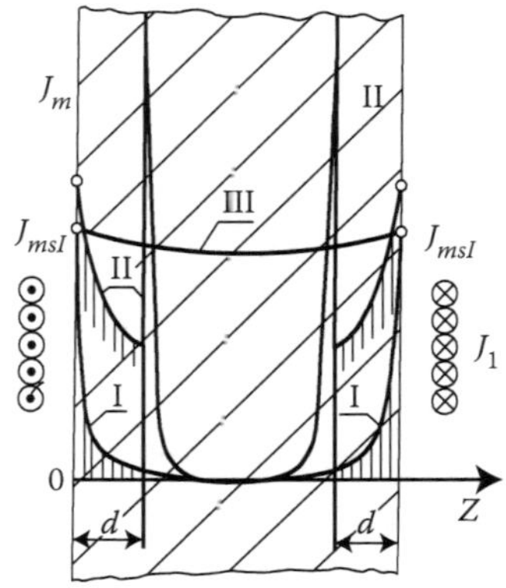

FIGURE 6.17 Distribution of the current density $J_m(z)$ at induction heating of a steel plate: I—initial, cold state, II—transient state with hot layer, III—hot, stationary state; J_1—the excitation alternating current (AC).

Distributions of eddy currents, field, and power loss in the initial and quasi-static state are represented by the same graphs as presented earlier (Figures 2.10 and 2.11) as well as by the same formulae (2.176) and (2.181)

$$H_m = H_{ms}\, \mathrm{e}^{-\alpha z},\ \mathrm{J}_m = \sigma E_m = \alpha\, H_{ms}\, \mathrm{e}^{-\alpha z} \tag{2.176}$$

$$\alpha = (1 + \mathrm{j})k,\ \delta = \sqrt{\frac{2}{\omega\mu\sigma}} = \frac{1}{k} \tag{2.181}$$

as well as formulae (3.9) and (3.10) for the power density (W/m^2), taking into account (3.10a)

$$S_p(z) = a_p\sqrt{\frac{\omega\mu_s}{2\sigma}}\frac{|H_{ms}|^2}{2}\mathrm{e}^{-2kz} \tag{3.9}$$

One should only substitute the corresponding constants σ_2, μ_2, or σ_3, μ_3. In the case of steel, the transition from the initial to the stationary state causes about a 90-fold increase of the equivalent depth-of-field penetration $\delta = \sqrt{2/(\omega\mu\sigma)}$ (2.181). This is why, when the thickness or diameter of the heated part is smaller than (2–3) δ_2, one should take into account the internal reflection of electromagnetic wave, using the methods and formulae given in Sections 4.3, 4.4, 4.9, and 4.10.

The volumetric distribution of the power density (in W/m^3) dissipated inside the heated body is described by the formula

$$P_{1\mathrm{V}} = \frac{1}{2\sigma}J_m^2 = \frac{\sigma}{2}E_m^2 \tag{6.73}$$

In the transient state, the surface layer of steel heated up to above the Curie point (about 800°C) can be in a first approach considered as a nonmagnetic metal screen (σ_2, μ_2) placed on the surface of the cold steel (σ_3, μ_3). The current density in the metal, $J_m = \sigma E_m$, can be then determined from the formula (4.18), $J_m(z) = \sigma E_m(z)$, where

$$J_m(z) = \alpha H_{ms1} \frac{e^{\alpha(d-z)} + M_2 e^{-\alpha(d-z)}}{e^{\alpha d} - M_1 M_2 e^{-\alpha d}} \tag{4.18}$$

$$M_2 = \frac{Z_3 - Z_2}{Z_3 + Z_2}; \quad M_1 = \frac{Z_1 - Z_2}{Z_1 + Z_2} \tag{4.13}$$

According to formulae (2.166) and (2.177), the wave impedances of particular media (Figure 4.9) are $|Z_1| \gg |Z_2|$ and $|Z_3|$, whereas

$$\frac{|Z_3|}{|Z_2|} = \sqrt{\frac{\mu_3\sigma_2}{\mu_2\sigma_3}} = \sqrt{\frac{c''}{c'}}$$

The wave reflection coefficients (4.13) take then the form $M_1 \approx -1$, $M_2 = \left(\sqrt{c'} - \sqrt{c''}\right)/\left(\sqrt{c'} + \sqrt{c''}\right)$.

According to Figure 1.28, the temperature coefficient of electric steel conductivity $c' = \sigma_3/\sigma_2 = 6\text{–}10$. Assuming next $c'' = \mu_3/\mu_2 \approx 1000$, we can accept $M_2 \approx -1$ and employ the simplified formulae (4.29).

Using formulae (2.176) and (4.29), we can find the relation of current densities on the boundary surfaces for particular thermal states (Figure 4.9):

At $z = 0$

$$\frac{J_{sII}}{J_{sI}} = \frac{\sigma_2 E_{II,z=0}}{\sigma_3 E_{I,z=0}} = \sqrt{\frac{\mu_2\sigma_2}{\mu_3\sigma_3}} = \frac{1}{\sqrt{c'c''}} \approx 0.01 \tag{6.74}$$

At $z = d$

$$\frac{J_{3III}}{J_{2III}} = \frac{\sigma_3 E_{III,z=d}}{\sigma_2 E_{III,z=d}} = c' = 6\text{–}10 \tag{6.75}$$

The ratio of the current density on the surfaces of cold metal J_{3III} and hot metal J_{sII} in the thermal transient state, according to Equation 4.18a, is

$$\frac{J_{3III}}{J_{sII}} = \frac{\sigma_3 E_{III,z=d}}{\sigma_2 E_{sIII,z=0}} = \frac{\sigma_3}{\sigma_2} = \frac{1}{|\mathrm{ch}\alpha_2 d|}$$
$$= \frac{c'}{\sqrt{\mathrm{ch}^2 k_2 d \cos^2 k_2 d + \mathrm{sh}^2 k_2 d sin^2 k_2 d}} \tag{6.76}$$

At $k_2d > 2$ one can accept $\text{sh}\,k_2d \approx \text{ch}\,k_2d \approx ½e^{k_2d}$, that is, $|\text{ch}\ \alpha_2 d| \approx ½e^{k_2d}$. Hence, $J_{3III} \geq J_{sII}$ when $e^{k_2d} \leq 2c'$, that is, $k_2d \leq \ln 2c' \approx 2.5–3.0$.

Although in reality the boundary between both layers is less evident than assumed here, nevertheless from formula (6.76) one can draw an important practical conclusion: until the thickness of the layer heated over Curie point is smaller than about two equivalent depths of field penetration $\delta_2 = \delta_{800°C}$ (more accurately, $d < \delta_2 \ln 2c'$), the highest density—hence the biggest amount—of the generated heat is concentrated not on the surface of the hardened product, but *on the boundary of heated and cool layer*. It helps to avoid overburning of the surface layer even at strong fields.

From analysis of formulae (2.181) and (2.188), it follows that at the beginning of heating the heat flux penetrates into the metal at the depth almost *100 times smaller than the electromagnetic wave*. At the same time, the heat flux penetration depth at which its amplitude decreases by 1/e (i.e., to 38% of its initial value) equals, in centimeters [6.18]

$$\text{For copper: } \delta_{T\,\text{Cu}} \approx 0.7\sqrt{t}$$

$$\text{For steel: } \delta_{T\,\text{Fe}} \approx (0.1\text{–}0.3)\sqrt{t} \tag{6.77}$$

where t is the time of the heat flux propagation, in seconds (bigger values apply to worse sorts of steel).

The phenomena described above is typical of the problems occurring in coupled electromagneticthermal fields; a good subject for analysis with the help of computer simulations. For example, these phenomena were investigated numerically by C. Emson and C. Riley (ISEF'87), but they did not carried out a full physical analysis of problem as the heating of steel tubes was analyzed by T. Skoczkowski and M. Kalus [6.13], [1.17].

6.9 HIGH-CURRENT LINES

In electrothermal engineering, in construction of large power transformers, high power switching stations, and so on, one of the fundamental problems consists of calculation of impedances of high power leads and their mutual influence and their impact on steel constructional elements.

6.9.1 Impedance

The resistance of high-current leads at AC currents is calculated with the formula

$$R_{ac} = k_n\, k_z\, k_w\, R_{dc} = k_{ad}\, R_{dc} \tag{6.78}$$

where R_{dc} is the resistance of the lead at direct current, k_n is the coefficient of skin effect, illustrated in Figures 6.18 through 6.20, in which f is the frequency (in Hz), R_{100} is the resistance (in Ω) of a conductor of length of 100 m, at direct current; k_z is the coefficient of *proximity* (Figures 6.21 through 6.23), and k_w is the coefficient of power losses in constructional elements coupled electromagnetically with the high-current lines, related to the so-called *contributed resistance* (Figures 6.24 and 6.25).

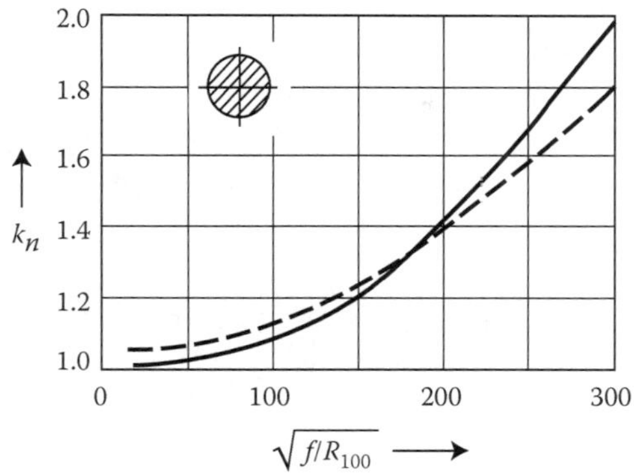

FIGURE 6.18 The coefficient of skin effect (k_n) of a cylindrical conductor: dashed line—M. Strzelecki [6.14] per Equation 6.80; continuous line—M. Smielanskiy [10.6] and T. Butkiewich. (Adapted from Strzelecki M.: Power losses and impedances of single-phase large-current paths. (in Polish). *XI SPETO*, Wisła, April 20–23, 1988, 245–253.)

The calculation of power losses and impedance of high-current lines has a long history and literature (Smielanskij [10.6]) where results are given only as graphs, without derivation of formulae. Computer methods allow for analysis of these problems in a more general way. Smielanskiy [6.14] reduced them to the solution of Fredholm's equations, expressing the current density in the form

$$J(x, y) = C - \frac{j\omega\sigma\mu_0}{2\pi} \iint_S J(x', y') \ln\frac{1}{r} dx' dy'$$

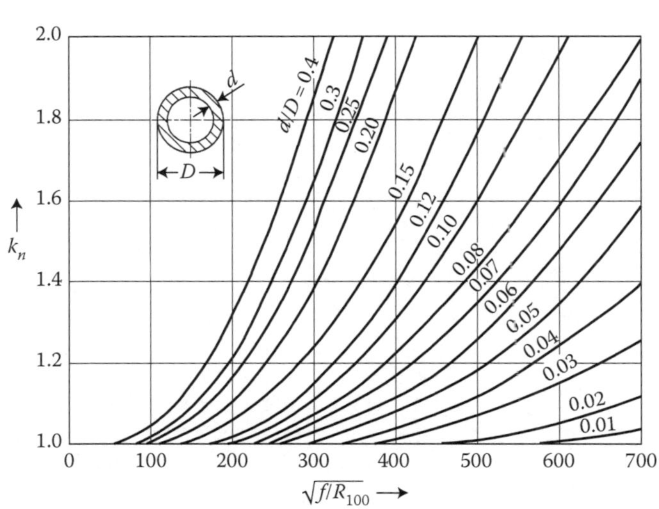

FIGURE 6.19 The coefficient of skin effect (k_n) of a tubular conductor. (Adapted from Smielanskiy M.J. and Bortnichuk N.I.: *Short Networks of Electric Furnaces*. (in Russian). Moscow: GEI, 1962.)

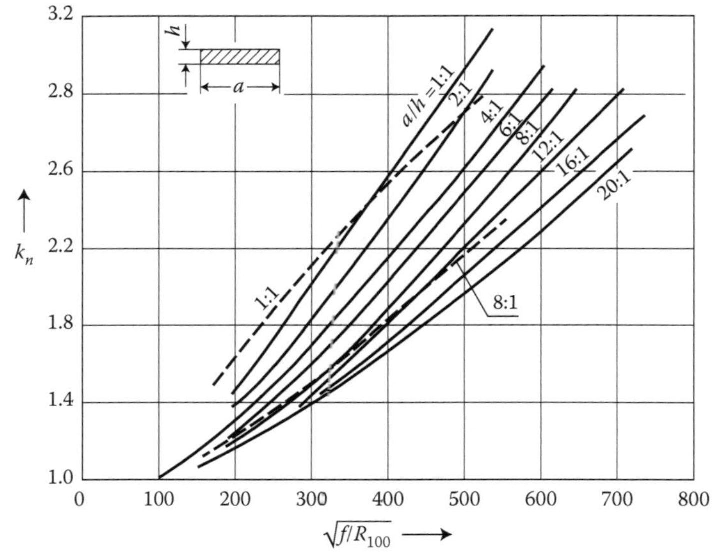

FIGURE 6.20 The coefficient of skin effect (k_n) of a rectangular conductor:——— Smielanskiy [10.6] - - - - Strzelecki [6.14], per Equation 6.80.

where the current of conductor is

$$I = \iint_S J(x', y')\, \mathrm{d}x'\mathrm{d}y' \tag{6.79}$$

and the coefficient of additional losses

$$k_{ad} = \frac{P_{AC}}{P_{DC}} = \frac{S}{I^2}\iint_S J(x,y)\cdot J^*(x,y)\,\mathrm{d}x\,\mathrm{d}y \approx \frac{S}{I^2}\sum_{i=1}^{N} |J \cdot J^*|\Delta S_i \tag{6.80}$$

The results of calculations in Ref. [5.14], for regions allowing for comparison, are in agreement with the results of other authors [10.6] (Figures 6.18 and 6.20).

6.9.2 Proximity Effect

In AC current-carrying conductors running nearby each other (Figure 6.21) there occurs a pushing out of the currents in the direction of higher density of the field H_{ms}, that is, toward the in-between gap at opposite currents, and outward at currents of the same direction. It is the so-called *proximity effect*.

According to the concept of P. Hammond [2.8], the power losses in two neighboring conductors (Figure 6.21) can be evaluated on the principle of superposition of antisymmetry of the own field H' (Figure 4.15) of power Equation 4.51, in W per m^2 of the width and length of the conductor

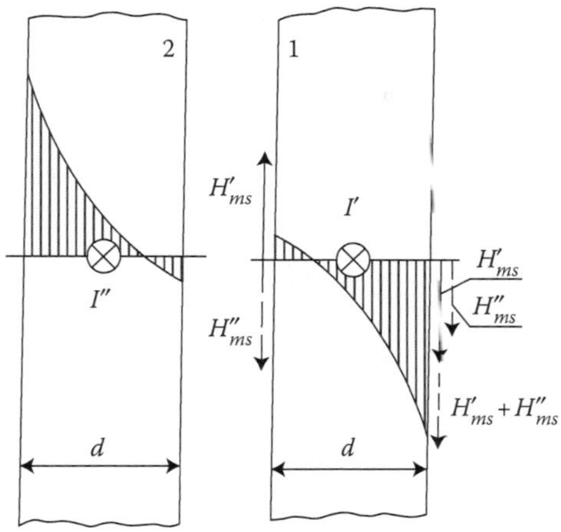

FIGURE 6.21 The proximity effect in nearby conductors with alternating currents $I' = I'' = I$.

$$P_1' = \eta \frac{k}{\sigma} |H'_{ms}|^2 \approx \frac{I_1^2 k}{2\sigma} \frac{\text{sh } kd + \sin kd}{\text{ch } kd - \cos kd} = \frac{I_1^2}{\sigma d} K_R \tag{6.81}$$

where $H_{ms} = I_1/\sqrt{2}$; I_1 is the current in A per m of width of the conductor and of the external field H'' of the neighboring bar (Figures 4.11 through 4.14) of power (4.47) in W/m²

$$P_1'' = \zeta \frac{k}{\sigma} |H''_{ms}|^2 \approx \frac{I_1^2 k}{2\sigma} \frac{\text{sh } kd - \sin kd}{\text{ch } kd + \cos kd} \qquad \left(H''_{ms} \approx I/\sqrt{2}\right) \tag{6.82}$$

The total power losses in the conductor 1 (Figure 6.21)

$$\begin{aligned} P_1 = P_1' + P_1'' &= \frac{I_1^2 \text{k}}{2\sigma} \left(\frac{\text{sh}\, kd + \sin kd}{\text{ch}\, kd - \cos kd} + \frac{\text{sh}\, kd - \sin kd}{\text{ch}\, kd + \cos kd} \right) \\ &= \frac{I_1^2 \text{k}}{2\sigma} \frac{\text{sh}\, 2kd + \sin 2kd}{\text{ch}\, 2kd - \cos 2kd} \end{aligned} \tag{6.83}$$

Considering the fundamental power losses at direct current $P_{1,0} = I_1^2(1/\sigma d \cdot 1)$, we get the *coefficient of resistance increase due to the proximity effect*

$$K_{R,\text{prox}} = \frac{R_{ac}}{R_{dc}} = kd \frac{\text{sh}\, 2kd + \sin 2kd}{\text{ch}\, 2kd - \cos 2kd} = K_R + K_{\text{prox}} \tag{6.84}$$

where

$$K_R = \frac{kd}{2} \frac{\text{sh}\, kd + \sin kd}{\text{ch}\, kd - \cos kd}; \quad K_{\text{prox}} = \frac{kd}{2} \frac{\text{sh}\, kd - \sin kd}{\text{ch}\, kd + \cos kd} \tag{6.85}$$

In literature, plots have been given for other arrangements (Figures 6.22 and 6.23).

The inductivity of two conductors carrying opposite currents and the *inductivity of three conductors* carrying three-phase currents is smaller the closer each other they are located. The inductivity of a circuit consisting of parallel conductors conducting identical currents can be reduced by increasing the distances between particular conductors. The self-inductance L of a single conductor of length l and of equivalent radius r equals to sum of the external inductance $L_{ext} = \frac{\mu_0 l}{2\pi}\left(\ln\frac{2l}{r} - 1\right)$ and the internal inductance $L_{int} = \frac{\mu l}{8\pi}$ (Neiman et al. [2.14], p. 131).

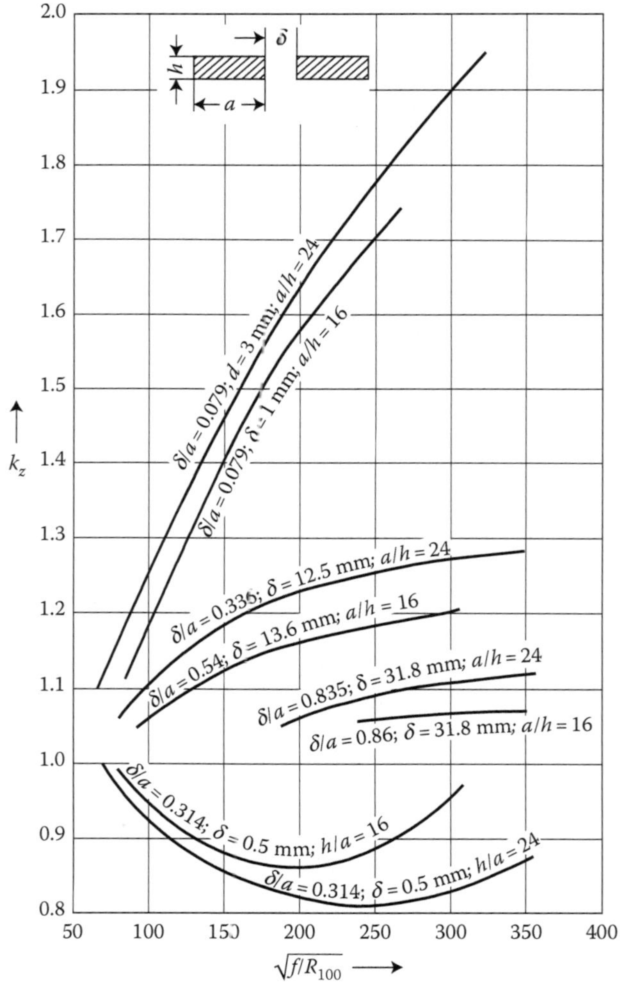

FIGURE 6.22 The proximity coefficient for rectangular bars. (Adapted from Smielanskiy M.J. and Bortnichuk N.I.: *Short Networks of Electric Furnaces.* (in Russian). Moscow: GEI, 1962 [10.6].)

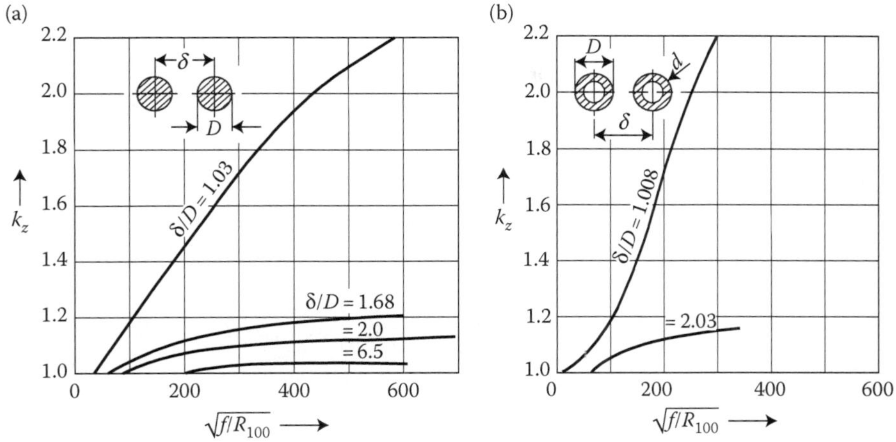

FIGURE 6.23 The proximity coefficients of bars: (a) cylindrical; (b) tubular at $d/D = 0.125$. (Adapted from Smielianskij M.J. and Bortniczuk N.I.: *Short Networks of Electric Furnaces.* (in Russian). Moscow: GEI, 1962.)

$$L = L_{ext} + L_{int} = \frac{\mu_0 l}{2\pi}\left(\ln\frac{2l}{r} - 1 + \frac{\mu_r}{4}\right) \tag{6.86}$$

where μ_r is the relative magnetic permeability of conductor material. In the case of a cylindrical conductor with the real radius R, we assume $r \approx 0.78\,R$, and in the case of a rectangular cross section, $r \approx 0.22\,(b + h)$ (Electrical Calendar SEP, 1959).

The mutual inductance M of two parallel conductors separated by the distance $D \ll l$

$$M = \pm\frac{\mu_0 l}{2\pi}\left(\ln\frac{2l}{D} - 1\right) \tag{6.87}$$

where " + " applies for identical directions of currents and " – " for opposite directions (Neiman et al. [2.14], p. 128).

At an asymmetrical, three-phase system of conductors, we have $D_{12} \neq D_{23} \neq D_{31}$. However, at a uniform transposition of the conductors, one can accept the average value

$$M = \frac{1}{3}(M_{12} + M_{23} + M_{31}) = \frac{l}{2\pi}\mu_0\left(\ln\frac{2l}{D'} - 1\right)$$

where $D' = \sqrt[3]{D_{12} \cdot D_{23} \cdot D_{31}}$.

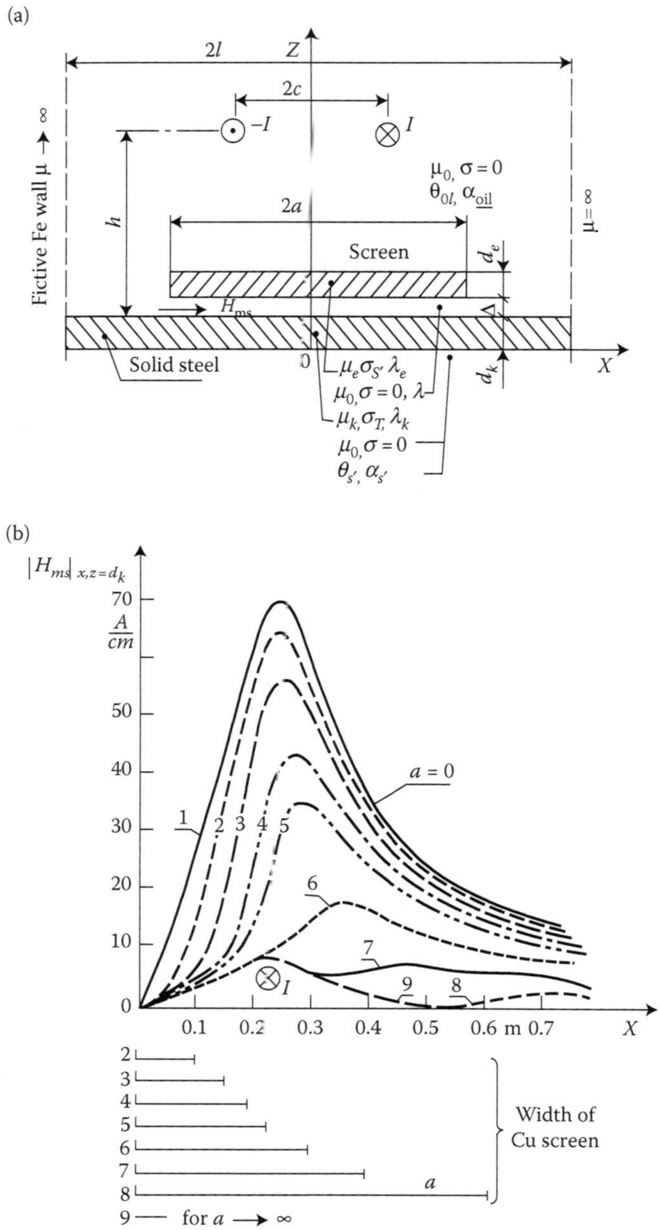

FIGURE 6.24 The calculation of power loss density $P_1 = \frac{a_p}{2}\sqrt{\frac{\omega\mu_s}{2\sigma}}\frac{|H_{ms}|^2}{2}$ (in W/m^2) in a steel wall screened partly from the field of a pair of conductors: (a) the calculation model; (b) the field on the steel surface, at $z = d_k$. (Adapted from Każmierski M. and Kersz I.: Effectiveness of longitudinal conducting screens. Part I. Electromagnetic field and stray losses. Part II. Temperature rise. (in Polish). *Rozprawy Elektrot.*, (2–3), 1989 [6.10].)

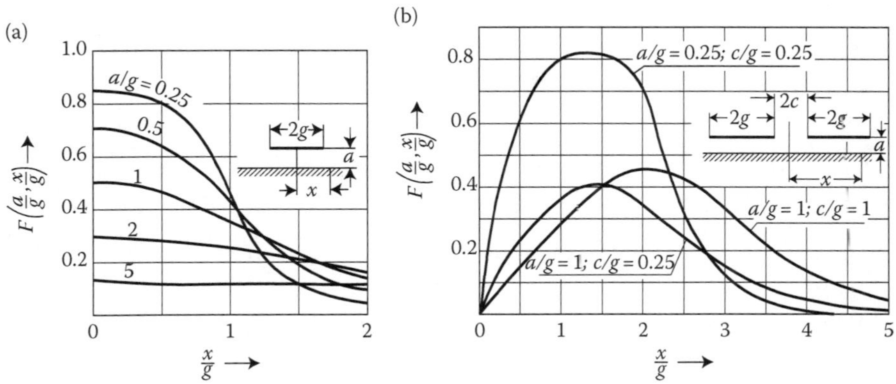

FIGURE 6.25 Distribution of the surface current density in an infinitely extended plate, excited by nearby strips with current: (a) single strip, (b) pair of strips. (Adapted from Lozinskiy M.G.: *Surface Hardening and Induction Heating of Steel.* (in Russian). Moscow: Mashgiz, 1949.)

6.9.3 Power Translocation

In the case of an asymmetric system of three-phase bars, a consequence of the inequality of mutual inductances of particular phase conductors ($M_{12} \neq M_{23} \neq M_{31}$) is the *power translocation* from one of the extreme bars (so-called "dead phase") to the second extreme bar of the flat R–S–T system. The self-inductance L and the resistance R of particular bars (jointly with receiver and supplying transformer) is usually the same. Therefore, for the voltages of particular bars we have the equations

$$\left.\begin{aligned} \underline{U}_1 &= (R + j\omega L)\underline{I}_1 + j\omega M_{12}\underline{I}_2 + j\omega M_{13}\underline{I}_3 \\ \underline{U}_2 &= (R + j\omega L)\underline{I}_2 + j\omega M_{23}\underline{I}_3 + j\omega M_{21}\underline{I}_1 \\ \underline{U}_3 &= (R + j\omega L)\underline{I}_3 + j\omega M_{31}\underline{I}_1 + j\omega M_{32}\underline{I}_2 \end{aligned}\right\} \tag{6.88}$$

If the currents in the bars create a symmetrical three-phase system, then

$$\underline{I}_2 = a^2\,\underline{I}_1,\ \underline{I}_3 = a\,\underline{I}_1, \quad \text{where } a = -\frac{1}{2} + j\frac{\sqrt{3}}{2} \text{ and } a^2 = -\frac{1}{2} - j\frac{\sqrt{3}}{2}$$

Considering that $M_{ik} = M_{ki}$ and $a^3 = 1$, we can represent the system of voltages in the following form:

$$\left.\begin{aligned} \underline{U}_1 &= [R + j\omega\,(L + a^2M_{12} + aM_{13})]\underline{I}_1 \\ \underline{U}_2 &= [R + j\omega\,(L + a^2M_{23} + aM_{12})]\underline{I}_2 \\ \underline{U}_3 &= [R + j\omega\,(L + a^2M_{31} + aM_{23})]\underline{I}_3 \end{aligned}\right\} \tag{6.89}$$

In the case of the symmetric distribution of conductors, or at transposition of the three-phase line conductors with equal distances, one can assume that $M_{12} = M_{23} = M_{31} = M$, and then the three voltages are equal.

In an asymmetrical system (Figure 4.33), instead, $M_{12} = M_{23} > M_{31}$. Introducing symbols:

$M_{12} = M_{23} = M$ and $M_{31} = M - 2N$, we can present Equation 6.89 in the form

$$\left.\begin{aligned} \underline{U}_1 &= [(R + \omega N\sqrt{3}) + j\omega (L - M + N)]\underline{I}_1 \\ \underline{U}_2 &= [R + j\omega (L - M + N)]\underline{I}_2 \\ \underline{U}_3 &= [(R - \omega N\sqrt{3}) + j\omega (L - M + N)]\underline{I}_3 \end{aligned}\right\} \tag{6.90}$$

From Equations 6.90 we can conclude that at symmetric currents the voltage drops on particular phase conductors are not equal. This is why, in the case of a symmetric load and its impedances comparable with the impedance of the supplying line, in order to maintain the symmetric current system it is necessary to create at the beginning of line the asymmetric phase voltage system, and vice versa—with the symmetric phase voltage system at the beginning of line, the current system will be asymmetric. From Equations 6.90 it follows that the resultant resistance of the first bar is bigger, whereas the resultant resistance of the third bar is smaller, by $\pm\omega N\sqrt{3}$, than the real resistance R. The active power related to this resistance is translocated from the third bar to the first one by the principle of a transformer coupling produced due to asymmetry of the busbar lines. The higher the power of installation, usually the smaller the symmetric impedance of network, and the bigger fraction of translocated power, the bigger asymmetry of the load of phases. This effect must be considered at the design stage of high-power lines. In the asymmetric system, the additional resistance of asymmetry is expressed by the formula (L.R. Neiman and P.A. Kalantarov, Vol. 3, 1948):

$$R' = \omega N\sqrt{3} = 1.2\frac{\mu_o}{4\pi} l\omega \tag{6.90a}$$

For instance, for a furnace of power of 7.5 MVA, current 40,000 A, frequency 50 Hz, and the supplying conductors of length 10 m, the power translocated from one extreme bar (*dead phase*) to another extreme bar, at the symmetry of currents, is

$$\begin{aligned} P' &= R'I^2 = 1.2\frac{\mu_O}{4\pi} l\omega I^2 = 1.2\frac{4\pi \times 10^{-7}}{4\pi} \times 10 \times 314 \times 40{,}000^2 \\ &= 605 \times 10^3 \text{ W} = 605 \text{ kW} \end{aligned}$$

6.9.4 Currents Induced in Steel Walls

Both at induction heating as well as at the design stage of busbar connections in high-current systems or as internal leads in power transformers and large synchronous turbogenerators, appears the need to determine eddy-current distribution induced in solid metal plates by high-current leads placed near these plates. The task is often accompanied by a corresponding electromagnetic screening or magnetic shunting. A good example here could be the solution of M. Kazmierski and I. Kersz [6.10] for a two-dimensional (2D) field of a pair of conductors (Figure 6.24a), described by the equations

$$\nabla^2 A_y = \begin{cases} 0 & \text{for dielectric} \\ \alpha_i^2 A_y & \text{for metal} \end{cases} \quad (\alpha_i = \sqrt{\mathrm{j}\omega\mu_i\sigma_i}) \tag{6.91}$$

having the solution (Figure 6.24b)

$$A_y = \begin{cases} \sum_{n=1}^{\infty} A_{in}\mathrm{e}^{-\nu_n(z-d_1)} \sin\nu_n x & (z > h) \\ \sum_{n=1}^{\infty} \left[A_n \mathrm{e}^{-\nu_n(z-d_1)} + W_n e^{\nu_n(z-d_1)}\right]\sin\nu_a x & (d_1 < z \le h) \end{cases} \tag{6.92}$$

where $d_1 = d_k + \Delta + d_e$, $\nu_n = (2n - 1/2l)\pi$.

From Equation 6.92, one can determine $H_{ms} = 1/\mu \text{ curl } A_y$ (Figure 6.24b), and next the density of power losses, in W/m²:

$$P_1 = a_p\sqrt{\frac{\omega\mu}{2\sigma}}\frac{|H_{ms}|^2}{2} \tag{3.10a}$$

as well as the local heating of the plate (Chapter 9).

M. Lozinskiy [6.18] presented, additionally, families of similar curves. In case of an infinitely long single bar, the ratio of current density in the plate to the current density in the bar can be, per Lozinskiy, determined from the formula

$$\frac{J_{x,\text{rms}}}{J_{b,\text{rms}}} = \frac{\delta_{\text{bar}}}{\delta_{\text{plate}}} F\left(\frac{a}{g}, \frac{x}{g}\right) \tag{6.93}$$

in which $J_{x,\text{rms}}$ is the density of rms current in the plate; $J_{b,\text{rms}}$ is the density of rms current in the bar, at the assumption that the thickness of the bar equals the equivalent depth of current penetration $\delta_b = \sqrt{2/\omega\mu_b\sigma_b}$, per (2.181); I_b is the current in the bar; and $F(a/g, x/g)$ is a function of the geometric proportions of the system (Figure 6.25a).

When in Figure 6.25a the width of strip ($2\ g$) is bigger than 8–10 times distance from plate (a), then almost all the current is concentrated in the part of the plate located under the strip. The linear density of eddy currents in the plate is then only slightly different from the linear density of the current in the strip. On the basis of Figure 6.25a, one can determine, from the principle of superposition, the distributions of the surface current density in a steel plate, induced by two strips conducting the same current in opposite directions—see Figures 6.25b and 6.26.

W. G. Deuring (Trans. AIEE, June 1957) published empirical formulae for power losses in a steel plate caused by a single bar parallel to the plate surface, conducting a current I (in A) of frequency 60 Hz. These formulae, after recalculation into the

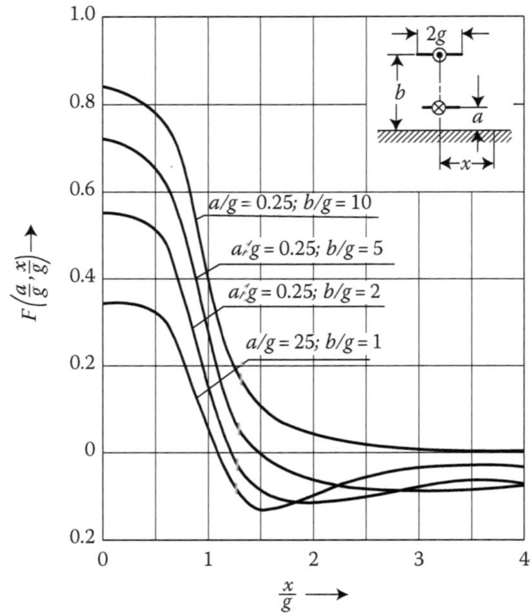

FIGURE 6.26 Distributions of the surface current density in a steel plate, induced by two nearby strips located one over each other and conducting currents in opposite directions. (Adapted from Lozinskiy M.G.: *Surface Hardening and Induction Heating of Steel.* (in Russian). Moscow: Mashgiz, 1949.)

metric international system of units "SI," provide the induced power losses in units of watts per meter (W/m) along the bar length, in the following form (where the current I is in A):

$$\left.\begin{array}{ll} P = 119 \times I^{1.9} \times 10^{-6}\ \text{W} & \text{at the bar distance from plate of 8.5 cm} \\ P = 111 \times I^{1.84} \times 10^{-6}\ \text{W} & \text{at the bar distance from plate of 17.4 cm} \\ P = 95.5 \times I^{1.8} \times 10^{-6}\ \text{W} & \text{at the bar distance from plate of 38.8 cm} \end{array}\right\} \quad (6.94)$$

In the case when the steel plate is screened by an aluminum sheet of a thickness 6.4 mm, the power losses at the frequency 60 Hz, in W/m, amount to

$$\left.\begin{array}{ll} P = 13.4 \times I^{2.03} \times 10^{-6}\ \text{W} & \text{at the bar distance from plate of 8.5 cm} \\ P = 11.6 \times I^{2.03} \times 10^{-6}\ \text{W} & \text{at the bar distance from plate of 9.8 cm} \\ P = 6.7 \times I^{1.96} \times 10^{-6}\ \text{W} & \text{at the bar distance from plate of 22.5 cm} \end{array}\right\} \quad (6.95)$$

See also Example under formula (7.76) in Chapter 7.

7 Electromagnetic Phenomena in Ferromagnetic Bodies

7.1 APPROXIMATION OF MAGNETIZATION CHARACTERISTICS

Due to complexity of the nonlinear electrodynamic processes in iron, various kinds of approximate and substitution methods have been broadly and effectively used in this area.

An *approximation*, as defined in this context, is an analytical representation of empirically determined magnetization characteristics of iron. Since an experimentally measured characteristic has always a limited number of points, its approximation is always connected with the problem of *interpolation*. Analytical approximation, however, contrary to *interpolation*, does not have to run through given points [7.3]. There exist many methods of interpolation and extrapolation [7.3], [7.4]. For example, one can use the popular Lagrange's interpolation formula. However, more convenient and more efficient is Newton's interpolation formula [7.3], [7.4]:

$$y(x) = y_0 + \Delta y_0(x - x_0) + \Delta^2 y_0(x - x_0)(x - x_1) + \cdots + \Delta^n y_0 (x - x_0)(x - x_1)\ldots(x - x_{n-1}) \tag{7.1}$$

in which $(x_0, y_0), (x_1, y_1), \ldots, (x_n, y_n)$ are the coordinates of given points, and Δy_0 are the differential coefficients determined correspondingly [7.3]. In this way, for example, for the curve of permeability of constructional steel (Figure 1.29, curve 4) an approximating parabolic polynomial of order n = 3 was obtained (J. Turowski [1.15/1], p. 251):

$$\mu_r(B) = 279(1.95 + 4.92B - 2.48B^2 - B^3) \tag{7.2}$$

On the basis of Weierstrass theorem, it can be proven [7.3] that with the help of a polynomial or trigonometric approximation one can obtain an arbitrarily small maximal error of approximation.

The approximating function of the magnetization characteristics should satisfy the following requirements:

- It should be as accurate approximation as possible.
- In cases when the function is used in operations containing differentiations, also its derivative $dy/dx = f(x)$ should be approximated as accurately as possible.
- The approximating function should not involve excessively complicated calculations and formulae.
- It should not contain too many constants.

TABLE 7.1
"Normal" (Standard) Magnetization Characteristic $E_u = f(i)$ and Its Derivative $dE_u/dt = f(i)$

i	0	0.5	1.0	1.5	2.0	2.5	3.0	3.5
E_u	0	0.58	1.0	1.2	1.3	1.4	1.45	1.5
$\frac{dE_u}{di}$	1.22	1.044	0.616	0.248	0.187	0.15	0.11	—

According to these indications, in practice one often resigns from arbitrarily accurate approximations by means of polynomials, in favor of simpler functions whose accuracy and form are dependent on specific needs and calculation capabilities. For example, in the range of flux densities from 0 to 1 T, for constructional steel one can utilize a better and more convenient approximation in the form of $\mu_r(B) = 1100 \cdot \sin(1.96B + 0.2)$ rather than the polynomial (7.2).

In the theory of electric machines, occasionally one employs the so-called "*normal*" (standard) *magnetization characteristic* (Table 7.1) which is kind of average from many real characteristics.

The standard magnetization characteristic is expressed in relative values, whereas the unit of emf (E_u) is typically based relative to the rated (nominal) voltage, and the unit of current (i)—to the magnetizing current corresponding to this value of emf. On the basis of this curve, one can evaluate and judge different approximation formulae (Table 7.2). These formulae can also be used for analytical approximation of the magnetization characteristic of steel, $B = f(H)$, after substituting E_u and i by B and H, respectively, as well as selecting new constants: a and b. According to Arkhangielskiy [7.5], the most accurate approximation, for both the magnetization curve and its derivative, is delivered by formula 9 from Table 7.2:

$$E_u = a \text{ arc tg } bi \tag{7.3}$$

while the minimum mean deviation $|\Sigma \Delta E_u / n|$ % for the magnetization curve is provided by formulae no. 6, 7, 8, and 12 from Table 7.2, whereas for the magnetization curve derivative—formulae no. 8, 10, and 11.

Formula 6 in Table 7.2 is, in fact, the so-called Froelich–Kenelly law (1.32)*

$$\frac{H}{B} = \frac{1}{\mu} = a' + b'H \quad \text{or} \quad \frac{1}{\mu - 1} = a + bH \tag{7.4}$$

while the first form concerns weaker fields (Table 7.3) and the second form—stronger fields,

$$b = \frac{d}{dH}\left(\frac{1}{\mu - 1}\right) = \frac{1}{B_{\text{sat}}},$$

where B_{sat} is the flux density of saturation.

* Adolf Froelich, Polish Army Officer 1920, an airplane researcher, the inventor. Arthur Kenelly American researcher and inventor 1881.

TABLE 7.2
Comparison of Approximating Equations of the Magnetization Characteristics of Steel with the "Normal" (Standard) Characteristic (Table 7.1)

No.	Equation	Constants a	Constants b	$\Delta E_{u,\,max}$(%)	*i* (Relative Units)
1	$E_u = a\sqrt[b]{i}$	0.914	2.14	13.7	3.5
2	$E_u = a\sqrt{i}$	0.884	—	16.0	3.5
3	$E_u = a\sqrt[3]{i}$	0.99	—	20.6	0.5
4	$E_u = a\sqrt{i} - bi$	0.985	0.068	10.3	3.5
5	$E_u = a\sqrt{i+b}$	0.82	0.088	12.1	3.5
6	$E_u = \dfrac{i}{a+bi}$	0.59	0.475	5.8	1.0 *i* 3.5
7	$E_u = a(1 - e^{-i})$	1.58	—	3.4	1.0
8	$E_u = a \text{ th } bi$	1.47	0.8	5.5	2.0
9	$E_u = a \text{ arc tg } bi$	1.12	1.2	2.6	1.0
10	$E_u = \dfrac{2}{\sqrt{\pi}}\int_0^{bi} e^{-t^2}dt$	1.44	0.7	7.4	2.0
11	$E_u = \dfrac{1}{\sqrt{2\pi}}\int_0^{bi} e^{-z^2/2}dz$	2.96	0.9	6.6	0.5
12	$E_u = \dfrac{1+a}{1+ai}i$	1.21	-	10.7	0.5

$\Delta E_{u,max}$—maximum deviation of curve calculated after given formula, from "normal" (standard) characteristic; *i*—current in relative values, at which these deviation appears.

TABLE 7.3
Values of the Exponents *n* in the Approximation Equation of Magnetization Characteristics in the Neiman's Formulae (7.9)

	Weak Fields		Strong Fields	
	B <	*n*	*B* >	*n*
Sort of Steel	T	–	T	–
Permalloy	0.35	0.32	0.70	20
Electrolytic iron	0.55	0.45	1.30	14
Transformer steel	0.90	0.55	1.20	10
Soft steel	0.60	0.55	1.30	9.5
Hard steel	0.90	0.68	1.15	7.5
Cast iron, annealed	0.32	0.65	0.60	4.5
Cast iron, casted	0.35	0.60	0.60	4.0

The form (7.4) enables a linear approximation of the nonlinear reluctivity $\nu = 1/\mu$ and the corresponding reluctance.

There exists an exponential variant (Govorkov [2.7]) of formula 7.4

$$B = \exp\left(\frac{H}{a + bH}\right) - 1 \tag{7.5}$$

The function 10 in Table 7.2, determined by the formula

$$\operatorname{erf}(x) = \Phi(x) = \frac{2}{\sqrt{\pi}} \int_0^x e^{-t^2} dt \tag{7.6}$$

and is called the integral of probability, Kramp's function, Gauss's integral of error distribution, or integral of error probability.

Despite its apparent complexity, the erf(x) function is convenient for calculations, because in the literature (e.g., [2.1], [2.11]) available are tables providing values of this function as well as of its derivative. Equation 11 in Table 7.2 is the Laplace's function.

Equations 1, 2, 3, 4, and 5 give $dE_u/di \to \infty$ at $i \to 0$, hence they cannot be used at small saturations. Equations 6, 7, and 12 do not have an inflexion point of the derivative $dE_u/di = f(i)$, as do the curves 8 through 11 and the "normal" curve, but $dE_u/di \neq \infty$ at $i = 0$.

Sometimes, it is more convenient to use inverse functions

$$H = \alpha \operatorname{sh} \beta B \tag{7.7}$$

$$i = E_u + e^{aE_u - b} \tag{7.8}$$

while in formula (7.8) for the "normal" magnetization characteristic the constants are $a = 5.12$ and $b = 6.73$ (V. Jenco, Electrichestvo 10/1951).

A convenient, simple, and for this reason often used, is the approximation proposed by Neiman [7.11] for alternating fields with amplitudes of the first harmonics H_m, B_m:

$$B_m = cH_m^{(1/n)} \quad \text{or} \quad |\mu| = cH_m^{((1-n)/n)} \tag{7.9}$$

where c is a coefficient fitted to specific magnetization curves, whereas the exponent n takes different values for the flux density B higher than the values given in Table 7.3.

Postnikov ([6.5], p. 585) adapted formula (7.9) to calculation of nonlinear impedances (6.65) of steel rotors of induction motors.

The *hysteresis loop* can be approximated (Bozorth [1.2]) with the help of formulae

$$\left.\begin{aligned} B &= B_{sat} \operatorname{erf}\frac{H \pm H_c}{H_c} \\ B &= B_{sat} \operatorname{th}\frac{H \pm H_c}{H_c} \end{aligned}\right\} \tag{7.10}$$

in which B_{sat} is the flux density of saturation; H_c is the coercive force; the minus sign concerns the growing branch of hysteresis loop and the plus sign applies to the decreasing branch of the loop. Approximation formulae of the demagnetization curve 1.32a are discussed in (Zycki [7.29]).

7.1.1 Approximation of Recalculated Characteristics

In technical calculations, often instead of using an approximation of the typical magnetization characteristic, $B = f(H)$, it is much more convenient to approximate its other form obtained from recalculation corresponding to the form appearing in the final formulae. For instance, in the formulae for power losses from tangential field Equations 3.10, 4.54, and so on, there occurs the value $\sqrt{\mu}\ H^2$. This value, for constructional steel, in the range of H from 0 to about 180 A/m, can be approximated by the function introduced by the author in 1958 [7.16]:

$$\sqrt{\mu}H^2 = c_1 H + c_2 H^2 \tag{7.11}$$

It makes the integration of total power losses significantly easier. By setting up, on the basis of corresponding magnetization curves (Figure 7.1), a system of two equations of the type

$$\Sigma\sqrt{\mu}H^2 = c_1 \Sigma H + c_2 \Sigma H^2$$

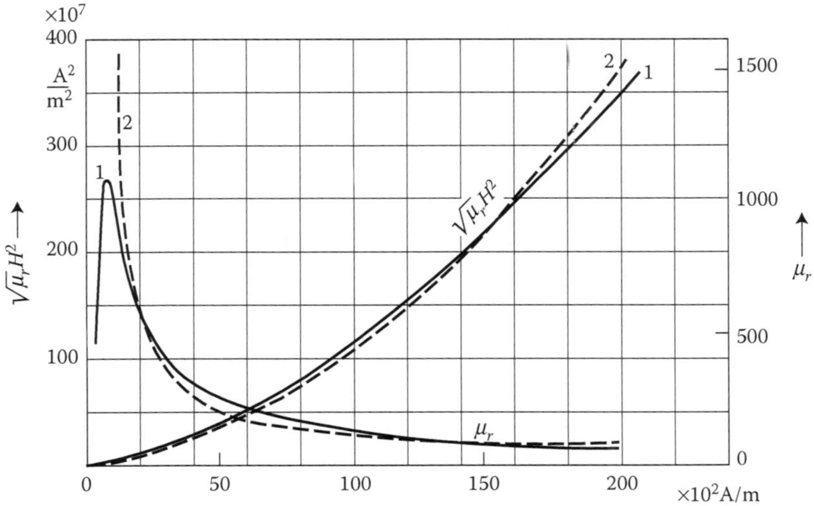

FIGURE 7.1 Recalculated magnetization curves of constructional steel: (1) analytical, (2) approximation: $\sqrt{\mu_r}H^2 = c_1 H + c_2 H^2$, $c_1 = 310 \times 10^2$ A/m; $c_2 = 7.9$. (Adapted from Turowski J.: Electromagnetic field and power losses in covers of transformers. PhD thesis. Technical University of Lodz, Poland. December 1957, pp. 1–142. (Published in parts in [1.15/1], [2.31], [4.16], [5.15], and [6.15]).)

in the work (Turowski [6.17]) there were found the constants $c_1 = 310 \times 10^2$ A/m and $c_2 = 7.9$.

In the formulae for the power loss from normal field (7.54), there occurs the dependence $1/\sqrt{\mu}$, which is a value difficult to assess. So, it is convenient to eliminate it with the help of the function $1/\sqrt{\mu} = f(\sqrt{\mu}H)$.

The real magnetization curve recalculated to that form (Figure 7.2) has a course that can be approximated by the straight line

$$\frac{1}{\sqrt{\mu}} = A_1 + A_2\sqrt{\mu}H \tag{7.12}$$

introduced by the author in 1962 [2.32], where for constructional steel

$$A_1 = 14\sqrt{\frac{\text{Am}}{\text{Vs}}} \quad \text{and} \quad A_2 = 0.13\frac{\text{m}^2}{\text{Vs}}$$

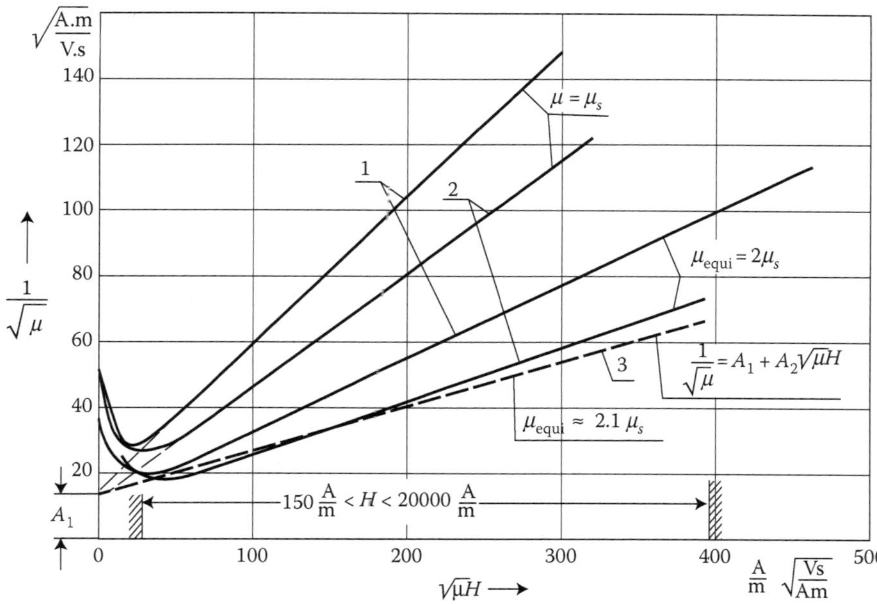

FIGURE 7.2 Approximations of recalculated magnetization curves of constructional steel [2.31]: 1—curve corresponding to the permeability determined for $H = H_{max}$; 2—curve corresponding to the permeability determined for $H = H_{rms}$; 3—approximating curves $\frac{1}{\sqrt{\mu}} = A_1 + A_2\sqrt{\mu}H$, where $A_1 = 14\sqrt{\frac{\text{Am}}{\text{Vs}}}$ and $A_2 = 0.13\ \frac{\text{m}^2}{\text{Vs}}$. μ_{equi}—equivalent permeability. (From Turowski J.: Losses and local overheating caused by stray fields. (in Polish). *Scientific Archives of the Technical University of Lodz "Elektryka,"* No. 11, 1963, pp. 89–179 (*Habilitation (DSc) dissertation; 1st Ministry Award)* [2.41].)

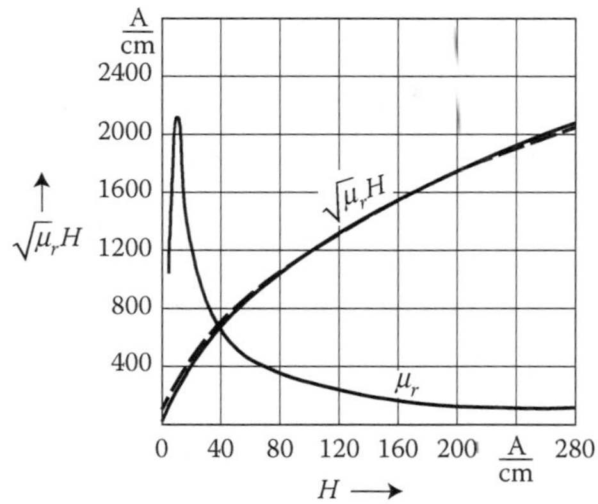

FIGURE 7.3 Approximation of the magnetization characteristic of constructional steel: — the real curve, - - - - - the analytical approximation $H = 1.05 \times 10^{-5}\left(\sqrt{\mu}H\right)^{1.77}$. (Adapted from Turowski J., Pawłowski J., and Pinkiewicz I.: The model test of stray losses in transformers. (in Polish) "*Elektryka*" *Scientific Papers of Techn. University of Lodz*, (12), 1963, 95–115; Kozlowski M. and Turowski J.: Stray losses and local overheating hazard in Transformers. *CIGRE. Plenary Session*, Paris 1972. Report 12-10 [5.16 and 7.9].)

A variant of formula (7.9), convenient for analytical investigations, was the approximation introduced in 1962 (Turowski et al. [5.16], [7.9]):

$$H = c\left(\sqrt{\mu_r}H\right)^n \tag{7.13}$$

where $c = 1.05 \times 10^{-5}\left(\frac{\text{A}}{\text{m}}\right)^{1-n}$, $n = 1.77$ (Figure 7.3).

7.2 METHODS OF CONSIDERING A VARIABLE MAGNETIC PERMEABILITY

The variability of magnetic permeability $\mu = \mu(H)$ into the depth of solid metal influences essentially the electromagnetic field as well as the active and reactive power loss distribution inside ferromagnetic metals. The task of the full numerical solution of this problem for plane wave inside a solid steel, starting from nonlinear Maxwell's equations, was formulated by the author in 1956 and 1962 [2.32] for electrical engineering, by Bozorth for physics [1.2], and by Govorkov for telecommunication ([2.7], pp. 339–343), and it was resolved only after advent of computers, in 1964 by K. Zakrzewski [7.26] and next developed by S. Wiak [7.25].

Steel linearization. The problem of iron linearization appeared when, due to the post-war's lack of copper (Rosenberg 1923 [7.13], Neiman 1949 [7.11]), it was necessary to use iron conductors. Due to numerical calculation difficulties, especially for 3D field problems, there are currently, simple, substitute methods in which the nonlinearity is considered as a kind of correction to the linear theory.

One of the best engineering approaches was born from the semiempirical ideas of Rosenberg [7.13] and Neiman [7.11, 7.12]—for the impedance of iron conductors—subsequently adopted and developed by J. Turowski [2.31, 6.17] as the linearization *coefficients* for electromagnetic field, the active (a_p) and reactive (a_q) losses, and the Poynting vector. After addition of Zakrzewski's numerical analysis [7.22, 7.27], a detailed list of corresponding coefficients was formulated (Table 7.4).

In the following sections, we present, in chronological order, the most important known linearization methods for one-dimensional sinusoidal electromagnetic field (i.e., a plane, polarized wave).

7.2.1 Rosenberg's Method for Steel Conductors (1923)

Assuming, that factual distribution of sinusoidal alternating magnetic field strength inside a solid iron half-space attenuates as the exponential function $H_m(z) = \mathrm{H}_{ms}e^{-\alpha z}$ (Figure 2.10) and considering that the decrease of $H_m(z)$ at strong fields is accompanied by the increase of the permeability $\mu = \mu(H)$ (Figure 1.29), Rosenberg [7.13] postulated that the flux density $B(z) = \mu(z)\ H(z)$ should have in metal a steep front (Figure 7.4a) and, hence, he substituted it by a rectangular distribution $B'(z)$ into the depth a. The active power-per-unit surface of the metal body, in W/m^2, deduced by the author [1.15/1] from the Rosenberg's conductor theory, was then

$$P_1 = S_{\mathrm{p}} = \int_0^a J_{\mathrm{rms}}^2 \frac{\mathrm{d}z}{\sigma} = \frac{2}{3}\frac{k_s}{\sigma} H_{ms}^2 = 1.34\sqrt{\frac{\omega\mu_{\mathrm{s}}}{2\sigma}}\frac{H_{ms}^2}{2} \tag{7.14}$$

which is almost the same as in formula (7.27) deduced from the Neiman's theory for steel conductors. The ratio of resistances is the same.

However, Rosenberg in 1923 [7.13] did not analyze the reactance of the iron conductors or the reactive power. But, as it was shown by Lasocinski in 1967 [7.10], strict application of the Rosenberg's assumptions led to results *not agreeing* with experiments, because he did not consider the phase shifts of the current densities $\boldsymbol{J}$.

7.2.2 Method of Rectangular Waves

This method, known for many years, shortly presented in [7.10] and (Turowski [1.15/1]), concerns strong fields and steels with a steep magnetization curve. The magnetic field intensity $H_m(z)$ in steel, at strong fields, decreases inwards a metal almost linearly, as one can see in Figures 2.10 and 7.4b. It means that the wave of flux density $B_{\mathrm{m}}(z)$ has a steep front, of the form similar to the magnetization curve $B_{\mathrm{m}}(H) \approx B_{\mathrm{m}}(z)$ (Figure 7.4a). If we substitute this characteristic with a rectangle (1 in Figure 7.4a), we shall obtain a rectangular wave of flux density, which penetrates the metal with a finite speed v. This penetration continues to the moment until the total flux $\Phi_1 = B_{\mathrm{sat}}\delta$ enters into the metal, that is, during 1/2 of period. It corresponds to the maximal depth of field penetration δ_{sat} (7.16). The layer of thickness δ_{sat} is immediately magnetized to the saturation B_{sat}, independently of the value of magnetic field

TABLE 7.4
Basic Formulae for a Solid Steel Half-Space Attacked by a Plane, Sinusoidal, Monochromatic Electromagnetic Wave

No.	Value	Metal of Magnetic Permeability $\mu = \text{const}$	Steel of $\mu = \mu_s(H_{ms}) = \text{var}$; $a_p \approx 1.5$; $a_q \approx 0.85$
1	Equivalent depth of penetration, m	$\delta = \sqrt{\frac{2}{\omega\mu\sigma}} = \frac{1}{k}$	$\delta = \frac{1}{a_p}\sqrt{\frac{2}{\omega\mu_s\sigma}} = \frac{1}{k}$
2	Length of electromagnetic wave in metal, m	$\lambda = 2\pi\delta = 2\pi\sqrt{\frac{2}{\omega\mu\sigma}}$	$\lambda = 2\pi\delta = \frac{2\pi}{a_p}\sqrt{\frac{2}{\omega\mu_s\sigma}}$
3	The magnetic flux per unit length, Wb/m, crosswise to the magnetic field intensity H lines, A/m	$\Phi_{m1} = (1-j)\sqrt{\frac{\mu}{2\omega\sigma}}H_{ms}$ $H_{ms} = (1+j)\sqrt{\frac{\omega\sigma}{2\mu}}\Phi_{m1}$ $\lvert\Phi_{m1}\rvert = \sqrt{\frac{\mu}{\omega\sigma}}\lvert H_{ms}\rvert$ $\lvert H_{ms}\rvert = \sqrt{\frac{\omega\sigma}{\mu}}\lvert\Phi_{m1}\rvert$	$\Phi_{m1} = a_p\frac{a_q - ja_p}{\sqrt{a_p^2 + a_q^2}}\sqrt{\frac{\mu_s}{\omega\sigma}}H_{ms}$ $H_{ms} = \frac{a_q - ja_p}{a_p\sqrt{a_p^2 + a_q^2}}\sqrt{\frac{\omega\sigma}{\mu_s}}\Phi_{m1}$ $\lvert\Phi_{m1}\rvert = a_p\sqrt{\frac{\mu_s}{\omega\sigma}}\lvert H_{ms}\rvert$ $\lvert H_{ms}\rvert = \frac{1}{a_p}\sqrt{\frac{\omega\sigma}{\mu_s}}\lvert\Phi_{m1}\rvert$
4	The active power loss density per unit surface, W/m²	$P_1 = \sqrt{\frac{\omega\mu}{2\sigma}}\frac{\lvert H_{ms}\rvert^2}{2} = \omega\sqrt{\frac{\omega\mu}{2\mu}}\frac{\lvert\Phi_{m1}\rvert^2}{2} = \frac{\omega}{2\sqrt{2}}\lvert H_{ms}\rvert\cdot\lvert\Phi_{m1}\rvert$	$P_1 = a_p\sqrt{\frac{\omega\mu_s}{2\sigma}}\frac{\lvert H_{ms}\rvert^2}{2} = \frac{\omega}{a_p}\sqrt{\frac{\omega\sigma}{2\mu_s}}\frac{\lvert\Phi_{m1}\rvert^2}{2} = \frac{\omega}{2\sqrt{2}}\lvert H_{ms}\rvert\cdot\lvert\Phi_{m1}\rvert$

continued

TABLE 7.4 (continued)

Basic Formulae for a Solid Steel Half-Space Attacked by a Plane, Sinusoidal, Monochromatic Electromagnetic Wave

No.	Value	Metal of Magnetic Permeability $\mu = \text{const}$	Steel of $\mu = \mu_s(H_{ms}) = \text{var}$; $a_p \approx 1.5$; $a_q \approx 0.85$
5	The consumption of reactive power per unit surface, var/m^2	$Q_1 = \sqrt{\frac{\omega\mu}{2\sigma}}\frac{\lvert H_{ms}\rvert^2}{2} = \omega\sqrt{\frac{\omega\sigma}{2\mu}}\frac{\lvert \Phi_{m1}\rvert^2}{2}$	$Q_1 = a_q\sqrt{\frac{\omega\mu_s}{2\sigma}}\frac{\lvert H_{ms}\rvert^2}{2} = \frac{a_q}{a_p}\omega\sqrt{\frac{\omega\sigma}{2\mu_s}}\frac{\lvert \Phi_{m1}\rvert^2}{2}$
6	Resistance and reactance per unit length along and crosswise the current $\frac{\Omega \cdot m}{m} = \Omega$	$R_1 = X_1 = \sqrt{\frac{\omega\mu}{2\sigma}}$ $\underline{Z}_1 = R_1 + jX$	$R_1 = a_p\sqrt{\frac{\omega\mu_s}{2\sigma}}$; $X_1 = \frac{a_q}{a_p}R_1 = a_q\sqrt{\frac{\omega\mu_s}{2\sigma}}$
7	Reluctance per unit $\frac{A \cdot m}{Wb \cdot m} = \frac{1}{H}$	$R_{m1z} = \frac{H_{ms}}{\Phi_{m1}} = R_{m1} + jR_{m1b}$ $R_{m1} = R_{m1b} = \sqrt{\frac{\omega\sigma}{2\mu}}$	$R_{m1} = \frac{a_q}{a_p\sqrt{a_p^2 + a_q^2}}\sqrt{\frac{\omega\sigma}{\mu_s}} = 0.37\sqrt{\frac{\omega\sigma}{\mu_s}}$ $R_{m1b} = \frac{1}{\sqrt{a_p^2 + a_q^2}}\sqrt{\frac{\omega\sigma}{\mu_s}} = 0.61\sqrt{\frac{\omega\sigma}{\mu_s}}$

Source: After Turowski J.: (1) *Calculations of Electromagnetic Components of Electric Machinery and Equipment*. (in Polish). Warsaw: WNT, 1982; and (in Russian). Moscow: Energoatomizdat, 1986; Neiman L.R: *Surface Effect in Ferromagnetic Bodies*. (in Russian). Leningrad-Moscow: Gosenergoizdat, 1949; Turowski J. and Zakrzewski K.: Problems of creation of strong magnetic fields. (in Russian). *World Electrotechnical Congress "WELC'77"*. Moscow, June 1977. Invited plenary paper, No. 1.59, pp. 1–15.

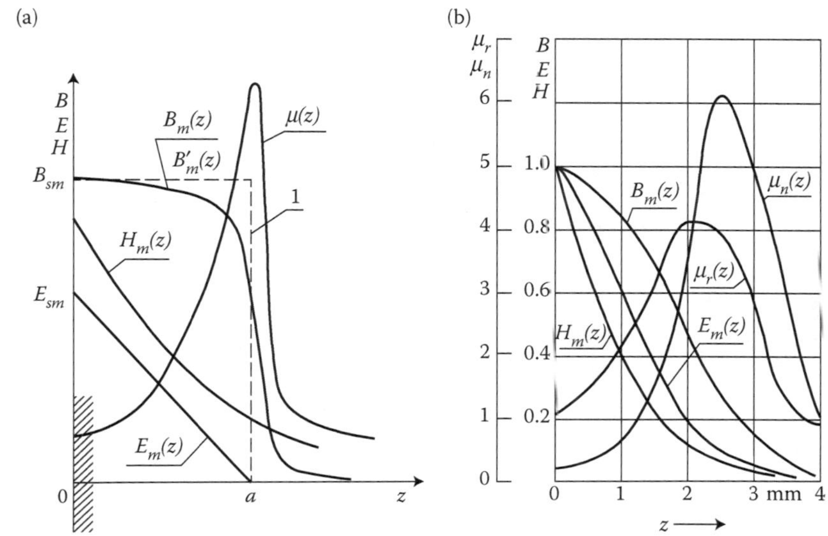

FIGURE 7.4 The field distribution inside a solid steel of variable permeability: (a) at the steep magnetization characteristic (Figure 7.5); (b) calculated numerically by Zakrzewski [7.26] for a constructional steel, after Figure 7.6, in relative units referred to the surface values: $H_{ms} = 4000$ A/m, $B_{ms} = 1.64$ T, $E_{ms} = 0.785$ V/m, $\mu_{rs} = 326$.

intensity on the surface. This situation continues until the beginning of the next half-period of the excitation field. It will generate a wave of opposite sign with amplitude $2B_{\text{sat}}$, which cancels the previous state of magnetization and magnetizes the steel in the opposite direction (Figure 7.5). The Maxwell's equations in this case can be substituted with the pair of equations:

$$\left.\begin{aligned} \frac{H}{\delta_t} &= \sigma E \\ E &= 2B_{\text{sat}} \frac{\mathrm{d}\delta_t}{\mathrm{d}i} \end{aligned}\right\} \quad 0 < \omega t < \pi \tag{7.15}$$

in which, for a periodic field on the surface $H_s = H_{ms} \sin \omega t$, we obtain the maximal depth of penetration

$$\delta_{\text{sat}} = \sqrt{\frac{2}{\omega\sigma B_{\text{sat}}/H_{ms}}} \tag{7.16}$$

The active Poynting vector on the surface (P.D. Agarwal, 1959), in W/m^2

$$S_{\text{p}} = \frac{8}{3\pi}\sqrt{\frac{3}{4}}\sqrt{\frac{\omega\mu_s}{2\sigma}}H_{ms}^2 = 1.47\sqrt{\frac{\omega\mu_s}{2\sigma}}\frac{H_{ms}^2}{2} \tag{7.17}$$

is almost the same as the one according the Neiman (7.27) [1.15/1], [7.10].

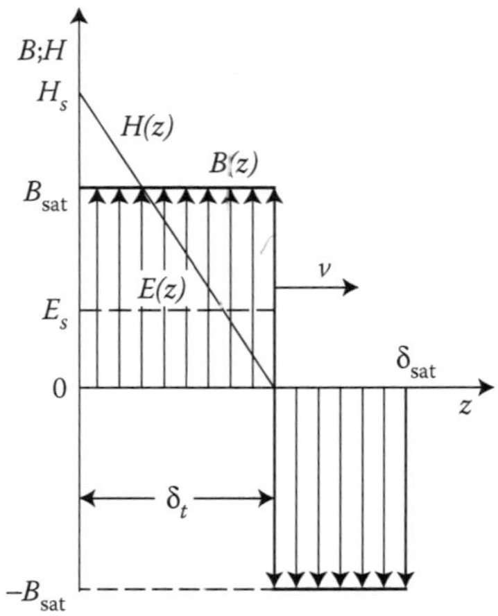

FIGURE 7.5 Distributions of the flux density $B_m(z)$, the magnetic field intensity $H_m(z)$, and the electric field $E_m(z)$ inwards a solid steel, at any time instant t, with the rectangular magnetization curve, and neglecting the hysteresis.

Also, the reactive component of the Poynting vector, in var/m^2

$$S_q = 0.74\sqrt{\frac{\omega\mu_s}{2\sigma}}\frac{H_{ms}^2}{2} \tag{7.18}$$

is approximately in agreement with that obtained from the Neiman's method.

Taking into account the hysteresis loop [7.10] causes a small reduction of the active power losses, which makes the expression (7.17) even closer to the Neiman's formula (7.27).

Agarwal obtained also a formula for the active power losses from eddy currents in an electrotechnical steel sheet, as per surface unit (in SI unit system):

$$P_{eddy} = \frac{8}{3\pi}\frac{H_{ms}^2}{\sigma\delta_{sat}}\left[1-\left(1-\frac{d^2}{\delta_{sat}^2}\right)^{3/2}\right] \tag{7.19}$$

with a much higher exponent for the thickness d than one can meet it in classical formulae of type (6.16). It was also confirmed by the author's investigation (compare curves 2 and 3 with curve 1 in Figure 6.5).

7.2.3 Neiman's Method (1949)

The method of Neiman [7.11], in spite of its limited accuracy, especially in the region of weak fields corresponding to a maximal permeability, has considerable advantages in comparison to other approximate methods, thanks to its simplicity, clear physical interpretation, and accuracy satisfactory for engineering calculations in regions of strong fields on the surface of investigated steel bodies.

In this method, the nonlinear dependence $B = f(H)$ for instantaneous $B(t)$ and $H(t)$, which can distort the forms of these curves, is omitted. However, the nonlinear dependence of the amplitudes $B_m = f(H_m)$ is taken into account as the basic magnetization characteristic. The relation between the instantaneous values $B(t)$ and $H(t)$ is considered by introduction of the complex magnetic permeability (1.49), where

$$\mu = \frac{B_m}{H_m} = \mu e^{-j\psi} \quad \text{and} \quad \sin\psi = \frac{\mu P_v}{\pi B_m^2 f} \tag{7.20}$$

where P_v is the hysteresis as per volumetric unit loss (W/m^3).

The relation between the amplitudes B_m and H_m is considered with the analytical approximation (7.9), that is, $B_m = cH_m^{1/n}$ for soft iron, and $B_m > 1.3$ T, $n = 9.5$ (Table 7.3), and c is a constant depending on materials [7.11]. The functional dependence $\mu = \mu(z)$ can be predicted and formulated (Figure 7.4) in such a way that after introducing it into the equations of plane wave

$$-\frac{dH_m}{dz} = \sigma E_m \quad \text{and} \quad \frac{dE_m}{dz} = -j\omega\mu H_m \tag{7.21}$$

one can obtain the solution $H_m = H_m(z)$, which together with $\mu = \mu(z)$ could give the dependence $B(z) = \mu(z)\, H(z)$, which satisfies Equation 7.21. This condition is fulfilled by the function

$$\mu(z) = \frac{\mu_s e^{-j\psi'}}{(1 - z/z_k)^2} \tag{7.22}$$

where μ_s is the permeability on the metal surface, determined from the basic magnetization characteristic (Figure 1.29) for the root mean square value $H_{s,rms} = (1/\sqrt{2})H_{ms}$ on the steel surface; z is the distance from the surface (Figure 7.4), z_k is the substitution depth at which the field practically disappears completely, and ψ' is the mean value of argument of the complex magnetic permeability (7.20), which corresponds to the hysteresis losses on the whole thickness of medium (usually $\psi = 0$–$10°$). Moreover

$$H_m = H_{ms}\left(1 - \frac{z}{z_k}\right)^{\beta} \tag{7.23}$$

$$E_m = E_{ms}\left(1 - \frac{z}{z_k}\right)^{\beta-1} \qquad J_m = \sigma E_m \tag{7.24}$$

$$E_{ms} = \frac{\beta}{\sigma z_k} \tag{7.25}$$

where $\beta = \beta + j\beta'' = f(n,\psi)$ and $z_k = 1/k_1\ \phi(n,\ \psi)$ are the complicated functions of metal magnetic properties, presented in papers [7.10], [7.11].

After investigation of many magnetic materials with different properties (Table 7.3), from cast iron to electrolytic iron, Neiman came to the conclusion that both the variability of magnetic permeability and the hysteresis loss, with the fields relevant to the heating of steel elements, can be considered for all ferromagnetic metals, with an accuracy satisfactory for technical needs, if one adopts the following constant values [7.11]:

$$\left.\begin{aligned} \mathrm{a_p} &= 1.3 \text{ to } 1.5 \approx 1.4 \\ \mathrm{a_q} &= \mathrm{a_p} tg\varphi = 0.8 \text{ to } 0.9 \approx 0.85 \end{aligned}\right\} \tag{7.26}$$

where $tg\varphi = 0.6$.

As a result, we obtain the simple formulae for a solid steel:

- For the *active* Poynting vector S_p on the steel surface, W/m²:

$$S_{\mathrm{p}} = P_1 \approx 1.4\sqrt{\frac{\omega\mu_s}{2\sigma}}\frac{|H_{ms}|^2}{2} \tag{7.27}$$

- For the *reactive* Poynting vector, var/m²:

$$S_{\mathrm{q}} = Q_1 \approx 0.85\sqrt{\frac{\omega\mu_s}{2\sigma}}\frac{|H_{ms}|^2}{2} \tag{7.28}$$

- For the surface *impedance* of a steel conductor:

$$Z_{St} \approx (1.4 + \mathrm{j}\,0.85)\frac{1}{\sigma\delta_s} \tag{7.29}$$

- The equivalent depth of penetration (2.181) into the steel

$$\delta_{St} = \frac{\delta}{\mathrm{a_p}}, \quad \delta = \sqrt{\frac{2}{\omega\mu_s\sigma}} = \frac{1}{k} \tag{7.30}$$

As can be seen, a strong alternating field in the steel with $\mu = \mu(H)$ extinguishes faster than at

$$\mu = \text{const}$$

The Neiman's method was created, in principle, for calculation of electric and magnetic impedances of steel conductors and busbars [7.11], [7.12]. Turowski [2.31] implemented this method in 1956 at the development of formulae for the Poynting vector and power of currents in solid steel parts. Formulae (7.27) through (7.30) were given, applied, and verified experimentally in the author's works [2.32], [2.33], [4.16], [4.24], [4.31], [4.29], [5.15], [5.16], [6.15], [7.15], [7.16], [9.10], and others. Later, they were also confirmed by the works of K. Zakrzewski in the years 1964–1975 [7.26], [7.27], and others, based on computer FDM analysis.

The formulae given above are sufficiently accurate in the case of fields whose value on the surface (H_{ms}) is bigger than the critical value H_{crit} corresponding to the maximum permeability (Table 7.3). At weaker fields, accuracy of these formulae is much lower, while in such cases it is recommended to apply rather $a_p \approx 1$ and $a_q \approx 0.6$.

The permeability μ_s occurring in the above equations, according to Neiman, should be determined from the primary magnetization curve at the rms value of magnetic field intensity $H_{s,rms}$ on the body surface. The author (J. Turowski, jointly with T. Janowski) proved experimentally [4.24] that for the active power (7.27) and for the apparent power a better agreement with measurements is obtained by taking μ_s for $H = H_{ms}$. Only for the reactive power (7.28) it is recommended to use $\mu_s(H_{s\,rms})$.

As per the author's investigations [2.41], these objections, due to small changes of permeability at strong fields (Figure 7.3), have the character of a small correction contained within natural variability of steel properties.

The Neiman's approach concerns *quasi-static sinusoidal processes* and does not consider a time deformation of the field in iron, which in most cases has a marginal significance. These disadvantages do not occur in the more modern numerical methods of Zakrzewski [7.26] and Wiak [7.25], but these methods are much more time and work consuming. The Neiman's approach, thanks to its simplicity, is largely popular for application in engineering jobs, because it allows one to apply classical differential equations at $\mu = \text{const}$ with simple corrections (Table 7.4) developed by J. Turowski and K. Zakrzewski.

7.2.4 Substitute Permeability

Comparing formulae (3.10) for metals with constant permeability

$$S = (1 + j)\sqrt{\frac{\omega\mu_s}{2\sigma}}\frac{|H_{ms}|^2}{2}; \quad \delta = \sqrt{\frac{2}{\omega\mu\sigma}}$$

with formulae (7.27) and (7.30) for steel, we can see that the power losses, the equivalent depth of penetration, and the steel impedance can be calculated from the classical formulae (3.10), (2.181), (2.185), and (2.185a), with a *constant substitute permeability* μ_{subst}, introduced by the author in 1958 [7.16] and equal

- For the active power, resistance, and the equivalent depth of penetration and flux in solid steel

$$\mu_{subst} = a_p^2\,\mu_s = (1.8 \text{ to } 2.1)\mu_s \tag{7.31}$$

- For the reactive power and reactance of solid steel

$$\mu_{subst,q} = a_q^2\,\mu_s = (0.66 \text{ to } 0.76)\mu_s \tag{7.32}$$

The permeability μ_{subst} is bigger than the surface value because it corresponds to some mean value of the magnetic field intensity inside the solid steel. With strong fields, in fact, a bigger permeability corresponds to a smaller magnetic field intensity. The permeability (7.31) is therefore some averaging value. However, in case of the reactive power (7.32), opposite conclusions occur. This is why the notion of the substitute permeability and the coefficients a_p and a_q should be used carefully and utilized only in verified formulae, such as collected in the Table 7.4, and one should also consider the way of field excitation (Section 7.3).

In the works by Kazmierski [7.7] and J. Gieras [2.6], the notion of substitute permeability was extended to screened ferromagnetics, which allowed to simplify calculations of such systems.

7.2.5 Computer Method

Starting from the general Maxwell's Equations 2.1 and 2.2, curl $\boldsymbol{H} = \sigma \boldsymbol{E}$ and curl $\boldsymbol{E} = -\mu_n\, \partial \boldsymbol{H}/\partial t$, and from the permeability curve, $\mu = f(H)$, determined experimentally for a given material (Figure 7.6), we can express the second Maxwell's equation with the use of the *nonlinear magnetic permeability* (1.48) as curl $\boldsymbol{E} = -\mu_n\, \partial \boldsymbol{H}/\partial t$, where $\mu_n = \partial B/\partial H = \mu(H) + H \cdot \partial \mu(H)/\partial H$, $H = H(z, t)$.

Penetration of a plane, polarized electromagnetic wave into a solid metal half-space (Figure 2.10) is described by the diffusion equation

$$\frac{\partial^2 H_y}{\partial z^2} = \sigma \mu_n \frac{\partial H_y}{\partial t} \tag{7.33}$$

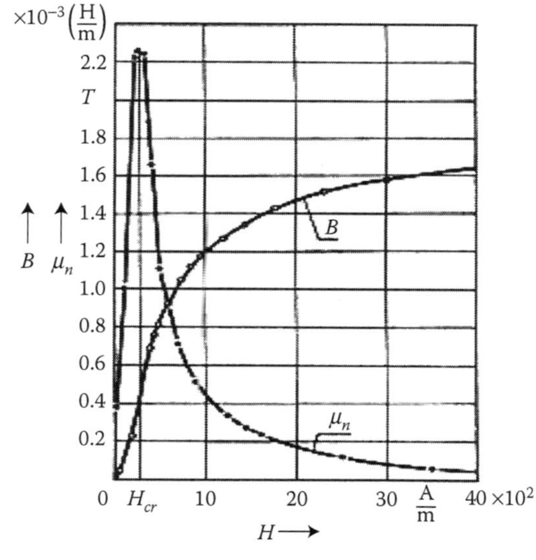

FIGURE 7.6 Magnetic characteristics of a steel sample (0.18 C, 0.52 Mn, 0.22 Si, 0.027 P, 0.032 S; $\sigma_{20°C} = 5.44 \times 10^6$ S/m). Vertex magnetization characteristic and characteristic of dynamic, magnetic permeability $\mu_d(H)$. (Adapted from Zakrzewski K.: Analysis of the electromagnetic field in solid iron by using numerical methods. (in Polish). *Archiwum Elektrot.*, XVIII(3), 1969, pp. 569–585 [7.26].)

The task and need of solving this equation was formulated by the author in 1956 [2.31], [2.32] as well as by Bozorth ([1.2], p. 621). However, only after implementation of computer technology *K. Zakrzewski** in 1964 took this task and resolved it [7.26], [7.27] transforming Equation 7.33 to the differential form, where $H_y = H$:

$$\frac{H_{i+1,k-1} - 2H_{i+1,k} + H_{i+1,k+1}}{(\Delta x)^2} = \sigma \mu_n (H_{i,k}) \frac{H_{i+1,k} - H_{i,k}}{\Delta t} \tag{7.34}$$

and using a discretization mesh method, where $\mu_n = \partial B / \partial H$.

The experimental magnetization curve $B = f(H)$ (Figure 7.6) was introduced directly into the computer program in the form of tables where necessary values of nonlinear permeability for each determined value of magnetic field intensity at the depth z were calculated, with consideration of the corresponding hysteresis loop.

In this way, the distribution of $H(z, t)$ in space and time inside steel was determined, at an optionally selected form of the excitation field on the surface of body, in the form

$$H_y(x = 0) = H_0 + \sum_{l=1}^{r} H_i \sin(\omega_l t + \psi_l)$$

On the basis of the calculated distribution $H(z, t)$, the remaining values of electromagnetic field, $B(z, t)$, $E(z, t)$, $\Phi(z, t)$ (Figure 7.4b), and power loss per surface unit of body, were calculated.

The described numerical method belongs to the most accurate physical and mathematical descriptions of the alternating field penetration into a solid steel, although it is still a simplified method. Results of the calculations were compared with the classical method of calculation (formulae (2.173) and (3.9)) for the permanent permeability $\mu = \mu_s = \text{const}$ and with the Neiman's method for cast steel and constructional steel.

Those investigations confirmed the study [7.11], [2.31] that considering the variability of the permeability causes

a. A faster attenuation of magnetic field intensity in comparison with that obtained with the classical method at $\mu = \mu_s = \text{const}$ at strong fields, and a smaller attenuation—at weaker fields.
b. The equivalent depth of penetration δ of an electromagnetic wave into solid steel is smaller at stronger fields and bigger at weaker fields, from the depth of penetration obtained by the classical formula (2.181) for constant permeability which exists on the surface.
c. Decreasing of the permeability toward the inside of metal at weak fields (lower than H_{critical} – peak of μ_n in Figure 7.6) as well as an initial increase and next a strong decrease at strong fields (Figure 7.4).
d. Small changes of flux density toward the inside of metal at strong fields, but its steep decrease at weak fields, starting from some depth (Figure 7.4).

* It was the PhD thesis under Janusz Turowski's supervision.

e. Deformation of time-dependent waveforms of the field value inside metal due to the medium nonlinearity.
f. The concepts of previous researchers relating to constant linearization coefficients for the active power and resistances ($a_p \approx 1.4$) as well as for the reactive power and reactance ($a_q \approx 0.85$) at stronger fields ($H_{ms} \geq 30$ A/cm) were confirmed in principle, with an appropriate extension (Zakrzewski [7.27]).
g. At weak fields, below μ_{max} (around 4–10 A/cm), $a_p \approx 1$ and $a_p \approx 0.5$ (J. Turowski [2.31], [6.17]). The computer program designed by K. Zakrzewski [7.27], after comparison with classical formulae at μ = const, allowed for accurate calculation of the linearization coefficients of particular parameters of solid steel (Figure 7.7), provided that its accurate magnetization characteristics of hysteresis loop vertex $B_{hist,vrtx} = f(H)$ and the steel conductivity σ_{steel} were well known. Accurate measuring of such parameters is a burdensome job. A method of calculating the corresponding coefficients a_{pi}, a_{qi} (i = w, H, Φ) was proposed in [7.27], with consideration

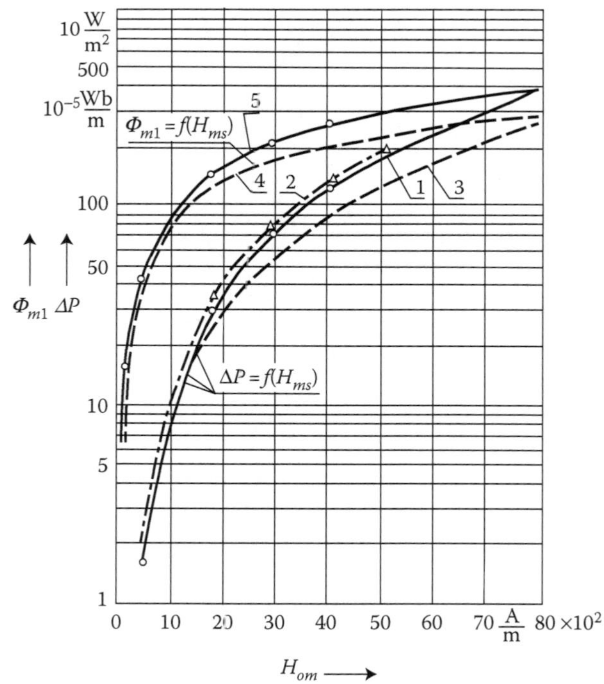

FIGURE 7.7 Characteristics of the power loss $\Delta P = f(H_{ms})$ and of the value of maximum flux $\Phi_{m1} = f(H_{ms})$ versus the amplitude H_{ms} of sinusoidal magnetic field intensity on the steel surface. Magnetization characteristic like in Figure 7.6: 1—the losses calculated numerically; 2—the losses measured on sample; 3—the losses calculated at $\mu_s(H_{ms})$ = const; 4—the magnetic flux calculated at $\mu_s(H_{ms})$ = const; 5—the flux calculated numerically. (Adapted from Zakrzewski K.: Analysis of the electromagnetic field in solid iron by using numerical methods. (in Polish). *Archiwum Elektrot.*, XVIII(3), 1969, 569–585 [7.26].)

of the magnetic field intensity H_{ms} or the flux Φ constraints. It is another inconvenience, caused by the high accuracy needed only for checking of the numerical program and coefficients.

Since, however, variation of the magnetization characteristics and σ for different constructional steels is high (Figure 1.27), and properties of these materials are normally not investigated, the authors of the work (Turowski, Zakrzewski [7.22]) decided to return to the primary Turowski's practical coefficients a_p and a_q from 1957 [2.31], [6.17], as proposed for iron wires by Rosenberg [7.13] in 1923 and Neiman (7.26) [7.11] in 1950. The practical novelty of the concept from [7.22] was the application of a_p and a_q for various nonlinear parameters (Table 7.4), according to gathered experience from joint researches, beginning from 1956 [2.31] including other collaborators.*

EXAMPLE

With the help of a coil of 10 turns, applied to the tank of a three-phase, model transformer, the biggest axial per unit flux in the tank was measured: $\Phi_{m1} = 14.9 \times 10^{-5}$ Wb/m. Assuming that the tank is built from the normal steel of common applicability St 4 s, with $\sigma_{20°C} = 6.2 \times 10^6$ S/m and assuming $\mu_{rs} = 700$, from formula 3 in Table 7.4, or Equation 3.11a, we calculate

$$|H_{ms}| = \frac{1}{a_p}\sqrt{\frac{\omega\sigma}{\mu_s}}|\Phi_{m1}| = \frac{1}{1.4}\sqrt{\frac{2\pi 50 \times 6.2 \times 10^6}{700 \times 0.4\pi \times 10^{-6}}} \times 14.9 \times 10^{-5} = 158.4 \text{ A/m}$$

Since it is less than $H_{ms,crit} = 450$ A/m (cf. Figure 1.29, curve 3) we adopt $a_p = 1$ and $\mu_{rs}(200 \text{ A/m}) = 620$, wherefrom, as the *first iteration*, we have $|H_{ms}| = 158.4 \times 1.4\sqrt{\frac{700}{620}} = 236$ A/m.

From calculations with the 3D program RNM-3D (Turowski et al. [7.21]), it was obtained for the same transformer $|H_{ms}|$ = <u>245 A/m</u> (difference of 4%). It is an excellent agreement and one should not expect a higher accuracy, due to variations of physical parameters of steel and limited accuracy of measurements. In large transformers, the field H_{ms} is much stronger and μ depends less on steel admixtures.

7.3 DEPENDENCE OF STRAY LOSSES IN SOLID STEEL PARTS OF TRANSFORMERS ON CURRENT AND TEMPERATURE

Equations 7.13, 3.10a, and 7.74a confirm the author's very important finding (J. Turowski [6.17], [10.17]) for load loss measurements and overloading hazards (Figure 9.2) and that in steel elements of transformers, the stray losses (7.74), due to the iron nonlinearity, can rise locally faster than the square of current I^2 (Figure 10.4). However, a full solution of Maxwell's equations with the nonlinear magnetic permeability $\mu(H)$ and nonsinusoidal excitation is too complicated to be used in

* T. Janowski [4.24], J. Pawłowski, I. Pinkiewicz [5.16], J. Lasocinski, A. Przytuła, M. Kazmierski, A. Ketner [10.30] J. Turowski, M. Kopeć, M. Turowski [7.21], and others, mainly at that time author's talented students.

regular engineering computations. Luckily, the author has shown [6.17], [10.17] that the solutions can be linearized and carried out, satisfactorily for engineering applications, at a constant surface value of $\mu_s(H_{ms}) = \text{const}$. Careful theoretical and experimental analysis shows (7.26) that at stronger fields ($H_{ms} > 5$ A/cm) changes of $\mu(H,z)$ along the z axis, inwards into the metal (Figure 2.10), can be taken into account by means of the linearization coefficients $a_p \approx 1.4$ for the active power and $a_q \approx 0.85$ for the reactive power. The magnetic nonlinearity $\mu(H, x, y) = var$ along the steel surface (X, Y) may be considered with the help of the author's specific analytical approximations (Figures 7.1 through 7.3; Equations 7.11 through 7.13):

$$\sqrt{\mu_s}H^2 \approx c_1 H + c_2 H^2 \approx cH^b, \qquad \frac{1}{\sqrt{\mu}} = A_1 + A_2\sqrt{\mu}H, \qquad \left(\sqrt{\mu_r}H\right)^n \approx H$$

Using this approach, we can settle that the per-unit power loss (in W/m²) in a solid steel half-space (Table 7.4, item 4) can be, according to Equations 3.10a and 3.11a, expressed by the formula

$$P_1 = \frac{a_p}{2}\sqrt{\frac{\omega\mu_0}{2\sigma}}\sqrt{\mu_s}\left|H_{ms}\right|^2 = \frac{\omega}{2\sqrt{2}}\left|H_{ms}\right|\cdot\left|\Phi_{m1}\right| \tag{7.35}$$

After applying the analytical approximation of the recalculated magnetization curve, according to Figure 7.3, in the form

$$H_{ms} = C\left(\sqrt{\mu_r}\left|H_{ms}\right|\right)^n = C\left(\sqrt{\frac{\omega\sigma}{2\mu_0}}\left|\Phi_{m1}\right|\right)^n \tag{7.36}$$

where $C = 1.05.10^{-5}\left(\frac{\text{A}}{\text{m}}\right)^{1-\text{n}}$ and $n = 1.77$,

then, after substituting Equation 7.36 into Equation 7.35, we get

$$P_1 = \frac{a_p}{2C^{\frac{1}{n}}}\sqrt{\frac{\omega\mu_0}{2\sigma}}\left|H_{ms}\right|^{1+\frac{1}{n}} = \frac{a_p}{2C^{0.62}}\sqrt{\frac{\omega\mu_0}{2\sigma}}\left|H_{ms}\right|^{1.565} \tag{7.37}$$

or for the magnetic flux:

$$P_1 = \frac{C}{2\sqrt{2}}\frac{\omega^{1+\frac{n}{2}}\sigma^{n/2}}{a_p^n\,\mu_0^{n/2}} = \frac{c}{2^{1.5}a_p^{0.89}\mu_0^{0.89}}\omega^{1.89}\sigma^{0.89}\left|\Phi_{m1}\right|^{2.77} \tag{7.38}$$

An important practical problem at the investigation of large electrical machines and transformers is identification of the dependence of additional losses in constructional elements made of solid steel on the excitation current I, frequency f, and temperature (conductivity σ).

Formulae (7.37) and (7.38) give a principal answer to these questions. It means that

1. When *the magnetic field intensity* H_{ms} on the steel surface is *proportional to the excitation current* ($H_{ms} = cI$), as it is, for instance, on the cover of a transformer tank (5.15), the power losses will follow the proportionality

$$P_1 \sim I^{1.6} \cdot f^{0.5} \cdot \sigma^{-0.5} \tag{7.39}$$

2. In a case when *the magnetic flux* penetrating the steel *is proportional to the current I* ($\Phi_{m1} = cI$), the losses will follow the proportionality

$$P_1 \sim I^{2.8} \cdot f^{1.9} \cdot \sigma^{0.9} \tag{7.40}$$

Such dependence is manifested approximately by the stray flux in cylindrical windings of transformers, because the main magnetic resistance on the pass of this flux is the narrow interwinding gap δ (Figure 4.18). Experiments carried out by the author showed that the fraction of the leakage flux penetrating the tank wall shows the dependence $\Phi_{m1} = c\,I^{\beta_1}$, where $\beta_1 = 0.9–0.8$ [5.16], [7.9]. Due to that, one can conclude that in this case occurs the proportionality

$$P_1 = kI^{\beta}, \quad \text{where } \beta = 2.77\beta_1 = 2.5 \text{ to } 2.2 \tag{7.41}$$

that is, the *losses in a solid steel wall can be locally increasing with the current faster than proportionally to* I^2 (Figures 7.22 [later in the chapter] and 10.4). However, since the losses (7.41) are only a part of the total load loss ΔP_{load} in transformers, the big value of the exponent β cannot show a remarkable discrepancy of the ΔP_{load} from the quadrature dependence, especially when measurements are carried out at currents smaller than the rated (nominal) one. This phenomenon is additionally obscured by the part of losses showing the dependence (7.39). However, one should take into account that *during overloading, local losses may grow faster than proportionally to* I^2. It can lead to an additional overheating hazard.

After the author's contribution at the CIGRE'64 Plenary Session [10.17], the International Study Committee No. 12 (Transformers) immediately increased the recommended short-circuit testing current from 25% to 50% of the rated current [10.27]. After further author's warnings (CIGRE'73 Norway and CIGRE'81 USA), in spite of industry objections, this current was increased even to 100% or more, due to fears of the possibility of thermal hazards at transformer overloading.

After such accidents as the "biggest power outage in history"—the Great North-East Blackout of 1965 in New York and Canada*—the reliability of transformers

* IEEE PES, January 1991.

has become one of the most important issues. The blackouts happened again in New York (1977), France (1978), U.S. Pacific Region (1996*) and later again in Canada, New York† (13.08.2003), France, and Northern Italy (27.09.2003).

7.4 POWER LOSSES IN STEEL COVERS OF TRANSFORMERS

In the case of a steel cover (Figure 5.21) with variable magnetic permeability and of significant thickness (8–10 mm), losses in the cover should be calculated [6.17, 7.16] with the help of formula (4.54). Since the integration is carried out on one side of the cover, this formula should be doubled and one should introduce the linearization coefficient a_p of 1.3–1.5 ≈ 1.4, which, according to Section 7.2, takes into account the variable permeability inside steel (J. Turowski [7.16])

$$P \approx a_p \sqrt{\frac{\omega\mu_0}{2\sigma}} \iint_A \sqrt{\mu_r}\, |H_{ms}|^2\, dA \tag{7.42}$$

In order to solve this integral, it is necessary to substitute the integrand expression by an analytical function of the magnetic field intensity H, as simple as possible. Using the approximation (7.11), we can split the double integral in (7.42) into two integrals

$$\iint_A \sqrt{\mu_r}\, |H_{ms}|^2\, dA = c_1 \iint_A |H_{ms}|\, dA + c_2 \iint_A |H_{ms}|^2\, dA,$$

in which the second one was already resolved in Section 6.3, for both single- and three-phase bushings. Solution of the first integral is carried out using the same approach. However, it is more difficult, due to the presence of a root of polynomial in the denominator of the integrand—see formulae (5.15) and (5.17). Hence, the investigated integrals were reduced to elliptic integrals. Then, we applied some simplifications based on an approximation of Legendre's tables for the cases utilized in practice, corresponding to the range $0.05 < R/a < 0.40$ (see Figure 5.21) in the works by J. Turowski [6.17], [7.16], [4.16], which allowed to develop the following formula for the power losses in steel covers:

$$P = a_p \frac{Ia}{2} \sqrt{\frac{\omega\mu_0}{\sigma}} \left(f_1 + f_I + \frac{1}{a} f_2 \right) \tag{7.43}$$

where $f_1 = f_1(c,(d/a)), f_2 = f_2(c,(d/\mathrm{R})), f_I = f_I(I/a)$ are complex functions of the geometric dimensions and the current I in the bushings. The functions are different for single- and three-phase transformers (Figures 5.20 through 5.22 and 7.8).

* IEEE Spectrum, April 1997.

† *Time Magazine*, August 25, 2003, pp. 30–30.

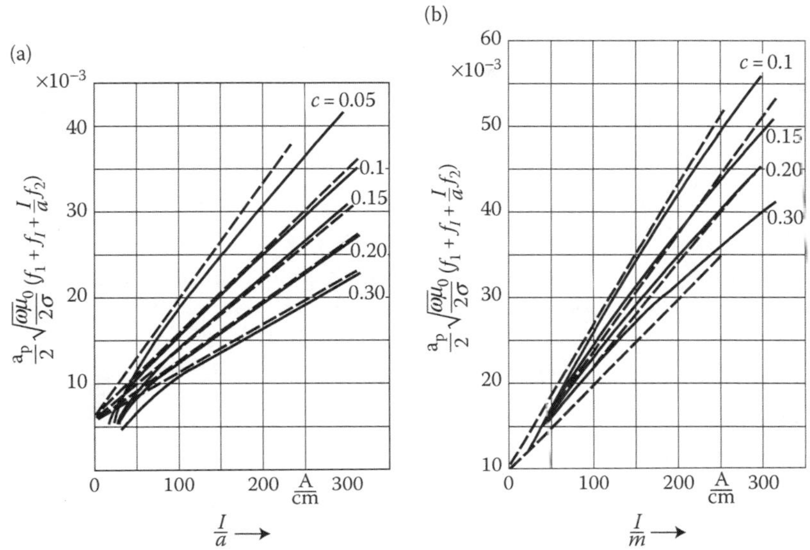

FIGURE 7.8 Comparison of the accurate formula (7.43)—continuous lines, with the simplified formula (7.44)—dashed lines, for (a) single-, and (b) three-phase transformer. (After Turowski J.: Losses in single and three phase transformers covers. (in Polish). *Rozprawy Elektrotechniczne*, (1), 1959, 87–119 [4.16].)

Using present computer techniques, the functions f_1, f_2, and f_I in formula (7.43) can be calculated fast and with accuracy. However, a high accuracy is not necessary here because in practice the designer does not have at his disposal accurate physical data of constructional steel. The conductivity σ of such steel can vary in the range $(4.8 \text{ to } 8) \times 10^6$ S/m (Table 1.6), which *a priori* limits the accuracy of calculation. Luckily, a typical constructional steel usually features an average value of about $\sigma_{steel} \approx 7 \times 10^6$ S/m.

The occurrence of σ under the root smoothes to a certain degree these discrepancies, from $(4.8–8) \times 10^6$ to $(2.2–2.8) \times 10^6$ S/m. Owing to that, one can accept for calculations the intermediate value of $\sigma = 7 \times 10^6$ S/m. Additionally, in the equations for f_1 and f_2 we can introduce average corrections with respect to the sheet thickness. As a result, a straight-line approximation of the analytical formulation (7.43) was obtained (J. Turowski [4.16]), which led to a satisfactory, simple formula for the power losses in transformer covers made of solid steel (for one side set of bushings):

$$P = m \times 5.5 \times 10^{-3} Ia \left(1 + \frac{0.0056}{c}\frac{I}{a}\right) \tag{7.44}$$

where $m = \sqrt{3}$ for three-phase transformers (Figure 5.22), and $m = 1$ for single-phase transformers (Figure 5.23); I is the current of one bushing (in A); a is the distance between bushings axes (in cm); $c = R/a$; and R is the radius of hole for bushing (in cm).

Formula (7.44) is valid in the range of $I/a > 28$ A/cm for three-phase transformers, and in the range of $I/a > 15$ A/cm for single-phase transformers. It is also valid in the range $0.1 < c = R/a < 0.3$. Within these limits are comprised practically all cases in which power losses can play any visible role. With some lower accuracy, formula (7.44) could be also applied beyond these limits. It is confirmed in Figure 7.8. In Table 7.5 given are exemplary values of losses in covers of several typical transformers [4.16]. They can serve also as indicators for an evaluation of losses in other steel parts crossed by electric current.

Influence of "transparency" of sheet. This effect takes place in thin plates or at a strong saturation. In such cases, a part of the electromagnetic power penetrates into the opposite side of the sheet, and power losses in the sheet get smaller. This can be calculated approximately on the basis of the following considerations. From Figure 4.14, we can see that at kd ≥ 1.7, the losses in a steel plate practically do not depend on the sheet thickness. Assuming $f = 50$ Hz and $\sigma_{20C} = 7 \times 10^6$ S/m, we get

$$k = \sqrt{\frac{\omega\mu\sigma}{2}} = 37.5\sqrt{\mu_r}\,\frac{192}{172 + t}\,1/\text{m}$$

Hence, at a thickness of sheet lower than the "critical" thickness

$$d_{\text{crit}} = \frac{1.7}{k} = \frac{45}{\sqrt{\mu_r}} \cdot \frac{172 + t}{192}\,\text{mm} \tag{7.45}$$

the power calculated with formula (7.43) or (7.44) should be multiplied by the coefficient $\varsigma_{\text{Fe}} < 1$ (Figure 4.14), which takes into account the reduction of losses in the result of "transparency" of sheet.

On the basis of dependence (7.45) and magnetization characteristic of steel (Figure 1.29), one can determine a graph of critical thickness over which the power losses in the cover do not depend on its thickness (Figure 7.9). The magnetic field intensity necessary for determination of d_{crit} can be evaluated from the formula $H_m \approx 0.9\, I/a$ (J. Turowski [4.16]).

From the above observation follows a possibility of reduction of the losses in cover by reducing its thickness. All formulae presented in this section were checked experimentally and were presented in the cited works (References) of the author J. Turowski.

7.5 CALCULATION OF STRAY LOSSES IN SOLID STEEL WALLS BY MEANS OF FOURIER'S SERIES

7.5.1 General Method

Calculation of power losses and local heating in inactive constructional parts made of solid steel [2.41], [7.17], and placed in strong electromagnetic fields consists of two basic tasks:

- Evaluation of the magnetic field H_{ms} on the body surface
- Calculation of the magnetic field $H_{ms}\,e^{-\alpha z}$ and the active and reactive power dissipation in the mass of body

TABLE 7.5
Losses in Steel Covers of Transformers without Nonmagnetic Gaps between Holes

Rated Power and Low Voltage of Transformer	Current I	a	$2R$	d	$c = \frac{R}{a}$	$\frac{I}{a}$	P_{load}	Losses in the Cover, at Temperature 95°C, Calculated by the Formula... Simplified	Accurate	
(kVA/kV)	(A)	(cm)	(cm)	(cm)	—	(A/cm)	(kW)	(W)	(W)	(%) P_{load}
500/0.231	1250	14	7	0.8	0.25	89.3	8.0	390	398	5.0
630/0.231	1580	14	7	0.8	0.25	113	9.45	558	582	6.2
1600/0.4	2310	14	7	1.0	0.25	165	19.5	1050	1080	5.5
16,000/3	3070	24	9	1.6	0.19	128	123	2180	2260	1.9

Source: After Turowski J.: Losses in single and three phase transformers covers. (in Polish). *Rozprawy Elektrotechniczne*, (1), 1959, 87–119.

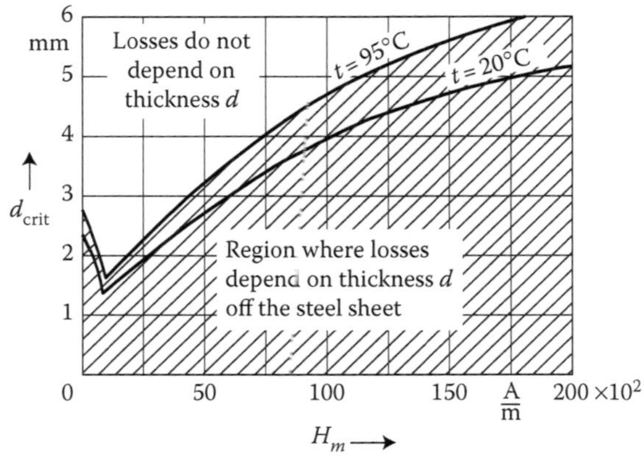

FIGURE 7.9 The critical thickness d_{crit} of the sheet made of constructional steel, below which the power losses depend on the sheet thickness *d*. (Adapted from Turowski J.: Losses in single and three phase transformers covers. (in Polish). *Rozprawy Elektrotechniczne*, (1), 1959, 87–119 [4.16].)

The first task can be resolved relatively easily, with an accuracy satisfactory for practice, with the help of the method of mirror images, Roth's method, Rogowski's method (Chapter 5), or numerical methods such as FEM-2D, FDM-2D, and RNM-3D [1.17]. We can find in this way a distribution of the normal component of magnetic field intensity H_{z0} (x,y) on the surface of body (Figure 5.26) from the side of air, at the assumption that the magnetic permeability of iron (μ_{Fe}) in comparison with that of air (μ_0) is infinitely large. More difficult, in general, is the solution of the electromagnetic Equations 2.1 through 2.6 inside metal with the consideration of variable permeability. This problem was elaborated in the publications by J. Turowski [2.41], [2.32], and [7.17].

7.5.1.1 Three-Dimensional Field

We assume that all components of a field are harmonic functions of time (2.136). Due to continuity of the flux entering constructional parts (2.4), the function of field distribution on the surface has, as a rule, a periodic character (Figure 7.10). On the basis of the principles presented in Section 7.2, we assume initially a constant permeability of steel, $\mu = \text{const}$, and similarly as in Section 2.10, we obtain the conductivity equation

$$\frac{\partial^2 H_{mz}}{\partial x^2} + \frac{\partial^2 H_{mz}}{\partial y^2} + \frac{\partial^2 H_{mz}}{\partial z^2} = \alpha^2 H_{mz} \tag{7.46}$$

where $\alpha = (1 + \text{j})k$ (2.139) characterizes the physical properties of steel. After applying Fourier's method $H(x, y, z) = X(x) \cdot Y(y) \cdot Z(z)$ to Equation 7.46, we can obtain a general solution in the form (2.144) and (2.148). Immediately, one can assume $C_{8n} = 0$, because otherwise the magnetic field would grow to infinity along the z axis, which is impossible. The other constants are estimated from boundary conditions on the metal surface. If the field distribution on the surface, $H_{z0}(x, y)$, is an even

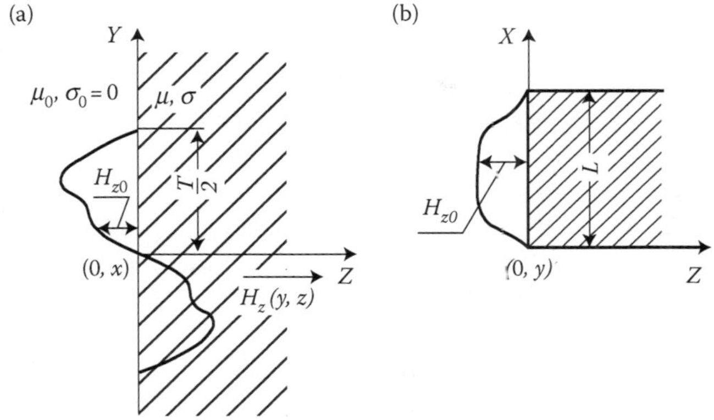

FIGURE 7.10 The periodic field on the surface of a steel half-space.

(cosinusoidal) function with respect to the z axis, then the constants $C_{2n} = C_{5n} = 0$. If it is an odd (sinusoidal) function, then $C_{1n} = C_{4n} = 0$.

Let us put the origin of the coordinate system in the nodal point (Figure 7.10). Then, on the basis of Equations 2.148, we get $C_{1n} = C_{4n} = C_{8n} = 0$ and

$$X(L) = C_{2n}\sin\beta_n L = 0 \quad \text{and} \quad Y\left(\frac{T}{2}\right) = C_{5n}\sin\eta_n\frac{T}{2} = 0$$

It means that theoretically there exists an infinite number of the β_n and η_n values

$$\beta_n = m\frac{\pi}{L} \quad \text{and} \quad \eta_n = n\frac{2\pi}{T}; \quad (m = 1, 2, 3,\ldots; \ n = 1, 2, 3,\ldots)$$

Hence, we have an infinite number of particular solutions of Equation 7.46

$$H_{m,zn} = \mathrm{C}_{mn}\sin m\frac{\pi}{L}x \cdot \sin n\frac{2\pi}{T}\mathrm{y}\mathrm{e}^{-\alpha_{mn}z} \tag{7.47}$$

where

$$\alpha_{mn} = \sqrt{\alpha^2 + \left(m\frac{\pi}{L}\right)^2 + \left(n\frac{2\pi}{T}\right)^2} \tag{7.48}$$

Equation 7.46 will also be satisfied by a linear combination of all particular solutions (7.47), which will be a general solution of this equation

$$H_{mz}(x,y,z) = \sum_{m=1}^{\infty}\sum_{n=1}^{\infty} C_{mn}\sin m\frac{\pi}{L}x \cdot \sin n\frac{2\pi}{T}y \in \mathrm{e}^{-\alpha_{mn}z} \tag{7.49}$$

The solution (7.49) is, in fact, nothing else than an expansion of the function $H_{mz} = f(x, y, z)$ in metal into a *double Fourier's series*. Particular components (7.47) of this sum represent particular space harmonics of the field distribution in the 0xy plane and their attenuation and phase shift during the field penetration into the metal, along the z axis.

As we can see, the harmonics of the higher order disappear faster than the harmonics of the lower order. It means that the field in the process of penetration into the metal is more and more similar in its form to the first harmonic of spatial distribution of the field. On the surface, from the iron side, at $z = 0$, we have

$$H_{mz}(x,y,0) = \sum_{m=1}^{\infty}\sum_{n=1}^{\infty} C_{mn} \sin m\frac{\pi}{L}x \cdot \sin n\frac{2\pi}{T}y \tag{7.50}$$

where

$$H_{mz}(x,y,0) = \frac{\mu_0}{\mu} H_{mz0}(x,y) \tag{7.51}$$

On a corresponding scale it is, in fact, an expansion into a *double Fourier's series* of the given distribution function of magnetic field $H_{mz0}(x, y)$ on the surface of metal from the air side.

Comparison of Equations 7.49 and 7.51 indicates the way to an accurate mathematical solution. However, integration of the function $H_{mz0}(x, y)$ in order to determine the constants C_{mn} of distribution (7.50) and (7.49) is too difficult. Therefore, in practice, one can apply the following simplifications.

In most cases, it is sufficient to consider in analysis only a few harmonics of the spatial distribution—from 1 to m_1 and from 1 to n_1. In this case, for instance, for the average values $T/2 = L = 20$ cm and $m_1 = n_1 = 1$, we get

$$\nu = \left(m_1\frac{\pi}{L}\right)^2 + \left(n_1\frac{2\pi}{T}\right)^2 = 0.4 \times 10^4 \frac{1}{\mathrm{m}^2}$$

In the case of steel with physical parameters $\mu_r = 500$–1000 and $\sigma = 7 \times 10^6$ S/m, at frequency 50 Hz, we obtain

$$\omega\mu\sigma = (140 \text{ to } 280) \times 10^4 \frac{1}{\mathrm{m}^2}$$

No doubt, therefore, that at not too high an order of harmonics (e.g., $m_1 \le 5$ and $n_1 \le 5$) and the object dimensions bigger than a few centimeters, one can adopt, with accuracy satisfactory for technical purposes

$$\alpha_{mn} = \frac{1}{\sqrt{2}}\left[\sqrt{\sqrt{(\omega\mu\sigma)^2 + \nu^2} + \nu} + \mathrm{j}\sqrt{\sqrt{(\omega\mu\sigma)^2 + \nu^2} - \nu}\right] \approx (1+\mathrm{j})\sqrt{\frac{\omega\mu\sigma}{2}} = \alpha \tag{7.48a}$$

Owing to this assumption, the exponential operator $e^{-\alpha_{mn}z}$ can be shifted in front of the symbols of the sum, which allows one to obtain on the basis of Equations 7.49 through 7.51, the formula for the magnetic field intensity in metal

$$\mathrm{H}_{mz}(x,y,z) \approx \frac{\mu_0}{\mu} H_{mz0}(x,y)\mathrm{e}^{-\alpha_{mn}z} \tag{7.52}$$

7.5.1.2 Two-Dimensional Field

Many technical problems can be resolved with the help of the 2D field theory, in which components of the field are varying in space only along the y and z axes (Figure 7.10a).

From Maxwell's Equations 2.2, at $H_x = 0$, $E_y = E_z = 0$, we obtain

$$\mathrm{E}_{mx} = \mathrm{j}\omega\mu\int \mathrm{H}_{mz}(y,z)\,\mathrm{d}y + \mathrm{C}(z) = -\mathrm{j}\omega\mu_0\,\mathrm{e}^{-\alpha z}\,F_y \tag{7.53}$$

$$H_{my} = -\frac{1}{\mathrm{j}\omega\mu}\frac{\partial E_{mx}}{\partial z} = -\frac{\mu_0}{\mu}\alpha\,\mathrm{e}^{-\alpha z}F_y = -(1+\mathrm{j})\mu_0\sqrt{\frac{\omega\sigma}{2}}\frac{1}{\sqrt{\mu}}F_y\mathrm{e}^{-\alpha z} \tag{7.54}$$

where

$$F_y = -\int H_{mz0}(y,0)\,\mathrm{d}y + C_0 \tag{7.55}$$

The constant C_0 in Equation 7.55 should be determined from the boundary conditions for the given field distribution on the surface.

The permeability in the form $1/\sqrt{\mu}$ appearing in formula (7.54) can be eliminated by introducing an analytical approximation (7.12), from which

$$H_{my} = \frac{H_{my}}{1/\sqrt{\mu}}\left(A_1 + A_2\sqrt{\mu}\left|H_{my}\right|\right)$$
$$= -(1+\mathrm{j})\mathrm{e}^{-\alpha z}\mu_0\sqrt{\frac{\omega\sigma}{2}}F_y\left(A_1 + A_2\mu_0\sqrt{\omega\sigma}\left|F_y\right|\mathrm{e}^{-kz}\right) \tag{7.54a}$$

The active and reactive power losses per surface unit equal to the Poynting vector (3.7) on the surface, at $z = 0$

$$P_{x1} = \frac{1}{2}\boldsymbol{E}_{mz}\boldsymbol{H}^*_{my} = (1-\mathrm{j})\frac{\mu_0^2\,\omega\sqrt{\omega\sigma}}{2\sqrt{2}}\left(A_1\left|F_y\right|^2 + A_2\mu_0\sqrt{\omega\sigma}\left|F_y\right|^3\right) \tag{7.56}$$

The active and reactive power losses per length unit of the investigated element along the x axis equal

$$P_{x1} = \int_{-\infty}^{+\infty} S_{z=0}\,\mathrm{d}y = (a_p - \mathrm{j}a_q)\omega\sqrt{\omega\sigma}\left(a_1 + a_2\sqrt{\omega\sigma}\right) \tag{7.57}$$

where

$$a_1 = \frac{A_1 \mu_0^2}{2\sqrt{2}} \int_{-T/2}^{T/2} |F_y|^2 \, dy \tag{7.57a}$$

$$a_2 = \frac{A_2 \mu_0^3}{2\sqrt{2}} \int_{-T/2}^{T/2} |F_y|^3 \, dy \tag{7.57b}$$

The linearization coefficients $a_p \approx 1.4$ and $a_q \approx 0.85$ were introduced with the purpose to consider the nonlinear permeability and hysteresis in solid steel, according to (7.26). In case of a nonferromagnetic metal, $1/\sqrt{\mu}$ = const and $a_p = a_q = 1$.

The coefficients a_1 and a_2 are easy to calculate at a simple distribution of field on the surface [7.20]. At a more complicated form of the curve $H_{mz0}(y)$, the integration with classical methods is too difficult to have practical usefulness. On the other hand, graphical integration, carried out in the works by Turowski [2.32], [2.41], is too cumbersome and not very accurate. Application of a computer program can help to calculate the coefficients a_1 and a_2 rapidly and with any desired accuracy (Figure 7.16 [later in the chapter]).

7.5.2 Analytical Formulae in Case of Sinusoidal Distribution of a Field on the Steel Surface

Let us assume that the distribution of the normal component of magnetic field intensity on the surface [7.20] of steel plate (Figure 7.11) is expressed by the formula

$$H_{mz0}(y,0) = \hat{H}_{mz0} \sin\frac{\pi}{\tau} y \tag{7.58}$$

According to formula (7.55)

$$\begin{aligned} F_y &= -\int H_{mz0}(y,0)dy + C_0 = -\hat{H}_{mz0} \int \sin\frac{\pi}{\tau} y dy + C_0 \\ &= \frac{\tau}{\pi} \hat{H}_{mz0} \cos\frac{\pi}{\tau} y + C_0 \end{aligned} \tag{7.59}$$

Value of the constant component C_0 in formula (7.59) depends on boundary conditions. One can consider here four characteristic cases (Figure 7.11):

Case (a). The analyzed element has a finite length in the direction of the y axis (steel bar), equal to the assumed period of space distribution of field 2τ, and all the flux Φ entering into the steel is closed within this period (Figure 7.11a). According to this assumption, for the outside region of $y \geq \tau$ and $y \leq -\tau$, one can assume that the components of field equal there $H_{mz0}(y, 0) = H_{my} = 0$. Hence, from formula (7.54) it follows for this region $F_y = 0$, and from formula (7.55) $C_0 = 0$. At such an approach,

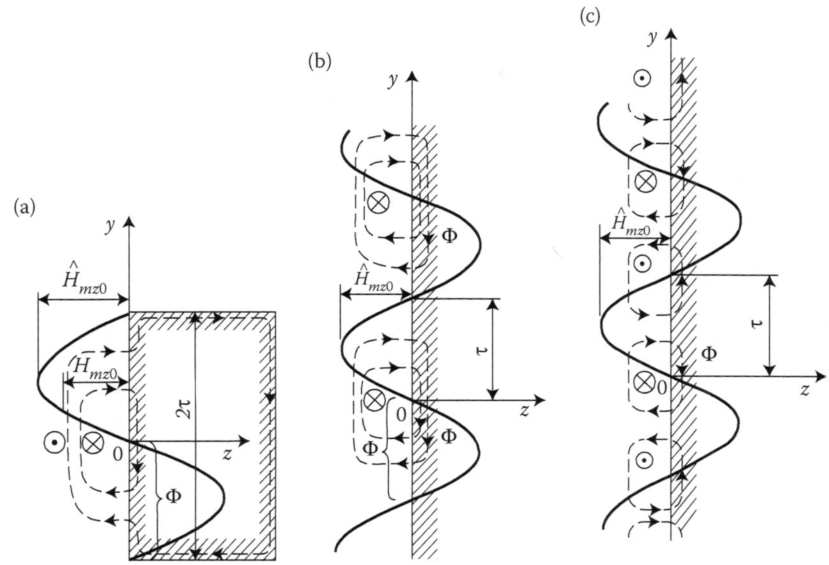

FIGURE 7.11 Typical cases of distribution of the normal component of field H_{mz0} on the surface of a steel half-space.

the flux entering into the metal in the section $0 < y < \tau$ is

$$\Phi_z(z = 0) = \mu_0 \int_0^{\tau} H_{mz0}(y,0)\mathrm{d}y = \frac{2\mu_C \tau}{\pi} \hat{H}_{mz0}$$

The alternating flux, instead, in the steel in the place $y = 0$ is

$$\Phi_y(y,0) = \mu \int_0^{\tau} H_{my}(y,0)\mathrm{d}y = \mu_0 \alpha F_y(y = 0) \int_0^{\tau} e^{-\alpha z}\mathrm{d}z = \frac{\mu_0 \tau}{\pi} \hat{H}_{mz0}$$

The flux $\Phi_y(y, 0)$ is therefore 2 times smaller than the flux $\Phi_z\,(z = 0)$ entering perpendicularly to the iron. It means that 1/2 Φ_z must be closed by other ways, for instance, by encircling the steel part (Figure 7.11a) or by other parts. At more complicated (nonsinusoidal) systems, the distribution of both fluxes can be very different.

Substituting expression (7.59) into formulae (7.57a,b), plus a simple integration, yields

$$a_1 = \frac{A_1 \mu_0^2 \tau^2}{2\sqrt{2}\pi^2} \hat{H}_{mz0}^2 \int_{-\tau}^{\tau} \cos^2 \frac{\pi}{\tau} y\,\mathrm{d}y = \frac{A_1 \mu_0^2 \tau^3}{2\sqrt{2}\pi^2} \hat{H}_{mz0}^2 \tag{7.60a}$$

$$a_2 = \frac{A_2 \mu_0^3 \tau^3}{2\sqrt{2}\pi^3} \hat{H}_{mz0}^3 \int_{-\tau}^{\tau} \cos^3 \frac{\pi}{\tau} y^3\,\mathrm{d}y = \frac{\sqrt{2} A_2 \mu_0^3 \tau^4}{\pi^3} \hat{H}_{mz0}^3 \tag{7.60b}$$

Next, by substituting the obtained coefficients into formula (7.57), one can obtain the apparent power consumed by the steel element (Figure 7.11a) per unit of its length:

$$P_{s1} = (1.4 - \mathrm{j}\,0.83)\frac{\mu_0^2\tau^3}{2\sqrt{2}\pi^3}\omega\sqrt{\omega\sigma}\,\hat{H}^2_{mz0}\left(A_1 + \frac{4\mu_0\tau}{\pi}\sqrt{\omega\sigma}A_2\hat{H}_{mz0}\right) \tag{7.61}$$

Case (b). A steel surface extends infinitely in the direction of the y axis, the field distribution is repeated periodically, and the flux is closed within the period 2τ (Figure 7.11b). The constant C_0 can be determined by equalization of the flux Φ_z entering into the steel within the space of one half-period (τ) with the flux Φ_y in the steel at $y = 0$.

$$\Phi_z(z=0) = \mu_0\int_0^{\tau} H_{mz0}(y,0)\,\mathrm{d}y = \frac{2\mu_0\tau}{\pi}\hat{H}_{mz0} \tag{7.62}$$

The alternating flux, instead, in steel in the place $y = 0$:

$$\Phi_y(y,0) = \mu\int_0^{\tau} H_{my}(y=0)\mathrm{d}z = \mu_0\alpha F_y(y=0)\int_0^{\tau} e^{-\alpha z}\mathrm{d}z = \mu_0\left(\frac{\tau}{\pi}\hat{H}_{mz0} + C_0\right) \tag{7.63}$$

After equalizing Equation 7.62 to Equation 7.63

$$\frac{2\tau}{\pi}\hat{H}_{mz0} = \frac{\tau}{\pi}\hat{H}_{mz0} + C, \quad \text{from where } C = \frac{\tau}{\pi}\hat{H}_{mz0}$$

Hence, $F_y = \frac{\tau}{\pi}\hat{H}_{mz0}\left(\cos\frac{\pi}{\tau}y + 1\right)$.

According to Equation 7.57a,b, we get now

$$a_1 = \frac{A_1\mu_0^2\tau^2}{2\sqrt{2}\pi^2}\hat{H}^2_{mz0}\int_{-\tau}^{\tau}\left(\cos\frac{\pi}{\tau}y + 1\right)^2\mathrm{d}y = \frac{3A_1\mu_0^2\tau^3}{2\sqrt{2}\pi^2}\hat{H}^2_{mz0} \tag{7.64a}$$

$$a_2 = \frac{A_2\mu_0^3\tau^3}{2\sqrt{2}\pi^3}\hat{H}^3_{mz0}\int_{-\tau}^{\tau}\left(\cos\frac{\pi}{\tau}y^* + 1\right)^3\mathrm{d}y = \frac{5A_2\mu_0^3\tau^4}{2\sqrt{2}\pi^3}\hat{H}^3_{mz0} \tag{7.64b}$$

After substitution of Equation 7.64a,b into Equation 7.57, we obtain the formula for the apparent power consumed by a solid steel surface (Figure 7.11b), per the period 2τ and unit length along the x axis:

$$P_{s1} = (1.4 - \mathrm{j}\,0.85)\frac{3\mu_0^2\tau^3}{2\sqrt{2}\pi^2}\omega\sqrt{\omega\sigma}\,\hat{H}^2_{mz0}\left(A_1 + \frac{5\mu_0}{3\pi}\sqrt{\omega\sigma}A_2\tau\sqrt{\omega\sigma}\,\hat{H}^2_{mz0}\right) \tag{7.65}$$

Case (c). A steel surface extends infinitely in the direction of the y axis, the field distribution is repeated periodically, and the flux entering the steel plate surface spreads symmetrically in both directions (Figure 7.11c). The constant C_0 in formula (7.59) can be determined by equalization of the flux Φ_z entering the steel within the space of a quarter of period ($\tau/2$) to the flux Φ_y in the steel in point $y = 0$. Changing then the upper limit of the integration in the integral (7.62) into $\tau/2$ and equalizing it to (7.63), we get

$$\hat{H}_{mz0}\left[-\frac{\tau}{\pi}\cos\frac{\pi}{\tau}y\right]_0^{\tau/2} = \frac{\tau}{\pi}\hat{H}_{mz0} + C, \text{ from where } C = 0$$

Proceeding further analogically as in case (a), we conclude that the consumption of apparent power by the surface of steel wall (Figure 7.11c) per one period is 2τ and per unit length along the x axis is expressed by formula (7.61). It appears that at theoretical calculations the real system from Figure 7.11a can be substituted by the theoretical system from Figure 7.11c.

Case (d). This system is analogical to that in Figure 7.11c, with only one difference; the wave of a normal component of magnetic field intensity on the steel surface H_{mz0} runs in the direction of the y axis with a uniform velocity equal $(\omega\tau/\pi)$.

An instantaneous value of the magnetic field intensity in an arbitrary place on the y axis, at the running field, is expressed by the formula

$$H_{z0}(y,0) = H_{mz0}\cos\left(\omega t + \frac{\pi}{\tau}y\right) \tag{7.66}$$

An instantaneous value, instead, of the field stationary in space, but oscillating in time according to Equation 7.58 is expressed by the formula

$$H_{z0}(y,0) = H_{mz0\,\max}\cos\,\omega t\sin\frac{\pi}{\tau}y \tag{7.67}$$

Therefore, the ratio of the average active power losses per one time period $T = 2\pi/\omega$ and one space period (double pole pitch) 2τ, equals to

$$\frac{P_{\text{run}}}{P_{\text{osc}}} = \frac{\dfrac{1}{T}\displaystyle\int_0^T \frac{1}{2\tau}\int_0^{2\tau} H_{mz0}^2\cos^2\left(\omega t - \frac{\pi}{\tau}y\right)\mathrm{d}t\,\mathrm{d}y}{\dfrac{1}{T}\displaystyle\int_0^T \frac{1}{2\tau}\int_0^{2\tau} \hat{H}_{mz0}^2\cos^2\omega t\,\sin^2\frac{\pi}{\tau}y\,\mathrm{d}t\,\mathrm{d}y} \tag{7.68}$$

where P_{run} is the losses at the running field, and P_{osc} is the losses at the oscillating field.

Formula (7.61) can be utilized also for calculations of linear induction motors* which are driving a solid steel body (e.g., a rail), by multiplying by 2 the power calculated with formula (7.61). This formula can have the same application for the calculation of the power of a rotation field in induction motors with solid steel rotor (infinitely long).

7.5.3 Computer Calculation of Power Losses in a Steel Plate Placed in the Field of Parallel Bars

Let us now consider a more complicated system (Figure 7.12) in which the coefficients a_1 and a_2 for formula (7.57) are calculated with a computer method.

The first task—finding the distribution of the field $H_{mz0}(y)$ on the surface of steel—was resolved in Chapter 5 (Equation 5.21a) with the method of mirror images (J. Turowski [2.32]):

$$H_{mz0}(y,0) = \eta_Q \frac{\sqrt{2}I}{2\pi h}\left[\ln\frac{c_1^2 + (y+h/2)^2}{c_1^2 + (y-h/2)^2} - \ln\frac{c_2^2 + (y+h/2)^2}{c_2^2 + (y-h/2)^2}\right] \tag{7.69}$$

where $\eta_Q = \dfrac{1+M_Q}{2}$ is the coefficient (5.8) considering a nonideal mirror image of the AC current in the solid steel (Figure 5.26).

The second task—finding the field and losses in the steel—can be resolved on the basis of formulae developed in Section 7.5.1. On the surface of an infinitely extended steel plate, the field distribution (7.69) initially reminds a distorted sinusoid (Figure 7.12). However, the distribution curve does not have next node on the y axis, but only asymptotically approaches the axis. Therefore, the power losses in the steel plate should be calculated with the help of (7.57), with integrating them not in the limits $\pm T/2$, but $\pm\infty$. After substituting Equation 7.69 into Equation 7.55, and applying the *de l'Hospitale* rule, we shall see (J. Turowski [2.32]) that at $y \to \infty$ the first component of the right side of Equation 7.55 approaches zero. Since, at the same time, at $y \to \infty$, E_{mx} (7.55) should be approaching zero as well, then $F_y \to 0$, and therefore $C_0 \to 0$. After a series of mathematical transformations (J. Turowski [2.41], [2.32]), we get

* Turowski J.: Magnetic field in the air-gap of linear induction motor. *Proc. Internat. Conference on Electrical Machines ICEM'80*. 15-17.09.1980. Athens. Part I, Paper LM1/8, pp. 60–67.
Turowski J., Wiak S.: Material aspects of a design of linear electromechanical converters. *Internat. Symposium ISEM,* Kanazawa, Japan. January 1990 (Invited paper).
Pudlowski Z.J., M., Savini A., Turowski J., Klaassen O.: A Comparison of Different Numerical Solutions for the Magnetic Field in a Linear Reluctance Self-Oscillating Motor. *Proc. of the 6th IEEE Biennial Conference on Electromagnetic Field Computation-CEFC'94*. 5–7 July, 1994. Aix-les-Bains (Lab.d'El., Grenoble), France, p. 146.
Rizzo M., Savini A., Turowski J.: A 3-D Investigation on the Non-Linear Magnetic Field of Linear Motors. *Proc. of 7th Biennial IEEE Conference on Electromagnetic Field Computation–CEFC'96,* Paper PC-13. March 18–20, 1996. Okayama, Japan.
Mendrela E.A., Turowski J.: Rotary-linear induction motor. *IEEE Winter Power Meeting*. New York 29.01-4.02.1978. IEEE Power Eng. Soc. Catalogue No 78, CH 1295-5-PWR. *IEEE Transactions PAS-97,* No 4, 1978, p. 1006.

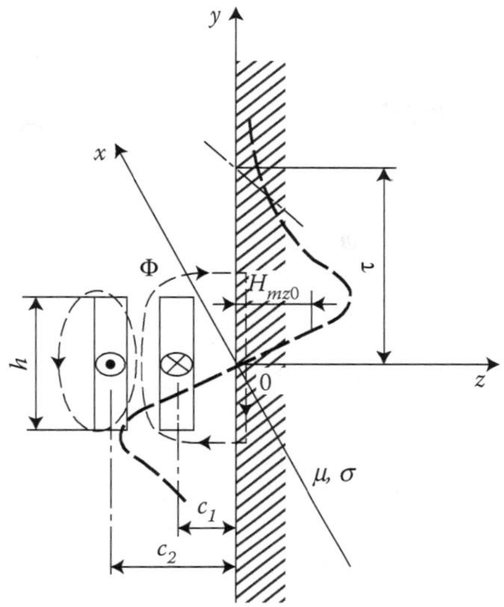

FIGURE 7.12 Scheme of distribution of the magnetic field of parallel bars on the surface of an infinitely extended steel plate.

$$\left.\begin{aligned} F_y &= \eta_Q \frac{\sqrt{2}I}{2\pi h} c_1[f(\zeta - h/c_1) - f(\zeta)] \\ f(\zeta) &= \ln\frac{1+\zeta^2}{(c_2/c_1)^2 + \zeta^2} + 2\,\text{arc tg}\,\zeta - 2\frac{c_2}{c_1}\,\text{arc tg}\,\frac{c_2}{c_1}\zeta \end{aligned}\right\} \tag{7.70}$$

$$\zeta = \frac{y + h/2}{c_1} \tag{7.71}$$

Considering the change of variables (7.71), after transformations (J. Turowski [2.32], [2.41]) we obtain *the total apparent power consumed by the steel plate, per unit length along the x axis*, calculated by means of the formula

$$P_{s1} = (1.4 - \text{j}\,0.85)\frac{\mu_0^2\omega\sqrt{\omega\sigma}}{2\sqrt{2}\pi^2} h\eta_Q^2 I^2 \left(A_1 k_1 + A_2 \frac{\mu_0\sqrt{\omega\sigma}}{\sqrt{2}\,\pi}\eta_Q I k_2 \right) \tag{7.72}$$

where

$$\left.\begin{aligned} k_1 &= \frac{1}{(h/c_1)^3}\int_{h/2c_2}^{\infty}\left[f\left(\zeta - \frac{h}{c_1}\right) - f(\zeta)\right]^2 \mathrm{d}\zeta \\ k_2 &= \frac{1}{(h/c_1)^4}\int_{h/2c_1}^{\infty}\left[f\left(\zeta - \frac{h}{c_1}\right) - f(\zeta)\right]^{-3} \mathrm{d}\zeta \end{aligned}\right\} \tag{7.73}$$

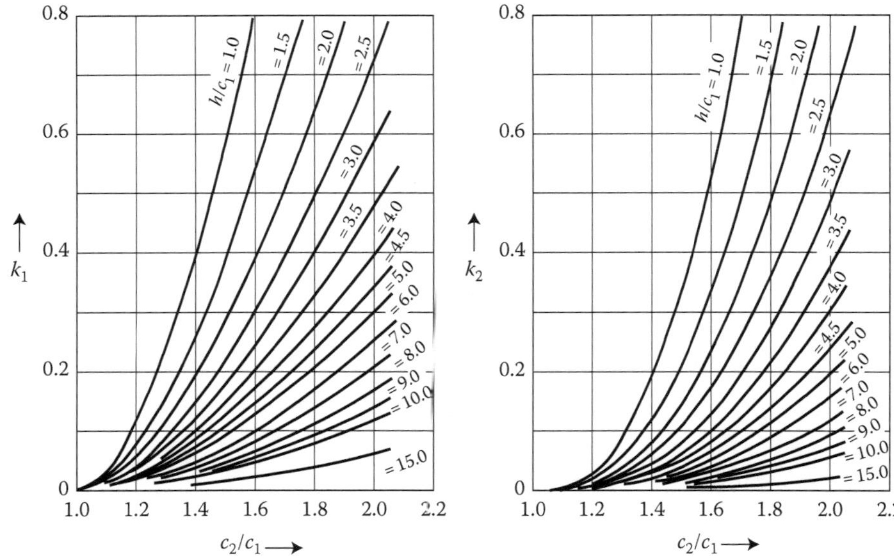

FIGURE 7.13 The coefficients of geometrical proportionality, k_1 and k_2, of bar pairs nearby a steel wall; Equation 7.73.

The coefficients of geometrical proportionality, k_1 and k_2, due to complexity of their integrand functions, were calculated with the help of computers* by numerical integration (Figures 7.13 and 7.14). These graphs represent an extension and supplement to the graphs published in (J. Turowski [2.32], [2.41]).

After substitution of numerical values of the constants A_1 and A_2 (7.12), $\mu_0 = 7 \times 10^6$ H/m, the correction coefficients of steel linearity a_p and a_q (7.26), as well as the coefficient of mirror "opacity" $\eta_Q = 0.8$ (Figure 5.26), we obtain the final approximate formulae for

- The *active power losses* in solid steel plate (in W/m) along the current I axis

$$P_1 = 1.12 \times 10^{-11} h f \sqrt{f\sigma} I^2 \left(k_1 + 5.3 \times 10^{-9} k_2 \sqrt{f\sigma} I\right) \tag{7.74}$$

- The *consumption of reactive power* by steel plate (in var/m)

$$Q_1 = 0.64 \times 10^{-11} h f \sqrt{f\sigma} I^2 \left(k_1 + 5.3 \times 10^{-9} k_2 \sqrt{f\sigma} I\right) \tag{7.75}$$

---

* Calculations were carried out by Dr. L. Byczkowska-Lipinska, next by Dr. G. Sobiczewska, and finally by Mr. M. Kopeć (MSc, Electronics Eng.), to whom the author (J. Turowski) expresses his acknowledgments.

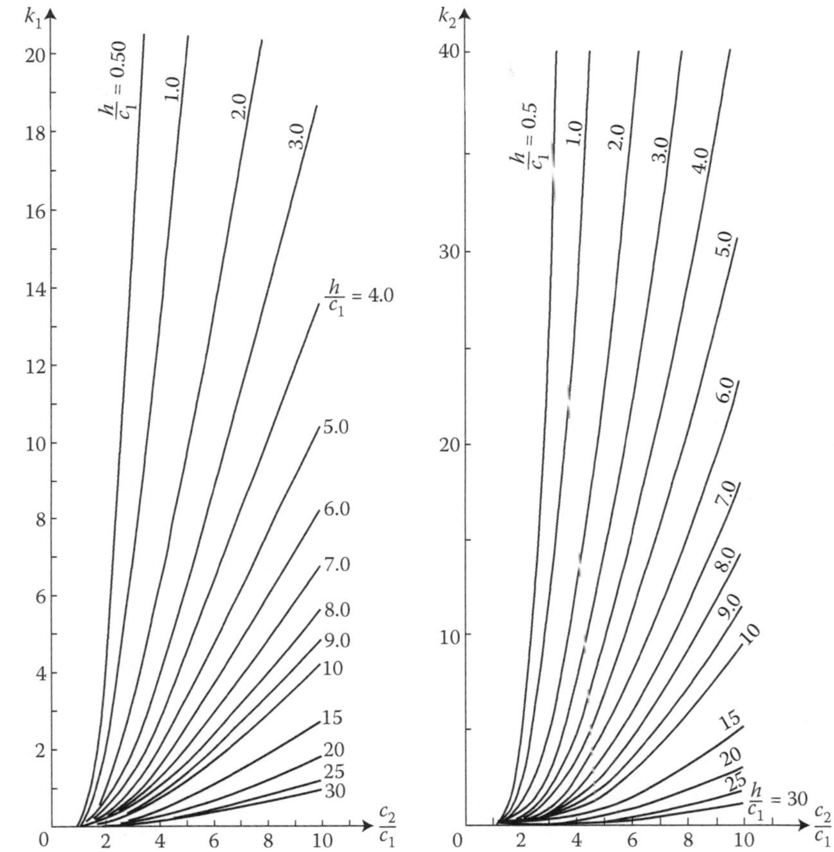

FIGURE 7.14 The coefficients k_1 (left) and k_2 (right) of geometrical proportionality of current bar pairs nearby a steel wall; Equation 7.73.

where all the variables, illustrated in Figure 7.12, are in the SI units. Formula (7.74) was verified experimentally (Figure 10.4) in J. Turowski's works [2.32] and [2.41] and several others. It can be utilized for evaluation of the *leakage losses* in steel tank of transformer (J. Turowski [2.34] and [7.18]). Then, instead of the current I, one should substitute the ampere-turns of one of the two windings. Formula (7.75), instead, can serve for the calculation of the *added reactance of bars placed nearby steel masses* in the form of flat surfaces. This reactance (in ohms per meter) amounts to

$$X_{r1} = \frac{Q_1}{I^2} = 0.7 \times 10^{-11} h f \sqrt{f\sigma}\left(k_1 + 5.3 \times 10^{-9} k_2 \sqrt{f\sigma} I\right) \tag{7.76}$$

and as we can see, it depends on the current. The range of application and accuracy of formulae (7.74) through (7.76) is defined mainly by the approximation (7.12), see (Figure 7.2) and η_Q.

EXAMPLE

A model of infinitely long busbars composed of 144 parallel wires (Figures 7.12 and 10.3) has parameters $h = 9.9$ cm, $c_2 = 8.3$ cm, $c_1 = 1.8$ cm; $\sigma_{St} = 7 \times 10^{-6}$ S/m; $f = 50$ Hz, $IN = 49 \times 144 = 7050$ A. Find the active power losses in a steel plate located in parallel next to the busbars.

SOLUTION

For $\frac{c_2}{c_1} = \frac{8.3}{1.8} = 4.6$ and $\frac{h}{c_1} = \frac{9.9}{1.8} = 5.5$, it can be determined from Figure 7.14: $k_1 = 2.4$ and $k_2 = 4.4$. The active power losses per unit length, according to Equation 7.74, amount to

$$P_1 = 1.12 \times 10^{-11} 9.9 \times 10^{-2} \times 50\sqrt{50 \times 7 \times 10^6}\, 7050^2$$
$$(2.4 + 5.3 \times 10^{-9}\, 4.4\sqrt{50 \times 7 \times 10^6}\, 7050) = 51.5(2.4 + 3.09) = 282 \text{ W/m}$$

From the measurements on the physical model (J. Turowski [2.41], Figures 10.3 and 10.4), 270 W/m was obtained, in spite of unavoidable, but justified, approximations and simplifications.

Using a PC, the above analytical formulae can be solved within microseconds or less, whereas with a numerical FEM-3D model it could take many hours.

A higher accuracy of calculation should not be expected (and is indeed not needed, although possible), because usually the parameters σ and μ of constructional steel are not measured with high accuracy, and their dispersion, especially in weaker fields, can be remarkable (cf. Figure 1.29 and Table 1.6), depending on chemical ingredients (Figure 1.9, Table 1.8; [5.9] and [1.20] p. 67) as well as on thermal (Figure 1.26) and plastic treatment. It should be also emphasized that the mirror-image coefficient η_Q (5.8) in some conditions can be smaller than the assumed value of $\eta_Q = 0.8$, and then the losses calculated with formula (7.74) can be a little smaller than the measured one, not to mention the inevitable measurement error for such a complicated method (cf. Figures 10.16 and 10.18).

As it follows from formulae (7.61), (7.65), and (7.72), the exponent at the current or magnetic field intensity, as was mentioned in Section 7.3 and proven in the CIGRE Reports (Jezierski, Turowski, et al. [10.17], [7.9], [10.18]), depends on the configuration of the field on the surface and *can be in some cases higher than 2* (compare Equations 7.40, 7.74, 7.74a and measurements Figure 10.4).

The total power loss in a solid steel wall with magnetic and electromagnetic screens can then be calculated from formulae (7.69) through (7.75), with the semiempirical correction x like in Equation 4.71:

$$P = \frac{1}{2}\sqrt{\frac{\omega\mu_o}{2\sigma_{St}}} \times$$
$$\times \left[p_e \iint_{A_e} \sqrt{\mu_{\text{rse}}}\, |H_{ms}|^2\, dA_e + p_m \iint_{A_m} \sqrt{\mu_{\text{rsm}}}\, |H_{ms}|^2\, dA_m + \text{a}_\text{p} \iint_{A_{St}} \sqrt{\mu_{\text{rs}}}\, |H_{ms}|^{2x}\, \text{d}A_{St} \right]$$
$$\approx \text{I}^2(a + bI) \tag{7.74a}$$

where $p_e = \sqrt{\dfrac{\mu_2\sigma_3}{2\mu_3\sigma_2}}$; $H_{ms}^2 = H_{\text{msx}}^2 + H_{\text{msy}}^2$ is the magnetic field intensity on the metal surface; μ_{rse}, μ_{rsm}, μ_{rs}, are the relative surface permeabilities of steel wall, where μ_{rse}, μ_{rsm} are the permeabilities behind the electromagnetic (e) or magnetic (m) screen, respectively, (in the first approach, one can assume they all are equal and much higher than 1); p_e, p_m are the coefficients of electromagnetic screening (4.42) or (4.67) or magnetic screening (shunting) (4.64), respectively; $x = 0.8$–1.3 is the coefficient of type of field excitation, according to the exponential dependencies (7.37) or (7.38), respectively, specific for particular construction (the x value is selected in a semiempirical way, for instance, for a transformer tank wall $x = 1.14$–1.15); A_e, A_m are the areas of surfaces screened electromagnetically (Cu, Al) or magnetically (with shunts packaged of laminated transformer iron); and A_{St} is the surface area of the nonscreened solid iron.

Formula (7.74a) is used for fast 3D calculation of power losses (Figure 7.22 [later in the chapter]) with the help of the rapid, interactive program RNM-3D (J. Turowski and M. Turowski [4.28], [7.15]). It is an alternative for very expensive and labor-intensive FEM-3D computational packages, illustrated in Figure 7.20 (later in the chapter).

7.6 POWER LOSSES IN A TRANSFORMER TANK

The great significance of a correct, 3D calculation of stray losses in the tank and inactive parts of power equipment construction is supported by the following facts:

1. The cost of capitalized load losses, depending on the location of the transformer, reaches from 3000 to 10,000 U.S. $/kW of these losses. For example, by using RNM-3D, an easy reduction of such losses by 50–100 kW can give a saving of 0.5–1 million U.S. $ for one unit.
2. Stray power losses could reach large local concentration with exceeding permitted temperature limit, which may damage the unit and the reliability of the entire energy system (the so-called Blackout).
3. Faulty shielding without credible and reliable 3D calculations sometimes may not decrease, but rather even *increase* these losses. 2D calculations give erroneous results as a rule, ignoring the large component of the circumferential magnetic field on the surface of the tank.

7.6.1 Two-Dimensional Numerical Solution

In contemporary large transformers, screened steel tanks are used, whereas at smaller powers (from 70–80 MV · A, Figure 4.16) the steel tanks are not screened. Calculation of power losses starts from calculation of the leakage field. Since it is difficult, especially with considering the steel saturation, various linearizations and simplifications are used, for instance, RNM-2D solutions. One of such solutions was the Hermitian 2D numerical calculation carried out at the Technical University of Lodz by K. Komęza and G. Sobiczewska-Krusz [7.8] (Figure 7.15) using a hierarchical FEM of higher order. This method enables a field calculation faster than with the classical FEM, with

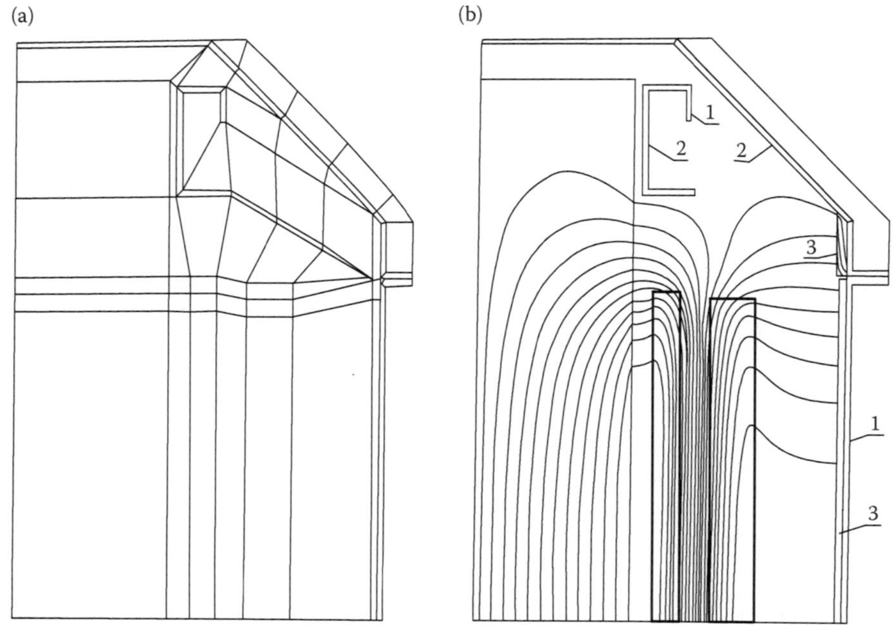

FIGURE 7.15 Two-dimensional (2D) calculation of leakage field of power transformer with the help of a high-order FEM: (a) discretization of the area; (b) the flux density lines, 1—constructional solid steel; 2—nonmagnetic solid steel; 3—package of transformer steel (magnetic screen = shunts). (Adapted from Komęza K. and Krusz G.: *International Symposium ISEF'89*, Lodz 1989, pp. 109–112 [7.8].)

a smaller number of elements, and a local increase of the order of Hermit's interpolation polynomial; for instance, from 0 to 2 (Figure 7.15b). It gives smooth and not broken lines of the flux density. In spite of plenty possibilities of this method, the important problem of three-dimensionality of the field is, however, still open. It has still to be considered with other approximate methods (see Section 7.5.2) or with numerical 3D methods (Section 7.5.3).

7.6.2 Three-Dimensional Analytical Calculations of a Stray Field and Losses in Tanks at Constant Permeability

Analytical-numerical methods (ANM) are inferior to numerical methods such as FEM and FDM in that, that ANM need simple boundary conditions. An advantage, however, of ANM is the possibility of considering of the third dimension, good physical clarity, and easily perceptible parametric dependencies, which simplifies a quick qualitative evaluation of phenomena.

7.6.2.1 Field on the Tank Surface

An asymmetric cylindrical winding with different zonal specific current loadings A_1, A_2, A_3,... (Figure 7.16) can be in general substituted by two fictitious windings superimposed on each other (E. Jezierski et al. [5.2]). One of the windings is symmetrical,

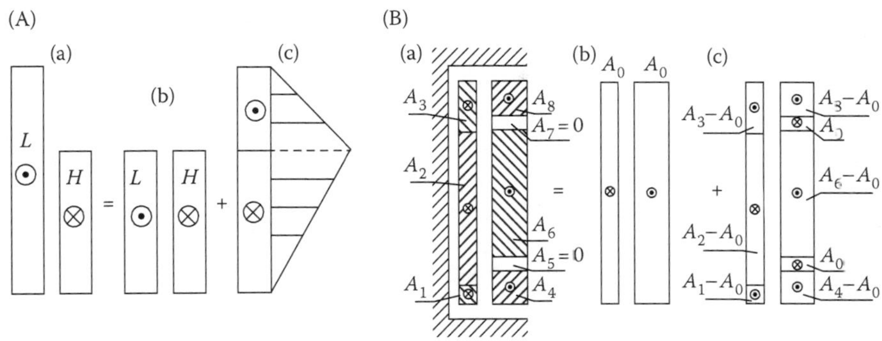

FIGURE 7.16 Division of ampere-turns of a transformer. (a) Asymmetric cylindrical winding, (b) fictitious symmetric component, and (c) fictitious sandwich component $L_a = I_2N_2$; $H_a = I_1N_1$. (A) A simplified model with a graph of radial ampere-turns; (B) with double asymmetry.

with uniformly distributed ampere-turns equal to the average ampere-turn of the real winding A_0 (Figure 7.16b), and the second one is a sandwich winding which produces a radial field (Figure 7.16c). The sum of the ampere-turns of the second fictitious winding group equals zero and after superposition upon the first group gives the real ampere-turn distribution in any chosen part of windings.

At normal axial asymmetry in cylindrical windings, not exceeding 5–10%, the power losses in the tank are determined mainly by the symmetric component of ampere-turns. The leakage flux Φ_{Tank} of symmetric group (b) of the transformer flow (ampere-turns) induces in the tank eddy currents that have a periodic distribution in the tank wall plane (Figure 7.17). Conforming with this, components of the magnetic

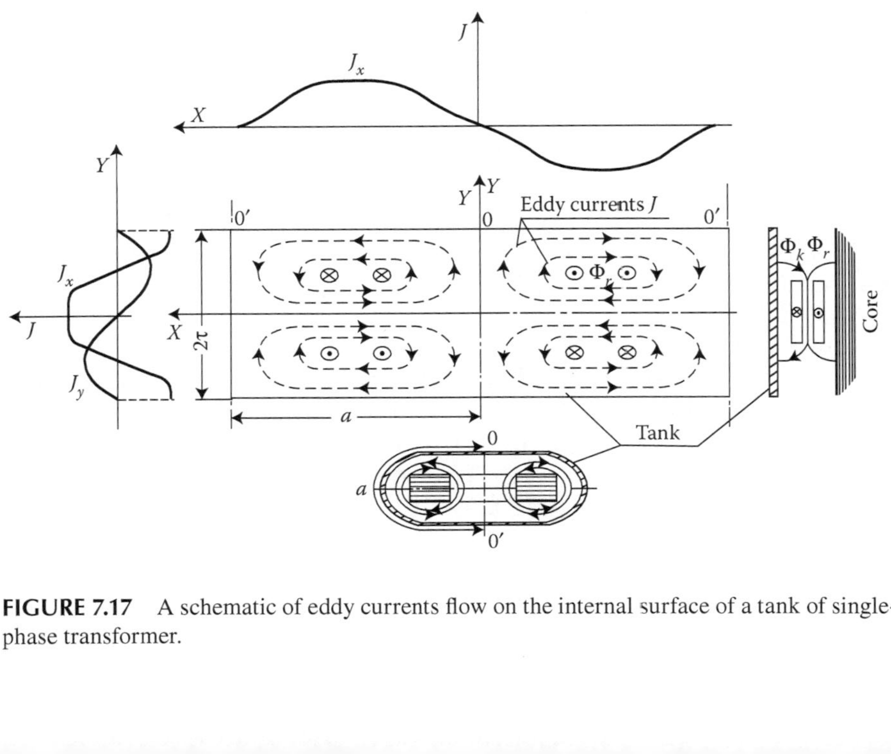

FIGURE 7.17 A schematic of eddy currents flow on the internal surface of a tank of single-phase transformer.

vector potential $\boldsymbol{A}$ (2.50), for an arbitrary mth, nth space harmonic of the field, can be expressed, after E. Markvardt* [1.15/1], in the form ($J_z = 0$)

$$\left.\begin{aligned} A_{mnx} &= \varphi \cos m\nu y \cdot \sin n\eta x \cdot e^{j\omega t} \\ A_{mny} &= \psi \sin m\nu y \cdot \cos n\eta x \cdot e^{j\omega t} \\ A_{mnz} &= 0 \end{aligned}\right\} \tag{7.77}$$

where $\nu = (\pi/\tau)$, $\eta = (\pi/a)$. The functions φ and ψ (of the z coordinate) unknown at this time, can be determined from the condition: div $\boldsymbol{A} = 0$ [1.15/1].

Substituting the components of vector potential (7.77) into (2.54a)

$$\frac{\partial^2 A_{mnx}}{\partial z^2} = \mu\sigma \frac{\partial A_{mnx}}{\partial t}, \quad \frac{\partial^2 A_{mny}}{\partial z^2} = \mu\sigma \frac{\partial A_{mny}}{\partial t} \tag{7.78}$$

we obtain two ordinary differential equations:

$$\frac{d^2\varphi}{dz^2} - (m^2\nu^2 + n^2\eta^2 + j\omega\mu\sigma)\varphi = 0$$

$$\frac{d^2\psi}{dz^2} - (m^2\nu^2 + n^2\eta^2 + j\omega\mu\sigma)\psi = 0$$

The solution of these equations (Turowski [1.15/1], p. 279), at $\mu = \text{const}$, leads to a double harmonic series of the space field

$$\left.\begin{aligned} A_x &= -\sum_{m=1}^{\infty}\sum_{n=1}^{\infty} \frac{B_{mnys}}{\alpha} \frac{\operatorname{ch}\alpha(d-z)}{\operatorname{sh}\alpha d} \cos m\nu y \cdot \sin n\eta x \cdot e^{j\omega t} \\ A_y &= \sum_{m=1}^{\infty}\sum_{n=1}^{\infty} \frac{B_{mnxs}}{\alpha} \frac{\operatorname{ch}\alpha(d-z)}{\operatorname{sh}\alpha d} \sin m\nu y \cdot \cos n\eta x \cdot e^{j\omega t} \\ A_z &= 0 \end{aligned}\right\} \tag{7.79}$$

These formulae completely describe the electromagnetic field in tank steel at given values of the field on its internal surface and at $\mu = \text{const}$.

7.6.2.2 Power Losses in a Tank

Power of electromagnetic field penetrating the tank was determined from Poynting's theorem after formula (3.7) concerning its normal component S_z. The power of mth, nth field harmonics per unit surface is determined from the formula

$$S_{mnz} = \frac{1}{2}\left(E_{mnxs}H^*_{mnys} - E_{mnys}H^*_{mnxs}\right) \tag{7.80}$$

* Markvardt E. G.: Electromagnetic calculations of transformers. Moscow, Russia, 1938, ONTI.

Corresponding components of E and H can be found easily from Equations 7.79 using formulae (2.50) and (2.51), $\boldsymbol{B}$ = curl $\boldsymbol{H}$ and $\boldsymbol{E} = -\,\partial \boldsymbol{A}/\partial t$, from which

$$S_{\text{mnz}} = \frac{j\omega\mu}{2\alpha}\left(H^2_{\text{mnys}}\cos^2 mvy\cdot\sin^2 n\eta x + H^2_{\text{mnxs}}\sin^2 mvy\cdot\cos^2 n\eta x\right)\text{cth}\,\alpha d \quad (7.81)$$

The average apparent power losses per unit surface, instead, are calculated from the formula

$$P_{s1\,\text{aver}\,mn} = \frac{\mathrm{j}\omega\mu}{8\alpha}\left(H^2_{\text{mnys}} + H^2_{\text{mnxs}}\right)\text{cth}\,\alpha d = \frac{\mathrm{j}\omega\mu}{8\alpha}H^2_{\text{mns}}\,\text{cth}\,\alpha d \quad (7.82)$$

where H_{mns} is the space and time amplitude of the resultant mth and nth harmonics of the tangential component of magnetic field intensity on the internal surface of tank.

At normal thicknesses of steel tanks walls (several to a dozen of millimeters) the dimensionless coefficient kd is on the order of 8–10 [line 5 under (4.20)], one can assume, therefore, with rough approximation: $c\text{th}\,\alpha d \approx c\text{th}\,(1+\mathrm{j})kd \approx \mathrm{j}$, and for a few first harmonics: $\alpha \approx \sqrt{\mathrm{j}\omega\mu\sigma}$ (7.48a). With these assumptions, the *average active power losses per surface unit* (W/m^2) *of tank* can be calculated from the formula

$$P_{s1\,\text{aver mn}} = \sqrt{\frac{\omega\mu}{2\sigma}}\frac{H^2_{\text{mns}}}{8} \quad (7.83)$$

The reactive power losses will have the same numerical value. Whereas the per-unit power for all harmonics (var/m^2) can be evaluated from the formula

$$P_{1\text{aver}} = \sqrt{\frac{\omega\mu}{2\sigma}}\sum_{m=1}^{\infty}\sum_{n=1}^{\infty}H^2_{\text{mns}} \quad (7.84)$$

The dependence of power losses in tank on the magnetic flux Φ_T penetrating the tank can be found considering, as per Equation 4.8, that the biggest flux in the tank, per unit length of its circumference (Wb/m), can be calculated from the formula

$$\Phi_{mT} = \frac{\mu H_{ms}}{\alpha}$$

from which, after substitution into Equation 7.82, we obtain ($c\text{th}\,\alpha d \approx \mathrm{j}$)

$$P_{s1\,\text{aver}\,mn} = \frac{\omega}{8}\sqrt{\frac{\omega\sigma}{2\mu}}\Phi^2_{mT} \quad (7.85)$$

Since the flux in the tank is only a part of the total flux leakage $\Phi_{m\delta\text{air}}$ in the inter-winding gap δ_{air} (Figure 4.18), which is proportional to the current, hence from formula (7.85) obtained at a constant permeability, it follows that the stray losses in the tank should be proportional to I^2. However, the author's investigations [10.17] (see also Section 7.3) have shown that the exponent at the current is rather in the range between 2 and 2.3. It is a result of *decreasing permeability $\mu(H)$ (occurring in the denominator) with an increase of the tank flux* $\Phi_{\text{m,tank}}$. Only at very weak fields in the tank, the permeability can be almost constant or slightly growing. This is why the exponent appearing at the current I can be smaller or equal to 2. This has caused *mistakes and erroneous conclusions* of some researchers who, while carrying out measurements at factory testing stations with too low currents, did not notice this phenomenon.

The tank stray field presented above and the loss analysis is relevant to the fields in the form of a standing wave of period equal to the length of circumference of the tank (i.e., single-phase transformers).

In three-phase transformers, due to the phase and space shift of ampere-turns of each of the legs, the stray field of a tank has the character of a wave deformed from sinusoid, traveling along the tank circumference with varying amplitude (Figure 7.22a and b [later in the chapter]). At the same time, the tangential component of this wave can be expressed by a double trigonometric series. On the basis of analysis of the field with solid metal rotor (6.72), it can be concluded that at similar values of amplitudes of particular harmonics, which occur at stationary field, the power losses in three-phase field are a little bigger. From the author's investigations, it follows that they are bigger by about 20–30% [1.16], depending on geometric proportion of the transformer and the tank (Figure 7.22a and b [later in the chapter]).

Ms. D. Przybylak carried out broader analytical investigations of 3D field on the surface of three-phase transformer tanks (PhD thesis, Instytut Energetyki, Poland, 1987).

7.6.2.3 Influence of Flux in a Tank

The ratio of the stray flux penetrating the tank, Φ_{tank}, to the total stray flux $\Phi_{\delta\text{air}}$ in the gap δ_{air} is determined mainly by the distance of the winding from the tank (a_T) and from the core (a_C) (4.60), as well as from the yoke beams (Figure 10.30). According to Markvardt [1.15/1], the flux $\Phi_{m,\text{tank}}$ in an unscreened tank can be calculated approximately with the equation

$$\Phi_{m,\text{tank}} = 1.13\frac{IN\delta'}{h}\left(0.06 + \frac{\text{sh}\pi\dfrac{a_C}{h}}{\text{sh}\pi\dfrac{a_C + a_T}{h}}\right) \tag{7.86}$$

where $\delta' = \delta_{\text{air}} + ((a_1 + a_2)/2)$ is the equivalent leakage gap. Yoke beams have a relatively small influence on the value of the $\Phi_{m,\text{tank}}$ flux. After substituting Equation 7.86 into Equation 7.85, the Markvardt's formula was developed [1.15/1]. In a similar way, many semiempirical formulae were developed, for instance, the "American" formula published in *Tikhomirov's* book, and the formulae by M. Lazarz (*ASEA Sweden; now ABB*), J. Kreuzer (*Elin, Austria*), Berezovski-Niznik-Kravtschenko (*Ukrainian Academy of Sci.*), Z. Valkovits (*"Rade Kontsar", Croatia*), and others. A comparative analysis of these formulae, presented in the Turowski's papers [4.18] and [4.28], showed that these formulae were relatively correct in a specific limited range and for transformer models typical for the specific company. This is understandable, due to significant simplifications and 2D calculations. A typical example is the Lazarz* formula for losses (in W) in tank and solid yoke beams in large transformers:

* Lazarz Mieczysław: "Design measures to reduce the power loss in power transformers". (in German) XI International Congress of Electricians (ITE), Berlin, 1965. Group. Ref. Nr. 3.1.

$$P_{\text{Lazarz}} = \text{k}(f\Phi_{m\delta\text{air}})^{\alpha}\,(\alpha = 1.5 \text{ to } 1.9); \qquad \Phi_{m\delta\text{air}} = \sum_{1}^{n}(H_{m,\text{aver}}t\, l_{\text{average}})n; \tag{7.87}$$

where $H_{\text{m,aver}}$ is the average in space magnetic field intensity on the width of windings including the gap, and l_{average} is the average length of winding. Some remnants of this formula passed probably together with ASEA to ABB, where it is still used with various coefficient modifications (Turowski et al. [4.30]). M. Lazarz confirmed that transformers with a large leakage field (e.g., autotransformers of high voltage) these losses could be large. As one can deduce from the small value of the coefficient α, formula (7.87) concerns transformers with screened tanks.

An advantage of such "statistic" formulae is their simple, parametric form, enabling a rapid assessment of stray losses. Being continuously corrected, they can give results close to the real ones to some extent. They are of course not scientific methods. They tell nothing about localization of losses, and fail when a nonconventional case appears.

During detailed investigations of stray losses in transformer tanks, the author in his works [7.18], [4.18], [4.30] endeavored to consider as much as possible the physical and geometric parameters as well as other factors influencing the power losses in a transformer tank, which resulted in the development of the equations presented below. They have a very useful form for physical understanding, or a rapid, expert design or reduction of 3D stray losses and hot-spots. A full 3D analysis possibility, however, has appeared since the authors developed the numerical model of equivalent magnetic circuit, based on the 3D reluctance network method (RNM-3D), and the interactive computer program, RNM-3D.

7.6.3 Parametric Analytical-Numerical (ANM-3D) Calculation of Stray Losses in a Tank of a Transformer

ANM make it easier to examine the impact of various parameters (J. Turowski [7.18]) on the test object. The analytical derivation of an accurate parametric model for the power losses in the transformer tank, however, is difficult and laborious. In the scientific and technical literature from 1955 to 1965, except the losses in covers (Sections 5.3, 6.3, and 6.4), there was almost nothing about the calculation and reduction of eddy currents (Figure 7.18), although the threat was growing [7.31].

As early as 1983 at the *CIGRE Conference*, it was alarmed that "dispersion of heat caused by the stray flux produces a very large concern, especially in large power transformers" and that now "it is becoming increasingly clear that the failures of large transformers are very expensive and the cost of their repair increasingly growing." Today, after over a quarter of century, modern computers not available then, and the underestimation of Maxwell's field theory, plus mechatronic approach, have helped the author (JT) to solve this problem. In constructing power transformers, however, most important are the details.

Derivation of accurate analytical, parametric formulae for the losses in transformer tanks is a difficult and cumbersome task (Figure 7.18). Simplifications, permanent semiempirical corrections, and computer modeling are inevitable (Figures

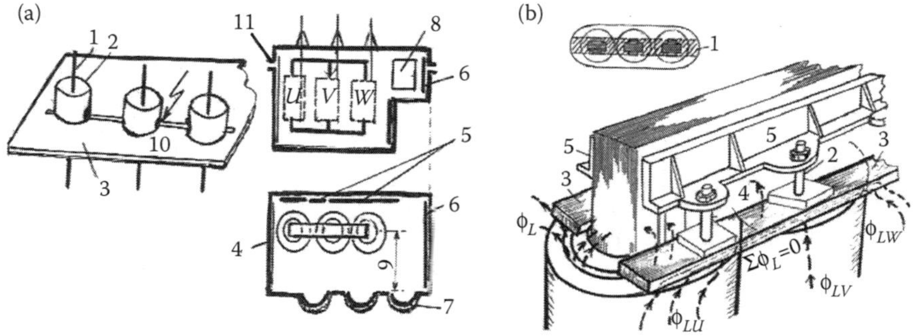

FIGURE 7.18 Power transformer steel parts in 3D field, at risk of excessive stray loss and heating: (a) 1—bushing, 2—turret, 3—cover, 4—tank, 5—Fe screen (shunts), 6—Cu screens, 7—oil pockets, 8—tap changer, 9—tank asymmetry, 10—hot-spot, 11—bolted joints; (b) 1—acceptable region of 2D calculation, 2—bolt and Al/Cu screen, 3—stray flux collectors, 4—stray flux towards metal plane, 5—yoke beam. (Adapted from Turowski J.: *Fundamentals of Mechatronics* (in Polish). AHE-Lodz, 2008 [1.20].)

7.21 through 7.24 [later in the chapter]). For this purpose, however, formula (7.74) can be used for the power loss in a metal wall placed in the field of a pair of bars (Figure 7.12) with AC current:

$$P_1 = 1.75 \times 10^{-11} hf\sqrt{f\sigma}\eta_Q^2 I_k^2 (k_1 + 6.6 \times 10^{-9} k_2 \sqrt{f\sigma}\eta_Q I_k) \tag{7.88}$$

where $I_k = IN$ is the rms value of the current flow in winding, in A.

The power losses calculated with formula (7.88) significantly depend on the assumed value of the average mirror-image coefficient η_Q (in Equation 7.74, $\eta_Q = 0.8$ was assumed). However, in result of subsequent mathematical transformations (J. Turowski [7.18]) it occurred that the coefficient η_Q significantly reduces itself, and in the end, in the formula for power losses in the tank, the η_Q plays a smaller role.

In order to apply formula (7.88) to calculate the power losses in tank, it had to be adjusted to accommodate the following considerations:

a. Split of the stray flux $\Phi_{\delta air}$ between the core flux Φ_{core} and the tank flux Φ_{tank} (4.59)
b. Influence of the 2D, looping shape of the eddy current flow paths in real transformer tank (Figure 7.17)
c. Variable distance of tank walls from winding, in result of the curvature of windings and three-phase core structure (Figures 7.19 and 7.23a and b [later in the chapter])
d. Superposition of fields of adjacent phase windings
e. Curvatures of the system

These factors were consecutively investigated both theoretically and experimentally in the work by Turowski [7.18] and others.

The influence of a split of the stray flux $\Phi_{\delta air}$ between the core flux Φ_{core} and the tank flux Φ_{tank} ($\Phi_{\delta air} = \Phi_{core} + \Phi_{tank}$) has been considered by assuming that the tank losses are generated only by the flux Φ_{tank}. Since the interwinding gap δ_{air} is the main reluctance in the way of these fluxes; hence, in the first approach one can presume that these fluxes are proportional to the current I, and more accurately $\Phi_{\text{mtank}} = cI^{\beta_1}$, where $\beta_1 = 0.9$–0.8 (7.41). Then, if I is the current of the winding, the power losses in tank wall (Figure 7.19) can be calculated approximately from formula (7.88), if in place of this current we substitute the computational current

$$I_k = I\frac{\Phi_{m,\text{tank}}}{\Phi_{m\text{Fe}}} \tag{7.89}$$

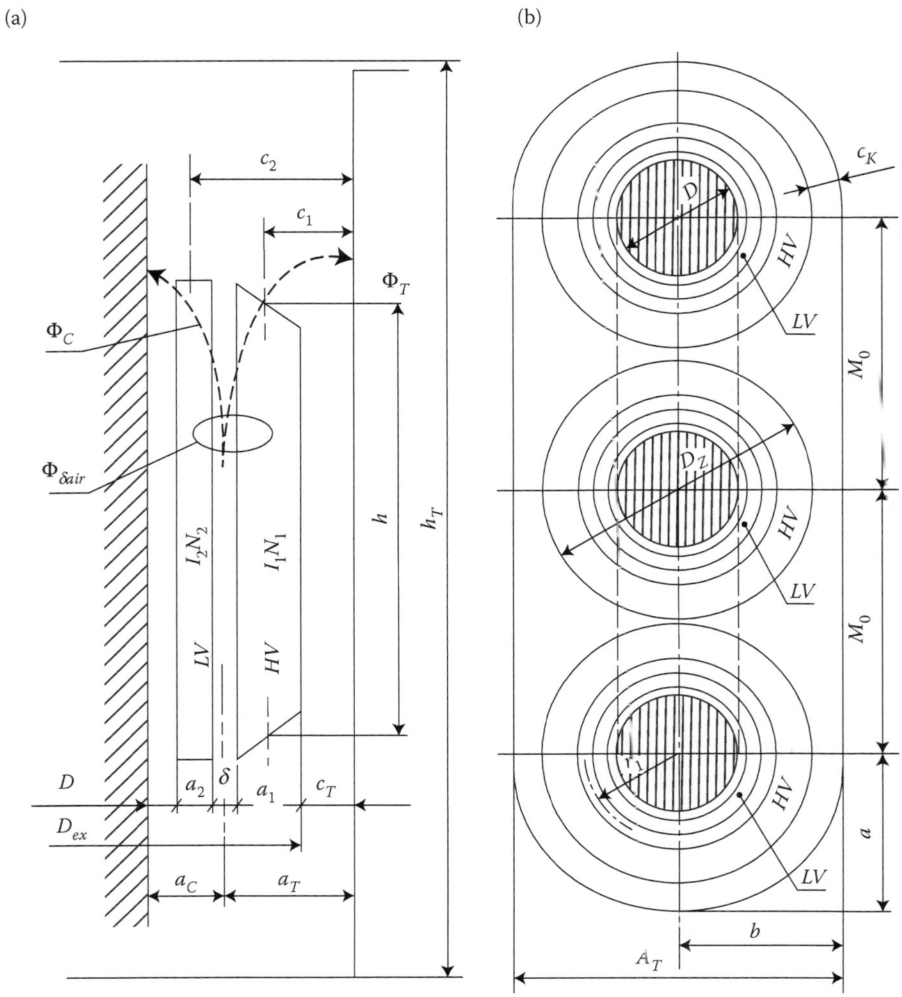

FIGURE 7.19 Computational dimensions of a three-phase transformer. (Adapted from Turowski J.: The formula for power loss in unscreened tanks of large single-phase transformers. (in Polish) *Rozprawy Elektrotechniczne*, (1), 1969, 149–176 [7.18].)

At geometrical proportions typical for large power transformers, one can assume that the flux $\Phi_{m\delta\text{air}}$ in δ_{air} gap is divided between the core (Φ_{core}) and the tank wall (Φ_{tank}) inversely proportional to the distance of the core (a_C) and of the tank (a_T) from the gap axis (4.59). Moreover, considering the shunting of the tank wall by the adjacent oil layer (c_T in Figure 7.19a), we accept the formula

$$I_k = I\frac{k_p a_C}{a_C + a_T}\frac{\Phi_{m\delta\text{air}}}{\Phi_{m\text{Fe}}} \tag{7.90}$$

where

$$k_p \approx \frac{1}{1+\dfrac{c_T}{\mu_r}\sqrt{\omega\mu\sigma}} \approx \frac{1}{1+0.38\,c_T\sqrt{f}} \tag{7.91}$$

$$\Phi_{m\delta\text{air}} = \sqrt{2}IN\mu_0\delta'/h_R \tag{7.92}$$

$$\delta' = \delta_{\text{ins}} + \frac{a_1 + a_2}{2} \quad \text{and} \quad h_R = \frac{h}{K_R} \tag{7.93}$$

K_R is the Rogowski's coefficient (4.58).

The flux $\Phi_{m\text{Fe}}$ penetrating a steel plate in a one-sided system (Figure 7.12) was calculated by integration, from 0 to ∞, of the normal component H_{mz0} of the field from the air side

$$\Phi_{m\text{Fe}} = \int_0^\infty \mu_0 H_{mz0}\,\text{d}y = \eta_Q\mu_0\frac{\sqrt{2}IN}{2\pi h}\left| h\cdot ln\frac{h^2+4c_1^2}{h^2+4c_2^2} + 4\left(c_1\text{arc tg}\frac{h}{2c_1} - c_2\text{arc tg}\frac{h}{2c_2}\right)\right| \tag{7.94}$$

However, the initial formula (7.88) concerns a single-dimensional field on the surface of steel plate. In the work (Turowski [7.18]), on the basis of a similar analysis as in Section 7.5.1, it was shown that power losses at a 2D field and looping flow of eddy currents (Figure 7.17) are about 50% of the losses occurring in single-dimensional field (Figure 7.12). This conclusion was confirmed experimentally on the inversed model (Figure 10.3) built under the direction and design of the author [2.41] and [5.16].

The final formula for the power losses in a transformer tank consists of the two main components:

a. Big losses occurring in the rounded parts of tank, with a constant distance a_C between the windings and the tank wall.
b. A smaller, but more difficult to compute, component of losses existing in the rectilinear parts of tank circumference (Figures 7.19b and 7.22a and b [later in the chapter]), where the fields of two or three legs compensate each other to a significant degree.

In the rectilinear part of tank, the distance between the windings and the tank walls is variable, and in [7.18] it was substituted by the average computational distance $a_{T\text{aver}} = a_T + 0.108 \cdot D_{\text{aver}}$.

Finally, after considering other simplifications [7.18] and later investigations [1.16], we obtain the formula containing the highest number of constructional parameters among all the recent formulae of such type (section 7.5.1). Thus, the power losses in the tank (in watts) are equal, after (Turowski [7.18]):

$$P \approx 3.44 k_m p_e \times 10^{-11} A_k h f \sqrt{f\sigma} I_k^2 \left(k_1 + 6.1 \times 10^{-9} k_2 \sqrt{f\sigma} I_k\right) \tag{7.95}$$

where the reduced computational current flow of one phase winding is

$$I_k = (Iz)^{\beta} k_p \frac{2\pi a_c}{a_c + a_T \dfrac{D}{D_{\text{ext}}}} \frac{\delta'}{h_R} \left| \frac{1}{\ln \dfrac{1+(2c_1/h)^2}{1+(2c_2/h)^2} + 2\left(\dfrac{2c_1}{h} \text{arc tg} \dfrac{h}{2c_1} - \dfrac{2c_2}{h} \text{arc tg} \dfrac{h}{2c_2}\right)} \right| \tag{7.96}$$

and

- $\beta = 0.99$–0.98 is the influence of eddy-current reaction in the tank (it is small; see Figure 10.30).
- k_p is the coefficient of tank wall shunting by a parallel oil layer c_T, according to (7.91) [7.18], p. 156.
- k_m is the coefficient accounting for construction specifics and the number of phases (legs) of transformer, as well as other factors which were not considered in the adopted assumptions and simplifications, while for different power transformers the recommended k_m values are
  - Single-phase, two legs—$k_m \approx 1$
  - Single-phase, shell-type—$k_m \approx 0.2$ to 0.3
  - Three-phase, shell-type, five legs—$k_m \approx 0.4$ to 0.6
  - Three-phase, three legs—$k_m \approx 1.2$ to 1.3
  - Three-phase, three legs, with a tap-changer in the tank (one round end of tank shifted away)—$k_m \approx 0.9$ to 0.96
  - p_e is the screening coefficient of tank (4.63), (4.64), (4.65), (4.67).
  - k_1, k_2 is the coefficients of the geometrical proportions of the system, in functions of h/c_1 and c_2/c_1.

  according to Figure 7.13 and Figure 7.14, calculated with a computer program.

Formula (7.95) was programmed on a computer in the Power Engineering Institute, Warsaw (J. Gortatewicz et al.) as well as in the Technical University of Lodz (by L. Byczkowska, [4.23] and by G. Sobiczewska)—the computer code called "Mast. Straty '82," based on the author's formula (7.95) but with less simplifications, also contained comparisons with calculation results using the statistic factory formulae of Markvardt (Moscow), Tickhomirov (Moscow-USA),

M. Łazarz (ASEA Ludvika, Sweden) and Kreuzer (Elin, Austria) [4.18] (see Section 7.6.2).

EXAMPLE

A sample set of input data for the "Mast. Straty '82" computer program (Figure 7.20) by G. Sobiczewska*:

S = 127 × 10[6] VA—the apparent power of transformer
F = 50.0 Hz—frequency
U1F = 141.46 × 10[3] V—phase voltage
U2F = 11.00 × 10[3] V—phase voltage
I1F = 299.30 A—phase current
I2F = 3848.5 A—phase current; two parallel branches 1924.24 A × 72 turns
UZ = 16.4%—short circuit voltage
UX = 16.5%—reactive component of short circuit voltage
GAMMA = 7 × 10[6] S/m—conductivity of steel
M = 3—number of phases
MK = 3—number of wound legs
IKOL = 3—number of transformer legs
KADZ = 2—type of tank (1—wavy, 2—flat)
EKRAN = 0—type of screen (0—no screen, 1—electromagnetic screen, 2—magnetic (shunt)
Z1 = 926—number of turns of HV winding
Z2 = 144—number of turns of LV winding

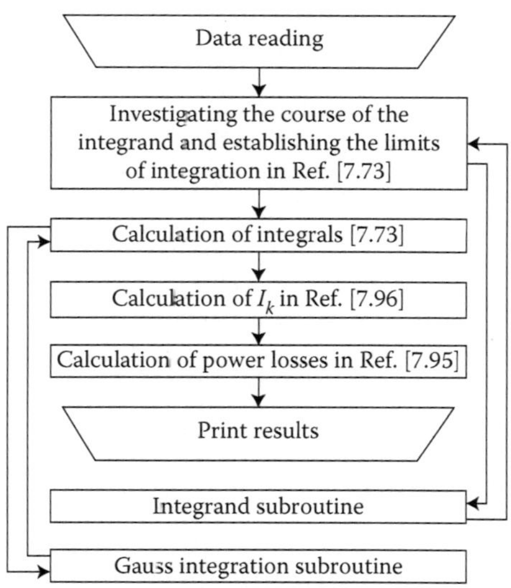

FIGURE 7.20 Flowchart of the numerical computation of power losses in a transformer tank, using formula (7.95), performed on a computer, by Turowski and Byczkowska [4.23].

* Sobiczewska G., Turowski J.: Program in Fortran 1900 for Odra-1305. ITMA PŁ No4213/211/1982, for ELTA Works.

H = 2,603 m—height of winding
HK = 4.7 m—height of tank
D = 0.76 m—diameter of core leg
DZ = 1.493 m—external diameter of windings
A1 = 0.1505 m—thickness of external winding
A2 = 0.12 m—thickness of internal winding
DELTA= 0.084 m—interwinding gap
MO = 1.61 m—the distance between the axes of legs
A = 1.02 m—the smaller radius of the rounded part of tank
B = 1.28 m—the bigger radius of the rounded part of tank
AFE = 0.385 m^2—the active cross section of leg
GAMMA2—conductivity of screen (S/m), if applicable
GAMMA3—conductivity of tank (S/m), behind the screen (if applicable)
MI3—permeability of tank material before screening (H/m)
DE—thickness of screen (m)
N—number of sheets (when the magnetic screen consists of flat layers, when EKRAN = 2)

Output printouts (results of calculations)

- Per Turowski formula (7.95): 159–236 kW (k_m = 1.2–1.3; β = 0.99–0.98)
- Per Markvardt formula: 176 kW
- Per Tikhomirov (US license) formula: 159 kW
- Per Kreuzer formula: 71–119 kW (*unknown screens or shunts*)
- Per Łazarz formula: 32–56 kW (*unknown screens or shunts*)

Note: The remarkable span of the above results is caused by differences in coefficients delivered by different authors. Therefore, the calculations were carried out for the smallest and the highest of possible values. Thus, these are extreme limits.

The example presented above proves that in order to obtain a smaller scatter of results, the parameters and structure of the transformer should be more accurately considered. Apart from experimental verification of all stages of the development of the final formula (7.95), it was also verified by the author (JT) on a large single-phase transformer (250/3 MVA, 242/√3 kV), thanks to the author's scientific collaboration with late Eng. *J. Kopeček* and Eng. *J. Klesa* from SKODA Transformer Works in Plzen (Czech Republic), as well as on a three-phase power transformer (127 MVA, 245/11 kV), thanks to the scientific collaboration with Eng. *M. Woolard* from Hawker Siddley Power Transformers in Walthamstow (London, UK), for which the author expresses his warmest thanks. In both cases, a good agreement of calculations with industrial measurements was obtained (J. Turowski [7.18]). A similar, good agreement has been obtained from the analysis and measurements in the Transformer Works ELTA (now ABB) in Lodz, Poland. However, a more accurate determination of the coefficient k_m is possible only by solution of a 3D model, as presented below (Figures 7.22 and 7.23a and b [later in the chapter]). Also, many other transformer works around the world confirmed it enthusiastically (see Table 11.2—list of users of RNM-3D).

7.6.4 Three-Dimensional Numerical Calculation of Stray Fields and Losses in Large, Three-Phase, Power Transformers

3D analytical-numerical ANM solutions (Sections 7.5.1. through 7.6.2) have the obvious advantages of showing the parametric dependences; however, they also have a fundamental defect such that the test object must be reduced to a relatively simple 2D model without the complex structure of screens, yoke beams, clampings, screws, and so on.

In recent decades, there has been enormous progress in the development of computational hardware and, in particular, sophisticated software for solving complicated, nonlinear, 3D fields. However, one should carefully select a method most suitable for a given case. For instance, the traditional Biot–Savart and Ampere's law (*law of flow*) has proven to be one of the best methods for calculating losses and hot-spots in transformer covers (Sections 5.3, 6.3, and 7.4). The most popular finite element method FEM-2D is also simple, but FEM-3D, which today is absolutely necessary, is very difficult, laborious, and costly. In terms of 3D solutions, a competitor to FEM-3D is the cheap, easy, and fast, hybrid reluctance network method RNM-3D.

7.6.4.1 FEM-3D

An example of a large-scale computation with the application of FEM-3D was presented by General Electric Co. (GEC) of Stafford, UK; see Figure 7.21 (Coulson et al. [7.6]).

However, the authors of [7.6] concluded that it was "not profitable for regular design use... and simpler methods of design of large generators and transformers should be searched". Since that time (1983), a tremendous progress in computational hardware and software has been made. However, the basic problems still remain. A typical challenge in the full 3D FEM analysis of power apparatus is given by the presence of large structural parts (several meters in length and height) with thicknesses on the order of centimeters, made of highly permeable and conductive material, while these parts have very small penetration depths, on the order

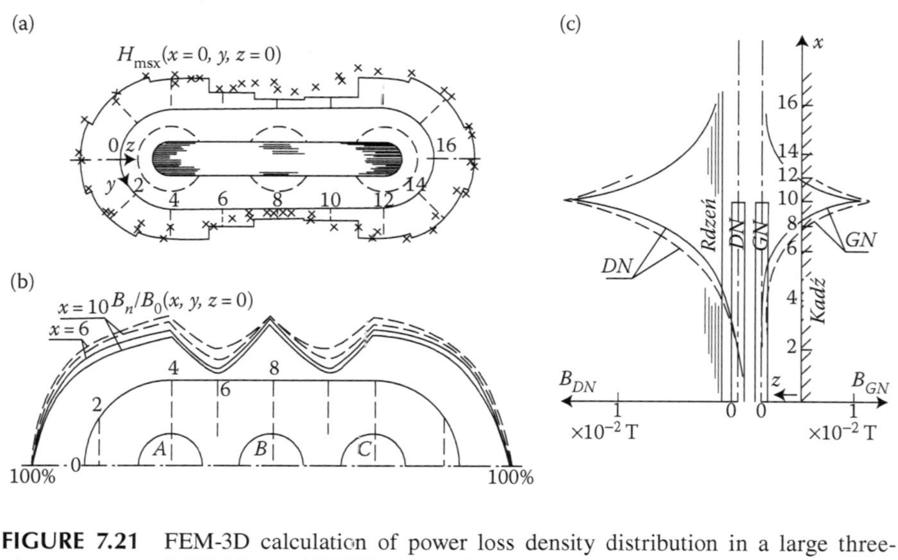

FIGURE 7.21 FEM-3D calculation of power loss density distribution in a large three-phase reactor (1680 Mvar, 400 kV). Reproduction with kind permission of GEC, Stafford Transformers, UK. (Adapted from Coulson M.A., Preston T.W., and Reece A.B.: *COMPUMAG'85 Conference*, Fort Collins, USA, 1985 [7.6].)

of couple of millimeters. The numerical treatment of such domains is a significant challenge for numerical analysis: the discretization needs to be very small, for example, with micrometer-size steps, in order to resolve the field penetration, which in turn translates into huge models, completely impracticable even on very powerful machines (e.g., with 64-bit processors) [9.13]–[9.15]. Simplified 3D FEM models may still take many hours of computation [7.36], [7.37]. Present FEM-3D simulations may need several dozens, if not hundreds of hours of preparatory work of a highly qualified specialist, and a complex, expensive software package. This is why, in transformer works, approximate or significantly simplified 2D models are still used, as well as semiempirical formulae, like those presented above.

The computational conflict of interfacing large (in tens of centimeters) finite elements in the leakage field region with very small (in μm) elements in the tank wall made of solid steel of ca. 15 mm thickness, with a few mm depth of exponential penetrating, was resolved by J. Turowski [1.15] by representing such a wall by a single reluctance element of thickness δ (Figure 2.11).

7.6.4.2 Three-Dimensional, Equivalent Reluctance Network Method: RNM-3D

In anticipation of the above-mentioned difficulties, the author of the book (J. Turowski [1.15]) already in 1960 commenced investigations in order to obtain a reliable, equivalent reluctance network method. In 1979, the author [7.19] created a 3D model and then, in early the 1980s, an algorithm (J. Turowski [4.19]) called "RNM-3D" (Figure 4.20a). In collaboration with the then young students of electronics engineering, Marek Turowski and Miroslaw Kopeć [4.27], the RNM-3D program soon reached such an efficiency level that it was able to calculate one constructional variant in a time shorter than 1 s (J. Turowski and M. Turowski [4.20] and [4.28] with M. Kopeć).

The RNM-3D calculations were successfully verified by experiment on numerous industrial cases, including a big industrial transformer (Figure 7.22), and many others (Table 11.2).

It is also possible to model five-leg transformers by introduction in Figure 4.20a zero-value reluctances at the coordinates of numbers (x = 1 to 21, y = 0 and 16, z = 2) as well as at (x = 20 and 22, y = 0 and 16, z = 3 and 5), respectively. Positive results of RNM-3D were also confirmed experimentally on 30 (thirty) ELTA-Lodz transformers (Turowski et al. [4.30]) by various specialists using several models, and many other large transformers in different times (e.g., Turowski et al. [2.34], [7.21], [4.28]), as well as by various researchers and industry designers, including more than 44 transformer works (Table 11.2) all over the world, from the United States, Canada, Europe, Turkey, Iran, India, China, Japan, Australia, Brazil, and so on.

The RNM-3D program and calculations are based on an expert system and interactive approach (Figures 1.1c, 1.1d, and 4.20c), with *pre-* and *postprocessor* and a special, dedicated Solver MSR-1100 (M. Turowski, M. Kopeć [4.27]), and recently Java, for EFACEC [4.31]. An important advantage of RNM-3D is its very fast operation, especially valuable for rapid modifications and optimization of the initial construction shape [4.28], [7.21]. It enables a multisided investigation of 3D fields, spatial screens, shunts, losses, and excessive heating hazards (Figures 4.20 through 4.22, 7.23, and 7.24).

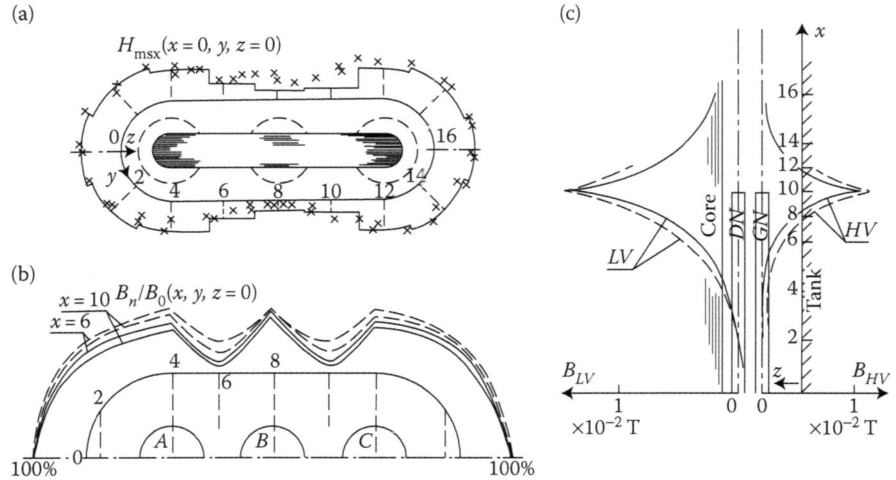

FIGURE 7.22 Experimental verifications of three-dimensional calculation of the field in tank of a large three-phase transformer: (a) by J. Turowski and S. Pietrzak 1968; (b), (c) M. Kazmierski 1970 [5.9]; — calculated with RNM-3D; × ××, - - - - measured. (Adapted from Turowski J., Kopeć M., and Turowski M.: *ISEF'89. Electromagnetic Fields in Electrical Engineering. COMPEL*. London: James & James, 1990, pp. 113–116; and *ITMA'91*. New Delhi, India [4.28].)

7.6.5 Industrial Implementation and Verification of the RNM-3D

7.6.5.1 Industrial Implementation of the RNM-3D Package

The main disadvantage of the above-mentioned, statistical "factory" formulae (Sections 7.6.2 and 7.6.3), besides remarkable simplifications, is their limitation to simple structures. A breakthrough in the possibility of rapid calculation of complex

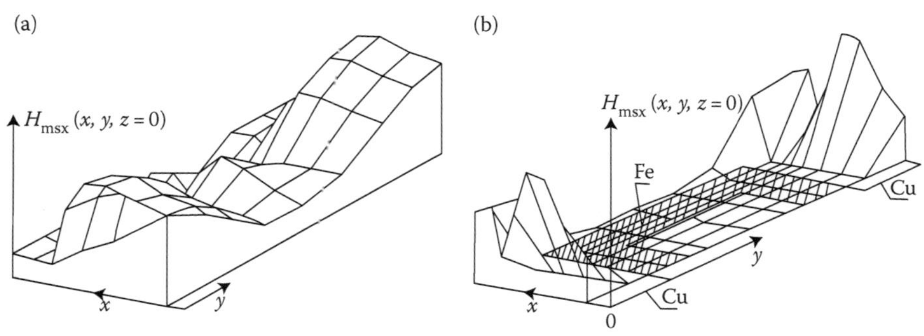

FIGURE 7.23 An interactive study of the field H_{max} on the surface of tank of a 150 kV three-phase transformer, with the help of the RNM-3D program: (a) without any screens; (b) with partial, combined Cu screens and Fe frame (Figure 4.20b). (Adapted from Turowski J., Kopeć M., and Turowski M.: *ISEF'89. Electromagnetic Fields in Electrical Engineering. COMPEL*. London: James & James, 1990, pp. 113–116; and *ITMA'91*. New Delhi, India [4.28].)

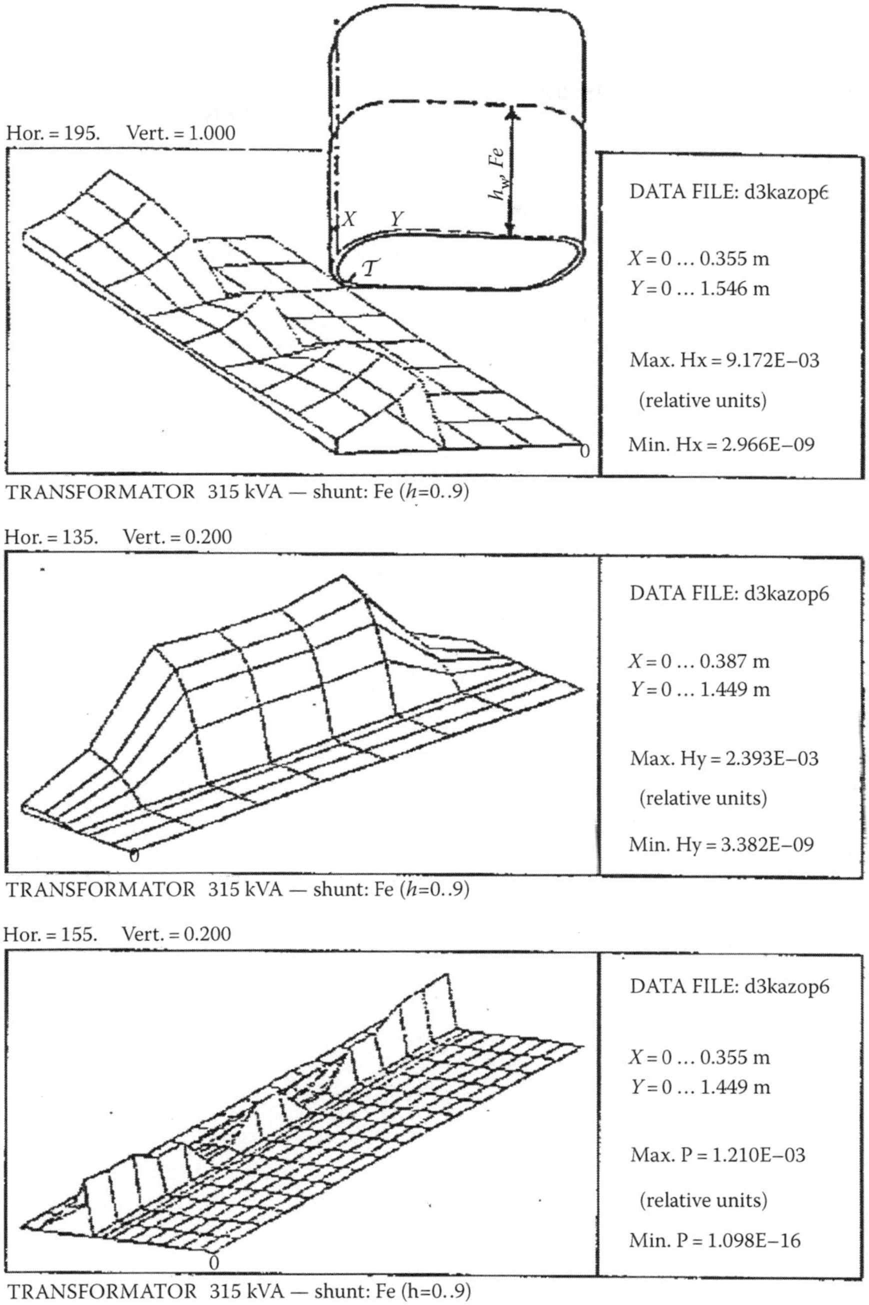

FIGURE 7.24 Example output results of RNM-3D three-dimensional calculations of tank losses with various wall screening. (Adapted from Turowski J., Kopeć M., and Turowski M.: *ISEF'89. Electromagnetic Fields in Electrical Engineering. COMPEL.* London: James & James, 1990, pp. 113–116; and *ITMA'91.* New Delhi, India [4.28].)

3D fields appeared in 1979 (J. Turowski [7.19]) when a 3D, reduced network model of transformer and the RNM-3D program was implemented. It was first verified in Transformer Works in India: "*GEC of India*" (Allahabad) and *Crompton Greaves* (Bombay = Mumbai), then in the United States: "*North American Transformers*" (Milpitas, CA), and then applied with success in more than 40 transformer works and research centers (Table 11.2) all over the world [1.8], [1.20]. The RNM-3D results agreed with factory tests accurately within 0–4% [2.24] of load losses, which was recognized as satisfactory or very good. A typical opinion has been as follows: "We are using your RNM-3D package extensively and for most of the jobs stray loss figure is closely matching with the tested value. The package is very much helping to understand the dependence of stray losses on various parameters of transformer design and shunt/screen arrangement" (D. A. Koppikar, CGL, Bombay 1993).

In the ABB-Elta Transformer Works in Lodz (Poland), there were carried out validations of about 30 large transformer units of power from 25 MVA/115 kV to 330 MVA/410 kV.

From among the main software packages developed at the Technical University of Lodz:

a. The simplest, RNM-3Dexe, for transformers with an oval tank, quarterly symmetrical [4.28]
b. The extended RNM-3Dasm for tanks with extensive asymmetry and [2.24], [2.39], [7.30], [7.31]

it was decided to use the simplest program RNM-3Dexe (a) as long as it allows an accuracy of computed results. The RNM-3Dexe program, thanks to its simplicity, allows a rapid and easy modeling of many cases.

With a clever approach, one can use it also for more complicated, asymmetric structures, using the symmetrization method [2.24]. With the use of RNM-3Dexe, all parameters can be calculated automatically within a time of less than 1 s [4.28], [4.29]. If it is necessary, one can also enter into the internal structure of the model, for instance, to consider differences between oval and rectangular shapes of tank. In such a case, the boundary reluctances should be adjusted with additional parallel reluctances and additional radial reluctances of the shaded cross section in Figure 7.25.

7.6.5.2 Transformers without Screens, Almost Symmetric

Smaller transformers usually have almost symmetrical (Figures 7.26 and 7.27) tanks. The short time for the investigation of large number of units prevented authors from applying the modifications mentioned above according to Figure 7.25. Calculations

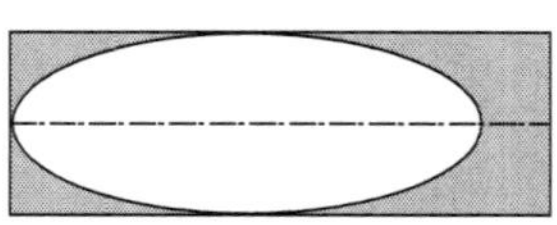

FIGURE 7.25 The principle of modification of RNM-3Dexe oval, symmetric model of transformer tank to a rectangular and/or asymmetric model. (Adapted from Turowski J., Kraj I., and Kulasek K.: *International Conference Transformer'01*, Bydgoszcz, Poland, June 2001 [4.30].)

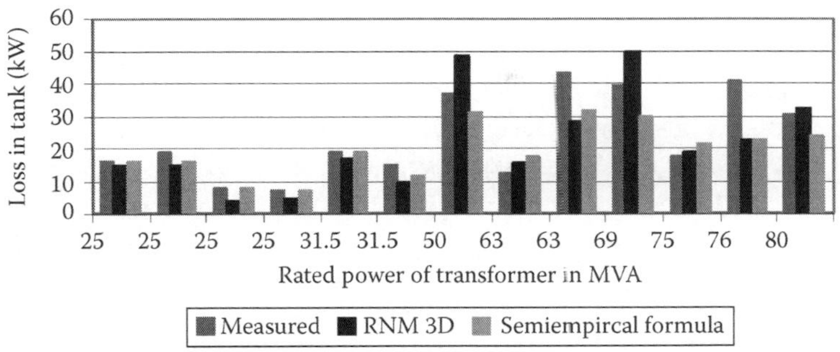

FIGURE 7.26 Stray losses in ABB-ELTA transformers with no screens. (Adapted from Turowski J., Kraj I., Kulasek K.: Industrial Verification of Rapid Design Methods in Power Transformers. *International Conference Transformer'01*, Bydgoszcz, Poland, June 2001 [4.30].)

were carried out automatically, in purpose to evaluate in a first approach, without consideration of constructional details, to what extent one can use the symmetrical, oval model of RNM-3Dexe, and when the intervention into a model structure is necessary or when the extended program RNM-3Dasm is absolutely necessary. In the results, for the group of transformers of 25–80 MVA (Figure 7.26), a relatively

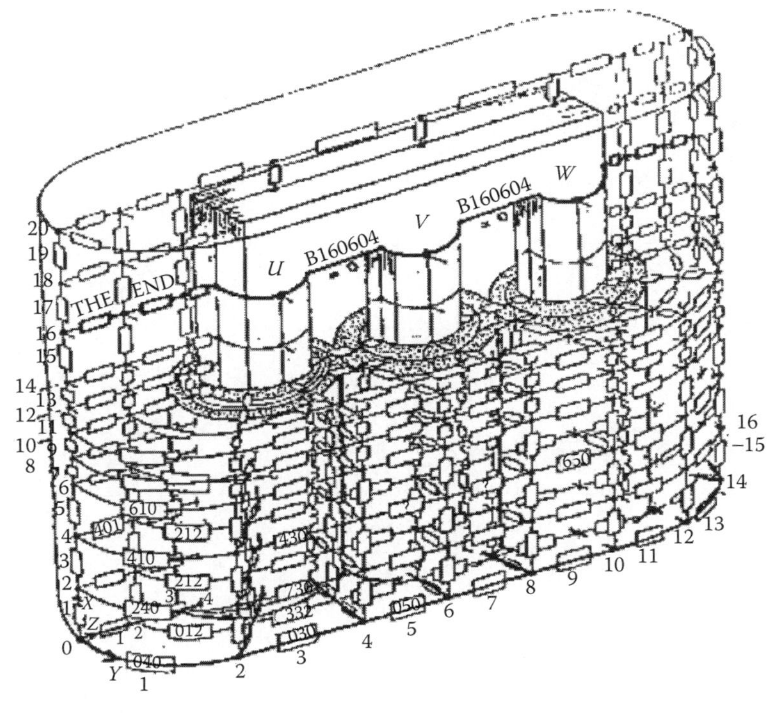

FIGURE 7.27 Symmetric model RNM-3Dexe.

good accordance was obtained between RNM-3Dexe calculated results and those measured at a test plant, as well as with a semiempirical statistical ABB "factory" formula. Single cases of larger discrepancies, as in the 76 MVA unit (Figure 7.26), need separate investigations of design details in order to determine what could have been the reason of that. Perhaps, there were bigger differences with respect to the symmetric model, or some other reasons.

For instance, at not a very large asymmetry with respect to the $Y = 0$ and $X = 0$ planes (Figure 7.27) we can, as in the paper [2.24], divide the transformer into four quarters and for each of them carry out a calculation as for a symmetric transformer. In the example in [2.24] of a 315 MVA/440 kV transformer with partial magnetic screens (shunts) on a tank wall, manufactured by *Crompton Greaves Co, Bombay,* four different partial values of stray losses were calculated.

The sum of the component quarters gave: 11.84 kW + 10.31 kW + 7.975 kW + 6.849 kW = 36.97 kW of total losses in tank. The total calculated load losses were 537.2 kW and those measured at the testing plant were 524.5 kW. It gave a good accordance of 2.4%, not achievable with other methods.

7.6.5.3 Large Transformers with an Extensive Asymmetry

In the second group of screened transformers (Figure 7.28), up to the rated power of 160 MVA the accordance of calculation, test results, and semiempirical formula was still good.

Disregarding single cases of 187 and 300 MVA, which need a separate investigation, all larger units show significantly higher measured power losses in a tank than the losses computed with the RNM-3Dexe program. Similarly higher are the losses calculated with the factory formula. It is not a surprise because the latter one is obtained from the testing data. It may mean that tanks of these transformers are not sufficiently screened, or the simplified RNM-3Dexe is not able to cover a strongly asymmetrical structure of the transformer. The latter conclusion was confirmed by a single tentative recalculation of the case of 300 MVA (Figure 7.28) using the full "asymmetric" program RNM-3Dasm.

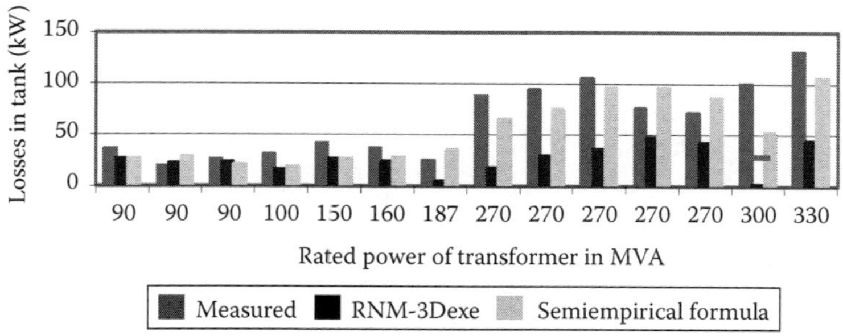

FIGURE 7.28 Stray loss in ABB-ELTA transformers with magnetic screens (shunts). The short horizontal line for the 300 MVA transformer indicates the power loss calculated with the full program RNM-3Dasymmetric. (Adapted from Turowski J., Kraj I., and Kulasek K.: Industrial verification of rapid design methods in power transformers. *International Conference Transformer'01*, Bydgoszcz, Poland, June 2001 [4.30].)

In order to examine the influence of asymmetry, the worst case of 300 MVA was recalculated with the help of the full "asymmetric" program, called *RNM-3Dasm*. As a result, 32.43 kW was obtained, as expected, instead of the wrong value of 1.58 kW obtained with the simplified RNM-3Dexe package. It was, of course, not a calculation error, but the application of a too simplified program to a more complex structure.

At more extensive asymmetries, application of the RNM-3Dasm program is inevitable. Taking into account that capitalized cost of load losses can reach from 3000 to 10,000 US$/kW, these useless superfluous power losses, between 30 kW and 80 kW (which may be worth up to 800,000 US$) *per one transformer*, are worthwhile to be investigated in detail with the use of an "asymmetric" model and relevant computer program. Even the example presented above from the paper [4.30] demonstrates that neglecting even small asymmetry can lead to errors of up to 60%.

7.6.5.4 Influence of the Structure and Screens Configuration

An additional source of discrepancies between calculations and measurements can be the selection of the model of the structure and location of screens. In the RNM-3Dexe model it was assumed, in fact, that screens were ideal. It means that they are so laminated that the leakage flux lines always pass along the sheets plane (curve 4 in Figure 7.29a, [4.30]).

In an opposite case, the reluctance (magnetic impedance) $R_{screen,\perp}$ of a package situated perpendicularly, across field lines (curve 2 in Figure 7.29a) will be expressed by the formula as for a solid steel:

$$\underline{R}_{my} = \Delta y(R_{m1} + jR_{m1r}) \quad R_{m1} = a_1\sqrt{\frac{\omega\alpha}{\mu_s}}; \quad R_{m1r} = \frac{a_2}{a_1}R_{m1} \tag{7.97}$$

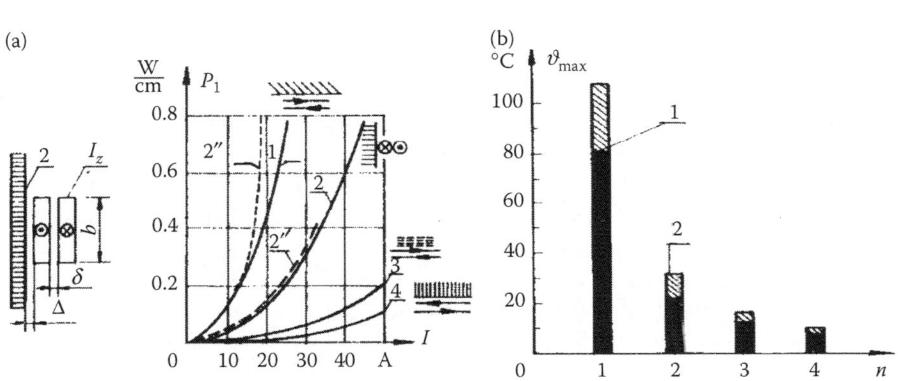

FIGURE 7.29 The influence of structure of laminated shunt package of transformer iron sheets on power loss and local overheating hazard: (a) power losses measured in steel walls laminated in different position and loss in solid constructional steel. (From Turowski J., Pawłowski J., and Pinkiewicz I.: The model test of stray losses in transformers. (in Polish) *"Elektryka" Scientific Papers of Techn. University of Lodz*, (12), 1963, 95–115.) (b) Influence of subdivision of laminated external package of core into *n* strips. (From Kazmierski M. et al. *CIGRE 1984 Plenary Session*. August 29—September 6, 1984. Report 12-12, 1–6 [5.16], [10.18].)

where $a_1 \approx 0.37$, $a_2 \approx 0.61$ are the linearization coefficients of steel [1.15], [4.30], σ is the conductivity of solid steel, and $\mu_s = \mu_s(H_{ms})$ is the surface permeability of solid steel [1.15].

7.6.5.5 Screening Mistake Risk

The value $R_{screen,\perp}$ for such a wrong screen, after Equation 7.97, would then vary from the reluctance of solid steel (curve 1, Figure 7.29a) exclusively by conductivity σ, which means

$$R_{screen,\perp} = R_{steel}\sqrt{\frac{\sigma_{screen}}{\sigma_{steel}}} = R_{steel}\sqrt{\frac{2.1}{7}} = 0.55 R_{steel} >> R_{screen,\equiv} \approx 0.$$

Whereas reluctances of the screen along the sheet plane $R_{screen,\equiv}$ (curve 4, Figure 7.29a) are almost equal zero.

At the same time, the power loss (in W/m²)

$$P_1 = a_p\sqrt{\frac{\omega\mu_s}{2\sigma}}\frac{|H_{ms}|^2}{2} = \frac{\omega}{a_p}\sqrt{\frac{\omega\sigma}{2\mu_s}}\frac{|\Phi_{m1}|^2}{2} \tag{7.98}$$

at $R_{screen,\perp}$, depending on way of excitation (Φ_m, or H_m), can amount from 55% (curve 2′, Figure 7.29a) to even $\sqrt{(7/2.1)} = 183\%$ of the losses in solid steel (curve 2″, Figure 7.29a).

Note: Erroneous (e.g., perpendicular to the field) positioning of screens (shunts) of laminated steel, may not only reduce losses, but even increase them significantly!

Since the losses in the rectilinear part (Figure 7.27) of the tank are 20–30% of the total loss in the tank, it means that an improper installation of screens, like at the curve 2 (Figure 7.29a), can cause superfluous, unwanted loss in the order of 30–40 kW.

Correspondingly, the unnecessary capitalized cost of power losses can reach from US$ 90,000 to 400,000 per one transformer unit.

As we can see, even old screen structures, such as in [10.18] and Figure 7.29b, made of strips "flatwise" (curve 3, Figure 7.29a) can be better than the error in packages situated perpendicular.

The RNM-3D program enables a rapid evaluation of effectiveness of screens and eliminates redundant power losses and their costs. It is in the interest of both parties—sellers and buyers of transformers. In addition to losses, one can investigate exact localization of hot-spots, excessive crushing forces, and so on. In a similar way, the influence of other geometric, material, and electric parameters on power losses in tank,* cover,† clamping, yoke beams, and so on were investigated (Turowski et al. [4.30]).

* I. Kraj: "Influence of tank screening on load loss in transformers", MSc thesis (in Polish). Tech. Univ. of Lodz, 2001, supervised by J. Turowski.

† N. Owczarek: "Influence of structure of transformer cover on additional losses due to bushing currents". MSc thesis (in Polish). Tech. Univ. of Lodz, 2001, supervised by J. Turowski.

Since, after Equation 7.98, the value of $H_{ms}(x, y, z)$ is responsible for the losses, from these graphs, we can easily localize the place of stray field concentration and the hot-spot. Hot-spots appears where magnetic field intensity exceeds the value $H_{ms,\mathrm{perm}}$, permitted from the point of view of the permitted temperature t_{perm} (Figures 9.2 and 9.3). At a remarkable nonuniformity of power loss distribution, like for instance on the surface of a transformer tank cover (Figure 6.8), the heat is transferred to colder parts of the cover. In such a case, the permissible value of $H_{ms,\mathrm{perm}}$ can be much higher. If, for example, the power loss distribution on a surface can be approximated with the exponential curve $P_1 = A\,\mathrm{e}^{-B|y|}$, the permissible field is

$$H_{ms,\mathrm{perm}K} = \frac{H_{ms,\mathrm{perm}}}{\sqrt{K}}, \quad \text{where } K \approx \frac{1}{cB\sqrt{d}+1} \ll 1 \qquad (7.99)$$

At a serious unevenness or nonuniformity, like on a transformer cover, K ≈ 0.5–0.79 (Figure 9.5).

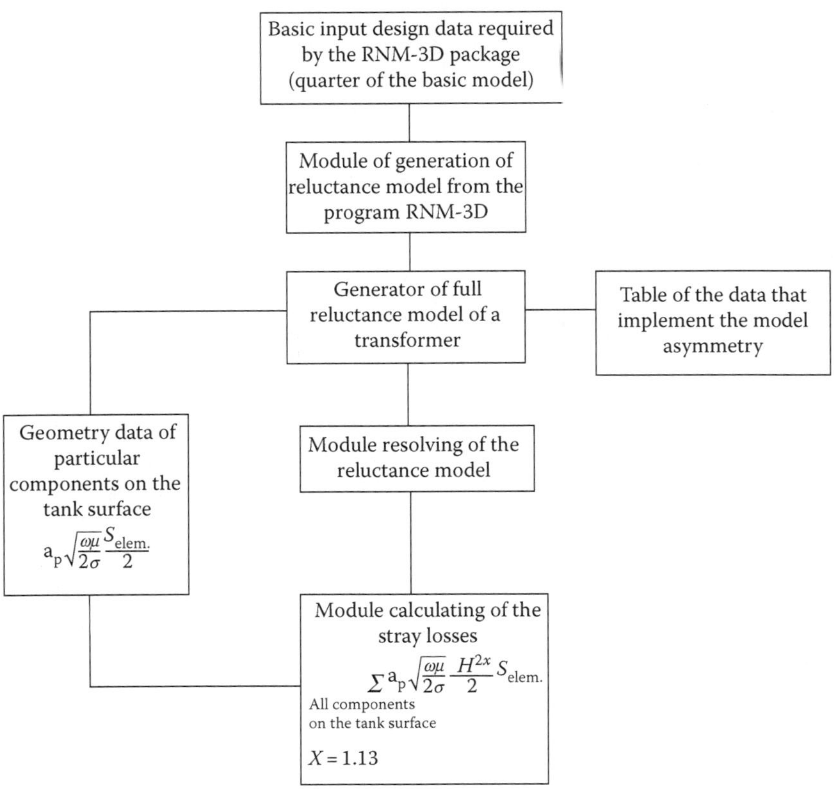

FIGURE 7.30 Block diagram of the RNM-3Dasm program. (From Turowski J. and Zwoliński G.: *Proc. of 5th Polish-Japanese Joint Seminar on Electromagnetics in Science and Technology. Gdansk'97*, Poland, May 19–21, 1997, pp. 165–168 [7.24].)

TABLE 7.6
Validation of the Total Package RNM-3Dexe for Eddy- and Stray-Loss Estimation in Transformers with Partial Magnetic Screens (Shunts) on the Tank Surface

	Calculated Eddy + Stray Losses								
Transformer	FEM Eddy (kW)	RNM-3D Tank-Loss (kW)	High-Current Loss (kW)	Frame and Other Stray Losses (kW)	Total (kW)	Copper Loss (kW)	Calculated Load Loss (kW)	Tested Load Loss (kW)	Difference (%)
31.5 MVA 132/33 kV	13.9	8.16	0.78	0.45	23.29	95.46	118.75	118.00	–0.3
80 MVA 138/10.5 kV	33.87	16.27	6.40	1.2	56.74	225.72	282.46	289.0	2.3
150 MVA 230/10.5 kV	36.26	13.75	9.9	3.5	63.31	377.4	440.81	448.0	1.6
160 MVA 220/132/11 kV	35.5	60.61	1.0	10.8	107.91	242.7	350.61	342.0	–2.5
160 MVA 220/145/12.1 kV	22.91	23.11	0.9	5.1	52.02	194.9	246.92	250.0	1.2
315 MVA 440/220/33 kV	57.43	36.97	1.2	15.8	111.4	425.8	537.2	524.5	–2.4

Source: Koppikar D.A., Kulkarni S.V., and Turowski J.: Fast 3-dimensional interactive computation of stray field and losses in asymmetric transformers. *IEE Proceedings. Generation, Transmission, Distribution*. 147(4), July 2000, 197–201.

In large power transformers, there are many dangerous places (Figure 7.18), which should be carefully investigated.

In some cases, it would be most convenient to combine FEM-3D and RNM-3D into one computational system, as it was shown in J. Turowski's papers [9.9], [10.29]. In some systems (Figure 7.15), remarkable computational advantages can be gained by application of hierarchical FEM with elements of higher order of Hermite's type (Komeza and Sobiczewska-Krusz [7.8]), which permits local refinement of mesh and the application of a smaller number of elements than in the classical FEM techniques.

Among many types of computational packages, FEM-2D belongs to the most popular ones, and FEM-3D belongs to the ones that are most expensive and require the most laborious preparatory work, while the mathematically equivalent RNM-3D is the easiest and most economical.

Figure 7.30 shows a block diagram of the source code of the significantly modernized and much extended "RNM-3Dasym" package for rapid calculation of core-type power transformers with extended structure asymmetry.

A comprehensive industrial verification of the stray-loss computation by RNM-3D is presented in Table 7.6, compiled at *Crompton Graves Transformer Works Bombay* (India) for six of the biggest three-phase core transformers. Thanks to RNM-3D, the *calculated load loss (kW)* in Table 7.6 is equal to the *Tested load loss (kW)*, with accuracy from −0.6% to 2.5%. It is the highest accuracy known in the literature. Sometimes, at wrong screening, the stray (RNM-3D) loss can reach 100% of $3I^2R$ or more. Application of "RNM-3Dasym" package could give even better results.

8 Forces in Electrodynamic Systems

8.1 PRINCIPLES OF CALCULATION OF FORCES ACTING ON BUSES AND WINDINGS OF TRANSFORMERS

The force d$\boldsymbol{F}$ of the magnetic field of flux density $\boldsymbol{B}$, acting on an element d$\boldsymbol{l}$ of a conductor carrying the current I, according to Ampere's law, is expressed by the vector product

$$\mathrm{d}\boldsymbol{F} = I\,(\mathrm{d}\boldsymbol{l} \times \boldsymbol{B}) = \boldsymbol{n}I\,\mathrm{d}l \cdot B \sin(\mathrm{d}\boldsymbol{l},\boldsymbol{B}) \tag{8.1}$$

The electromagnetic force acting on a body of volume V, in which flows the electric current of density $\boldsymbol{J} = \sigma\boldsymbol{E}$, is expressed by Ampere's law

$$F = \int_V (\boldsymbol{J} \times \boldsymbol{B})\mathrm{d}V \quad \text{or} \quad \mathrm{d}\boldsymbol{F} = (\boldsymbol{J} \times \boldsymbol{B})\,\mathrm{d}V \tag{8.2}$$

where for an alternating field, $\underline{E}_m = E_m\,\mathrm{e}^{j\omega t}$, $\underline{B}_m = B_m\,\mathrm{e}^{j\omega t}$, the force components are

$$\left.\begin{aligned} F_x &= ½\,\sigma\,\mathrm{Re}(E_{my}B_{mz}{}^* - E_{mz}B_{my}{}^*) \\ F_y &= ½\,\sigma\,\mathrm{Re}(E_{mz}B_{mx}{}^* - E_{mx}B_{mz}{}^*) \\ F_z &= ½\,\sigma\,\mathrm{Re}(E_{mx}B_{my}{}^* - E_{my}B_{mx}{}^*) \end{aligned}\right\} \tag{8.2a}$$

The most popular, four basic methods of calculating electrodynamic forces include (1) Ampere's law, (2) Virtual forces, (3) Maxwel's Stress Tensor, and (4) Hamilton's Principle with its Euler–Lagrange differential equation—for transient, dynamic processes (Turowski [1.20]), which can be used in Mechatronic design, instead of the complex and expensive numerical simulator, such as Saber.

The first method based on Ampere's law (8.1), and (8.2)—consists of integrating interactive forces in infinitely smaller elements.

The second method follows from the *principle of virtual forces* (Turowski [1.20], [1.16], [1.17]). In this method, it is assumed that the *displacements* (dg) *of current-carrying circuits* (Figure 8.1) *in the direction of the acting force* (f_{eg}) *are so small that the currents* i_k *or the linkage fluxes* ψ_k *in the interacting circuits do not undergo any change*. Since, at the same time, the work of field forces ($f_{eg} \cdot \mathrm{d}g$) is equal to the increment dW_m of the energy of a magnetic field, that is, $f_{eg} \cdot \mathrm{d}g = \mathrm{d}W_m$, hence the

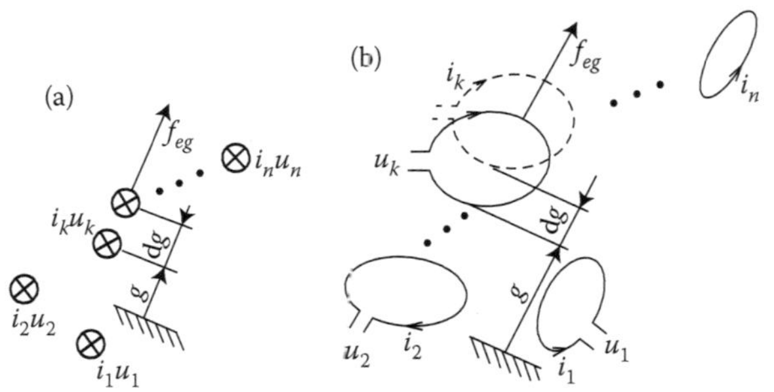

FIGURE 8.1 The virtual work $f_{eg} \cdot \mathrm{d}g$ in a system of: (a) conductors, (b) current circuits. (Adapted from Turowski J.: *Fundamentals of Mechatronics* (in Polish). AHE-Lodz, 2008 [1.20]; Turowski J. (1) (Editor and author 35%) et al.: *Analysis and Synthesis of Electromagnetic Fields*. (in Polish). Wroclaw: OSSOLINEUM - PAN, 1990; where §3 Reluctance network method RNM; §5 Coupled fields. (2) Computational magnetics. (in English). (J. Sykulski, editor), London: Chapman & Hall Stellenbosch, 1995 where: §4 Turowski J.: Reluctance networks RNM. and §6 Coupled Fields [1.17].)

instantaneous value of electromagnetic force in the direction of the coordinate x_j is expressed by the formula

$$f_{ej} = -\left(\frac{\partial W_m}{\partial x_j}\right)_{\psi_k=\text{const}} = \left(\frac{\partial W'_m}{\partial x_j}\right)_{i_k=\text{const}} \tag{8.3}$$

where (Turowski [1.16], [1.18])

$$W_m = \frac{1}{2}\int_v \boldsymbol{H}\cdot \mathrm{d}\boldsymbol{B}\,\mathrm{d}V = \frac{1}{2}\int_v \boldsymbol{A}\cdot\boldsymbol{J}\,\mathrm{d}V = \frac{\mu}{8\pi}\int_{v_i}\int_{v_k}\frac{\boldsymbol{J}_i\boldsymbol{J}_k}{\vec{\boldsymbol{r}}_{ik}}\mathrm{d}V_i\mathrm{d}V_k$$

$$= \sum_{i=1}^{n}\sum_{k=1}^{n}\frac{1}{2}M_{ik}(x_j)i_i i_k = \sum_{k=1}^{n}\int_0^{\psi_k} i'_k \mathrm{d}\psi' k \tag{8.4}$$

is the magnetic energy of the electrodynamic system; and

$$W'_m = \sum_{k=1}^{n} i_k\psi_k - W_m = \sum_{k=1}^{n}\int_0^{i_k}\psi_k\left(i_1,\ldots,i_{i-1},i'_i,0,\ldots,0;x_1,\ldots,x_m\right)\mathrm{d}i'_i \tag{8.5}$$

is the *magnetic coenergy* of the electrodynamic system (Figure 8.3 [later in the chapter]), developed with the help of Legendre transformation (Turowski [1.20]), introduced to replace the difficult determination value of ψ_k, by the easy-to-measure current i_k of kth circuit (Figure 8.2).

Formula (8.3) is able to calculate transient forces and stresses with the help of Hamilton's Principle and Euler–Lagrange differential equation (Turowski [1.16], [1.20], [1.18]). Both methods are, in fact, identical and only the basis of a user's

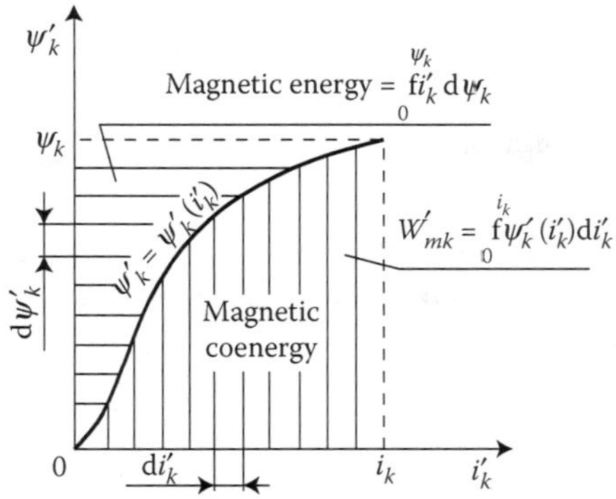

FIGURE 8.2 The magnetic coenergy, $W'_{mk} = i_k \psi_k - W_{mk}$, of the kth nonlinear circuit (From Turowski J.: *Fundamentals of Mechatronics* (in Polish). AHE-Lodz, 2008 [1.20].)

experience dictates the choice. However, the method of virtual works follows immediately from Hamilton's Principle and Euler–Lagrange equations

$$\frac{\partial L}{\partial q_k} - \frac{\mathrm{d}}{\mathrm{d}t}\left(\frac{\partial L}{\partial \dot{q}_k}\right) - \frac{\partial F}{\partial \dot{q}_k} = -G_k \tag{8.6}$$

where $k = 1, 2, 3, \ldots, l$ is the number of *degrees of freedom*, that is, the number of basic coordinates of a system diminished by the number of constraint equations (Turowski [1.18]); $L = T–V$ is the *Lagrange's state function*; F is the Raleigh's loss function; G_k is the exciting force of electric (voltages) and mechanical (forces and torques) origin; q_k is the *generalized* coordinates (Q_k or x_k), equal to the number of degrees of freedom; $\dot{q}_k = \mathrm{d}q_k/\mathrm{d}t$.

From Equation 8.6, after calculating the *kinetic energy* T and the *potential energy* V of the system, one can find immediately all the *equations of motion* (8.6a) of electromechanical and electromagnetic systems (Turowski [1.16], pp. 134, 229–233 and 254–256):

$$\begin{aligned} \boldsymbol{u} &= \mathbf{Ri} + \frac{\mathrm{d}}{\mathrm{d}t}\mathbf{Mi} \\ \boldsymbol{m}\frac{d^2y}{dt^2} + D\frac{\mathrm{d}}{\mathrm{d}y} + K_y &= \frac{1}{2}\boldsymbol{i}^T\frac{\mathrm{d}}{\mathrm{d}y}\mathbf{Mi} + \boldsymbol{mg} \end{aligned} \tag{8.6a}$$

where $\boldsymbol{u}, \boldsymbol{i}$ are the vectors of voltages and currents in the individual windings, respectively; $\boldsymbol{R}$, $\boldsymbol{M}$ are the resistance plus self- and mutual-inductance matrices of these windings; D is the mechanical damping factor (e.g., friction), K is the elasticity constant (spring, spacers), and g the acceleration of gravity. The first component on the right-hand side of Equation 8.6a shows the electromagnetic force acting on the test coil, where $\boldsymbol{i}^T$ is a transposed matrix. The second component represents an external force, for example, the gravity force, $\boldsymbol{mg}$.

Since the per unit energy of a magnetic field is $w_m = \mu H^2/2$, then from Equation 8.3 it follows the vector equation of *volume force*

$$f = k \text{ grad } |H|^2 \tag{8.7}$$

whereas its components, for instance, acting on ferromagnetic particles σFe_2O_3 in magnetic separators of chimney dust of power stations (K. Adamiak EA&EM Mendrela [8.4]), are represented by the expressions

$$f_x = k\left(H_x \frac{\partial H_x}{\partial x} + H_y \frac{\partial H_y}{\partial x}\right), \quad f_y = k\left(H_x \frac{\partial H_x}{\partial y} + H_y \frac{\partial H_y}{\partial y}\right) \tag{8.7a}$$

On a ferromagnetic particle of volume V and permeability μ acts the force $F = (\mu - \mu_0) /(2\mu_0 \mu)\, V \text{ grad } B^2$. The force F_1 acting in the direction u_1 on the charge Q can be moreover expressed by Green's function $G(u_1, u_2, u_3)$—see Ref. [4.3], p. 65.

The third method allows the calculation of mechanical forces in a magnetic field [8.10], [1.16] with the help of Maxwell's Stress Tensor (2.75). For instance, in an electromagnetic reversing linear motor (Rais, J. and M. Turowski [8.13]), an actuator or vibrator, the driving force acting on the surface S of the left column (Figure 8.3) from the air side is

$$\boldsymbol{F} = \oiint_S \vec{\boldsymbol{T}}^n \, dS = \iint_S [\mu_0 (\boldsymbol{n} \cdot \boldsymbol{H})\boldsymbol{H} - \frac{1}{2}\mu_0 |\boldsymbol{H}|^2 \boldsymbol{n}] \, dS \tag{8.8}$$

Considering the vector sum $\boldsymbol{H} = H_n\boldsymbol{n} + H_t\boldsymbol{t}$, we get

$$F = \frac{1}{2}\mu_0 \, \boldsymbol{n} \iint_S (H_n^2 - H_t^2) \, dS + \mu_0 \, \boldsymbol{t} \iint_S H_n H_t \, dS = \boldsymbol{n}F_n + \boldsymbol{t}F_t \tag{8.8a}$$

As can be seen, two forces, $\boldsymbol{F}_n = F_n\boldsymbol{n}$ and $\boldsymbol{F}_t = F_t\boldsymbol{t}$, act on the moving armature of the electromagnet. If the magnetic core is laminated with thin sheets and not

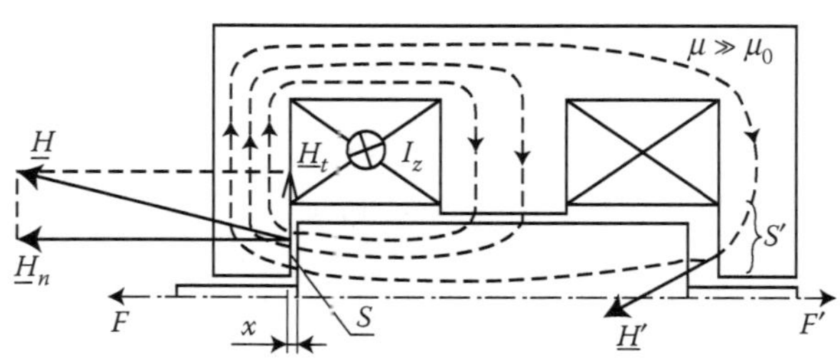

FIGURE 8.3 Calculation of the thrust force F_n with the help of *Maxwell's Stress Tensor* and the field H from the air side.

too strongly saturated, one can assume $H_t = 0$, from which $F_t = 0$, and then on the armature acts only the force

$$F_n = \iint_S \frac{BH}{2} \mathrm{d}S$$

In another case, of armature made of solid steel, high frequency, or steepness of control pulses, a high speed of reversal motion of armature, and small gap x, causing a big armature saturation, the component H_t on the steel surface grows and H_n decreases. In such a case, it can happen that $H_t < H_n$ and a repulsion of the armature can occur. At calculation of the resultant force, all its components should be summarized over the whole active surface of armature, including the braking force $F' = \iint_S (1/2)\, B'_n\, H'_n\, \mathrm{d}S < 0$.

Electromagnetic forces can also be considered as internal forces of magnetic fields, emerging on the surface of bodies placed in the field (*ponderomotive forces*). These forces tend to shorten distances between bodies in the direction of field lines, that is, to shorten the length of the field tubes and lines, as well as they tend to increase a distance in the direction crosswise to the lines (Figure 8.4a). These forces, per the unit surface of a cross-section or side of tube, equal to the volume density of energy, J/m³, of magnetic field ([4.3], p. 318):

$$p_{1V} = \frac{\boldsymbol{BH}}{2} = \frac{B^2}{2\mu_0} \tag{8.9}$$

The *ponderomotive* pull and stress forces in a magnetic field include: the force acting on a conducting wire in a magnetic field (Figure 8.4a), the pulling or pushing interaction force of parallel conductors (Figure 8.4b,c), attraction forces of conductors to steel surfaces (Figure 5.2a), as well as repulsion of conductors from the surface of good conductors (Figure 5.2b).

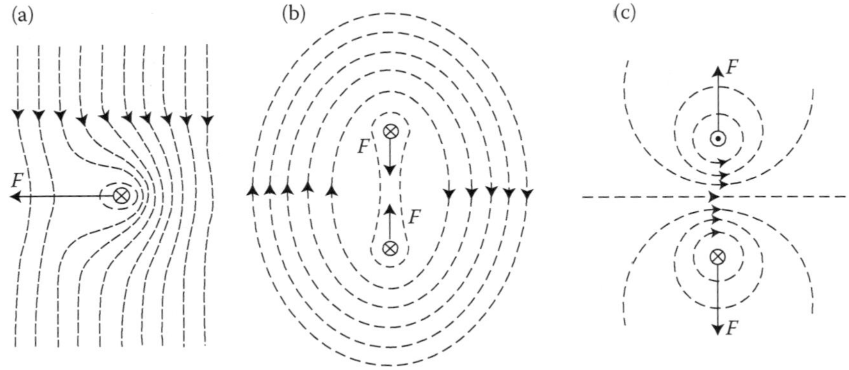

FIGURE 8.4 Pulling, attractive, and repulsive ponderomotive forces in a magnetic field.

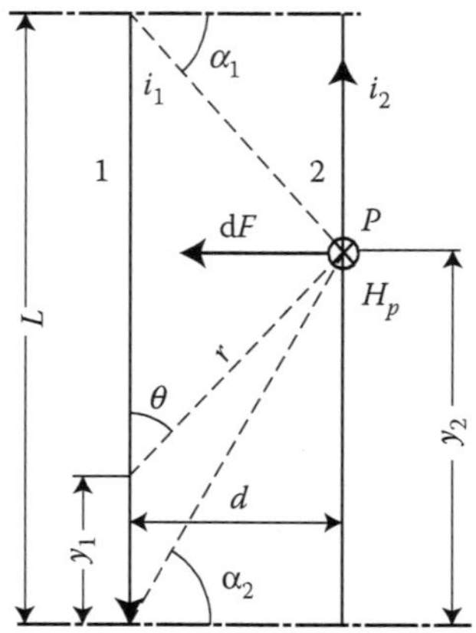

FIGURE 8.5 Determination of interactive force between two parallel conductors.

8.1.1 Interaction Force of Parallel Conductors

To illustrate both described methods, we shall determine the interaction force of two parallel conductors of the dimensions as in Figure 8.5, carrying currents i_1, i_2. We assume that the thickness of both conductors is negligibly small in comparison with the distance d between them.

In accordance with formula (2.16a), the magnetic field intensity in an arbitrary point P of the conductor 2 will have the following form:

$$H_P = \frac{i_1}{4\pi}\int_0^l \frac{\sin\theta\, dy_1}{r^2} = \frac{i_1 d}{4\pi}\int_0^l \frac{dy_1}{[(y_2 - y_1)^2 + d^2]^{3/2}}$$

$$= \frac{i_1}{4\pi d}\left[\frac{l - y_2}{\sqrt{(l-y_2)^2 + d^2}} + \frac{y_2}{\sqrt{y_2^2 + d^2}}\right] = \frac{i_1}{4\pi d}(\sin\alpha_1 + \sin\alpha_2)$$

and is directed toward the figure plane. The force acting on the element dy_2 of the conductor 2 equals $\mu_0 \cdot i_2 \cdot dy_2 \cdot H_P$ and is directed to the left.

Hence, the resulting force attracting the conductors carrying currents of the same direction is

$$F = \mu_0 i_2 \int_0^l H_p dy_2 = \frac{\mu_2 i_1 i_2}{2\pi d}\left(\sqrt{l^2 + d^2} - d\right)$$

The force per unit length of conductors, in N/m:

$$F = \frac{\mu_2 i_1 i_2}{2\pi d}\left(\sqrt{1 + \frac{d^2}{l^2}} - \frac{d}{l}\right) \approx \frac{\mu_0}{2\pi d} i_1\, i_2 \qquad (8.10)$$

A simplified form of formula (8.10) is broadly used. However, it should be remembered that it is valid only at $l \gg d$. At short conductors, application of the simplified formula can lead to significant errors.

Now, let us determine the force of repulsion of conductors shown in Figure 8.5 in the case when they create sides of rectangular loop of length l, that is, when $i_2 = -i_1 = i$. In this case, we apply the method of increment of field energy at small (virtual) deformation of system (8.3). Self-inductance of such a loop equals to

$$L = \frac{l}{\pi}\left(\mu_0 \ln\frac{d}{R} + \frac{\mu}{4} - \frac{\mu_0 d}{l}\right)$$

where R is the conductor radius and μ is the permeability of its material.

If we slightly increase the distance d between the conductors, then L will increase. Therefore, the derivative $\partial L/\partial d$ is positive, hence it delivers a force in the direction of the change, which means—repulsion.

From the differentiation of Equations 8.3 and 8.4, the repulsion force per unit length of conductors equals

$$\begin{aligned} F_1 &= \frac{F}{l} = \frac{\mathrm{d}W_1}{\mathrm{d}x_i} = \frac{\mathrm{d}}{\mathrm{d}x_i}\left(\frac{1}{2}L_1 i_1^2 + M i_1 i_2 + \frac{1}{2}L_2 i_2^2\right) \\ &= \frac{1}{2l} i^2 \frac{\partial L}{\partial d} = \frac{\mu_0 i^2}{2\mu}\left(\frac{1}{d} - \frac{1}{l}\right) \approx \frac{\mu_0}{2\pi d} i^2, \quad (l \gg d) \end{aligned} \tag{8.11}$$

The result is, obviously, in accordance with formula (8.10) obtained using Ampere's method for two different currents.

Short-circuit forces in switching stations are calculated in two steps: *first step*—concerning the determination of short-circuit currents—is described in Publication 909 of International Electrotechnical Commission (IEC) [8.7] and its modifications, and the *second step*—consisting of calculating forces between conductors, is described in the publication CIGRE* Sc-23-WG02 and its modifications.

Forces in transformer windings, at sudden short circuits on the transformer output terminals, are often very large and need careful calculation early to avoid possible damage of windings during operation. Classical investigations in this field include, in Section 5.5.2, the works of Roth [1.15]. The winding cross-section in the core window of a single-phase transformer from Figure 8.6 is a particular case of the system presented in Figure 5.27. Owing to the symmetry of system with respect to the vertical axis a–a, which is at the same time an equipotential line, the analysis can be limited to one-half of the window region. Thanks to it, the previous Equations 5.25 through 5.28 are also valid here.

Let us find the forces acting on jth coil of the winding I (Figure 8.6) in which flows a current of density J_j. Formula (8.2) determines the force acting on an element

* CIGRE = Conférence Internationale des Grand Réseaux Electriques à haute tension (International Conference of Large Electric Networks of High Voltage).

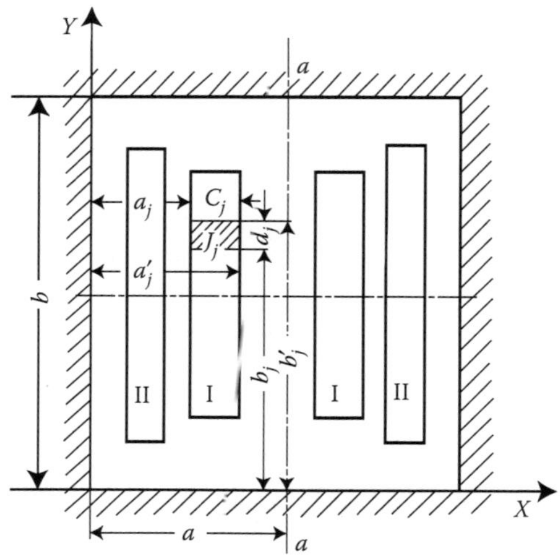

FIGURE 8.6 Determination of the forces in transformer winding, using the Roth method.

$dx \cdot dy$ of the coil cross-section. This element is in the field of components B_x, B_y which can be determined [1.15] from the magnetic vector potential A (5.25).

The components F_{1x} and F_{1y} of the resultant force on the jth coil of winding, per unit length along the Z-axis, are equal

$$F_{1xj} = \int_{a_j}^{a'_j}\int_{b_j}^{b'_j} J_j B_y \, dx \, dy = -J_j \int_{a_j}^{a'_j}\int_{b_j}^{b'_j} \frac{\partial A}{\partial x} dx \, dy$$

$$F_{1yj} = \int_{a_j}^{a'_j}\int_{b_j}^{b'_j} J_j B_x \, dx \, dy = J_j \int_{a_j}^{a'_j}\int_{b_j}^{b'_j} \frac{\partial A}{\partial y} dx \, dy$$

Substituting the values of $\partial A/\partial x$ and $\partial A/\partial y$ from Equation 5.25, we obtain (as per [3.3]) the components of forces, *radial*

$$F_{1xj} = -J_j \sum_{h=1}^{\infty}\sum_{k=1}^{\infty} C_{h,k} \frac{\left(\cos m_h a'_j - \cos m_h a_j\right)\left(\sin n_k b'_j - \sin n_k b_j\right)}{n_k} \tag{8.12}$$

and *axial*

$$F_{1xj} = J_j \sum_{h=1}^{\infty}\sum_{k=1}^{\infty} C_{h,k} \frac{\left(\cos n_k b'_j - \cos n_k b_j\right)\left(\sin m_h a'_j - \sin m_h a_j\right)}{m_k} \tag{8.13}$$

where $C_{h,k}$ is a trigonometric series of the current density distribution ([1.15], p. 217) in the transformer windings. Owing to the system symmetry also with respect to the horizontal axis (Figure 8.6), the resultant axial force F_{1y} acting on the whole winding equals zero. However, the axial forces and stresses acting on particular elements of windings can be high and dangerous for the transformer. On the basis of formulae (8.12) and (8.13), one can calculate forces in any place of the windings. Such calculations are however burdensome, due to slow convergence of the series. The application of computers (J. Kulikowski, 1964) accelerated this process and gave the formulae (8.12) and (8.13) practical significance. Practical methods and examples are given in fundamental books of E. Jezierski [1.5], [5.2].

However, it should be noticed that on photos of damaged windings the damages are mainly outside the core window, whereas calculations are usually limited to the window interior. The reason for these discrepancies is clear when one considers that the axial forces are due to the radial component of flux density, B_r, which is bigger outside the window. However, outside exists a 3-D field, which can be easily calculated with the authors' RNM-3D programs. RNM-3D is also convenient for calculating the force in winding (B_r in Figures 4.21a and 4.22) [4.28]. Both the author's conclusion and the usefulness of RNM-3D were confirmed by the Polish–Italian research team (Rizzo et al. [8.19]).

The classical analytical–numerical method of Roth (5.25), described above, works only for simplified geometric forms of investigated models, whereas modern numerical methods, such as FDM, FEM, and the authors' hybrid computer packages RNM-3D (Figures 4.21 and 7.21), permit calculation of more complicated forms, in particular, considering important influences of screens, shunts, yoke beams, flux collectors (Figure 7.29), and so on. Programs of the RNM-3D class are broadly used in plenty of transformer works all over the world, including Poland, Russia, the United States, Canada, India, Italy, Spain, Portugal, China, Japan, Australia, Mexico, Brazil, Iran, and others.

The electrodynamic transient forces can be investigated rapidly and cheaply with the help of the author's "Hamilton-84" program (J. Turowski [1.20], p. 138) based on Euler–Lagrange's differential equations (8.6). It was also discussed in Turowski's works [1.16] (pp. 134, 229–233, and 254–256), as well as in Ref. [1.20].

8.2 FORCES ACTING ON BUS BARS LOCATED NEAR STEEL CONSTRUCTIONAL ELEMENTS

The force acting on a current-carrying conductor placed near a metal mass can be conveniently calculated using the method of mirror images (Chapter 5).

Let us consider a conductor of length l in the distance h from a steel wall of permeability μ. Such a system, according to Figure 5.1, can be substituted with two parallel currents: i and $[(\mu_r - 1)/(\mu_r + 1)]i$. Next, if we substitute into formula (8.10) the value $d = 2\,h$ and the above-mentioned currents, we shall obtain the attraction force of the conductor to the steel surface, in N/m of the conductor length:

$$F_1 = \frac{\mu_0}{4\pi h}\frac{\mu_r - 1}{\mu_r + 1}i^2\left(\sqrt{1+\frac{d^2}{l^2}} - \frac{d}{l}\right) \approx \frac{\mu_0}{4\pi h}\frac{\mu_r - 1}{\mu_r + 1}i^2 \quad \text{at} \quad l \gg d \tag{8.14}$$

In the case when the conductor passes nearby a steel beam, the dimension l can be taken as approximately equal to the dimension of the beam along the conductor length. In the case of a nonmagnetic metal beam, with $\mu_r < 1$, formula (8.14) represents the repulsion force of the conductor from the metal. This phenomenon can occur at alternating currents (AC) flowing nearby a nonmagnetic metal mass, for instance, copper screens.

8.3 FORCES ACTING ON CONDUCTOR SURFACES

The pulling and pressure forces acting on the surface of metal bodies placed in a magnetic field can be calculated either with the help of *Maxwell stress tensor* (2.75) and (8.8), or directly from formula (8.9) ([1.15], p. 289). These forces, independent of how the field is directed in both media, act perpendicular on any surface element [1.15]. They are always directed from the medium of higher magnetic permeability toward the medium of lower permeability.

Surface forces occur practically only on the iron surface. At $\mu_2 = \mu_r\mu_0$ and $\mu_1 = \mu_0$, we obtain the force per unit surface

$$p = \frac{\mu_r - 1}{2\mu_0 \mu_r}\left(B_{1n}^2 + \mu_0^2 \mu_r H_{1t}^2\right) = \frac{\mu_r - 1}{2\mu_0 \mu_r}\left(B_{1n}^2 - \mu_0 \mu_r B_{1t} H_{1t}\right) \tag{8.15}$$

directed toward the exterior of the iron. In formula (8.15), the symbols B_{1n}, B_{1t}, and H_{1t} relate to the field in the air, where usually the last two components can be omitted in comparison with the first one. Hence, in practice, formula (8.9) is used. For instance, the *force of attraction (transportation) of horseshoe electromagnet* is calculated with the formula

$$F = 2A\frac{B^2}{2\mu_0} \tag{8.16}$$

in which A is the surface of one pole of the electromagnet.

The force of *magnetic tension* (in N) of one pole of electric machine, per machine unit length is calculated also using the formula (8.9) and equals

$$F_N = \int_0^\tau \frac{B^2(x)}{2\mu_0}\,\mathrm{d}x \tag{8.17}$$

in which x is the abscise along the machine circumference, and $\tau = \pi D/2p$ is the pole pitch.

If the magnetic field is parallel to the steel surface ($B_n = 0$), hence the force remains continuously perpendicular to the surface, though it will be significantly weaker

$$p = \frac{(\mu_r - 1)\mu_0 H_{1t}^2}{2} = \frac{(\mu_r - 1)\mu_0 B_2^2}{2\mu_0^2 \mu_r^2} \approx \frac{B_2^2}{2\mu_0 \mu_r} \tag{8.18}$$

where B_2 is the flux density in steel, parallel to the boundary surface.

The surface forces explain the lightening of current-carrying conductors from the mechanical charge, with the help of iron screens or slots surrounding conductors in electric machines.

8.4 FORCES IN SLOT PARTS OF WINDINGS OF ELECTRIC MACHINES

Forces and stresses appearing in windings of electric machines can be divided into three fundamental groups: electromagnetic forces in *end windings*, electromagnetic forces in *slotted parts* of windings, and *dilatation forces* of thermal elongation of bars in slot (Bytnar [8.5]).

The first type of these forces were until recently considered the most dangerous and therefore they have been better investigated and in principle resolved [1.15], [8.2], [8.16], [8.17]. The main assumptions at field calculations in end winding are given in Section 5.2.

Dilatation forces are not of electromagnetic nature, but their action can be amplified by superposition of these forces onto the electromagnetic forces in slotted part of windings.

We shall limit, therefore, our considerations to the forces operating in the *slotted part of windings* of electric machines, investigated by the author since 1960 (J. Turowski [8.17]) through subsequent joint publications [8.14]–[8.16], [8.18], and in the summary [8.8]. These forces, in addition to stresses exerted on the bottom insulation and wedges of slots, can cause the vibration of conductors in slots, which can lead to consecutive damaging and erosion of insulation and crushing of hollow conductors, and so on. These problems require a special winding fastening.

It is generally known that the crosswise flux density in a slot, in the critical moment of a three-phase turbogenerator short circuit at $e_m = 0$ (Figure 8.7) in the subtransient state (J. Turowski [1.18], p. 159) can reach high values and cause crosswise teeth saturation.

Since 1931, the use of J. F. Calvert's formulae (*Transactions AIEE,* 1931, p. 178) for $\mu_{Fe} = \infty$ in contemporary large turbogenerators have led to absurdly large crashing forces in slots.

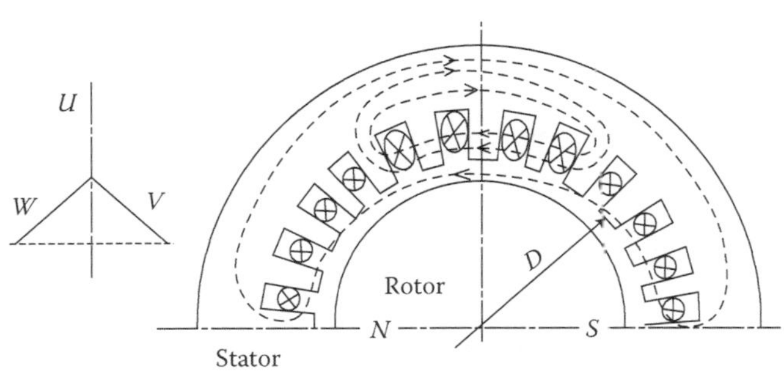

FIGURE 8.7 The course of flux density lines in slots, in the worst time instant of the three-phase short-circuit; subtransient state ($q = 4$).

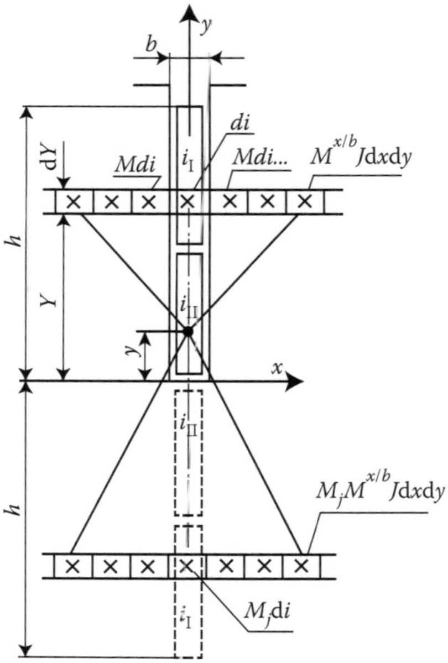

FIGURE 8.8 Calculation of the field and forces in slots of a turbogenerator, using the method of mirror images. (Adapted from Turowski J. *Archiwum Elektrotechniki*, (4), 1960, 677–691 [8.17].)

In 1960, the author, starting from the method of mirror images (Figure 8.8), developed an analytical formulae (J. Turowski [8.17], [1.15]) in which the saturation of teeth and slots proportion was taken into account (Figure 8.9) with the help of an analytical coefficient $k = k(\beta) \leq 1$. So calculated forces acting on the slot bottom, in N per m of slot length, are expressed by the following formulae.

a. For a single-layer winding (Figure 8.9a) or two equiphase layers in slots of two-layer windings

$$F_{b1} = -2\frac{\mu_o}{b}I_m^2 \cdot k(\beta) \tag{8.19}$$

where

$$k(\beta) = \frac{2}{\pi\beta}(0.577 + \ln\beta + \text{ci}\beta\cos\beta - \text{si}\beta\sin\beta) \tag{8.20}$$

$\beta = (2h/b)\ln(\mu_r + 1)/(\mu_r - 1)$, ci$\beta$ and siβ are cos and sin integrals, respectively.* At $2 < \beta < 12$, $k(\beta) \approx 1/\beta$.

* cos and sin integrals are: $\text{ci}\,x = -\int_x^\infty \frac{\cos x}{x}\text{d}x$; $\text{si}\,x = -\int_x^\infty \frac{\sin x}{x}\text{d}x$.

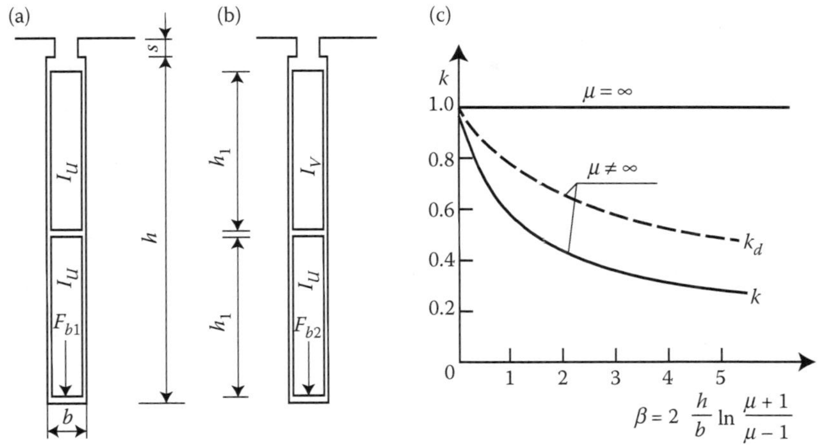

FIGURE 8.9 Forces in a slot for one slot pitch, after the author's formulae [8.17], [1.15]: (a) layers with the same phase currents; (b) layers with the currents of adjacent phases; (c) influence of iron saturation and slot proportion on the forces in the slot bottom, calculated analytically at $\mu = \text{const} \neq \infty$.

b. For a slot of double layer, shortened winding, when currents of the upper and bottom layer belong to different phases

$$F_{b2} = -\frac{1}{2}\frac{\mu_o}{b} I_m^2 k_d(\beta) \tag{8.21}$$

where $1.5 < \beta < 10, k_d \approx 1/\sqrt{\beta}$.

In both formulae (8.19) and (8.21) the I_m is the maximum impulse current i_{imp} in a layer (rod). In reality, the force dependences on μ and h/b are more complicated, but the analytical description (8.19), (8.21), and others, given in the author's work [8.17], are a much better basis than the old ones for a rapid-design expert system and to help evaluate more correct values, as well as to create a fast computer program.

Formulae (8.19) and (8.21) apply to one, separated slot pitch and a constant average permeability $\mu_{\text{aver}} = \mu(I_m, h/b)$ in the whole iron region. A method of selecting the μ_{aver} was given by the author in his work [8.17], then enhanced analytically in 1980 ([1.15], p. 205). Next, in a joint work (Rizzo et al. [8.19]) characteristics of μ_{aver} were calculated by comparison of the analytical method [8.17] with nonlinear FEM of variable permeability $\mu = \mu(H)$—see Figure 8.10.

In works [8.14], [8.15], the Polish–Italian team of M. Rizzo, A. Savini, and J. Turowski resolved this job with the finite element method FEM-2D (Figure 8.11) for the whole pole pitch, considering variable steel permeability $\mu = \mu(H)$ and displacement of stator flux from rotor to gap in the generator subtransient state. These investigations confirmed the discovery in 1960 (J. Turowski [8.17]) of significant dependence of the forces in slots on saturation and the proportion of slots, according to the dependence in Figure 8.9c. The maximal radial force (Turowski [1.15], p. 296)

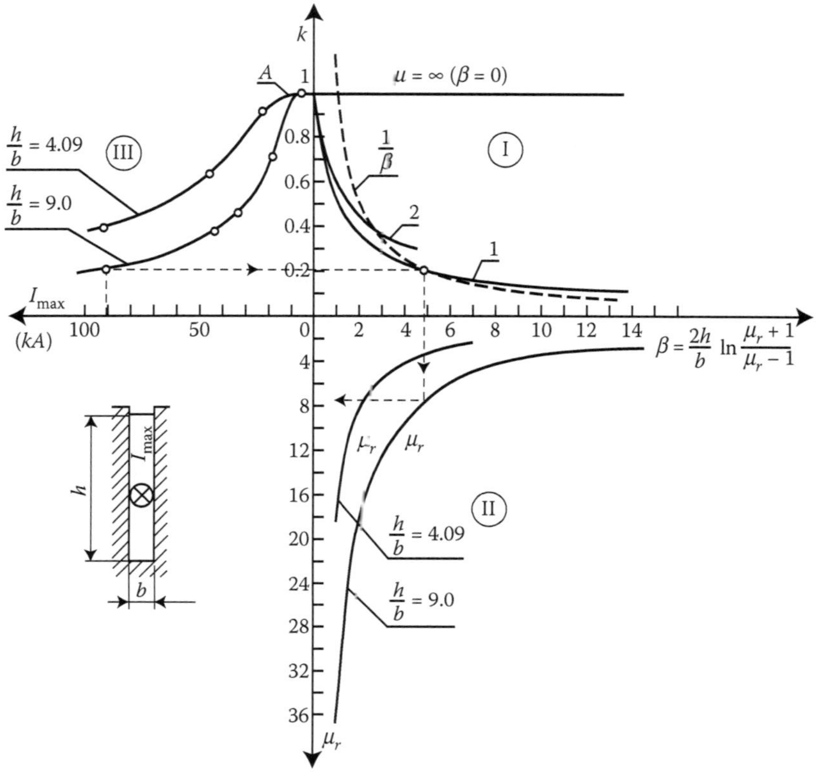

FIGURE 8.10 Influence of iron saturation on the forces at the slot bottom of turbine generator: I—characteristics $k(\beta)$; 1—FEM calculation at $\mu = \mu(H)$; 2—analytical formula (8.20); II—the corresponding average permeabilities $\mu_r = \mu_r(\beta)$, constant for the whole iron region, at h/b = const; III—$k = f(I_{\text{Max}})$ characteristics calculated at $\mu = \mu(\beta)$ for h/b = const; *A*—the zone of rated (nominal) currents. (Adapted from Rizzo M., Savini A., and Turowski J. *IEEE Transactions on Magnetics*, 25(4), July 1989, 2977–2979 [8.14].)

$$F_{y,mx} = -\frac{\mu_o}{2} J_m^2 b\, k(h^2 - y^2) \qquad \left(J_m = I_m \frac{2}{bh} \right) \tag{8.22}$$

is proportional to the circumferential flux density B_Θ (Figure 8.12). Figure 8.12 shows the circumferential $B_\Theta = (1/I_m)\, f_y(y)$ in a slot of one of the largest turbogenerators, with the rated (nominal) specifications: 1450 MVA, $U_N = 24$ kV, $I_N = 35$ kA, 2p = 4, 36 slots, winding shorting coefficient $S/\tau = 7/9$, $X_d'' = 0.144\ \Omega$, $\kappa^* = 1.7$. The author (J. Turowski) expresses here his warmest gratitude to *Kraftwerk Union Mülheim* company for their kind approval to use this generator data (*IEEE Trans. PAS-94*, 4/1975). Abscesses $B_\Theta > 0$ mean the force $f_y(y)$ directed toward the slot bottom, and $B_\Theta < 0$—toward the slot top. At the assumption of $\mu = \infty$, the characteristics $B_\Theta(y)$ do not cross the vertical axis and start from the origin of the coordinate system. In all

* Damping of subtransient short-circuit inrush current multiplication factor; usually $\kappa = 1.6$–1.8 [1.18].

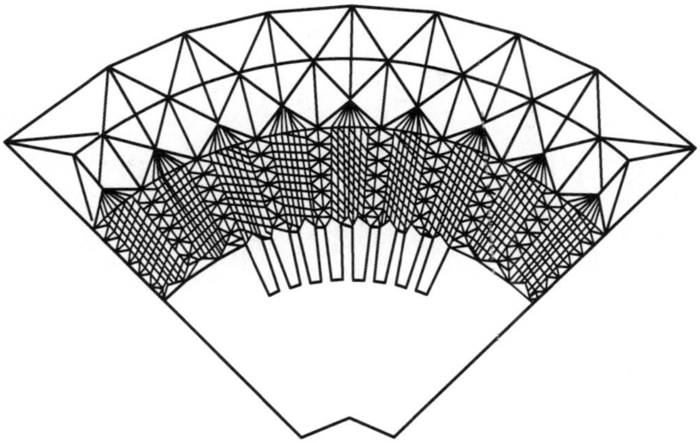

FIGURE 8.11 A nonlinear 2-D FEM model for calculation of field and forces in slots of turbogenerator in the region of full pole pitch. (Adapted from Rizzo M., Savini A., and Turowski J.: *IEEE Transactions on Magnetics*, 25(4), July 1989, 2977–2979 [8.14]; Rizzo M., Savini A., and Turowski J.: *Electromagnetic Fields in Electrical Engineering*. James & James, London 1990, *ISEF'89*, pp. 193–196 [8.15].)

three cases in Figure 8.12, the same impulse current of a three-phase short-circuit was assumed: $I_{m1} = I_{m2} = 232$ kA, calculated from the formula

$$I_m = \kappa\sqrt{2}\,\frac{U_N/\sqrt{3}}{X_d''} \tag{8.23}$$

Comparison of the nonlinear FEM calculation for full pole pitch (Figure 8.11) and for the single slot pitch (Figure 8.8) at $\mu = \mu(H)$ and at nonsaturated iron

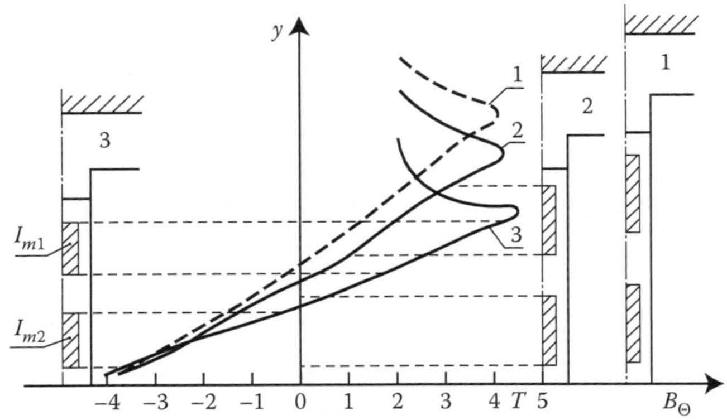

FIGURE 8.12 The circumferential flux density B_Θ in a slot of a 1450 MVA turbogenerator, at impulse short-circuit, at various proportions of slots: 1—$h/b = 5.2$; 2—$h/b = 4.5$; 3—$h/b = 3.9$. (Adapted from Rizzo M., Savini A., and Turowski J. *Electromagnetic Fields in Electrical Engineering*. London: James & James, 1990, ISEF'89, pp. 193–196 [8.15].)

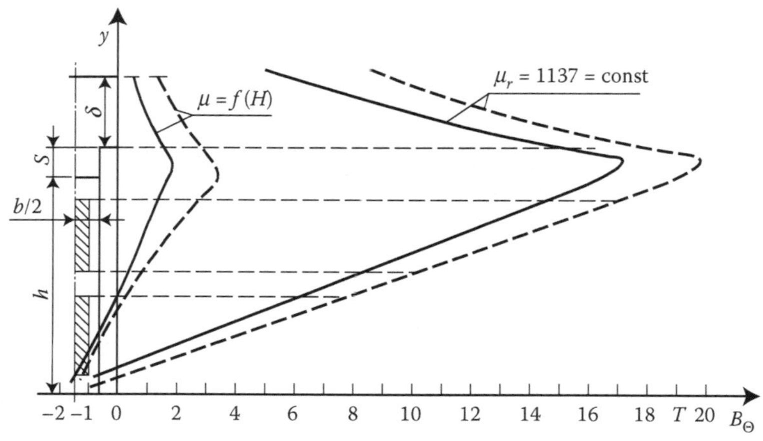

FIGURE 8.13 The flux density B_Θ (and force f_y) in a slot of 1450 MVA turbogenerator, $h/b = 4.5$: — — full pole pitch (Figure 8.11); - - - - single slot pitch (Figure 8.10). (Adapted from Rizzo M., Savini A., and Turowski J *IEEE Transactions on Magnetics*, 25(4), July 1989, 2977–2979 [8.14].)

$\mu_r = 1137 = \text{const}$ (Figure 8.13), confirmed that the forces in slots are, in fact, many times smaller than those calculated with the assumption of nonsaturated iron, as well as that calculation of forces for a single pole pitch gives results somewhat bigger than that for full pole pitch. It ensures an indispensable safety margin for inaccuracy of calculations and simplifications, heating, and so on.

This allows a return to analytical formulae [1.15], [8.17], which are convenient for rapid analysis of functional dependences, discovery of extremes, and so on. The constant average permeability needed for such calculations can be taken from graphs as in Figure 8.10.

As it follows from Figures 8.9 and 8.10, the proportion of the slot dimensions, h/b, has a big influence on the forces in slots. In small generators, an average ratio of the slot h/b is around 3.5–5, for instance, for 12 MVA $h/b = 90/22$ mm = 4.9. In bigger generators, for instance, 1500 MVA, cos $\varphi = 0.8$; 27 kV, $2p = 4$, this ratio reaches $h/b = 268.5/29$ mm = 9.26. Since at very big saturation (of order 3T), μ_r can be around 5, the biggest possible value of the argument β (see Equation 8.20) can be

$$\beta = 2(h/b)\ \ln\left[(\mu_r + 1)/(\mu_r - 1)\right] \approx 0.8(4.09 \text{ to } 9.26) = 3.3 \text{ to } 7.5$$

At very small saturation ($\mu \to \infty$) β is very small and $k \approx 1$.

The forces acting on the bottom and wedge of the slot containing conductors of different phases (in a double layer, shortened winding) are from 4 to 2.5 times smaller, depending on saturation and the proportion h/b of the slot, from the forces acting in a single-layer winding. From the viewpoint of forces existing in the slotted part of winding, deeper and narrower slots are more beneficial. So it is done in modern generators.

EXAMPLE 8.1

Task: Determine the influence of crosswise saturation of teeth on the short-circuit forces in the slotted part of windings in a generator of 12 MVA, 6.5 kV, *Y*, cos $\varphi_N = 0.7$; 50 Hz; 3000 rev/min, with a two-layer, notshortened winding of stator. Semi-closed slots, $h = 90$ mm, $b = 22$ mm ($h/b = 4.09$). Solid iron rotor without damping winding ($\kappa = 1.7$). The leakage reactance of armature, taking into account slot heads saturation in short-circuit instance (open slot), $X_r = 0.39\ \Omega$.

SOLUTION

The impulse short-circuit current $I_{max} = \kappa\sqrt{2}(E_m/X_r) = 1.7(6300/\sqrt{3})/0.39 =$ 22.350 A. The total current flow of a slot at short circuit $I_{Max} = 2I_{max} = 44{,}700$ A. The crosswise (quadrature) flux density in the slot, immediately over the winding

$$B_{quadr} = \mu_0 \frac{I_{Max}}{\mathrm{b}} 0.4\pi \times 10^{-6} \frac{44{,}700}{22 \times 10^{-3}} = 2.55\ \mathrm{T}$$

From the curve in Figure 8.10, for $h/b = 90/22 = 4.09$ and $I_{Max} = 44.7$ kA, we obtain $k \approx 0.65$. Hence, on the slot bottom, according to Equation 8.19c, will act the force

$F_{bott} = -2\ \mu_0\ (I_{max}^2/b)k = -2 \cdot 0.4\pi \times 10^{-6}(22{,}350^2/0.022)\ 0.65 = -37{,}092$ N/m $\approx$ −3.8 tons per meter of slot length.

EXAMPLE 8.2

Task: Determine the influence of saturation on short-circuit forces in a slot of dimensions 90 mm × 25 mm. The total current flow of slot at short circuit is 100,000 A.

SOLUTION

The biggest flux density in a slot is $B_{quadr} = 0.4\pi\ 10^{-6}\ (100{,}000/0.025) = 5.03$ T. From Figure 8.10, for $h/b = 90/25 = 3.6$ and $I_{Max} = 100$ kA, we read $k = 0.45$. Hence, the force acting on the bottom isolation, per meter of the slot length, will be in this case

$F_{bott} = -2 \cdot 0.4\pi \times 10^{-6} \cdot (50{,}000^2/0.025) \cdot k = -251{,}327 \cdot 0.45 = -113{,}097$ N/m = −11.5 tons per meter of slot length.

If the impact of saturation was not considered, we would have obtained the force acting on the bottom of the slot equal to $F_{bott} = 25.6$ tons/m. The bottom insulation would be quickly eroded and the large turbogenerator destroyed, with all the consequences for a power system hazard. Luckily, the saturation of iron mitigates this threat.

Taking into account that current flow in the slots of contemporary turbogenerators can reach up to 150 kA and that these are alternating forces of 100 Hz, it is reasonable to pay attention to the strength of the insulation and hollow conductors on the slot bottom (Patel, IEEE [8.12]) and to fastening of the windings in the slot (Kranz CIGRE [8.8]).

A full analysis of the consequences of the forces acting in slots needs, in addition, an investigation of the *dilatation forces* of thermal elongation of bars in a slot (Bytnar [8.5]).

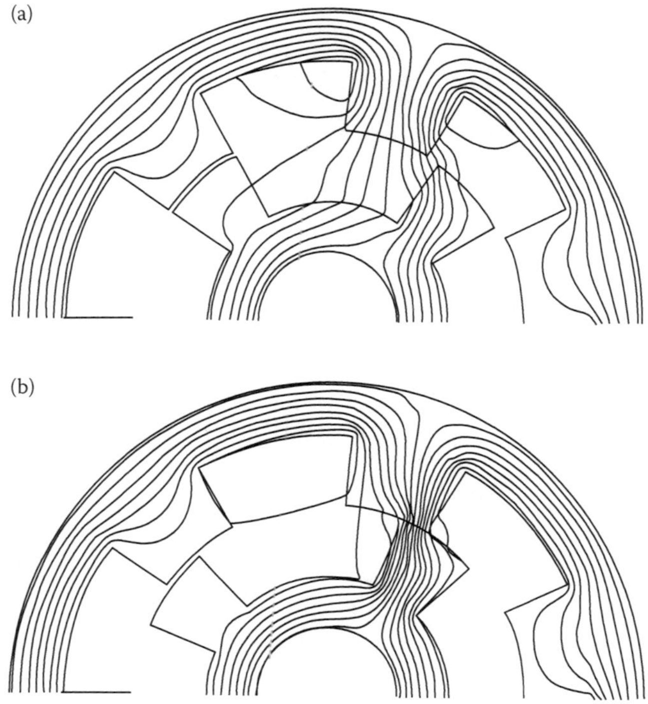

FIGURE 8.14 Fields and forces in a switched reluctance motor, calculated with FDM-2D, at $\mu = \mu(B)$. (A. Pelikant, PhD thesis 1989, supervisor J. Turowski.)

8.5 RELUCTANCE FORCES AND TORQUES

In systems in which the paths of flux distribution (flux density lines) are relatively easy for initial determination (Figures 4.18, 4.20, and 8.15) it is convenient to apply the reluctance network method (RNM) which is the fastest and the simplest (Turowski et al. [2.34] [7.21], [8.13]), especially for three-dimensional fields (Figure 7.21). In two-dimensional systems, especially where an initial assessment of field distribution is difficult for unambiguous determination, it is sometimes more convenient to apply FEM of FDM (Figure 8.14). In systems with a regular structure (Figure 8.15) both FDM and RNM can be used [8.13], [8.21]. RNM delivers fewer details on the accurate course of field lines. However, RNM-3D permits easier consideration of influence of the third dimension of space (Figure 8.15).

8.5.1 Dynamics of Thyristor-Controlled Reversible Motors

Many of their variants have been produced at the Novosibirsk Institute of Electrical Engineering (Figure 8.15) for various application—from small actuators and micro-vibrators to large and heavy pile driving and seismic hammers. From Hamilton's

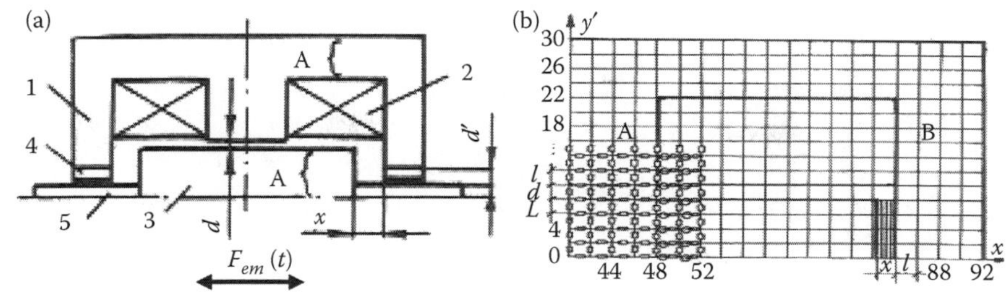

FIGURE 8.15 Field and electrodynamic forces $F_{em}(t)$ in an electronically controlled reversible electromagnetic pump: (a) 1—iron laminated stator, 2—excitation coil $Ni(t)$, 3—movable, laminated armature, 4—nonmagnetic sleeves, 5—solid iron drawing plug; x—working gap, A—cross-section area; (b) an efficient, cost-effective RNM-3D model, with 400 nodes. (Adapted from Rais V.R., Turowski J., and Turowski M.: *Proc. of Internat. Symposium on Electromagnetic Fields in Electrical Engineering—ISEF'87.* 23–25 Sep. 1987. Paper No. VI-8, pp. 291–294, Pavia, Italy (Plenary Report) [8.13].)

Principle and Euler-Lagrange (8.6) differential equation (J. Turowski [1.20]), we get [8.13] the equations of motion for:

- Voltage:

$$u(t) = Ri + \frac{\mathrm{d}i}{\mathrm{d}t} + i\frac{\mathrm{d}L}{\mathrm{d}x}\frac{\mathrm{d}x}{\mathrm{d}t}$$

  and
- Force:

$$F_e(t) = M\frac{\mathrm{d}^2x}{\mathrm{d}t} + \alpha\frac{\mathrm{d}x}{\mathrm{d}t} + K(x - D) - \frac{1}{2}i^2\frac{\mathrm{d}l}{\mathrm{d}x} \tag{8.24}$$

  and $\Psi(x, i) = L(x, i) \cdot i(x)$.

Verification of the RNM-3D calculations are presented in Figure 8.16.

Analysis of such a structure and an electronic control program are only possible with the helpful simulation of electromagnetic fields (Figure 8.17) and calculating the reluctance thrust force $\boldsymbol{F}$ (8.8) with the help of Maxwell's stress tensor $\vec{\mathrm{T}}^{\mathrm{n}}$ as a surface integral:

$$\boldsymbol{F} = \iint_S \vec{\boldsymbol{T}}^{\mathrm{n}}\,\mathrm{d}S = \iint_S \left[\mu_0\left(\vec{\mathrm{n}}\vec{H}\right)\vec{H} - \frac{1}{2}\mu_0\left|H\right|^2\vec{n}\right]\mathrm{d}S = \mathbf{1}_n F_n + \mathbf{1}_t F_t \tag{8.25}$$

In this way, it is possible to simulate [6.16] electromechanical transient processes at various cycles (Figure 8.17) of electronic control, plunger rebounds at impact on limiting bumper, and so on.

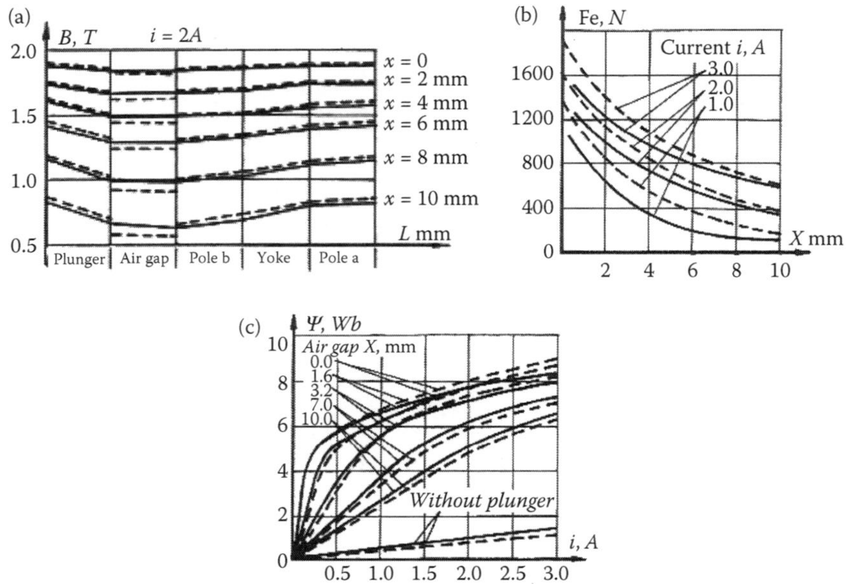

FIGURE 8.16 Results of analysis of the actuator of Figure 8.15: (a) comparison of $B(l)$ calculated with the RNM model of: — 400 nodes and – – – 95 nodes; (b) experimental verification: — measured, – – – calculated $f(x)$; (c) experimental verification of $\psi(i)$. (Adapted from Rais V.R., Turowski J., and Turowski M. *Proc. of Internat. Symposium on Electromagnetic Fields in Electrical Engineering—ISEF*'87. 23–25 Sep. 1987. Paper No. VI-8, pp. 291–294, Pavia, Italy (Plenary Report).)

8.5.2 Torque of Hybrid Stepping Motors with Permanent Magnets in Slots

Stepping motors (J. Turowski [1.19], pp. 118–134) belong to the most popular, electronically controlled actuators in modern mechatronics. The *hybrid* ones (Figure 8.18a) are a combination of variable-reluctance and permanent magnet motors. The effect of permanent magnets, inserted into stator slots, is such that they focus magnetic field lines in corresponding teeth. As a result, the reluctance stepping torque, depending on the material used (Figure 8.18b), can be increased even twice. FEM-2D and 3D analyses were performed [8.25] for linear and nonlinear demagnetization curves (Figure 1.22). The analyses showed that the linearization of demagnetization curves, used widely in the design of all types of electric machinery, may be nonadequate to the particular application.

In the author's (JT) team, a number of other small motors were tested (Figure 8.19). For their modeling and simulations, various methods were used, including RNM, FEM-2D, and -3D, and the hybrid FEM Hermitian method (Figure 7.16).

Two-phase DC brushless, permanent magnet *motor with uneven gap* (Figure 8.19f), according to the Muller's US Patent, is one of the simplest structures with self-starting and positioning at no current. After a nonlinear 2-D FEM analysis [8.27],

A simplified form of formula (8.10) is broadly used. However, it should be remembered that it is valid only at $l \gg d$. At short conductors, application of the simplified formula can lead to significant errors.

Now, let us determine the force of repulsion of conductors shown in Figure 8.5 in the case when they create sides of rectangular loop of length l, that is, when $i_2 = -i_1 = i$. In this case, we apply the method of increment of field energy at small (virtual) deformation of system (8.3). Self-inductance of such a loop equals to

$$L = \frac{l}{\pi}\left(\mu_0 \ln\frac{d}{R} + \frac{\mu}{4} - \frac{\mu_0 d}{l}\right)$$

where R is the conductor radius and μ is the permeability of its material.

If we slightly increase the distance d between the conductors, then L will increase. Therefore, the derivative $\partial L/\partial d$ is positive, hence it delivers a force in the direction of the change, which means—repulsion.

From the differentiation of Equations 8.3 and 8.4, the repulsion force per unit length of conductors equals

$$\begin{aligned} F_1 &= \frac{F}{l} = \frac{\mathrm{d}W_1}{\mathrm{d}x_i} = \frac{\mathrm{d}}{\mathrm{d}x_i}\left(\frac{1}{2}L_1 i_1^2 + M_{21} i_1 i_2 + \frac{1}{2}L_2 i_2^2\right) \\ &= \frac{1}{2l} i^2 \frac{\partial L}{\partial d} = \frac{\mu_0 i^2}{2\mu}\left(\frac{1}{d} - \frac{1}{l}\right) \approx \frac{\mu_0}{2\pi d} i^2, \quad (l \gg d) \end{aligned} \tag{8.11}$$

The result is, obviously, in accordance with formula (8.10) obtained using Ampere's method for two different currents.

Short-circuit forces in switching stations are calculated in two steps: *first step*—concerning the determination of short-circuit currents—is described in Publication 909 of International Electrotechnical Commission (IEC) [8.7] and its modifications, and the *second step*—consisting of calculating forces between conductors, is described in the publication CIGRE* Sc-23-WG02 and its modifications.

Forces in transformer windings, at sudden short circuits on the transformer output terminals, are often very large and need careful calculation early to avoid possible damage of windings during operation. Classical investigations in this field include, in Section 5.5.2, the works of Roth [1.15]. The winding cross-section in the core window of a single-phase transformer from Figure 8.6 is a particular case of the system presented in Figure 5.27. Owing to the symmetry of system with respect to the vertical axis a–a, which is at the same time an equipotential line, the analysis can be limited to one-half of the window region. Thanks to it, the previous Equations 5.25 through 5.28 are also valid here.

Let us find the forces acting on jth coil of the winding I (Figure 8.6) in which flows a current of density J_j. Formula (8.2) determines the force acting on an element

* CIGRE = Conférence Internationale des Grand Réseaux Electriques à haute tension (International Conference of Large Electric Networks of High Voltage).

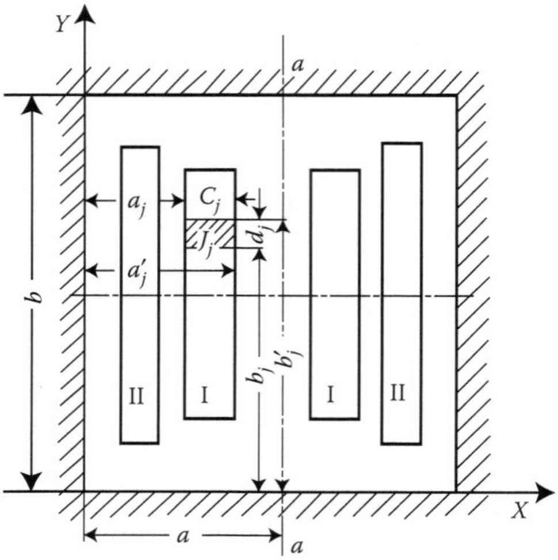

FIGURE 8.6 Determination of the forces in transformer winding, using the Roth method.

$\mathrm{d}x \cdot \mathrm{d}y$ of the coil cross-section. This element is in the field of components B_x, B_y which can be determined [1.15] from the magnetic vector potential A (5.25).

The components F_{1x} and F_{1y} of the resultant force on the jth coil of winding, per unit length along the Z-axis, are equal

$$F_{1xj} = \int_{a_j}^{a'_j}\int_{b_j}^{b'_j} J_j B_y \,\mathrm{d}x\,\mathrm{d}y = -J_j \int_{a_j}^{a'_j}\int_{b_j}^{b'_j} \frac{\partial A}{\partial x}\,\mathrm{d}x\,\mathrm{d}y$$

$$F_{1yj} = \int_{a_j}^{a'_j}\int_{b_j}^{b'_j} J_j B_x \,\mathrm{d}x\,\mathrm{d}y = J_j \int_{a_j}^{a'_j}\int_{b_j}^{b'_j} \frac{\partial A}{\partial y}\,\mathrm{d}x\,\mathrm{d}y$$

Substituting the values of $\partial A/\partial x$ and $\partial A/\partial y$ from Equation 5.25, we obtain (as per [3.3]) the components of forces, *radial*

$$F_{1xj} = -J_j \sum_{h=1}^{\infty}\sum_{k=1}^{\infty} C_{h,k} \frac{\left(\cos m_h a'_j - \cos m_h a_j\right)\left(\sin n_k b'_j - \sin n_k b_j\right)}{n_k} \tag{8.12}$$

and *axial*

$$F_{1xj} = J_j \sum_{h=1}^{\infty}\sum_{k=1}^{\infty} C_{h,k} \frac{\left(\cos n_k b'_j - \cos n_k b_j\right)\left(\sin m_h a'_j - \sin m_h a_j\right)}{m_k} \tag{8.13}$$

where $C_{h,k}$ is a trigonometric series of the current density distribution ([1.15], p. 217) in the transformer windings. Owing to the system symmetry also with

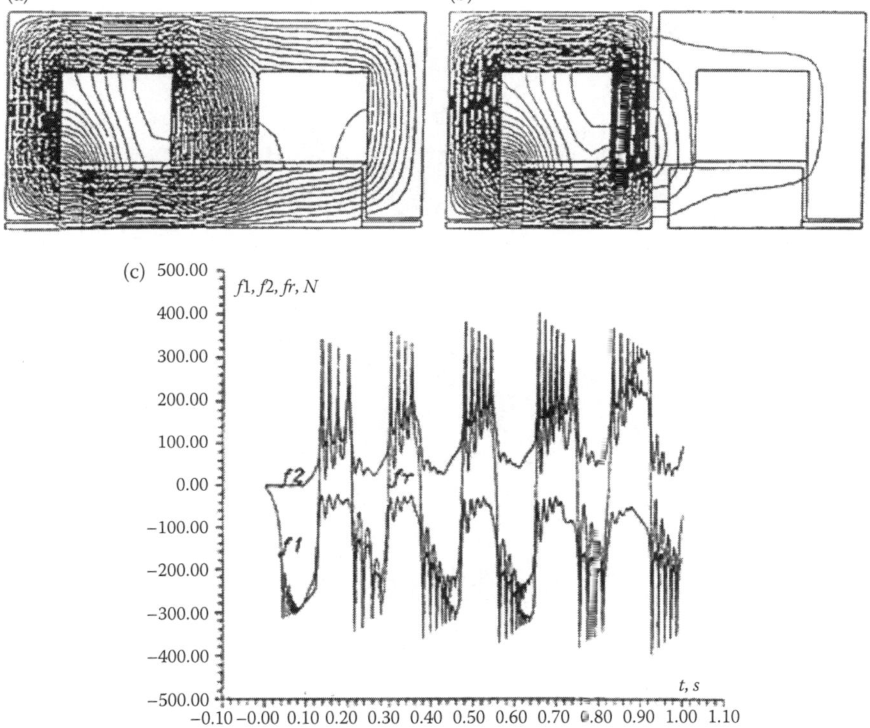

FIGURE 8.17 Computer simulation of the field in a linear reversible actuator: (a) single stator, (b) double stator, (c) thrust component forces $f_1(t)$, $f_2(t)$ and the resultant force $f_r(t)$, in N, versus control time (s). (Adapted from Wiak S. and Turowski J. International Seminar on Linear Electric Machines. September 1–6. 1990, Odessa, Ukraine; Turowski J. et al. *Proc. of Internat. Aegean Conference on Electr. Machines and Power Electronics-ACEMP'92*. May 27–29, 1992. Turkey: Kusadasi. (Invited paper). Plenary Session. *Proceedings*, Vol. 2/2, pp. 430–437 [6.16].)

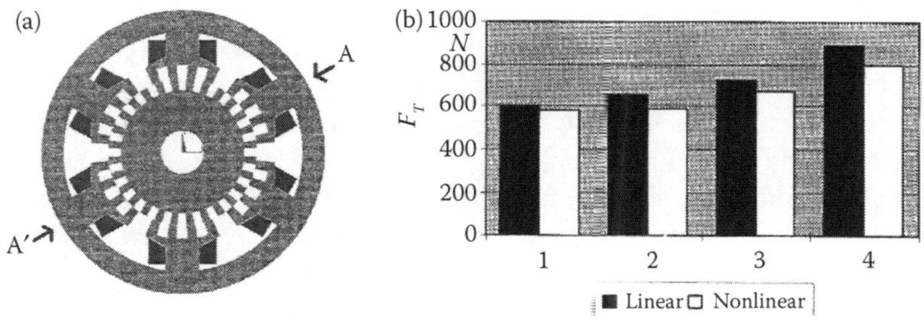

FIGURE 8.18 Choice of permanent magnet sort in slots: (a) cross-section, (b) stepping force: 1—empty slot, 2—slot with Alnico-8, 3—slot with Sr Ferrite, 4—slot with Sm_2CO_{17}. (From Rizzo M., Savini A., and Turowski J. *COMPEL*, 17(1–3), 1998, pp. 318–323 [8.25].)

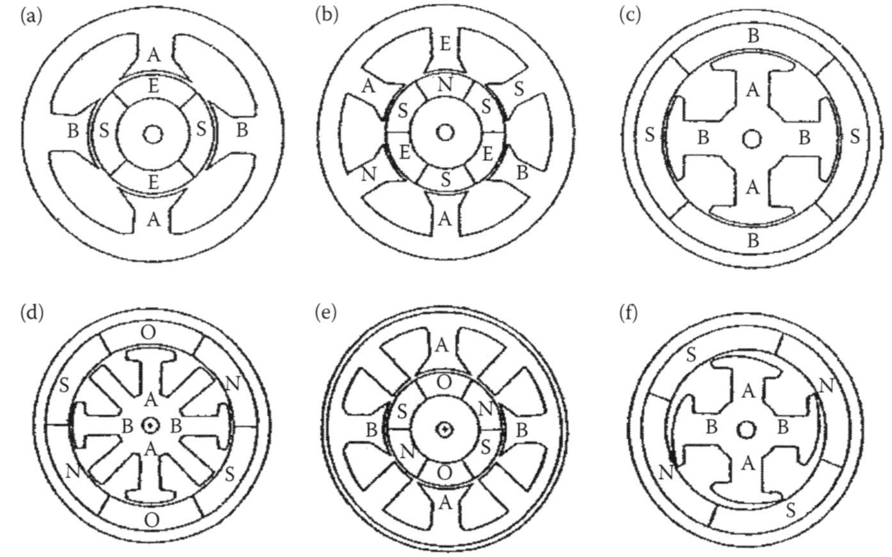

FIGURE 8.19 Cross-sections of various permanent magnet motors: (a) brushless inside field 4-pole, 2-phase, (b) brushless inside field 6-pole, 3-phase, (c) brushless outside field 4-pole, 2-phase, (d) brushless outside field 4-pole, 2-phase motor with auxiliary poles, (e) brushless inside field 4-pole, 2-phase motor with auxiliary poles, (f) brushless outside field 4-pole, 2-phase with variable gap.

using Maxwell's stress tensor (8.25), it was concluded that the increase of nonuniformity of the gap increases the synchronizing torque and decreases the driving torque. The latter decrease is a disadvantage of the uneven-gap solution. However, the disadvantage is not great in the sense of decreasing the efficiency of the motor because the load has even greater influence on it. Applying the Hermitian FEM [8.28] was helpful in achieving a smoother analysis of the torque-position characteristics.

Two-phase dc brushless motors *with auxiliary salient poles* (Figure 8.19e). According to Wessels US Patent, although more expensive, this solution delivers a better possibility in self-starting and positioning of these motors [8.27], [8.29]. From a nonlinear FEM analysis, using Equation 8.25, it was concluded that the location of a permanent magnet, inside or outside (Figure 8.19d), and the presence of unwound auxiliary poles, may bring about quite different results. Two kinds of motors were investigated: (2-phase, 4-pole) and (3-phase, 6-pole), in both versions—inside and outside field. The results show that for the inside-field motor the torques are of the same order of magnitude, whereas for the outside field motor the 4-pole configuration gives a smoother and bigger torque than the 6-pole one (Turowski J., Komęza K., Pelikant A., Wiak S., M., Savini A. [6.16]).

Effects of pole structure in small permanent magnet DC motors, especially pole shoes (Rizzo et al. [8.30]) and permanent magnet material [8.31], have crucial influence on the performance of these machines.

Reluctance stepping motors performance can also be improved by adding permanent magnet inserts into the slots (Rizzo et al. [8.32]).

9 Local Heating of Structural Parts

9.1 ELECTROMAGNETIC CRITERIA OF LOCAL EXCESSIVE HEATING

In large electric machines and power transformers, elements of switching stations, and so on, a local *excessive heating hazard* of constructional parts can appear, due to eddy currents induced by concentrated alternating magnetic field. This is one of the most important factors of *reliability* of such equipment, and because of their strong interconnection with power systems the consequences of their failure can be often *extremely expensive* [9.5], [9.13]–[9.15]. This is why it was necessary to develop a simple method, as general as possible, for *prediction* and *examination* of possibility of appearance of such overheating hazards. In 1960, the author proposed [2.41], [9.10] such a method based on *electromagnetic criteria*. It proved very useful, especially for contemporary interactive computer methods of design and assessment of reliability of large power equipment (Figures 4.20 through 4.22, and Figure 9.2 [later in the chapter]).

As a criterion deciding whether a given element or its part will not be overheated over the permitted temperature during its operation, we have adopted the value of H_{ms} of the tangential component of alternating magnetic field strength on the surface of the body. A table of such permitted values and corresponding graphs can be created on the basis of the ability of heat transport by different bodies and materials (see Table 9.1 and Figure 9.2 [later in the chapter]). It is a preliminary, but important indicator of the necessity of a more detailed analysis of the problem and localization of endangered spots. Only after exceeding or approaching these critical values, one should apply additional protection means, for example, screens, ventilators, fans, temperature sensors, and so on, as well as more accurate calculations, which are not too easy.

In the case of solid steel elements placed in an external alternating electromagnetic field, one can assume that all the power losses are created at the surface of these bodies (3.9). In the case when the thickness of these parts is so small that we can neglect the temperature drop across this thickness (what usually takes place), and when the per-unit losses are the same on the entire body surface and there is no evacuation of heat through the metal, all the heat P_1, in W per m^2 of the surface, will be transferred by the body surface to the surrounding ambient (Figure 9.1), according to the dependence (at $t_{01} = t$).

$$P_1 = \alpha' \theta \tag{9.1}$$

where α' is the coefficient of heat dissipation by thermal convection and radiation; $\theta = t - t_0$ is the temperature rise of metal with respect to the surrounding ambient.

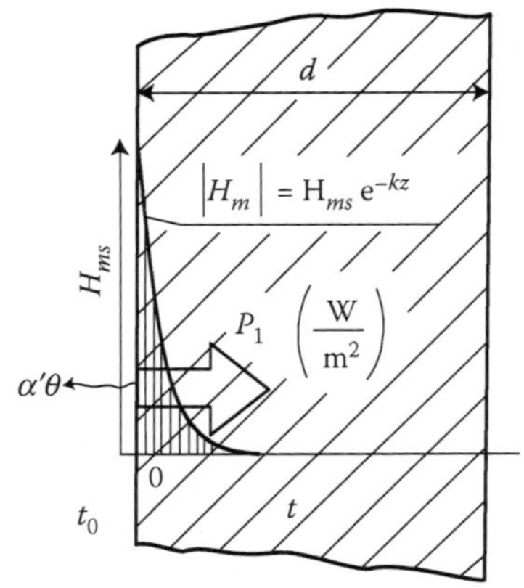

FIGURE 9.1 Thermal dissipation of the eddy-current power P_1 ; t_0, t_{01}—the ambient temperatures on both sides.

The variation of the coefficient α' as a function of temperature, in the limits of the temperature rise permitted for electric machines and transformers, can be approximately expressed by the relation (J. Turowski [9.10])

$$\alpha' = \alpha_0' \left(\frac{\theta}{\theta_0}\right)^{0.25} \tag{9.2}$$

It needs an *additional discussion*. The dependence (9.2) has been widely adopted for vertical surfaces (Jezierski [1.5]). Regarding horizontal surfaces (metal covers, beams, bottoms, etc.), there are often different, sometimes controversial, opinions. As the most justified ones, we can take the generalized formulae obtained on the basis of *criteria of similarity of:* Nusselt (Nu), Grashoff (Gr), and Prandtl (Pr).

After having used this way to generalize numerous data, M. Micheiev [9.2] gave applied formulae for the coefficient of heat removal by *convection*

$$\alpha_c = A_2\,(\theta/l)^{0.25} \quad \text{at} \quad (\mathrm{GrPr})_m = 5\times 10^2 \ldots 2\times 10^7 \tag{9.3}$$

$$\alpha_c = A_3\,\theta^{\,0.33} \quad \text{at} \quad (\mathrm{GrPr})_m > 2\times 10^7 \tag{9.4}$$

where Pr = 0.722 ([9.2], Table 35); the wall temperature $t_w = 110°C$; the air temperature $t_f = 35°C$, the average temperature $t_m = 73°C$; $\beta = 1/T = 1/(346°C)$; $\theta = 75°C$; $g = 9.81\ m/s^2$; $A_2 = 1.12$, $A_3 = 1.21$ ([9.2], Table 6); the coefficient of kinematic viscosity $\nu = 1.96\times 10^{-5}\ m^2/s$ ([9.2], Table 35), and l is the characteristic dimension ("the smaller side of the heated plate"). In the case of heat concentration in a relatively narrow region (e.g., around a bushings on transformer cover—Figure 6.8), the characteristic dimension l should be assumed as much smaller than the width of the cover plate itself and bigger than the region of generation of major part of the power losses.

In the work [9.10], the author adopted $0.1\text{ m} \le l \le 0.7\text{ m}$, hence

$$(\mathrm{GrPr})_{\mathrm{m}} = \beta\frac{gl^3}{\nu^2} = \theta\cdot\mathrm{Pr} = \frac{1}{346}\cdot\frac{9.81\cdot(0.1^3\text{ to }0.7^3)}{1.96^2\cdot10^{-10}}\cdot 75\cdot 0.722 = 4\cdot10^6\text{ to }1.37\cdot10^9$$

From Equation 9.3, for $l = 0.1$ m, we get $\alpha_c = 5.86$ kcal/h m² K = 6.82 W/m² · K and according to Equation 9.4, for $l > 0.7$ m we get $\alpha_c = (1.21 \times 75^{0.33})\cdot 1.163$ = 5.94 W/m² · K independently on the dimension l.

For further calculation it was then adopted (J. Turowski [9.10]) $l > 0.7$ m and $\alpha_c \approx 6.0$ W/m² · K.

The coefficient of heat removal by radiation was adopted as $\alpha_r = 8.2$ W/m² · K and the *total coefficient of heat removal**:

$$\alpha_{110°\mathrm{C}} = \alpha_c + \alpha_r = 6.0 + 8.2 = 14.2\frac{\mathrm{W}}{\mathrm{m}^2\cdot\mathrm{K}} = \alpha_0' \tag{9.2a}$$

Then, the problem of calculation of temperature reduces itself to determination of the per unit power losses (in whole plate) in W/m² of a one-sided surface (4.55)

$$P_1 = k_d\cdot \mathrm{a_p}\sqrt{\frac{\omega\,\mu_0}{2\,\sigma}}\cdot\sqrt{\mu_r}\,|H_{ms}|^2 \tag{9.5}$$

where, depending on the way of excitation and the wall thickness d, $k_d = \kappa$, ζ, or η (Figures 4.13 through 4.15), as well as on the selection of possible corrections to the coefficient of heat removal α', considering the heat removal by the metal conductivity in the case of nonuniform loss distribution on the considered surface or in the case of nonuniform distribution of temperature in the metal mass. Let us now consider several typical cases to which we can reduce various constructional jobs we can encounter in practice.

Thick metal plates ($d > \lambda$) with constant magnetic permeability, excited from two sides (Figures 4.12, 4.14, and 4.15). The power (9.5), according to Equation 9.1, is transferred to the surrounding ambient in the form of heat by convection and radiation directly by both surfaces. Hence, substituting Equations 9.5 and 9.2 into Equation 9.1, at $k_d = \mathrm{a_p} = 1$, $t_{01} = t_0$ ($P_1 = 2\alpha'\theta$), we obtain the highest *permitted* (critical) *value of magnetic field strength* $H_{ms,\,\mathrm{perm}}$† on the plate surface, over which the temperature rise can exceed the permitted value θ_{perm}

$$H^2_{ms,\mathrm{perm}} \le 2\sqrt{\frac{2\,\sigma}{\omega\,\mu}}\,\frac{\alpha_0'}{\theta_0^{0,25}}\,\vartheta^{1.25} \tag{9.6}$$

* In the Polish edition of this book the coefficients of convection α_c and radiation α_r have denotation α_k and α_p, respectively.

† In the Polish edition, $H_{ms,\,\mathrm{perm}} = H_{ms,\,\mathrm{dop}}$.

The plate can be cooled by air or oil. Let us safely assume that the highest permissible temperature of the plate is 110°C,* and the highest permitted temperature rise of plate over the temperature of the cooling medium is: for air $\theta_{01} = 75$ K and for transformer oil $\theta_0 = 15$ K, whereas the coefficients of heat removal at natural convection—for air the $\alpha_{01}' = 15$ W/m² · K (vertical surface in the air [9.10]) and for the oil $\alpha_0' = 90$ W/m² · K, then the products

$$\alpha_{01}'\theta_{01} = 15 \times 75 = 1125 \frac{\text{W}}{\text{m}^2} \text{ for the air}$$

and

$$\alpha_0'\theta_0 = 90 \times 15 = 1350 \frac{\text{W}}{\text{m}^2} \text{ for oil} \tag{9.7}$$

show very little difference in values. We can assume, therefore, that the calculated values $H_{ms,\text{perm}}$ will be practically the same for both cases of cooling and correspond to the lower of them.

For a plate made of copper ($\sigma_{\text{Cu},20°\text{C}} = 56 \times 10^6$ S/m, $\mu = \mu_0$), at 50 Hz and cooling conditions as above, we have

$$H_{ms,\,\text{perm}} \leq 320 \text{ A/cm} \tag{9.6a}$$

For a plate made of austenitic, *nonmagnetic steel*, this number would be 200 A/cm.

Thick plates made of steel, with two-sided cooling, with symmetrical, plane, polarized waves incident from both sides. In this case, to the formula (9.5) we can apply J. Turowski's approximation (7.11) of magnetization curve $\sqrt{\mu_r}\,|H^2| = c_1H + c_2H^2$, from where

$$P_1 = \frac{\text{a}_\text{p}}{2}\sqrt{\frac{\omega\,\mu_0}{2\,\sigma_t}}(c_1H_{ms} + c_2H_{ms}^{\;2}) \tag{9.8}$$

where $\text{a}_\text{p} \approx 1.4$; $c_1 \approx 310 \times 10^2$ A/m ; $c_2 = 7.9$ (for constructional steel and cast steel); $\sigma_t = \sigma_{20°\text{C}}\,192/(172 + t)$, t is the temperature of the investigated steel part.

Taking into account formulae (9.1) and (9.2), we obtain the equation

$$\frac{\text{a}_\text{p}}{2}\sqrt{\frac{\omega\,\mu_0(172 + t)}{2\,\sigma_{20°\text{C}} \cdot 192}}\left(\text{c}_1H_{ms} + \text{c}_2H_{ms}^{\;2}\right) = \alpha_0'\,\frac{\theta^{1.25}}{\theta_0^{0.25}} \tag{9.9}$$

from which the tangential field H_{ms} on the surface of the steel wall, excited and cooled two-sided, symmetrically, should not exceed the permitted value $H_{ms,\text{perm}}$ for the maximum permitted temperature t_{perm} Figure 9.2)

$$|H_{ms}| \leq H_{ms,\text{perm}} = \left(\sqrt{1 + 3.29 \times 10^{-8}\,c} - 1\right) \tag{9.10}$$

* Discussions at CIGRE and recommendations of ANSI (USA) standards suggest the possibility to permit locally, at overloading, the temperature of hot spot outside of winding up to 140°C, and exceptionally even up to 200°C (Partyga, Turowski [9.6]).

where

$$c \approx 2\alpha_0' \frac{\theta_{perm}^{1.25}}{\theta_0^{0.25}} \sqrt{\frac{\sigma_0}{\omega\mu_0} \cdot \frac{192}{172 + t_{perm}}}$$

θ_{perm}, t_{perm} are the temperature rise and the body temperature assumed as permitted.

For instance, at $f = 50$ Hz, with the average steel conductivity $\sigma_{20°C} = 7 \times 10^6$ S/m, in the conditions of air cooling, at the body temperature $t = 110°C$, $\theta = \theta_0 = 75$ K, $\alpha_0' = 15$ W/m$^2 \cdot$ K, we obtain for the steel

$$H_{ms,\,perm} \leq 40 \text{ A/cm} \tag{9.10a}$$

This value can be taken as an approximate value at the first assessment of possibility of local excessive heating, and then its accuracy may be improved if necessary.

Calculation of the tangential component $H_{ms}(x, y)$ on the surface of the steel wall, with the help of the interactive program RNM-3D (Figures 4.22, 4.21a, 7.21), together with Figure 9.2, gives the possibility of fast, initial detection of spots endangered by local overheating. In these spots one should do more detailed calculations and analysis, apply additional preventing means (screens, change of construction and/or materials, cooling), and place monitoring temperature sensors.

For instance, a change of positioning of the bolted joints of the cover with the tank (1, 2, 3,... in Figure 9.3) changes the overheating hazard of joint bolts in a remarkable way. The dashed line in Figure 9.3 represents the $H_{ms}(x)$ distribution on the surface of the tank without any joints. A full analysis, however, should consider three-dimensional model (Figure 7.22) and 3-D distribution of lines of eddy currents.

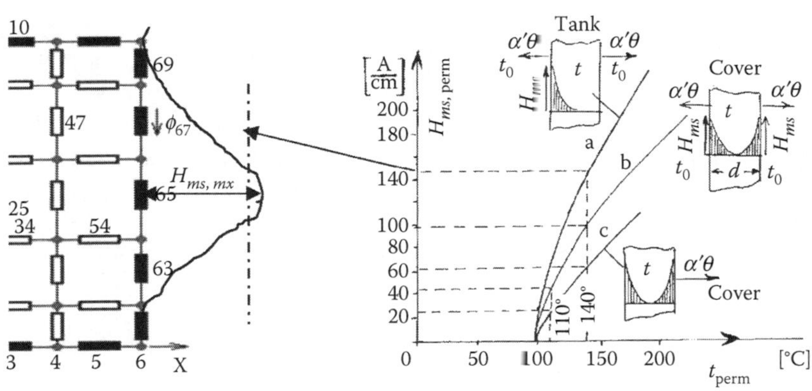

FIGURE 9.2 The permitted values (criterion of overheating hazard) of the tangential component of field $H_{ms,perm}$ on the surface of solid ($d \geq \lambda$) steel plates for assumed permitted temperatures t_{perm}. For example, a—tank wall excited from one side and cooled from both sides; b—cover-plate excited and cooled from both sides; c—cover-plate with hot oil under it, or a clamping plate of the core.

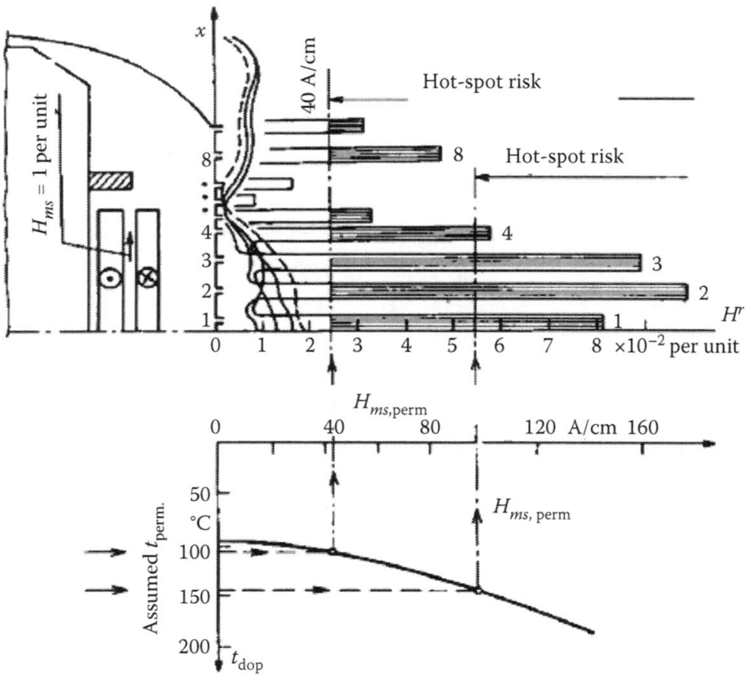

FIGURE 9.3 Detection of local overheating hazard of tank-cover bolted joints in a power transformer. (Adapted from Turowski J. *Proc. Internat. Symposium on Electromagnetic Fields in Electrical Engineering—ISEF'85*. Warsaw, Poland, 26–28 September 1985. Paper No. 63, pp. 271–274 (Plenary lecture) [9.9].)

We shall obtain almost the same value as Equation 9.10a when we assume an oil cooling, according to the permitted conditions (9.6). From comparison of Equations 9.6a and 9.10a, it follows that on the surface of steel elements the magnetic field should be 5–8 times weaker than in the case of parts made of nonmagnetic metals.

In the case of analogical thick plates in which the heat is evacuated from one side only, as it occurs, for instance, when under a transformer cover has gathered oil heated to the temperature of the steel plate (Figure 9.2c), the computational coefficient α_0' of heat removal in formulae (9.7) and (9.9) should be reduced roughly by half. In this case, we get the most unfavorable conditions:

$$\begin{aligned} &\text{for Cu:} \quad H_{ms,\text{Cu,perm}} \le \frac{320}{\sqrt{2}} = 226\ \text{A/cm} \\ &\text{for steel:} \quad H_{ms,\text{perm}} \le \frac{40}{\sqrt{2}} = 28\ \text{A/cm} \end{aligned} \tag{9.11}$$

In the other case when on a thick, two-sided cooled plate impinges a plane wave from one side only (Figure 9.2a), we get the most advantageous conditions, corresponding to doubling of the coefficient α_0' of heat removal. Hence,

$$\text{for Cu plate:} \quad H_{ms,\text{Cu,perm}} \leq 320 \cdot \sqrt{2} = 453 \text{ A/cm}$$

and

$$\text{for solid steel:} \quad H_{ms,\text{perm}} \leq 40 \cdot \sqrt{2} = 56.5 \text{ A/cm} \tag{9.12}$$

Electromagnetically transparent magnetic screens, at *one-sided* incidence of electromagnetic waves, can be analyzed with the help of formula (4.32).

Therefore, in the case of *two-sided cooling* we obtain:

$$H^2_{ms,\text{perm}} \leq \frac{4}{\kappa}\sqrt{\frac{2\sigma}{\omega\mu}}\,\frac{\alpha'_0}{\theta_0^{0.25}}\,\theta^{1.25} \tag{9.13}$$

where κ is expressed by formula (4.33) and plotted in Figure 4.13. At $kd \geq 1$ we can accept $\kappa = 1$ and formula (9.13) becomes identical with Equation 9.12. On the other hand, at $kd < 1$ the permitted magnetic field intensity on the surface decreases very fast with the decrease of the thickness—almost linearly ($1/\kappa \approx 1/kd$). This means that thin, one-sided screens, especially made of steel, can be easily overheated. In the case of steel with the parameters $\mu_r = 500$–1000 and $\sigma_{20°C} = 7 \times 10^6$ S/m, $k = 830$ to 1180 1/m. In the case of Cu, instead, $k \approx 100$ 1/m. Therefore, at $kd < 1$, that is, for the thicknesses of steel sheets smaller than $d = 1/(0.83 \div 1.18) \simeq 1$ mm, and/or for copper sheets thinner than 1 cm, we obtain (in A/m)

$$\begin{aligned} &\text{for Cu:} \quad H_{ms,\text{Cu,perm}} \leq 4530 \times 10^2 \sqrt{d} \qquad (d < 1\text{cm}) \\ &\text{for solid steel:} \quad H_{ms,\text{perm}} \leq (1620 \ldots 1940) \times 10^2 \sqrt{d} \qquad (d < 1\text{mm}) \end{aligned} \tag{9.14}$$

where d is the thickness of sheet, in meters.

$$\text{For example, if } d = 0.5 \text{ mm}, \ H_{ms,\text{Cu,perm}} \leq 4530 \times 10^2 \sqrt{5 \cdot 10^{-4}} = 101 \times 10^2 \text{ A/m, and}$$

$$\text{for steel: } H_{ms,\text{perm}} \leq (36 \ldots 43) \cdot 10^2 \text{ A/m.}$$

Tank walls of small transformers can be approximately considered as the single-sided transparent sheets.

In the case of transparent sheets, two-sided cooled, with a plane wave incident from both sides (e.g., electrostatic screens, thin covers of transformers, etc.), the per unit losses are expressed by formula 4.47. Hence, similarly as before, the highest permissible magnetic field intensity on the metal surface is

$$H^2_{ms,\text{perm}} \leq \frac{4}{\zeta}\sqrt{\frac{2\sigma}{\omega\mu}}\,\frac{\alpha'_0}{\theta_0^{0.25}}\,\theta^{1.25} \tag{9.15}$$

At $kd > 2.3$ (Figure 4.14), $\zeta \simeq 1$ and formula 9.15 becomes identical with formula 9.12. On the other hand, at the sheet thicknesses smaller than 2 … 3 mm in the case of steel, and less than 2.3 cm in the case of copper, that is, at $kd < 2.3$, the coefficient ζ can be approximately substituted by the parabola $0.2(kd)^2$, hence

$$\left.\begin{aligned} &\text{for Cu:} \quad H_{ms,\text{Cu,perm}} \le \frac{453\times 10^2}{0.45\,kd} \approx \frac{10^3}{d} \\ &\text{for solid steel:} \quad H_{ms,\text{perm}} \le \frac{56.5}{0.45\,kd} = \frac{0.15 \text{ to } 0.11}{d} \end{aligned}\right\} (kd < 2.3) \tag{9.16}$$

where d is the thickness of sheet, in meters.

For instance, if $d = 0.5$ mm,

$$\text{for Cu:} \quad H_{ms,\text{Cu,perm}} \le \frac{10^3}{0.0005} \approx 20{,}000 \times 10^2 \text{ A/m}$$

and

$$\text{for solid steel:} \quad H_{ms,\text{perm}} \le \frac{0.15 \text{ through } 0.11}{5.10^{-4}} = 300 \text{ to } 220 \text{ A/m}$$

From comparing the two last examples, it follows that a two-sided screen is much less endangered by local overheating than a screen exposed to one-sided field penetration. It has a confirmation in practice, because a single sheet of transformer core placed in a uniform AC electromagnetic field, or a flat conductor with AC current (Figures 4.11 and 4.12) operate in such conditions of two-sided symmetric excitation.

A steel plate screened by copper from the side of incidence of a plane electromagnetic wave (e.g., screens of tank wall, etc.). At the analysis of formula 4.40 it was stated that even at small thicknesses of screens the losses in the screen–steel system are practically the same as in the screen itself. Therefore, the active power losses in such systems can be expressed approximately with the formula (4.32). At the assumption of two-sided cooling of the system, we can determine hence $H_{ms,\text{perm}}$ from the formula (9.13), obtaining for the corresponding thicknesses (in A/m):

$$H_{ms,\text{perm}} \le 453 \times 10^2 \text{ A/m} \quad \text{at} \quad d \ge 0.01 \text{ m}$$

and (9.17)

$$H_{ms,\text{perm}} < 4530 \times 10^2 \sqrt{d} \quad \text{at} \quad d < 0.01 \text{ m}$$

Perpendicular penetration of field lines into steel surface. This is the case often occurring in design practice. All the permitted values determined above have a full application also here. One should remember, however, that they concern the tangential component H_{ms} on the surface of the metal. In this case, it is most convenient to apply the method of *mirror images* (Chapter 5). For instance, when a current-carrying bus bar passes through a cavity of solid iron in the form of slot (Figure 9.4),

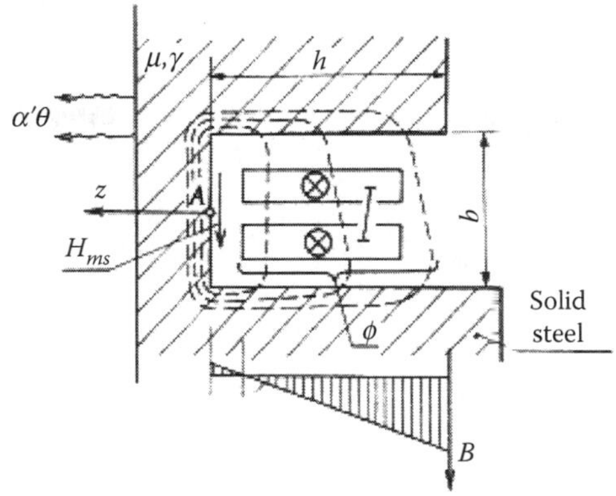

FIGURE 9.4 Determination of the permitted current I of a bus-bar in a steel cavity, from the criterial values $H_{ms} < H_{ms,\mathrm{perm}}$ (Figure 9.2).

the highest value of magnetic field and the biggest thermal hazard will occur at the bottom of the slot. The flux penetrating into the slot walls, $\Phi_m = (\mu_0\sqrt{2}\,I/b)(h/2)$ (formula 5.29), closes the loop through the bottom and equals to the module (4.8). After equalization of both values, we find the strongest magnetic field intensity on the bottom surface, in the point A (Figure 9.4):

$$H_{ms,A} = \frac{\mu_0}{\mu_w}\frac{h}{\mathrm{a_p}\,b}\,k\,I = \mu_0\frac{h}{2\,b}\sqrt{\frac{\omega\sigma}{\mu_w}}I; \quad (\mathrm{a_p} \approx \sqrt{2}) \tag{9.18}$$

where $\sigma_{20°\mathrm{C}} = 7 \times 10^6$ S/m, $\mu_w = \mu\,(H_{ms,A})$—according to Figures 1.27 or 1.29.

The value (9.18) cannot exceed the permitted values $H_{ms,\mathrm{perm}}$ determined above (Figure 9.2). Otherwise, one should apply a screening, another material, cooling, and/or other means to reduce the possibility of local overheating hazard.

In the case when bus-bars run near a metal wall, in order to check the possibility of overheating hazard, one can use the method given in Section 7.4 and the next ones.

In Table 9.1, we have collected a summary of the permitted magnetic loads of constructional parts, from the point of view of possible local excessive heating. All values should be considered as first approximations. Table 9.1 can also be used for experimental checking of local overheating hazard on the basis of the *temperature method* of measurement of the per unit losses (Chapter 10). The *temperature method* gives the possibility to measure the magnetic field intensity on the body surface in a very short time without the need to heat the object.

The presented method can also serve for an inverse synthesis design. If, for instance, we wish to design a hardening coil to heat a given steel element to the hardening temperature, we can start in Figure 9.2 from the desired temperature $t_{\mathrm{perm}} = t_{\mathrm{hardening}}$ and then find the necessary field—on the $H_{ms,\mathrm{perm}}$ axis. From this value of $H_{ms,\mathrm{perm}} = H_{ms,\mathrm{hardening}}$, we can easily design the necessary coil. The same operation can be done for annealing or melting temperature and fields.

TABLE 9.1
Permitted (Criterial) Values of the Tangential Component $H_{ms,\text{perm}}$ of Magnetic Field Intensity on a Surface of Metal Plates from the Point of View of Local Heating over Permitted Temperature $t_{\text{perm}} = 110°C$, according to Turowski's Overheating Criteria

Kinds of Excitation Fields	Cooling System	Material of the Part	$H_{ms,\text{perm}} \times 10^2$ A/m	Formula Number
Plane-wave incident symmetrically from both sides	Air from both sides $\theta_{\text{perm}} = 75$ K or trafo oil $\theta_{\text{perm}} = 15$ K	Copper ($d > 23$ mm)	320	9.6a
As above	As above	Aluminum ($d > 30$ mm)	280	—
As above	As above	Nonmagnetic steel ($d > 65$ mm)	200	—
As above	As above	Constructional steel ($d > 3$ mm)	40	9.10a
As above	As above, one-sided cooling	As above	28	9.11
As above	As above	Copper ($d > 23$ mm)	226	9.11
Plane wave from one side	As above, two-sided cooling	Copper ($d \geq 10$ mm)	453	9.12
As above	As above	Steel ($d \geq 1$ mm)	56.5	9.12
As above	As above	Copper ($d < 10$ mm)	$4530\sqrt{d}$[a]	9.14
As above	As above	Steel ($d < 1$ mm)	$1800\sqrt{d}$[a]	9.14
As above	As above	Hot rolled transformer steel 0.35 mm	17	—
Plane wave, from both sides	As above	Copper ($d < 23$ mm)	$10^3/d$[a]	9.16
As above	As above	Steel ($d < 3$ mm)	$0.13/d$[a]	9.16
Plane wave incident from the screen side	As above	Steel screened by Cu: –of thickness $d \geq 10$ mm –of thickness $d < 10$ mm	 453 $4530\sqrt{d}$[a]	 9.17

Source: Turowski J. *Scientific Archives of the Technical University of Lodz "Elektryka*, No. 11, 1963, pp. 89–179 (*Habilitation (DSc) dissertation; 1st Ministry Award*) [2.41]; Kozlowski M. and Turowski J. *CIGRE. Plenary Session*, Paris 1972. Report 12-10 [7.9]; Turowski J. *Izvestia Vysshikh Uchebnykh Zavedenij. SSSR "Elektromekhanika,"* 8/1961, pp. 91–114 [9.10].

[a] d—in meters, $\theta = t - t_0$.

9.2 METHODS OF PREVENTION OF LOCAL OVERHEATING OF A METAL CONSTRUCTION

Elimination of the possibility of local excessive heating can be achieved by the removing of both causes and effects of heating. To the first group belong the above-mentioned methods of weakening of magnetic fields on the body surface, with the help of magnetic screens (Chapter 4), cutting slits (Section 6.4), as well as reduction of power losses by application of nonmagnetic materials. Thermal effects of local concentration of power losses can be decreased by corresponding heat evacuation to less endangered spots. Quantitatively, this effect can be considered with the help of the coefficient of irregularity of heat distribution, *K*, derived by the author in 1960 [9.10].

9.2.1 Coefficient of Irregularity of Heat Distribution, K (J. Turowski [9.10])

If the amplitude of the magnetic field intensity H_{ms} exceeds the permitted values, given in Table 9.1, one should take into account the possibility of excessive heating of the part. These values in power transformers and electrical machines can be easily exceeded; however, irregular distribution of per-unit power losses mitigates in general this hazard.

A typical example is the uneven loss distribution on a plate surface, accompanied by heat evacuation along the plate plane from hotter to cooler spots, which may occur, for example, on transformer cover plates (Figure 6.8), [4.16], [6.17] tanks, yoke beams, clamping frames at the end of turbine generators, and so on. Let us assume that irregularity of loss distribution on such a plate can be represented with the help of an exponential curve (Figure 9.5):

$$P_1 = Ae^{-B|y|} \tag{9.19}$$

In the above formula (9.19), the constant *B* represents the degree of nonuniformity of the space distribution of losses, and the constant *A* is its highest value. At $B \to 0$ the curve of loss distribution tends to a uniform distribution, discussed in Section 9.1.

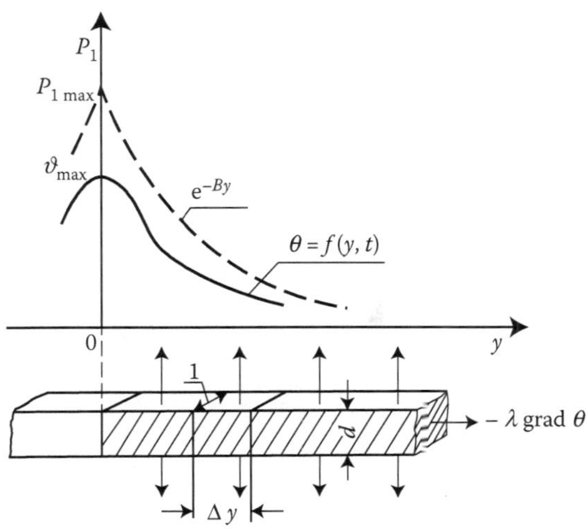

FIGURE 9.5 Nonuniform distribution of power losses and temperature along the *Y*-axis. (Adapted from Turowski J. *Izvestia Vysshikh Uchebnykh Zavedenij. SSSR "Elektromekhanika,"* 8/1961, pp. 91–114 [9.10].)

Since the sheet thickness is usually small in comparison with the remaining dimensions, and the thermal conductivity is high, the temperature drop across the sheet thickness can be neglected. Due to this, we can assume that temperature varies only in the y direction.

The exchange of heat between the sheet surface and the surroundings occurs on the basis of convection and radiation. A part of the heat is carried away by heat conduction from hotter to cooler spots. Assuming a constant ambient temperature, we can write the equation of thermal conductivity in the form

$$c\rho\frac{\partial\Theta(y,t)}{\partial t} = \lambda\frac{\partial^2\Theta(y,t)}{\partial y^2} + q \tag{9.20}$$

where c is the thermal capacity, ρ the mass density, λ the thermal conductivity of metal, q the heat amount per unit volume, consisting of power losses ($2P_1/d$) as the positive source, and of the heat carried away to the surroundings ($\alpha'\theta$) as the negative source of heat (the *symbol* Δ *below means an element of length in* Figure 9.5, *and not the Laplacian*)

$$q = \frac{2P_1\,\Delta y}{d\,\Delta y} - \frac{2\alpha'\,\Delta y}{d\,\Delta y} = \frac{2A\mathrm{e}^{-B|y|}}{d} - \frac{2\alpha'\Theta}{d} \tag{9.21}$$

Let us assume that the heat removal is symmetrical from both sides of the sheet, which is natural at vertical sheets. At horizontal sheets, we make a justified assumption that the coefficient of heat removal from the upper surface is bigger than the calculation value by the same amount by which the bottom surface coefficient is smaller [9.10]. In the considered case, the transient state is not interesting, and we shall resolve Equation 9.20 directly for the steady-state conditions, taking into account Equation 9.21

$$\lambda d\frac{\partial^2\Theta(y)}{\partial y^2} - 2\alpha'\theta = -2A\mathrm{e}^{-B|y|} \tag{9.22}$$

A general solution of Equation 9.22 takes the form (Mikheiev [9.2]):

$$\Theta = \frac{-2A}{B^2\lambda d - 2\alpha'}\mathrm{e}^{-B|y|} + C_1\exp\left(\sqrt{\frac{2\alpha'}{\lambda d}}\,y\right) + C_1\exp\left(-\sqrt{\frac{2\alpha'}{\lambda d}}\,y\right) \tag{9.23}$$

At $y \to \infty$, $\theta = 0$, hence $C_1 = 0$.

Since in point $y = 0$ the body temperature is maximum, the following condition for extremum should be satisfied

$$\left(\frac{\partial\Theta}{\partial y}\right)_{y=0} = 0 \tag{9.24}$$

After substitution of Equation 9.23 into Equation 9.24, we can determine the second constant:

$$C_2 = \frac{2AB}{\left(B^2\lambda d - 2\alpha'\right)\sqrt{\frac{2\alpha'}{\lambda d}}}$$

From it, the temperature rise in the distance y from the hot spot is

$$\Theta(y) = \frac{2A}{B^2\lambda d - 2\alpha'}\left(\frac{B}{\sqrt{\frac{2\alpha'}{\lambda d}}}\mathrm{e}^{-\sqrt{\frac{2\alpha'}{\lambda d}}\,y} - \mathrm{e}^{-B|y|}\right) \tag{9.25}$$

and the maximum temperature rise:

$$\Theta_{\max} = \frac{2A}{B^2\lambda d - 2\alpha'}\left(\frac{B}{\sqrt{\frac{2\alpha'}{\lambda d}}} - 1\right) = \frac{A}{\alpha'}K \tag{9.26}$$

where the coefficient of nonuniformity of heat distribution is

$$K = \frac{1}{B/\sqrt{(2\alpha'/\lambda d)} + 1} \approx \frac{1}{cB\sqrt{d} + 1} \le 1 \tag{9.27}$$

where in the simplified formula it was assumed: $\alpha' = 14.2$ W/(m^2 K) and λ = 45 W/(m K)—the thermal conductivity of steel, wherefrom $c = 1.26$ 1/m$^{1/2}$.

The coefficient of nonuniformity K (9.27) takes into account the influence (on the maximal temperature) of the heat transfer to cooler places. In the case of uniform losses on the whole surface of sheet ($B = 0$), or in the case of neglecting of the sheet thermal conductivity along the plane ($\lambda = 0$), the coefficient $K = 1$ and the maximal temperature is conditioned only by the heat evacuation through the entire plate surface, like in the cases considered in Section 9.1.

At a big nonuniformity of loss distribution, which occurs, for example, on the transformer cover (Figure 6.8), this coefficient's values are $K = 0.5$–0.7. At a more uniform distribution, this coefficient approaches 1. Hence, the values given in Table 9.1 can be considered as *criteria of overheating*, and the coefficient K—as an indispensable in such cases *factor of safety*, that makes values $H_{ms,\mathrm{perm},K}$ more accurate owing to its division by $\sqrt{K}$ that is,

$$H_{ms,\mathrm{perm},K} = \frac{H_{ms,\mathrm{perm}}}{\sqrt{K}} \tag{9.28}$$

By applying additional means (cooling, change of material, screens, smaller thickness d, etc.), which decrease the value of the coefficient K, one can significantly increase the value of the permitted field $H_{ms,\text{perm},K}$ and thus reduce the thermal hazard of parts. The decrease of K can be achieved by influence on parameters in formula (9.27). These considerations did not take into account the influence of metal parts which are contiguous to the investigated object, the heat transfer in the directions of remaining axes, x and z. This requires more accurate, time-consuming research and calculations, which make sense to undertake when the field value on the surfaces approaches the permitted 1. More important, however, is the elimination of this hazard rather than its very accurate calculation.

9.3 HEATING OF TRANSFORMER COVER PLATES

Cover plate of power transformer is a typical example of the part endangered by the local overheating [1.48], [9.10], [9.11]–[9.15]. According to Table 9.1, the highest permitted magnetic field intensity $H_{ms,\text{perm}}$ for the cover adjacent to a layer of hot oil is 28 A/cm (9.11). However, a big nonuniformity of loss distribution, represented by the coefficient $K = 0.5$–0.7 (9.27), allows to increase this limit to the value of $H_{ms,\text{perm},K} = \dfrac{28\ \text{A/cm}}{\sqrt{0.5 \text{ to } 0.7}} = (40 \text{ to } 34)\ \text{A/cm}.$

The magnetic field intensity H_{ms} on the surface of cover of a three-phase transformer is expressed by formula (5.16), and the loss distribution on the surface, at a constant permeability, is presented in Figure 6.8. In reality, the steel saturation smoothes the peaks of field distribution in the bushings vicinity. Owing to it, the task can be simplified by assuming that in the x direction (Figure 6.8) the loss distribution is uniform and equals to its value at $x = 0.5$ a. On the other hand, in the y direction the losses decrease according to the exponential rule (9.19). The first assumption is justified by the fact that the heat flow in the x direction, between the bushings, is difficult because of the presence of insulators as well as due to some increase of losses around the bushings. The constants A and B for the second assumption can be evaluated by comparison of the curve obtained from formulae (9.8), (5.16) at $x = 0.5$ a with the curve (9.19). From that, we obtain

$$B = \frac{1.4}{a}, \quad A = 0.7\sqrt{\frac{\omega\mu_o}{2\sigma}}\,(310 \cdot 10^2 H_{m0} + 7.9\, H_{m0}^2) \tag{9.29}$$

where $H_{m0} \simeq 0.8$ I/a.

By comparing the expression with H_{m0} in Equation 9.29 with the value $H_{ms,\text{perm}}/\sqrt{K}$, we can find the *highest permissible current in bushings*, or check if at a given current the cover will not locally heat excessively. In the J. Turowski's works [9.10] and [2.41], hot-spot temperatures for different covers were calculated with the formula (9.26), and the obtained results showed good agreement with measurements made on several full-scale models.

9.4 PERMISSIBLE CURRENT IN BUSHINGS

In design practice, there is often a need to pass an AC high current bus-bars through metal barriers (transformer tank walls and covers, steel beams, electric machine housings, etc.). Formulae (9.26) through (9.29) allow to determine the maximal current in bushings (Figure 5.21), over which there is a hazard of excessive heating of the wall, above the permissible temperature.

Substituting expressions (9.29) and (9.27) into Equation 9.26 and assuming, in accordance with [9.10], the coefficient of heat removal by convection and radiation, at natural cooling, to be

$$\alpha' = \alpha'_{110°C}\left(\frac{\Theta}{110°C}\right)^{0.25}; \quad \alpha'_{110°C} = 14.2\frac{W}{m^2 \cdot K}$$

the coefficient of heat conductivity of steel $\lambda = 45$ W/(m · K), $f = 50$ Hz, and $\mu_0 = 0.4\pi \cdot 10^{-6}$ H/m, we get

$$\Theta_{max}^{1.25} = 0.7\sqrt{\frac{\omega\mu_0}{2\sigma_t}\frac{\sqrt[4]{110°C}}{\alpha'_{110}}}\left(388 \cdot 10^2\frac{a}{I} + 7.9\right)K =$$

$$= 1.05\sqrt{\frac{172+t}{\sigma_{20°C}}}\left(3.88\frac{I}{a} + 7.9 \cdot 10^{-4}\frac{I^2}{a^2}\right)K \qquad (9.30)$$

where

$$K = \frac{1}{1.4/a\sqrt{\left(2\,\alpha'_{110}/\lambda d\right)} + 1} = \frac{1}{1.76\sqrt{d}/a + 1}$$

We shall not make a significant error if we omit for simplification, the dependence on the thermal coefficient α' which is under the root of the expression for K.

In the case of a hot, oil-cooled transformer, at the most unfavorable conditions when there is practically no heat transfer to the hot oil, the coefficient α' of heat removal through the cover should be multiplied by 1.1/2 (the heat removal through the upper surface is about 10% higher than the mean calculated value).

Taking the same permitted heating as in Equation 9.8, we obtain from Equation 9.30 the condition for the current, in the form

$$75^{1.25} > \frac{2.1}{1.1}\sqrt{\frac{172+110}{\sigma_{20°C}}}\left(3.88\frac{I}{a} + 7.9 \cdot 10^{-4}\frac{I^2}{a^2}\right)\frac{1}{\sqrt{\frac{2}{1.1}} \cdot \frac{1.76\sqrt{d}}{a} + 1} \qquad (9.31)$$

After solution of the inequality (9.31), at $\sigma_{20°C} = 7 \times 10^6$ S/m, we finally obtain the *Permitted current in the bushings*:

$$I < 24.6 \cdot 10^2 a\left[\sqrt{1 + 3.9\left(1 + \frac{2.4\sqrt{d}}{a}\right)} - 1\right] \qquad (9.32)$$

where I is the rms value of current (in A); a the distance between the axes of bushings (in m) (Figure 5.21); and d the thickness of cover (sheet) (in m).

For example, at $a = 0.2$ m and $d = 7$ mm, the current in bushings should not exceed 966 A.

At the same time, the highest magnetic field strength, according to Equation 9.29, reaches the value $H_{m0} = 38.6$ A/cm, and the coefficient $K = 0.58$ (9.27). The corresponding overheating criteria value is $\frac{H_{ms,\text{perm}}}{\sqrt{K}} = \frac{28\text{ A/cm}}{\sqrt{0.58}} = 37$ A/cm. Since this is already the limit value, in practice it is safer to accept currents smaller than this limit, using rather the values $H_{ms,\text{perm}}$ from Table 9.1.

At currents higher than permitted, one should use one of the several preventing methods (Section 6.4) mentioned at the beginning of Section 9.2.

High-current leads placed near the steel covers and walls pose similarly acute problems for design and reliability of large power transformers, by inducing in these walls high eddy currents and causing strong local heating as well. Such leads usually run parallel to these walls. The work by Bereza et al. [9.4] presented a method and results of investigations of such an arrangement, with taking into account inserts from nonmagnetic steel and magnetic screens, perpendicular to the current axis.

In the work by Apanasewicz and Kazmierski [9.3] the influence of the field of high current leads on the field and power losses in the region of bolted joints of tank cover of transformer was investigated.

9.4.1 Eddy-Current Loss and Hot Spots in Bushing Turrets

The so-called bushing turrets (2 in Figure 7.19a), after flat covers (3 in Figure 7.19a), belong to the power transformer's inactive parts which are the most endangered by local overheating.

The simplest way of their calculation is with the help of the Ampere and Biot–Savart laws (Figures 5.20, 5.21, 5.22, 6.7, 6.8, and Section 7.4).

A simple and rapid computer program with the formulae (5.15) through (5.17) can provide a solution in microseconds, which could be also done by hand. In this way, one can determine the distribution of the magnetic field intensity $H_{ms}(r, \Theta)$ or $H_{ms}(x, y)$ on the internal surface of a turret of an arbitrary form, similar to Figure 6.8, and similarly the hot spot from Figure 9.2, similar to Figure 9.3, and finally the per unit loss—from Table 7.4. The total losses in the whole turret can be obtained by integration over its internal surface, according to formula 7.74a. The accuracy of this result depends on consideration of particular details. But usually even a simplest calculation is satisfactory for a technical application.

9.4.2 Computer Calculation

Using the above described simple method, an automatic program was designed for calculation of the fields, losses, and hot spots in flat covers without nonmagnetic inserts (Figure 9.6a,b), covers with nonmagnetic inserts (Figure 9.6c), as well as covers with turrets.

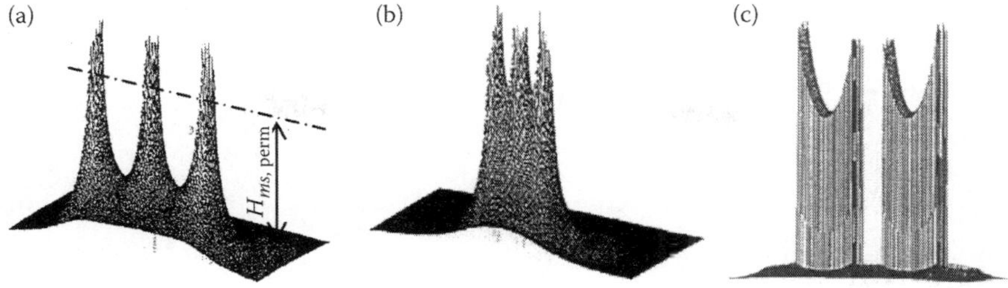

FIGURE 9.6 Distributions of the tangential field $H_{ms}(x, y)$ on the surface of a flat cover made of constructional steel: (a, b) influence of the distance between bushing axes, (c) the magnetic field concentration at the nonmagnetic insert ($e = 47$ mm, see Figure 6.9), per (Turowski, Pelikant [9.12]); $H_{ms,\text{perm}}$ hot-spot localization as in Figure 9.2.

9.4.3 Single-Phase Turrets

A hybrid, analytical–numerical method was applied to simulation and calculation of turrets of a very large single-phase transformer (J. Turowski [9.11]) (Figure 9.7). Here the author's theory of electromagnetic and magnetic screening was applied, as well as the method of mirror images to reduce bushings to one conductor (Figure 9.8) as in Figures 5.1 and 9.4. In the case considered, a model of the box will be similar

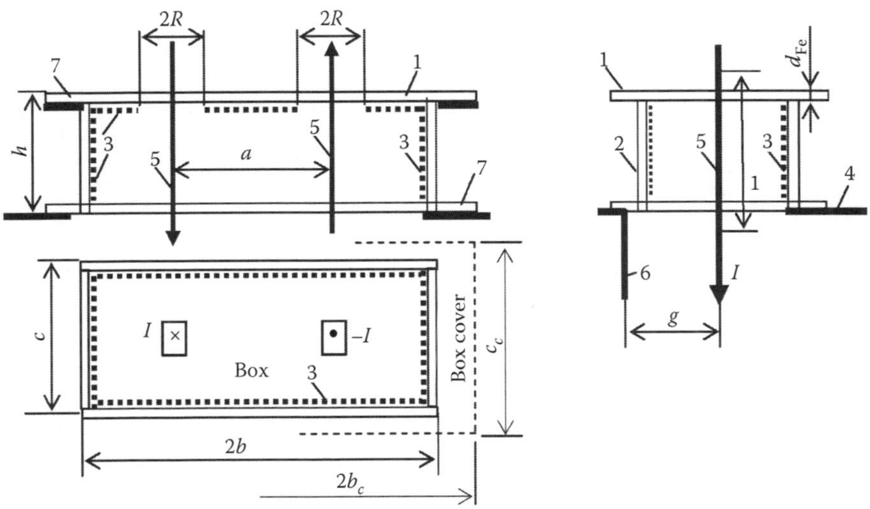

FIGURE 9.7 Dimensions of a single- phase high-current bushing turret: 1. Cover of the bushing box of thickness d_{FeC}, 2. Steel walls of the bushing box of thickness d_{Fe}, 3. Copper screens of thickness d_{Cu}, 4. Cover of tank of thickness d_{Fe}, 5. Copper loss in the bushing conductors of cross-section s_{bCu}, 6. Eddy current loss induced in tank by the leads, 7. Flange bolted joints, 8. the losses in tank walls induced by transformer windings. (Adapted from Turowski J.: *Rapid Evaluation of Excessive Local Heating Hazard in Bushing Turrets of Large Power Transformers. XVI International Conference on Electrical Machines—ICEM'04, Cracow, Poland, Sept. 2004* [9.11].)

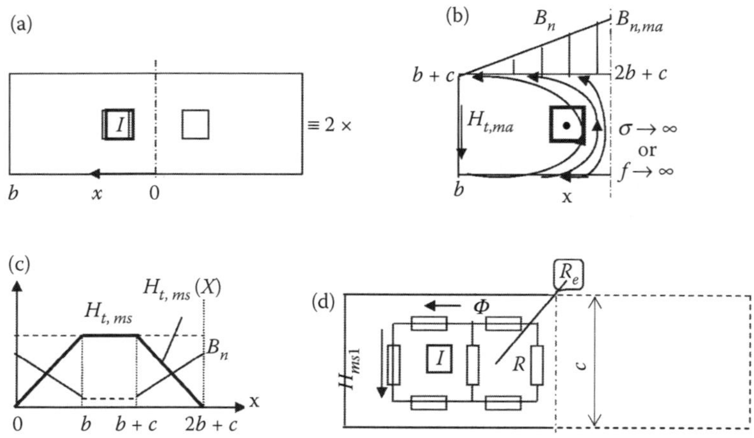

FIGURE 9.8 Simulation of single-phase bushing turret: (a) cross-section, (b) equivalent mirror image, (c) field distribution, (d) simplified RNM model to evaluate boundary value H_{ms1} on the internal surface of the box. $R_e = 3/4\,R = 0.75\,R$; $\sqrt{2}\,\sqrt{2}\,I = \Phi(R + 2.75)\,R$; $H_{ms1} \cdot c = \Phi R = \sqrt{2}\,I(R/(3.75\,R)$. (Adapted from Turowski J.: *Rapid Evaluation of Excessive Local Heating Hazard in Bushing Turrets of Large Power Transformers. XVI International Conference on Electrical Machines—ICEM'04, Cracow, Poland, Sept. 2004* [9.11].)

to the calculation of a hot spot in the cavity bottom (Figure 9.4). Only the H_{ms} on the cavity bottom will be different.

It can be evaluated in the simplest way with the help of RNM (equivalent Reluctance Network Method—Figure 9.8). Full analysis can be done with the help of the authors' "MSR1100" program (see Ref. [1.17/2], pp. 145–178, and [4.27]). However, the first approach from J. Turowski [9.11] can be used, that is, a simplest model (Figure 9.8) for the evaluation of an approximate hot-spot value H_{ms1}:

$$H_{ms1} = \frac{\sqrt{2}\,I}{3.75c} = \frac{\sqrt{2} \cdot 17{,}000}{3.75 \cdot 0.3455} = 18.6 \times 10^3\,\frac{\text{A}}{\text{m}} \tag{9.33}$$

9.5 THREE-PHASE TURRETS, SIMULATED BY A RAPID, EQUIVALENT-CIRCUIT RNM MODEL

The "Ampere-turns shifting method (ASM)" was used in Figure 5.36 to the RNM model of thick winding by equivalent magnetic circuit (J. Turowski [1.17/2], p. 153). This new author's concept (Figure 9.9) includes substitution of cumbersome field analysis by a simple and rapid RNM circuit model, equivalent to electromagnetic field calculations as in Figures 5.35 and 9.8.

In Figure 9.9a, all the currents, real and fictitious, are exactly superimposed on each other, respectively. As a result, all three meet in the mesh cell (square) $xy = 73$ (i.e., $x = 7$, $y = 3$) and they all disappear as the three-phase vector sum $\mathrm{I}_{Um} + \mathrm{I}_{Vm} +$

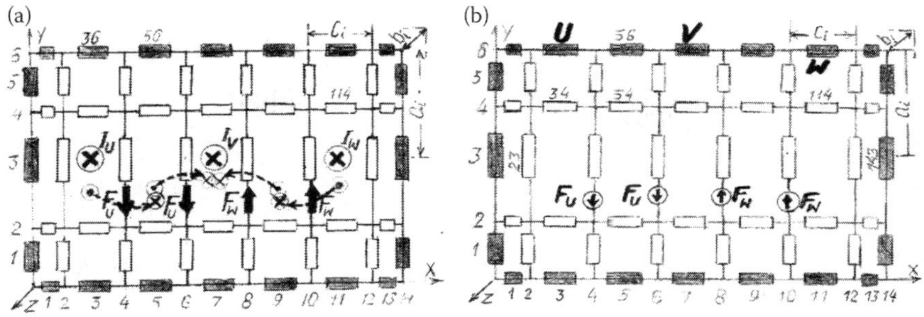

FIGURE 9.9 RNM model of three-phase bushing system in a cover (Figure 5.21) or a common rectangular turret (Figure 9.7): (a) ⊗ $\boldsymbol{I_{Um}}$, ⊗I_{Vm}, ⊗I_{Wm}—three-phase bushings; ⊙ → ⊗—two fictitious coils with currents I_{Um}; ⬇⬇—two black arrows are two fictitious mmf $F_U = \sqrt{2}I_U$ compensating the flow of two I_{Um} coils; ⊗← ⊙—two fictitious coils with currents I_{Wm}; ⬆⬆—two black arrows are two fictitious mmf $F_W = \sqrt{2}I_W$ compensating the flow of two fictitious I_{Wm} coils; (b) the final RNM model.

$I_{Wm} = 0$. There remains only a multi-node circuit (Figure 9.9b) with four mmf sources: two F_U and two F_W.

9.5.1 Calculation of Reluctances for RNM

The entire space is divided into boxes in Figure 9.10, where the black boxes ■ are complex reluctances of metal, with reluctance $R_{mi} = l_{ix}/\mu_{ix}\,\delta \cdot l_{iy}$, $\delta = 1/a_p\sqrt{\omega\mu\sigma/2}$, and the white boxes □ represent a dielectric inside the box (Figure 9.11), with the

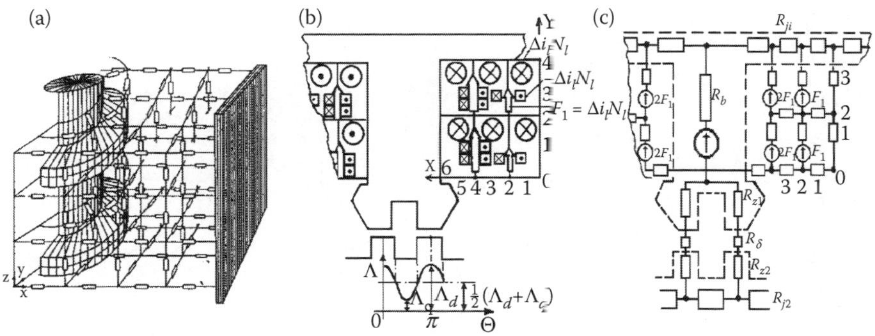

FIGURE 9.10 RNM-3D modeling of distributed field by an equivalent reluctance circuit: (a) magnetic separator of chimney smoke (Adapted from Ciosk K., Gierczak E., Mendrela E.A. *Magnetic and Force Field in the Gap of Magnetic Screw Separator* (in Polish). Wiadomosci Elektrotechniczne. LXII, No 6, 1994, p. 236–238.); (b) pole winding of stepping motor; (c) equivalent reluctance scheme for a wide winding. (Adapted from Turowski J.: Computational magnetics. (in English). J. Sykulski (Editor), London: Chapman & Hall Stellenbosch, 1995 where: §4) Turowski J.: Reluctance networks RNM [1.17/2].)

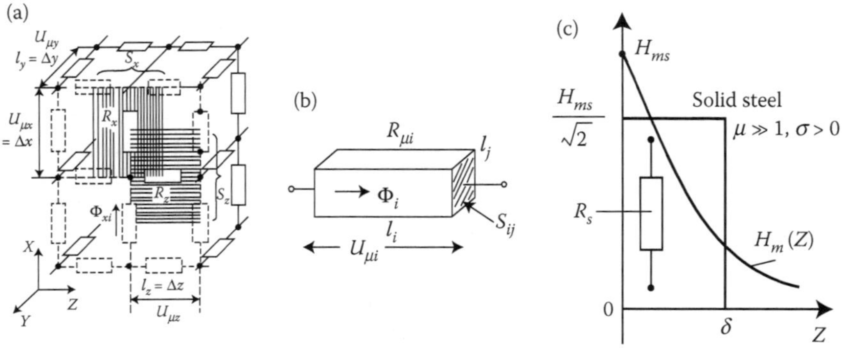

FIGURE 9.11 Calculation of reluctances for Figures 9.9 and 9.10: (a) a fragment of the RNM-3D network; (b) reluctance in a dielectric; (c) reluctance of a solid conductor, for example, for steel $H_m(z) = H_{ms}\, e^{-\alpha z}$, $\alpha = (1 + j)k$; $k = 1/\delta$—attenuation constant in metal Figure 2.10.

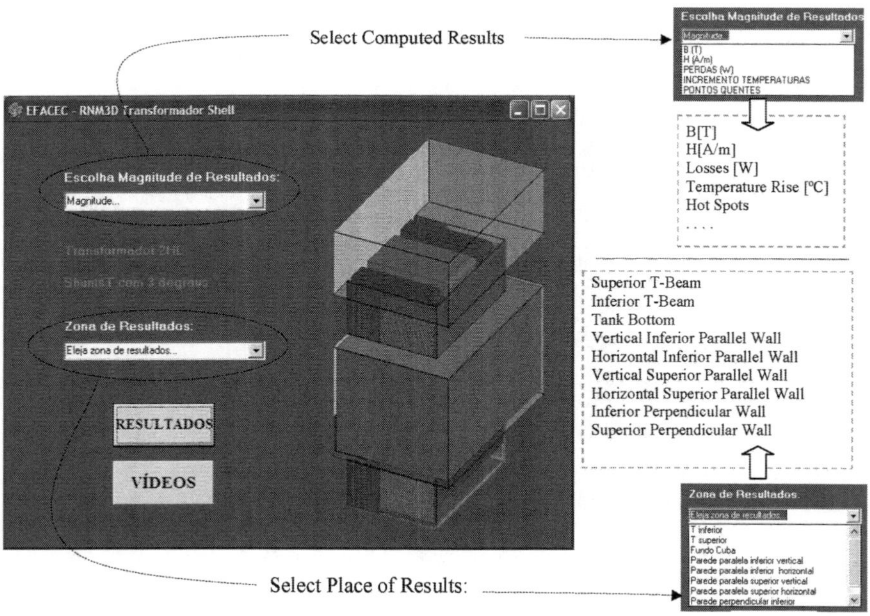

FIGURE 9.12 Automatic calculation of the special magnetic shunts in large, single-phase, shell-type transformers by the interactive program "RNM-3D$_{\text{shell}}$" during design in EFACEC—Porto (Portugal) and Georgia (USA) Transformer Works. (Adapted from Turowski J. et al.: *(EFACEC Trafo Works, Porto, Portugal) International Advanced Research Workshop on Transformers – ARWtr2007* X.M. Lopez-Fernandez (editors), Baiona, Spain, October 2007, pp. 56–68.)

reluctance $R_i = \dfrac{l_{ix}}{\mu_{ix}\ S_{iyz}}$. When Cu or Al electromagnetic screens are applied on the wall, then $R_{Cu} \approx (\alpha_2 \sinh\alpha_2 d)/(\mu(\cosh \alpha_2 d - 1)) \approx 2/\mu_0 d$.

In a first approach, one can accept $R_{Cu} \to \infty$. When a magnetic screen (shunt) is applied, one can accept $R_{Fe} \to 0$.

In the work by Shenyang Polytechnic University* (China), the authors presented results of calculation of field and eddy currents induced in the so-called turrets through which high current bushings of LV side in a large transformer are passing. They used the boundary element method (BEM-3D) with three-vertex polar elements. The calculated power losses from the fields of bushings were 2.62 kW without screens, 2.16 kW—with an Al screen, and 1.86 kW—with a Cu screen, respectively. The same results can be easily and quickly calculated using the RNM-3D (Figure 9.8d). A more accurate analysis can be done by using a higher density grid, and by using "MSR-1100" or "Java" code as the Solver in the RNM-3D program (Turowski et al. [4.31]).

A similar calculation of the special magnetic shunts in shell-type transformers is given in Figure 9.12.

The "RNM-3D$_{shell}$" program was designed according to the principles of Mechatronics and Interactive Expert System (Figure 1.1b,c), where the author of this book was the *Area* and *Domain Expert* (J. Turowski [1.20]). Professor X. M. Lopez-Fernandez was a consultant, and the Spanish industrial engineers, Ms Andrea Soto and Mr David Souto, who were at the same time graduate students of the author (J. Turowski) at the Technical University of Lodz, were the *Knowledge Engineers* [4.36].

* Yang Junyou, Chen Yongbin, Tang Renyuan, Li Yan: Application of high precision boundary element method to 3-D eddy current fields due to leads of transformers. *International Conference on Electromagnetic Field Problems and Application—ICEF'92*. Hangzhou. China, 14–16 Oct. 1992, Paper OB2-2, pp. 97–98.

10 Methods of Experimental Investigations

10.1 EXPERIMENTAL VERIFICATION OF THEORETICAL CALCULATIONS

Owing to the complexity of real electromagnetic processes, especially those in constructional steel elements, along with various heating problems, it is difficult to develop a trustworthy theory and formulae for rapid and cost-effective industrial calculations and design methods, without permanent and versatile experimental verification of both initial assumptions and final results, like for instance measuring of rotational core losses in machine lamination [10.35].

Experimental verification of assumptions and results should be understood in a broader sense, that is, not only as laboratory tests. Often, an expensive laboratory experiment can be substituted by a mental experiment, which is often applied on the basis of experience and knowledge, when one wishes to resolve something quickly. For instance, a theory of screens of a certain type can be verified by finding an analogy to power transformer with short-circuited secondary winding, or to induction motor with solid rotor, deep slot in foil winding (Section 5.7), or other examples, well known from engineering practice. In the case when we are forced to apply simplifying assumptions, an experiment plays a primary role. It applies especially to nonlinear systems, in particular with steel, for which lack of permanent experimental verification of assumptions and theoretical analysis can easily lead to wrong conclusions. In cases when we can manage to develop a theory without introduction of simplifications (e.g., for some simple linear systems), an experiment plays rather only the role of inspection and does not have to be very accurate. It serves only for detection of rough accidental errors. In such cases, the theory can give results more accurate than experiment. When investigating new objects or phenomena, except of a system approach (J. Turowski [1.20]), it is often more advantageous to begin from experiments. It makes it easier to select proper assumptions and direct the theoretical investigation to a proper route. In contrast with experimental research, the initial simple calculations cost almost nothing (Figure 10.1). However, subsequently these expenses can grow quickly; reaching the *break-even point* for the *cost of risk* (*CR*), for example, the three-dimensional, three-phase FEM-3D simulations (Figure 7.15). Then, one should pass to simpler, more advantageous methods of calculation, like for instance the three-dimensional, three-phase RNM-3D (Figures 7.20 and 7.22). These differences are apparent from the comparison presented in Figure 10.2a.

Sometimes it pays not to push through theoretical analysis, but to pass as soon as possible from theory to experimental studies. From a certain moment, the mathematical investigations lose their sense; for example, due to the material and processing

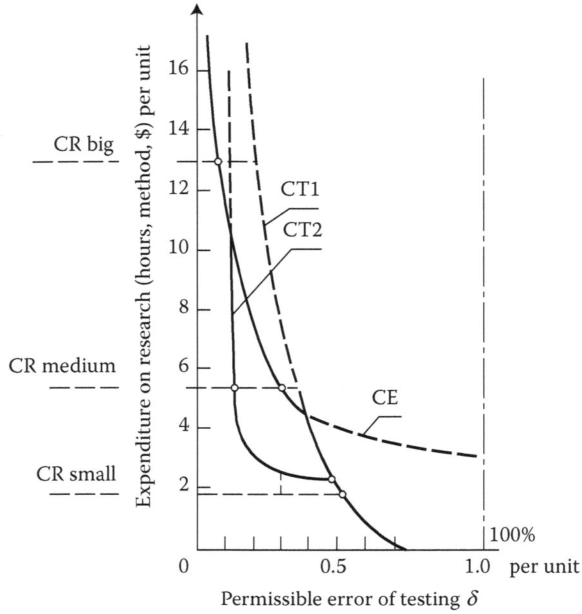

FIGURE 10.1 Profitability of industrial investigations. CR—(cost of implementing risk) limit, above which one should proceed to implementation; CT—cost of theoretical investigation, depending on effectiveness of method: for example, CT1—FEM 3D method (Figure 7.20); CT2—RNM-3D (Figure 7.21); CE—cost of experimental investigations. (Adapted from Turowski J.: (1) *Calculations of Electromagnetic Components of Electric Machinery and Equipment.* (in Polish). Warsaw: WNT, 1982; and (in Russian). Moscow: Energoatomizdat, 1986.)

discrepancies of the parameters like σ, μ, ε, which may be so big (7.26) that further refining of purely mathematical model and calculations is an unjustified waste of time and effort.

In such a case, one should proceed to experimental and prototype investigations. However, one should always remember that "time is money" and both theoretical and experimental investigations should be accurate, but rapid. There are many methods theoretically equivalent, however only some of them are rapid enough (Figures 9.10 and 10.2a). They need, for instance, clever reduced models (e.g., analytical–numerical, mirror images, equivalent circuit, linearization) or applications of other analogous fields.

Discrepancies in the cost and effort of selecting the research methods after Figure 10.1, between theoretical, experimental, and prototype, can be as great as the example, in Figure 10.2a.

The cost of prototypes, especially of large and dangerous ones, can be very expensive. The Cost of implementing Risk (CR), equal to possible losses in the case of implementation failure, can be very large (e.g., New York 1977, Chernobyl 1986). Overlapping of different phenomena during normal operation of the object makes their separated research and scientific generalization of results more difficult. Not less important, apart from reliability, is the examination of the *cost of abandoning*

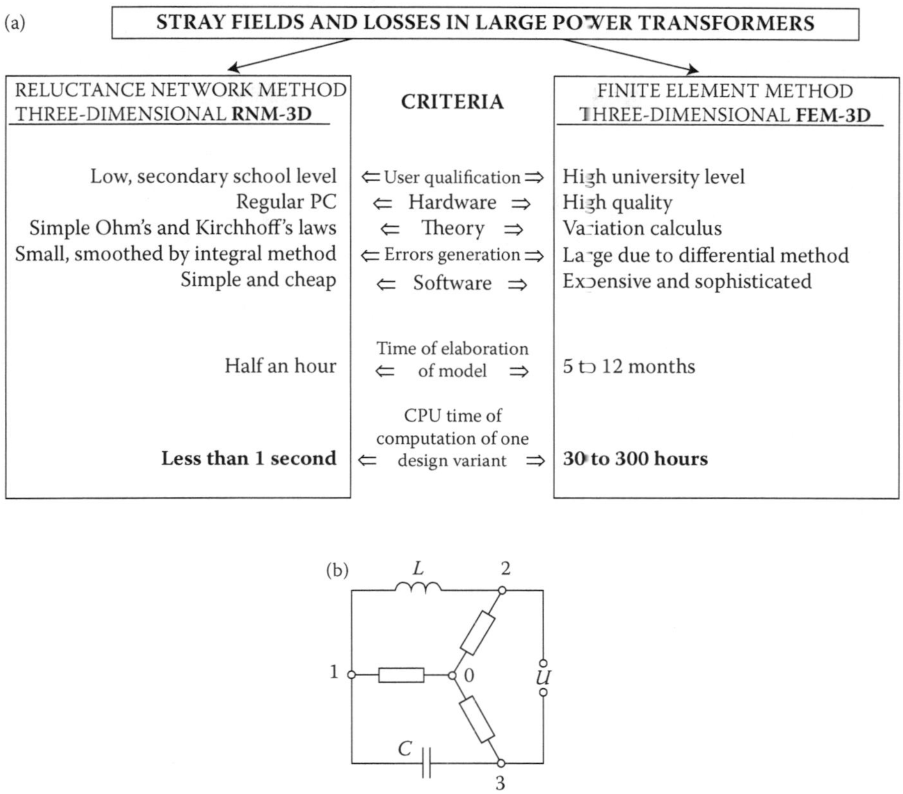

FIGURE 10.2 (a) Dependence of cost-effectiveness of modeling and simulation on the method selected. (Adapted from Turowski J.: *Proceedings of Internat. Aegean Conference—ACEMP'04*. May 26–28, 2004. Istanbul, Turkey, pp. 65–70.) (b) Simplest layouts of static converters of single-phase current into three-phase current.

(e.g., not complete repair of a system after its failure, abandonment of protection expenditure, etc.).

This is why the design of effective methods for experimental research on simplified and reduced models plays a big role here. This problem belongs to the *theory of similarity* (Section 10.2) [1.20], [2.41], [9.2]. A correctly designed model, supported by profound theoretical investigations and an experienced Knowledge Expert, can often bring necessary results in a faster and more cost-effective manner (Figure 10.1, CE).

There are several theories and principles of modeling, which generally can be divided into two basic groups:

- *Physical modeling*, whose objective is identification of physical essence of phenomenon and laws ruling its behavior, in order to give an adequate mathematical form of the model, or—an experimental verification of correctness of assumptions and results of theoretical investigations, prototype investigations, and so on.

- *Mathematical modeling*, the purpose of which is an exact or approximate solution of a problem described with mathematical equations (mathematical simulation of phenomena and processes).

The first group includes physical models containing real elements of investigated objects (cores, coils, etc.), where one physical parameter can be substituted by another (for instance, the change of model scale at a change of frequency [2.41], [10.3], [10.34]).

For example, the values g and f_{eg} in Figure 8.1 can be considered not only as the space displacement and force, but broader—as *generalized coordinates* and *generalized forces* [1.18], [1.20]. The number of such coordinates, g, equals at the same time the number of degrees of freedom, whereas:

When g_i is	Then the force f_{eg} is
Linear displacement x_i	Mechanical force F_i
Rotation angle φ_i	Rotational torque of force T_i
Electric charge Q_i	Electromotive force E_i
Surface S_i	Surface tension F_{si}
Volume V_i	Pressure q_i

Some tools of mathematical modeling with the help of theories of physical fields and theory of motion and control (mainly Hamilton principle [1.16], [1.20], [1.18], [1.15]), as well as approximate numerical, mesh-based programs (Turowski [1.20]) ANM, FDM, RNM, FEM, BEM, and so forth, are presented in Table 10.1.

10.2 PRINCIPLES OF THEORY OF ELECTRODYNAMIC SIMILARITY

The theory of similarity has gained a broad application in model investigations in electrothermal technology, hydrodynamics, aerodynamics in electrodynamics, and others. Its objective is to deliver the following basic indications:

- How one should build and test a reduced (scaled) model of the investigated structure to obtain similar processes in the model and original device.
- How to process the data/results obtained from the reduced model investigations, to be able to extend them to the original device.

To obtain the model phenomena similar to the original device, it is necessary to fulfill defined conditions, called *criteria of similarity*. To establish the criteria of similarity, one can use the method of *dimension analysis* [2.8] or the method of *analysis of equations*, which is considered more reliable ([10.3], p. 118).

Criteria of electrodynamic similarity [2.41], [1.20]. Various systems and phenomena of the same nature are called *similar*, when their analogous quantities can be described with the help of analogical equations connected mutually by defined constants, called *criteria of similarity*.

TABLE 10.1
Two Complementary Domains of Rapid Design of Electromagnetic and Electromechanical Coupled Systems

FIELD APPROACH [1.15]
Design of Machines and Apparatus

The most general description is the Navier–Stokes equation for elastic fields[a]

$$\frac{\partial \mathrm{v}}{\partial t} + (\mathrm{v}\,\mathrm{grad})\,\mathrm{v} = -\frac{1}{\rho_{\mathrm{m}}}\,\mathrm{grad}\,p + \mathrm{v}\nabla^2\mathrm{v} + \frac{\mathrm{v}}{3}\,\mathrm{grad}(\mathrm{div}\,\mathrm{v}) + \mathrm{P}$$

+ Fourier

Eqs. for thermal fields + Duhamel – Neumann Eq. for thermo–elasticity are hard for joint solution. Then, thanks to big differences in time constants T_t and materials, they are investigated as independent fields:

ELECTR	MAGNT	ELE-MAG	TEMPR	MECHN
$D = \varepsilon E$	$B = \mu H$	$T_t \ll$	T_t	$\gg T_t$
$\sigma \approx 0$	$\partial/\partial \approx 0$		*Thermokinetics*	*Newton' laws*

Maxwell Equations

$$\mathrm{curl}\,\boldsymbol{H} = \boldsymbol{J}_{\mathrm{total}}; \quad \mathrm{curl}\,\boldsymbol{E} = -\partial\boldsymbol{B}/\partial t$$

$$\boldsymbol{J}_{\mathrm{total}} = \sigma\boldsymbol{E} + \frac{\partial \mathbf{D}}{\partial t} + \sigma(\boldsymbol{v}\times\boldsymbol{B}) + \rho\boldsymbol{v}_{\rho} + \mathrm{rot}(\boldsymbol{D}\times\boldsymbol{v}) + $$
$$+ J_{\mathrm{external}}$$

CIRCUIT APPROACH [1.18], [1.20]
Design of Dynamic Systems Motion

The most general description of a systems in motion is the Hamilton's Principle for coupled electromechanical systems

$$I = \int_{t_1}^{t_2} L\,\mathrm{d}t = \min, \quad L = T - V \quad (*)$$

Dynamics of coupled, dissipative electromechanical systems needs solution of Euler–Lagrange partial differential equations

$$\frac{\partial L}{\partial q_k} - \frac{\mathrm{d}}{\mathrm{d}t}\left(\frac{\partial L}{\partial \dot{q}_k}\right) - \frac{\partial F}{\partial \dot{q}_k} = -\mathrm{G}_k,$$

This is necessary condition of existence of minimum of integral (*) obtained as variation $\delta I = 0$. $k = 1,2,\ldots, n$; $L(q_k, \dot{q})= T - V$—Lagrange's state function, T', V—kinetic (co) energy and potential energy, respectively, $G(t)$—forces of constraints, F—Raleigh loss function $q_k = q_k(t)$—general coordinate–extreme

DIFFERENTIAL EQUATIONS [2] → LUMPED PARAMETERS → FORMALIZED DIAGRAMS FOR EQUATIONS OF MOTION [3.55]

APPROXIMATE COMPUTER METHODS

	FEM-3D	FDM-3D	BEM-3D	RNM-3D	TABLE OF ENERGY EQUATIONS +
ANM	**Finite**	Finite	Boundary	**Reluctance**	+ MATLAB or "**Hamilton**" **Package**
	Element	Difference	Element	**Network**	*Almost as simple and rapid as RNM-3D*
	Method	Method	Method	**Method**	*programs.*
	Differential		Integral		*Differential*

[a] It is an equation describing heat exchange between a solid body surface and a fluid passing by, with a velocity **v**, where $\rho_{\mathrm{m}} = \rho_{\mathrm{m}}(x, t)$ is the fluid density, $p = p(x, t)$ is the hydrodynamic pressure, υ is the (constant) dynamic viscosity of incompressible fluid, υD**v** is the density of the internal friction force, **P** represents other forces (per unit volume), such as gravity, or the density of the Lorentz force $f_{\mathrm{L}} = \boldsymbol{J}\times\boldsymbol{B}$.

Then, the criteria of similarity are dimensionless numbers being invariants, that is, being the same for both systems. Criteria of similarity have identically equal values for the *original* and the *model* and they are the condition of existence of similarity.

The method of modeling of electromagnetic processes with the change of object scale is based on the principle that all electromagnetic phenomena which occur in the model and in the original are subordinated to the same Maxwell Laws (2.1) through (2.6). Hence, we have for the original:

$$\left.\begin{aligned} \mathrm{curl}_{\mathrm{or}} H_{\mathrm{or}} &= \sigma_{\mathrm{or}} E_{\mathrm{or}} + \frac{\partial D_{\mathrm{or}}}{\partial t_{\mathrm{or}}} \\ \mathrm{curl}_{\mathrm{or}} E_{\mathrm{or}} &= -\frac{\partial B_{\mathrm{or}}}{\partial t_{\mathrm{or}}} \end{aligned}\right\} \tag{10.1}$$

$$\left.\begin{aligned} \mathrm{curl}_{\mathrm{mod}} H_{\mathrm{mod}} &= \sigma_{\mathrm{mod}} E_{\mathrm{mod}} + \frac{\partial D_{\mathrm{mod}}}{\partial t_{\mathrm{mod}}} \\ \mathrm{curl}_{\mathrm{mod}} E_{\mathrm{mod}} &= -\frac{\partial B_{\mathrm{mod}}}{\partial t_{\mathrm{mod}}} \end{aligned}\right\} \tag{10.2}$$

The similarity of fields and electrodynamic processes is based on the assumption of geometric similarity of the model and the original, that is, the same scale for all linear dimensions is assumed:

$$m_{\mathrm{l}} = \frac{l_{\mathrm{mod}}}{l_{\mathrm{or}}} = \frac{\mathrm{curl}_{\mathrm{or}}}{\mathrm{curl}_{\mathrm{mod}}}$$

Moreover, assuming for the remaining values the scales denoted by the corresponding indexes

$$\left.\begin{aligned} m_{\mathrm{H}} &= \frac{H_{\mathrm{mod}}}{H_{\mathrm{or}}}, \quad m_{\mathrm{E}} = \frac{E_{\mathrm{mod}}}{E_{\mathrm{or}}}, \quad m_{\mathrm{f}} = \frac{f_{\mathrm{mod}}}{f_{\mathrm{or}}} = \frac{H_{\mathrm{mod}}}{H_{\mathrm{or}}} = \frac{t_{\mathrm{or}}}{t_{\mathrm{mod}}} \\ m_{\sigma} &= \frac{\sigma_{\mathrm{or}}}{\sigma_{\mathrm{mod}}}, \quad m_{\mu} = \frac{\mu_{\mathrm{or}}}{\mu_{\mathrm{mod}}}, \quad m_{\varepsilon} = \frac{\varepsilon_{\mathrm{or}}}{\varepsilon_{\mathrm{mod}}} \end{aligned}\right\} \tag{10.3}$$

and substituting them into Equations 10.2, we obtain for the model

$$\left.\begin{aligned} \frac{m_{\mathrm{H}}}{m_{\mathrm{l}}} \mathrm{curl}_{\mathrm{or}} H_{\mathrm{or}} &= m_{\sigma} m_{\mathrm{E}} \sigma_{\mathrm{or}} E_{\mathrm{or}} + m_{\varepsilon} m_{\mathrm{E}} m_{\mathrm{f}} \frac{\partial D_{\mathrm{or}}}{\partial t_{\mathrm{or}}} \\ \frac{m_{\mathrm{E}}}{m_{\mathrm{l}}} \mathrm{curl}_{\mathrm{or}} E_{\mathrm{or}} &= -m_{\mu} m_{\mathrm{H}} m_{\mathrm{f}} \frac{\partial B_{\mathrm{or}}}{\partial t_{\mathrm{or}}} \end{aligned}\right\} \tag{10.4}$$

Identity of Equations 10.4 for the model and Equations 10.1 for the original can be obtained when we fulfill the conditions

$$m_{\mathrm{H}} = m_{\sigma} m_{\mathrm{E}} m_{\mathrm{l}} = m_{\varepsilon} m_{\mathrm{E}} m_{\mathrm{f}} m_{\mathrm{l}} \quad \text{and} \quad m_{\mathrm{E}} = m_{\mu} m_{\mathrm{H}} m_{\mathrm{f}} m_{\mathrm{l}} \tag{10.5}$$

Henceforth, after substitution, we obtain the *two similarity criteria* of electromagnetic field:

$$m_f m_\mu m_\sigma m_l^2 = 1 \tag{10.6}$$

$$m_\varepsilon m_\mu m_f m_l^2 = 1 \tag{10.7}$$

which may be also presented in the form

$$f_{mod}\mu_{mod}\sigma_{mod}l_{mod}^2 = f_{or}\mu_{or}\sigma_{or}l_{or}^2 = \Pi_1 = \text{idem} \tag{10.6a}$$

$$\varepsilon_{mod}\mu_{mod}f_{mod}^2 l_{mod}^2 = \varepsilon_{or}\mu_{or}f_{or}^2 l_{or}^2 = \Pi_2 = \text{idem} \tag{10.6a}$$

or shortly

$$\Pi_1 = f\mu\sigma l^2 = \text{idem} \tag{10.6b}$$

$$\Pi_2 = \varepsilon\mu f^2 l^2 = \text{idem} \tag{10.7b}$$

Joint fulfillment of both criteria Π_1 and Π_2 is difficult in practice. For instance, at a reduction of the model scale by increasing the supply frequency (at $m_\varepsilon = m_\mu = 1$) we obtain the difficult-to-satisfy technical condition of $m_\sigma = m_f = 1/m_l$, because it is difficult to find a material of so big conductivity σ, not taking into account the extremely costly cryogenic superconductivity. So, one usually does not model simultaneous processes occurring in conducting and dielectric media, especially in the cases when the displacement currents can be neglected ($\partial D/\partial t = 0$) it is sufficient to take into account only the criterion Π_1, considering the similarity as approximate. In the case of modeling fields in dielectrics, the basis of modeling should be the criteria Π_2, whereas Π_1 can be omitted.

The theory of similarity, as seen in Equations 10.4, needs a constant scale of the material parameters m_σ, m_ε, and m_μ. In the case of linear media, this condition is usually satisfied as $m_\sigma = m_\varepsilon = m_\mu = 1$. Hence, there are no obstacles in selecting an arbitrary scale of the magnetic field intensity m_H and of the current density $m_J = m_\sigma m_E$; one should only satisfy the condition obtained from Equations 10.5 and 10.6, namely:

$$\frac{m_J m_l}{m_H} = \frac{m_I}{m_H m_l} = 1, \quad \text{that is} \quad \frac{Jl}{H} = \frac{I}{Hl} = \text{idem} \tag{10.8}$$

where

$$m_I = m_J m_l^2$$

If no other conditions were determined, then the *scale of current density* m_J in formula (10.8) can be chosen arbitrarily.

The *scale of power* is defined in various ways, depending on the character of the object and method of investigation. In conductors without skin effect, the scale of power losses per unit volume (W/m³) is defined by the formula

$$m_p' = \frac{m_E^2}{m_\sigma} \tag{10.9a}$$

In solid metal parts of thickness larger than the electromagnetic wavelength λ in the metal, the scale of power loss per unit surface of the body (W/m²), according to formula (3.10a), is expressed by the formula (Turowski [2.41])

$$m_{\mathrm{P}}'' = \sqrt{\frac{m_f\, m_{\mu_s}}{m_\sigma}} \cdot m_H^2 \tag{10.9b}$$

The scale (10.9b) applies also to steel, because at sufficiently strong fields H_{ms} on the body surface, $a_{p,\mathrm{or}} \approx a_{p,\mathrm{mod}}$ [10.34]. At studying local (point-wise) power loss in a transient case (Section 10.5) one can sometimes assume the condition of equality of point-wise unit losses $m_p' = 1$ or $m_p'' = 1$ [2.41]. However, at model investigations of a total object in quasistatic state, it is more convenient to start from corresponding points of solid model parts and of the original that have the same power dissipated on similar geometrically ΔA surfaces, that is,

$$|S_{\mathrm{P}}|\ \Delta A = a_{\mathrm{P}}\sqrt{\frac{\omega\,\mu_{\mathrm{S}}}{2\sigma}}\,\frac{|H_{ms}|^2}{2}\,\Delta A = \mathrm{idem}$$

and then, considering (10.6), we have the condition

$$m_{\mathrm{p}}''' = \sqrt{\frac{m_f m_\mu}{m_\sigma}} \cdot m_H^2 m_l^2 = \frac{m_H^2 m_l}{m_\sigma} \tag{10.9c}$$

If a steel wall is *screened*, then according to (4.39)

$$P_{1\mathrm{scr-St}} \approx \frac{1}{k_{\mathrm{scr}}\, d_{\mathrm{screen}}}\sqrt{\frac{\omega\mu_S}{2\sigma}}\,\frac{|H_{ms}|^2}{2} = \frac{1}{2d_{\mathrm{scr}}\sigma_{\mathrm{scr}}}|H_{ms}|^2$$

The scale of surface loss in a *screened* wall is

$$m_{p\ \mathrm{scr}} = \frac{m_H^2}{m_{\sigma\,\mathrm{scr}} m_l} \tag{10.9d}$$

The scale of loss in the magnetic core and magnetic screen (shunt), laminated from transformer sheets (1.40), (6.16), at the assumption of $\Delta p_{\mathrm{St}} = C(B_m^a)f_b$ [10.34], has the form

$$m_{p,\mathrm{Fe}} = m_B^a m_f^b \tag{10.9e}$$

where in cold rolled (anisotropic) sheets: $B_m \geq 1.4$ T; $a = 1.9$; $b = 1.63$.

For conductors of average skin effect, the scale of resistance is determined starting from the assumption that similar phenomena in the model and in the original the ratio of DC resistances (R_{DC}) and the ratio of AC resistances (R_{AC}) are the same in both systems:

$$m_R = \frac{R_{\text{mod, DC}}}{R_{\text{or, DC}}} = \frac{R_{\text{mod, AC}}}{R_{\text{or, AC}}} = \frac{l_{\text{mod}}/S_{\text{mod}}\sigma_{\text{mod}}}{l_{\text{or}}/S_{\text{or}}\sigma_{\text{or}}} = \frac{l_{\text{or}}\sigma_{\text{or}}}{S_{\text{mod}}\sigma_{\text{mod}}} = \frac{1}{m_l m_\sigma} \tag{10.10}$$

where s_{mod} and s_{or} are the cross-sections of the corresponding segments of conductors of the model and of the original, respectively, $s_{\text{mod}}/s_{\text{or}} = l_{\text{mod}}^2/l_{\text{or}}^2$.

If we consider the skin effect, by means of the coefficient K_R (Figure 5.29), then at AC current:

$$m_R' = m_{KR}\frac{1}{m_l m_\sigma} \tag{10.10a}$$

Owing to similar ratios R/X in both systems, the reactance scale also is

$$m_X = \frac{X_{\text{mod}}}{X_{\text{or}}} = \frac{1}{m_l\, m_\sigma}, \quad \text{or} \quad m_X' = m_{KX}\frac{1}{m_l m_\sigma} \tag{10.10b}$$

The scale of slot flow for geometrically similar circuits l_{mod} and l_{or} can be determined from the law of flow (Ampere's law)

$$m_{IN} = \frac{I_{\text{mod}}N_{\text{mod}}}{I_{\text{or}}N_{\text{or}}} = \frac{\oint_{l_{\text{mod}}}(\boldsymbol{H}_{\text{mod}}\mathrm{d}\boldsymbol{l}_{\text{mod}})}{\oint_{l_{\text{or}}}(\boldsymbol{H}_{\text{or}}\mathrm{d}\boldsymbol{l}_{\text{or}})} = m_H m_l \tag{10.11}$$

For widely used systems containing nonlinear and anisotropic elements of variable permeability, $\mu = f(H)$, the author in 1962 [2.41] proposed the additional third criterion (at application of the same materials)

$$m_H = 1, \text{ that is, } \Pi_3 = H = \text{idem} \tag{10.12}$$

which is indispensable for fulfillment of the basic condition for Equation 10.4: $m_\mu = \text{const.}$

In such a case, we obtain:

$$\left.\begin{array}{l} \text{the scale of permeability: } m_\mu = 1 \\ \text{the scale of flux density: } m_B = m_\mu m_H = 1 \\ \text{the scale of current (10.8): } m_I = m_l \\ \text{the scale of current density: } m_J = 1/m_l \\ \text{the scale of power: } m_P'\text{—without change} \end{array}\right\} \tag{10.13}$$

$$m_P'' = \sqrt{\frac{m_f m_{\mu_s}}{m_\sigma}} = \frac{1}{m_l m_\sigma}$$

$$m_p''' = \frac{m_l}{m_\sigma}$$

The criterion Π_3 was formulated in Russia ([10.3]) and [10.6]) in a somewhat different approach—as a requirement that the relative characteristics of μ of the model and of the original be identical:

$$\mu^r = \frac{\mu_k}{\mu_{k+1}} = \varphi\left(\frac{H_k}{H_{k+1}}\right) = \text{idem} \tag{10.12a}$$

where k and $k + 1$ are pair of points in space (of the model or of the original); μ_k and H_k are the permeability and the magnetic field intensity in these points, respectively. Condition (10.12a) is, however, very difficult to realize, because it would require a special selection of the magnetic materials.

In 1977, Zakrzewski [10.34], using the parabolic approximation (7.9) $H = C_1 \cdot B^n$ of the magnetization curve, developed a set of scale coefficients for different sorts of steel (Table 7.3), where

$$m_\mu = m_H^{\frac{1-n}{n}} \tag{10.12b}$$

10.3 PRINCIPLE OF INDUCTION HEATING DEVICE MODELING

With electro-thermal device modeling, due to their operation at high, but relatively stable temperature, the iron nonlinearity can be neglected and taken simply as $m_\mu = m_\sigma = m_\varepsilon' = 1$. Hence, the similarity criteria (10.6) and (10.7) take the form

$$\Pi_1 f l^2 = \text{idem} \tag{10.14}$$

$$\Pi_2 f^2 l^2 = \text{idem} \tag{10.15}$$

Reduction of model dimensions can be then obtained by an *increase* of the supply frequency.

It causes at the same time the increase of the displacement currents flowing through inter-turn capacitances and ground leakage capacitances. This is why, considering the criterion Π_1, it is not recommended (Dershwartz, Smielanskiy [10.11]) to reduce the model dimensions too much and simultaneous increase the frequency, because it can cause deformation of the research results. Usually, for devices of the rated frequency 50–500 Hz, for modeling one uses sources of frequency 2500–10,000 Hz, which gives sufficient economy by several-fold reduction of the scale (Smielanskiy [10.6], [10.11]).

The objective of modeling can be [10.11]: determination of the coefficient for recalculation of the voltage supplied to the model inductor into corresponding voltage of the original; determination of the dependence between powers consumed by charge, screens, tanks, and so on; finding cos φ_{or} if cos φ_{mod} is known; and determination of the number of turns of inductor.

If, for instance, the rated frequency of the object is 500 Hz, whereas for modeling a source of frequency 8000 Hz was applied, the linear dimensions of the model will be

$$\frac{1}{m_l} = \frac{l_{or}}{l_{mod}} = \sqrt{\frac{f_{mod}}{f_{or}}} = \sqrt{16} = 4 \tag{10.14a}$$

that is, 4 times smaller than the dimensions of the original.

The scale m_H of the magnetic field intensity is determined from the condition (10.9c)

$$m_H = \frac{1}{\sqrt{m_l}} = \sqrt{m_f} \quad \text{or} \quad H_{mod} = \sqrt[4]{16}\, H_{or} \tag{10.16}$$

From formula (10.11), at the assumption $N_{mod} = N_{or}$, and considering Equations 10.14a and 10.16, we get the current of excitation winding

$$I_{mod} = \mathrm{m}_H \mathrm{m}_l I_{or} = \sqrt{m_l}\, I_{or} = \frac{I_{or}}{\sqrt[4]{16}} \tag{10.17}$$

Next, by demanding the equality of power in the excitation windings

$$I_{mod}^2 R_{mod} = I_{or}^2 R_{or} \tag{10.18}$$

we obtain the resistance of the model winding

$$R_{mod} = \frac{1}{m_H^2 m_l^2} R_{or} = \frac{1}{m_l} R_{or} = \sqrt{16}\, R_{or} \tag{10.19}$$

The same value can be obtained from Equation 10.10 at considering condition (10.16).

Thanks to the similarity of the model and the original, the same dependence as Equation 10.19 is valid for the reactance

$$X_{mod} = \sqrt{16} X_{or} \tag{10.20}$$

Since $\cos\varphi = \dfrac{R}{\sqrt{R^2 + X^2}}$, from Equations 10.19 and 10.20 it follows, that

$$\cos\varphi_{mod} = \cos\varphi_{or} \tag{10.21}$$

From the power equality conditions (10.9c) and (10.18), we get

$$U_{\text{mod}} I_{\text{mod}} \cos \varphi_{\text{mod}} = U_{\text{or}} I_{\text{or}} \cos \varphi_{\text{or}}$$

Considering Equations 10.17 and 10.21, we obtain the voltage

$$U_{\text{mod}} = \sqrt[4]{16}\, U_{\text{or}} \tag{10.22}$$

An investigation is carried out in such a way that the inductor of the model is supplied by an arbitrary, known voltage U'_{mod}. After stabilization of the model temperature, one determines the power losses in the inductor, in the housing, and in the screens, on the basis of the temperature of cooling water and the temperature of particular parts of housing. Then, by subtracting the determined losses from the wattmeter indicator, one can evaluate the power consumed by the heated charge.

The losses and the effective power determined in this way equal to the real losses and power existing in the original, at the voltage

$$U'_{\text{or}} = 0.5 U'_{\text{mod}}.$$

The effective (output) power and power losses at the rated voltage are obtained from the recalculation

$$\frac{P_N}{P'_{\text{or}}} = \left(\frac{U_N}{U'_{\text{or}}} \right)^2 \tag{10.23}$$

If the obtained value is not equal to the assumed power, then one should change the number of turns in the inductors. In case of too large power losses in the tank or in screens, these parts should be shifted away by some distance, according to Equation 10.6a. This is one of the possible examples of the design of heating units by modeling, proposed by Russian specialists [10.11].

10.4 MODELING OF HIGH CURRENT LINES

Modeling is one of the best ways to design and evaluate secondary high current networks for electric furnaces, because calculation and measurement on the completed large objects are too difficult and burdened by many errors. A correct network should have, among others, as high symmetry as possible, as well as the smallest possible reactance and power losses. The basis for such calculation are formulae (10.14a), (10.19), and (10.20). Contact resistances, which play an important role in real systems, cannot be modeled. Because of it, in the model all the connection should be welded. One important difficulty here is the measurement of resistances, reactances, and power in the system. For this purpose, one applies special meter circuits, bridges (Turowski [10.26]), discussed in Sections 10.6 and 10.8.

Since generators of higher frequency (2.5–10 kHz) are usually single-phase, for a supply of three phase models, one sometimes applies static converters of single-phase currents into three-phase current. Figure 10.2b shows two simple examples of such converters.

In the case of a three-phase symmetric receiver, with impedances $Z_1 = Z_2 = Z_3 = Z = |Z|\ e^{j\varphi}$ Figure 10.2a), we select the reactances X' and X'' in such a way that the impedances $Z_1 + \mathrm{j}\ X'$ and $Z_3 + \mathrm{j}\ X''$ have the same modules and arguments, equal respectively to $+\pi/3$ and $-\pi/3$. Then the currents I_1, I_2, I_3 and the voltages U_1, U_2, U_3 on the terminals of the impedances Z_1, Z_2, Z_3 will create symmetric three-phase systems. At $|\varphi| < \pi/3$ one of the additional reactances should be inductive ($X' = \omega L$), and the second one capacitive ($X'' = 1/\omega C$). In the case of a receiver of active character, that is, at $Z = R(\varphi = 0)$, one should assume $\omega L = 1/\omega C = \sqrt{3}R$. The same condition applies to the system in Figure 10.2b.

10.5 MODELING OF TRANSFORMERS AND THEIR ELEMENTS

Modeling of whole transformers can be carried out analogically as described in Section 10.3. However, in a case where we wish to take into account the influence of the permeability variation, we should take into account formulae (10.12) and (10.13) which, however, will complicate the modeling due to the change of scale.

In the case of modeling the local power losses in transient states (Section 10.6), instead of the relation (10.9c) one sometimes introduces the condition $m_p' = 1$, which at the assumption of $m_\mu = m_\sigma = 1$ gives the scales $m_H = m_l, m_f = 1/m_l^2$, and $m_{IN} = m_l^2$. This method, however, is burdened by an error resulting from nonlinearity of the permeability, and can be used in principle only for steel elements saturated very weak or very strong.

In a direct relation to the theory of similarity are the so-called "model laws" or "laws of growth" of transformers (E. Jezierski [1.5]), [4.23], in which electromagnetic, thermal, and other parameters are expressed by linear dimensions of the object. This method facilitates an estimated dimensioning of new transformers, on the basis of previous, already fabricated and tested ones.

In addition to modeling based on the theory of similarity, an important significance may have modeling investigation of particular parts by transformation of their systems into *simpler forms*, per analogy to eddy-current distribution, without change of scale or frequency of excitation currents. One such solution was the designed by the author in 1961 [2.41]: "*inverted*" *model of transformer tank,* enabling measurement of power losses in real transformer walls (Figure 10.3). This model acts like an overturned glove. It provides the possibility to measure power losses in solid steel walls and screened steel walls, with the help of external excitation windings. The measurements are similar to *Epstein* apparatus. The essence of this approach consists in the investigated wall made of solid steel 1 (Figure 10.3), which simulates the real tank wall, is placed inside the excitation windings, and not outside of them as it normally is. This makes it easier to investigate the wall, to change the paths of induced currents from 2-D to 3-D distribution, to change distances, and so on. The laminated magnetic core,* less important here, is simulated by the external package of six transformer sheets.

* The modeling concerns the fields and losses at short-circuit test, when the main magnetic flux in the core is negligibly small.

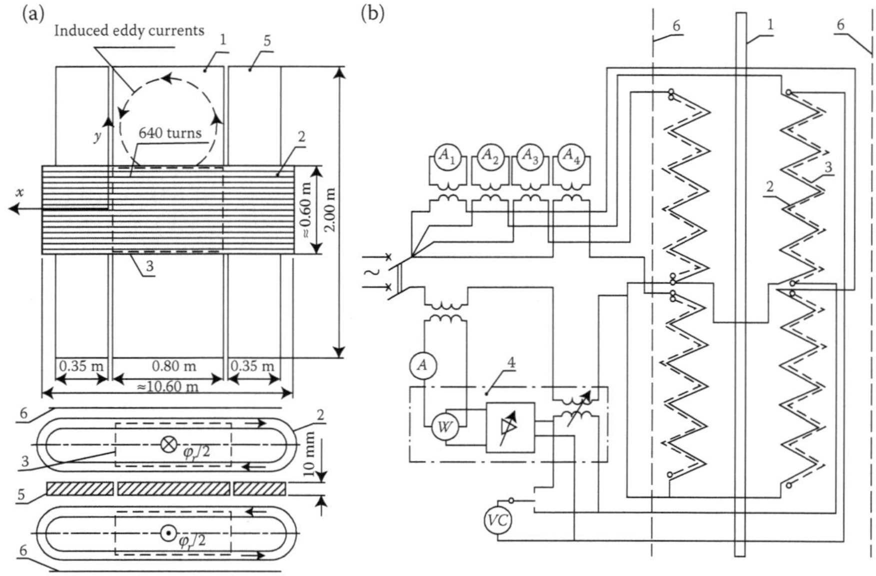

FIGURE 10.3 The J. Turowski's inverted model of transformer tank: (a) dimensional sketch; (b) scheme of the measurement system; 1—the investigated steel wall, or "tank"; 2—the excitation current winding; 3—the measuring potential winding; 4—electronic compensation wattmeter, after J. Turowski's patent [10.24] to [10.26]; VC—digital voltmeter; 5—steel plates equalizing the stray field distribution; 6—packages of transformer sheets, simulating the transformer core; Φ_r—stray flux. (Adapted from Turowski J.: Losses and local overheating caused by stray fields. (in Polish). *Scientific Archives of the Technical University of Lodz "Elektryka,"* No. 11, 1963, pp. 89–179 *(Habilitation (DSc) dissertation; 1st Ministry Award)* [2.41].)

The author's "*inverted model*" of the transformer tank was first built at a smaller scale, 500 × 880 mm (Turowski et al. [5.16]), under extended Mr Spuv's (MTZ—Moscow) concept, and next on a bigger scale* (Figure 10.3). For many years, these models delivered a lot of experimental information, for instance shown in Figures 10.4 and 10.5. They served as important experimental verifications of the theoretical works of the J. Turowski and his former students and assistants, including Zakrzewski, Kazmierski, Ketner, Janowski, Sykulski, Wiak, and others, at the Technical University of Lodz, Industrial Power Institute and ELTA (Now ABB) Transformer Works in Lodz [2.32], [2.41], [4.18], [5.9], [5.14], [5.19], [7.7], [7.17], [7.18], [7.20], [7.22], and numerous PhD and MSc theses. In 1982, Zakrzewski and Wiak built a reduced model of the system from Figure 10.3, in a reduced scale, after Equation 10.6b, for the frequencies of 200–500 Hz. The full model from Figure 10.3, of frequency 50 Hz, permitted, among others, for modeling of tank walls of single-phase transformers (Figure 7.18) with the help of a thick steel plate placed between excitation coils.

* The team: J. Turowski, T. Janowski, and K. Zakrzewski, on a contract from ELTA Transformer Works, Lodz.

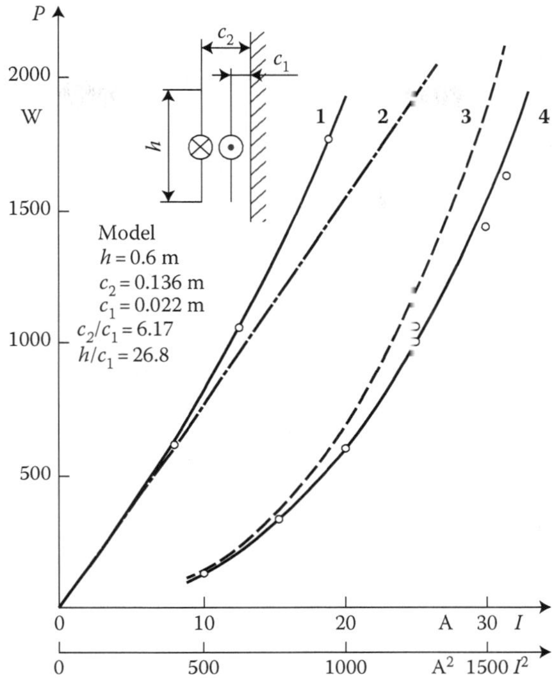

FIGURE 10.4 Experimental verification of formulae (7.74) and (7.95) with the J. Turowski's *inverted model* (Figure 10.3), and of the dependence of losses in steel wall on current (after J. Turowski [7.18]) where the losses are: 1—the measured $P_{meas} = f(I^2)$, 2—straight line; 3—calculated with formula (7.95) $P_{calc} = f(I)$; 4—measured $P_{meas} = f(I)$ on the inverted model from Figure 10.3.

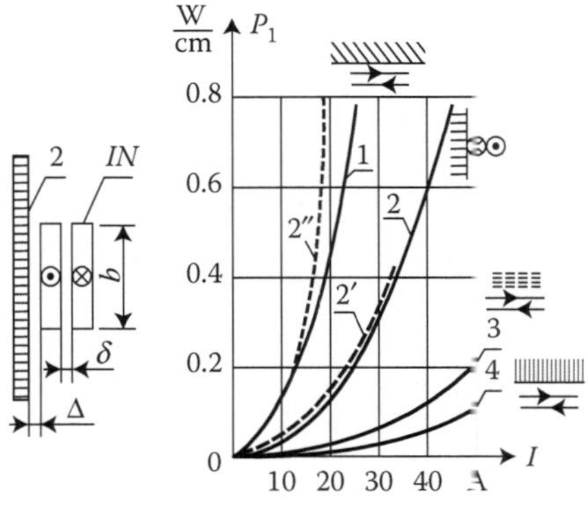

FIGURE 10.5 Influence of arrangements (2, 3, 4) of Fe sheets packages (shunts), made of stripes of *Epstein* apparatus, versus the exciting current (I) on the power losses in a steel wall: 1—P_{St} in solid steel wall. The measurements were made on the small inverted model (Figure 10.3) by (Turowski, Pawłowski, Pinkiewicz [5.16]), [2.41]. At wrong (2) shunts (see Section 7.6.5) [4.30]: 2′—$P'_{St}(H) = P_{St}(\sigma_{tr}/\sigma_{St}) \approx 55\% P_{St}$, 2″—$P''_{St}(\Phi) = P_{St}(\sigma_{St}/\sigma_{tr}) \approx 183\% P_{St}$.

From comparison of Figure 7.17 with Figure 10.3, it follows that the eddy currents induced in both surfaces of the central plate *1* in Figure 10.3 have the same character of distribution as on the internal surface of tank wall of a single-phase transformer. The external steel plates 5 serve only for elimination of edge deformation of the leakage flux (Φ_l) distribution on the steel plate surface. If the steel plate of the model is of sufficient thickness (≥10 mm), then the processes occurring on one of its surfaces do not have practically any influence on the eddy-currents flowing on the opposite surface.

After appropriate over-switching of the exciting winding 2 in Figure 10.3, one can obtain a model of a plate infinitely extended in the X direction, with the induced currents flowing only in this one direction. Such a system was investigated in Section 7.5. For the measurement of power losses in the central plate of the model, the *method of separate voltage coil* (3 in Figure 10.3) was applied. The voltage coil 3 exactly coincides spatially with the exciting current winding 2 and has exactly the same number of turns as the current winding. A similar method is used in the popular *Epstein apparatus*. The method permits eliminatation from the measurements, the losses in copper of the excitation winding (Section 10.6).

A full analysis of magnetic screening (shunting) should be carried out with the help of a computer program, like RNM-3D or a similar effective and rapid design tool. The designer must remember about the harm that may be caused by wrong location of shunts (see Section 7.6.5 and curve 2″ in Figure 10.5).

10.6 THERMOMETRIC METHOD OF PER-UNIT POWER LOSSES MEASUREMENT

A natural complement to Chapter 9 and the method of computational checking of local overheating hazard is the laboratory “thermometric method” of determination of the local losses and the magnetic field intensity H_{ms} on the metal surface. It is carried out by measurement of the initial change of temperature in the investigated spot in function of time, at switching on or off of the load. This method permits a direct, point-wise measurement of power losses and their concentration on complete electric machines, transformers, and other electric devices of large power, without the need of their overall heating. This methods allows also a discovery of possible local overheating by comparing the measured magnetic field intensity H_{ms} on the surface of the considered region with the criterial permitted value $H_{ms,\mathrm{perm}}(t_{\mathrm{perm}})$ (Figure 9.2). The temperature T is normally measured with the help of a thermocouple fastened to the surface of the investigated bodies (in the case of solid metal parts) or inside (in the case of a laminated iron core). A short measurement duration (several seconds) provides the possibility to carry out the measurements at a large load without heating up the whole object.

10.6.1 Method of Initial Rise of Body Temperature

For an arbitrary point of a metal body of rectangular anisotropy (2.83a) (Kącki [9.1], pp. 222, 226), which contains a volumetric source of heat $P_{1\mathrm{V}}$ (in W/m^3) one can write the equation of thermal equilibrium

$$P_{1V} = c\rho_m \frac{\partial T}{\partial t} - \text{div}\,(\ddot{\lambda}_i\,\text{grad}\,T) \tag{10.24}$$

where

$$\text{div}\,(\ddot{\lambda}_i \text{grad}\,T) = \lambda_x \frac{\partial^2 T}{\partial x^2} + \lambda_y \frac{\partial^2 T}{\partial y^2} + \lambda_z \frac{\partial^2 T}{\partial z^2} \tag{10.24a}$$

where $\ddot{\lambda}_i = \lambda_i(x, y, z)$ is the diagonal tensor of thermal conductivity of this metal; c the specific heat (Ws/(kg · K)); ρ_m the mass density (kg/m³) T the body temperature (°C).

In the case when the investigated element has the form of a sheet in which in its entire thickness (the same per unit power P_{1V} (W/m³)) is dissipated, the equation of thermal equilibrium (10.24) should be completed by the heat dissipated to ambient temperature T_0

$$P_{1V} = cr_m \frac{\partial T}{\partial t} - \text{div}\,(\ddot{\lambda}_i\,\text{grad}T) + \alpha'(T - T_0) \tag{10.25}$$

where T_0 the ambient temperature (°C); t the time (s); α' the resultant coefficient of heat dissipation to ambient, by convection and radiation [W/(m² · K)].

If the body temperature T is measured so fast from the initial instant $t = 0$ of field excitation, that the temperature of all points of the body and ambient can be assumed to be practically the same, that is

$$\text{grad}\,T = \mathbf{i}\frac{\partial T}{\partial x} + \mathbf{j}\frac{\partial T}{\partial y} + \mathbf{k}\frac{\partial T}{\partial z} = 0 \quad \text{and} \quad T = T_0 \tag{10.26}$$

then from Equations 10.24 and 10.25 we will obtain (Figure 10.6)

$$P_{1V} = c\rho_m \left(\frac{\partial T}{\partial t}\right)_{t=0} = c\rho_m \left(\frac{\partial \Theta}{\partial t}\right)_{t=0} = c\rho_m\,\text{tg}\,\alpha_0 \tag{10.27}$$

where $\Theta = T - T_0$ is the temperature rise of the body (T_0 = const, ambient temperature).

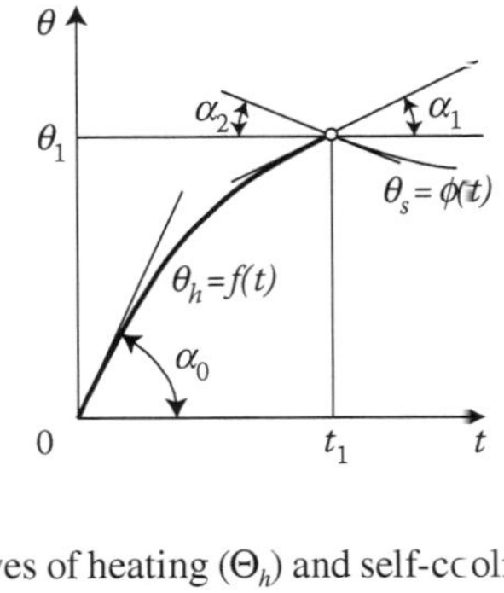

FIGURE 10.6 The ideal curves of heating (Θ_h) and self-cooling (Θ_s) of a homogeneous body; $\Theta = T - T_0$, T_0 = const (ambient temperature).

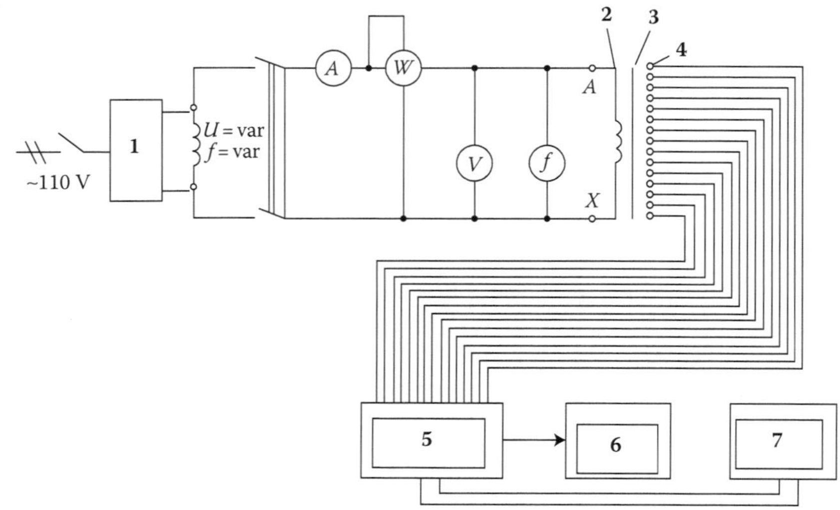

FIGURE 10.7 A scheme of the system for measurement of local losses in a transformer core by the thermometric method: 1—frequency and voltage converter; 2—winding of the investigated transformer; 3—iron core of the transformer; 4—thermocells; 5—electronic commutator; 6—computer/printer; 7—digital voltmeter. (Adapted from Turowski J., Komęza K., and Wiak S.: Experimental determination of the distribution of power losses in the cores and construction elements of transformers. *Przegląd Elektrotechniczny*, (10), 1987, 265–268 [10.31].)

Therefore, the measure of the active power (10.27) dissipated in a given point of a body is expressed by tangent of the angle α_0 of inclination of the tangent to the heating curve in the initial point.

The body should have a stationary initial temperature, close to the ambient temperature. The conclusion drawn from Equation 10.27 is valid both for isotropic and anisotropic bodies, for instance, for laminated iron cores of transformers and electric machines.

Limitations of the method of initial rise of body temperature arise from the necessity to completely cool down the object, in case if we need to repeat the measurement in another point as well as the necessity of sudden switching on of the investigated object to full power. At large objects, for example, transformers, this is not recommended. The former drawback can be overcome by application of an automatic measurement in many points simultaneously (Figure 10.7) (J. Turowski, Komeza, Wiak [10.31]). The latter drawback can be avoided by using an object previously loaded and heated to a stationary temperature.

10.6.2 Method of Switching On or off Of the Investigated Object

Measurements of per unit power loss in a given point can be in principle performed also at an arbitrary temperature of the investigated body, with the help of measurement of the speed of temperature rise in the time instant of switching on or off the investigated object, for example, a power transformer (method of two tangents). It is necessary, however, to eliminate the heat dissipation to ambient (thermal screens, deeper layers of body, etc.).

For instance, let the functions $\Theta_h = f(x, y, z, t)$ and $\Theta_s = \varphi(x, y, z, t)$ represent heating and self-cooling, respectively, of the selected point of body, before and after switching off the heat source. If at the time instant $t = t_1$ (Figure 10.6) we switch off the current, then the power P_{1V} will immediately disappear, but the field of temperature in the first moment will still be without change

$$\text{grad } \Theta_{h1} = \text{grad } \Theta_{s1} = \text{grad } \Theta_1 \tag{10.28}$$

The temperature increments as a function of time, before the switching off $(\partial\Theta_h/\partial t)_{t=t_1}$ and after the switching off $(\partial\Theta_s/\partial t)_{t=t_1}$ fulfill for the instant $t = t_1$ the equations

$$\left.\begin{aligned} c\rho_m \left|\frac{\partial\Theta_h}{\partial t}\right|_{t=t_1} &= P_{1V} - \lambda_i \nabla^2\theta_1 \\ -c\rho_m \left|\frac{\partial\Theta_s}{\partial t}\right|_{t=t_1} &= 0 - \lambda_i \nabla^2\theta_1 \end{aligned}\right\} \tag{10.29}$$

After adding by sides both these equations and replacing the absolute values of the time derivatives by tangents of respective angles, α_1 and α_2 (Figure 10.6), we obtain:

$$P_{1V} = c\rho_m(\text{tg } \alpha_1 + \text{tg } \alpha_2) \tag{10.30}$$

One can prove that in the case of homogeneous body, $\text{tg } \alpha_0 = \text{tg } \alpha_1 + \text{tg } \alpha_2$.

The described method of two tangents is more rarely used than the previous method (of initial temperature rise), because it is burdened by a bigger number of side-effect factors, which results in errors that in practice are bigger than in the method of initial temperature rise.

10.6.3 Accuracy of the Method

Equations 10.24, 10.27 show that the method error, and therefore measurement difficulties, are the greater, the more difficult to fulfill in practice are the basic requirements (10.26), that is, per (J. Turowski, Kazmierski, Ketner [10.30]):

1. The smaller are the per unit losses P_{1V} (weaker field, smaller permeability, big electric conductivity, etc.).
2. The higher is the thermal conductivity of the body (e.g., copper, big cross-sections).
3. The more nonuniform is power loss distribution (e.g., into a thick steel plate).

4. The higher is the gradient of power loss distribution in the point of measurement.
5. The higher is the coefficient of heat transfer to ambient.
6. The higher is temperature in the point of measurement on the body surface, in comparison with the ambient temperature.
7. The longer is the initial time interval between the instant of switching on the object and the first correct recording of temperature (thermal inertia of the measuring system).

Moreover, the errors depend on such factors as

8. Recording frequency, the kind and sensitivity of the measuring apparatus and thermoelements (thermocouples).
9. Method of fixing the thermoelements (deformation of temperature field by the thermocouple, as well as the heat capacity of measuring system).
10. Subjective reading and drawing of tangents (subjective error).
11. Necessity of gradual increasing of load of large objects, for example, at short-circuit test of power transformers and generators.

It should be also remembered that the distribution of power loss $P_{1V}(x, y)$ determined by thermometric method in cold conditions can be somewhat different from the distribution obtained at a heated operation.

The measurement error, depending on the mentioned factors, can reach from 1% to 15%.

At measurements of the power loss in laminated core, with more or less uniform loss distribution, one can reach accuracy of 1–2% or even better [10.27], [10.31]. Measurements carried out by the author et al. [10.31] showed a high repeatability of results and, a convenient to objective evaluation, linearity of heating curves (Figure 10.8). Owing to it, a good for technical purposes agreement (3.7–5.3%) between the measured results and the average loss measured by wattmeter was reached. Particularly troublesome, however, are the measurements in case of *thick steel plates*, with big nonuniformity of the power loss distribution. An intensive heat evacuation from the thin, in comparison with the whole plate thickness, active layer of steel, by the adjacent large mass of cold metal, causes a large discrepancy between the measured and real losses. In such a case, it is necessary to introduce the corrective increasing coefficients k_t (Figure 10.9) which reaches 200–500% or more—depending on the thickness of the heated plate and the way of its excitation. These correction coefficients were calculated with the help of a complicated computer program (Niewierowicz, J. Turowski [10.20]).

10.6.4 Computer Recalculation of the Measured Power Losses in Solid Steel

The per unit power losses in solid steel equal the measured value (10.27) multiplied by the coefficient k_t (Figure 10.9)

$$P_{1V} = k_t P_{1V\text{meas}} \tag{10.31}$$

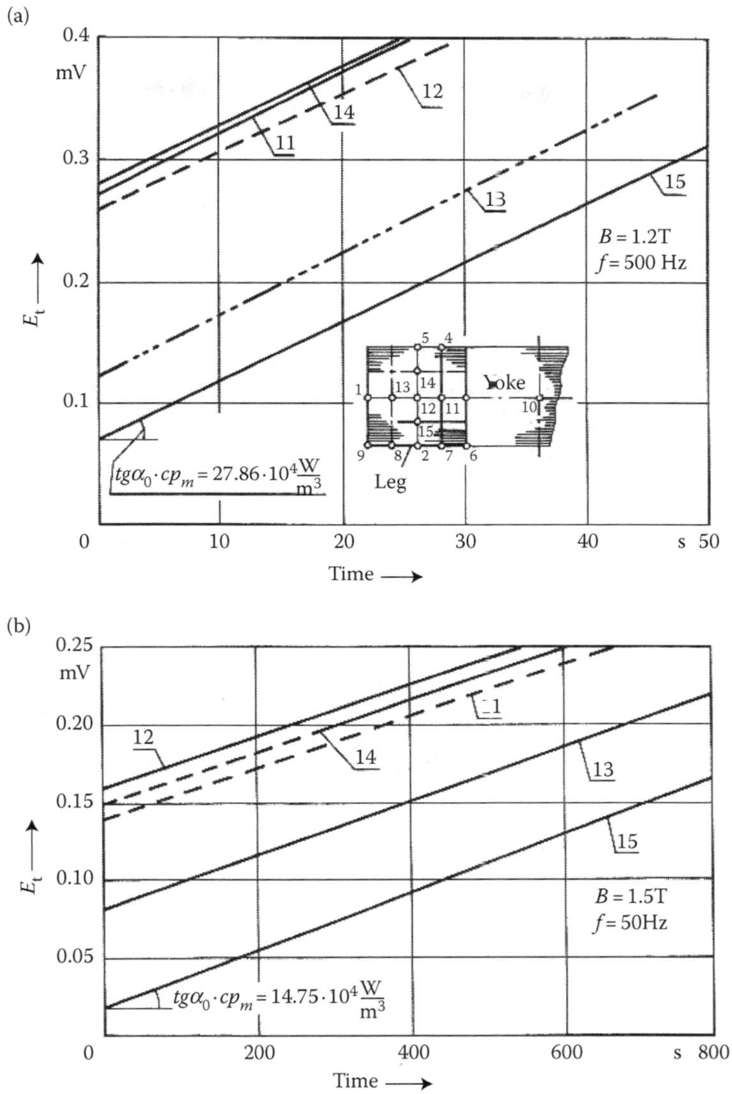

FIGURE 10.8 Initial heating curves of transformer core, after [10.31]: (a) short time scale and (b) longer time scale. 1, 2, … 14, 15 are the numbers of thermoelements; their positions are depicted in the inset in figure (a); E_t—electromotive force (emf) of the thermoelement.

where

$$k_t = \left(\frac{P_{1\mathrm{VS}t}}{P_{1\mathrm{Vmeas}}}\right)_{t=\Delta t} = \left(\frac{\theta_{id}}{\theta_\lambda}\right)_{t=\Delta t} \geq 1 \tag{10.32}$$

$\Theta_{id} = P_{1V}(z)\ \Delta t/c\rho_m$—the temperature rise of the investigated point of the body after the time $t = \Delta t$ from the excitation, at adiabatic heating (with no heat evacuation,

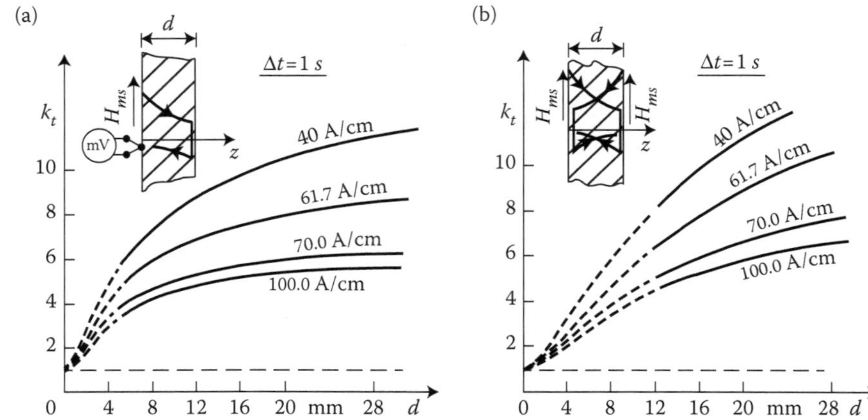

FIGURE 10.9 The enlarging coefficient $k_t(d)$ by which one should multiply the per unit losses $P_1 = c\rho_m \mathrm{tg}\alpha_0$ measured on the surface of a solid metal wall (After Niewierowicz N. and Turowski J.: *Proceedings IEE*, 119(5), 1972, 626–636 [10.20]) at the excitation: (a) one-sided, (b) double-sided; after the time of 1 s from the instant of switching on.

$\Delta\Theta = 0$); $\Theta_\lambda = \Theta_\lambda(t)$—the real temperature rise in the investigated point of body, at the time instant $t = \Delta t$, calculated with a full numerical (computer-based) solution of the diffusion Equation 10.24.

This task was solved under the author's supervision in (Niewierowicz, J. Turowski [10.20]) for ferromagnetic and nonferromagnetic thick plates on which a plane electromagnetic wave impinges from one side (Figure 4.8) or from both sides (Figure 4.12). In such a case, Equation 10.24 takes the form:

$$\lambda_z \frac{\partial^2 \theta_\lambda}{\partial z^2} - c\rho_m \frac{\partial \theta_\lambda}{\partial t} = -P_{1V}(z) \tag{10.33}$$

where $P_{1V}(z) = \frac{1}{2}\sigma E_m^2(z)$ is the distribution of the volumetric power losses (W/m^3) resulting from the eddy currents induced inside the solid metal plate, with considering the skin effect. These values, for various ways of excitation, at $\mu = \text{const}$, are given in Chapter 3.

For the case of a solid steel plate of variable permeability, $\mu_{\text{var}} = \mu[H(z)]$, to the right-hand side of Equation 10.33 one has applied the per unit power loss $P_{1V}(z, \mu_{\text{var}})$ calculated with the computer program of K. Zakrzewski [7.26] (Figure 7.4b)—under J. Turowski's supervision. In this way, graphs of the coefficient k_t (Figure 10.9) were obtained, as well as a *new, extended thermometric method* which consist in it that instead of drawing of the heating curves (Figure 10.6 and Figure 10.8) one reads only one temperature increment after an assumed time, for example, at $\Delta t = 30$ s, and then from the numerically calculated graph (Figure 10.10) one reads the per unit surface losses P_1 and the per unit volume losses P_{1V} as well as the magnetic field intensity H_{ms} on the body surface. The H_{ms} is the direct indicator of overheating hazard, after Figure 9.2.

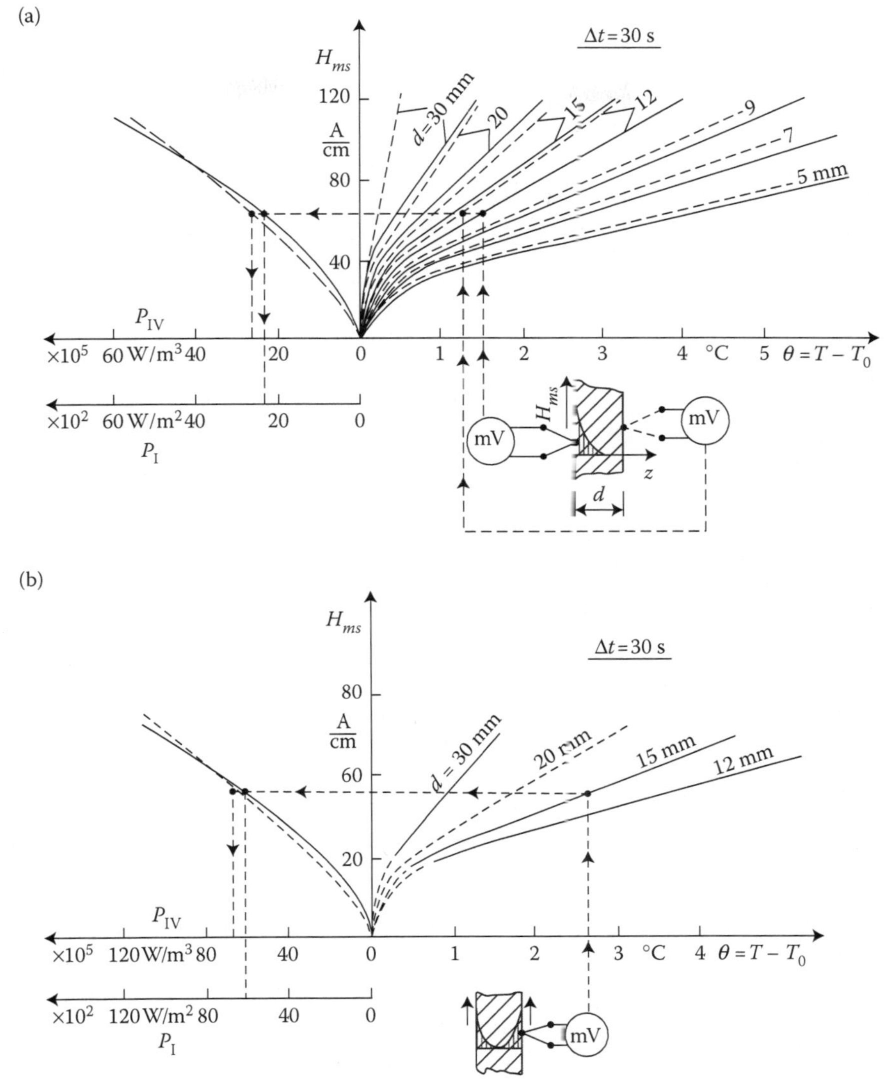

FIGURE 10.10 The extended thermometric method [10.20] of determining H_{ms} and the per unit power loss P_1, P_{1V}, by taking only one temperature reading $\Theta_t(T)$ after the time interval $\Delta t = 30$ s: (a) At one-side excitation, (b) at double-side excitation. — surface losses P_1; – – — volume P_{1V}; ← — measurement of temperature (and loss) at the wall side opposite to excitation (a), for example, external surface of transformer tank.

In Figure 10.10a provided are also plots (dashed lines) from which one can read the power loss in a steel wall by measuring temperature on the side of the wall opposite from excitation, for example, outside transformer tank, which is very convenient in practice.

In the work (J. Turowski, Kazmierski, Ketner [10.30]), results of measurements of losses in the cover of a three-phase transformer (Figure 10.11), by the thermometric method, were presented.

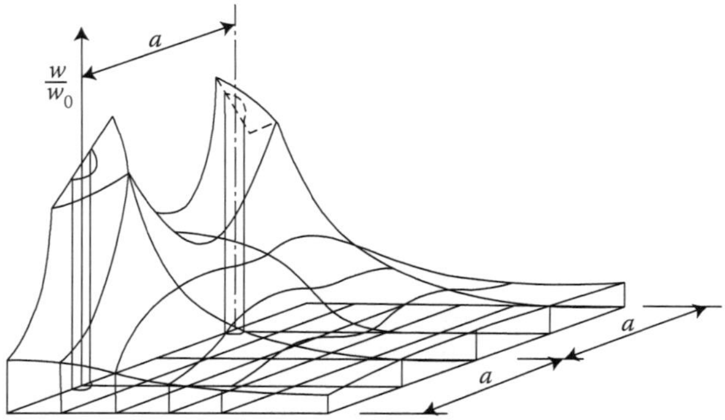

FIGURE 10.11 Distribution of power loss density (W/m²) in one half of a transformer cover, measured by the thermometric method in the work by Turowski J. et al. [10.30] and calculated by the author in [4.16].

10.6.5 Measurement Techniques

The task of loss measurement in both cases (10.27) and (10.30) is reduced to the measurement of small temperature increments, in small time intervals, as close as possible to the initial point of the heating curve. These conditions can be satisfied by a semi-automatic system (Figures 10.7 and 10.8). It can be computerized by introduction of a buffer (interface) memory which enables automatic computer processing in real time. It is also possible to build a system and a computer program which would use simultaneously the method from Figure 10.10.

At measurements of bodies made of copper, high sensitivity measuring devices are necessary, enabling measurement of small temperature increments in small time intervals—on the order of 0.5–1 s. In the case of bodies made of steel, with magnetic fields over 30×10^2 A/m at frequency of 50 Hz, the measurement can last even 20–30 s. It should not be, however, in any case prolonged over 1.5 min. A big significance here has a correct graphic post-processing of the obtained results [10.30].

To find the correct value of tg α_0 (Figure 10.6), it is necessary to find on the graph a rectilinear part, rejecting the initial deflection caused by inertia of the measuring system, and the final deflection caused by outflow of heat from the investigated point due to the loss of balance of the condition (10.26).

10.6.6 Approximate Formulae

In the case of uniform loss distribution in the whole volume of body (laminated cores, conductors, etc.) and at comparison of relative loss distribution, it is sufficient to determine only the volume losses P_{1V} (10.27). If, however, on the basis of measurement we wish to determine the total losses in solid steel bodies, by the method of integration of losses measured on the surface, then the volume losses P_{1V} (in W/m³) should be recalculated into the surface losses P_1 (in W/m²). A similar recalculation is needed when from measured losses we wish to find the magnetic field intensity H_{ms}, or vice versa.

The *average* volume losses P_{1V} at the depth of penetration to solid steel can be determined from Equations 2.181, 7.30, and 3.12 [10.30]:

$$P_{1V} = P_1/\delta \approx \frac{\pi}{\sqrt{2}} f\mu_s, H_{ms}^2, \quad \text{from which: tg } \alpha_0 \approx \frac{\pi}{\sqrt{2}c\rho_m} f\mu_s H_{ms}^2 \tag{10.34}$$

where, after Equation 3.12

$$P_1 = a_p \sqrt{\frac{\omega\mu_s}{2\sigma}} \frac{H_{ms}^2}{2} \tag{3.12}$$

Assuming for the case of steel the average values $c = 500$ Ws/(kg · K). $\rho_m = 7.85 \times 10^3$ kg/m³, $\sigma_{20°C} = 7 \times 10^6$ S/m, and $f = 50$ Hz, we get the formulae dedicated to steel

$$P_1 \approx 3.76 \times 10^{-6} \sqrt{\mu_{rs}} \cdot H_{ms}^2, \quad \text{W/m}^2 \tag{10.35}$$

$$P_{1V} \approx 1.4 \times 10^{-4} \mu_{rs} H_{ms}^2, \quad \text{W/m}^3 \tag{10.36}$$

$$\text{tg } \alpha_0 \approx 35.7 \times 10^{-12} \mu_{rs} H_{ms}^2, \quad \text{°C/s} \tag{10.37}$$

in which μ_{rs} is the relative permeability and H_{ms} is the maximal value of magnetic field intensity (in A/m) on the surface.

The above formulae can be utilized for determining the permitted per unit losses or the permitted tg α_0, if we know the permitted magnetic field intensity $H_{ms,\text{perm}}$ on the surface of body (Table 9.1, Figure 9.2).

If, instead, on the basis of experimentally determined tg α_0 we wish to determine the per-unit losses P_{1V} and P_1 or the tangential value of magnetic field intensity H_{ms} on the surface of body, then we can use the formulae for solid steel:

$$P_{1V} \approx 3.92 \times 10^6 \text{ tg } \alpha_0 \tag{10.38}$$

$$P_1 \approx 1.05 \times 10^5 \frac{\text{tg}\alpha_0}{\sqrt{\mu_{rs}}} \tag{10.39}$$

$$\sqrt{\mu_{rs}} \cdot H_{ms} \approx 1.69 \times 10^5 \sqrt{\text{tg}\alpha_0} \tag{10.40}$$

To facilitate calculations with the help of the above formulae, in Figure 10.12 are given graphs of correspondingly recalculated magnetization characteristics for an average constructional steel and for cast iron. On the basis of formulae (10.38) through (10.40), the data for the curves in Figure 10.13 were calculated. The formulae

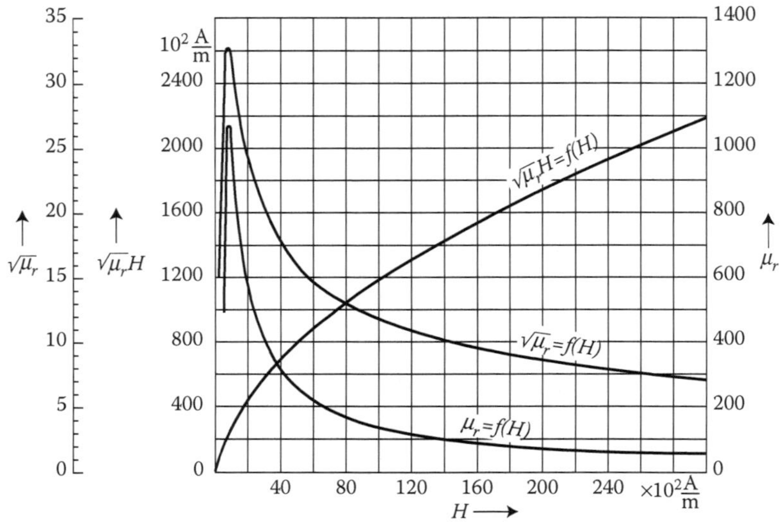

FIGURE 10.12 Plots of the recalculated magnetization curve of constructional steel and cast iron [10.30]. *Analytical approximations*—see Figure 7.3 (Adapted from Kozlowski M. and Turowski J.: *CIGRE. Plenary Session*, Paris 1972. Report 12-10 [7.9].)

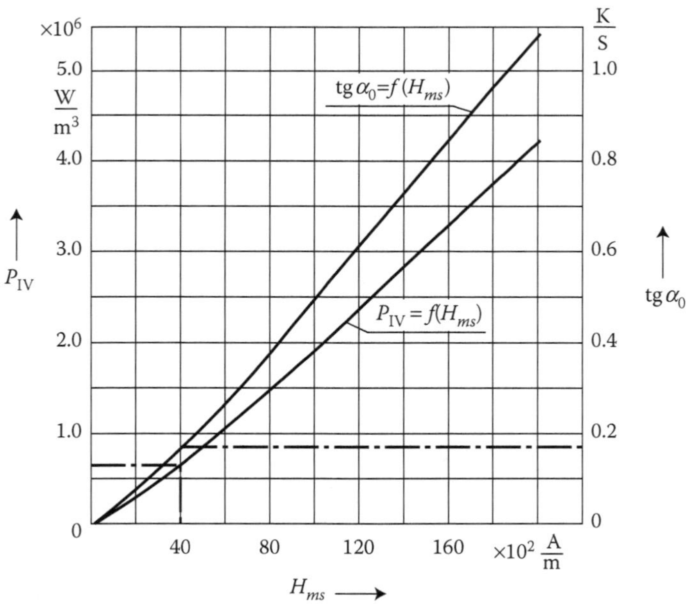

FIGURE 10.13 The dependence of tg α_0 and the per unit losses P_{1V} on the magnetic field intensity H_{ms} on the surface of solid steel parts. (Adapted from Turowski J., Kazmierski M., and Ketner A.: Determination of losses in structural parts by the thermometric method. *Przegląd Elektrotechniczny*, (10), 1964, 439–444 [10.30].)

(10.35) through (10.40) as well as Figure 10.13 do not consider the corrections for the heat evacuation into interior of the metal mass (Figure 10.9). This can be done with the help of Figure 10.9 or Figure 10.10. At thin sheets (cores), these corrections are not necessary.

The highest volume losses on the solid body surface itself, in the case of an ideal, dimensionless thermoelement, are

$$P_{1V,\max} = \frac{a_p}{2}\sigma E_{ms}^2 = \frac{a_p}{\sigma}k^2 H_{ms}^2 = \sqrt{2}\frac{\pi}{\Delta_s}H_{ms}^2 = 2P_{1V} \tag{10.41}$$

10.7 INVESTIGATION OF PERMISSIBLE OVER-EXCITATION OF POWER TRANSFORMERS

In turbo-generators and generator transformers, at the so-called "*power drop*", that is, at a sudden disappearance or turn-off of the active power load, due to, for example, a short-circuit in the power network, a sudden jump of the voltage on the terminals of generator can happen, to the value of $E_{mN} = (1 + \Delta U)U_N$, (where $\Delta U = 0.4\ldots0.5$, as a relative value is the voltage regulation* at a simultaneous increase of the rotational speed of turbine, $n_s = f/p$. It causes a further increase of the electromotive force of turbo-generator ($E = cpn_s$) and a simultaneous increase of the frequency $f = pn_s$. In response to this, the speed controller can cause a drop of speed, and hence also of the frequency f. Similar swinging of U and f can emerge in other states of power unit, for instance, at the start-up, and so on. Then, per the relation $U \approx 4.44\, fN\, (B_m\, s)$, the flux density in the magnetic core, $B_m = k(U/f)$, can easily exceed the permitted rated (nominal B_{mN}) value, even up to 140% (Nowaczyński [10.21]). As a result of a strong magnetic core saturation (over-excitation) the main flux can by displaced from the core and penetrate the solid constructional parts (screws fastening the core, clamping plates, yoke beams, rings, and even windings themselves). This over-excitation can cause inadmissible local overheating of these parts, especially from the side adjacent to the core. This requires an investigation of the asymmetric skin-effect in solid steel elements, in order to determine the permissible time of duration of the over-excitation.

According to the international recommendation of IEC (Publication 76-1, 1976), the over-excitation cannot be higher than $B_m/B_{mN} = 1.4$ within 5 s. Maintenance needs, however, require more accurate indications. Unfortunately, every transformer type can be different, and this is why standards of different countries provide their own recommendations (USA, Germany, Russia, etc.) which sometimes are contradictory [3.2], [10.21], [10.29].

In 1977, J. Nowaczynski [10.21] in cooperation with the author (JT), carried out extended theoretical and experimental investigations of coupled electromagnetic and thermal fields. As a result, characteristics were obtained which were also different from other sources (Figure 10.14). These results from specific transformers of local

* The voltage regulation $\Delta U = \frac{E_0 - U_N}{U_N}\,100\%$, where U_N is the voltage at the rated load, and E_0—at no load and rated I_f and n_s....

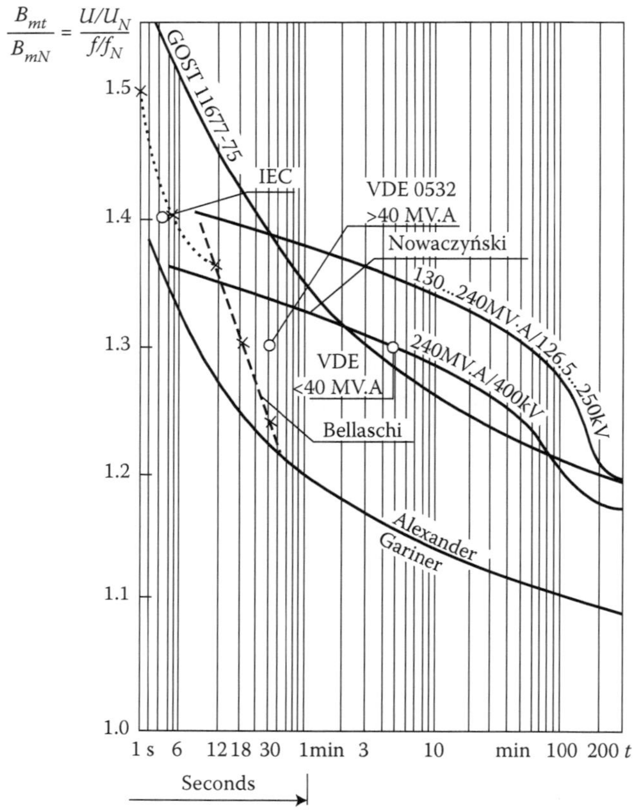

FIGURE 10.14 The permitted duration of over-excitation of power transformers, according to various recommendations [10.21], [10.29]; GOST—a Russian standard, VDE—a German standard. Other sources are from J. Nowaczyńsk: [10.21].

manufacturers provided evidence for the necessity to use three-dimensional modeling and calculation (Figures 7.20 through 7.22), which is a serious challenge for calculation and design technology on the international scale.

After a critical analysis of the measuring methods with the help of Rogowski's straps, the thermometric method and coil probes, Nowaczynski [10.21] proposed a new method of coil probes (Figure 10.15), which enabled independent measurements of field in the X and Y axes on both sides of a solid steel wall, at its asymmetric excitation, appearing strongly at over-excitation.

10.8 MEASUREMENT OF POWER AT VERY SMALL POWER FACTORS (COS φ) AND/OR SMALL VOLTAGES

At model investigations or testing of electrodynamic systems, such as elements of electric machines and transformers (Turowski [5.15]), bus-bar systems [4.24], [5.16], [5.19], as well as at investigation of transformers [10.21], [10.27] and

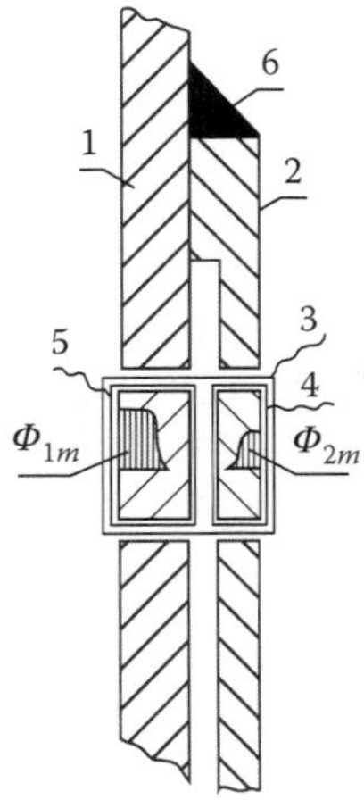

FIGURE 10.15 Probe for measurement of asymmetric fields in steel plates: 1—the investigated plate; 2—magnetic shunt made of solid steel; 3—coil for measurement of resultant magnetic flux ; 4, 5—coils for measurement of component fluxes Φ_{1m} and Φ_{2m} in the X and Y axes; 6—girth (circumferential) welding around the measuring plate 2. (After Nowaczyński J.: (1) Local overheating of construction elements of transformers, during overload and over-excitation. (in Polish). PhD Dissertation under the direction of J. Turowski. Inst. of El. Machines and Transformers, Technical University of Lodz, Sepember 1977; and (2) *Sci. Reports of Power Inst.* (7) 1979, 1–130 [10.21].)

reactors for power systems [10.12], the power factors can be very small, down to 0.001 or less. Significant measuring errors can then appear (Figure 10.17 [later in the chapter]), due to the angle error of wattmeters and instrument transformers, which causes that usually wattmeters are not suitable for measurement of device power. Such measurements are usually carried out with the help of special bridge or compensation methods, or with special wattmeters. Bridge methods, due to their complicated structure and labor-intensive service, have not found a broader application in industrial practice [10.27]. In consideration of it, the author proposed and patented a special *compensation wattmeter* (Turowski [10.24], [10.25], [10.26], Figure 10.16) and next, jointly with T. Janowski, the Wattmeter Attachment [10.14] acting on the same principle, but simpler in manufacturing. This wattmeter and the attachment are described in J. Turowski's book [1.15].

The scheme of a wattmeter (Figure 10.16) permits technical power measurements at an arbitrary small power factor (cos φ) or arbitrary small voltage. The system ensures operation of the normal, basic wattmeter 1 always at the power factor practically equal 1 (cos $\varphi = 1$), which means basically at full scale of the wattmeter, independently on the power factor of the receiver. The principle of operation of the system is explained by the phasor* diagram for the first harmonic of current and voltage of the load (Figure 10.16b). Analogical phasor diagrams can be developed for any other harmonics. The voltage circuit of the basic wattmeter 1 is connected

* Phasor = vector in a complex plane.

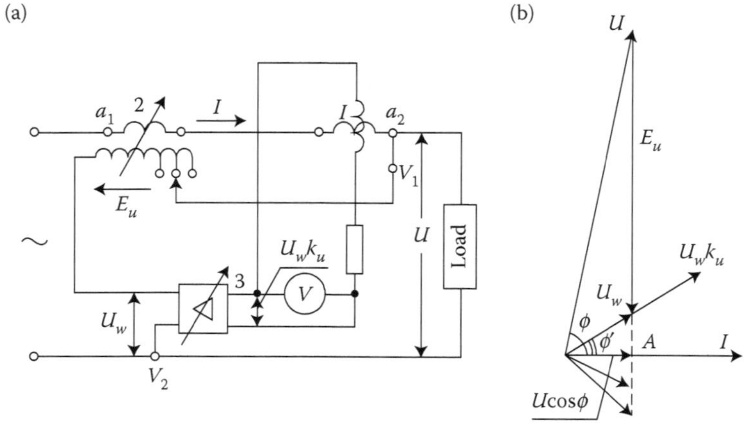

FIGURE 10.16 The compensation wattmeter, according to J. Turowski's Patent [10.24], [10.25], [10.26], for measuring power at arbitrary small power factors and/or arbitrary small voltages: (a) scheme; (b) phasor diagram of voltages; 1—a normal wattmeter of rated cos $\varphi_N = 1.00$; 2—an air-core control transformer; 3—amplifier with controllable amplification.

not to the receiver voltage (as it is normally done), but to the vector difference of the receiver voltages and the controlled secondary voltage of the air-core transformer 2. This enables a reduction of the voltage measured by the wattmeter to the value U cos φ being in phase with the current in the current coil. The power readout is done after reducing the voltmeter V indicator to minimum, and after matching amplification of amplifier 3 to the voltage range of wattmeter 1.

Accurate reduction of the voltmeter readout to minimum is not indispensable, because the basic wattmeter 1 always reads the active power anyway. This feature enables measurements of power at deformed waveforms of currents and voltages. Amplifier 3 serves for amplification of the resultant voltage U_w in case of very small power factors of load (e.g., at heating test of large power system reactors).

The system from Figure 10.16 can also serve to measure power at very small voltages, when a conventional wattmeter cannot be applied at all. It is called "wattmeter attachment" and can be used as an independent apparatus (Janowski [10.14]). Application of amplifier 3 brings additional important properties to the wattmeter. Thanks to high input resistance of the amplifier, the power measurement is realized practically without absorption of current by the voltage circuit of wattmeter. The power necessary for displacement of the moving element of wattmeter is delivered to the voltage coil from external source, which is the power supply of the amplifier. This feature is especially valuable at measurements of small powers with small voltages. Thanks to it, also the angle error of air-core transformer can be minimal.

Accuracy of the whole apparatus depends mainly on the accuracy of air-core transformer 2 and not much on the accuracy of amplifier 3 [1.15]. The most stringent limitations are imposed to the angle error of *air-core transformer* 2.

An important advantage of the compensation wattmeter is the possibility to compensate the angle error of the current measurement transformer (J. Turowski [10.26]). At small power factors, this error (in percent) is expressed by the formula

$$\Delta_{\%} = 0.0291\delta \operatorname{tg}\varphi\% \approx 0.0291\frac{\delta}{\cos\varphi}\% \quad (10.42)$$

where δ is the angle error of the measuring transformer, in angular minutes. The error (10.42) at small values of cos φ can reach significant values (Figure 10.17).

The resultant angle-error at small angles δ and δ_k comes out here multiple times smaller than the error (10.42) of classical method

$$\Delta_{\%} = \frac{|U||I|\cos\varphi - |U||I|\cos\varphi(\cos\delta \mp \sin\delta \operatorname{tg}\delta_k)}{|U||I|\cos\varphi}$$
$$= 1 - \cos\delta \pm \sin\delta \operatorname{tg}\delta_k \approx 0 \quad (10.43)$$

because the angles δ and δ_k are as a rule very small.

The measuring current transformer should not be connected in such a way that its secondary winding is connected in series with the primary winding of the air-core transformer, because in such a case the wattmeter would measure the power $|U||I|$ $\cos(\varphi - \delta)$ and the angle-error would be the same as in formula (10.42).

The described instrument can be used as a normal wattmeter, that is, as a *single* one, or in a *system of two or three wattmeters,* with measuring transformers, and so on. In the case of measuring power at small voltages (to ca. 100 V) the scheme from

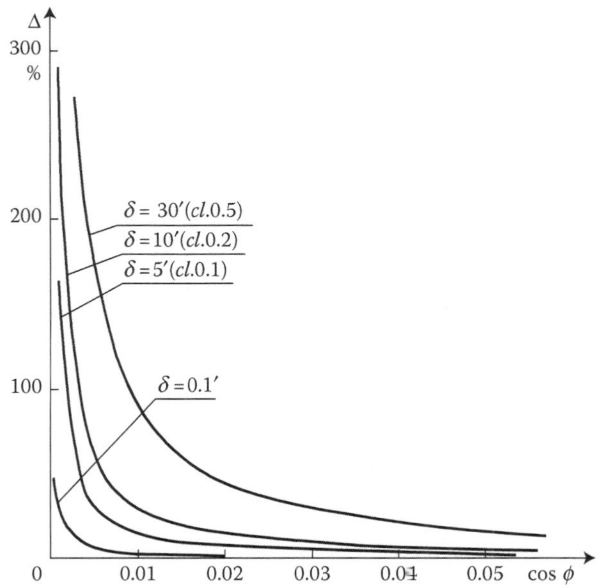

FIGURE 10.17 Percentage angle-errors of measuring current transformer at measurement of active power, at different angle error δ (the wattmeter class, *kl.*), as a function of the power factor (cos φ) of load. In the parentheses are given example classes (*kl.*) of the measuring current transformer.

Figure 10.16a can be directly applied. In the case of higher voltages, the electromotive force (emf) E_u for compensation would be too big, which would require building a huge air-core transformer 2. In such a case, the terminals v_1 and v_2 of wattmeter could be connected to the voltage U through an appropriately chosen voltage divider. The J. Turowski's wattmeter is especially suitable for short-circuit tests of large power reactors (Figure 7.15) versus bridges [10.12], power transformers (Figure 10.14), modeling investigations (Figure 10.3), and so forth.

The J. Turowski's semi-electronic, compensation wattmeter, patented in 1962 [10.24], has been used for many years, taught during lectures, and further developed in technical universities (e.g., Gotszalk [10.1], pp. 131–138). It has been a convenient technical tool, cited, among others, in university textbooks of magnetic measurements (Nałęcz et al. [10.5], p. 274). It has capabilities competitive with many electronic apparatuses and bridges, since it exceeds them in simplicity, low price, easy application in three-phase systems, and so on. The "Wattmeter Compensation Attachments" built on this principle by T. Janowski [10.14] enables the measurement of receiver power with cos $\varphi \geq 0.0005$ or smaller, at voltages $U \geq 0.1$ V, with accuracy between 0 and less than 2% [1.15/1].

10.8.1 Bridge Systems

The measurements of power and induction impedances at small power factors are sometimes performed using various bridge systems. Such systems, at the application of elements with minimal angle error, permit to obtain a significant measuring accuracy. Their drawbacks include high labor-demand, sensitivity to of curve deformation of current or voltage, and big difficulties at power measurement in three-phase systems. These drawbacks do not occur in the author's compensating wattmeter described above (Figure 10.18). As an example, in Figure 10.19, is shown a scheme of

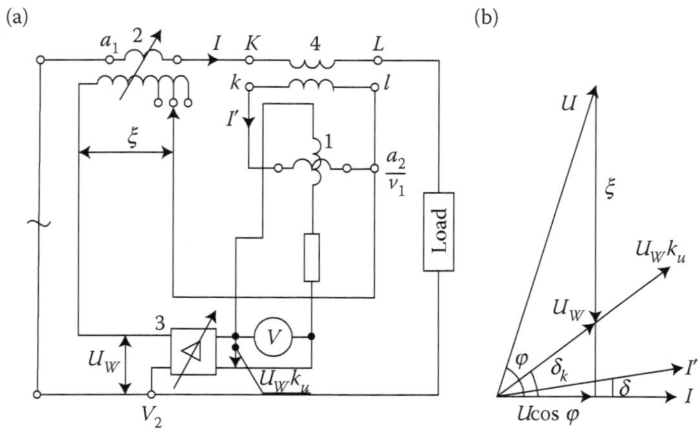

FIGURE 10.18 The Turowski's compensation wattmeter: (a) the way of connecting the measuring current transformer 4; (b) a phasor diagram of voltages and angle-errors. (Adapted from Turowski J.: *Proceedings IEE*, 112(4), 1965, 740–741 [10.26].)

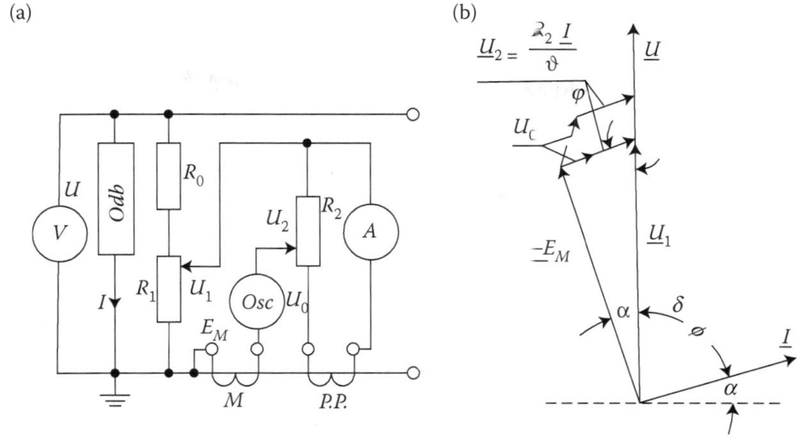

FIGURE 10.19 A scheme of measuring bridge with mutual inductance (a), and a phasor diagram of voltages (b). (Adapted from Specht T.R., Racemacher L.B., and Moore H.R.: *Electr. Eng.*, (5) 1958 [10.23].)

a bridge [10.23] which consists of a resistive voltage divider, R_0, R_1, a coil of mutual inductance M, a resistive potentiometer R_2, a measuring current transformer PP, and an oscilloscope Osc as an indicator of equilibrium state of the bridge.

The zero oscilloscope is connected to the voltage U_0 which is a geometric sum of the voltages—E_M, U_1, and U_2. Since the emf E_M is perpendicular to the receiver current I, and the voltage U_2 is parallel to I (Figure 10.19), by controlling alternatively resistances R_1 and R_2 one can bring the oscilloscope voltage U_0 to zero. The control process should be started from setting the voltage U_1 to be close to the value of emf E_M. After reaching equilibrium ($U_0 = 0$), we determine the power factor from the ratio:

$$\cos\varphi = \frac{|U_2|}{|U_1|} = \frac{R_2(R_1 + R_0)|I|}{R_1 U \vartheta} \approx \operatorname{tg}\alpha = \frac{R_2}{\vartheta \omega M} \tag{10.44}$$

where $\vartheta = N_1/N_2$ is the transformation ratio.

If elements of the system have the angular error not exceeding 1′ (*one minute*) then the measuring error will probably be about 0.5%.

Another bridge system (Figure 10.20) was used by Deutsch [10.12] at a precise power measurement at currents up to 150 A and voltages up to 500 kV and power factors in the range of 0.1–0.001, which occur in high-voltage reactors and at short-circuit tests of high-voltage power transformers. High voltage is supplied to the bridge by a high-voltage lossless (with gas insulation) capacitor. In this system, an accurate measuring current transformer, with transformation ratio error not higher than 0.001% and the angle error less than 0.2′ at loadings 5–200%, was applied. The secondary winding of the measuring current transformer was connected to the system in such a way that the voltage U_2 over the resistance R_2 is reversed by

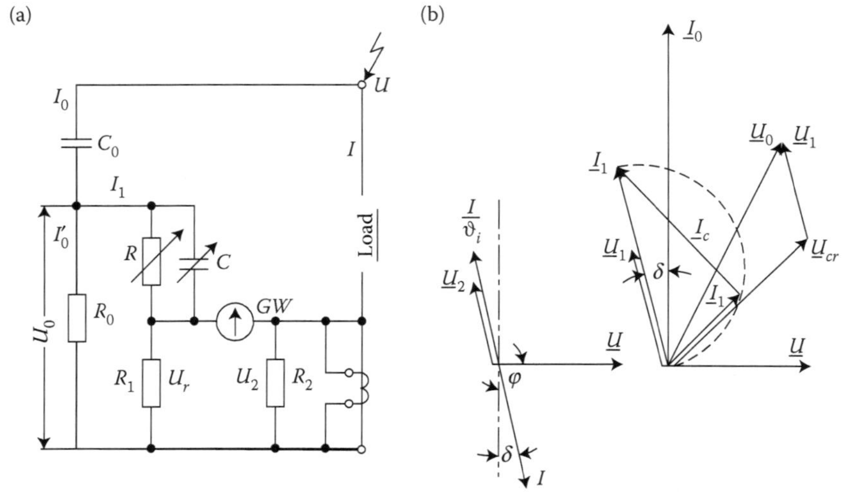

FIGURE 10.20 A high-voltage measuring bridge with a measuring current transformer and calibration capacitor. (a) Schematic, (b) phasor diagram. (Adapted from Deutsch G.: Measurement of active power losses of large reactors. (in German). *Brown Boveri Mitt.*, 47(4), 1960.)

180° with respect to the receiver current. The searched phase shift φ is evaluated with the help of the complementary angle $\delta = 90° - \varphi$ at the instant of equalization of the voltages $U_1 = U_2$. Value and phase of the voltage U_1 is regulated by varying the resistance R and the capacitance C (Figure 10.20). The current I_0 at the same time remains invariable and its phase is ahead with regard to the supplying voltage exactly by an angle of $\pi/2$, because capacitive reactance of the capacitor C_0 (100 pF, 30 MΩ) is incomparably greater than impedance of the circuit R_0, R_1, C, R (no more than 1 kΩ).

At voltage balance, the angle between voltage U_1 and current I_0 equals $\delta = 90° - \varphi$. From relations between the circuit parameters of the system one can obtain:

$$\text{tg}\delta = \omega RC \frac{R}{(R_0 - R_1)(1 + \omega^2 C^2 R^2) + R} \approx \frac{\omega C R^2}{R_0 + R + R_1} \approx \cos\varphi \quad (10.45)$$

Assuming that the resistance of a given object is much smaller than its reactance, that is, $R_x \ll \omega L_x$ (which gives an error smaller than 1% at $\cos\varphi = 0.1$ and smaller than 0.01% at $\cos\varphi = 0.01$), Deutsch in [10.12] derived the formulae for calculation of

- Reactance of the investigated object

$$\omega L_x = \frac{R_2(R_0 + R_1 + R)}{\vartheta_1 \omega C_0 R_0 R_1} \quad (10.46)$$

- Resistance of the object

$$R_x \approx \omega L_x \cos \varphi \tag{10.47}$$

- Voltage supplied to the object

$$|U| \approx \omega L_x |I| \tag{10.48}$$

- Apparent power

$$|S| \approx \omega L_x |I|^2 \tag{10.49}$$

- Active power losses

$$P = |S| \cos \varphi \tag{10.50}$$

At direct grounding of the screen, an additional measuring error at the evaluation of power factor appears:

$$\Delta = \frac{\cos\varphi - (\cos\varphi)'}{(\cos\varphi)'} = -\frac{\omega C_e}{(\cos\varphi)'} K\left(1 + \frac{R}{R_1}\right) \tag{10.51}$$

where C_e is the capacitance of screen to ground, $(\cos \varphi)'$ is according to Equation 10.45, and

$$K = R_0 R_1/(R_0 + R + R_1) = \text{const.}$$

The corrected value of power factor:

$$\cos \varphi = (\cos \varphi)' (1 - \Delta) \qquad \text{where error } \Delta \approx C_0/C \tag{10.52}$$

Other guidelines and an example of applying measurement in a parallel reactor of 8120 kvar, 275 kV, one can find in the Deutsch's work [10.12] or in the previous edition [1.15] of this book.

10.8.2 Compensatory Measurement of Additional Losses

At model investigations, we often wish to measure additional losses in a system with exclusion of the fundamental *Joule* losses $R\,|I|^2$ in the winding exciting the field. In such a case, we can apply the method of a separate voltage (potential) winding overlapping in space exactly with the main excitation winding (Turowski et al. [5.16]). The excitation winding 1 (Figure 10.21) produces a leakage flux. With this flux is partly linked a short-circuited secondary circuit 2, representing active and reactive losses and the resultant flow produced by eddy currents induced in solid metal parts of the investigated object (power transformer). Along with the winding 1, a voltage

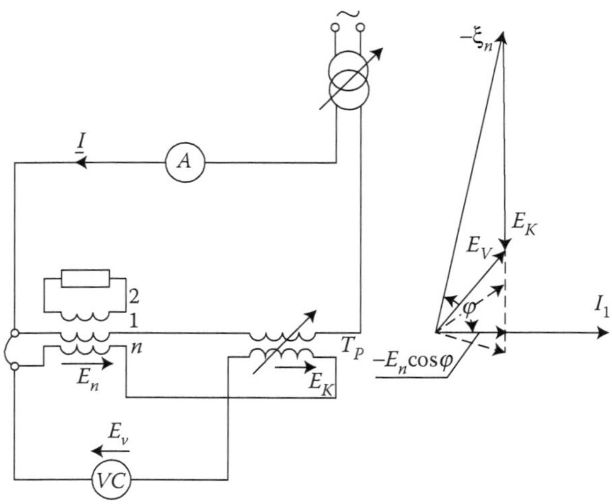

FIGURE 10.21 Measurement of additional losses with the help of the compensatory method, with application of a separate additional potential winding (n): 1—the main exciting winding, 2—the investigated object, T_p—regulatory air-core transformer, VC—digital voltmeter (harmonics analyzer) which measures the first harmonic of voltage. (Adapted from Turowski J., Pawłowski J., and Pinkiewicz I.: The model test of stray losses in transformers. (in Polish) *"Elektryka" Scientific Papers of Techn. University of Lodz*, (12), 1963, 95–115.)

measuring winding n is wound. This winding should be spatially aligned with the exciting winding 1 as much as possible. This is due to the following considerations.

When the winding 1 is supplied by a sinusoidal alternating voltage, and at $I_n = 0$, we get (Figure 10.21):

$$\left.\begin{aligned} U_1 &= I_1(R_1 + j\omega L_1) - j\omega M_{12} I_2 \\ 0 &= I_2(R_2 + j\omega L_2) - j\omega M_{12} I_1 \\ E_n &= j\omega M_{n1} I_1 + j\omega M_{n2} I_2 \end{aligned}\right\} \tag{10.53}$$

where R, L, M are, respectively, the resistance, self-inductance, and mutual inductance of corresponding circuits denoted with the given subscripts, correspondingly.

After solving the system of Equation 10.53, by making a few transformations we obtain the apparent power developed in the winding 2, expressed by the measured quantities I_1 and E_n

$$S_1 = E_n I_1^* = -\frac{M_{n2}}{M_{12}} R_2 |I_2|^2 + j\omega\left(M_{n1} |I_1|^2 + \frac{M_{n2} L_2}{M_{12}} |I_2|^2 \right) \tag{10.54}$$

from where the measured active power is

$$P = |E_n||I_1|\cos\varphi = \frac{M_{n2}}{M_{12}} R_2 |I_2|^2 \tag{10.55}$$

Therefore, if the potential winding n has the same number of turns and in space is situated exactly like the exciting winding 1, that is, $M_{n2} = M_{12}$, then an ideal wattmeter whose voltage circuit is connected to the measuring potential winding n, whereas its current circuit is in series with the excitation winding 1, will show the power consumed by the investigated external object 2 as the result of eddy currents induced in it. Since such systems have exceptionally low power factors (0.0001–0.01), the power measurement has been done with the J. Turowski's semi-electronic, compensating circuit [5.14] (Figure 10.21). Minimal readings of the digital voltmeter VC were found by regulating the voltage E_k with the help of the regulating air-core transformer T_p. The voltage E_k is immediately equal to the active component of electromotive force $E_{v,\min} = E_n \cos \varphi$. A conventional digital voltmeter can be applied only to sinusoidal processes. In models with solid steel elements, at sinusoidal alternating current in excitation winding 1, the emf in the potential winding n is deformed. Since power equals to a sum of products of corresponding harmonics of current and voltage, and at the same time the current is sinusoidal, for evaluation of the power suffices to measure only the first harmonic of emf, $E_{v,\min}$. For this purpose, the voltmeter we used was a harmonic analyzer that directly read the first harmonic of voltage, in volts [5.14]. An improvement of this method consists in the application of the J. Turowski's semi-electronic, compensating wattmeter described above (Figure 10.18) or the author's compensating attachment to a conventional wattmeter.

A similar method was applied by a former J. Turowski's PhD student and assistant, T. Janowski [10.13], for measurement of stray losses occurring outside of transformer windings.

10.9 OTHER METHODS OF MEASUREMENTS

There exists many *measuring converters of magnetic field,* described in literature [1.2], [10.5]. In the domain of AC fields, these can be measuring coils (Figure 10.15), Rogowski's strip (Figure 10.22), hallotrones (Figure 1.13), fluxmeters, and so on. In the DC domain: coils with ballistic galvanometer or fluxmeter, and so on. [10.5], spherical coils to point field measurements, Rogowski's strips, rotating and oscillating coils (Groszkowski's magnetometer), rotating disk, compensatory schemes with

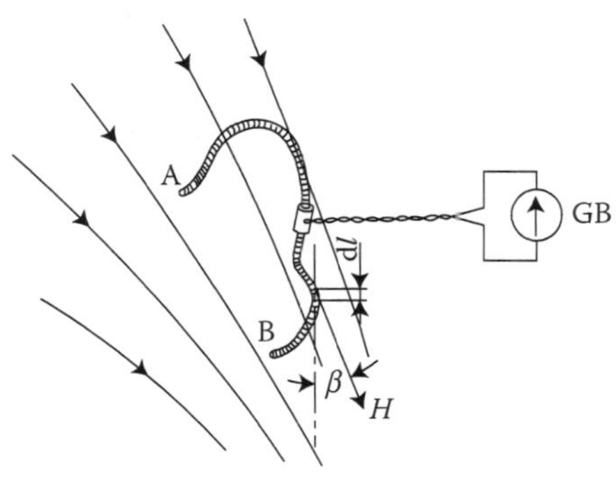

FIGURE 10.22 Measurement of *magnetic voltage* with the help of *Rogowski's strap*; GB—ballistic galvanometer or a digital voltmeter.

inductive converters and ballistic galvanometer, magnetron and electro-radiation converter (electronic elements), electrodynamic coil, permeameter, magnetoresistive elements (gaussotron), transducers, yoke permeameters, and others.

Magnetization curves are measured, among others, by the classic ballistic method ([1.2], p. 669), which still is considered as one of the best. It requires, however, to make samples in the toroidal form. *Power losses* in cores of laminated sheets are measured with the conventional Epstein's apparatus ([1.2], p. 677). See also "Plotting Magnetization Curves" on http://info.ee.surrey.ac.uk/Workshop/advice/coils/BHCkt/index.html.

Application of measuring systems based on modern semiconductor elements [10.4] made it possible to measure field and power losses in magnetic and electric circuits, temperature [9.7] as well as the dynamic hysteresis loop, with much higher accuracy than it was possible with the help of classical methods, Gaussmeters, for example, www.maurermagnetic.ch.

Measuring coils (Figure 10.15) belong to the simplest and cheapest measuring methods of alternating field, on the basis of the formula

$$E = 4\sigma_k f N B_m s, \qquad \sigma_k = E/E_{\text{average}} \tag{10.56}$$

whereas for a sinusoidal curve the shape coefficient $\sigma_k = 1.11$.

At the direct current (DC) field, measurements are possible with the help of a ballistic galvanometer, more or less automated, by removing the coil from the field and inserting the coil into the field.

At weak fields, one should consider the *terrestrial magnetic field* (about 0.4 A/cm).

Rogowski's strap. The magnetic voltage U_m or flow θ_{AB} between points A and B of space, equal to the linear integral of magnetic field intensity along an arbitrary path between these points (Figure 10.22):

$$\int_{AB} \boldsymbol{H} \cdot \mathrm{d}\boldsymbol{l} = \int_{AB} H \cdot \cos\beta \,\mathrm{d}l = \theta_{AB} \tag{10.57}$$

can be measured with the help of a so-called *magnetic strap*, called also *Rogowski's strap*. This strap consists of a flexible core made of insulation material, on which is wound a double-layer of thin isolated wire (both layers in the same direction). The ends of this winding meet in the center of strap and are bifilarly led to a digital voltmeter—in case of alternating field. In case of magnetostatic field, the terminals of strap are connected to a ballistic galvanometer and the measurement is performed by a rapid removal of the strap from the region occupied by field, or by switching off the current exciting the magnetostatic field.

The flux coupled with the winding of strap (Figure 10.22), after considering (10.57), can be expressed as

$$\Psi = \int_A^B \mu_0 H \cos\beta \,\mathrm{A}\, N' \mathrm{d}l = \mu_0 \mathrm{A} N' \theta_{AB} \tag{10.58}$$

wherefrom:

$$\theta_{AB} = k\,\Psi \tag{10.59}$$

where $k = 1/\mu_0 AN'$ is the constant of strap, which is calculated on the basis of a known number of turns N' per unit length of strap and of the average cross-section of strap, A. Finally, the magnetic voltage at alternating field is determined on the basis of the measured electromotive force (emf) E_u

$$\theta_{AB} = k\frac{E_u}{4.44f} \tag{10.60}$$

At DC field, instead,

$$\theta_{AB} = k\,C_\psi\,\alpha \tag{10.61}$$

where C is the constant of the ballistic galvanometer at measurement of flux (specified in catalogues).

By closing the strap around current conducting bars, one can measure on the basis of (Ampere) law of flow (10.57) the current in the bar, what is especially valuable at investigations of not easily accessible bars and conductors in parallel branches.

10.9.1 Measurement of Magnetic Field Intensity

In the case when the investigated field is uniform, one can measure the magnetic field intensity H with the help of *Rogowski's strap* on the basis of the dependence $H = \theta_{AB}/l_{AB}$, where l_{AB} is the length of AB section. At such measurements, most convenient is to use a rectilinear strap of length l_{AB}.

10.9.2 Measurement of Electric Field Intensity

A measurement of electric field intensity E in a metal consists in determination of the difference between two potentials of the points in the distance of a unit length and placed on one line of the same current density $\boldsymbol{J}$.

In an electrostatic field, the voltage U_{AB} measured in such a way does not depend on the shape of curve connecting both points A and B. In a rotational field of alternating electric field, instead, the reading of an ideal voltmeter connected between the same points A and B (Figure 10.23) will be varied, depending on the shape of curves connecting these points with the voltmeter. In each different case, 1, 2, or 3 in Figure 10.23, the voltmeter will display a different value of voltage along the corresponding curve.

To measure the potential difference U_{AB} between the points A-B on a selected surface, the current filament conductors leading to the voltmeter should be placed in such a way that they are aligned with this current line, and the terminal conductor leads should be bifilarly twisted (Figure 10.23).

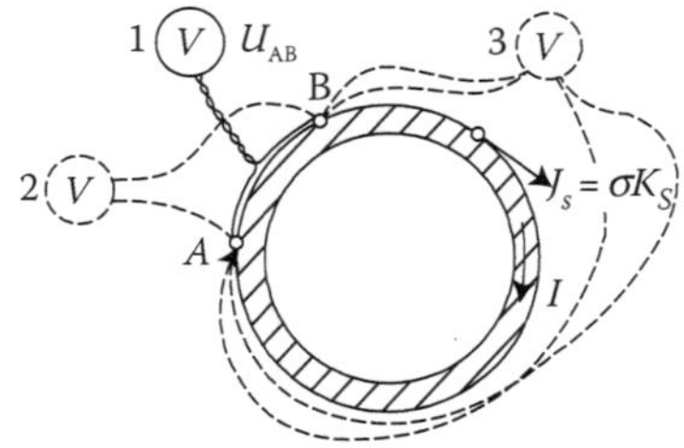

FIGURE 10.23 Measurement of the electric field intensity on the surface of a metal tube or ring: 1—correct positioning of conductors, 2, 3—not correct.

The electric field intensity, the current density, and the power loss density (in W/m³) on the metal surface one can determine from the dependences:

$$\left.\begin{aligned} E_{mx} &= \frac{\sqrt{2}U_{\text{AB}}}{l_{\text{AB}}} \\ J_{mx} &= \sigma E_{mx} = \frac{\sqrt{2}\sigma U_{\text{AB}}}{l_{\text{AB}}} \\ P_{1\text{V}} &= \frac{J_{ms}^2}{2\sigma} = \frac{\sigma U_{\text{AB}}^2}{l_{\text{AB}}^2} \end{aligned}\right\} \tag{10.62}$$

The voltage U_{AB}, measured in a similar way, can also be a basis for determination of the internal impedance of a conductor on its section of length l_{AB}

$$Z_{\text{intern}} = R + \text{j}\, X_{\text{intern}} = \frac{U_{\text{AB}}}{I}$$

or

$$Z_{\text{intern}} = \sqrt{R^2 + X_{\text{int}}^2} = \frac{|U_{\text{AB}}|}{|I|} \tag{10.62a}$$

where R is the resistance, X_{intern} the internal reactance of the section AB of conductor.

For conductors of a profile cross-section, a filament of current should be selected for measuring that is not encircled by flux lines originating from inside the conductor. Such a filament can be recognized by this that it has the biggest current density and is situated on protruded edges of the cross-section profile.

For measurements of the voltage U_{AB} one can use all kinds of compensators of AC currents [1.15], [7.12], however more modern digital voltmeters and oscilloscopes perform similar measurements in a much simpler way, including the distribution of current density along the height of motor deep-slot [10.9].

In the simplest probes (Figure 10.24) for measurement of difference U_{AB} of potentials, the most important principle is to obtain very good contacts in points A and

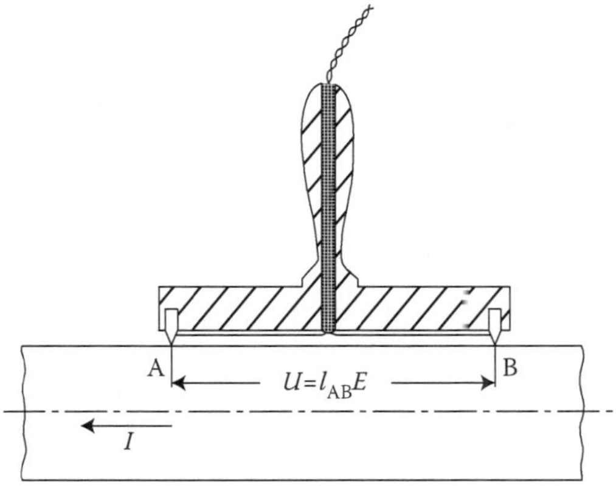

FIGURE 10.24 Probe for measurement of the electric field intensity on metal surface.

B (hard blade) and a good adherence of the horizontal conductors to the investigated current filament. The output terminal conductors should be very well mutually twisted.

10.9.3 Measurement of the Poynting Vector with the Help of Probes

A direct measurement of the Poynting vector on the metal surface, that is, measurement of the energy entering into a conductor volume is in principle possible on the basis of formulae (3.7). For that purpose, we need to add to the probe shown in Figure 10.24 a probe in the form of Rogowski's strap (in a horseshoe core with coil) at an angle of 90°. The span of both probes should be identical. The emf from the blade probe should be connected through a voltage amplifier to the voltage coil of wattmeter, and the emf from the strap probe, after amplification by a current amplifier, should be connected to the current coil of the wattmeter. The wattmeter connected in this way, after corresponding re-scaling, should display the active power per surface unit of metal, that is, the active power component of the Poynting vector.

If the field on the metal surface has more than one of the components H and E in the chosen system, then from the power measured in this system one should subtract, according to Equation 3.7, the power measured after turning the probe by an angle of 90°. A wattmeter acting on the same principle is described in the R. E. Tompkins work (*J. Appl. Phys.* 3, 1958).

10.9.4 Measurement of Power Flux*

The power of losses in some *dissipative* region, for example, in a section of magnetic core, can be determined by integration according to Poynting theorem (3.1) of the normal component of the Poynting vector on the surface of this section. If we initially assume that a section of the core has a cylindrical form (Figure 10.25) and

* Morozov D.N. *Additional Losses (in Russian) Transformatory*. 8th Edition. Moscow, GEI, 1962.

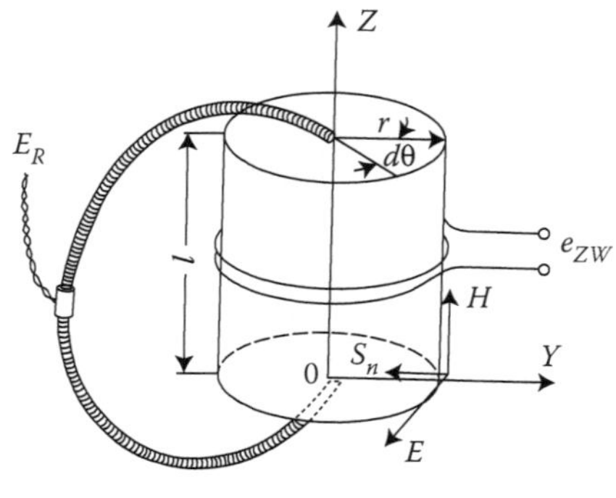

FIGURE 10.25 Integration of the *Poynting* vector $\boldsymbol{S} = \boldsymbol{E} \times \boldsymbol{H}$ on the surface of a core section.

on its surface exist only the axial component of vector H and the circumferential component of vector E, then the power dissipated in the whole volume, at sinusoidal curves of E and H, according to Equation 3.1 will be

$$P = \oint_A \frac{1}{2}(E_m \times H_m^*) n \mathrm{d}A = \frac{1}{2}\int_0^{2n} E_m r \mathrm{d}\theta \int_0^l H_m \mathrm{d}z \quad (10.63)$$

Since the first integral on the right–hand side of Equation 10.63 equals the voltage of one turn encircling the core,

$$e_{\text{turn}} = \frac{1}{\sqrt{2}}\int_0^{2\pi} E_{\text{m}} \cdot r\mathrm{d}\theta,$$

and the second integral equals to the magnetic voltage, $\left(1/\sqrt{2}\right)\int_0^l \boldsymbol{H}_{\text{m}} \cdot \mathrm{d}z = U_m$, corresponding to the section l of the core, therefore

$$P = e_{\text{turn}} U_m \cos \varphi \quad (10.64)$$

where φ is the angle of phase shift between the curves of flow of the magnetic voltage U_m and of the emf of turn e_{turn}. Application of current amplifier to the turn emf e_{turn} and of voltage amplifier to the emf obtained from *Rogowski's* strap (Figure 10.25) can enable using a wattmeter for a direct measurement of power loss in the investigated section of core. The method is valid for an arbitrary shape of core cross-section.

The described measurements can be carried out also with the help of vector meters of various kinds or compensatory systems that measure values of the components of formula (10.64). At nonsinusoidal functions, the measurements should be done separately for every harmonic.

Measurement of the Poynting vector with the help of a Hall effect sensor*. In order to measure the electromagnetic power of a plane wave incident perpendicularly

* Barlow H.L.M. *Proc. IEE*, Pt. B, No 2, 1955.

onto a metal surface, a *Hall sensor* plate should be positioned in such a way that the direction of the control current (along the plate) be identical with the direction of the vector $\boldsymbol{E}$, and value of the current should be proportional to $\boldsymbol{E}$. At the same time, the vector $\boldsymbol{H}$ of magnetic field intensity should be perpendicular to the plane of the sensor. In such an arrangement, the Hall voltage will be proportional to the power passing through the plate.

The Hall probe enables, for instance, investigations of loss distribution in transformer core (Allen [10.8]). For this purpose, the Hall plate should be affixed by its edge to the core surface so that it is penetrated by the tangential component of magnetic field intensity existing on the surface. The output terminals of the control current are connected to the circuit of measuring turns enclosing the part of core adhering to the plate. Since the emf induced in these turns equals $e = \oint \boldsymbol{E} \cdot d\boldsymbol{l}$, hence the induced by it control current of sensor is proportional to the electric field intensity on the surface of core.

Computer-aided digital measurement methods are enabled to obtain significant accuracies of measurements, with a simultaneous processing of results. An analog signal from a *measuring sensor*, for example, hallotron, is converted by an *analog-to-digital converter* (ADC) into a *digital signal of a unified block system*. Such an approach can support, for example, complex measurements of magnetic field $\boldsymbol{H}$ and its gradient, temperature, and other quantities in superconducting magnets (Janowski et al. [10.15]), in stationary and transient states, for example, for rapid detection of the decay of superconductivity in particular elements of a cryomagnet. In this case, the measurement and analysis of results is performed in real time of transient changes of field, temperature, current, velocity of propagation of normal zone, and so on, in the difficult conditions of very low (cryogenic) temperatures, broad range of changes of parameters, and complicated structure of the cryomagnet. At the Technical University of Lublin (Poland) for this purpose a system of hallotron and temperature sensors were used and spaced on a cryomagnet, connected by amplifiers to an analog or digital millivoltmeter. The system for steady-state measurements enabled measurements of the magnetic field with an accuracy of 0.005 T or better. At measurements of transient states, the sensors were connected through an analog-to-digital converter to a microcomputer with output to monitor and plotter. The described system enabled a rapid concurrent recording of many analog values from numerous sensors, their real-time analysis, and graphic display or illustration. The short time of the analog-to-digital conversion (27 μs or less) and the frequency of recording (8–30 kHz) enabled a rapid registration and analysis of all the experimental data. The time interval between consecutive registrations, set up by programming, was within the limits of 100 μs to 100 ms, and the range of input voltages 2 mV to 1000 V. Data from sensors were recorded to hard disc with period between registrations of 100 ms to 10 s, or to the computer memory at 100 μs to 100 ms intervals.

10.10 DIAGNOSTICS OF METAL ELEMENTS

The nondestructive testing of materials and products, essential for studies of reliability of machines and devices, are divided into *radiological*, *ultrasonic*, and *electromagnetic*. The electromagnetic investigations are traditionally divided into *magnetic* and *inductive* (eddy currents). According to [10.2], about 50 million parts a day were investigated in a global scale with the help of magnetic and inductive methods. They can be bars, tubes, balls for ball bearings, aircraft design parts, automobile parts, railway rails, ropes, welds, and so on.

These measurements are divided into: *structurescopy* (methods of structure inspection), *defectoscopy* (discontinuity detection), and methods of *thickness measurements* of layers, walls, and coatings.

Foerster's probe (1939) is a very sensitive probe enabling measurement of stationary fields in the range of 10^{-5}–10^{-3} A/cm. In this probe primary windings of two cores are connected in a series in a push-pull system, and the secondary windings are connected in a series in accordance. A superposition of the measured stationary field on the inductance of both cores causes an appearance of a resultant inductance and an output electromotive force (emf) proportional to the measured field. Turning the probe with regards to the investigated field, one can determine its value and direction [10.2]. Differential probes help eliminate the influence of the terrestrial field from the measurement.

Foerster's probes are used, for instance, to investigate thick-walled tubes (cannon tubes). The measured tube is magnetized by a DC current with the help of an axial conductor introduced into the tube. After switching off the current, the tube is slowly rotated while the internal surface is being checked with the help of Foerster's probe introduced into the tube by a long rod. Results are registered in a self-recorder [10.2].

Hallotrons—Figure 1.13, owing to their small dimensions, are applied to pointwise measurements in places not easily accessible, in the range from about 0.8 A/cm to very high field intensities, for both stationary and alternating fields.

Measuring yokes [10.2], in the form of a horseshoe core (compare Figure 10.15), are used for measurement of coercive force connected with the structure and chemical composition of steel.

Powder method—permits, according to [10.2], to discover with a high efficiency heterogeneity, cracks, inclusions, etc., in ferromagnetic materials. It is one of the most often used in practice methods of defectoscopy. Dry and wet methods are used. The method consists in the initial magnetization of the object and then inspection of a concentration of a poured dry powder or a powder floating in liquid.

Methods of eddy currents are utilized with alternating fields induced with the help of the so-called *contact coils* affixed to the object surface, or *passage coils*, external and internal, applied to tubes, cylinders, and so on. Calculation of fields of these coils and of the field deformations due to defects in the investigated object belongs to difficult, specific jobs of the Technical Electrodynamics [10.2], [1.15].

One of the important tasks of defectoscopy is an automation of the investigation processes and diagnostics.

10.11 CRITICAL DISTANCE OF TANK WALL FROM TRANSFORMER WINDINGS

The state of a power transformer tank wall and the wall distance from windings have essential importance for field configuration in the leakage field region. This configuration has in turn an influence on the distribution and values of short-circuit forces, additional losses in windings from the radial field component and from the stray loss in the tank itself, as well as on a possibility *to measure stray losses* at short circuit tests.

To approach this problem scientifically, the author in 1965 [10.28] introduced the notion of relative *critical distance of a tank wall* from the axis of interwinding gap, called the *critical tank distance.* It is the ratio of the distance (a_T) of tank surface from the axis of the interwinding gap to the distance (a_C) of the core surface from this axis. Above the critical distance, the tank has no influence anymore on the leakage field distribution between the core and tank. It means that the withdrawal in such a case of a transformer from its tank does not remarkably change the flux turning to the core.

This criterion enables them to answer the question, whether and when the withdrawal of a removable part of a transformer from the tank causes a regrouping of leakage flux, force localization, additional losses in windings and tank, and so on. At the tank distance smaller than the critical one, it will be impossible, for instance, to evaluate the power losses in the tank wall with the help of short circuit tests "*with tank and without tank*," as it was proposed before without a proper understanding. *At distances equal or larger than the critical one, this test can be fully reliable.* Since in large transformers the relative distance of tank walls is often near to this border, an investigation of it is absolutely necessary.

An indicative (approximate) division of leakage flux between core and tank, at geometric proportions appearing in large transformers, can be expressed with the formula (4.59), rephrased here:

$$\frac{\Phi_C}{\Phi_T} \approx \frac{a_T}{a_C} \quad \text{or} \quad \frac{\Phi_T}{\Phi_\delta} = \frac{a_C}{a_C - a_T} \tag{4.59'}$$

If we remove one of the iron walls surrounding both sides the winding (e.g., the tank), the component Φ_T of the gap leakage flux Φ_δ ($\Phi_\delta = \Phi_T + \Phi_C$) will not disappear entirely, but will be substituted by the component $\Phi_{\text{air}} = \text{const}$ closing the loop through the remaining free air space. The reluctance (magnetic resistance) of this free space is constant and due to its large "cross-section"—small.

Hence, the reluctance of the free air space can be substituted by a reluctance in the direction of *fictitious iron surface* F_{air} placed in the distance a_{air} from the axis of the interwinding gap (Figure 10.26). Such an arrangement will again satisfy the dependence (4.59), hence

$$\Phi_\delta \approx \Phi_C + \Phi_{\text{air}}, \quad \left|\frac{\Phi_C}{\Phi_\delta}\right| = \frac{a_{\text{air}}}{a_C + a_{\text{air}}}$$

$$\left(\frac{a_T}{a_C}\right)_{\text{CR}} = \frac{a_{\text{air}}}{a_C} \approx \frac{\left|\dfrac{\Phi_C}{\Phi_\delta}\right|}{1 - \left|\dfrac{\Phi_C}{\Phi_\delta}\right|} \tag{10.65}$$

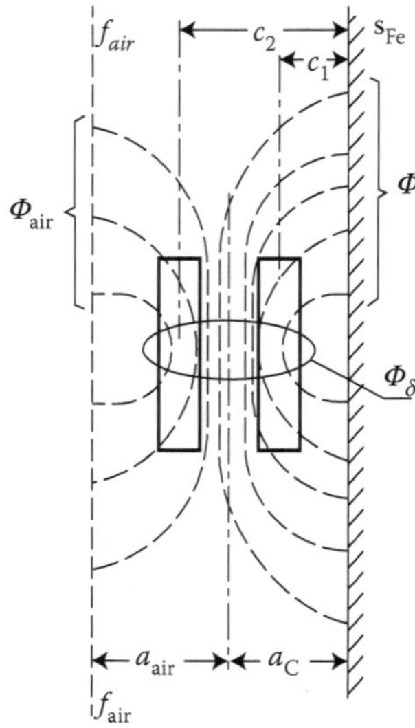

FIGURE 10.26 Determination of the critical distance $a_{air} = a_{CR}$ of steel wall; F_{air}—a fictitious iron surface which gives the same resultant reluctance as the unlimited air space; S_{Fe}—the iron core surface.

This distance a_{air}, equivalent with respect to reluctance to the iron surface F_{air} (Figure 10.26), is now called the *critical distance* a_{CR}. In normal power transformers, the tank distance $a_T \le a_{CR}$ is still in the scope of the approximate dependence (4.59). At $a_T > a_{CR}$, the flux Φ_T remains practically constant and equal Φ_{air}. Hence, also the component Φ_C turning to the core does not change.

The right-hand side of Equation 10.65 can be calculated with computer methods such as FEM or RNM (Figures 10.31 and 10.32 [later in the chapter]), but as an approximate assessment one can use the simple analytical formulae derived in the book (J. Turowski [1.15/1], p. 343):

$$\left(\frac{a_T}{a_C}\right)_{CR} = \frac{\eta\left|\ln\frac{1+(2c_1/h)^2}{1+(2c_2/h)^2} + 2\left(\frac{2c_1}{h}\alpha_1 - \frac{2c_2}{h}\alpha_2\right)\right|}{2\pi\frac{\delta'}{h_R} - \eta\left|\ln\frac{1+(2c_1/h)^2}{1+(2c_2/h)^2} + 2\left(\frac{2c_1}{h}\alpha_1 - \frac{2c_2}{h}\alpha_2\right)\right|} \tag{10.66}$$

where, after (Figure 5.5), η = 0.7 to 0.8; α_1 = arctg $h/2c_1$ and α_2 = arctg $h/2c_2$—angles resulting from Figure 10.27.

In large transformers, the values of $h/2c_1$ and $h/2c_2$ are in the range of 0.1–0.2. We can therefore, at the assumption that $(2c_1/h)^2 \ll 1$ and $(2c_2/h)^2 \ll 1$, neglect in Equation 10.66 the logarithmic components, and hence we get approximately [1.15]

$$\left(\frac{a_T}{a_C}\right)_{CR} \approx \frac{\pi}{8(a_C/h)} - 1 \tag{10.67}$$

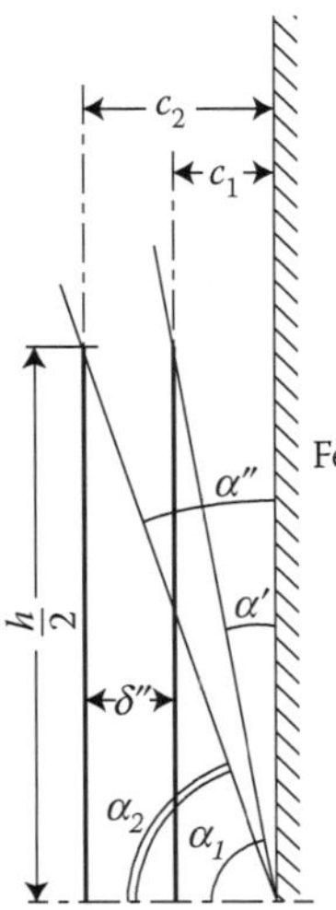

FIGURE 10.27 Geometric proportions of linearly concentrated ampere-turns of transformer, nearby a solid iron wall.

In large power transformers, the ratio a_C/h usually does not exceed the limits of 0.08–0.1. One can, therefore, in view of the approximate character of these considerations, adopt in average that the *relative critical tank distance* $(a_T/a_C)_{Cr}$, *above which the tank has no influence on the leakage flux distribution between the core and the tank surface, is about* 3–3.5.

At more accurate calculations, one should use the full formula (10.66), especially at higher values of the ratios $2c_1/h$ and $2c_2/h$, and take into account the Rogowski's coefficient $k_R < 1$, as well as $h_R = h/k_R > h$.

An experimental verification and application of a_{cr} as a criterion of selection of measurement method of power losses in tank are given in the work [10.16].

The above considerations concern non-screened tanks. It is necessary, therefore, to investigate the influence of a screens.

10.11.1 Critical Distance of Electromagnetically Screened Tank Walls

The author (JT) defined and derived corresponding formulae in 1969 [1.15/1], which next were confirmed experimentally [10.22], on a model transformer, according to Figure 10.3.* Due to the dominating role of the gap reluctance, $R_\delta = h_R/(\mu_0\delta')$ (Figure 4.18), the total leakage flux is almost constant, $\Phi_{m\delta} \approx$ const, independently on changes of a_C and a_T. Hence,

$$\frac{\Phi_{T,\text{scr}}}{\Phi_{C,\text{scr}}} \approx \frac{R_C}{R_{T,\text{scr}}}, \quad R_C \approx R_{C,\text{scr}} \approx \frac{2a_{C,\text{scr}}}{\mu_0\beta h} \tag{10.68}$$

* Turowski J., Sykulski J., and Sykulska E.: Evaluation of critical distance of a screened tank from windings (in Polish). Internal Report of ITMAE, Tech. Univ. of Lodz, No. 52, 1974.

where $\beta = 0.6–0.8$ is the coefficient of the cross-section of the flux path $\Phi_{m,C,\mathrm{scr}}$ beyond the gap δ $\Phi_{T,\mathrm{scr}}$, $\Phi_{C,\mathrm{scr}}$ is the leakage flux components beyond gap, at a screened tank, turning to the tank and core, respectively, R_C and $R_{T,\mathrm{scr}}$—reluctances for these flux components, $a_{C,\mathrm{scr}}$—the tank distance at a screened tank.

Assuming that a Cu screen entirely repulses the leakage flux from the tank wall and forces it to pass through the gap c_T (Figure 10.28), we can write

$$R_{T,\mathrm{scr}} \approx \frac{h + a_{T,\mathrm{scr}}}{\mu_0 c_T} \quad \text{and} \quad \frac{\Phi_{C,\mathrm{scr}}}{\Phi_\delta} \approx \frac{R_{T,\mathrm{scr}}}{R_C + R_{T,\mathrm{scr}}} \tag{10.69}$$

The new critical distance of a (screened) tank $(a_{T,\mathrm{scr}}/a_{C,\mathrm{scr}})_{\mathrm{CR}}$ can be evaluated by searching for such a special relation

$$\frac{a_{\mathrm{T,scr}}}{a_{\mathrm{C,scr}}} = \left(\frac{a_{\mathrm{T,scr}}}{a_{\mathrm{C,scr}}}\right)_{\mathrm{CR}} = x \tag{10.70}$$

at $a_{T,\mathrm{scr}} = a_C$, for which flux $\Phi_{m,C,\mathrm{scr}}$ turning to the core will be after screening the same as the flux Φ_C existing in the non-screened transformer at a distance not smaller than the critical one (10.67). It means that considering (4.59) and (10.67) we get:

$$\frac{\Phi_{C,\mathrm{scr}}}{\Phi_\delta} \approx \frac{\Phi_C}{\Phi_\delta} = \frac{1}{1 + \left(\dfrac{a_T}{a_C}\right)_{\mathrm{CR}}} = \frac{8}{\pi}\frac{a_T}{h} \tag{10.71}$$

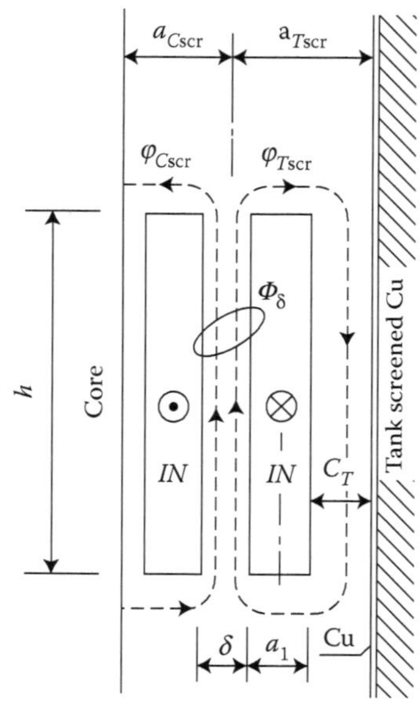

FIGURE 10.28 Distribution of the leakage flux Φ_δ between core and tank wall screened with Cu screens.

After equalization of (10.69) and (10.71), we obtain an equation in the form $\frac{a+x}{b+cx} = d$, from which: $x = \frac{a-bd}{cd-1}$ (Savini, Turowski [10.22]), that is,

$$\left(\frac{a_{T,\mathrm{scr}}}{a_C}\right)_{\mathrm{CR}} \approx \frac{\frac{h}{a_C} - \left(\frac{h}{a_C} - \frac{a_1+\delta}{\beta h}\right)\frac{8}{\pi}\frac{a_C}{h}}{\left(1+\frac{2a_C}{\beta h}\right)\frac{8}{\pi}\frac{a_C}{h} - 1} \tag{10.72}$$

In large transformers, where $\frac{h}{a_C} \gg \frac{a_1+\delta}{\beta h}$ and $\frac{2c_C}{\beta h} \ll 1$, Equation 10.72 can be simplified to the form

$$\left(\frac{a_{T,\mathrm{scr}}}{a_C}\right)_{\mathrm{CR}} \approx \frac{h}{a_C} \tag{10.73}$$

It means that at the ratio

$$\frac{a_{T,\mathrm{scr}}}{a_C} > \frac{h}{a_C} \tag{10.74}$$

the presence of an *electromagnetically screened tank wall has no influence on the flux* Φ_C turning toward the core (Figure 10.28). In large transformers, the ratio $h/a_C = 10$ to 12, which gives $(a_{T,\mathrm{scr}}/a_C)_{\mathrm{CR}}$ also equal 10–12. Since in practice $a_{T,\mathrm{scr}}/a_C$ is in the range of 1–5, we can conclude that *screening of tank walls always has an influence on the change of the stray flux distribution.* A result of superimposition of Cu screens on a tank wall is an increase of the stray flux component Φ_C directed to the core, and reduction of the stray flux component Φ_T directed to the tank walls (Figure 4.17b).

The above considerations concern a two-dimensional (2-D) model. Experimental verification confirmed (Figures 10.29, 10.30) the above conclusions (10.73). However, in order to perform a full investigation of a real transformer, one should resolve a three-dimensional (3-D), three-phase model (Figure 7.20) and investigate the influence of yokes, yoke beams, and so on.

In the work (Savini and Turowski [10.22]), on the basis of the computer program RNM-3 (Turowski, Kopec [4.27]) (Figures 4.21 and 4.22), the leakage flux distribution in more complex structures was calculated (Figure 10.30) and compared with similar calculations done jointly with Professor A. Savini at the University of Pavia (Italy) with the help of the finite element method FEM-2D. From the curve A in Figure 10.30 one can see that the repellent reaction of eddy-currents induced in the steel wall is balanced by an attraction of flux, which keeps the flux Φ_C almost constant. This cannot be said for systems with Cu/Al screens or Fe shunts, which can be clearly seen in the RNM solution (Figure 10.31) where a presence of ideal screens was assumed.

The influence of yoke beams can be evaluated on the principle of mirror images (Figure 10.32). Until recently, it was thought that this influence was not big. However,

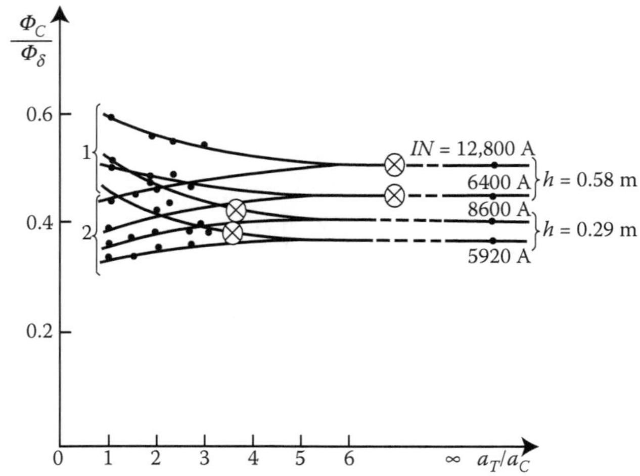

FIGURE 10.29 Experimental verification of simplified formula (10.73) for critical distance of a tank wall screened by copper: 1—with a copper (Cu) screen, 2—steel without any screen, ⊗– $(a_{T,\text{scr}}/a_C)_{\text{Cr}}$ calculated with (10.73) (*IN*, *h*—per Figure 10.28, measurements per Figure 10.3).

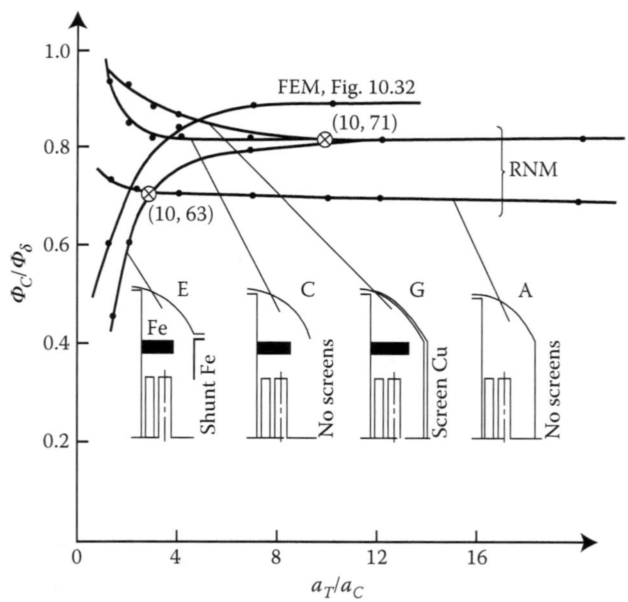

FIGURE 10.30 Influence of tank screening and yoke beams on the leakage flux distribution in a 240 MVA transformer. (Adapted from Savini A. and Turowski J.: Chapter 3.6 in the book "*Electromagnetic Fields in Electrical Engineering*". New York: Plenum Press, 1988, pp. 119–127 [10.22].) ⊗—the "critical" distance of tank.

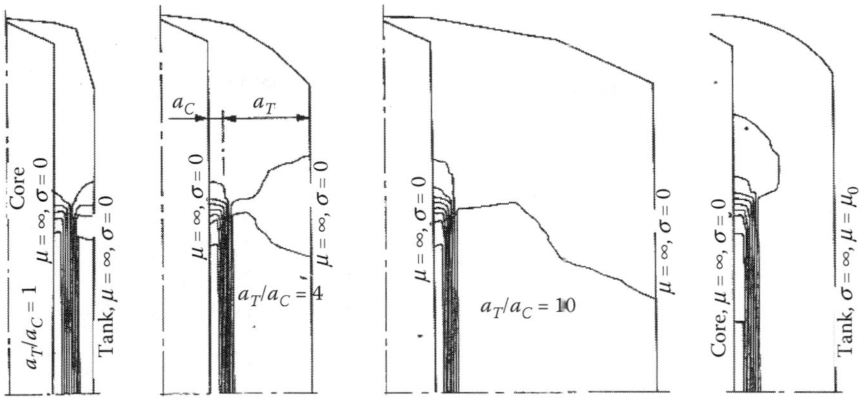

FIGURE 10.31 FEM-2D calculations of the effect of ideal screening of tank walls and the relative distances a_T/a_C on the stray flux distribution in a large transformer. (Adapted from Savini A. and Turowski J.: Chapter 3.6 in the book "*Electromagnetic Fields in Electrical Engineering*". New York: Plenum Press, 1988, pp. 119–127 [10.22].)

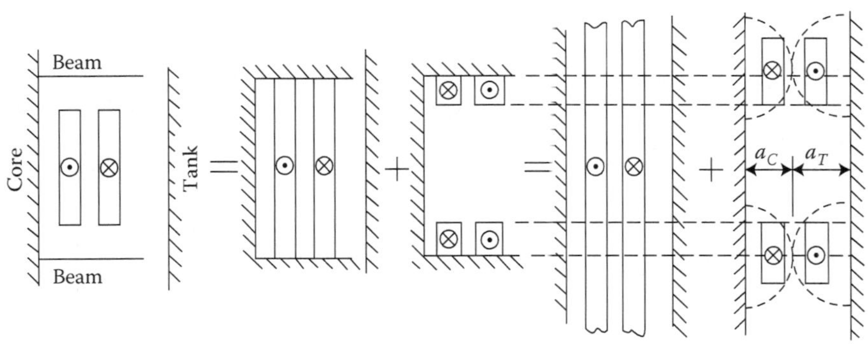

FIGURE 10.32 Assessment of the influence of yoke beams on distribution of leakage field; Beam = yoke beams.

contradictory opinions from large transformer works in England and Ukraine led J. Turowski to the conclusion that the source of discrepancy is the difference in the values of the upper voltage and, thus, the insulating clearance between the beams and the winding ends, and the problem must be investigated in 3D. Only a 3D examination of the role of flux collectors demonstrated that the issue was important, and the work [4.12] by J. Turowski et al. explained it evidently.

10.12 INFLUENCE OF FLUX COLLECTORS

For a long time, the role of the so-called top and bottom flux collectors (Figure 4.18a—e_2, Figure 7.19b—3) and their influence on stray losses and electrodynamic forces in transformers has been an object of conflicting opinions. In some publications it has been suggested [10.36] that such collectors—in the form of laminated shunts—will straighten and redirect the stray magnetic lines towards the collectors as a three-phase magnetic node. There, they vanish according to the rule $\Phi_U + \Phi_V + \Phi_W = 0$.

Although theoretically correct, this effect has been sometimes negated by other specialists. The answer lies in the different features of transformers, depending on the insulation clearances, "critical distances" of tank walls (Figure 10.26) [10.37], screening systems, etc. To give a full, quantitative answer to these conflicting opinions, a detailed 3-D analysis was carried out [4.12] with the help of the rapid RNM-3Dexe package. In the RNM models, the whole sophisticated theory, a complicated geometry and many physical phenomena (magnetic nonlinearity, solid iron electromagnetic processes, skin effects, eddy current reaction, electromagnetic and magnetic screening, laminated iron shunt effects, etc.) are hidden within the analytical representation of reluctances and special composition of the program source code. Thanks to this, the hybrid RNM-3D programs are very simple and rapid when performing such 3-D analysis. The answer was presented using a graph and it was confirmed that **there is no controversy**. The effect of flux collectors is clear and depends on their clearance h from the winding edges. For instance, for a 240 MVA transformer, at the collector clearances h of 0, 116, 232, and 464 mm, the power loss in the tank wall ΔP was 0, 6.9, 78, and 307 kW, respectively. The zero clearance will, of course, never occur due to insulation requirements, but it was helpful for interpolation purposes. The ΔP values depend additionally on the whole 3-D configuration of screens and may vary. Nevertheless, generally **the dependence $\Delta P = f(h)$ is almost linear.**

11 Conclusion

This book is mainly aimed to deliver theoretical and expert knowledge base (Figure 1.1c) for expert systems and tools for knowledge engineers to create adequate models of electromagnetic and electromechanical high-power equipment, suitable for rapid design. Another important component of this methodology is the implementation of modern methods of Mechatronics and the Hamilton's Principle and Euler–Lagrange equation (8.6) [1.20] to the study of the motion and dynamics of machines. This knowledge is inseparably linked with the applied electromagnetics theory and practice, electromechanical and other theories, the use of various construction materials, as well as heating processes and methods of tests and measurements that determine the reliability of objects and their accompanying energy systems.

There were many engineering problems discussed and resolved, from small, complicated models, proposed by international group of specialists grouped in the "TEAM Workshop" (Figure 11.1), to large machines (Sections 4.7, 8.4) and transformers (Table 11.2 [later in the chapter]).

Further investigations of the main field and leakage fields and their consequences have been carried out on three-dimensional models (Figures 4.20, 4.21, 7.22–7.24, etc.), which have already been implemented in Transformer Works ELTA (now ABB) in Lodz, Poland, Crompton Greaves in Mumbai, India, and more than 44 transformer works and universities all over the world (see Table 11.2 [later in the chapter]), from the United States, Canada, Europe, Asia, and Australia. These methods have gained an international recognition, among others, at the IMACS'88 Congress in Paris [2.34], 3DMAG Symposia in Japan [11.1], ISEF'85 [9.9], ISEF'87 [8.13], ISEF'89 [4.28], and so on, through ISEF'2009 [11.4], ISEF'2013 [1.53], and others. New enhancements were recently implemented in the new versions of the programs "RNM-3Dshell," successfully designed and industrially implemented by the author together with Vigo University, Spain (Prof. X. M. Lopez-Fernandez), EFACEC Transformer Works in Porto, Portugal, and Savannah, Georgia, USA [4.31], [1.8], as well as the WEG Company Brazil, and others.

RNM-3D was applied to resolve rapidly and successfully a group of the "TEAM Workshop Models"—Nos. 7, 10, and 13. Figure 11.1 presents a comparative solution of the Problem No. 7, which shows a predominant effectiveness of the RNM-3D approach over the FEM-3D. The FEM-3D solution, in Figure 11.1a, was made in cooperation with Japanese partners [2.25] using a Japanese supercomputer NEC SX-1E [11.3]. The same problem, in Figure 11.1b, was resolved on a PC using the Reluctance Network Method RNM-3D (Wiak, Pelikant, Turowski [11.2], [11.3], [11.4]), in a much easier, faster, and greatly cheaper manner.

Another proof, even more impressive than in Figure 11.1, of application of RNM-3D and mechatronic approach effectiveness, are the conclusions shown in Tables 5.1 and Table 11.1, which show criteria of best selection among five most popular methods and programs.

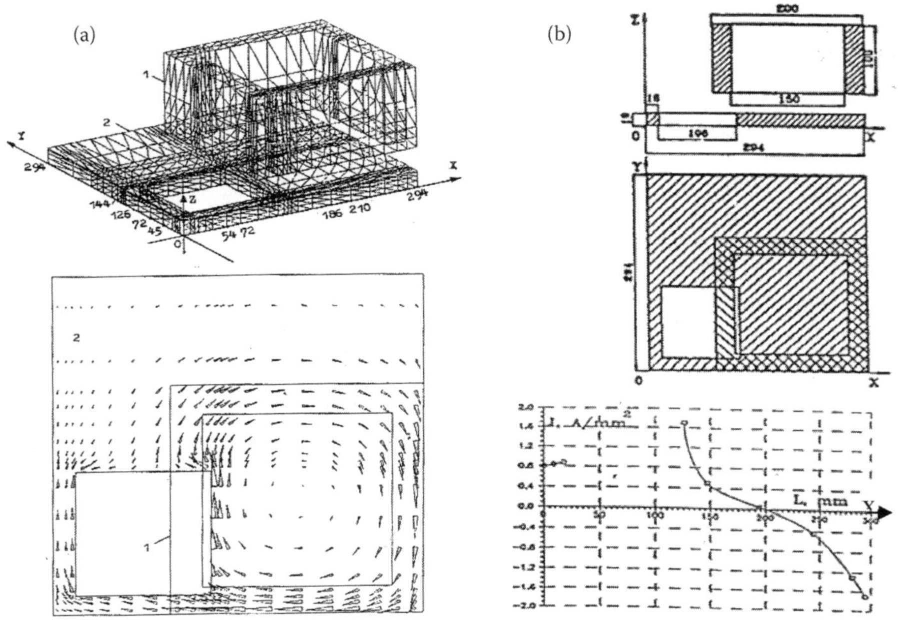

FIGURE 11.1 Comparative solutions of the Problem No. 7 in the international project "TEAM Workshop," with two different numerical methods and computer programs: (a) FEM-3D and the NEC SX Supercomputer-1E, the number of nodes: 11832, CPU time: 499–3903 sec. (Olszewski P. and Fujiwara K.: Results of test Problem 7. *Proc. of International Symposium and TEAM Workshop on 3-D Electromagnetic Field Analysis (3DMAG)*, September 11–13, 1989, Okayama, Japan, Paper IV-17, pp. 55–56.); FEM-3D grid (top) and distribution of the currents (bottom) induced on the upper surface of the metal plate—difficult and labor-intensive job; (b) RNM-3D [11.3]: a scheme of the model (top) and the current density distribution on the plate (bottom), along the *X*-axis—a simple, quick, and easy solution.

In many practical applications, the equivalent reluctance network method (RNM) approach has proven to be the simplest and the most user-friendly method. It is the clearest method for busy engineers, because they can very easily grasp the analogy between the water flow in pipes, the current flow in electric circuits, and the magnetic flux flow in magnetic circuits. The fundamental circuit-theory laws, such as Ampere's (law of flow), Ohm's and Kirchhoff's laws, as well as their magnetic analogies, can easily be used to analyze such circuits, as they are well taught in secondary schools. However, the secret of success is hidden in sophisticated pre-calculation of parameters, which nevertheless is mostly not required from the RNM-3D regular users.

It is important to follow the principles of mechatronics [1.20], presented briefly in Section 1.1. Especially important is the point "(13) Employment of Expert systems," which means, to apply, wherever possible, all the simplified, synthetic formulae and analytic approaches created by past generations of specialists. A recent spectacular confirmation of effectiveness of such approach are the fundamental works [1.53],

TABLE 11.1
Criteria and Comparison of Cost, Usefulness, and Simplicity of Different Approximate Numerical Methods for Fast 3D Interactive Modeling, Computation, and Design of Electromagnetic Fields, Forces, and Losses in Electromechanical Converters, Reactors, and Transformers

No.	Criteria	Method	Specifics: Low — Medium — High
1	Degree of elaboration in literature	FDM	==============
		FEM	==3D===q3D==============2D
		BEM	=====
		RNM	========
		ANM	=============
2	Cost of hardware and 3D software	FDM	=================
		FEM	====================
		BEM	========================
		RNM	=====
		ANM	========
3	Cost and duration of a complex 3D computation of one design variant	FDM	====================
		FEM	========================
		BEM	========================
		RNM	====
		ANM	=======
4	Cost and time of training of user	FDM	===================
		FEM	======================
		BEM	======================
		RNM	====
		ANM	===================
5	Cost, difficulty, and time to fulfill 3D convergent conditions	FDM	=====================
		FEM	======================
		BEM	======================
		RNM	====
		ANM	========
6	Cost and time to fulfill boundary conditions	FDM	====================
		FEM	======================
		BEM	======================
		RNM	====
		ANM	=====
7	Stability of solution	FDM	====
		FEM	====
		BEM	=====
		RNM	========================
		ANM	============

continued

TABLE 11.1 (continued)
Criteria and Comparison of Cost, Usefulness, and Simplicity of Different Approximate Numerical Methods for fast 3D Interactive Modeling, Computation, and Design of Electromagnetic Fields, Forces, and Losses in Electromechanical Converters, Reactors, and Transformers

No.	Criteria	Method	Specifics: Low	Medium	High
8	Errors of approximation and rounding of numbers relative to computation time	FDM			
		FEM			
		BEM			
		RNM			
		ANM			
9	Self-generation of errors and discontinuities (e.g., as differentiation effect)	FDM			
		FEM			
		BEM			
		RNM			
		ANM			
10	Cost and time of solution of internal 3D problems in nonlinear, conducting media, considering local heating	FDM			
		FEM			
		BEM			
		RNM			
		ANM			
11	Cost and time of solution of external 3D problems in nonlinear, conducting media	FDM			
		FEM			
		BEM			
		RNM			
		ANM			
12	Cost and duration of computation of 2D and quasi-3D models	FDM			
		FEM			
		BEM			
		RNM			
		ANM			
13	Possibility of hybrid cooperation with other methods	FDM			
		FEM			
		BEM			
		RNM			
		ANM			
14	Quality of industrially confirmed verifications	FDM			
		FEM			
		BEM			
		RNM			
		ANM			

TABLE 11.1 (continued)
Criteria and Comparison of Cost, Usefulness, and Simplicity of Different Approximate Numerical Methods for fast 3D Interactive Modeling, Computation, and Design of Electromagnetic Fields, Forces, and Losses in Electromechanical Converters, Reactors, and Transformers

No.	Criteria	Method	Specifics: Low	Specifics: Medium	Specifics: High
15	Usefulness of output data for direct, fast, interactive industrial design	FDM			
		FEM			
		BEM			
		RNM			
		ANM			
16	Implementation and sales of applicable programs (Reference list)	FDM			
		FEM	3D	quasi 3D	2D
		BEM			
		RNM			
		ANM			

Source: Adapted from Turowski J.: *Fundamentals of Mechatronics* (in Polish). AHE-Lodz, 2008.

[4.31], and others, carried out for big and modern Transformer Works: EFACEC in Portugal (Porto) and in Georgia, USA, WEG (Blumenau—Brazil) in cooperation with specialists from Vigo University (Spain) and Design Office of EFACEC.

This appreciation of the RNM-3D approach is by no means a tendency to reject, at least for power specializations, the other methods mentioned in Table 11.1, especially FEM-3D. Every of the 15 basic methods listed in Section 2.2, and many more of their variants, modifications, and *hybrid* combinations, are nothing but one of the tools which should be chosen (Table 11.1) as most convenient for the given task or model. For example, for covers (Section 5.3), the "old" Ampere's and Biot–Savart laws have proven to be the best [4.13], [7.16], [9.12], for Shell-type transformers—RNM-3Dshell with JAVA Solver [4.31], for antenna and external fields—the BEM, and so on.

The secret of the RNM-3D success is that these programs are based on an *Expert System*, with deep scientific fundamentals of a *Knowledge Base* [1.20]. The interactive approach (Figure 4.20c) is supported by *Machine Intelligence* which is hidden inside the Source Code. Thanks to it, the *user can operate from outside with simple tools, in the form of Ohm* and *Kirchhoff* laws, *similarly like a sophisticated, intelligent automobile can be controlled from outside by a driver using a simple, traditional steering wheel.*

And last, but not least, the hard industrial verification (Table 11.2), with the enthusiastic opinions of users, are the best proof of the RNM-3D tremendous industrial effectiveness.

TABLE 11.2
List of Selected Users of the J. Turowski's Methods and Programs of the RNM-3D Class

1	Transformer Works in Żychlin, Poland (Power loss and overheating of the cover. JT PhD thesis) 1957
2	Transformer Works MTZ in Moscow, Russia 1960. Hot-Spot in remote transformers 160 MVA in Siberia
3	Transformer Works *SKODA-Plzen*. Czech Rp. 1969. *J. Kopecek, J. Klesa. Industrial verification.*
4	Transformer Works *ELTA*, Lodz, Poland (1987–1988)
5	Transformer Works *SHENYANG*, China (1991–1992)
6	Transformer Works *GEC-ALSTOM* OF INDIA, Allahabad, India (1991–1992)
7	University of Sydney, Department of Electrical Engineering, Australia (1992)
8	Transformer Works *BAODING*, China (1993), and other Universities and Works
9	Transformer Works *CROMPTON GREAVES*, Mumbai, India (1993)
10	Repair Works of Power Board *ENERGOSERWIS, Lubliniec*, Poland (1993)
11	Transformer Works NGEF-*ABB*, Bangalore, India (1994)
12	Transformer Works *WILSON TRANSFORMER*, Australia (1994)
13	Transformer Works CROMPTON GREAVES, Mumbai, India (1994). Source Code
14	Transformer Works *WILSON TRANSFORMER*, Australia (1994). Testing calculations
15	Transformer Works *NORTH AMERICAN TRANSFORMERS*, Milpitas CA, USA,1995 r.
16	Transformer Works *IRAN-TRANSFO (SIEMENS)*, Teheran (1995).
17	Transformer Works HACKBRIDGE-HEWITTIC AND EASUN, Chennai, India (1995)
18	Monash University, Department of Electrical Engineering, Melbourne, Australia (1995)
19	Transformer Works TRANSFORMERS AND ELECTRICALS KERALA (*HITACHI*), Kerala, India (1996)
20	Transformer Works BHARAT HEAVY ELECTRICALS, Bhopal, India (1996)
21	Transformer Works CROMPTON GREAVES, Mumbai Program "Bolt heatings" 1996 r.
22	Transformer Works CROMPTON GREAVES, Mumbai Program "Clampings" 1998
23	Transformer Works CROMPTON GREAVES, Mumbai Program "Turret"
24	Transformer Works CROMPTON GREAVES, Mumbai Program RNM-3Dasm 1999
25	Transformer Works *EMCO* TRANSFORMERS LIMITED, Mumbai, India, 1997
26	Power Institute, Warszawa-Mory, Dr. J. Kulikowski
27	*University of Palermo, Italy*, Department of Electrical Engineering, Italy, 1998
28	*University of Pavia, Italy*, Department of Electrical Engineering, Italy, 1998
29	Repair Works of Power Board ZREW S.A., Janów, Poland (1998)
30	*Pouwels Trafo Belgium*, Mechelen (1999). RNM-3Dexe
31	Ingenieria y Desarrolo Transformadores de Potencia, Tlalnepantla, Mexico. Testing (2000)
32	Delta Star Inc. San Carlos, CA. USA. Testing (2000)
33	Transformer Works *ABB*-ELTA Lodz (2000)
34	*Pouwels Trafo Belgium,* Mechelen (2000). RNM-3Dasm, Source Code
35	GE-PROLEC. Power Division. Transformer Manufacturer. Monterrey. Mexico. Testing (2000)
36	Delta Star Inc. San Carlos, CA. USA. Package RNM-3Dexe (2001)
37	*ABB* Lodz. Counting over more than 30 power transformer units from 25 to 330 MVA (I. Kraj- 2001)
38	*ABB* Lodz Development of automatization of stray loss modeling and calculation (M. Swiatkowski-2002)

TABLE 11.2 (continued)
List of Selected Users of the J. Turowski's Methods and Programs of the RNM-3D Class

39	Power Transformers Division *ABB* Inc. 1600 Lionel-Boulet Blvd. Varennes, QC CANADA, J3X 1S4 Power loss and overheating hazard in high current bushing turrets in 440 MVA 1phase trafo. July 2003
40	University of Vigo – Spain
41	Transformers Works *EFACEC* Porto, Portugal 2006
42	Technical University of Lodz, Poland—for numerous PhD and MSc theses. Since 1950 till now
43	Transformers Works *EFACEC* Savannah, Georgia, USA 2009
44	*WEG* Equipamentos Elétricos S.A. Brazil 2010; RNM-3Dexe

Source: Adapted from Turowski J.: Electromagnetic field and losses in the transformer housing. (in Polish). *"Elektryka" Science Papers*, Technical University of Lodz., No. 3, 1957, pp. 73–63; Turowski J.: Electromagnetic field and power losses in covers of transformers. PhD thesis. Technical University of Lodz, Poland. December 1957, pp. 1–142. (Published in parts in [1.15/1], [2.31], [4.16], [5.15], and [6.15].)

Here are some users' and experts' opinions (the author (JT) has their names, positions, and addresses):

User 1: "… We are using your RNM-3D package extensively and for most of the jobs the computed stray loss figure is closely matching the tested value. The package is very much helping us to understand the dependence of stray losses on various parameters of transformer design and shunt/screen arrangement." (India)

User 2: "… Dear Prof. J. Turowski, … we received the modified version of the computer program, which gave us good estimation of stray losses, and I also simulated provision of shunts on tank, clamp, and got good results …" (USA)

User 3: "… Dear Prof. J. Turowski, … The method that you use for studying the leakage field appeals to me greatly and you have now fashioned it into a reliable engineering tool. I am convinced that this method will be widely used in industry and will become the preferred method of calculation…." (UK)

User 4: "… Monsieur le Professor Turowski … Cher Monsieur, Votre notice sur le logiciel RNM-3D m'est bien parvenue. Toutes mes félicitations pour le beau succes du travail que vous lui avez consacré." (Paris, France)

User 5: "… I had used RNM-3D very extensively and found it very useful and reliable …" (Australia. From an e-mail to Prof. J. Turowski, dated February 8, 1999).

User 6: "Dear Professor Turowski, My name is Maria E. M., I worked with Prof. S. and Prof. B. on the paper about screen design. I would like to tell you that your program is very useful and complete and I liked very much to work with it. I am very grateful for the program that you gave us and the opportunity to work with it."

User 7: "Respected Prof. Turowski, ... the Reluctance Network Method can fulfill the requirements of very fast estimation and control of the tank stray loss" (Mumbai, India)

Some impressions, especially those in Table 11.1, may be considered to be more or less subjective. However, as one can see, the conclusions in this chapter are based on J. Turowski's extensive professional experience, profound scientific basis, numerous researches and design cases, observations of publications and conferences throughout many years, as well as on many experiments and broad industrial implementations in over 50 major transformer works and research institutions around the world. All of this has been enriched by the experience, research and numerous publications of M. Turowski in the field of computer science, numerical methods, and physics phenomena, both in the power electronics and modern nanotechnology as well as integrated circuits.

This book is the result of many years of harmonious cooperation between both authors, with utilization of their vast industrial and academic experience.

11.1 FINAL COMPLEX EXAMPLE

Material gathered in this book provides several more or less complex tools to relatively simple engineering solutions based on theoretical fundamentals of electrodynamics, electro-mechanics, and mechatronics [1.20]. These tools are based on many years of scientific, educational, and industrial experience of the authors and their collaborators. But the real and very difficult challenge arises when you need to explain dynamic processes in complex multidisciplinary incidents that occasionally occur in the contemporary world of modern technology. It is good if a serious threat can be resolved, like in the recent example that occurred with perfect plane "Dreamliner," where the producer managed to remove the failure. But in recent years there have been many accidents with a tragic final. For example, there was an air crash [11.5] that was among others thoroughly examined by the methods of engineering electrodynamics. It was a most complicated combination of all dynamic processes, and at the same time, many various physical phenomena. Above it, these processes are often influenced decisively by an automatic and manual control. These phenomena are carefully examined by experts of all specializations, both theoretical and experimental [11.6].

Appendix

A.1 BASIC QUANTITIES, UNITS, AND PHYSICAL CONSTANTS

TABLE A.1
Units of Measurement of Selected Quantities, in SI System (Legally Approved)

Quantity	Symbol of the Quantity	SI Units of Measure, and Relation between Them	Units of Measure Temporarily Admitted as Legal, and Relationships between Them
		Basic Units of Measure	
Length	l, L	m (meter)	1 μm = 10^{-6} m, 1 Å = 10^{-10} m
Mass	m	kg (kilogram)	1t = 1000 kg
Time	t, τ	s (second)	1 min; 24 h (day) = 86,400 s, (1 year ≈ 8760 h)
Electric current	i, I	A (ampere)	
Temperature	T, t	K (kelvin)	1°C = 1 K; 0 K = −273.15°C
Luminous intensity (power)	I	cd (candela)	
		Complementary Units of Measure	
Plane angle	α, β, γ	rad (radian)	1° = $2\pi/360$ rad 1 revolution = 2π rad
Solid angle	ω	sr (steradian)	1° solid round = 1 sr = 4π
		Derived Units of Measure (Selected)	
Temperature difference	$\Delta T. \Delta t, \theta$	1 K	1°C = 1 K
Heat, energy	Q	1 J = 1 N m	1 cal = 4.1868 J; 1 J = 2.38884×10^{-4} kcal
Frequency	f	1 Hz = 1/s	
Linear speed	v	m/s	1 km/h = 1/3.6 m/s
Angular speed	$\omega, n\ (\Omega)$	rad/s	1 rev/s = 2π rad/s 1 rev/min = $2\pi/60$ rad/s
Mass density	ρ	kg/m^3	1 g/cm^3 = 1 kg/dm^3 = 10^3 kg/m^3
Force	$\boldsymbol{F} = m\boldsymbol{a}$	1 N = 1 kg m/s^2 = Ws/m	1 dyne = 10^{-5} N (1 kG = 1 kp = 9.80665 N)
Moment of a force	$\boldsymbol{M} = \boldsymbol{F} \times \boldsymbol{l}$	1 N m = 1 kg · m^2/s^2	1 kGm = 9.80665 N m
Pressure, stress	p	1 Pa = 1 N/m^2	1 bar = 10^5 N/m^2 1 kG/cm^2 = 1 at (technical) = 98066.5 N/m^2; 1 kG/m^2 = 1 mm H_2O = 1 Tr (tor)
Work, energy	W, A	1 J = 1 N m = 1 Ws	1 Wh = 3600 J; 1 kGm = 9.80665 J

continued

TABLE A.1 (continued)
Units of Measurement of Selected Quantities, in SI System (Legally Approved)

Quantity	Symbol of the Quantity	SI Units of Measure, and Relation between Them	Units of Measure Temporarily Admitted as Legal, and Relationships between Them
Power	$S = P + jQ$	VA; 1 W = 1 J/s; var	1 VA (appearent) = 1 var (*reactive*); (1 HP = 0.7355 kW = 75 kGm/s)
Specific heat	c	J/(kg · K)	1 cal/(g · K)
Thermal conductivity	λ	W/(m · K)	1 cal/(cm · s · K) = 418.68 W/(m · K)
Current density	$\boldsymbol{J}$	A/m^2	$1\ A/mm^2 = 10^6\ A/m^2$
Electric charge	Q	1 C (Coulomb) = 1 As	1 Ah = 3600 C
Electric voltage	U or V	1 V = 1 W/A = 1 J/C	
Electric field intensity	$\boldsymbol{E}$	1 V/m = 1 N/C	
Electric capacitance	C	1 F = 1 C/V = 1 As/V	
Electric resistance, reactance, impedance	R X, Z -respct.	1 Ω = 1 V/1 A	
Resistivity, specific resistance	ρ	$1\ \Omega\ m = 1\ \Omega \cdot (1\ m^2/1\ m)$	$(1\ \Omega\ mm^2/m = 10^{-6}\ \Omega\ m)$; reactivity, impeditivity, respectively
Electric conductance	G	1 S = 1/Ω	
Electric conductivity, specific conductance	$\sigma = 1/\rho$	1 S/m = 1/Ω m	
Magnetic flux	$\boldsymbol{\Phi}$	1 Wb = 1 Vs	$1\ Mx = 10^{-8}\ Wb$
Magnetic flux density	$\boldsymbol{B}$	$1\ T = 1\ Wb/m^2 = 1\ Vs/m^2$	$1\ Gs = 10^{-4}\ T = 10^{-8}\ Vs/cm^2$
Magnetic field intensity	$\boldsymbol{H}$	1 A/m	$1 Oe = \frac{1}{4\pi} 10^3\ A/m = 79.58\ A/m$
Inductance	L, M	1 H = 1 Wb/A = 1 Ωs	
Magnetomotive force	F_m, θ	1 A (ampere-turn)	$1\ Gb = 10/4\pi\ A$
Reluctance, magnetic resistance, reactance, impedance, respectively	R_m	1/H = 1/Ωs	

A.I.1 More Important Physical Constants

Propagation velocity of electromagnetic wave in vacuum $c_o = (2.997930 \pm 0.000003)\ 10^8$ m/s

Acceleration due to gravity in normal conditions $g_n = 9.80665\ m/s^2$

Electric constant, permittivity of vacuum $e_o = (8.85416 \pm 0.00018)10^{-12}$ F/m $= 1/\mu_0 c_0^2 = 1/(4\pi \cdot 9)\ 10^{-9}$ F/m

Magnetic constant, magnetic permeability of vacuum $\mu_0 = 4\pi \times 10^{-7}$ H/m = 1.25664×10^{-6} H/m

A.II FUNDAMENTAL FORMULAE OF VECTOR CALCULUS

A.II.1 Systems of Coordinates

A.II.1.1 System of Rectangular (Cartesian) Coordinates

x, y, z—rectangular coordinates (Figure A.1)

i, j, k—unit vectors (versors) in the directions of corresponding coordinate axes

A.II.1.2 System of Cylindrical Coordinates (Figure A.2)

r, θ, z—cylindrical coordinates

$$\left.\begin{aligned} x &= r\cos\theta \\ y &= r\sin\theta \\ z &= z \end{aligned}\right\} \quad \text{connection with rectangular (Cartesian) coordinate system}$$

$\mathbf{1_r}$, $\mathbf{1_\theta}$, $\mathbf{1_z}$—unit vectors (versors), mutually perpendicular

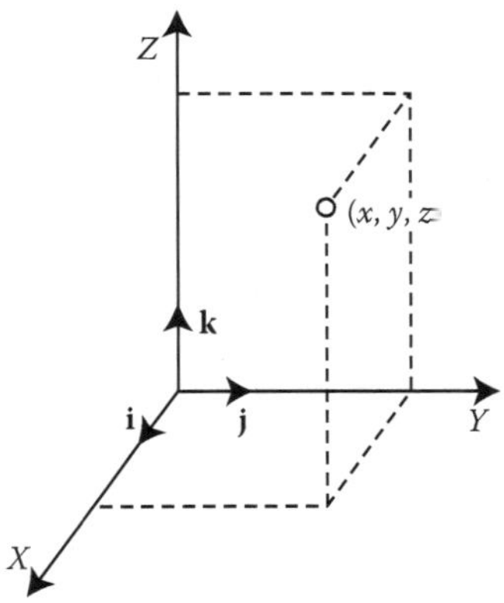

FIGURE A.1 Rectangular (Cartesian) coordinate system.

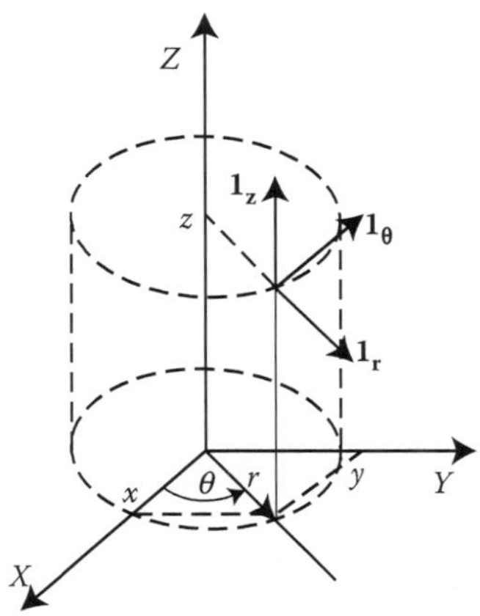

FIGURE A.2 Cylindrical coordinate system.

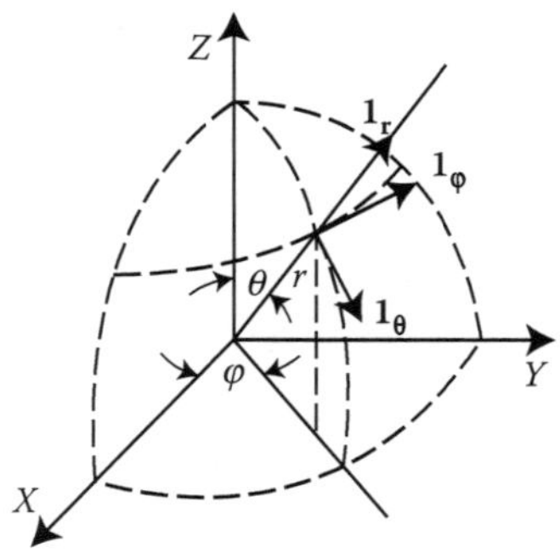

FIGURE A.3 Polar (spherical) coordinate system.

A.II.1.3 System of Polar (Spherical) Coordinates (Figure A.3)

r, θ, φ—spherical (polar) coordinates

$x = r \sin \theta \cdot \cos \varphi$; $y = r \sin \theta \cdot \sin \varphi$; $z = r \cos \theta$—connection with rectangular coordinate system.

$\mathbf{1_r}$, $\mathbf{1_\theta}$, $\mathbf{1_\varphi}$—unit vectors (versors), mutually perpendicular.

A.II.2 Formulae of Vector Algebra

A.II.2.1 Vector

- In rectangular (Cartesian) system: $\boldsymbol{A} = \mathbf{i}A_x + \mathbf{j}A_y + \mathbf{k}A_z$
- In cylindrical system: $\boldsymbol{A} = \mathbf{1_r}A_r + \mathbf{1_\theta}A_\theta + \mathbf{1_z}A_z$
- In polar system: $\boldsymbol{A} = \mathbf{1_r}A_r + \mathbf{1_\theta}A_\theta + \mathbf{1_\varphi}A_\varphi$

where $\boldsymbol{A}_i$ is the measure or modulus of projections of the vector $\boldsymbol{A}$ on the corresponding axes of coordinates;

$$i = x, y, z, r, \theta, \varphi.$$

A.II.2.2 Additions of Vectors (Figure A.4)

A.II.2.3 Scalar Product (Figure A.5)

$W = \boldsymbol{A} \cdot \boldsymbol{B} = AB \cos \alpha = (\mathbf{i}A_x + \mathbf{j}A_y + \mathbf{k}A_z) \cdot (\mathbf{i}B_x + \mathbf{j}B_y + \mathbf{k}B_z) = A_xB_x + A_yB_y + A_zB_z$

$\boldsymbol{A} \cdot \boldsymbol{B} = \boldsymbol{B} \cdot \boldsymbol{A}$ *Example*: $\boldsymbol{A}$ – road, $\boldsymbol{B}$ – force, W – work $= \boldsymbol{A} \cdot \boldsymbol{B}$

$\mathbf{i} \cdot \mathbf{j} = \mathbf{i} \cdot \mathbf{k} = \mathbf{j} \cdot \mathbf{k} = 0, \quad \mathbf{i} \cdot \mathbf{i} = \mathbf{j} \cdot \mathbf{j} = \mathbf{k} \cdot \mathbf{k} = 1$

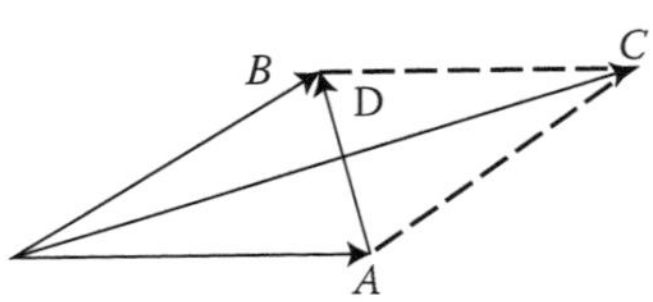

FIGURE A.4 Addition of vectors.

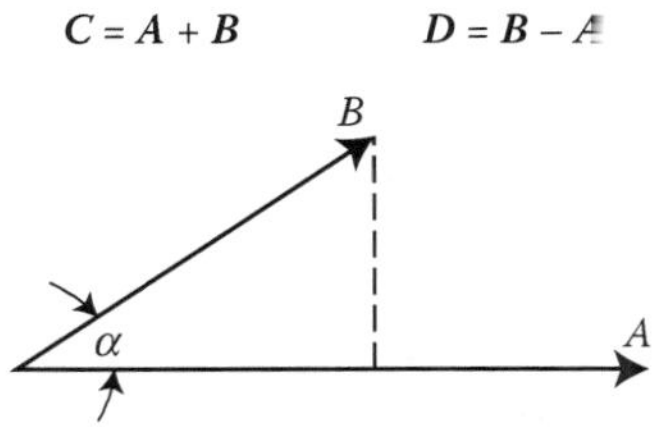

FIGURE A.5 Scalar product.

A.II.2.4 Vector Product (Figure A.6)

$$\boldsymbol{F} = \boldsymbol{A} \times \boldsymbol{B} = \boldsymbol{n}\, AB \sin \alpha = (\mathbf{i}A_x + \mathbf{j}A_y + \mathbf{k}A_z) \times (\mathbf{i}B_x + \mathbf{j}B_y + \mathbf{k}B_z)$$

$$= \begin{vmatrix} \mathbf{i} & \mathbf{j} & \mathbf{k} \\ A_x & A_y & A_z \\ B_x & B_y & B_z \end{vmatrix} = \mathbf{i}(A_yB_z - A_zB_y) + \mathbf{j}(A_zB_x - A_xB_z) + \mathbf{k}(A_xB_y - A_yB_x);$$

$$\boldsymbol{A} \times \boldsymbol{B} = -\boldsymbol{B} \times \boldsymbol{A}; \quad \mathbf{i} \times \mathbf{i} = \mathbf{j} \times \mathbf{j} = \mathbf{k} \times \mathbf{k} = 0; \quad \mathbf{i} \times \mathbf{j} = \mathbf{k} \quad \mathbf{j} \times \mathbf{k} = \mathbf{i}, \quad \mathbf{k} \times \mathbf{i} = \mathbf{j}.$$

Example: A force $\boldsymbol{F}$ acting on a conductor with current $\boldsymbol{A}$ in the magnetic field $\boldsymbol{B}$ equals to the area of parallelogram with sides $\boldsymbol{A}$ and $\boldsymbol{B}$.

A.II.2.5 Scalar Product, Mixed

$$(\boldsymbol{A} \times \boldsymbol{B}) \cdot \boldsymbol{C} = (\boldsymbol{B} \times \boldsymbol{C}) \cdot \boldsymbol{A} = (\boldsymbol{C} \times \boldsymbol{A}) \cdot \boldsymbol{B} = \boldsymbol{A} \cdot (\boldsymbol{B} \times \boldsymbol{C}) = \boldsymbol{B} \cdot (\boldsymbol{C} \times \boldsymbol{A});$$

For example, the volume of a parallelepiped with edges $\boldsymbol{A}, \boldsymbol{B}, \boldsymbol{C}$: $(\boldsymbol{D} \times \boldsymbol{A}) \cdot (\boldsymbol{B} \times \boldsymbol{C}) = \boldsymbol{D}[\boldsymbol{A} \times (\boldsymbol{B} \times \boldsymbol{C})]$

A.II.2.6 Vector Product, Mixed

$$(\boldsymbol{A} \times \boldsymbol{B}) \times \boldsymbol{C} = \boldsymbol{B}(\boldsymbol{C} \cdot \boldsymbol{A}) - \boldsymbol{A}(\boldsymbol{B} \cdot \boldsymbol{C})$$

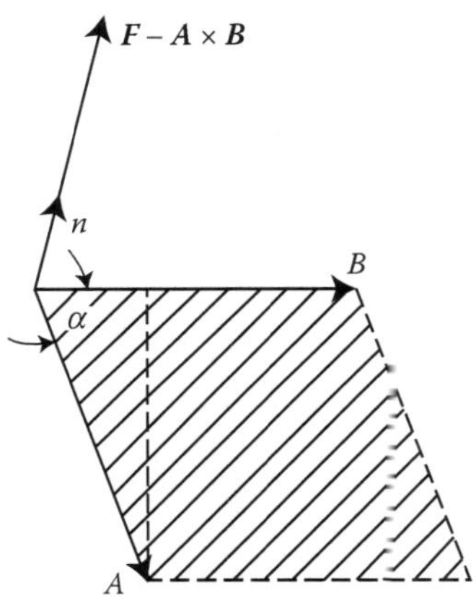

FIGURE A.6 Vector product.

$$A \times (B \times C) = (A \cdot C)B - (A \cdot B)C$$

$$(D \times A) \times (B \times C) = [(D \times A)C]B - [(D \times A)B] \cdot C$$

A.II.3 Formulae of Vector Analysis

A.II.3.1 Differential Operators

$$\nabla = \mathbf{i}\frac{\partial}{\partial x} + \mathbf{j}\frac{\partial}{\partial y} + \mathbf{k}\frac{\partial}{\partial z} \text{ — nabla operator (vector)}$$

$$\nabla^2 = \frac{\partial^2}{\partial x^2} + \frac{\partial^2}{\partial y^2} + \frac{\partial^2}{\partial z^2} \text{ — Laplace operator (scalar)}$$

A.II.3.2 Gradient (Vector, Operation on a Scalar)

In rectangular (Cartesian) coordinates: $\text{grad } P = \mathbf{i}\frac{\partial P}{\partial x} + \mathbf{j}\frac{\partial P}{\partial y} + \mathbf{k}\frac{\partial P}{\partial z} = \nabla P$

In cylindrical coordinates: $\text{grad } P = \mathbf{1_r}\frac{\partial P}{\partial r} + \mathbf{1_\Theta}\frac{1}{r}\frac{\partial P}{\partial \Theta} + \mathbf{1_z}\frac{\partial P}{\partial z} = \nabla P$

In polar (spherical) coordinates: $P = \mathbf{1_r}\frac{\partial P}{\partial r} + \mathbf{1_\Theta}\frac{1}{r}\frac{\partial P}{\partial \theta} + \mathbf{1_\varphi}\frac{1}{r\sin\Theta} \cdot \frac{\partial P}{\partial \varphi} = \nabla P$

A.II.3.3 Divergence (Scalar, Operation on a Vector)

Definition: $\text{div } \boldsymbol{A} = \lim\limits_{\Delta v \to 0} \frac{\oiint_s \boldsymbol{A} \cdot \mathrm{d}\boldsymbol{s}}{\Delta v}$

In rectangular (Cartesian) coordinates: $\text{div } \boldsymbol{A} = \frac{\partial A_x}{\partial x} + \frac{\partial A_y}{\partial y} + \frac{\partial A_z}{\partial z} = \nabla \cdot \boldsymbol{A}$

In cylindrical coordinates: $\text{div } \boldsymbol{A} = \frac{\partial \boldsymbol{A}_r}{\partial r} + \frac{\boldsymbol{A}_r}{r} + \frac{1}{r}\frac{\partial \boldsymbol{A}_\Theta}{\partial \Theta} + \frac{\partial \boldsymbol{A}_z}{\partial z}$

In polar (spherical) coordinates: $\text{div } \boldsymbol{A} = \frac{1}{r^2}\frac{\partial}{\partial r}(r^2 \boldsymbol{A}_r) + \frac{1}{r\sin\Theta}\frac{\partial}{\partial \Theta}(\boldsymbol{A}_\Theta \sin\Theta)$

$$+ \frac{1}{r\sin\Theta}\frac{\partial \boldsymbol{A}_\varphi}{\partial \varphi}$$

A.II.3.4 Curl (Rotation = Rot, Operation on a Vector)

$$\text{curl}_n \boldsymbol{A} = \lim_{\Delta S \to 0} \frac{\oint_L \boldsymbol{A} \cdot \mathrm{d}l}{\Delta S}$$

In rectangular (Cartesian) coordinates:

$$\operatorname{curl} \boldsymbol{A} = \begin{vmatrix} \mathbf{i} & \mathbf{j} & \mathbf{k} \\ \frac{\partial}{\partial x} & \frac{\partial}{\partial y} & \frac{\partial}{\partial z} \\ A_x & A_y & A_z \end{vmatrix} = \mathbf{i}\left(\frac{\partial A_z}{\partial y} - \frac{\partial A_y}{\partial z}\right) + \mathbf{j}\left(\frac{\partial A_x}{\partial z} - \frac{\partial A_z}{\partial x}\right) + \mathbf{k}\left(\frac{\partial A_y}{\partial x} - \frac{\partial A_x}{\partial y}\right)$$

In cylindrical coordinates:

$$\operatorname{curl} \boldsymbol{A} = \begin{vmatrix} \frac{\mathbf{1_r}}{r} & \mathbf{1_\theta} & \frac{\mathbf{1_z}}{r} \\ \frac{\partial}{\partial \mathbf{r}} & \frac{\partial}{\partial \theta} & \frac{\partial}{\partial z} \\ A_\mathbf{r} & rA_\mathbf{\theta} & A_\mathbf{z} \end{vmatrix} = \mathbf{1_r}\left(\frac{1}{r}\frac{\partial A_z}{\partial \theta} - \frac{\partial A_\theta}{\partial z}\right) + \mathbf{1_\theta}\left(\frac{\partial A_r}{\partial z} - \frac{\partial A_z}{\partial r}\right)$$

$$+ \mathbf{1_z}\left(\frac{A_\theta}{r} + \frac{\partial A_\theta}{\partial r} - \frac{1}{r}\frac{\partial A_r}{\partial \theta}\right)$$

In polar (spherical) coordinates:

$$\operatorname{curl} \boldsymbol{A} = \begin{vmatrix} \frac{\mathbf{1_r}}{r^2 \sin\theta} & \frac{\mathbf{1_\theta}}{r \sin\theta} & \frac{\mathbf{1_\varphi}}{r} \\ \frac{\partial}{\partial r} & \frac{\partial}{\partial \theta} & \frac{\theta}{\partial \varphi} \\ A_r & rA_\theta & r\sin\theta\, A_z \end{vmatrix}$$

A.II.3.5 Scalar Laplacian (Laplace Operator; Scalar—Operation on a Scalar)

In rectangular (Cartesian) coordinates:

$$\nabla^2 F = \operatorname{div} \cdot \operatorname{grad} F = \frac{\partial^2 F}{\partial x^2} + \frac{\partial^2 F}{\partial y^2} + \frac{\partial^2 F}{\partial z^2} \equiv \nabla \cdot \nabla F \equiv \Delta F$$

In cylindrical coordinates: $\nabla^2 F = \frac{\partial^2 F}{\partial r^2} + \frac{1}{r}\frac{\partial F}{\partial r} - \frac{1}{r^2}\frac{\partial^2 F}{\partial \Theta^2} + \frac{\partial^2 F}{\partial z^2}$

In polar (spherical) coordinates:

$$\nabla^2 F = \frac{\partial^2 F}{\partial r^2} + \frac{1}{r^2}\frac{\partial^2 F}{\partial \Theta^2} + \frac{1}{r^2 \sin^2\Theta}\frac{\partial^2 F}{\partial \varphi^2} + \frac{2}{r}\frac{\partial F}{\partial r} + \frac{\operatorname{ctg}\Theta}{r^2}\frac{\partial F}{\partial \Theta}$$

A.II.3.6 Vector Laplacian (Laplace Operator; Vector—Operation on a Vector)

In rectangular (Cartesian) coordinates:

$$\nabla^2 A = \mathbf{i}\nabla^2 A_x + \mathbf{j}\nabla^2 A_y + \mathbf{k}\nabla^2 A_z = \frac{\partial^2 \mathbf{A}}{\partial x^2} + \frac{\partial^2 \mathbf{A}}{\partial y^2} + \frac{\partial^2 \mathbf{A}}{\partial z^2}$$

In cylindrical coordinates:

$$\nabla^2 \boldsymbol{A} = \mathbf{1}_r \left(\nabla^2 \boldsymbol{A}_r - \frac{2}{r^2}\frac{\partial \boldsymbol{A}_\Theta}{\partial \Theta} \frac{\boldsymbol{A}_r}{r^2} \right) + \mathbf{1}_\Theta \left(\nabla^2 \boldsymbol{A}_\Theta + \frac{2}{r^2}\frac{\partial \boldsymbol{A}_r}{\partial \Theta} - \frac{\boldsymbol{A}_\Theta}{r^2} \right) + \mathbf{1}_z (\nabla^2 \boldsymbol{A}_z)$$

A.II.3.7 Gauss's Theorem (*Green's*)

$$\oiint_S D \cdot \mathrm{d}\boldsymbol{S} = \iiint_V \operatorname{div} D \cdot \mathrm{d}\boldsymbol{V}$$

A.II.3.8 Stokes's Theorem

$$\oint_L H \cdot \mathrm{d}\boldsymbol{l} = \iint_s \operatorname{curl} H \cdot \mathrm{d}s; \qquad (L\text{—edge curve of limited surface } S)$$

A.II.3.9 Relationships between the Vector Operators

$$\operatorname{grad}(UV) = V \operatorname{grad} U + U \operatorname{grad} V$$

$$\operatorname{div}(U\boldsymbol{A}) = \boldsymbol{A} \operatorname{grad} U + U \operatorname{div} \boldsymbol{A}$$

$$\operatorname{curl} U\boldsymbol{A} = (\operatorname{grad} U) \times \boldsymbol{A} + U \operatorname{curl} \boldsymbol{A}$$

$$\operatorname{div}(\boldsymbol{A} \times \boldsymbol{B}) = \boldsymbol{B} \operatorname{curl} \boldsymbol{A} - \boldsymbol{A} \operatorname{curl} \boldsymbol{B}$$

$$\operatorname{div} \operatorname{curl} \boldsymbol{A} = 0$$

$$\operatorname{curl} \operatorname{grad} U = 0$$

$$\operatorname{curl} \operatorname{curl} \boldsymbol{A} = \operatorname{grad} \operatorname{div} \boldsymbol{A} - \nabla^2 \boldsymbol{A}$$

$$\operatorname{div} \operatorname{grad} U = \nabla^2 U$$

EXAMPLE

If one substitutes $\boldsymbol{H} = -\operatorname{grad} V_{m1}$ instead of $\boldsymbol{A}$, then $\operatorname{div} \boldsymbol{B} = \operatorname{div}(\mu \boldsymbol{H}) = 0$ and $\boldsymbol{H}$ grad $\mu + \mu \nabla^2 V_m = 0$

$$(\boldsymbol{A} \operatorname{grad})\boldsymbol{B} = \boldsymbol{A} \operatorname{div} \boldsymbol{B} - \operatorname{curl}(\boldsymbol{A} \times \boldsymbol{B}) - \boldsymbol{B} \operatorname{div} \boldsymbol{A} + (\boldsymbol{B} \operatorname{grad})\boldsymbol{A}$$

Derivative of a scalar function V in the direction $\boldsymbol{l}$:

$$\mathrm{d}V = \operatorname{grad} V \cdot \mathrm{d}\boldsymbol{l}, \quad \frac{\mathrm{d}V}{\mathrm{d}\boldsymbol{l}} = \frac{\boldsymbol{l}}{l} \operatorname{grad} V.$$

$$\operatorname{ci} x = -\int_x^\infty \frac{\cos x}{x} \mathrm{d}x; \ \operatorname{si} x = -\int_x^\infty \frac{\sin x}{x} \mathrm{d}x \text{ are cos and sin integrals}$$

References

CHAPTER 1

1.1. Ashcroft N.W. and Mermin N.D.: *Solid State Physics*. New York: Rinehart and Winston, 1976.
1.2. Bozorth R.M.: *Ferromagnetism*. New York: Van Nostrand, 1951.
1.3. Ertan H.B., Üctug M.Y., Colyer R., and Consoli A. (Editors and co-authors): *Modern Electrical Drives*. Dordrecht: Kluwer Academic Publishers with NATO Scientific Affairs Division. May 2000. *Other authors*: Kutman T., Turowski J., Wiak S., Jack A.G., Hanitsch R., Parspour N., Rahman M.A., Zhou P., Ertan H.B., Jufer M., Bose B.K., Lipo T.A., Kükrer O., Ehsani M. Husain I., Capolino G.A., Üctug M.Y., Timàr P.L., Schmidt I., Retter G.I., Serrano-Iribarnegaray L., Rashid M.H., Akin E., Halsall C.L., Zingel R.G.W., Belmans R.J.M., Szentirmai L., Colyer R.E. and Trzynadlowski A.M.
1.4. Jasinski A.H.: 2002. *Transition Economies in the European Research and Innovation Area: New Challenges for their Science and Technology*. Department of Management, University of Warsaw, 2004, wherein the Chapter: Turowski J: Innovative challenges in technology management.
1.5. Jezierski E.: *Transformers. Theory*. Warsaw: WNT, 1975 (see also [5.2]).
1.6. Kolbinski K. and Słowikowski J.: *Electrical Materials*. (in Polish). Warsaw: WNT, 1988.
1.7. Kozłowski L.: *Elements of Atomic Physics and Solid*. Cracow: AGH, 1972.
1.8. Lopez-Fernández X.M., Ertan H.B., and Turowski J. (Editors and co-authors): *TRANSFORMERS: Analysis, Design and Measurements*. New York: Boca Raton, CRC Press, 2013, 610pp.
1.9. *Handbook of Electrical Materials*. (in Polish). Warsaw: PWT, 1959.
1.10. Savini A. and Turowski J. (Editors and co-authors): *Electromagnetic Fields in Electrical Engineering*. New York: Plenum Press, 1988.
1.11. Simonyi K.: *Theoretical Electrotechnics*. (in German). Berlin: VEB Deutscher Verlag der Wissenschaften, 1979.
1.12. Smolinski S.: *Superconductivity*. (in Polish). Warsaw: WNT, 1983.
1.13. *Handbook of Electrical Materials*. (in Russian) Vol. 2, Moscow: Gosenergoizdat, 1960.
1.14. Tarczyński W.: *Electrodynamics of Electrical Apparatus*. (in Polish). Ed. of Tech. Univ. of Lodz, 2007.
1.15. Turowski J.: (1) *Technical Electrodynamics*. (in Polish). Warsaw: WNT, 1969, 1st edition. (2) Turowski J.: *Tiekhnicheskaia electrodinamika*. (in Russian). *Izdatielstwo "Energia"*. Moscow, 1974, 2nd edition. (3) Turowski J.: *Technical Electrodynamics*. (in Polish). Warsaw: WNT, 1993, 3rd edition. Modified and updated.
1.16. Turowski J.: (1) *Calculations of Electromagnetic Components of Electric Machinery and Equipment*. (in Polish). Warsaw: WNT, 1982; and (in Russian). Moscow: Energoatomizdat, 1986.
1.17. Turowski J. (1) (Editor and author 35%) et al.: *Analysis and Synthesis of Electromagnetic Fields*. (in Polish). Wroclaw: OSSOLINEUM - PAN, 1990; where: §3 Reluctance network method RNM; §5 Coupled fields. (2) Computational magnetics. (in English). J. Sykulski (Editor),London: Chapman & Hall Stellenbosch, 1995 where: §4) Turowski J.: Reluctance networks RNM. and §6 Coupled fields.

1.18. Turowski J.: *The Theory of Electric Machines. AC Machines.* (in Polish) 3rd edition. Tech. Univ. of Lodz 1984.
1.19. Turowski J.: *Electromechanical Components of Automatics.* (in Polish) 2nd edition. Tech. Univ. of Lodz, 1989.
1.20. Turowski J.: *Fundamentals of Mechatronics* (in Polish). Lodz: AHE Press, 2008.
1.21. Urbański W.: *Lectures on Higher Physics* (in German: *Vorträge über höhere Physik*), Lemberg: Lwów, 1857.
1.22. Wyatt O.H. and Dew-Hughes D.: *Metals, Ceramics & Polymers: An Introduction to the Structure and Properties of Engineering Materials.* Cambridge: Cambridge University Press, 1974.
1.23. Wesolowski K.: *Materials Science.* Vols. I, II, III. (in Polish). Warsaw: WNT, 1966.
1.24. Wilkes P.: *Solid-State Theory in Metallurgy.* Cambridge: Cambridge University Press, 1973.
1.25. Wolfarth E.P.: *Ferromagnetic Materials.* Vol. 3. Amsterdam: North-Holland Publ. Co. 1982.
1.26. Another superconducting breakthrough. High temperature superconductor challenges critical field record. *High-Tech. Materials Alert*, 4(3), 1987, 10.
1.27. Anuszczyk J.: Power losses in electrotechnical sheet steel under rotational magnetization. *ISEF'87*, Pavia, Italy, September 23–25, 1987. Paper 1–1, pp. 1–4.
1.28. Atherton D.L., Szpunar B., and Szpunar J.A.: A new approach to Preisach diagram. *IEEE Trans. MAG*-23, 3, 1987, 1856–1865.
1.29. Bessho K., Yamada S., Furuta M., and Hayashigishi M.: Characteristics and application of a new multilayer coil with core for AC high magnetic field. *ISEF'89*, Lodz, Poland, Sept. 20–22, 1989, pp. 83–86.
1.30. *Silicon Electrical Sheets.* Bochnia: Catalogue of Metallurgical Processing Plant, 1982.
1.31. Kadowaki K. et al.: Superconductivity at 95 K in the single-phase $UBa_2Cu_3O_{9-y}$ system. *Physica*, 145 B, 1987, 1–4.
1.32. Kitada M.: Kindzoku, *Metals and Technol.* (Jap.), 57(5), 1987, 2–7.
1.33. Kordecki A.: Thermo-compensation of induction changes in permanent magnets of electric machines. *The Science Works of Inst. Electromechanical Systems in Wroclaw University of Technology.* Conference series No 39, 1987, pp. 5–19.
1.34. Kozłowska A.: Core transformers with amorphous band. (in German) *Bulletin SEV*, 78(19), 1987, 1201–1204.
1.35. Latek W.: Rational starting of turbogenerator. *Przegląd Elektrot.*, 5, 1964, 235–239.
1.36. METGLAS Electromagnetic Alloys. USA 1981. (http://www.metglas.com/products/).
1.37. Industry Branch Norm BN-85/3382-20. Components of electronic devices. Cores for transformers and reactors coils. Determination of the magnetic and electric sizes. *Inst. Tele-Radiotechn.* 1986, pp. 1–19.
1.38. Sabrie J.L.: Cryoelectricity materials. *General Report of Section 1. CIGRE Symposium on New and Improved Materials for Electrotechnology—Elactra*, October 1987, No. 114, pp. 29–33.
1.39. Sagawa M., Fujimura S., Yamamoto H., Matsuura Y., and Hiraga K.: Permanent magnet materials based on the rare-earth-iron-boron tetragonal compounds. *IEEE Trans. on Magnetics MAG-20*, 5, Sept. 1984, 1584–1589.
1.40. Tanaka Y. et al.: Characteristics of epoxy-impregnated AC superconducting coils. *INTERMAG'87*, Tokyo, Japan, April 14–17, 1987, Paper GE-09.
1.41. Technical Data on Domain Refined Orientcore. Hi-B. z DKH, Nippon Steel Corporation Catalogue No. EXE 706, Dec. 1984–Nov. 1987.
1.42. Tellinen J. and Järvik J.: Calculation of losses in electrical steel by longitudinal superposition of alternating and constant magnetic fields. *ISEF'85*, Warsaw, Poland, Sept. 1985, Paper 82, pp. 351–353.

1.43. Turowski J.: Mechatronics Impact upon Electrical Machines and Drives. *Proceedings of Internat. Aegean Conference—ACEMP'04*. May 26–28, 2004. Istanbul, Turkey, pp. 65–70.
1.44. Ueda A., Hirao T., Hatanaka H., Morita M., and (Mitsubishi El. Co).: Experimental results of field winding and concept of rotor component development of superconducting generator. *INTERMAG'87*. Tokyo, Japan. April 14–17, 1987, Paper GE-07.
1.45. Walter J.L. et al.: Magnetic properties of cube textured silicon-iron magnetic sheet. *Journal of Applied Physics*, 29(3), March, 1958.
1.46. Wiener G. et al.: Cube textured in body centered magnetic alloy. *Journal of Applied Physics*, 29(3), March, 1958.
1.47. Zhang X.: Nd-Fe-B magnets in China. *INTERMAG'87*. Tokyo, Japan. April 14–17, 1987, paper CC-01.
1.48. Lopez-Fernandez, X.M., Penabad-Duran, P., and Turowski, J. Three-dimensional methodology for the overheating hazard assessment on transformer covers, *IEEE Transactions on Industry Applications*, 48(5), September/October 2012, 1549–1555.
1.49. Halbach K.: Physical and optical properties of rare earth cobalt magnets. *Nucl. Instrum. Methods Phys Res.*, 187(1), Aug. 1981, 109–117.
1.50. Derlecki S., Kuśmierek Z., Szulakowski J., and Dems M.: Magnetic properties of amorphous materials used as corps of electric machines. *Przegląd Elektrotechniczny (Electrical Review)*, 88(5a), 2012, 10–13.
1.51. Dems M. and Komęza K.: *Static and Dynamic Modeling of Induction Motors Operating States*. (in Polish). Monograph released by the Techn. University of Lodz, 2011, 332pp.
1.52. Dems M. and Komęza K.: Performance characteristics of a high-speed induction motor with an amorphous stator core. *XX Intern. Confer. on Electric Machines ICEM 2012*, Marseille, France, Sept. 2012, paper FD-005207.
1.53. Pennabad-Duran P., Di Barba P., Lopez-Fernandez X.M., and Turowski J.: Electromagnetic and Thermal parameter identification method for best prediction of temperature distribution on transformer tank covers. *XVI International Symposium on Electromagnetic Fields in Mechatronics, Electrical and Electronic Engineering—ISEF'2013*, Ohrid, Macedonia, Sept. 12–14, 2013.
1.54. Sykulski J.K.: Superconducting transformers. Chapter 4, pp. 95–108, in: *TRANSFORMERS: Analysis, Design and Measurements* X.M. Lopez-Fernández, H.B. Ertan and J. Turowski (editors), New York: Boca Raton, CRC Press, 2013.
1.55. Bochnia Product Catalog 2006: Grain Oriented Transformer Sheets and Strips. According to EN 10 107 2005 Standards. www.stalprodukt.com.pl
1.56. Turowski J., Lopez-Fernandez X.M., Soto Rodriguez A., and Souto Revenga D.: Rapid design of sandwich windings transformers for stray loss reduction. *ISEF 2007—XIII International Symposium on Electromagnetic Fields in Mechatronics, Electrical and Electronic Engineering*. Prague, Czech Republic, September 13–15, 2007.

CHAPTER 2

2.1. (a) Abramowitz, M. and Stegun, I.A. (Eds.): *Handbook of Mathematical Functions with Formulas, Graphs, and Mathematical Tables*, 9th printing. New York: Dover, 1972. (b) Antoniewicz J.: *Tables of Functions for Engineers*. (in Polish). Warsaw: PWN, 1980.
2.2. Ciok Z.: *Methods for Calculating the Electromagnetic and Flow Fields*. (in Polish). Warsaw: University of Technology, 1981.
2.3. Dąbrowski M.: *Magnetic Fields and Circuits of Electric Machines*. (in Polish). Warsaw: WNT, 1971.
2.4. Galus Z.: *Theoretical Foundations of Chemical Electroanalysis*. (in Polish). Warsaw: PWN, 1973.

2.5. Gieras J.F.: *Advancements in Electric Machines*. Berlin: Springer, 2008.

2.6. Gieras J.F.: *Linear Induction Drives*. NJ, USA: Clarendon Press. 1994.

2.7. Govorkov W.A.: *Electric and Magnetic Field*. (in Polish). Warsaw: WNT 1962 (Moscow, 1968).

2.8. Hammond P.: *Applied Electromagnetism*. Oxford: Pergamon Press, 1979.

2.9. Hammond P. and Sykulski J.K.: *Engineering Electromagnetism. Physical Processes and Computation*. New York, Oxford: Oxford University Press, 1994.

2.10. Jackson J.D.: *Classical Electrodynamics*. John Wiley & Sons, 1975 (Warsaw: PWN, 1982).

2.11. Jahnke E. and Emde F.: *Tables of Functions*. Moscow: Fizmatgiz, 1959.

2.12. Kącki E.: *Partial Differential Equations in Physics and Technology Issues*. (in Polish). Warsaw: WNT, 1989.

2.13. Kulkarni S.V. and Khaparde S.A.: *Transformer Engineering. Design and Practice*. New York, Basel: Marcel Dekker Inc., 2004.

2.14. Neiman L.R. and Demerchian K.S.: *Theory of Electrical Engineering*. (in Russian). Vols. 1 and 2. Leningrad: Energoizdat, 1981.

2.15. Rawa H.: *Electricity and Magnetism in Technology*. (in Polish). Warsaw: PWN, 1994.

2.16. Wiak S. (Editor and co-author): *Mechatronics*. EXIT, Lodz: Technical University of Lodz, 2009.

2.17. Zienkiewicz O.: *The Finite Element Method*. Oxford: Butterworth-Heinemann, 6th edition, 2000.

2.18. Davey K.R. and Barnes W.J.: T-Ω technique applied to three-dimensional eddy current problems using the boundary integral equation. *IEE Proc*. 133, (Pt. B, 6), 1986, 359–364.

2.19. Fichtner W., Rose D.J., and Bank R.E.: Semiconductor device simulation. *IEEE Trans. on Electron Devices*. ED-30(9), 1983, 1018–1039.

2.20. Franz A. et al.: A generalization of the finite-difference method suitable for semiconductor device simulation. *IEEE Trans. on Electron. Devices*. ED-30(9)), 1983, 1070–1082.

2.21. Frolov S.I.: Determination of volume and surface densities of electromagnetic forces in non-linear, anisotropic magnetic systems. (in Russian). *Elektrichestvo*, 8, 1989, 62–66.

2.22. Gangyu X. and Nasar S.A.: Analysis of fields and forces in a permanent magnet of linear synchronous machine based on the concept of magnetic charge. *IEEE Trans. on Magnetics*, 25(3), 1989, 2713–2719.

2.23. Komęza K. and Krusz G.: Application of finite element method to analysis of the impact of eddy currents on distribution of stray field in transformer. (in Polish). *Rozprawy Elektrot*., 32(3), 1986, 785–799.

2.24. Koppikar D.A., Kulkarni S.V., and Turowski J.: Fast 3-dimensional interactive computation of stray field and losses in asymmetric transformers. *IEE Proceedings. Generation, Transmission, Distribution*. 147(4), July 2000, 197–201.

2.25. Nakata T., Takahashi K., Fujiwara K., Muramatsu K., and Cheng Z.G.: Comparison of various methods for 3-D eddy current analysis. *Proceedings of the 4th Joint Magnetism and Magnetic Materials-Intermag Conference (MMM-Intermag)*, 12–15 July 1988, Vancouver, British Columbia, Canada.

2.26. Pietruszka M. and Jabłoński M.: Flux distribution inside anisotropic cores of three phase transformers at no load. *Proc. ISEF'89*, Lodz, September 20–22, 1989, 117–120.

2.27. Rawa H.: About some problems of formulating equations describing the electromagnetic field in heterogeneous environments. The role of the volume density of electric charge. (in Polish). *Archiwum Elektrot*., 34(3–4), 1985, 765–774.

2.28. Superczynski M.: Superconducting electric propulsion test craft Jupiter II. *IEEE Trans. on Magnetics*, MAG-23(2), 1987, 348.

2.29. Sykulski J.K. and Hammond P.: Calculation of electric and magnetic fields by means of the method of tubes and slices. *ISEF'87*, Pavia, Italy, 1987, 49–53.
2.30. Turowska M.: Influence of polarogram nonlinearity on current features in electrochemical oscillator. *Soc. Sci. Lodz, Acta Chim.*, 18, 1973, 73–84.
2.31. Turowski J.: Electromagnetic field and losses in the transformer housing. (in Polish). *"Elektryka" Science Papers*, Technical University of Lodz., No. 3, 1957, pp. 73–63.
2.32. Turowski J.: Methods of calculating the additional losses caused by stray flux in a transformer. (in Polish). *Rozprawy Elektrot.*, (3/4), 1962, 563–599.
2.33. Turowski J.: Calculation of losses and temperatures in the rail pipe screen. (in Polish). *Archiwum Elektrotechniki*, 1, 1963, 61–79.
2.34. Turowski J.: Modeling and simulation of electromagnetic field in electrical machines and transformers with the help of equivalent reluctance network. *12th IMACS World Congress*, Paris, July 18–22, 1988, Paper T.18, PM2.H.04.
2.35. Selberherr S.: *Analysis and Simulation of Semiconductor Devices*. Vienna, Austria: Springer-Verlag, 1984.
2.36. Turowski M.: Two-dimensional semiconductor device modelling on personal microcomputer, *Proc. ISMM International Symposium "Mini and Microcomputers and Their Applications"*, Sant Feliu de Guixols, Spain, June 27–30, 1988, pp. 25–28.
2.37. Wiak S. and Zakrzewski K.: Numerical calculation of transients in electrical circuits containing elements with nonlinear eddy-current skin effect. *IEE Proc.*, 134(Pt A, 9), 1987, 741–746.
2.38. Zygielbaum P.S.: Magnetohydrodynamics: An engineering perspective. *IEEE Trans. PAS*, 100(5), 1981, 2529–2538.
2.39. Zwoliński G., Turowski J., Kulkarni S.V., and Koppikar D.A.: Method of fast computation of 3-D stray field and losses in highly asymmetric transformers. *Proc. 5th International Conference on Transformers. TRAFOTECH'98*. January 23–24, 1998. Mumbai, India, pp. I–51 to I–58.
2.40. Sze S.M. and Ng K.K., *Physics of Semiconductor Devices*. New York: Wiley and Sons, 2007.
2.41. Turowski J.: Losses and local overheating caused by stray fields. (in Polish). *Scientific Archives of the Technical University of Lodz "Elektryka*, No. 11, 1963, pp. 89–179 (*Habilitation (DSc) dissertation; 1st Ministry Award).*
2.42. Raman A. and Turowski M.: Mixed-Mode TCAD Tools, Chapter 31 in *Extreme Environment Electronics*. Boca Raton, FL: CRC Press, November 2012.
2.43. Cidronali A., Collodi G., Santarelli A., Vannini G., and Manes G.: Millimeter-wave FET modeling using on-wafer measurements and EM simulation. *IEEE Trans. on Microwave Theory and Techniques*, 50(2), Feb 2002, 425–432.
2.44. Labiod S., Latreche S., Bella M., and Gontrand C.: Combined electromagnetic and drift diffusion models for microwave semiconductor device, *Journal of Electromagnetic Analysis and Applications*, 3(10), 2011, 423–429. [available online: http://www.scirp.org/Journal/PaperInformation.aspx?paperID=8146].
2.45. Mirzavand R., Abdipour A., and Moradi G., Full-wave semiconductor devices simulation using ADI-FDTD method, *Progress in Electromagnetic Research M*, 11, 2010, 191–202.

CHAPTER 3

3.1. Bron O.B.: *Electromagnetic Field as a Form of Matter* (in Russian). Moscow: Gosenergoizdat, 1962.
3.2. Leites L.V.: *Electromagnetic Calculations of Transformers and Reactors*. (in Russian). Moscow: Energia, 1981.

3.3. Poynting T.: On the transfer of energy in electromagnetic field. *Trans. of Roy. Soc. Lo*, London. Part II, 1884, pp. 343–361.

3.4. Turowski J.: (1) Generalized equations for electromechanical and electrodynamic energy conversion in transformers. *Proc. of Internat. Symposium on "Electrodynamics. Forces and Losses in Transformers"*. PŁ-SEP-OŁIL-ELTA, Lodz, Poland, October 18–20, 1979, pp. 296–309; and (2) *Rozprawy Elektrotechniczne*, 26, 1980, pp. 911–927.

CHAPTER 4

4.1. Appolonskiy C.M.: *Calculation of the Electromagnetic Screening Shells*. (in Russian). Leningrad: Energoizdat, 1982.

4.2. Kaden T.: *Electromagnetic Shields in High-Technology and Telecommunications Mechanics*. (in Russian, *Translated from German*). Moscow: Gosenergoizdat, 1959.

4.3. Smythe W.R.: *Static and Dynamic Electricity*. New York, 1950 (and Moscow: Gosenergoizdat 1954), 3rd edition. New York: McGraw-Hill, 1968.

4.4. Wiszniewski N.J. et al.: *High-Pressure Apparatus with Hermetic Driving System*. (in Polish). Warsaw: WNT, 1963.

4.5. Davey K.R. and King E.I.: A three dimensional scalar potential field solution and its application to the turbine generator end region. *IEEE Trans*. PAS-100(5), 1981, 2302–2310.

4.6. Djurovič M. and Monson J.E.: Stray losses in the step of a transformer yoke with a horizontal magnetic shunt. *IEEE Trans*. PAS-101(2), 1982, 2995–3000.

4.7. Doi S., Ito K., and Nonaka S.: Three-dimensional thermal analysis of stator end-core for large turbinegenerators using flow visualization results. *IEEE Transactions*, PAS-104(7), 1985, 1856–1862.

4.8. Hołoga Z.: Enclosed heavy current busbars. *Polish Engineering*, (4), 1983, 7–10.

4.9. Jack A.G. and Mecrow B.C.: Computation and validation against measurement of stator end-region fields in synchronous generator. *IEE Proc*., 133, Pt. C(1), 1986, 26–32.

4.10. Mendrela E.A., Mendrela E.M., and Turowski J.: Flux density field, currents, forces and power losses and integral parameters in induction motors with rotary and linear multi-layer shielded rotor. (in Polish). *Report of IMET Tech. Univ. of Lodz*, No. 2687 CPBR 5.7, Phase I, October 1987.

4.11. Rizzo M., Savini A., and Turowski J.: Influence of slot dimensions on forces in turbine generators. *Proc. of Internat. Symposium on Electromagnetic Fields in Electrical Engineering—ISEF'89*, September 20–22, 1989, Lodz, Poland. Session "Synchronous Machines II", Paper II.7.

4.12. Rizzo M., Savini A., and Turowski J.: Influence of flux collectors on stray losses in transformers. *IEEE Transactions on Magnetics*, 36(4), July 2000, 1915–1918.

4.13. Rizzo M., Savini A., and Turowski J.: Destructive forces in large power transformers. *Proceedings of the International IGTE Symposium on Numerical Field Calculation in Electrical Engineering*, Graz, Austria, September 16–18, 2002.

4.14. Savini A. and Turowski J.: The effects of saturation on forces in slots of large synchronous generators. *Archiwum Elektrotechniki,* PAN. (3/4), 1985, 611–620.

4.15. Szymański G. and Patecki A.: (1) Eddy current and temperature of the sheath in three-phase pipe-sheathing systems. *IEEE Trans. on Magnetics*, Mag-20(5), September 1984, 2004–2006; and (2) ISEF'89, Lodz, pp. 97–100.

4.16. Turowski J.: Losses in single and three phase transformers covers. (in Polish). *Rozprawy Elektrotechniczne*, (1), 1959, 87–119.

4.17. Turowski J.: Development trends in the construction of large turbogenerators. (in Polish). *Przegląd Elektrotechniczny*, (11), 1965, 456–459.

4.18. Turowski J.: Comparative analysis of existing methods of calculation of losses in tanks and constructional parts of transformers. (In Polish). *Proc. of Institute of Elektrotechnics*, Warsaw, 1972, No. 72, pp. 17–35

4.19. Turowski J.: Reluctance model of stray field in transformer. (in Polish). *Rozprawy Elektrotechniczne*, (4), 1984, 1121–1144.

4.20. Turowski J. and Turowski M.: A network approach to the solution of stray field problems in large transformers. *Rozprawy Elektrotechniczne*, PAN.(2), 1985, 405–422.

4.21. Turowski J.: Wave method of calculating the screens and elements of electric machines. (in Russian). *Elektricziestwo*, (12), 1965, 53–58 (6 fig., 1 tab.).

4.22. Turowski J. and Brulikowski J.: Effect of distance of screen from the steel wall on the power losses in the system shielded. (in Polish). *Proc. of Jubilee Scientific Session of the 25th Anniversary of the Faculty of Electrical Engineering*, Technical University of Lodz. June 1970, No. 2, pp. 94–101.

4.23. Turowski J. and Byczkowska L.: Computer calculation of power losses in tanks of transformers. (in Polish). *Ibid*, June 1970, No. 2, pp. 102–111.

4.24. Turowski J. and Janowski T.: Model investigation of tube screens and the skin effect in iron. *Archiwum Elektrotechniki.*, (1), 1963, 81–103.

4.25. Turowski J. and Przytuła A.: Distribution of equiphase stray fluxes in three-winding transformers. *"Elektryka" Scientific Reports of the Tech. Univ. of Lodz*, No. 8, 1961, pp. 91–114.

4.26. Turowski J.: Influence of screens and shunt structure on stray field distribution. *Proc. of Jubilee Conference "The Transformer Production is 100 Years Old"*. Budapest, June 18–21, pp. 470–477.

4.27. Turowski M. and Kopeć M.: Dedicated network program for interactive design of screens and field distribution in electromagnetic devices. *Rozprawy Elektrot.*,32(3), 1986, pp. 835–844.

4.28. Turowski J., Kopeć M., and Turowski M.: Fast 3-D analysis of combined Cu & Fe screens in three-phase transformers. *ISEF'89. Electromagnetic Fields in Electrical Engineering. COMPEL.*London: James & James 1990, pp. 113–116; and *ITMA'91*. New Delhi, India.

4.29. Turowski J.: Application of reluctance network to estimate stray losses in transformer tank. *Proc. of Power Transformers Conference "Transformer'99"*. Kolobrzeg, Poland, April 1999, pp. 19–26.

4.30. Turowski J., Kraj I., and Kulasek K.: Industrial verification of rapid design methods in power transformers. *International Conference Transformer'01*, Bydgoszcz, Poland, June 2001.

4.31. Turowski J. *(TUL)*, Lopez-Fernandez X.M. *(Vigo Univ. Spain)*, Soto A., and Souto D. *(EFACEC Trafo Works, Porto, Portugal)*: Stray losses control in core- and shell type transformers. Part I: Upgrading of energy-saving and reliability of large transformers. Chapter in the book: *International Advanced Research Workshop on Transformers—ARWtr2007* X.M. Lopez-Fernandez (editors), Baiona, Spain, October 2007, pp. 56–68.

4.32. Turowski J.: Effect of saturation on the forces in the windings of electrical machines. (in Polish). *Archiwum Elektrotechniki*, (4), 1960, 677–691.

4.33. Zarębski W.: Calculation of eddy currents induced in the bus-bar screens. (in Polish). *Przegląd Elektrotechniczny*, XXXVIII(8), 1962, 334–339.

4.34. Leon O.C. and Lin P.-M.: *Computer-Aided Analysis of Electronic Circuits: Algorithms and Computational Techniques*. Englewood Cliffs, NJ: Prentice-Hall, 1975.

4.35. Turowski M. and Raman A.: Device-circuit models for extreme environment electronics. *Proc. of 19th Int. Conf. Mixed Design of Integrated Circuits and Systems (MIXDES 2012)*. Warsaw, Poland, May 24–26, 2012, pp. 350–355.

4.36. Lopez-Fernandez X.M., Soto A., Souto D., and Turowski J.: Stray losses control in shell type transformers. Part II: Method of three-dimensional reluctance network solution (RNM-3D$_{Shell}$). *Advanced Research Workshop on Transformers—ARWtr'2007*, Baiona, Spain, Oct. 2007, pp. 65–76.

4.37. Sevgi L.: Electromagnetic screening and shielding-effectiveness (SE) modeling. *IEEE Antennas and Propagation Magazine*, 51(1), 2009, pp. 211–216.

CHAPTER 5

5.1. Hague B.: *Electromagnetic Problems in Electrical Engineering*. London. Oxford University Press, 1929.

5.2. Jezierski E., Gogolewski Z., Kopczyński Z., and Szmit J.: *Transformers. Construction and Design*. (in Polish). Warsaw: WNT, 1963.

5.3. Sajdak Cz. and Samek E.: *Induction Heating*. Katowice: "Śląsk" Publishing House, 1985.

5.4. Śliwiński T. and Głowacki A.: *Starting Parameters of Induction Motors*. (in Polish). Warsaw: PWN, 1982.

5.5. Roters H.C.: *Electromagnetic Devices*. New York: J. Willey, 1961.

5.6. Althammer P.: Calculation of magnetic field in turbogenerators (in German). *Brown Boveri Mitt.*, (5), 1964, 281–292.

5.7. Carpenter C.J.: The application of the method of images to machine end-winding fields. *Proc. IEE—Part A: Power Engineering*, 107(35), October 1960, 487–500.

5.8. Hammond P.: Electric and magnetic images. *Proc. IEE—Part C*, 107(12), Sep. 1960, 306–313.

5.9. Kaźmierski M.: An estimation of approximate methods for the analysis of electromagnetic fields in transformers. *Rozprawy Elektrotechniczne*, 16(1), 1970, 3–26.

5.10. Lehmann T.: Methode graphique. *Lumière Electr.* 1909, 103–137; and *Rev. Gen. Electr.* 1933, 351.

5.11. Mullineux N., Reed J.R., and White I.J.: Current distribution in sheet and foil-wound transformers. *IEE Proceedings*, (1), 1969, 127–129.

5.12. Rizzo M., Savini A., and Turowski J.: Influence of saturation on a torque of DC traction motors. *Internat. Conference on Electric Machines. ICEM'88*. Pisa, Italy. September 12–13, 1988, Vol. 1, pp. 85–88.

5.13. Rizzo M., Savini A., and Turowski J.: The harmonic content of the flux distribution curves in DC traction motors. *Proc. of the Internat. Conference on Electric Drives ICED'88*. Polana Brasov, Romania, September 20–22, 1988, Paper 8.1–10, pp. 1–9.

5.14. Turowski J., Fleszar J., Gierczyk J., and Mendrela E.: Criteria for the applicability of the method of analytical extensions for computing of the field by Fourier series. *National Conference on Electromagnetic Field Theory*. Jadwisin, Poland, 1982, pp. 91–96.

5.15. Turowski J.: Power losses in transformer covers at various methods of their reduction. (in Polish) *Archiwum Elektrotechniki*, (4), 1959, 529–556.

5.16. Turowski J., Pawłowski J., and Pinkiewicz I.: The model test of stray losses in transformers. (in Polish) *"Elektryka" Scientific Papers of Techn. University of Lodz*, (12), 1963, 95–115.

5.17. Turowski J., Andryszczak W., and Pelikant A.: Application of the method of mirror images to calculation of the slot impedances, taking into account the saturation of the teeth. (in Polish). *"Elektryka" Scientific Papers of Techn. University of Lodz*, (74), 1983, 187–198.

5.18. Turowski J.: Additional losses in foil- and bar-wound transformers. *Transactions IEEE. Power, App. and Syst. Winter Power Meeting and Tesla Symposium*. New York, USA. Paper A-76151. *Catalog Power Eng. Soc.* No 76, CH 1075-1-PWR, January 1976, pp. 1–7.

5.19. Zakrzewski K. and Sykulski J.: Mirror image of AC currents in unilateral conductive screen, in the light of the of vector potential method. (in Polish). *Rozprawy Elektrotechniczne*, 23(1), 1977, 73–92.
5.20. Zakrzewski K. and Turowski J.: Methods of analysis of magnetic screening transformer tank. (in Polish) *Rozprawy Elektrotechniczne*, (3), 1972, 575–601.
5.21. North G.G.: Harmonic analysis of a short stator linear induction motor using a transformation technique. *IEEE Trans. on Power Apparatus and Systems*, PAS-92(5), 1973, 1733–1743.
5.22. Krishnan R.: *Permanent Magnet Synchronous and Brushless DC Motor Drives*. Boca Raton: CRC Press, 2009.
5.23. Sharifian M.B., Shaarbafi K., Faiz J., and Feyzi M.: Slot fringing effect on the magnetic characteristics of the electrical machines. *Proc. ICECS 2003*, 2, 778–781.
5.24. Harley R.G. and Duan Y.: Traditional design of cage rotor induction motors, *CEME-2009*, Georgia Institute of Technology, Nov. 2009.
5.25. Deepa V. and Bindu R.: Design of three phase induction motor in windows programming platform. *Proc. Third Biennial National Conference, NCNTE-2012*, 56–61.
5.26. Nietushil A.B. and Polivanov K.M.: *Bases of Electrical Engineering. Part 3. Theory of Electromagnetic Field.* (in Russian). Moscow: Gosenergoizdat, 1956.

CHAPTER 6

6.1. Di Barba P.: *Multiobjective Shape Design in Electricity and Magnetism.* London: Springer, 2010.
6.2. Gieras J., Piech Z., and Tomczuk B.: *Linear Synchronous Motors: Transportation and Automation Systems (Electric Power Engineering Series)*, CRC Press; 2nd edition, August 2011.
6.3. Gieras J.: *Advancement in Electric Machines*. London: Springer, 2010.
6.4. Courant R.: *Differential and Integral Calculus*. New York: Wiley-Interscience, 1992; Ishi Press, 2010.
6.5. Postnikov I.M.: *Design of Electric Machines*. (in Russian). Kiev: GITL, 1960.
6.6. Ryzhik I.M. and Gradstain I.S.: Tables of integrals, series and products. (in Russian). Moscow, Gos. Izd. Tech.-Teoret. Literat. 1951.
6.7. Stafl M.: *Electrodynamics of Electrical Machines*. Praha: Academia, 1967.
6.8. Turowski J.: *Electromachine Components of Automatics*. (in Polish). Techn. University of Lodz, 1989.
6.9. Hańka L.: Induced transient phenomena in planar bodies. (in Czech). *Elektrotechn.*, Obzor 8, 1962 (51), 378–382.
6.10. Kaźmierski M. and Kersz I.: Effectiveness of longitudinal conducting screens. Part I. Electromagnetic field and stray losses. Part II. Temperature rise. (in Polish). *Rozprawy Elektrot.*, (2–3), 1989.
6.11. Lasociński J.: The electromagnetic field in the gap of machine with a solid ferromagnetic rotor taking into account the finite length of the machine. (in Polish). *Rozprawy Elektrot.*, XII(1), 1966, 69–92.
6.12. Margolin S.D.: Calculation of magnetic skin-effect in sheet steel. (in Russian). *Journal Tech. Phys.* 10, 1948.
6.13. Skoczkowski M. and Kalus M.: The mathematical model of induction heating of ferromagnetic pipes. *IEEE Trans. on Magnetics*, 25(3), 1989, 2745–2750 (and *ISEF'87 Conf.*, Pavia).
6.14. Strzelecki M.: Power losses and impedances of single-phase large-current paths. (in Polish). *XI SPETO*, Wisła, April 20–23, 1988, 245–253.

6.15. Turowski J.: Selection of optimal thickness of electrotechnical sheet from the standpoint of electrical properties of the core. (in Polish). *Przegląd Elektrotechniczny*, 8, 1964, 361–368.

6.16. Turowski J., Komeza K., Pelikant A., Wiak S. M., and Savini A.: Electromagnetic field problems in small electronic controlled motors. *Proc. of Internat. Aegean Conference on Electr. Machines and Power Electronics-ACEMP'92*. May 27–29, 1992. Turkey: Kusadasi. (Invited paper). Plenary Session. *Proceedings*, Vol. 2/2, pp. 430–437.

6.17. Turowski J.: Electromagnetic field and power losses in covers of transformers. PhD thesis. Technical University of Lodz, Poland. December 1957, pp. 1–142. (Published in parts in [1.15/1], [2.31], [4.16], [5.15], and [6.15]).

6.18. Lozinskiy M.G.: *Surface Hardening and Induction Heating of Steel*. (in Russian). Moscow: Mashgiz, 1949.

CHAPTER 7

7.1. Chari M.V.K. and Silvester P.P.: *Finite Elements in Electrical and Magnetic Field Problems*. New York: J. Wiley, 1980.

7.2. Dąbrowski M.: *Analysis of Magnetic Circuits. Power Losses in the Circuits*. (in Polish). Warszawa-Poznań: PWN, 1981.

7.3. (a) Fortuna Z., Macukow B., and Wąsowski J.: *Numerical Methods*. (in Polish). Warsaw: PWN, 1982. (b) Faires J.D. and Burden R.L.: *Numerical Methods*. Pacific Grove, CA: Brooks Cole, 3rd edition, 2002.

7.4. (a) Golde W., Norek C., and Paszkowski S.: *Outline of Approximation Theory and its Applications in Electrical Engineering*. (in Polish). Warsaw: PWN, 1958. (b) Hamming R.W.: *Numerical Methods for Scientists and Engineers*. New York: Dover Publications; 2nd edition, 1987.

7.5. Arkhangielskiy B.I.: The analytical expression of the magnetization curve of electric machines. *Elektrichestvo*, (3), 1950, 30–32.

7.6. Coulson M.A., Preston T.W., and Reece A.B.: 3-Dimensional finite element solvers for the design of electrical equipment. *COMPUMAG'85 Conference*, Fort Collins, USA, 1985.

7.7. Kaźmierski M.: Replacement permeability of screened ferromagnetism. (in Polish). *Rozprawy Elektrot.* 22(2), 1976, 287–301.

7.8. Komęza K. and Krusz G.: Hermitian finite elements in numerical analysis of transformer leakage magnetic field. *International Symposium ISEF'89*, Lodz 1989, pp. 109–112.

7.9. Kozlowski M. and Turowski J.: Stray losses and local overheating hazard in transformers. *CIGRE. Plenary Session*, Paris 1972. Report 12–10.

7.10. Lasociński J.: Approximate methods for analyzing one-dimensional electromagnetic fields in solid iron. (in Polish). *Rozprawy Elektrot.*, (3), 1967, 555–577.

7.11. Neiman L.R: *Surface Effect in Ferromagnetic Bodies*. (in Russian). Leningrad-Moscow: Gosenergoizdat, 1949.

7.12. Neiman L.R. and Zaicev I.A.: Experimental verification of the surface effect in tubular steel bars. (in Russian). *Elektrichestvo*, (2), 1950, 3–8.

7.13. Rosenberg E.: Eddy current in solid iron. (in German). *E. u M.*, 41, 1923, 317–325.

7.14. Savini A.: Modelling hysteresis loops for finite element magnetic field calculation. *IEEE Trans. on Magnetics*, 18(2), March 1982, 552–557.

7.15. Turowski J.: Fast evaluation of 3-D eddy current loss distribution in transformer tanks. *Proc. of 13th IMACS World Congress on Computation and Applied Mathematics*. July 22–26, 1991. Dublin, Ireland. Session 12.2.5 (Invited paper).

7.16. Turowski J.: Losses in covers of three phase transformers caused by the electromagnetic field of bushings (in Polish). *Science Reports "Elektryka"*, No. 4, Tech. University of Lodz, 1958, pp. 79–101.
7.17. Turowski J.: A method for the automatic calculation of the eddy current losses in solid iron. (in German). *XI. International Scientific Colloquium of the Technical University Ilmenau.* 1966. Lecture series "Electrical Machines" Part I, pp. 51–56.
7.18. Turowski J.: The formula for power loss in unscreened tanks of large single-phase transformers. (in Polish) *Rozprawy Elektrotechniczne*, (1), 1969,149–176.
7.19. Turowski J.: The meshed model of stray field in three-phase transformer. (in Polish). *Science Reports "Elektryka"*, Techn. University of Lodz, No. 66, 1979, pp. 105–122.
7.20. Turowski J.: Generalized method of calculating the stray losses in metal parts of the construction. (in Polish). *Rozprawy Elektrotechniczne*,(1), 1968, 179–195.
7.21. Turowski J., Kopeć M., and Turowski M.: Method of fast analysis of 3-D leakage fields in large three-phase transformers. *COMPEL—The International Journal for Computation and Mathematics in Electrical Engineering.* London: James & James, 1990, Vol. 9, Supplement A, pp. 167–170 and pp. 302–304.
7.22. Turowski J. and Zakrzewski K.: Problems of creation of strong magnetic fields. (in Russian). *World Electrotechnical Congress "WELC'77"*. Moscow, June 1977. Invited plenary paper, No. 1.59, pp. 1–15.
7.23. Turowski J., Turowski M., and Kopeć M.: Method of three-dimensional network solution of leakage field of three-phase transformers. *IEEE Transactions on Magnetics*, 26(5), September 1990, 2911–2919.
7.24. Turowski J. and Zwoliński G.: Hybrid model for fast computation of 3-D leakage field and eddy current losses in highly asymmetric structures. *Proc. of 5th Polish-Japanese Joint Seminar on Electromagnetics in Science and Technology.* Gdansk'97, Poland, May 19–21, 1997, pp. 165–168.
7.25. Wiak S.: Analysis of transients in electrical circuits containing solenoid with initially magnetized cores for impulse excitations. *Archiv für Elektrotechnik*, 71, 1988, 1–7.
7.26. Zakrzewski K.: Analysis of the electromagnetic field in solid iron by using numerical methods. (in Polish). *Archiwum Elektrot.*, XVIII(3), 1969, 569–585.
7.27. Zakrzewski K.: Practical method for calculating active and reactive power in solid iron. (in Polish). *Rozprawy Elektrotechniczne*, 24(1), 1975, 213–233.
7.28. Zolotariov N.A.: Mathematic modeling of magnetic hysteresis. *Elektrichestvo*,(6), 1989, 75–79.
7.29. Życki Z.: The accuracy of calculation of magnetic circuits with permanent magnets. *Scientific Papers of Inst. of Electromachine Systems*, Wroclaw University of Technology, No. 39, 1987, pp. 61–80.
7.30. Zwoliński G.: Stray fields and losses in asymmetric transformer structures. PhD thesis. Technical University of Lodz, 1999 (supervisor *J. Turowski*).
7.31. Zwolinski G. and Turowski J.: Stray losses in multiwinding transformers on load. *International Conference on Power Transformers—Transformer 03*. Pieczyska, Poland, May 2003.
7.32. Kulkarni S.V. and Khaparde S.A.: Stray loss evaluation in power transformers—A review. *Proc. IEEE. PES Winter Meeting*, Jan. 2000, pp. 2269–2274, Paper 0-7803-5938-0/00.
7.33. Lopez-Fernandez X.M., Turowski J., Comesaña, Souto-Rodriguez A., Couto D., and Jácomo Ramos: Three-dimensional leakage magnetic field distribution in shell-form power transformer. *ISEF 2009—XIV Int. Symposium on Electromagnetic Fields in Mechatronics, Electr. and Electron. Engng*. Arras, France, September 10–12, 2009.

7.34. Turowski J.: Fast calculation and reduction of stray losses in power transformers. (Theory in the light of industrial practice). Updated for the Transformer Works WEG in Brazil. *Internat. "Brasil-Poland-Spain" Seminar*, Lodz, Poland, December 6–9, 2011, pp. 1–11.

7.35. Hargreaves P.A., Mecrow B. C., and Hall R.: Calculation of iron loss in electrical generators using finite element analysis. *IEEE Trans. on Industry Applications*, 48(5), Sept/Oct. 2012, 1460–1466.

7.36. Li Y., Eerhemubayaer, Sun X., Jing Y., and Li J.: Calculation and analysis of 3-D nonlinear eddy current field and structure losses in transformer. *2011 International Conference on Electrical Machines and Systems (ICEMS)*, pp. 1–4.

7.37. Zhu Z., Xie D., Wang G., Zhang Y., and Yan X.: Computation of 3-D magnetic leakage field and stray losses in large power transformer. *IEEE Trans. on Magnetics*, 48(2), 739–742, Feb. 2012.

CHAPTER 8

8.1. Kurdziel R.: *Action of Thermal and Dynamic Short-circuit Currents*. Warsaw: PWT, 1957.

8.2. Kuzniecov I.F. and Cicikian G.N.: *Electrodynamic Forces in the Current-carrying Parts of Electrical Equipment and Conductors*. Leningrad: Energoatomizdat, 1989.

8.3. Puchała A.: *The Dynamics of Electromechanical Equipment*. Warsaw: PWN, 1977.

8.4. Adamiak K., Mendrela E.A., and Mendrela E.M.: Force field distribution in the chamber of shaft magnetic separators. *ISEF'85*, Warsaw, Paper X.69, pp. 295–298.

8.5. Bytnar A.: Compensating abilities of turbogenerators working in an electrical power systems. *Scientific Archives of Warsaw University of Technology "Elektryka"*, (90), 1989, 21–35.

8.6. Drak B., Paszek S., and Paszek W.: Electrodynamic forces acting on the stator winding overhangs of the turbogenerator. *ISEF'89*, 229–232.

8.7. *IEC Publication 909*: Short-circuit Current Calculation in Three-Phase AC-Systems, 1988.

8.8. Kranz R.: Special Report of Group 11 (Rotating Machines) CIGRE 1984. Session Paris, Report 11.00.

8.9. Lefevre Y., Lajoie-Mazenc M., and Daval B.: Force calculating in electromagnetic devices. *ISEF'87*. Paper V-4, pp. 227–230.

8.10. Nowak L.: Optimization of acyclic transducers including the electromechanical dynamic states. Polit. Poznan, Dissertations. No. 191, 1988.

8.11. Patecki A. and Szymański C.: Electrodynamic forces in three-phase pipe-sheathing system. *ISEF'85*, Warsaw, Poland, Paper XI.78. pp. 335–337.

8.12. Patel M.R.: Crushing stresses on hollow conductors in large water-cooled generator stator slots. *IEEE Trans.* PAS-102(5), 1983, 1365–1370.

8.13. Rais V.R., Turowski J., and Turowski M.: Reluctance network analysis of coupled fields in reversible electromagnetic pump. *Proc. of Internat. Symposium on Electromagnetic Fields in Electrical Engineering—ISEF'87*. 23–25 Sep. 1987. Paper No. VI-8, pp. 291–294, Pavia, Italy (Plenary Report).

8.14. Rizzo M., Savini A., and Turowski J.: Force calculation in large electrical machines. *IEEE Transactions on Magnetics*. 25(4), July 1989, 2977–2979.

8.15. Rizzo M., Savini A., and Turowski J.: Influence of slot dimensions on forces in turbine generators. *Electromagnetic Fields in Electrical Engineering*. London: James & James, 1990, *ISEF'89*, pp. 193–196.

8.16. Savini A. and Turowski J.: The effects of saturation on forces in slots of large synchronous generators. *Archiwum Elektrotechniki,* PAN. No. 3/4, 1985, 611–620.

8.17. Turowski J.: Effect of saturation on the forces in the windings of electrical machines. *Archiwum Elektrotechniki*, (4), 1960, 677–691.
8.18. Rizzo M., Savini A., and Turowski J.: Evolution in calculating crushing forces and vibrations in slots of large turbo-alternators. *Proc. of International Conference on the Evolution and Modern Aspects of Synchronous Machines*, Zurich, Switzerland, 27–29 Aug. 1991, SM-100.
8.19. Rizzo M., Savini A., and Turowski J.: Dependence of forces, eddy currents, and stray losses on geometric structure and parameters of power transformers. *Proc. of the 4th International Conference on Electromagnetic Field Problems & Applications, ICEF-2000*, 18–20 September, 2000, Tianjin, China. International Academic Publishers. Beijing: World Publishing Corporation, 2000, pp. 351–354.
8.20. Sobiczewska G., Sosnowski S., and Turowski J.: The dynamics of maneuvering trucks and cranes with linear motors. *Proc. SEP-ITMA-PŁ-OBRME Conference*, Katowice-Chorzów, November 4–5, 1976. "Electric Machines". BOBRME Problem Papers, No. 24/1976, pp. 38–42.
8.21. Worotynski J., Turowski M., and Mendrela E.A.: The accuracy of calculation of magnetic fields, inductance and forces in electromagnetic devices using the reluctance network method (RNM), *ISEF'93. COMPEL*. London: James & James Science Publishers, 1994, pp. 159–163.
8.22. Pathmaathan M. and Soong W.L.: Nesimi Ertugrul: Maximum-torque-per-ampere control of phase advance modulation of an SPM wind generator. *IEEE Trans. on Industry Applications,* 48(5), September/October 2012, 1443–1451.
8.23. De Donato G., Capponi F.G., Rivellini G.A., and Caricchi F.: Integra-slot versus fractional-slot concentrated-winding axial-flux PM machines: Comparative design, FEA, and experimental tests. *IEEE Trans. on Industry Applications*, 48(5), September/October 2012, 1487–1495.
8.24. Weizhong Fei, Chi Kwong Luk P., Jian Xin Shen, Yu Wang, and Mengjia Jin.: A novel PM flux switching machine with an outer-rotor configuration for in-wheel light traction applications. *IEEE Trans. on Industry Applications*, 48(5), September/October 2012, 1496–1505.
8.25. Rizzo M., Savini A., and Turowski J.: Effect of the nonlinearity of demagnetization curves on the torque of hybrid stepping motors with permanent magnet in slots. *COMPEL*. 17(1–3), 1998, pp. 318–323.
8.26. Wiak S. and Turowski J.: Computer simulation of field and transient processes in linear electromechanical converter. *International Seminar on Linear Electric Machines*. September 1–6. 1990, Odessa, Ukraine.
8.27. Rizzo M., Savini A., and Turowski J.: Influence of the shape of nonuniform gaps on the torques of DC brushless motors. *4th Biennial IEEE Conference on Electromagnetic Field Computation*. Toronto, Canada, October 22–24, 1990.
8.28. Wiak S., Komęza K., Savini A., Turowski J., and Rizzo M.: Field and torque calculation of fractional DC permanent magnet motor by Hermitian finite element method. Book: *Electromagnetic Field Problems and Applications—ICEF'92*. International Beijing: Academic Publishers, 1992, pp. 480–484.
8.29. Rizzo M., Savini A., and Turowski J.: Influence of number of poles on torques of DC brushless motors with auxiliary salient poles. *Intermag'91 Conference*, Pittsburgh, USA.
8.30. Rizzo M., Savini A., and Turowski J.: FEM analysis and design of polar shoes in permanent magnet DC motors. *Book: Simulation and Design of Applied Electromagnetic Systems* (T. Honma editor). Amsterdam: Elsevier Science B.V., 1994, pp. 265–268.
8.31. Rizzo M., Savini A., and Turowski J.: The shaping of flux density at the air gap of small DC motors with different permanent-magnet poles. Book: *Electric and*

Magnetic Fields (A. Nicolet and R. Belmans editors). New York: Plenum Press, 1995, pp. 319–322.

8.32. Rizzo M., Savini A., and Turowski J.: Influence of permanent-magnet inserts on the reluctance torque of small electric machines. *Electromagnetic Field Problems & Applications ICEF'96*. Beijing: International Academic Publishers, 1996, pp. 144–146.

CHAPTER 9

9.1. Kącki E.: *Thermokinetics*. (in Polish) Warsaw: WNT, 1967.

9.2. Michejev M.A.: *Principles of Heat Transfer*. (in Polish). Warsaw: WNT, 1953.

9.3. Apanasewicz S. and Kazmierski M.: Electromagnetic field in cover-joint region due to AC through LV leads in generator transformers. *ISEF'89*. 20-22.9.1969, pp. 137–140.

9.4. Bereza V.L., Furman J.I., Ivankov C.F., and Nizhnik L.P.: Eddy currents and losses in the transformer tank cover induced by the field of heavy current taps. *ISEF'85*. Warsaw, Paper No. 57, pp. 245–248.

9.5. Long L.W.: Overloading and loading limitation of large transformers. *CIGRE, Electra*, No 86, 1983, pp. 33–51.

9.6. Partyga S. and Turowski J.: Current problems of exploitation and construction of transformer. (in Polish). *Przegląd Elektrotechniczny*, No. 8–9, 1982, 234–236.

9.7. Sofia, J.W.: Analysis of thermal transient data with synthesized dynamic models for semiconductor devices. *IEEE Transactions on Components, Packaging, and Manufacturing Technology Part A (CPMT)*, 18, March 1995, 39–47.

9.8. Turowski J.: Localization of stray losses and excessive temperature rises of structural elements by the reluctance network method. *Works of the Inst. of Elektrotech.*, (132), 1984, 7–19.

9.9. Turowski J.: Overheating hazard in flanged bolt joints of transformers. *Proc. Internat. Symposium on Electromagnetic Fields in Electrical Engineering - ISEF'85*. Warsaw, Poland, 26–28 September 1985. Paper No. 63, pp. 271–274 (Plenary lecture).

9.10. Turowski J.: Local overheating hazard in transformers. (in Russian). *Izvestia Vysshikh Uchebnykh Zavedenij. SSSR "Elektromekhanika"*, 8/1961, pp. 91–114.

9.11. Turowski J.: Rapid evaluation of excessive local heating hazard in bushing turrets of large power transformers. *XVI International Conference on Electrical Machines—ICEM'04*, Cracow, Poland, September 2004.

9.12. Turowski J. and Pelikant A.: Eddy current losses and hot-spot evaluation in cover plates of power transformers. *IEE Proceedings—Generation, Transmission and Distribution*, 144(6), November 1997, pp. 435–440, and *ISEF'97*, Gdansk, Poland, 25–27 September 1997.

9.13. Cranganu-Cretu B. and Schneider M.: Coupled electromagnetic-thermal analysis for ABB power transformers. *Proc. Int. Colloq. Transform. Research Asset Manage.* Cavtat, Croatia, November 2009, pp. 1–9.

9.14. Penabad-Duran P., Lopez-Fernandez X.M., Turowski J., and Ribeiro P.M.: 3D heating hazard assessment on transformer covers. Arrangement decision. *COMPEL Int. J. Comput. Math. in Elect. Electron. Eng.*,31(2), 2012, 703–715.

9.15. Lopez-Fernandez X.M., Penabad-Duran P., Turowski J., and Ribeiro P.M.: Nonlinear heating hazard assessment on transformer covers and tank walls. *Electrical Review*, (7b), 2012, 28–31.

CHAPTER 10

10.1. Gotszalk R.: *Laboratory Techniques and Measurements of Electrical Quantities*. (in Polish). Wroclaw University of Technology, 1985.

10.2. Heptner H. and Stroppe H.: *Magnetic and Inductive Testing of Metals*. (in Polish). Ed. "Śląsk", 1972.

10.3. Ivanov-Smolenskiy A.V.: *Electromagnetic Fields and Processes in Electric Machines and their Physical Modeling*. (in Russian). Moscow: Energia, 1969.

10.4. Bolshakova, I.A. et al.: Magnetic measuring instrumentation with radiation-resistant Hall sensors for fusion reactors: Experience of testing at JET. *IEEE Transactions on Nuclear Science*, 59, Aug. 2012, 1224–1231.

10.5. Nałęcz M. and Jaworski J.: *Magnetic Measurements Techniques*. (in Polish). Warsaw: WNT, 1968.

10.6. Smielianskij M.J. and Bortniczuk N.I.: *Short Networks of Electric Furnaces*. (in Russian). Moscow: GEI, 1962.

10.7. Trzaska Z., Krzemiński S., Wincenciak S., Sikora J., Skoczylas J., and Filipowicz F.: *Laboratory of Electromagnetic Field*. (in Polish). Warsaw: Warsaw University of Technology, 1979.

10.8. Allen P.H.O.: Loss measurement survey by Poynting vector wattmeter. *Electrical Journal*. 161(25), 1958.

10.9. Bolkowski S. and Karwat T.: The use of power-frequency multipliers in the laboratory of the electromagnetic field. (in Polish) *Proc. of XI Seminar SPETO'88*. Wisła, Poland, 1988, 143–145.

10.10. Cjchy J. and Dąbrowski M.: The study of magnetic properties of structural steel. *Archiwum Elektrot*. 44(2), 1975, 329–343.

10.11. Derszwarc G.B. and Smielanskij M.J.: Electrodynamic modeling of electrothermal devices. (in Russian). *Elektrichestvo*, (10), 1951, 47–50.

10.12. Deutsch G.: Measurement of active power losses of large reactors. (in German). *Brown Boveri Mitt.*, 47(4), 1960.

10.13. Janowski T.: The method of measurement of additional load losses occurring outside the windings. (in Polish). *Sci. Proc. of WSI-Lublin, Elektryka*, October 1969.

10.14. Janowski T.: Power measurements using wattmeter compensation attachment (in Polish). *Sci. Proc. of WSI-Lublin, Elektryka*, October 1970, 123–137.

10.15. Janowski T. and Malinowski H.: Superconducting solenoid model for transient studies. (in Polish). *Proc. SPETO'89*. Wisła, Poland, 1989.

10.16. Janowski T. and Turowski J.: The criterion for the selection of method for measuring power losses in tanks of transformers. (in Polish). *Rozprawy Elektrotechniczne*, (1), 1970, 205–226.

10.17. Jezierski E. and Turowski J.: Dependence of stray losses on current and temperature. *CIGRE, 20th Plenary Session*, Paris 1964, Report 102, 1–13.

10.18. Kazmierski M., Kozlowski M., Lasocinski J., Pinkiewicz I., and Turowski J.: Hot spot identification and overheating hazard preventing when designing a large transformer. *CIGRE 1984 Plenary Session*. August 29—September 6, 1984. Report 12-12, 1–6.

10.19. Klesa J. and Kopecek J.: Some problems of design of the tank of large transformers. (in Czech). *Elektrotechn. Obzor*, (2), 1967, 53–56.

10.20. Niewierowicz N. and Turowski J.: A new thermometric method of measuring power losses in solid metal elements. *Proceedings IEE*, 119(5), 1972, 626–636.

10.21. Nowaczyński J.: (1) Local overheating of construction elements of transformers, during overload and over-excitation. (in Polish). PhD Dissertation under the direction of J. Turowski, Inst. of El. Machines and Transformers, Technical University of Lodz, September 1977; and (2) *Sci. Reports of Power Inst*. (7) 1979, 1–130.

10.22. Savini A. and Turowski J.: Influence of structure geometry, screens, and eddy currents on the critical distance of tank wall in power transformers. Chapter 3.6 in the book "*Electromagnetic Fields in Electrical Engineering*". New York: Plenum Press, 1988, pp. 119–127.

10.23. Specht T.R., Rademacher L.B., and Moore H.R.: Iron and copper loss measurement in transformers. *Electr. Eng.*, (5) 1958.

10.24. Turowski J: Wattmeter for measuring power at practically any low power factor and any low voltage. Polish Patent No. 47975, 5 Nov. 1962, *Patent Office Bulletin*, No. 6, 1963. Online: tech.money.pl/przemysl/patenty/pl-47975-504288.html.

10.25. Turowski J.: Wattmeter system for measuring at very low power factor and small voltages. *Przegląd Elektrotechniczny*, (4), 1964, pp.176–177.

10.25a. Turowski J.: The instrument for measurement of losses at the low power factor. CIGRE, 20th Plenary Session, Paris 1964, *Report of Polish National Committee of the CIGRE*. Instytut Energetyki. Warsaw-Paris 1964, pp. 1–3. Supplement.

10.26. Turowski J.: Power measurement at very low power factor or low voltage. *Proceedings IEE*, 112(4), 1965, 740–741.

10.27. Turowski J.: Testing of large transformers on CIGRE'64 (in Polish). *Przegląd Elektrotechniczny*, (11), 1965, 460–464.

10.28. Turowski J.: The method of calculating the distribution of stray flux between the core and tank. (in Polish). Technical University of Lodz, Dept. of Electric Machines. 1965. Report No E6/1z/64, Part I, 9–36.

10.29. Turowski J.: Calculation of stray fields and the losses in large transformers (in Italian). Istituto di Elettrotecnica. Universita degli Studi di Pavia, Italy, February 25, 1985, pp.1–54.

10.30. Turowski J., Kazmierski M., and Ketner A.: Determination of losses in structural parts by the thermometric method. *Przegląd Elektrotechniczny*, (10), 1964, 439–444.

10.31. Turowski J., Komęza K., and Wiak S.: Experimental determination of the distribution of power losses in the cores and construction elements of transformers. *Przegląd Elektrotechniczny*, (10), 1987, 265–268.

10.32. Wandel & Golterman: *Electronic Precision Measurement*. (in German). Eningen: Drexler Druck, 1983.

10.33. Zakrzewski K.: Mathematical modeling of electromagnetic fields in massive iron. (in Polish). *Rozprawy Elektrot.*, 16(1), 1970, 27–43.

10.34. Zakrzewski K.: Physical modeling of fields and load losses in transformers. *Rozprawy Elektrot.*, 25(2), 1979, 401–418.

10.35. Alatawneh N. and Pragasen P.: Design of a novel test fixture to measure rotational core losses in machine lamination. *IEEE Trans. on Industry Applications,* 48(5), Sept/Oct. 2012, 1460–1466.

10.36. Darley V.: The practical application of FEM in transformer design & development. *COMPEL*, 11, 125–128, 1992.

10.37. Turowski J., Kopeć M., Rizzo M., and Savini A.: 3-D analysis of critical distance of tank wall in power transformers. *Proc. Internat. Symposium on Electromagnetic Fields in Electrical Engineering—ISEF'91*, Sep. 1991. Southampton, UK, pp. 229–232.

CHAPTER 11

11.1. Turowski J.: Method of fast analysis of 3-D leakage fields in large three-phase transformers. *Proc. of International Symposium and TEAM Workshop on 3-D Electromagnetic Field Analysis—3DMAG,* Okayama, Japan, 11–13 September 1989, Paper VII.10, 123.

11.2. Wiak S., Pelikant A., and Turowski J.: Solution of 3-D electromagnetic field in Problem No. 10 TEAM WORKSHOP. *Proc. of Seminar EDT-3D on Electrodynamics of Three-Dimensional Systems*. Lodz, Poland. Ministry of National Educ. and Poland Section of IEEE, 7 December 1990, 12–14.

11.3. Wiak S., Pelikant A., and Turowski J.: Solution of the TEAM WORKSHOP Problem No. 7 by Reluctance Network Method (Asymmetrical Conductor with a Hole). *COMPEL*, James & James. London, 13(1), March 1994, 59–62.

11.4. Wiak S., Pelikant A., and Turowski J.: Magnetostatic field calculation of 3-D non-linear model by reluctance network method. (Problem No. 13 of TEAM Workshop). *Proc. of Internat. Symposium on Electromagnetic Fields in Electrical Engineering—ISEF'91*, September 18–20, 1991, Southampton, UK, 267–270.

11.5. Gieras J.: Selected technical aspects of Tu-154 m Smolensk air crash on April 10, 2010. *Mathematical and Computational Forestry Natura-Resouce Sciences MCFNS*, 5(5), 2013, 37–70; ISSN 1946-7664.

11.6. *Proceedings of the Smolensk Air Crash Conference*, Oct. 22, 2012. (In English and Polish). Warsaw, 2013, pp. 1–188.

Index

A

B

C

F

G

H

N

T

U

V

W

Y